Gold Nanoparticles in Analytical Chemistry

Comprehensive Analytical Chemistry

Volume 66

Advisory Board

Gold Nanoparticles in Analytical Chemistry

Comprehensive Analytical Chemistry

Volume 66

Edited by

Miguel Valcárcel and Ángela I. López-Lorente

Department of Analytical Chemistry
University of Córdoba
Córdoba, Spain

ELSEVIER

AMSTERDAM • BOSTON • HEIDELBERG • LONDON • NEW YORK • OXFORD
PARIS • SAN DIEGO • SAN FRANCISCO • SINGAPORE • SYDNEY • TOKYO

Elsevier
Radarweg 29, PO Box 211, 1000 AE Amsterdam, Netherlands
The Boulevard, Langford Lane, Kidlington, Oxford OX5 1GB, UK
225 Wyman Street, Waltham, MA 02451, USA

British Library Cataloguing in Publication Data
A catalog record for this book is available from the British Library

Library of Congress Cataloging-in-Publication Data
A catalogue record for this book is available from the Library of Congress

ISBN: 978-0-444-63285-2
ISSN: 0166-526X

Contents

10. Determination of Gold Nanoparticles in Biological, Environmental, and Agrifood Samples

Ángela Inmaculada López-Lorente and Miguel Valcárcel

Part III
Gold Nanoparticles as Analytical Tools

11. Applications of Gold Nanoparticles in Electroanalysis

Marek Trojanowicz

13. Gold Nanoparticles as (Bio)Chemical Sensors

Miguel Peixoto de Almeida, Eulália Pereira, Pedro Baptista, Inês Gomes, Sara Figueiredo, Leonor Soares, and Ricardo Franco

14. Lateral Flow Biosensors Based on Gold Nanoparticles

Lourdes Rivas, Alfredo de la Escosura-Muñiz, Josefina Pons, and Arben Merkoçi

Contributors to Volume 66

María Jesús Almendral Parra, Departamento de Química Analítica, Nutrición y Bromatología, University of Salamanca, Plaza de la Merced s/n, Salamanca, Spain

Vincenzo Amendola, Department of Chemical Sciences, University of Padova, Padova, Italy

Pedro Baptista, CIGMH, Departamento de Ciências da Vida, Faculdade de Ciências e Tecnologia, Universidade Nova de Lisboa, Caparica, Portugal

Encarnación Caballero-Díaz, Department of Analytical Chemistry, University of Córdoba, Córdoba, Spain

Shaowei Chen, Department of Chemistry and Biochemistry, University of California, Santa Cruz, CA, USA

Han-Wen Cheng, Department of Chemistry, State University of New York at Binghamton, Binghamton, NY, USA

José M. Costa-Fernández, Department of Physical and Analytical Chemistry, University of Oviedo, Oviedo, Spain

Patricia Crespo, Instituto de Magnetismo Aplicado and Dpto. Física de Materiales, Universidad Complutense de Madrid, Madrid, Spain

Elizabeth R. Crew, Department of Chemistry, State University of New York at Binghamton, Binghamton, NY, USA

Jorge Ruiz Encinar, Department of Physical and Analytical Chemistry, University of Oviedo, Oviedo, Spain

Alfredo de la Escosura-Muñiz, Institut Catala de Nanociencia i Nanotecnologia (ICN2), Bellaterra (Barcelona), Spain

Sara Figueiredo, CIGMH, Departamento de Ciências da Vida, Faculdade de Ciências e Tecnologia, Universidade Nova de Lisboa, Caparica, Portugal

Ricardo Franco, REQUIMTE, Departamento de Química, Faculdade de Ciências e Tecnologia, Universidade Nova de Lisboa, Caparica, Portugal

Miguel Angel García, Instituto de Magnetismo Aplicado and Dpto. Física de Materiales, Universidad Complutense de Madrid, Madrid, Spain

Inês Gomes, REQUIMTE, Departamento de Química e Bioquímica, Faculdade de Ciências, Universidade do Porto, Porto, Portugal; Instituto de Medicina Molecular, Faculdade de Medicina da Universidade de Lisboa, Lisboa, Portugal

Yan Guo, School of Environmental Science and Engineering, Nanjing University of Information Science and Technology, Nanjing, Jiangsu, P. R. China; Department of Chemistry and Biochemistry, University of California, Santa Cruz, CA, USA

Antonio Hernando, Instituto de Magnetismo Aplicado and Dpto. Física de Materiales, Universidad Complutense de Madrid, Madrid, Spain

Dominik Hühn, Fachbereich Physik, Philipps Universität Marburg, Marburg, Germany

Christine Kranz, Institute of Analytical and Bioanalytical Chemistry, University of Ulm, Albert-Einstein-Allee 11, 89081 Ulm, Germany

Bernhard Lendl, Institute of Chemical Technologies and Analytics, Vienna University of Technology, Vienna, Austria

Ángela Inmaculada López-Lorente, Department of Analytical Chemistry, University of Córdoba, Córdoba, Spain

Jin Luo, Department of Chemistry, State University of New York at Binghamton, Binghamton, NY, USA

Moreno Meneghetti, Department of Chemical Sciences, University of Padova, Padova, Italy

Arben Merkoçi, Institut Catala de Nanociencia i Nanotecnologia (ICN2), Bellaterra (Barcelona), Spain; Institucio Catalana de Recerca i Estudis Avançats (ICREA), Barcelona, Spain

Boris Mizaikoff, Institute of Analytical and Bioanalytical Chemistry, University of Ulm, Albert-Einstein-Allee 11, 89081 Ulm, Germany

Sara Sánchez Paradinas, Departamento de Química Analítica, Nutrición y Bromatología, University of Salamanca, Plaza de la Merced s/n, Salamanca, Spain; Institut für Physikalische Chemie und Elektrochemie, Leibniz Universität Hannover, Schneiderberg, Hannover, Germany

Wolfgang J. Parak, Fachbereich Physik, Philipps Universität Marburg, Marburg, Germany

Lucia Pasquato, Department of Chemical and Pharmaceutical Sciences, University of Trieste, Trieste, Italy

Miguel Peixoto de Almeida, REQUIMTE, Departamento de Química e Bioquímica, Faculdade de Ciências, Universidade do Porto, Porto, Portugal

Paolo Pengo, Department of Chemical and Pharmaceutical Sciences, University of Trieste, Trieste, Italy

Eulália Pereira, REQUIMTE, Departamento de Química e Bioquímica, Faculdade de Ciências, Universidade do Porto, Porto, Portugal

Rosario Pereiro, Department of Physical and Analytical Chemistry, University of Oviedo, Oviedo, Spain

Josefina Pons, Inorganic Chemistry Unit, Chemistry Department, Science Faculty, Autonomous University of Barcelona, Barcelona, Spain

Georg Ramer, Institute of Chemical Technologies and Analytics, Vienna University of Technology, Vienna, Austria

Lourdes Rivas, Institut Catala de Nanociencia i Nanotecnologia (ICN2), Bellaterra (Barcelona), Spain; Inorganic Chemistry Unit, Chemistry Department, Science Faculty, Autonomous University of Barcelona, Barcelona, Spain

Alfredo Sanz-Medel, Department of Physical and Analytical Chemistry, University of Oviedo, Oviedo, Spain

Shiyao Shan, Department of Chemistry, State University of New York at Binghamton, Binghamton, NY, USA

Zakiya R. Skeete, Department of Chemistry, State University of New York at Binghamton, Binghamton, NY, USA

Leonor Soares, REQUIMTE, Departamento de Química e Bioquímica, Faculdade de Ciências, Universidade do Porto, Porto, Portugal; REQUIMTE, Departamento de Química, Faculdade de Ciências e Tecnologia, Universidade Nova de Lisboa, Caparica, Portugal

Mauro Stener, Department of Chemical and Pharmaceutical Sciences, University of Trieste, Trieste, Italy

Laura Trapiella-Alfonso, Department of Physical and Analytical Chemistry, University of Oviedo, Oviedo, Spain

Marek Trojanowicz, Department of Chemistry, University of Warsaw, Poland and Laboratory of Nuclear Analytical Methods, Institute of Nuclear Chemistry and Technology, Warsaw, Poland

Miguel Valcárcel, Department of Analytical Chemistry, University of Córdoba, Córdoba, Spain

Chuan-Jian Zhong, Department of Chemistry, State University of New York at Binghamton, Binghamton, NY, USA

Series Editor's Preface

"Nanotechnology has been defined as the technology of the twenty-first century, and it is expected that the broad range of nanomaterials together with their applications on the global market will constantly increase in the coming years." This sentence was written two years ago in the preface to Volume 59 of Comprehensive Analytical Chemistry, *Analysis and Risk of Nanomaterials in Environmental and Food Samples*, edited by myself and Dr M. Farre.

It is then obvious that there is a need for the Comprehensive Analytical Chemistry series to look for new books in the field of nanomaterials. This task was relatively easy. In one of my regular telephone conversations with Prof. Miguel Valcárcel, an old friend and well-known expert in analytical chemistry, he suggested editing a book on gold nanoparticles. I accepted immediately.

The book that you have in your hands contains 14 chapters. The first five cover general aspects such as an introduction to analytical nanoscience and nanotechnology, the synthesis, characterization, and toxicity of gold nanoparticles. In the second part, gold nanoparticles are considered as target analytes, with emphasis on their characterization and determination, including spectroscopic, mass spectrometric, and separation techniques. Part three describes the use of gold nanoparticles as analytical tools. They can be incorporated in electrodes, and used as (bio)chemical sensors as well as lateral flow biosensors.

With the comprehensive information on this type of nanoparticles, this multipurpose book with novel applications in biology, the environment, and food is a useful addition to the series and will be of great benefit to the broad nanoscience and nanotechnology community.

Finally I would like to thank both editors of this book, Miguel Valcárcel and Ángela I. López-Lorente, for the amount of work, time, and expertise that they devoted to it. I would like to acknowledge as well the various well-known authors for their contributions in compiling such a world-class and timely book that will be of help to newcomers, PhD students, and those senior researchers who consider nanotechnology as one of the emerging challenges in the years to come.

D. Barceló
IDAEA-CSIC, Barcelona, and ICRA, Girona
July 10, 2014

Volume Editor's Preface

Today we are immersed in a full expansion of Nanoscience and Nanotechnology (N&N). Analytical Science is an integral part of N&N since reliable information about the nanoworld is crucial in order to make well-founded scientific and technical decisions in this area. Two key facets of Analytical Nanoscience and Nanotechnology (AN&N) can be noted: on the one hand, the consideration of nanoparticles and nanostructured materials as tools for the innovation and improvement of (bio)chemical measurement processes, and, on the other hand, their consideration as objects (analytes). The use of nanomaterials as analytical tools is the more developed field, however, the balance is bound to change over the next few years due to the growing significance of the characterization of nanomaterials and the development of new instruments based on nanotechnological approaches.

Among the wide variety of nanoparticles commonly used in AN&N, namely carbon nanostructures such as carbon nanotubes, carbon dots, graphene, fullerenes, nanodiamonds, etc., semiconductor nanoparticles (quantum dots), or metallic nanoparticles (i.e., silver, titanium oxide, or magnetic nanoparticles), this book focuses on nanoparticles of a specific nature: gold. In this sense, the book is unique as it presents a systematic review on the different aspects of gold nanoparticles in analytical chemistry. Without doubt, gold nanoparticles are among the most relevant nanoparticles, having analytical connotations at a similar level to carbon nanotubes.

The aim of this book is to bring gold nanoparticles closer to the reader interested in AN&N, providing a comprehensive overview. Although the focus is on gold nanoparticles, many of the general conclusions can be extrapolated to other nanoparticles. Those professionals working not only in AN&N but also in different fields involving the use of gold nanoparticles, such as catalysis, biological and medical applications, can also benefit from the book since many of the exceptional properties of gold nanoparticles can be applied for different purposes.

The 14 chapters are classified into three sections. First, basic aspects of gold nanoparticles such as their synthesis, physicochemical properties, or derivatization procedures are described in order to envisage their potential. The second part of the book reviews the techniques employed for both the characterization and determination of gold nanoparticles. The last part is devoted to the improvement of analytical processes by using gold nanoparticles as tools in electrochemistry, spectroscopy, or biosensors.

The editors wish to express their gratitude to those who have helped to bring this book to completion. We would like to thank to all the authors for their contributions and the exhaustive revisions they have performed. We also like to thank the cooperation of Elsevier and, for his technical support, José Manuel Membrives.

Miguel Valcárcel
Ángela I. López-Lorente
July 2014

Part I

Generalities

Chapter 1

Analytical Nanoscience and Nanotechnology

Ángela Inmaculada López-Lorente and Miguel Valcárcel*
Department of Analytical Chemistry, University of Córdoba, Córdoba, Spain
**Corresponding author: E-mail: qa1meobj@uco.es*

Chapter Outline

Gold Nanoparticles in Analytical Chemistry. http://dx.doi.org/10.1016/B978-0-444-63285-2.00001-8

1. CONTEXTUALIZATION

1.1 Definitions

The common characteristic of nanoscience and nanotechnology (N&N) is the size of the target objects, which are comprised in the so-called "nanometric scale," typically between one and 100 nm.

Nanoscience has multiple complementary definitions, such as "the science of the synthesis, analysis and manipulation of materials at atomic, molecular, and macromolecular scales where physico-chemical properties may differ significantly from those at a larger particulate scale," [1] or, simply: "the science based on the diverse structures of materials which have dimensions of a billionth part of the meter" [2].

On the other hand, **Nanotechnology** "deals with the design, characterization, production and application of structures, devices and systems by controlling the shape and size at the nanometer scale" [1].

A substantial aspect of nanoscience and nanotechnology (N&N) is its multidisciplinary as well as transversal and convergent character. Physicists, chemists, and engineers are the scientists and professionals more directly involved, but their convergence with other areas such as information technology and communication, biotechnology, and materials science, in a first approach, and medicine, pharmacy, agrifood, and diverse types of industries such as textile or energetic, in another, has to be pointed out.

Analytical science cannot be left out of N&N [3] and, in fact, it is even present in many definitions of N&N since reliable information about the nanoworld is crucial to make well founded scientific and technical decisions in this area. Words belonging to the analytical discipline such as "analysis" or "characterization" and others shared with other disciplines such as "use" or "employment" summarize the two key facets of the relationship between analytical chemistry and nanoscience and nanotechnology, namely, (1) the consideration of nanoparticles and nanostructured materials as objects (analytes) or (2) tools for the innovation and improvement of the (bio) chemical measurements processes.

The major application areas of nanotechnology can be classified into four groups [3], namely, (1) nanobiotechnology and nanomedicine, (2) nanomaterials, (3) nanoelectronics, and (4) nanosensors/nanodevices, nanotechnological instrumentation, and nanometrology. The last area is directly related to analytical science, which also plays an essential role in the other three, for example, dealing with the monitoring of production processes or both the characterization and use of end products.

1.2 Classifications

There are several emerging possibilities when introducing nanoscience and nanotechnology in the analytical scope. Therefore, a multiple classification based on four complementary criteria has been created, which is shown schematically in Figure 1 and is described in the following text.

The first criterion (Figure 1(1)) considers the type of material analyzed, which can be conventional (macro or micro in size) or nanomaterials. In the first case, nanoparticles can be involved in the analytical process, conferring to it nanotechnological character. An example is the use of quantum dots functionalized with antibodies, which can be injected in organisms in order to detect carcinogenic processes [4]. In the second possibility, the target is the own nanoworld, which coincides with the consideration of nanomaterials as analytes. For example, the determination of nanomaterials such as gold nanoparticles [5] or carbon nanotubes [6–8] from environmental and biological matrices [9].

The second criterion (Figure 1(2)) relies on the analytical consideration of nanoparticles and nanostructured materials as objects (analytes) or tools involved in the analytical process. The extraction of chemical information

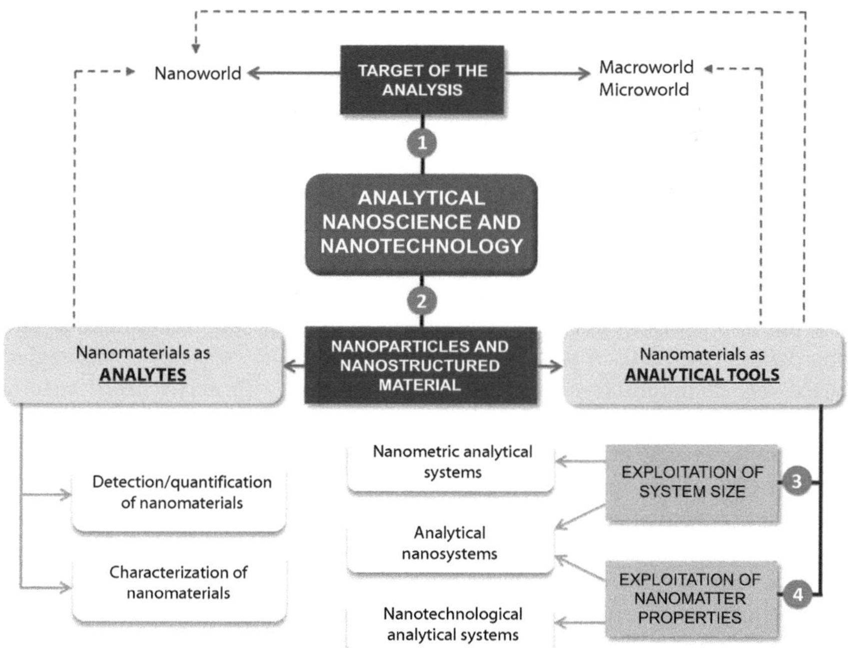

FIGURE 1 Inherent classifications of analytical nanoscience and nanotechnology take into account four criteria: (1) target of the analysis; (2) consideration of the nanomatter; (3) exploitation of the nanosize; and (4) exploitation of the nanomatter properties.

from the structured nanomaterials (composition, chirality, reactivity, etc.) is an indispensable complement to the physical characterization, which is more well-known (dimensions, topography, etc.) [10]. On the other hand, nanomaterials can be used as analytical tools in order to develop new analytical processes or to improve existing ones (i.e., development of optical sensors, development of stationary and pseudostationary phases in chromatography and capillary electrophoresis, mechanical sensors, etc.).

Criteria 3 and 4 (Figure 1(3) and (4)) are based on exploitation in the analytical scope of the exceptional properties of nanomaterials, in exploiting the nanosize, or both. This leads to the definition of three types of analytical systems related to nanoscience and nanotechnology: nanotechnological analytical systems, nanometric analytical systems, and analytical nanosystems [11]. *Nanotechnological analytical systems* exploit the exceptional physico-chemical properties of nanomaterials (although they are in micro/macro analytical systems) accounting for the most current uses of analytical nanoscience. *Nanometric analytical systems*, which are based exclusively on the nanosize of the devices involved, are exemplified by nanochip liquid chromatography systems [12] exploiting the advantages of working with flow rates as low as a few nanolitres per minute, a nanopipette [13], or levitated nanodrops as analytical containers [14]. Finally, *analytical nanosystems* successfully integrate the previous two types of systems by exploiting both the nanosize and nanomaterials properties (e.g., individual carbon nanotubes for use as electrodes [15], supramolecular systems that selectively recognize an analyte [16], and the so-called lab-on-a-particle [17]).

1.3 Synthesis of Nanoparticles

Nanomaterials can exist in the environment from a natural source, such as organic colloids, magnetite, aerosols, iron oxides, etc. The nanotechnological revolution is posed in a change of paradigm in the fabrication of products. Two approaches can be used to raise nanosize, namely, (1) "top-down" strategies, based on methodologies which achieve nanosize materials from macromaterials (nanoparticles are directly generated from bulk materials via the generation of isolated atoms usually involving physical methods such as milling or attrition, repeated quenching and photolithography [18]) and (2) "bottom-up" strategies, based on the creation of complex nanostructures from atomic or molecular functional elements. They comprise molecular components as starting materials linked with chemical reactions, nucleation, and growth processes to promote the formation of clusters. Numerous kinds of nanoparticles have been produced by liquid-phase synthesis, using techniques such as co-precipitation of sparingly soluble products by addition, exchange, and reduction reactions, oxidation, hydrolysis [19], sol—gel processing [20], microemulsions [21], etc. The latter approach is generally considered to be far

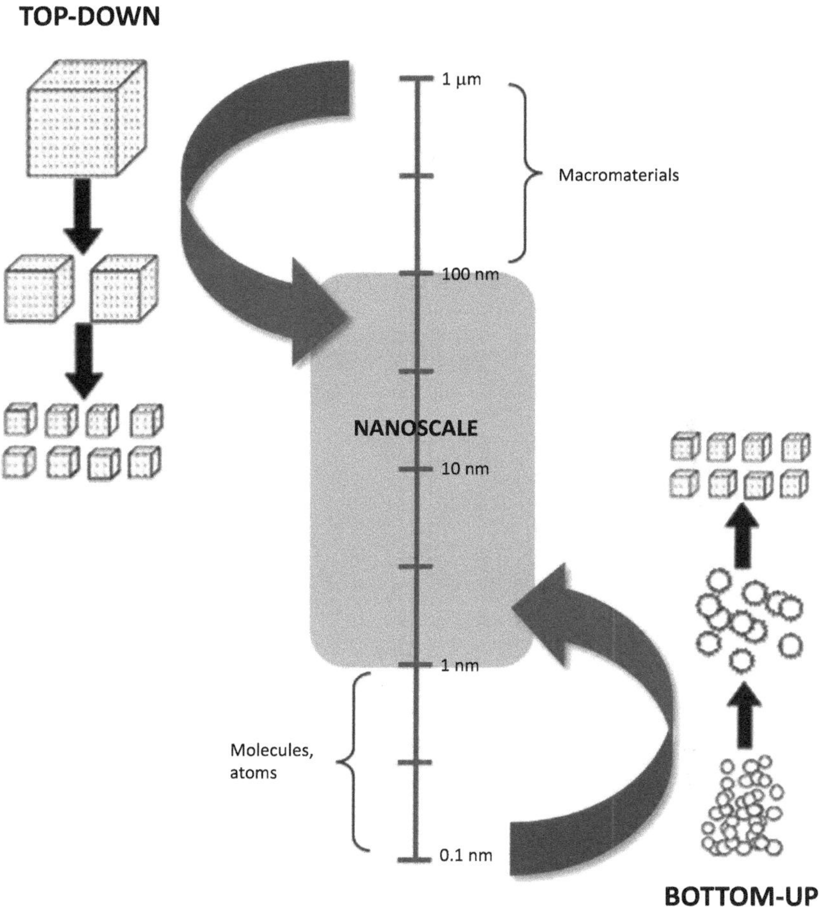

FIGURE 2 Scheme of the two approaches employed in the fabrication of nanomaterials: "top-down" and "bottom-up."

more promising due to the higher level of control offered. Figure 2 shows a scheme of the different strategies ("top-down" and "bottom-up") used to achieve the nanoscale.

1.4 Types of Nanoparticles

According to the IUPAC Glossary, a nanoparticle is a microscopic particle whose size is measured in nanometers, often restricted to so-called nanosized particles (NPs; < 100 nm in aerodynamic diameter), also called ultrafine particles [22]. With the expected increase in the applications of nanotechnology, more and more products will be manufactured containing components which will fit the commonly used definition of the nanoscale, as having a size

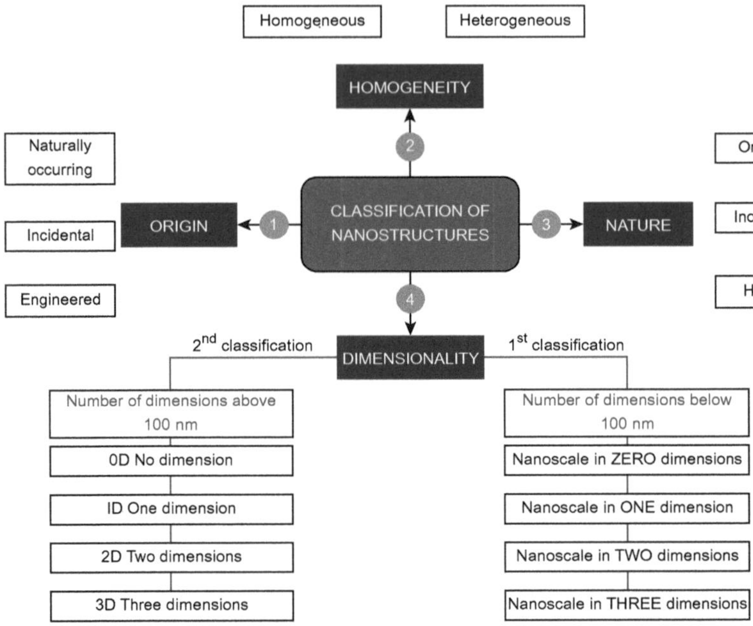

FIGURE 3 Classification of nanostructures according to their origin (1), homogeneity (2), nature (3), and dimensionality (4).

between approximately one and 100 nm. These wide varieties of nanostructures have been classified in multiple ways in the literature. Figure 3shows the most relevant types of nanostructures in analytical nanoscience and nanotechnology, classified according to four nonexclusive criteria.

Nanoparticles can be classified as natural, anthropogenic (incidental), or engineered in origin [9] (Figure 3(1)).

From a practical point of view, it is important to know the homogeneity of the nanostructured materials both for scientific studies as well as for industrial applications. Homogeneity can be referred to in terms of chemical composition or dimensionality (Figure 3(2)). Identical nanoparticles are those with the same chemical composition and dimensions. On the contrary, nanoparticles with the same chemical composition but different dimensions usually present different properties.

Concerning the nature or chemical composition of nanostructures, those can be classified (Figure 3(3)) as inorganic (e.g., noble metal nanoparticles, quantum dots, etc.), organic (fullerenes, carbon nanotubes, dendrimers, molecular imprinted polymers, etc.) or mixed (gold nanoparticles modified with calixarenes, carbon nanotubes functionalized with ferrocene, etc.). In this context, there is a growing interest in the development of hybrid nanoparticles, which can be defined as well-organized nanomaterials consisting of two or more types of individual nanocomponents [23].

The last classification of nanomaterials is based on dimensionality criteria (Figure 3(4)). As shown in the figure, two classifications may be done, taking into account both the strict dimensions (in the nanoscale) of the nanostructure that give rise to those exceptional properties and the dimensions of the material where nanostructures are present.

The Royal Society of Chemistry and the Royal Academy of Engineering classified nanostructures in function of the number of dimensions in the nanoscale (below 100 nm) [24], distinguishing three types of nanostructures: (1) nanoscale in one dimension, such as surfaces with nanometric thickness (e.g., graphene sheets); (2) nanoscale in two dimensions, such as carbon nanotubes, inorganic nanotubes, nanowires, etc.; (3) nanoscale in three dimensions, which includes metallic nanoparticles and their oxides, quantum dots, fullerenes, and dendrimers. Classification of nanoscale at zero dimension can also be added, such as materials composed by dispersed nanoparticles.

Other authors [25] have classified nanostructures depending on the number of dimensions which exceed 100 nm, being above the nanoscale, nanostructures being thus categorized as 0D, 1D, 2D, or 3D. A 0D nanostructure is a material with all its dimensions comprised in the nanometric scale (e.g., metallic nanoparticles, quantum dots, etc.). Carbon nanotubes are an example of 1D nanostructures, which have one dimension of micro/macrometric size, such as nanowires or nanorods. 2D nanostructures have two dimensions above nanoscale while one of them is below 100 nm. That is the case of surface nanocoatings or thin films of molecular monolayers. Finally, 3D nanostructures are those whose three dimensions escape from the nanoscale, but the material is comprised by a set of nanoparticles forming a block of micro/macrometric size (e.g., nanoporous materials, powders).

We will focus on engineered nanoparticles, which can be classified according to their nature as organic, inorganic, or hybrid.

1.4.1 Organic Nanoparticles

1.4.1.1 Carbon Nanomaterials

Graphitic forms include 0D fullerene, 1D CNT, and 3D graphite, and the 2D case comes to the graphene, a single layer of carbon atoms formed in a honeycomb lattice.

Graphene is an open, flat, two dimensional structure composed of carbon atoms organized in a network of hexagons attached to each other. This is possibly a result of sp^2 hybridization of the carbon atoms present in the sheet. It has a large specific surface area and can be easily modified with functional groups, especially via graphene oxide. Graphene quantum dots (GQDs), a new kind of quantum dots, have emerged and ignited tremendous research interest. GQDs are defined as graphene sheets with lateral dimensions less than 100 nm in single, double, and few (3 to <10) layers. GQDs show low cytotoxicity, excellent solubility, chemical inertia, stable photoluminescence, and better surface grafting. Therefore, they are promising in optoelectronic devices, sensors, bioimaging, etc. [26].

Another two examples of 0D carbon-based fluorescent nanomaterials are diamond nanocrystals (DNs) and carbon dots (CDs), which have also drawn much attention in recent years. In general, DNs consist of about 98% carbon with residual hydrogen, oxygen, and nitrogen; they possess a sp^3 hybridized core and have small amounts of graphitic carbon on the surface. Luminescent CDs comprise discrete, quasispherical carbon nanoparticles with sizes below 10 nm.

Fullerenes are closed-cage carbon molecules containing pentagonal and hexagonal rings. They comprise a wide range of isomers and homologous series, from the most studied C_{60} and C_{70} to the so-called higher fullerenes like C_{240}, C_{540}, and C_{720}. They possess relatively high electron affinity, hydrophobic surface, and high surface/volume ratio.

Carbon nanotubes have received special attention ever since their discovery by Iijima in 1991 [27]. They are tubular in shape and consist entirely of covalently bonded carbon atoms. They can be described as hollow graphitic nanomaterials comprising one (single-walled carbon nanotubes, SWNTs), two (double-walled carbon nanotubes, DWNTs) or multiple (multiwalled carbon nanotubes, MWNTs) layers of graphene sheets. They possess nonpolar bonds and high aspect ratios, which make them insoluble in water and facilitate their aggregation.

Carbon nano-onions (CNOs) are nanoparticles with a diameter between 20 and 100 nm composed of several layers of concentric fullerenes and constitute the spherical version of MWNTs. They were described by Ugarte [28] in 1992, although few applications have been reported.

Moreover, carbon nanofibers (CNFs) are solid carbon fibers with lengths in the order of a few microns and diameters below 100 nm. CNFs differ from CNTs in the absence of a hollow cavity, and the diameters of CNFs are generally higher than those of the corresponding CNTs [29].

Finally, carbon nanocones were first synthesized by vapor condensation of carbon atoms on a graphite substrate [30]. The disclination of each structure corresponds to the presence of a given number of pentagons in the seed from which it grew: disks (no pentagons), five types of cones (one to five pentagons), and open tubes (six pentagons). The unique electronic distribution, which is provided by these pentagonal rings to the carbon nanocones, results in enhanced local density at the cone apex. One major class of cone structures are single-walled carbon nanohorns (SWNHs) with the narrowest opening angle with five pentagonal rings in their apex (Figure 4).

1.4.1.2 Other Organic Nanomaterials

Organic polymeric nanoparticles are prepared from polymers and considered potential drug delivery devices. Dendrimers are one kind of polymeric nanoparticles constructed by the successive addition of layers of branching groups.

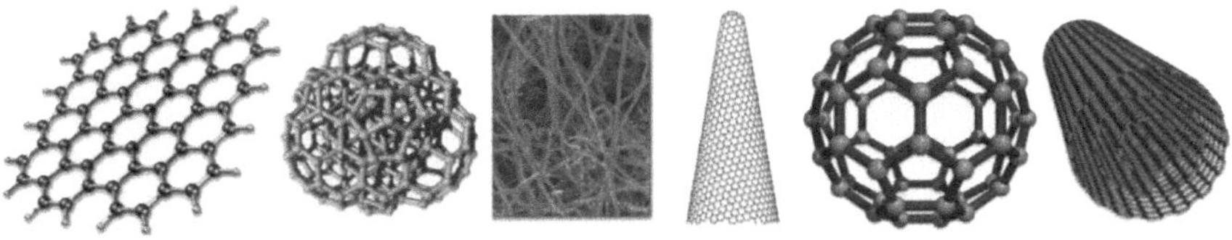

FIGURE 4 Main types of carbon-based nanomaterials. *Adapted with permission from Ref. [29].*

Molecularly imprinted polymers (MIPs) are a specific class of selective sorbents that have built-in functionality to achieve the complementary recognition of a given chemical compound [31]. Although MIP particles can be nanometric in size, they usually fall in the micrometric range.

In addition, an increasing number of diverse nanomaterials are emerging for biopharmaceutical applications, such as liposomes, nanomicelles, nanovesicles. nanoemulsions, etc. [32].

1.4.2 Inorganic Nanoparticles

Inorganic nanoparticles cover a broad range of substances including elemental metals, metal oxides, and metal salts. Silver nanoparticles (AgNPs) are used in many products as a bactericide, whereas gold nanoparticles (AuNPs) are explored for many possible applications because of their catalytic activity. Both silver and gold nanoparticles possess the so-called surface plasmon as a result of the collective oscillation of the electrons, which gives them excellent optical properties and a large enhancement of the electric field on their surface.

Quantum dots (QDs) are semiconductor nanocrystals with all three dimensions falling in the 1−10 nm size range. In many respects, these luminescent nanocrystals constitute a transitional stage between bulk semiconductors and single atoms. The QD core is made up of elements from the II−VI (e.g., CdSe, CdTe, CdS, and ZnSe), III−V (e.g., InP and InAs), or IV−VI (e.g., PbSe) group [33]. They have aroused widespread interest by virtue of their exceptional optical, electronic, electrochemical, photophysical, redox, and catalytic properties.

Nanoparticulate metal oxides are widely used, such as TiO_2, Al_2O_3, ZrO_2, MnO and CeO_2, as well as nanoparticles of iron oxides (FeOx). Attention should be paid to the increasingly used magnetic nanoparticles, which have been synthesized with a number of different compositions and phases, including iron oxides, such as Fe_3O_4 and γ-Fe_2O_3, pure metals, such as Fe and Co, and spinel-type ferromagnets, as well as alloys [34]. Moreover, silica nanoparticles (SiO_2) are characterized by presenting high surface areas and exhibit intrinsic surface reactivity, which allows chemical modifications [10].

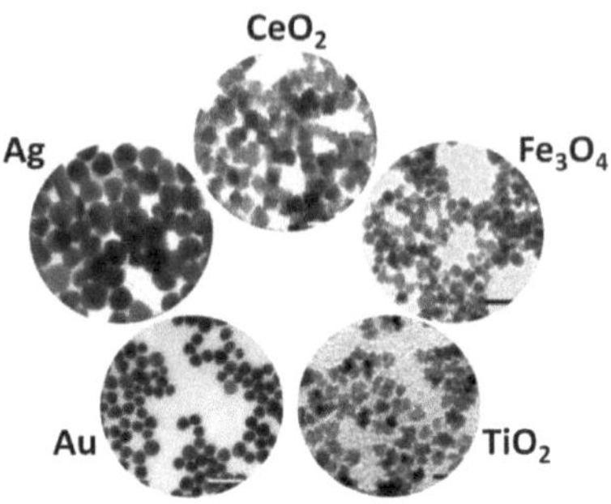

FIGURE 5 Transmission electron microscopy images of model inorganic nanoparticles, namely, Ag, Au, TiO_2, Fe_3O_4, CeO_2, and NPs. *Adapted with permission from Ref. [36].*

Nano-size zeolite, clays, and ceramics are other nanoparticles that have been proposed for various applications [35] (Figure 5).

1.4.3 Hybrid Nanoparticles

Hybrid nanoparticles can be defined as well-organized nanomaterials consisting of two or more types of individual nanocomponents [23]. Those nanocompounds can be bound via organic/inorganic molecular bridges or directly attached to one another. Apart from the special type of bonding between the nanocompounds, the exceptional properties of hybrid nanoparticles are due to the highly organized arrangement of the nanoconstituents.

Nanoparticles possess excellent properties that can be boosted or supplemented by combining two or more types of materials into a hybrid nanocomposite. In general, hybrid nanoparticles can be classified into two different types according to the combined properties [23], namely, (1) properties of the isolated nanoparticles are different but complementary, and (2) properties of the isolated nanoparticles are of the same nature, but their combination produces important synergistic effects.

1.5 Properties of Nanoparticles

The "nanoscale" has introduced a new scenario where impressive changes in physico-chemical principles, laws, and properties are observed as regards macro- and micro-materials.

Two especially significant differences have been described: the surface/volume ratios and chemical reactivities of nanostructured matter in comparison to macro- and micro-scale matter. In addition, quantum effects are enhanced by effect of the configuration of molecular orbitals (similar to that of atomic orbitals) and, as a consequence, chemical, optical, electrical, thermal, and magnetic characteristics are unique on the nanoscale. Such changes have shaped the impact of nanotechnology in a number of scientific, technical, and industrial fields.

2. INTRODUCTION TO ANALYTICAL NANOSCIENCE AND NANOTECHNOLOGY

2.1 Facets of Analytical Nanoscience and Nanotechnology

Generally, nanomaterials in the analytical nanoscience and nanotechnology (AN&N) scope are considered as analytical objects or analytical tools. Analytical nanoscience and nanotechnology currently provide one of the most promising avenues for developments in analytical science, derived from their two main fields of action, namely the analysis of nanostructured materials and their use as tangible tools [37]. The use of nanotechnological tools in analytical methods can improve the analytical properties and enables the development of new types of analysis. On the other hand, analysis of the nanoworld is an issue of analytical chemistry.

A third classification, as an interface between the two previous, can be introduced, which is the use of nanomaterials in analytical processes for the characterization and/or determination of other nanomaterials, see Figure 6. Examples of this facet are, for instance, the use of silver or gold nanoparticles as SERS substrates for the determination of other nanomaterials (e.g., carbon nanotubes [6]), or the use of membranes composed by multiwalled carbon nanotubes for the preconcentration and determination of single-walled carbon nanotubes [7].

2.2 Types of Analytical Systems

The use of nanomaterials as analytical tools is one of the facets of analytical nanoscience and nanotechnology. Nanotechnology-based analytical processes

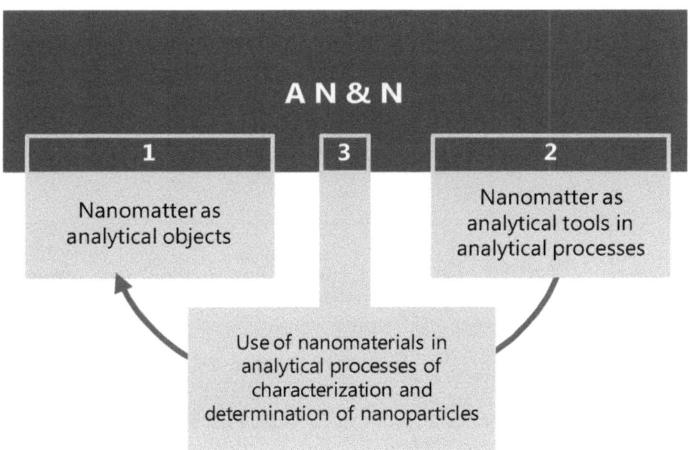

FIGURE 6 Different facets of nanoparticles in analytical nanoscience and nanotechnology (AN&N) scope.

can exploit both the nano size and exceptional properties of structured nanomatter. In this sense, as it was previously introduced (see Figure 1 and related text), analytical systems based on nanoscience and nanotechnology can be classified into three types [11]:

1. *Nanometric analytical systems* are based on nanosize and/or nanofluidics, for example, nanochip liquid chromatography. However, some authors have placed them outside the scope of nanotechnology.
2. *Nanotechnological analytical systems* exploit the exceptional physico-chemical properties of nanomaterials.
3. *Analytical nanosystems* integrate the previous two systems, for example, individual carbon nanotubes for use as electrodes (Figure 7).

2.2.1 Nanometric Analytical Systems

As previously defined, nanometric analytical systems are analytical systems that take advantage of the fact that some technical characteristics or elements of the analytical process have nanometric size of flow. Thus, their foundation is not based on the new physico-chemical scenario imposed by nanomaterials and their exceptional physical and chemical properties. They can be considered as a trend of miniaturization.

They cannot be completely considered inside the scope of nanotechnology, although because of the undoubted advantages they offer, many authors include them. Some representative examples of this kind of systems are reported as follows.

Some nanometric analytical systems are based on nanometric volumes employed. Such is the case of the use of levitated nanodrops [14]. This strategy has been employed for online monitoring of chemical reactions in

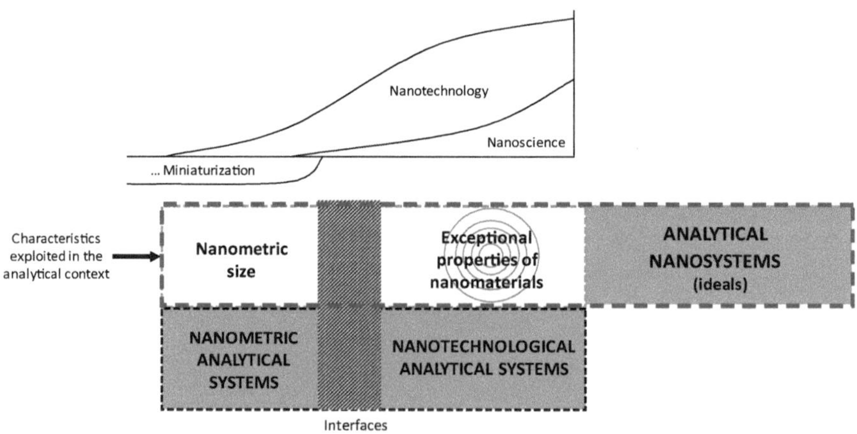

FIGURE 7 Difference between the three general types of analytical systems related with nanoscience and nanotechnology.

ultrasonically levitated, nanoliter-size droplets by Raman spectroscopy. A flow-through microdispenser connected to an automated flow injection system was used to dose picoliter droplets into the node of an ultrasonic trap. A sequence of reagents can be injected via the microdispenser into the levitated droplet. Thus, chemical reactions can be carried out and monitored online. The droplet suspended in the air enables the removal of interferences from the recipient and a rapid evaporation of the solvent. This has eased the direct analysis with Raman microscopy [14] as well as the development of interfaces in direct couplings in which sample solvent is a problem, such as in matrix-assisted laser desorption/ionization time of fight mass spectrometry (MALDI-TOF-MS) [38].

The chromatographic systems called "Nano-LC" are based on the use of liquid mobile phases with a flow rate of a few nanolitres per minute [39], in contrast with conventional ones in the range of microliters per minute. That implies a new design for the chromatography instrument, but it offers advantages such as ease of use, both the delay time for the gradient to arrive at the head of the column and the dead volume between the separation column and the ESI tip are minimized, resulting in a shorter running time and reduced band broadening.

The so-called "nanopipettes" [13] are integrated, carbon-based pipettes with nanoscale dimensions that can probe cells with minimal intrusion, inject fluids into the cells, and concurrently carry out electrical measurements. They allow the direct transference of nanovolumes in the range of nano/picoliters between micro/nanometric compartments. This transference is based on the electro-osmotic phenomena. Their applications in the field of biotechnology are very promising, as well as in high-throughput analysis, key in the field of the pharmacological industry to perform multisynthetic experiments in the development of new drugs.

2.2.2 Nanotechnological Analytical Systems

Nanotechnological analytical systems exploit the exceptional properties of the nanometric materials (nanoparticles and nanostructured materials) with analytical purposes. They posses micrometric dimensions, but once their dimensions can be reduced into the nanoscale, they could be called "analytical nanosystems."

In these systems, there is a micro-element which acts as a bridge between the nanocomponents and the signal transduction system through a simple physical bond. The active nanocomponent is responsible for the interaction and produces the analytical signal.

One illustrative example is micro-electromechanical systems ("nano-cantilevers") [40,41], which are based on the immobilization of a biochemical receptor in a probe of nanometric dimensions in a micromechanical silicon system that is sensitive to the environment, including chemical compounds and biological entities which affect their mechanical characteristics in such a manner that the change can be measured in terms of electrical or optical properties [5]. Although a micro-electromechanical system is micrometric in size, its sharp tip, which allows interaction forces at the atomic scale to be sensed, is a nanometric device. Their most remarkable advantages are their reduced dimensions, portability (in some cases), and their high levels of sensitivity and selectivity for the determination of a wide variety of analytes. Their scope is defined by the biochemical interaction analyte-receptor.

Other examples of nanotechnological analytical systems are the field effect transistors (FETs), which have been revitalized by the introduction of nano-structured materials [41]. Such is the case of the use of semiconducting carbon nanotubes or a network of nanotubes placed between two electrodes of micrometric dimensions. Carbon nanotubes are derivatized in order to achieve a more specific interaction with the analyte. When it is retained, a change in the electronic properties of the carbon nanotubes is originated, which leads to a variation in the potential between the electrodes, used as an analytical signal. Semiconducting carbon nanotubes have been employed for the determination of a wide variety of gaseous analytes, such as NH_3, CO, and CO_2 among others because they exhibit larges changes in electric conductivity in their presence [42].

Finally, another example of this type of system is a nano-electrode, consisting of a single multiwalled carbon nanotube (2—3 μm in length and 30 nm of diameter) bonded to the end of an etched tungsten tip [15], which acts as a bridge. This is a real nano-electrode since it has nanometric dimensions and exploits the exceptional properties of nanomatter, although it is coupled to a micrometric device, the tungsten tip. The carbon nanotube surface can be functionalized with a wide variety of (bio) chemical compounds in order to improve the selectivity and the field of application. It offers the advantage of compatibility with micro/nano-size samples (e.g., cells). This nano-electrode has been used to determine dopamine and glutamate (immobilizing glutamate oxidase enzyme in CNT surface) in a physiological medium with a quantification limit of 100 μm.

2.2.3 Analytical Nanosystems

These systems can be defined as instruments or devices that have a nanometric size and are controlled by the physico-chemical laws of Nanoscience. In addition to the nanometric size, the exceptional properties of nanomatter are also exploited. This is an ideal situation, since microcomponents are needed to connect the nanoworld with the macroworld [3].

A nanochemical approach to a real analytical nanosystem is the so-called "lab-on-a-particle" [17], which aims to emulate the micrometric scale developments ("lab-on-a-chip"). This nanosystem consists on a supramolecular nanoarchitecture that incorporates chemical entities that can be employed as a "gate" to allow controlled access to a specific site on the supramolecular complex. An example is mesoporous silica nanoparticles functionalized with switchable molecules, whose inner pores can be used to entrap chemical species. The gate will open upon the application of physico-chemical external impulses, such as photochemical or electrochemical, and can release confined guests or allow molecular species of the solution to be incorporated.

2.3 Evolution and Limit of Analytical Nanoscience and Nanotechnology

Nanoscience and nanotechnology unarguably have promising prospects. A report of the US National Science Foundation [43] predicted that a new revolution based on the bio-nano-info triangle will shortly surpass the present evolution of the computer-info binomial.

The scientific and industrial impact of nanoscience and nanotechnology has grown dramatically in recent years. The exponential growth of the number of papers on this topic published in the last 10 years in the SciFinder and the vast amount of economic resources invested in industrial technological developments each year [44] constitute the best support for the brilliant present and promising future of nanoscience and nanotechnology. It should be pointed out that there are more than one million scientific papers published on this topic (Figure 8).

The impact of nanotechnology leaves no shadow of doubt. Hassan [45] emphasized that the most crucial aspect in this respect is the small things-big changes binomial. Analytical chemistry should play a major role in the coming advances in N&N, particularly in these aspects: (1) the environmental and toxicological impacts of nanotechnology, (2) the need for homogeneous, pure,

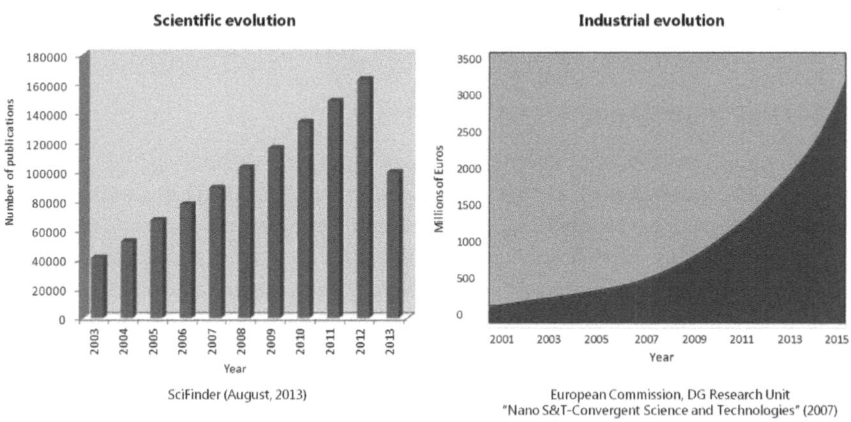

FIGURE 8 Scientific and industrial evolution of nanoscience and nanotechnology: (left) number of publications per year, (right) millions of euros involved in European nanotechnological industry.

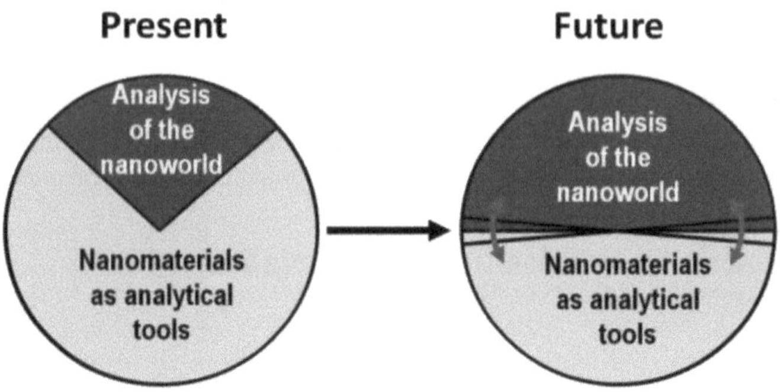

FIGURE 9 Present situation and trends in the two facets of analytical nanoscience and nanotechnology: nanomaterials as analytical tools and analysis of the nanoworld.

well-characterized nanoparticles, (3) the need to address nanometrology, and (4) nanomedicine, in the development of biomeasurement nanosystems and the characterization and monitoring of nanostructured pharmaceuticals [3].

Regarding analytical nanoscience and nanotechnology, the use of nanomaterials as analytical tools is the most developed field. Although more than one half of reported applications relate to the use of nanoparticles, the balance is bound to change over the next few years due to the growing significance of the characterization of nanomaterials and the development of new instruments based on nanotechnological approaches (Figure 9).

Nowadays, the prefix "nano" is opening many doors, which leads to many authors misusing it to their advantage. The prefix "nano" should be used in connection to nanomaterials, nanostructures, and nanodevices that exploit not only the size but also the exceptional properties of the nanoworld. The same happened in the past with other overexploited keywords such as "sensor." Fortunately, this terminological abuse has vanished over time.

2.4 Ethical and Social Implications

Nanotechnology has been deemed as a key emerging technology for fulfilling the "grand challenges of our time" (Lund Declaration [46]) in areas such as health care, energy production, environmental protection, and potable water procurement. However, it shows two contradictory connotations, namely, (1) the production of new materials with outstanding mechanical, optical, electric, and magnetic properties for a wide variety of uses, which is highly positive, and (2) its uncertain effects on human health and the environment, which is highly negative [47].

There are two major deficiencies regarding the impact of nanotechnology. One is that the risks and ethical implications of nanotechnology in the production and industrial domains have not been considered. The other is that, although the potential hazards of nanotechnological products has been described, their toxicity remains uncertain due to the lack of scientific consensus in many cases [48].

Nano-ethics, which is the term used to designate ethics in the development of nanosize objects, comprises three different fields of study (the first two are closely related): ecotoxicity, nanotoxicity, and violation of privacy. Ecotoxicity is related to the potential environmental damage of nanomaterials, whereas nanotoxicity focuses on the effects of nanoparticles on health. Assessing the risks of nanomaterials is especially important in nanomedicine, where they are used to prevent, diagnose, and treat specific diseases [49]. The violation of privacy field is associated with the production of nanosensors. The development of new communication technologies may result in an increasing invasion of privacy and violation of personal rights [47].

Government's concern about this issue is raised, as well as of various prestigious scientific and technical organizations. No doubt that this is a pending issue of nanotechnology, which is the key for future consolidation. Recently, an integral approach to the social responsibility of nanoscience and nanotechnology [47] has been reported, which proposes a framework for conducting responsible research in nanoscience.

3. USE OF NANOPARTICLES AS TOOLS IN ANALYTICAL PROCESSES

3.1 Objectives

One of the main goals of analytical chemistry is the development of new methodologies which improve existing ones and meet new demands of (bio) chemical information posed by the present social and economic problems. The growing demand of (bio) chemical information requires the development of new tangible and intangible tools to support analytical processes. Improvements in analytical processes can be measured in terms of analytical properties: capital, basic, and productivity-related (see Figure 10).

The new scene created by the growth of nanoscience and nanotechnology has had an impact in the field of analytical chemistry. Nanotechnological tools have been used in analytical methods by exploiting the excellent properties of nanoparticles in order to improve well-established analytical methods or to develop others for new analytes or matrices. The use of nanoparticles should lead to improved selectivity, sensitivity, rapidity, miniaturizability, or portability of the analytical system. This facet of analytical nanotechnology is the most extensively explored [3].

The most widely used nanoparticles in analytical sciences include (1) silica nanoparticles, (2) metallic nanoparticles (quantum dots, gold nanoparticles, etc.), (3) carbon nanoparticles (mainly fullerenes and carbon nanotubes), (4) organic polymer nanoparticles (e.g., molecular imprinted polymers), and (5) supramolecular aggregates (nanomicelles, nanovesicles), as depicted in

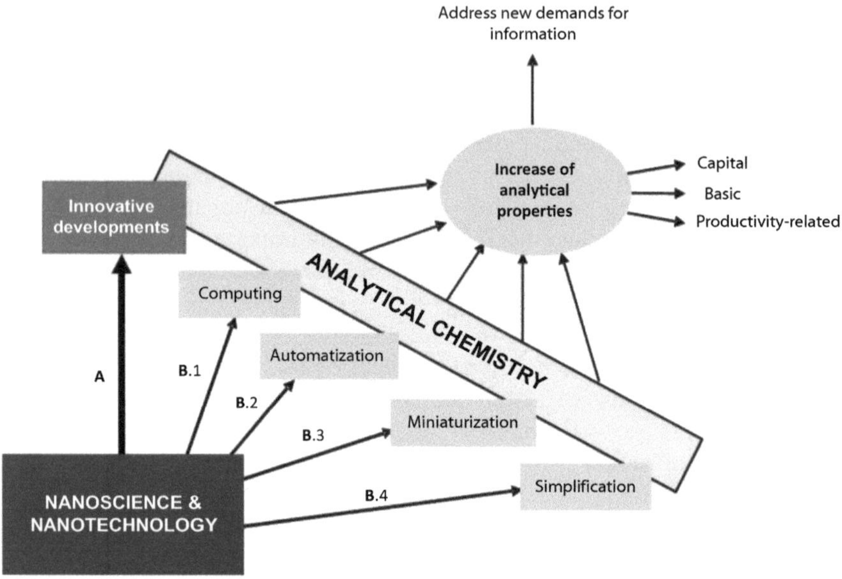

FIGURE 10 The multiple roles of nanotechnology in analytical science: (A) direct impact, and (B) indirect impact through their role as support for other scientific-technical advances.

Figure 11. The explored nanoparticle property can be electrical, optical, thermal, magnetic, or chemical. Frequently, however, two or more properties are explored at once.

An analytical process can be divided into several steps, as represented in Figure 11, namely, sample preparation (including sampling), chromatographic (LC, GC) and capillary electrophoretic techniques, detection and data handling and treatment to offer the results as required. The role of nanoparticles differs between the different steps of the analytical process. Figure 11 shows the proportion of described analytical procedures in which nanoparticles are involved.

3.2 Sample Treatment: Purification and Preconcentration of Analytes

Sample preparation has also profited from the use of nanomaterials, albeit to a lesser extent than detection. The incorporation of nanoparticles in the sample treatment step, in general, helps to simplify it. For example, conventional sorbents for solid-phase extraction and solid-phase micro-extraction have been replaced by nanomaterials. Nanoparticles, according to their participation and role in the sample treatment step can be classified as follows [10]: (1) nanoparticles acting as sorbent agents [50], where direct interaction between the analyte and the nanoparticle takes place; (2) nanoparticles acting as an inert

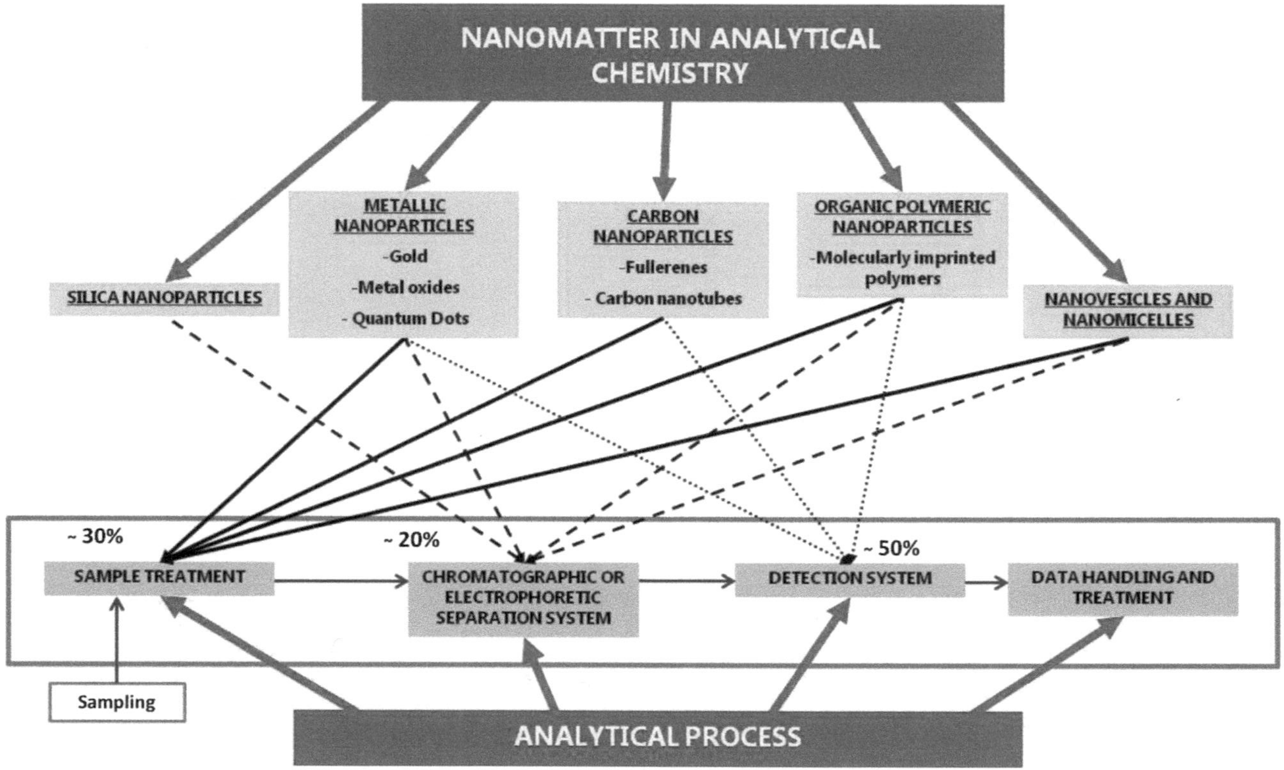

FIGURE 11 The role of nanoparticles in different steps of the analytical process.

support, such as a silica nanoparticle functionalized with a complexation agent; (3) nanoparticles having special magnetic properties, which can either directly adsorb the analyte or can be functionalized with organic groups (the use of a magnetic field can simplify the analytical procedure); and (4) nanoparticles acting as ionization agents for the direct analysis of samples by ion secondary mass spectrometry.

Carbon nanomaterials have become the focus of attention thanks to their singular π-π electron configuration, as well as metal oxides by virtue of their high surface area. In this sense, carbon nanotubes have been widely used as sorbents for solid-phase extraction [51]. Packed nanotubes tend to aggregate to some extent, their inclusion as a nanoscience application requires that aggregation be avoided, and interactions between analytes and isolated nanoparticles be favored [52]. Conical carbon nanoparticles, such as single-walled nanohorns (SWNHs) [53] or carbon nanocones/disks [54] have been also used for solid-phase (micro) extraction. In addition to solid-phase extraction, nanoparticles have been used in other sample treatments such as membrane filtration, for example, with membranes composed of or modified by carbon nanotubes [7,55].

3.3 Improvement of Chromatographic and Electrophoretic Separations

Furthermore, nanostructured materials are being increasingly used in *separation science* as stationary phases in gas and liquid chromatography or electrochromatography, and also as pseudo-stationary phases in electrokinetic chromatography. The use of nano-size materials in electro-driven separation systems has received more attention than that of laminar flow liquid and gas chromatographic (LC and GC) systems [56]. As novel stationary phases and dynamic coatings, NPs greatly enhance separation performance in relation to resolution, selectivity, and efficiency. They display a large surface area-to-volume ratio, which is ideal for low mass-transfer effects in chromatography [57].

So far, nanoparticles have found extensive use in capillary electrophoresis (CE), packed capillary electrochromatography (CEC), open tubular CEC formats, and microchip CE. Several nanoparticles have been employed as pseudostationary phases in capillary electrophoresis [58], such as SWNHs, CNTs, and fullerenes [59] or graphene [60]. A new electrophoretic mode designated as micellar nanoparticle dispersion electrokinetic chromatography (MiNDEKC) has been developed which proved to be effective in the separation of aromatic compounds [61] or even chiral compounds [62].

There are also some reports of the use of nanoparticles in LC, ion chromatography (IC), and GC. Nanoparticles for GC are usually packed into a tubular column. As an alternative to packing, CNTs can be self-assembled into a gas chromatographic column [63]. NPs have been also inserted into a monolithic column in LC [64].

3.4 Improvement of Detection Processes

Detection is the analytical step in which nanoparticles have been most widely used by virtue of their ability to replace conventional materials, as well as the advantages of electrochemical biosensors. In this sense, biomolecules are stabilized by the nanoparticles, which increase active surfaces and facilitate electron transfer. In addition, the development of optical nanosensors has been fostered by the exceptional optical properties of metallic nanoparticles derived from their plasmon resonance, as well as photofluorescent properties in the case of semiconducting nanoparticles (quantum dots) [65], carbon dots [66] (CDs), or graphene quantum dots [67].

Quantum dot-based sensors for chemical and biological detection have been widely developed [68] by virtue of the special optical and electronic properties of the component QDs plus the possibility to functionalize them relatively easily with a wide variety of biological as well as relevant molecules for other important applications [69].

Gold nanoparticles have been also widely used by virtue of their optical properties. For example, mercury traces have been detected by ultraviolet—visible spectroscopy using gold nanoparticles functionalized with complementary DNA sequences, which are intensely colored and are present as colloidal dispersion [70].

Carbon nanotubes and gold nanoparticles are the types of nanoparticles most widely used for developing electrochemical (bio) sensors [71]. CNT-based sensors can be used to detect changes in their electronic properties resulting from the sorption of molecules on their surface. The electronic properties of CNTs encourage their use as electrodes to mediate electron—transfer reactions with electroactive species in solution [72]. Combining gold nanoparticles and carbon nanotubes was found to enhance some electrocatalytic properties of electrodes [73].

4. ANALYSIS OF NANOPARTICLES AND NANOSTRUCTURED MATERIAL

4.1 Information from the Nanoworld

The development of analytical methodologies for extracting quality information from nanoparticles or nanostructured compounds is of high interest due to the poor knowledge available about their actual composition and characteristics, as well as their presence in environmental, biological, or agrifood matrices. Developing effective analytical methods for the rapid, accurate characterization of nanoparticles is mandatory with a view to advancing nanotechnology [3].

This topic has clear connotations of interface between physics and chemistry, as can be inferred from Figure 12, where the physical (size, properties, and topography) and chemical characteristics (composition,

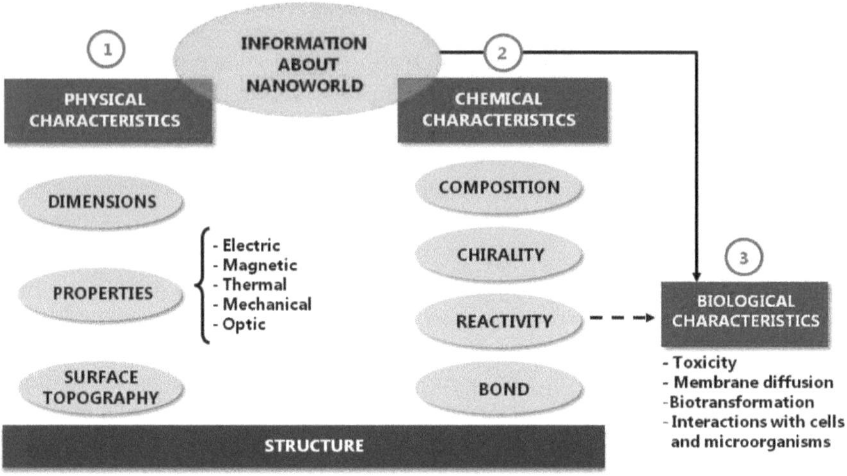

FIGURE 12 Types of information about nanomatter according to the nature of its characteristics.

chirality, reactivity, and type of bond) are reflected. It should be pointed out in this context that there is an increasing interest in biological characteristics (diffusion through membranes, toxicity, biotransformations, nanoparticles interaction with cells and microorganisms) derived from the chemical reactivity of the nanomatter. Among the types of characterization that are needed, perhaps none is more challenging than chemical characterization [74].

Extracting information from the nanoworld entails measuring dimensions of only a few nanometers. So-called nanometrology is, therefore, a hot topic. Although nanometrology is of special concern to physicists, analytical chemists can make substantial contributions to its advancement.

Several types of techniques are available to characterize NPs [75] and their surface chemistry [76]. Herein, the different techniques have been classified as microscopic, spectroscopic, and separation.

4.2 Determination and Characterization of Nanoparticles

As can be seen in Figure 13 characterization encompasses determination, which has analytes as targets. Objects or systems are characterized, being related with knowledge in the primary data-information (provided by analytes)-knowledge hierarchy. In the analytical-chemical context, information is related to qualitative and quantitative results, while knowledge is related with reports. Characterization can be defined as the identification and/or quantification of specific properties or characteristics of the sample or material [77]. On the other hand, determination can be considered as the application of the complete analytical process for detecting, identifying, and quantifying an analytical parameter [78].

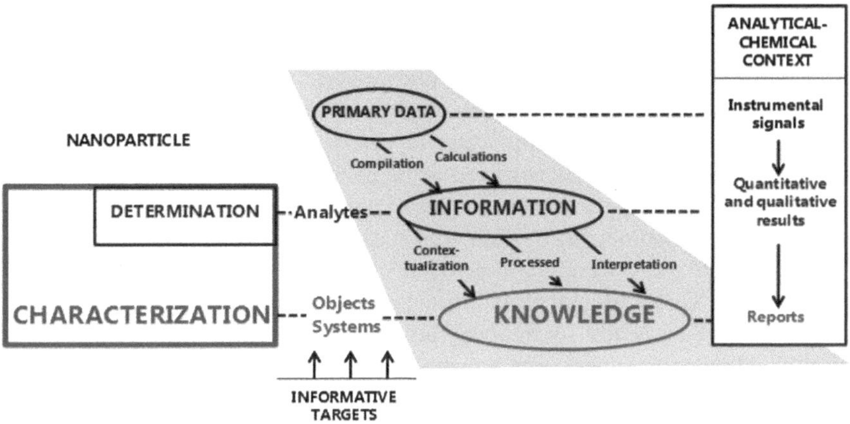

FIGURE 13 Differentiation between characterization and determination in the analytical-chemical context and their relationship with primary data-information-knowledge hierarchy.

4.3 Microscopic Techniques

Microscopic techniques include optical, electron, and scanning probe microscopy. The most used techniques for the characterization of NPs are electron microscopies (EM), transmission electron microscopy (TEM), high-resolution TEM, and scanning electron microscopy (SEM), and scanning probe microscopies (SPM), namely, atomic force microscopy (AFM) and scanning tunneling microscopy (STM). Depending on the technique, resolutions down to the subnanometer range can be achieved. Besides the imaging of nanoparticles, these methods enable the determination of the aggregation, dispersion, size, structure, and shape of NPs. In addition, SEM and AFM offer three-dimensional images of nanoparticles.

However, microscopic techniques present shortcomings in sample preparation as well as the statistical uncertainty due to the human subjectivity implied when deciding which parts of the grid are photographed. In order to measure an accurate size distribution of nanoparticles, it is necessary to count and measure thousands of particles in order to obtain reliable counting statistics. Regarding sample preparation, SEM and TEM have to operate in a vacuum and, therefore, only dry, solid samples can be investigated. The transfer of the sample from dispersion to dried state changes the size distribution, as well as the aggregation state of the nanoparticles, or leads to the precipitation of salts. Furthermore, electron microscopy is usually a destructive method, meaning that the same sample cannot be analyzed twice or by another method for validation. Also, biological samples often need treatment, such as heavy metal staining, for improved contrast. In this case, a scanning transmission electron microscope (STEM) belonging to the group of TEMs can be of use.

In order to image NPs in more natural conditions, environmental scanning electron microscopy [79] (ESEM) has been developed. Although its sample chamber and detector cannot achieve atmospheric pressure, nanoparticles can be visualized under almost natural conditions [80], with residual hydration water still on the particles. This water layer also serves as a conductor on the surface so the sample does not need to be conductively coated [81].

AFM is one of the most common nanometrology methods and has numerous applications [82]. With this technique, it is possible to analyze samples under moist conditions or even in liquids, which affords minimum perturbation. Under liquid conditions, particles can eventually stick to the cantilever, which leads to imaging artifacts. This smearing effect can be minimized by using a noncontact scanning mode where the tip is not touching the nanoparticles but only feels its forces.

When combined with analytical methods, additional information can be gained about the elemental composition of the sample. Electron microscopy methods can be combined to perform energy dispersive X-ray spectroscopy (EDX or EDS) and electron energy loss spectroscopy (EELS) for the elemental analysis of the nanoparticles, whereas selected area electron diffraction (SAED) gives information about the crystalline properties of NPs [83].

X-ray microscopy (XRM) can provide spatial resolution imaging of a specimen in the aqueous state without the need for sample preparation [84]. A variation of the XRM is the scanning transmission X-ray microscopy (STXM), which has been used, for example, to characterize metallic Fe particles for remediation purposes [85].

Moreover, near-field scanning optical microscopy (NSOM), with a resolution between 50 and 100 nm, can be used to image aggregates of nanoparticles [86]. A modification of confocal microscopy—confocal laser scanning microscopy (CLSM)—is able to detect fluorescent samples and can be used to characterize colloids since it is able to image thick samples [80].

4.4 Separation Techniques

Separation techniques such as size-exclusion chromatography (SEC), hydrodynamic chromatography (HDC), capillary electrophoresis (CE), dielectrophoresis (DEP), and field-flow fractionation (FFF) can also be used to characterize NPs. These methods are sensitive and enable further analysis of the sample.

The most employed method is SEC [87], which, combined with detection techniques such as voltammetric, ICP-MS, DLS, and multiangle laser light scattering (MALLS), also called static light scattering (SLS), can be applied to the characterization of AuNPs, QDs, and SWNTs, for example. A drawback of SEC is that irreversible adsorption onto the stationary phase may occur.

Separation and characterization of NPs with hydrodynamic chromatography is based on their hydrodynamic radius. Liquid chromatography coupled with voltammetric detection has been also employed for the separation of metal nanoparticles [88].

Capillary electrophoresis (CE) has emerged as a useful technique, with fewer surface effects than in other techniques (e.g., SEC) and with low consumption of sample and reagents, for the characterization of nanoparticles [89]. CE has been used to separate a variety of differently sized materials, including gold and silver nanoparticles, CNTs, and QDs, among others [90]. Different modalities have been employed such as isoelectric focusing (IEF) or gel electrophoresis.

Dielectrophoresis (DEP) has also emerged as a powerful technique in order to manipulate nanoparticles [90]. Special emphasis is given to the dielectrophoretic separation of CNTs, due to its usefulness in discriminating metallic ones from semiconductor [91], since it is capable of separate species not only by size but also by dielectric constant. However, DEP is further from routine analytical application than gel or capillary electrophoresis.

Field-flow fractionation separates nanoparticles according to particle size in terms of their diffusion coefficients in a very thin, open channel. Separation relies on the combination of an applied field and a longitudinal carrier flow. It has been successfully used to analyze a wide range of NPs [92,93] including SiO_2, CNTs, QDs, gold, and silver.

Other separation techniques used in this field are centrifugation [94], ligand-assisted extraction [95], solid-phase extraction [96], microextraction with ionic liquids [5,8], microfiltration [97], dialysis [98], and cloud-point extraction [99] of different nanomaterials (Table 1).

4.5 Spectroscopic Techniques

Spectroscopic methods also have a wide range of applications for the analysis of nanomaterials. Fluorescence-based techniques are used for the analysis of nanoparticles with a strong native fluorescence, such as quantum dots (QDs), or more recently, carbon dots (CDs) or graphene quantum dots, which can be characterized by their absorption or fluorescence spectra. The position of emission bands can be correlated with particle size and is commonly used to monitor size in the synthesis of QDs [100], for example. In describing the UV−Vis absorption spectra of metal NPs, the term surface plasmon is used, which describes the oscillating electron clouds present at the metal-solution interface. Particle concentrations can be quantified from absorption measurements provided the optical constants for the nanoparticles are known [100,101].

Vibrational spectroscopies, both Raman and infrared, are useful for the analysis of nanoparticles, especially carbon nanoparticles. The two techniques differ in the excitation source used, the way signals are monitored, and the selection rules employed on vibrational modes. At present, Raman spectroscopy is the most important technique for the analysis of carbon nanotubes [102] because Raman measurements are simple, can be done at room

TABLE 1 Specifications of Methods for Analysis and Characterization of Nanoparticles

Method	Approximate Size Range (nm)	Limit of Detection[a]	Level of Sample Perturbation
AFM	0.5 to >1000	ppb-ppm	Medium
Centrifugation	10 to >1000	(d.d.)[b]	Low
Dialysis	0.5–100	(d.d.)[b]	Low
DLS	3 to >1000	ppm	Minimum
Electrophoresis	3 to >1000	ppm	Minimum
EM-EELS/-EDX	Analysis spot size: ~1 nm	ppm	High
ESEM	40 to >1000	ppb-ppm	Medium
FFF	1–1000	(d.d.)[c]	Low
HDC	5–1200	(d.d.)[b]	Low
ICP-MS	Depends on fractionation	ppt-ppb	
LIBS	5 to >1000	Ppt	Minimum
Microfiltration	100 to >1000	(d.d.)[b]	Low-medium
SEC	0.5–10	(d.d.)[b]	Medium
SEM	10 to >1000	ppb-ppm	High
SLS/MALLS	50 to >1000		Minimum
TEM/HR-TEM	1 to >1000	ppb-ppm	High
TEM-SAED	Analysis spot size: 1 nm		High
Turbidimetry/ nephelometry	50 to >1000	ppb-ppm	Minimum
XRD	0.5 to >1000	Dry powder	High

[a]*For comparison mass concentration limit of detection for 100 nm particles are estimated.*
[b]*(d.d.)= detection dependant*
[c]*(UV: ppm, Fluo/ICP-MS: ppb)*
Adapted with permission from Ref. [81].

temperature and under ambient pressure, are quick, nondestructive, and noninvasive [103], and it is possible to find sophisticated laboratory equipment as well as less expensive, portable equipment. The widespread acceptance of Raman spectroscopy relies on its usefulness to provide information about vibrational properties that can be correlated with the structure and electronic properties of the nanotubes. Tip-enhanced Raman scattering (TERS) has been also used for carbon nanotubes [104], for example. In the case of SWNTs,

NIR-fluorescence [105] is a technique widely used for their characterization according to their chirality [106] as well as their quantitative analysis in an aquatic environment [107].

The average bulk chemical composition of a sample can be determined by using spectroscopic techniques such as atomic absorption spectroscopy, inductively coupled plasma with atomic emission spectroscopy (ICP-AES), optical emission spectroscopy (ICP-OES) [108], and mass spectrometry (ICP-MS) [109,110], especially for the analysis of metallic nanoparticles. These techniques feature good limits of detection, providing compositions results which are sample averages, and they also afford multi-element analysis. However, they are destructive and subject to matrix interferences [111]. Mass spectrometry techniques are gaining growing importance thanks to their compatibility with any type of sample, their extremely high sensibility, and easy coupling with separation techniques to obtain real-time information [112].

Light scattering is a very commonly used method to determine particle size. Among the scattering techniques, dynamic light scattering (DLS) is widely used for sizing of NPs and determining their aggregation in suspensions [113]. The advantages of DLS are the rapid and simple operation, readily available equipment, and minimum perturbation of the sample. The limitations are the interpretation and critical review of the data obtained, especially for polydisperse systems [81]. Turbidimetry and nephelometry have been also employed to measure particle concentration [114].

Laser-based methods worth mentioning are the small angle X-ray scattering (SAXS), which is able to characterize mono- and poly-disperse systems, and laser-induced breakdown spectroscopy (LIBS), which has a very low detection limit and is suitable for the size and concentration analysis of colloids. Nuclear magnetic resonance (NMR) spectroscopy [115] is used to determine the dynamics and three-dimensional structure of the samples, whereas X-ray spectroscopy provides crystallographic information which can be used for the characterization of NP surfaces and coatings [116].

4.6 Other Techniques

In addition to those techniques, electrochemical methods [117] of analysis of nanoparticles have been also employed for their analysis such as voltammetry [118,119]. Thermogravimetric analysis (TGA) has been also utilized, for example, as a bulk characterization method for determining carbon nanotube quality after manufacturing [120] and/or after oxidation processes [121].

Nowadays, there is a tendency to combine microscopic and spectroscopic techniques in order to obtain information about both the size and chemical composition of materials. Examples are the combination of AFM-ATR-IR [122], AFM-Raman [123], or AFM-SECM (scanning electrochemical microscopy) [124]. These techniques may also be useful in the case of the analysis of nanomaterials.

4.7 Nanometrology

Nanotechnology constitutes a large industrial sector and is expected to continue to grow at a very fast rate. Such a rapid development has raised the need for accurate, precise control of the dimensions of nanomaterials. Reliable and precise measurement in the nanoscale is a bottleneck for the full development of nanoscience and nanotechnology.

Nanometrology can be defined as the science of measurement at the nanoscale level. Nanometrological measurements include not only length or size (shape, aspect ratio, and size distribution) but also chemical composition, nanoparticle concentrations, and optical, force, mass, electrical, and various other types of properties [125].

The complexity of nanometric measurements arises from the small dimensions of nano-objects, which are usually less than 100 nm in size. Measurements at this scale require a high precision (frequently about 0.1 nm) and, hence, effective methods of measurement. Moreover, a feature of nanomaterials is the exceptional properties they exhibit by virtue of their small size. However, efficiently exploiting such properties requires a sound knowledge of their nature and the ability to characterize the physics or chemistry of very small objects [125].

To measure is to compare, and for this purpose new standards and reference materials are needed, both for instrumental and metrological calibration. There is a lack of reference nanomaterials, only a few size, but not mass or number concentration certified nanoparticle reference materials are available (e.g., NIST reference materials 8011–8013 [126], ERM-FD100 and ERM-FD304 [127], or BAM-N001 [128]). The development of chemical nanometrology requires extensive specific infrastructure, in which standards constitute its Achilles heel.

5. FINAL REMARKS

Analytical nanoscience and nanotechnology constitutes a subdiscipline that is not diverted from general exponential growth of almost all branches of nanoscience and nanotechnology in the second decade of the twenty-first century. Frequently, it is forgotten that the nanometric size is not the key of this growth, but the exceptional physico-chemical properties of nanostructured materials that have conferred them such scientific and technological importance.

In fact, the rapid growth of nanotechnology may alter the natural sequence of any scientific-technical evolution: research + development + innovation (R + D + I). The intermediate step, which is indispensable, is being alarmingly reduced in nanotechnology, with the aim of rapidly introducing into the market the nanotechnological products.

Two key facets of analytical nanoscience and nanotechnology can be defined, (1) the consideration of nanoparticles and nanostructured materials as tools for the innovation and improvement of (bio) chemical measurement processes, which is the aspect most developed to date, and (2) their consideration as objects, with the aim of extracting physico-chemical information from the nanoworld. A third option exists as an interface between both of them, which consists of nanoparticles being employed as tools in analytical processes for the characterization/determination of other nanomaterials.

ACKNOWLEDGMENTS

The authors wish to thank Spain's Ministry of Innovation and Science for funding Project CTQ2011-23790 and Junta de Andalucía for Project FQM4801.

REFERENCES

[1] The Royal Society and the Royal Academy of Engineering, Nanoscience and Nanotechnologies: Opportunities and Uncertainties, The Royal Society, London, UK, 2004.
[2] P. Poole, F.J. Owens, Introduction to Nanotechnology, Hoboken, NJ, Wiley, 2003.
[3] M. Valcárcel, B.M. Simonet, S. Cárdenas, Anal. Bioanal. Chem. 391 (2008) 1881–1887.
[4] A. Zajac, D. Song, W. Qian, T. Zhukov, Colloids Surf. B Biointerfaces 58 (2007) 309–314.
[5] A.I. López-Lorente, B.M. Simonet, M. Valcárcel, Analyst 137 (2012) 3528–3534.
[6] A.I. López-Lorente, B.M. Simonet, M. Valcárcel, B. Mizaikoff, Anal. Chim. Acta 788 (2013) 122–128.
[7] A.I. López-Lorente, M.L. Polo-Luque, M. Valcárcel, Anal. Chem. 85 (2013) 10338–10343.
[8] A.I. López-Lorente, B.M. Simonet, M. Valcárcel, Talanta 105 (2013) 75–79.
[9] A.I. López-Lorente, B.M. Simonet, M. Valcárcel, Front. Biosci. E4 (2012) 1024–1042.
[10] R. Lucena, B.M. Simonet, S. Cárdenas, M. Valcárcel, J. Chromatogr. A 1218 (2011) 620–637.
[11] M. Valcárcel, B.M. Simonet, Anal. Bioanal. Chem. 399 (2011) 1–2.
[12] R.A. Brennen, H. Yin, K.P. Killeen, Anal. Chem. 79 (2007) 9302–9309.
[13] M.G. Schrlau, E.M. Falls, B.L. Ziober, H.H. Bau, Nanotechnology 19 (2008) 15101.
[14] N. Leopold, M. Haberkorn, T. Laurell, J. Nilsson, J.R. Baena, B. Lendl, Anal. Chem. 75 (2003) 2166–2171.
[15] H. Boo, R.A. Jeong, S. Park, K.S. Kim, K.H. Am, Y.H. Lee, J.H. Han, H.C. Kim, T.D. Chung, Anal. Chem. 78 (2006) 617–620.
[16] A. Burns, H. Ow, U. Wiesner, Chem. Soc. Rev. 35 (2006) 1028–1042.
[17] A.B. Descalzo, R. Martínez-Mañez, F. Sancenon, K. Hoffmann, K. Rurack, Angew. Chem. Int. Ed. 45 (2006) 5924–5948.
[18] G. Cao, Nanosctructures and Nanomaterials. Synthesis, Properties & Applicationd, Imperial College Press, London, 2004.
[19] B.L. Cushing, V.L. Kolesnichenko, C.J. O'Connor, Chem. Rev. 104 (2004) 3893–3946.
[20] B.G. Trewing, I.I. Slowing, S. Giri, H.T. Chen, V.S.Y. Lin, Acc. Chem. Res. 40 (2007) 846–853.
[21] J. Eastoe, M.L. Hollamby, L. Hudson, Adv. Colloids Interf. Sci. 128 (2006) 5–15.
[22] http://sis.nlm.nih.gov/enviro/iupacglossary/glossaryn.html (accessed 18.05.13).
[23] A.I. López-Lorente, B.M. Simonet, M. Valcárcel, Anal. Bioanal. Chem. 399 (2011) 43–54.
[24] http://www.nanotec.org.uk (accessed 18.05.13).

[25] A. Daniszewska, W. Lojkowski, H. Fecht, K.J. Kurzydlowski, U. Narkiewicz, G.A. Salishchev, M.J. Zehetbauer, M. Kulczyk, M. Chmielecka, D. Kuzmenko, Solid State Phenomena 114 (2006) 345–391.
[26] L. Li, G. Wu, G. Yang, J. Peng, J. Zhao, J.-J. Zhu, Nanoscale 5 (2013) 4015–4039.
[27] S. Iijima, Nature 354 (1991) 56–58.
[28] D. Ugarte, Nature 359 (1992) 707–709.
[29] B.-T. Zhang, X. Zheng, H.-F. Li, J.-M. Lin, Anal. Chim. Acta 784 (2013) 1–17.
[30] M.H. Ge, K. Sattler, Chem. Phys. Lett. 220 (1994) 192–196.
[31] F. Kolarov, K. Niedergall, M. Bach, G.E.M. Tovar, G. Gauglitz, Anal. Bioanal. Chem. 402 (2012) 3245–3252.
[32] T. Elbayoumi, V. Weissig, J. Biomed. Nanotechnol. 5 (2009) 620–633.
[33] C. Carrillo-Carrión, S. Cárdenas, B.M. Simonet, M. Valcárcel, Chem. Commun. (2009) 5214–5226.
[34] A.-H. Lu, E.L. Salabas, F. Schüth, Angew. Chem. Int. Ed. 46 (2007) 1222–1244.
[35] B. Nowack, T.D. Bucheli, Environm. Pollut. 150 (2007) 5–22.
[36] A. Sánchez, S. Recillas, X. Font, E. Casals, E. González, V. Puntes, Trends Anal. Chem. 30 (2011) 507–516.
[37] M. Valcárcel, B.M. Simonet, Anal. Bioanal. Chem. 399 (2011) 1–2.
[38] T. Laurell, G. Marko-Varga, J. Nilsson, Lab-on-a-chip (2003) 271–284.
[39] K. Cottingam, Anal. Chem. 77 (2005) 51A.
[40] L. Carrascosa, M. Moreno, M. Álvarez, L. Lechuga, Trends Anal. Chem. 25 (2006) 196–206.
[41] J. Riu, A. Maroto, F.X. Rius, Talanta 69 (2006) 288–301.
[42] O.K. Varghese, P.D. Kichambre, D. Gong, K.G. Ong, E.C. Dickey, C.A. Grimes, Sen. Actuators B 81 (2001) 32–41.
[43] NSF, Societal Applications of the Nanoscience and Nanotechnology, US National Science Foundation (NSF), Arlington, VA, 2001.
[44] European Commission, DG Research A. Hullmann, The Economic Development of Nanotechnology. An Indicator Based Analysis, CORDIS, Bruselles, 2006.
[45] M.H.A. Hassan, Science 309 (2005) 65–66.
[46] Lund declaration (2009): http://www.vr.se/download/18.7dac901212646d84fd38000336/ (accessed 30.07.13).
[47] E. Caballero-Díaz, B.M. Simonet, M. Valcárcel, J. Nanopart. Res. 15 (2013) 1534.
[48] G. Hunt, M. Mehta, Nanotechnology Risks, Ethics and Laws, Routledge, Oxford, 2006.
[49] M. Mahmood, D. Casciano, Y. Xu, A.S. Biris, J. Appl. Toxicol. 32 (2012) 10–19.
[50] I. Márquez-Sillero, E. Aguilera-Herrador, S. Cárdenas, M. Valcárcel, J. Chromatogr. A 1217 (2010) 1–6.
[51] B. Suárez, B.M. Simonet, S. Cárdenas, M. Valcárcel, J. Chromatogr. A 1159 (2007) 203–207.
[52] M. Valcárcel, S. Cárdenas, B.M. Simonet, Y. Moliner-Martínez, R. Lucena, Trends Anal. Chem. 27 (2008) 34–43.
[53] J.M. Jiménez–Soto, S. Cárdenas, M. Valcárcel, Anal. Chim. Acta 714 (2012) 76–81.
[54] J.M. Jiménez–Soto, S. Cárdenas, M. Valcárcel, J. Chromatogr. A 1216 (2009) 5626–5633.
[55] A.I. López-Lorente, B.M. Simonet, M. Valcárcel, Anal. Chem. 82 (2010) 5399–5407.
[56] E. Guihen, Trends Anal. Chem. 46 (2013) 1–14.
[57] C. Nilsson, S. Birnbaum, S. Nilsson, J. Chromatogr. A 1168 (2007) 212–224.
[58] C. Nilsson, S. Nilsson, Electrophoresis 27 (2006) 76–83.
[59] Y. Moliner-Martinez, M. Barrios, S. Cárdenas, M. Valcárcel, J. Chromatogr. A 1194 (2008) 128–133.

[60] S. Benítez-Martínez, B.M. Simonet, M. Valcárcel, Electrophoresis 34 (2013) 2561–2567.
[61] B. Suárez, B.M. Simonet, S. Cárdenas, M. Valcárcel, Elecctrophoresis 28 (2007) 1714–1722.
[62] Y. Moliner-Martínez, S. Cárdenas, M. Valcárcel, Electrophoresis 28 (2007) 2573–2579.
[63] C. Saridara, S. Mitra, Anal. Chem. 77 (2005) 7094–7097.
[64] Y. Li, Y. Chen, R. Xiang, D. Ciuparau, L.D. Pfefferle, C. Horvath, J.A. Wilkins, Anal. Chem. 77 (2005) 1398–1406.
[65] L. Wang, W. Ma, L. Xu, W. Chen, Y. Zhu, C. Xu, N.A. Kotov, Mater. Sci. Eng. R 70 (2010) 265–274.
[66] L. Cao, X. Wang, M.J. Meziani, F. Lu, H. Wang, P.G. Luo, Y. Lin, B.A. Harruff, L.M. Veca, D. Murray, S.-Y. Xie, Y.-P. Sun, J. Am. Chem. Soc. 129 (2007) 11318–11319.
[67] S. Zhu, J. Zhang, C. Qiao, S. Tang, Y. Li, W. Yuan, B. Li, L. Tian, F. Liu, R. Hu, H. Gao, H. Wei, H. Zhang, H. Sun, B. Yang, Chem. Commun. 47 (2011) 6858–6860.
[68] Z. Yue, F. Lisdat, W.J. Parak, S.G. Hickey, L. Tu, N. Sabir, D. Dorfs, N.C. Bigall, ACS Appl. Mater. Interfaces 5 (2013) 2800–2814.
[69] W.J. Parak, D. Gerion, T. Pellegrino, D. Zanchet, C. Micheel, S.C. Williams, R. Boudreau, M.A. Le Gros, C.A. Larabell, A.P. Alivisatos, Nanotechnology 14 (2003) 15–27.
[70] J.S. Lee, M.S. Han, ChA. Mirkin, Angew. Chem. Int. Ed. 46 (2007) 4093–4096.
[71] G.A. Zelada-Guillén, A. Tweed-Kent, M. Niemann, H.U. Göringer, J. Riu, F.X. Rius, Biosens. Bioelectr. 41 (2013) 366–371.
[72] M. Trojanowicz, Trends Anal. Chem. 25 (2006) 480–489.
[73] J. Manso, M.L. Mena, P. Yáñez-Sedeño, J. Pingarrón, J. Electroanal. Chem. 603 (2007) 1–7.
[74] M.R. Winchester, R.E. Sturgeon, J.M. Costa-Fernández, Anal. Bioanal. Chem. 396 (2010) 951–952.
[75] R. Zenobi, Anal. Bioanal. Chem. 390 (2008) 215–221.
[76] K.A. Wepasnick, B.A. Smith, J.L. Bitter, D.H. Fairbrother, Anal. Bioanal. Chem. 396 (2010) 1003–1014.
[77] E. Prichard, Analytical Measurement Terminology, LGC/RSC, 2001.
[78] L. Cuadros, L. Gámiz, A. Carrasco, C. Ruiz, Glossary of Analytical Terms, first ed., GRASEQA, 2013.
[79] D.J. Stokes, Philos. Trans. R. Soc. London, Ser. A 361 (2003) 2771–2787.
[80] K. Tiede, A.B.A. Boxall, S.P. Tear, J. Lewis, H. David, M. Hassellöv, Food Addit. Contam. 25 (2008) 795–821.
[81] M. Hassellöv, J.W. Readman, J.F. Ranville, K. Tiede, Ecotoxicology 17 (2008) 344–361.
[82] J.H. Hafner, C.L. Cheung, A.T. Woolley, C.M. Lieber, Prog. Biophys. Mol. Biol. 77 (2001) 73–110.
[83] D. Mavrocordatos, W. Pronk, M. Boller, Water Sci. Technol. 50 (2004) 9–18.
[84] J. Thieme, I. McNulty, S. Vogt, D. Paterson, Environ. Sci. Technol. 41 (2007) 6885–6889.
[85] J.T. Nurmi, P.G. Tratnyek, V. Sarathy, D.R. Baer, J.E. Amonette, K. Pecher, C.M. Wang, J.C. Linehan, D.W. Matson, R.L. Penn, M.D. Driessen, Environ. Sci. Technol. 39 (2005) 1221–1230.
[86] A.D. Maynard, Philos. Trans. R. Soc. Lond. A 358 (2000) 2593–2609.
[87] G.T. Wei, F.K. Liu, J. Chromatogr. A 836 (1999) 253–260.
[88] Y. Song, M.L.A.V. Heien, V. Jimenez, R.M. Wightman, R.W. Murray, Anal. Chem. 76 (2004) 4911–4919.
[89] A.I. López-Lorente, B.M. Simonet, M. Valcárcel, Trends Anal. Chem. 30 (2011) 58–71.
[90] A.I. López-Lorente, B.M. Simonet, M. Valcárcel, Compr. Anal. Chem. 59 (2012) 33–89.

[91] R. Krupke, F. Hennrich, H. v. Löhneysen, M.M. Kappes, Science 301 (2003) 344–347.
[92] F. von der Kammer, S. Legros, T. Hofmann, Trends Anal. Chem. 30 (2011) 425–436.
[93] M. Baalousha, B. Stolpe, J.R. Lead, J. Chromatogr. A 1218 (2011) 4078–4103.
[94] O. Akbulut, C.R. Mace, R.V. Martinez, A.A. Kumar, Z. Nie, M.R. Patton, G.M. Whitesides, Nano Lett. 12 (2012) 4060–4064.
[95] L. Li, K. Leopold, Anal. Chem. 84 (2012) 4340–4349.
[96] B. Suárez, Y. Moliner-Martínez, S. Cárdenas, B.M. Simonet, M. Valcárcel, Environ. Sci. Technol. 42 (2008) 6100–6104.
[97] S. Bandow, A.M. Rao, K.A. Williams, A. Thess, R.E. Smalley, P.C. Eklund, J. Phys. Chem. B 101 (1997) 8839–8842.
[98] S.F. Sweeney, G.H. Woehrle, J.E. Hutchison, J. Am. Chem. Soc. 128 (2006) 3190–3197.
[99] J.-B. Chao, J.-F. Liu, S.-J. Yu, Y.-D. Feng, Z.-Q. Tan, R. Liu, Y.-G. Yin, Anal. Chem. 83 (2011) 6875–6882.
[100] W.W. Yu, L. Qu, W. Guo, X. Peng, Chem. Mater. 15 (2003) 2854–2860.
[101] W. Haiss, N.T.K. Thanh, J. Aveyard, D.G. Ferning, Anal. Chem. 79 (2007) 4215–4221.
[102] M.S. Dresselhaus, G. Dresselhaus, A. Joiro, J. Phys. Chem. C 111 (2007) 17887–17893.
[103] A. Jorio, M.A. Pimenta, A.G. Souza Filho, R. Saito, G. Dresselhaus, M.S. Dresselhaus, New J. Phys. 5 (2003) 1.1–1.17.
[104] N. Hayazawa, T. Yano, H. Watanabe, Y. Inouye, S. Kawata, Chem. Phys. Lett. 376 (2003) 174–180.
[105] R.B. Weisman, Anal. Bioanal. Chem. 396 (2010) 1015–1023.
[106] S. Maruyama, Y. Miyauchi, Y. Murakami, S. Chiashi, New J. Phys. 5 (2003) 149.1–149.12.
[107] A. Schierz, A.N. Parks, K.M. Washburn, G.T. Chandler, P.L. Ferguson, Environ. Sci. Technol. 46 (2012) 12262–12271.
[108] S. Elzey, D.-H. Tsai, S.A. Rabb, L.L. Yu, M.R. Winchester, V.A. Hackley, Anal. Bioanal. Chem. 403 (2012) 145–149.
[109] A. Scheffer, C. Engelhard, M. Sperling, W. Buscher, Anal. Bioanal. Chem. 390 (2008) 249–252.
[110] R. Allabashi, W. Stach, A. de la Escosura-Muñiz, L. Liste-Calleja, A. Merkoçi, J. Nanopart. Res. 11 (2009) 2003–2011.
[111] B.M. Simonet, M. Valcárcel, Anal. Bioanal. Chem. 393 (2009) 17–21.
[112] A.R. Montoro-Bustos, J. Ruiz-Encinar, A. Sanz-Medel, Anal. Bioanal. Chem. 405 (2013) 5637–5643.
[113] A. Ledin, S. Karlsson, A. Ker, B. Allard, Water Res. 28 (1994) 1539–1545.
[114] J.A. Sene, M.V.B. Pinheiro, K. Krambrock, P.J.S. Barbeira, Talanta 78 (2009) 1503–1507.
[115] D.J. Nelson, C.N. Brammer, Anal. Bioanal. Chem. 396 (2010) 1079–1086.
[116] M. Farré, K. Gajda-Schrantz, L. Kantiani, D. Barceló, Anal. Bioanal. Chem. 393 (2009) 81–95.
[117] Y.-G. Zhou, N.V. Rees, R.G. Compton, Angew. Chem. Int. Ed. 50 (2011) 4219–4221.
[118] D. Hernández-Santos, M.B. González-García, A. Costa-García, Electrochim. Acta 46 (2000) 607–615.
[119] M. Pumera, M. Aldavert, C. Mills, A. Merkoçi, S. Alegret, Electrochim. Acta 50 (2005) 3702–3707.
[120] E. Mansfield, A. Kar, S.A. Hooker, Anal. Bioanal. Chem. 396 (2010) 1071–1077.
[121] L.S.K. Pang, J.D. Saxby, S.P. Chatfield, J. Phys. Chem. 97 (1993) 6941–6942.
[122] M. Brucherseifer, C. Kranz, B. Mizaikoff, Anal. Chem. 79 (2007) 8803–8806.

[123] C.E. Harvey, E.M. v.S. Lantman, A.J.G. Mank, B.M. Weckhuysen, Chem. Commun. 48 (2012) 1742–1744.

[124] J. Wiedemair, J.-S. Moon, D.E. Eaton, B. Mizaikoff, C. Kranz, IFMBE Proc. 25 (2010) 372–375.

[125] R. Herrera-Basurto, B.M. Simonet, Encyclopedia of Analytical Chemistry, John Wiley and Sons, 2013. http://dx.doi.org/10.1002/9780470027318.a9177.

[126] National Institute of Standards & Technology (NIST), http://www.nist.gov/mml/csd/inorganic/goldnanorms.cfm (accessed 10.01.14).

[127] Institute for Reference Materials and Measurements, http://irmm.jrc.ec.europa.eu/reference_materials_catalogue/Pages/index.aspx (accessed 12.02.14).

[128] Federal Institute for Materials Research and Testing. http://www.bam.de/en/fachthemen/referenzmaterialien/rm_aktuell.htm (accessed 12.02.14).

Chapter 2

Synthesis of Gold Nanoparticles

Han-Wen Cheng, Zakiya R. Skeete, Elizabeth R. Crew, Shiyao Shan, Jin Luo and Chuan-Jian Zhong*
Department of Chemistry, State University of New York at Binghamton, Binghamton, NY, USA
**Corresponding author E-mail: cjzhong@binghamton.edu*

Chapter Outline

1. INTRODUCTION

Gold nanoparticles less than ~100 nm in size are perhaps one of the most extensively studied systems in the last decade, as reflected by a surge of interests not only in the field of analytical chemistry but also in nanoscience and nanotechnology in general. This stems largely from the unique optical, electrical, catalytic, and molecular recognition properties of nanoparticles in different sizes, shapes, and assemblies, which can be exploited for a wide

Gold Nanoparticles in Analytical Chemistry. http://dx.doi.org/10.1016/B978-0-444-63285-2.00002-X

range of technological applications [1]. This chapter focuses on the synthesis of gold nanoparticles, with emphasis on understanding the growth mechanism in terms of sizes and properties and their relevance to analytical chemistry. Historically, the synthesis of gold nanoparticles traces back to ancient times (many centuries ago) when the synthesis of gold colloids was used for color decoration in staining glasses. It was until about 100 years ago when Michael Faraday performed the first scientific experiment in the synthesis of pure gold colloids by chemical reduction of a solution of gold chloride [2]. However, there has been a surge of interests in the synthesis of gold nanoparticles in the last two decades, centering on the ability to control the size, shape, structure, and surface properties of gold nanoparticles for a wide range of nanoscience and nanotechnology. Since there have been numerous reviews covering this subject area [3], especially in the last few years [4–7], this chapter has no intention to provide a comprehensive review on the subject of synthesis of gold nanoparticles, rather, will focus on selected recent studies from the authors′ laboratories in terms of understanding the size and surface control.

For the synthesis of gold nanoparticles in a solution (wet-chemical synthesis method), the general approach involves the chemical reduction of chloroauric anions ($AuCl_4^-$) as the gold precursor. The Turkevich method reported in 1951 [8,9], i.e., citrate reduction of $AuCl_4^-$ in aqueous solution, and the Brust method reported in 1994 [10], i.e., borohydride reduction of $AuCl_4^-$ in an organic solution transferred from an aqueous solution by a phase-transfer agent (tetraoctylammonium bromide (TOABr)), are perhaps two of the most well-known methods for the synthesis of gold nanoparticles in terms of size control (Figure 1). In general, the reduction of $AuCl_4^-$ by a reducing agent

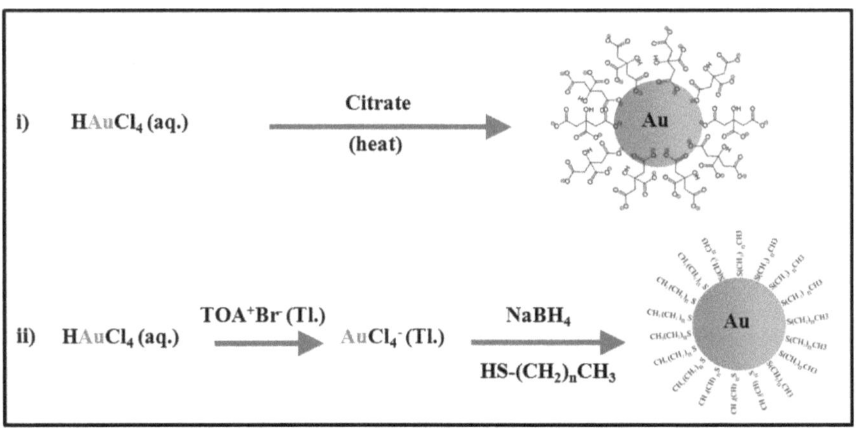

FIGURE 1 Illustration of two synthesis methods: (i) citrate reduction of $AuCl_4^-$ in aqueous solution and (ii) borohydride reduction of $AuCl_4^-$ in organic solution transferred from aqueous solution by a phase-transfer agent (tetraoctylammonium bromide, TOABr).

produces neutral gold atoms, which undergo nucleation and growth processes to form gold nanoparticles (AuNPs) in the presence of the capping molecules or ions (e.g., citrate ($C_3H_5O(CO_2)_3^{3-}$)) or alkanethiols ($HS-(CH_2)_nCH_3$), which protect the nanoparticles from further growth or aggregation.

Because the size, shape, and surface properties of Au NPs resulting from the wet-chemical synthesis processes depend on a combination of parameters with different weights, including speed of reduction, temperature, concentration, stirring condition, capping molecules, etc., the ability to achieve a controlled synthesis has been a constant topic of current research. In addition to modifications or refinements of the two well-known methods, there have been many new methods for the control of size, shape, and surface properties for Au NPs [4–7], including recent efforts in the synthesis of gold nanoclusters with 25 or less Au atoms [11]. For example, the modification of Turkevich method by introducing $NaBH_4$ at room or lower temperature could lead to smaller sized Au NPs (<3–5 nm). Smaller sized Au NPs (3 or 12 nm) could be used as seeds for further reduction of $AuCl_4^-$ to form larger sized particles [4].

In general, the use of wet-chemical processes to synthesize Au NPs is a "bottom up" approach [5], in contrast to a "top down" approach where bulk gold is broken down by a strong ion irradiation or arc force. Various functional Au NPs are prepared by two-step process: (1) nucleation and (2) successive growth. For size-controlled growth of spherical Au NPs, seed-mediated growth of Au NPs is carried out under the mild condition of reducing agent and Au precursor [4,12,13]. The control of the experimental conditions is important to ensure that the Au atoms formed by the reduction reaction are deposited on the surface of the seed particles without further nucleation. In view of the finding that the fast addition of reducing agent favors nucleation whereas the slow addition of reducing agent suppresses nucleation, a step-by-step addition of reducing agent is found to be more effective than a one-step seeding process. Shape-controlled growth of Au NPs often involved adding additional surfactants such as cetyltrimethylammonium bromide (CTAB), sodium dodecylsulfonate, and poly(vinylpyrrolidone) (PVP) [13–15]. Examples include Au nanorods, the formation of which depends on many control parameters such as concentration, seed particle size, temperature, pH, etc. [16–18], and many other shapes (e.g., nanobelts, nanoprisms, octahedron, etc.). Different surface passivation and in situ functionalization have also been used for the control of the surface properties [19].

For the synthesis of Au NPs in organic phase [6], a key process in the Brust–Schiffrin Method for thiol-protected Au NPs, reported first in 1994, is to transfer $AuCl_4^-$ from aqueous phase to toluene using the TOABr where it is reduced by sodium borohydride ($NaBH_4$) in presence of dodecanethiol [10]. The resulting particles are in the range of 2–5 nm. A further modification of this method by manipulating the reaction temperature and other parameters was reported by Murray et al. demonstrating the ability to control size down to about 1 nm with relatively high monodispersity [20]. There are also many other modifications in terms of introducing different capping ligand structures [21–26].

Some of the recent studies of the synthesis of Au NPs have been thoroughly reviewed by Murphy and El-Sayed [7] in terms of size, shape, and surface chemistry. Examples include phosphine-stabilized gold particles produced from the reduction of PPh_3AuCl by diborane in benzene, yielding $Au_{55}(PPh_3)_{12}Cl_6$ (1.4 ± 0.4 nm diameter) [27], gold nanorods based on seeded growth of small (ca. 1.5 nm diameter) single-crystal seed particles by borohydride reduction in the presence of CTAB, and ascorbate reduction in the presence of Ag^+ and CTAB [16,28] and gold nanorods by other methods [29,30], gold nanocages and nanoframes [31,32], hollow gold nanospheres [33], gold tetrahedra, cubes, octahedra, and icosahedra [34,35], gold nanocubes [36], and gold nanostars [37]. For spherical Au NPs, a number of synthetic methods have been reported for the preparation using seeds in aqueous solutions [38–46]. For example, Au NPs with size ranging from 5 to 40 nm were synthesized using Au seeds prepared by borohydride reduction of Au salt in the presence of citrate as a capping agent, followed by growth in the presence of CTAB and ascorbic acid [12,38]. Other examples involve the synthesis of Au NPs of 10–20 nm sizes at elevated temperature using acrylate as both reducing and capping agent [39,40], and photochemical seeded growth of Au NPs of 5–100 nm sizes in the presence of TX-100 (Triton) X-100 Poly-(oxyethylene)iso-octylphenyl ether, and ascorbic acid [41]. Seeding methods have often been used for growing larger sized Au NPs (20–100 nm) in the presence of ascorbic acid or hydroxylamine [4,17,29,42–49]. In the seeded growth, the basic perception for the growth mechanism was that Au(III) was reduced on the surface of a smaller sized nanoparticles by citrate or hydroxylamine as reducing agent at room temperature. This type of growth mechanism has been considered for synthesis of Au nanorods (20–100 nm) in the presence of trace silver ions [17,29].

The importance of the development of refined or new synthesis methods for making Au NPs stems from the rapid growth of interest in exploration of Au NPs as building blocks toward multifunctional materials. This requires not only the ability to tune the size, shape, and composition at the nanoscale but also the ability to engineer the interparticle properties as multifunctional assemblies for chemical and biological recognition, sensing, imaging, etc. [50]. Two of the important questions in such applications include how the unique nanoscale properties of building blocks are retained in the assembly, and how the nanostructured assembly can be engineered to display the targeted functional properties. In comparison with studying nanoscale chemistry at individual nanoparticle level, a challenge to the exploration of an ensemble of nanoparticles is the understanding of the interparticle interactions of the nanoscale building blocks either in dispersed state (e.g., in solutions) or in solid state (e.g., in thin films). Molecularly mediated assembly of nanoparticles through size-controlled assembly of nanoparticles via molecules or biomolecules with defined sizes and structures can be produced in the format of solution-dispersed and finite (e.g., submicron) dimensions to display collective properties [51]. Thin-film assembly of nanoparticles using multifunctional molecular mediators is another approach for the creation of multifunctional and collective properties of the nanoscale building blocks in continuous and macroscopic dimensions. The exploration of nanoparticle-structured molecular recognition is an important front of many emerging chemical and biological sensor technologies [51–60]. Important attributes of metal nanoparticle-structured sensing materials include interparticle electronic conduction and framework affinity, large surface area-to-volume ratio, multidentate ligating microenvironment, and molecularly defined nanoporosity. The nanoparticle-structured sensing materials tunable by size, shape, composition, and spatial properties [61] not only sequester small molecules into the interior with porous molecular access or specific binding, but also have the advantage of a high surface activity due to the predominance of interfacial phenomena. Indeed, the use of Au NPs as building blocks toward functional materials has generated tremendous interests in microelectronics, catalysis, sensor, medicine, etc. [61]. Some of these attributes have been extensively discussed in several recent reviews covering topics ranging from electronic properties of nanoparticle assemblies to chemical sensors [59,60]. In addition to the viability of charging a single electron on a single nanoparticle or hopping/tunneling electrons in a collective ensemble of nanoparticles as highly sensitive materials, the enrichment of ligands and voids in a high surface area-to-volume ratio microenvironment plus a 3D assembly provides framework nanoporosity for signal amplification.

In addition to shape control as highlighted above, the control of size and surface properties of Au NPs is perhaps the most basic element to any exploration of Au NPs as building blocks in the field of analytical chemistry.The formation of Au NPs involves two basic processes, nucleation and growth, which require a good understanding of the mechanistic details in order to establish the controllability of size, shape, and surface properties. As described in the next two sections, some of the important findings in our studies of the synthesis of Au NPs in aqueous and organic solutions, respectively, demonstrate that the understanding of the growth mechanistic details is essential for harnessing the unique nanoscale properties of Au NPs.

2. SYNTHESIS OF GOLD NANOPARTICLES IN AQUEOUS PHASE

2.1 Seeded and Aggregative Growth

For the synthesis of Au NPs in aqueous phase, seeded growth method have been reported for controlling particle size, most of the methods have been based on the modification of Turkevich method through a two-step process (nucleation and growth). For example, under the mild condition of reducing agent and Au precursor, smaller sized Au NP (3 or 12 nm) was first formed by reduction using $NaBH_4$ as seeds followed by further reduction of $AuCl_4^-$ to form larger sized particles [4]. This process, while effective in particle size growth, depends heavily on the speed and timing for the addition of reducing agent for the control of size monodispersity, which is often difficult. To understand the issues related to controllability, a highly effective method was developed in our laboratory for the growth of highly monodispersed Au NPs in the size range of 10–100 nm diameter [62]. This method involves "aggregative growth mechanism" for the controlled growth of Au NPs, where the intermediate process involves the coalescence of the smaller sized particles on the larger sized particles. Using $AuCl_4^-$ as a starting Au precursor and acrylates as both reducing and capping agents, the aqueous seed formation and seeded growth are shown to produce highly monodispersed Au NPs in the size range of 10–100 nm diameters.

In addition to considering the reduction of Au(III) on the surface of Au seeds by a reducing agent for the growth of larger sized particles in most of the previous reports, traditionally Ostwald ripening has often been considered to be another general driving force for the growth of particle sizes. In Ostwald ripening, smaller particles dissolve preferentially with subsequent crystallization onto larger particles [63], which involves nucleation and growth processes for the growth of larger sized particles from smaller sized particles. However, little has been known for addressing the fundamental question on how the formation of larger sized nanoparticles correlates with the initial, intermediate, and final size distributions in terms of the growth parameters. For initial and final size distributions, our earlier and recent studies demonstrated that the interparticle aggregative coalescence was an important factor operative for the size growth of gold [64] and copper nanoparticles [65] under thermally activated growth processes in organic or molten solvents. Buhro et al. [66] have carried out a detailed experimental and theoretical study of the thermal coarsening kinetic process of thiolate-capped Au NPs in the presence of TOABr, which led to a major new insight into the aggregative nucleation and growth mechanism. The aggregation parameters are shown to control the final particle size and size distribution, which are substantiated by the observation of polycrystalline particles, the observation of early-stage bimodal size distributions, the observation of the sigmoidal growth kinetics, and the close correlation of the nucleation parameters with the particle sizes and distributions. This insight also ruled out the operation of Ostwald ripening at the early stage of the particle growth.

Insights into the growth mechanism were gained by determination of the size and optical properties as a function of the growth parameters such as the reaction time and the seed/precursor concentrations [67,68], demonstrating the operation of the aggregation growth mechanism in this simple and reproducible growth process. For example, Au NPs with average sizes of 60 and 90 nm can be easily grown using 30 and 60 nm particles as seeds, respectively (Figure 2(A)). The size of the nanoparticles can be obtained over 10–100 nm range, whereas the monodispersity for the nanoparticles is very high. While the overall morphology of these nanoparticles appears spherical, faceted edges along the outlines of many nanoparticles can be observed upon close examination, which is consistent with the polycrystalline properties of the Au NPs. The size and monodispersity of these nanoparticles are further evidenced by the size distributions of the nanoparticles (Figure 2(B)). In addition

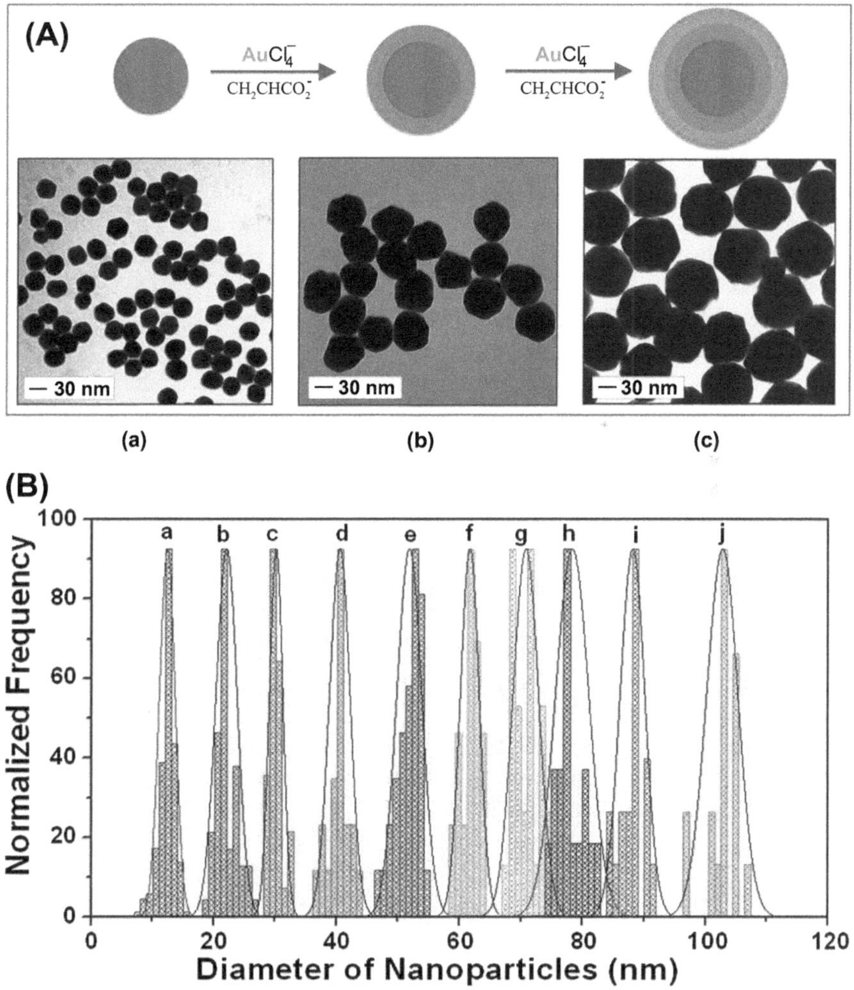

FIGURE 2 (A) TEM micrographs for 30 nm seeds (a), and the seed-grown 60 nm (b) and 90 nm (c) Au NPs. (B) Size distributions of Au NPs of different sizes. *Reproduced with permission from Ref. [62]. Copyright 2010 American Chemical Society.*

to showing that the average sizes are controllable in the range of 10–100 nm, these distribution curves exhibited remarkably high monodispersity, with an average standard deviation of ±1.6 nm [67]. The relative standard deviations (RSD) are ~8% for the smaller sized (<30 nm), ~4% for the medium-sized (30–50 nm), and ~3% for the larger sized (>50 nm) nanoparticles.

The size–concentration relationship was assessed by considering a theoretical model based on an initial assumption of 100% conversion efficiency for the seeded growth [62]. On the basis of a spherical model for a Au NP with radius r (cm) and growth thickness d (cm), the mass balance between the seed-grown particles and the concentration of $AuCl_4^-$ (C_{mM}) yields the following d–r correlation:

$$\frac{d}{r} = \sqrt[3]{1 + 2.49 \times \frac{C \times V}{N \times r^3}} - 1 \tag{1}$$

where N is the number of Au NPs in the total volume; C is the concentration of Au precursor (mol/cm^3); and V is the total reaction volume. Note that this equation is consistent with those reported previously [41,69], where the final diameter of the particles is related to the diameter of seeds and concentration of Au used.

In Figure 3, the calculated and the theoretical fitting results are compared with the experimental data. While values for the reaction volume (V_{expt}), and the particle radius (r_{expt}) can be experimentally determined, the actual (or effective) number of nanoparticle seeds that have participated in the seeded growth (N_{eff}) may not be 100% of the initial number of seeds. Considering this factor for the evaluation of the efficiency for the seeded growth based on values of V_{expt}, r_{expt}, and $N_{initial}$, a coefficient,

$$\alpha = 2.49 \times \frac{V_{expt}}{N_{eff} \times r_{expt}^3} \tag{2}$$

was introduced. For example, for growing particles of 32–61 nm using 32.8 nm seeds (1.1×10^{11} NPs/mL), the addition of 25 mL seeds in a 125 mL synthesis solution (i.e., 2.8×10^{12} NPs) would theoretically yield $\alpha = 23.3$, which corresponds to the solid line in Figure 3(A)(a). By fitting the experimental data in the full concentration range and in the concentration range of <0.1 mM, i.e., the dashed lines in Figure 3(A)(b) and B(e), the fitting results yielded $\alpha = 33.2$ and 40.6, respectively. In comparison with the theoretical $\alpha = 23.3$, there are two important findings: (1) $N_{eff} = 0.85 \times 10^{11}$ NPs/mL seeds (i.e., 70%) are effectively grown to larger size particles for the full concentration range, and (2) $N_{eff} = 0.69 \times 10^{11}$ NPs/mL seeds (i.e., 57%) are effectively grown to larger size particles for the <0.1 mM concentration range. In the case of using 61 nm seeds (2.4×10^{10} NPs/mL) to grow sizes up to 100 nm, the use of 25 mL seeds (i.e., 6.1×10^{11} NPs) in a 125 mL synthesis solution would yield $\alpha = 18.0$, which corresponds to the solid line in

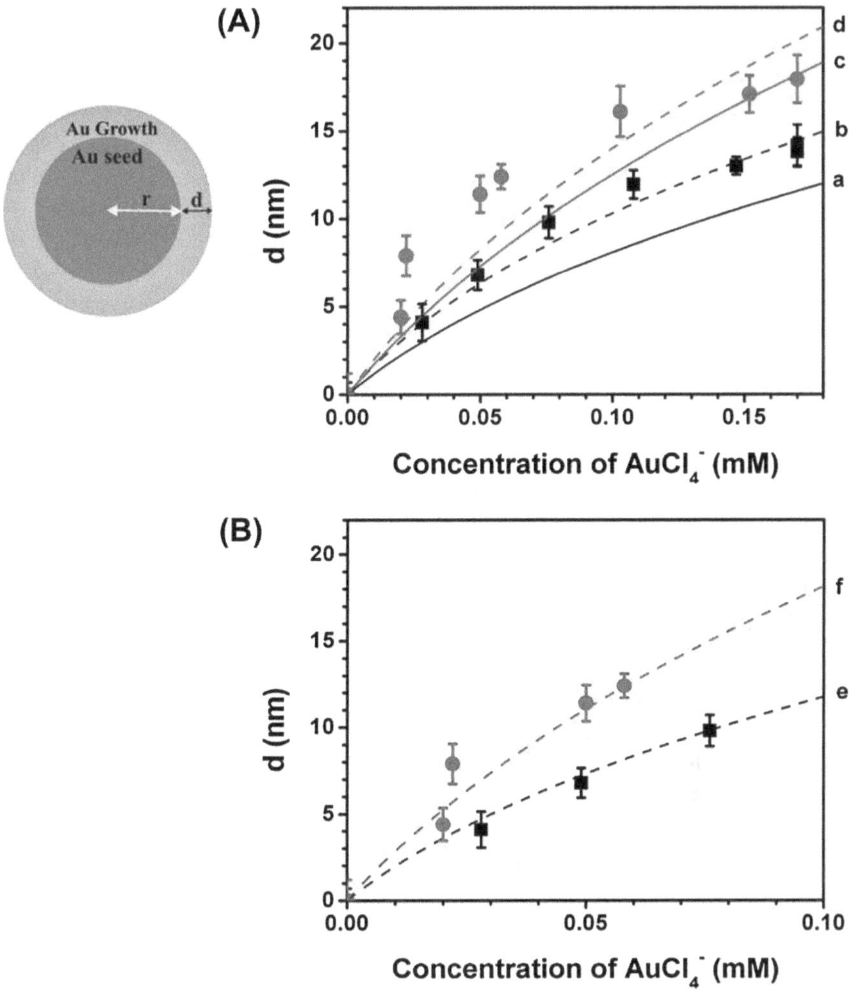

FIGURE 3 Comparison between theoretical (A) curves a and c and experimental (A) curves b and d; (B) curves e and f correlation between the growth thickness and the concentration of $AuCl_4^-$ for using 32-nm and (a, b, and e) and 61-nm sized particles (c, d, and f) as seeds. (A) curve fitting to the full concentration range (dashed lines); (B) fitting to the concentration range of <0.1 mM (dashed lines). Solid lines represent theoretical lines based on Eqn (1). *Reproduced with permission from Ref. [62]. Copyright 2010 American Chemical Society.*

Figure 3(A)(c). By fitting the experimental data in the full concentration range and in the concentration range of <0.1 mM, i.e., the dashed lines in Figure 3(A)(d) and B(f), the fitting results yielded $\alpha = 21.0$ and 30.5, respectively. In comparison with the theoretical $\alpha = 18.0$, the value of N_{eff} is derived as 2.07×10^{10} NPs/mL (i.e., 86%) for the entire concentration range, and as 1.43×10^{10} NPs/mL (i.e., 59%) for the concentration range of <0.1 mM.

The seed formation and the seeded growth process can also be monitored in terms of the surface plasmon resonance (SP) band characteristics and the sizes of the resulting nanoparticles. The Avrami's theoretical model for crystallization and growth [70], which was shown to be useful for assessing the crystallization kinetics of macromolecules or polymers [71,72], was considered in our initial data analysis. This model was also previously found viable for describing the formation of thin-film assemblies of Au NPs mediated by thiol-based linker molecules [54]. The absorbance (A) of the SP band can thus be related to the volume fraction of the nanoparticles grown in the solution as a function of the reaction time (t),

$$A = C\left(1 - e^{-kt^n}\right) \tag{3}$$

where k is the apparent rate constant, n is the critical growth exponent, and C is the proportionality constant. Based on monitoring the SP band (~524 nm) as a measure of the formation of gold seeds (31.2 nm), the fitting result to the experimental data yielded an apparent rate constant $k = 1.45 \times 10^{-10}$ (the unit is hour^{-n} for all rate constants) with a critical growth exponent $n = 6.5$. While there are subtle differences in k and n, the results are indicative of the controllability of the particle growth by controlling a combination of reaction time and seed size. The size growth of the nanoparticles was also determined by sampling particles at different times according to the kinetic change. The fitting result of the particle size data using Eqn (3) yielded an apparent rate constant of 1.8×10^{-9} with a critical growth exponent of 6.0, which are quite close to the result obtained from the fitting the optical data ($k = 1.1 \times 10^{-9}$ ($n = 6.0$)).

To gain further insights into the growth mechanism, the kinetics of the size growth was followed systematically by measuring both SP band absorbance and the particle sizes at a series of specified reaction times. Figure 4(A) shows the representative set of data for the kinetics of the particle size growth. Indeed, the kinetic trend for the particle size growth is similar to the SP band absorbance data, both displaying a similar sigmoidal function. The fitting result of the data using Eqn (3) yielded $k = 3.6 \times 10^{-4}$ and $n = 2.5$. In general, the kinetics for the size growth of the nanoparticles under constant initial concentrations of the seed and $AuCl_4^-$ is characteristic of a sigmoidal shape.

On the basis of the work by Buhro et al. [66], there are several basic criteria for the operation of the aggregative growth mechanism, including (1) the polycrystalline particles, (2) the early-time bimodal size distributions, (3) the sigmoidal growth kinetics, and (4) the close correlation of the nucleation parameters with the particle sizes and distributions. By examining the detailed sizes and morphologies of the nanoparticles obtained at earlier and final stages of the size growth, bimodal distributions were indeed observed. By further examining the detailed morphologies of the nanoparticles grown from this seeded growth process, the observation of the multiple crystal facets and boundaries is

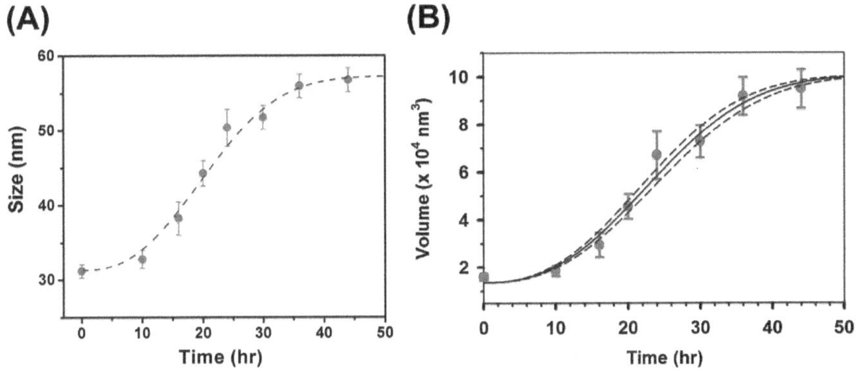

FIGURE 4 (A) The kinetic plot for particle size growth from 31 nm seeds. The curve fitting is based on Eqn (3), producing an apparent rate constant of $k = 3.6 \times 10^{-4}$ ($n = 2.5$). (B) Theoretical fitting based on Eqn (4) to the kinetics for the size growth data. Solid line represents the theoretical fit to the experimental data with k_g value of $2.56 \times 10^{-4}\,h^{-n}$ and n value of 2.5. The dashed upper and lower dashed lines represents the theoretical curves as a result of a change in k_g by +10% and −10% (both using the same value of n). *Reproduced with permission from Ref. [62]. Copyright 2010 American Chemical Society.*

clearly indicative of the polycrystalline nature for the as-grown nanoparticles as a result of the interparticle coalescence, which serves as an important piece of evidence in support of the "aggregative growth mechanism". The presence of the crystalline domains in the nanoparticle is similar to the observation reported recently for Au NPs grown from thermal evolution [66].

A further analysis of the correlation of nucleation parameters with the particle sizes and distributions was carried out based on the theoretical and experimental work by Buhro et al. for a size growth condition with an elevated temperature and organic solvents [66]. It must be emphasized that this analysis was preliminary at this stage because of the limited data sets as shown in Figure 4(A). To assess the viability of the model of the aggregative growth mechanism, the diameters of the formed particles at different intervals (Figure 4(A)) were converted to volume distributions by assuming spherical morphologies. In the sigmoidal size (volume) versus time profile for nanoparticle growth, the initial induction-like period is associated with nucleation, which is the formation of critical aggregates in the case of aggregative growth. The regime is followed by an active growth regime, in which the critical aggregates grow by aggregative coalescence with primary nanoparticles, until the primary nanoparticles are consumed. This type of growth was examined by plotting the particle volume at time (V_t) against the growth time,

$$V_t = V_i + \left(V_{lim} - V_i\right)\left(1 - \exp\left(-k_g t^n\right)\right) \quad (4)$$

where k_g is the growth-rate parameter, n is the Avrami exponent, and V_i is the primary nanoparticle mean volume. By fitting the data using Eqn (4), the fitting results yielded k_g value of $2.56 \times 10^{-4}\ h^{-n}$ and n value of 2.5. As shown in Figure 4(B), the kinetic data were compared between the experimental and the theoretical fitting results. The two additional curves, one in which k_g was increased by 10% from the original value, and the other in which k_g was decreased by 10% (both using the same value of n), demonstrate that the fits to the initial induction period, associated with nucleation, were not affected by the variations in the k_g value. However, the fits in the rising portion of the sigmoidal data, the active growth regime, are strongly affected by the variations in the k_g value. In addition, the fraction of nanoparticles with a given volume was further analyzed as a function of the volume of the formed nanoparticles for the estimates of the nucleation and aggregation parameters. The maximum aggregative nucleation rate Γ_{max} and the time taken for reaching that rate τ_n were determined from the nucleation function scaled as $\Gamma(t)$ [66]. The results demonstrated the feasibility of this theoretical approach for developing the correlation of the nucleation parameters with the particle sizes and distributions. This "seeded and aggregative growth" mechanism can be described by a quantitative correlation between the size growth and the seed/precursor concentrations for the precise control of the size of the nanoparticles.

2.2 Optical and Spectroscopic Properties of Au NPs in Correlation with Particle Size

Studies of the optical properties of Au NPs have shown that a change in absorbance or wavelength of the SP band provides a measure of particle size, shape, concentration, and dielectric medium properties [73]. For example, the study of the optical properties of Au NPs prepared by reverse micelles showed a red shift with the size of gold colloids, which is expected from Mie theory. The enhancement of surface-enhanced Raman scattering (SERS) intensity was shown to be tunable by varying the particle size, particle−substrate separation, and substrate material. Au NPs of various shapes and sizes self-assembled on gold have been shown to create a large localized electromagnetic field enhancement. Gold nanoshells have been shown to provide SERS enhancement comparable to gold nanosphere dimers. In recent studies of molecularly mediated assemblies of Au NPs [74,75], the SP band and the SERS properties were shown to display clear dependence on the size and interparticle spatial properties of the nanoparticle assemblies.

The size monodispersity of the Au NPs permitted an explicit comparison between the maximum wavelengths of the SP bands measured and the theoretical ones using the highly monodispersed nanoparticles (10−100 nm), which is important especially for the consideration of using the maximum wavelength as a measure of the particle sizes in this size range [67]. Optically,

Au NPs in solutions are well known to display visible colors and SP bands in the visible spectral region. From 30 to 90 nm, the color exhibits a smooth transition from dark red to pink. Figure 5(A) shows a typical set of normalized UV–vis spectra for the Au NPs of different sizes. The wavelength of the SP bands is clearly dependent on the particle size, as evidenced by the increase of the wavelength (λ_{max}) at maximum absorbance of the SP band from 519 to 569 nm for the nanoparticle of the different sizes. This red shift is also accompanied by a small broadening of the SP band in the long-wavelength region.

The correlation between the particle sizes and the λ_{max} of the SP band can be analyzed based on Mie theory. The spectral simulation was based on Mie theory (Eqn (5)),

$$\frac{\gamma}{NV} = \frac{18\pi\varepsilon_{\alpha}^{3/2}}{\lambda} \times \frac{\varepsilon_2}{(\varepsilon_1 + 2\varepsilon_{\alpha})^2 + \varepsilon_2^2} \tag{5}$$

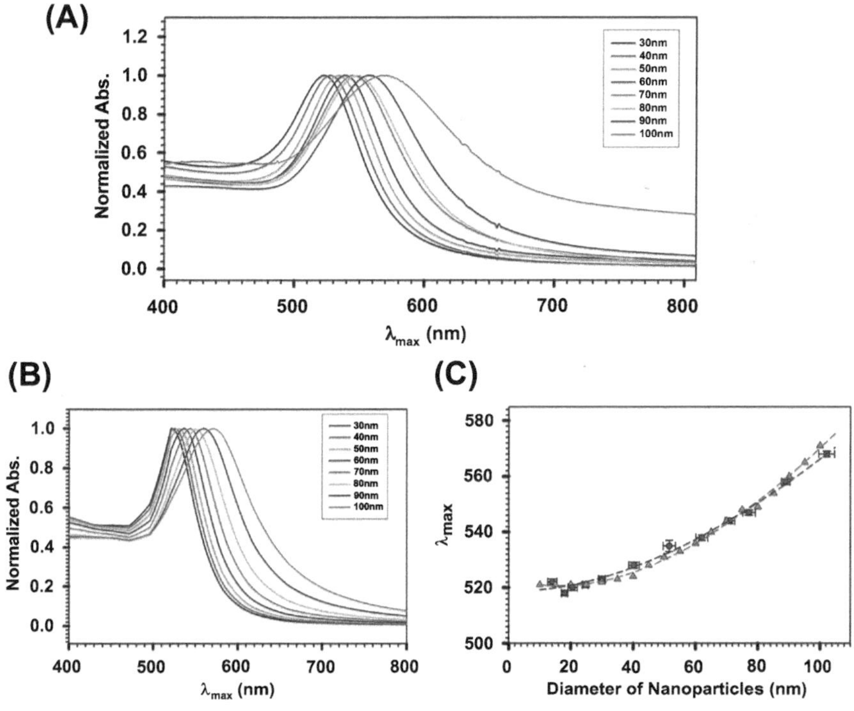

FIGURE 5 (A) UV–vis spectra (normalized) for Au NPs with different particle sizes in aqueous solution. (B) The simulated spectra of the SP band (normalized) for Au NPs of different sizes. Refractive index of water was used. (C) Comparison between experimental and theoretical relationships for SP band wavelength (λ_{max}) versus diameter of the nanoparticles (nm). Dashed lines are fitting curves. *Reproduced with permission from Ref. [67]. Copyright 2007 American Chemical Society.*

γ, the extinction coefficient, *NV*, the volume concentration of the assembled nanoparticles (*N*, the number of spheres per unit volume; *V*, the volume of each sphere), $\varepsilon\epsilon$, the dielectric medium constant, ε_1 and ε_2, the real and complex part of Au), and a modified Mie theory. Since the dielectric medium constant (ε) is related to refractive index, i.e., $(n_{eff} + ik_{eff})^2 = \varepsilon_{eff}$ (ε_{eff} is the effective dielectric constant, n_{eff} is the real part of the effective complex index of refraction for the nanocomposite, and k_{eff} is the imaginary part of the effective complex index of refraction), the dependence of the SP band for Au NPs of different sizes can be considered a result of the increase in effective refractive index for nanoparticles [76].

Figure 5(B) shows a set of simulated spectra (normalized) for Au NPs of different sizes in aqueous solution. The simulated spectra show a clear red shift of the SP band as the particle size increases. The increase of the λ_{max} value of the SP band displays a gradual increase with the particle size. In Figure 5(C), the experimental and theoretical results for the particle size dependence of the λ_{max} value of the SP band are compared. The polynomial fitting curves to both cases are also included for comparison. While there are subtle differences in the absolute values of λ_{max}, the trends are in general very comparable between the experimental and the theoretical data. It appears that the size correlation with the surface plasmon resonance properties is almost in complete agreement between the experimental and theoretical data in the medium size range of the nanoparticles, whereas slight deviations are shown in both smaller and larger sizes of the nanoparticles. The close agreement between the experimental and theoretical results provides insights into the validity of using the maximum wavelength of the SP band as a close measure of the particle size.

Insights have also been developed into the size correlation of the SERS properties. The size monodispersity of the Au NPs also permitted the correlation between the particle size and the SERS characteristics for the nanoparticles in the solution, which is in contrast to most of the other works focusing primarily on SERS correlation for nanoparticles on solid substrates. Gold and silver nanoparticles have extensively been studied for SERS. The correlation among particle size, SP band, and SERS intensity has mostly been limited to particles immobilized on a certain substrate. The particle size–SERS correlation was examined by comparison of SERS spectra of Raman labels on Au NPs in solution and on a solid substrate [77]. In view of the importance of size correlation in these optical and spectroscopic properties for nanoparticles in solution, it is also important to understand how the coupling of the localized fields between nanoparticles in solution influences the SERS effect.

Using the Au NPs, the SERS signals between samples in the solution and on the substrate are compared. Upon addition of 4-mercaptobenzoic acid (MBA, a Raman reporter molecule) to solutions of Au NPs of different sizes capped with acrylate, the particles were gently shaken overnight and concentrated by centrifugation. A sample of the preconcentrated solution was then cast onto a Au thin-film substrate (Figure 6(A)). Upon drying, the surface samples were examined using SERS. Two intense diagnostic peaks of the MBA are observed at 1081 and 1591 cm^{-1}. As shown in Figure 6(A) inserts, the intensity of the MBA peak at 1591 cm^{-1} shows a maximum for nanoparticles approximately ranging from 60 to 80 nm.

Experimentally, the nanoparticles capped with MBA in solution were first subjected to centrifugation before taking Raman spectra. Figure 6(B) shows a representative set of SERS spectra for MBA adsorbed on Au NPs of different sizes in aqueous solution. Note that the MBAs in the absence of Au NPs did not exhibit any Raman signals for concentrations up to 2 mM. For MBAs adsorbed on Au NPs in the solution, the MBA peaks are observed at 1078 and 1594 cm^{-1}, which are shifted slightly in comparison with those at 1081 and 1591 cm^{-1} for the substrate samples. These two peaks correspond to υ(CC) ring-breathing modes of MBA. The other weaker/less-intense bands correspond to δ(CH) (1130–1180 cm^{-1}) and $\upsilon(COO^-)$ (1380–1400 cm^{-1}) modes.

The above comparison led to several important observations. First, the peak intensity for MBA on the nanoparticles (30–90 nm) in aqueous solution increases with particle size. When the particle size is >40 nm, the SERS signals are clearly detectable. This finding is remarkable because it has been known to be difficult to detect SERS signals in solution phase. The ability to detect SERS signals in solutions provides not only a means for probing the SERS mechanism but also an alternative way for expanding SERS applications for chemical analysis. Second, the MBA bands observed for nanoparticles >50 nm in solution are quite similar to those on a solid substrate. The size correlation of the SERS intensity suggests the existence of a critical size range of the nanoparticles in the solution beyond which the particle–particle interaction is operative and responsible for the SERS effect. For particle sizes >50 nm, the SERS intensity of the adsorbed MBAs is detectable and it increases with the particle size; for particle sizes <50 nm, the SERS signal is not detectable within the detection limit.

An important question is what factors are responsible for the observation of the SERS effect of the nanoparticles in solution. It is important to note that no peak was observed prior to centrifugation. An explanation is then linked to the effect of centrifuging the MBA–nanoparticle solution on the interparticle properties. It is likely that the centrifugation brought the nanoparticles closer together. The decrease in the interparticle distance between the particles allowed more effective coupling of the adsorbed MBAs to occur. Another possible factor is the formation of small clusters of nanoparticles capped with MBA in the solution, such as dimers or trimers of nanoparticles. The adsorbed MBA molecules confined in such clusters are thus the "hot spots" responsible for the observed SERS effect. This possibility is consistent with previous observations of SERS signal produced by particle–substrate coupling and particle–particle coupling.

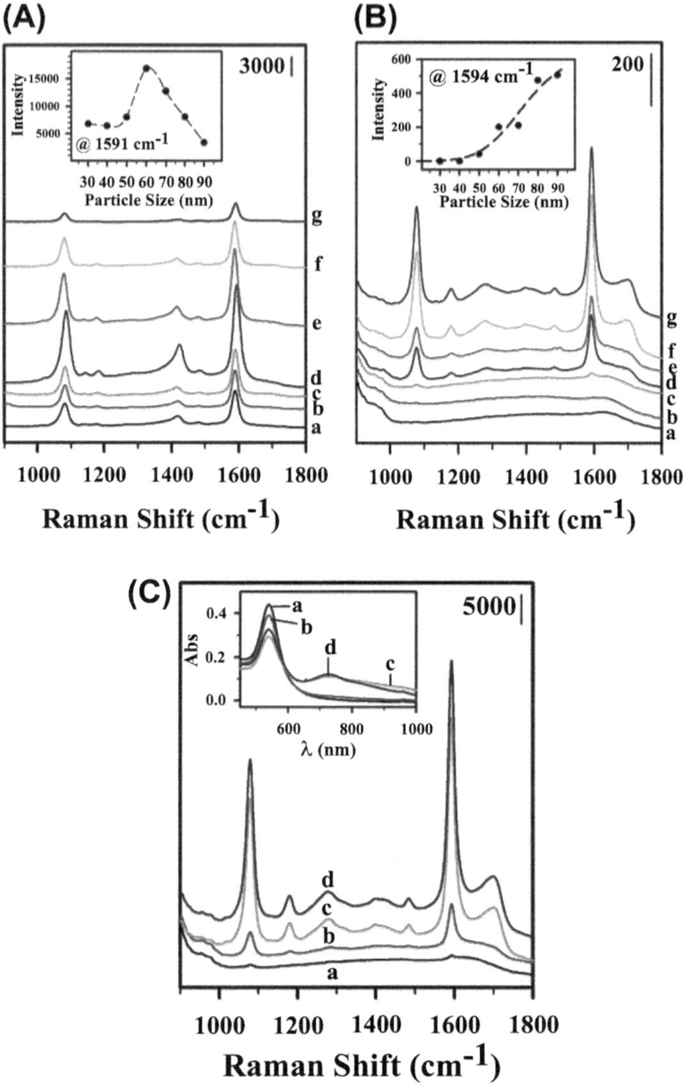

FIGURE 6 (A) SERS spectra for MBA-labeled Au_{nm} of different sizes on an Au-film substrate (insert: SERS intensity at 1591 cm^{-1} versus particle size). (B) SERS spectra of MBA-labeled Au_{nm} of different sizes in aqueous solution (insert: SERS intensity at 1594 cm^{-1} versus nanoparticle size). Particle sizes: a = 30; b = 40; c = 50; d = 60; e = 70; f = 80; g = 90 nm. (Intensity = counts/s). (C) SERS spectra for solutions of MBA in the presence of Au_{60nm} after centrifuging at different speeds: (a) 2000, (b) 5000, (c) 10,000, and (d) 14,000 rpm. Insert: UV−vis spectra for the corresponding nanoparticle solutions. *Reproduced with permission from Ref. [77]. Copyright 2008 IOP science.*

To further understand the exact mechanism responsible for the formation of dimers/trimers, we examined several questions related to the dependence of the SERS intensity on centrifugation speed, surface coverage, and salt concentration. The first question is how the dimer/trimer formations are affected by different centrifuging speed and surface coverage. As shown in Figure 6(C), the intensities of the two bands increase as the speed of the centrifugation increases from 2,000 to 14,000 rpm, demonstrating that the degree of particle aggregation increases as the speed increases. This assessment is supported by UV—vis data (insert), which showed that an SP band [67,74,75,78] is observed around ~750 nm as the centrifugation speed increased, which is a characteristic of the SP band change as a result of the particle aggregation. Furthermore, the coverage of MBAs on the nanoparticle surface is also an important parameter pertaining to the degree of aggregation between MBA-capped particles. Results showed that the Raman peak intensity increases with the increase of the MBA molecules added, which is indicative of the increase in surface coverage of MBAs. In addition, while the dimer/trimer formation is also dependent on the salt concentration, the salt-induced aggregation of the MBA-capped nanoparticles is irreversible, whereas the centrifugation-induced aggregation of nanoparticles can lead to small clusters, which are dispersible and stable in the solution.

In summary, the controllable size and the high monodispersity of Au NPs with 20—100 nm diameters in aqueous solutions have permitted a meaningful correlation of the particle size with the surface plasmon resonance and spectroscopic properties. The close agreement between the experimental and theoretical results provides insights into the validity of determining the maximum wavelength of the SP band as a measure of the particle size. The size correlation with the intensity of SERS from the adsorption of MBA on the nanoparticles in aqueous solutions reveals that they demonstrate the existence of a critical size of the nanoparticles in the solution beyond which the particle—particle interaction is operative and responsible for the SERS effect. These findings serve as the basis of size correlation for exploiting the optical and spectroscopic properties of Au NPs of different sizes in aqueous solutions, which has become increasingly important in analytical or bioanalytical applications.

3. SYNTHESIS OF GOLD NANOPARTICLES IN ORGANIC PHASE

3.1 Thermally Activated Size Evolution via Aggregative Growth

Au NPs with smaller size and higher monodispersity are often synthesized in organic phases. The smaller size and higher monodispersity are perhaps some of the main attributes in comparison with Au NPs synthesized in aqueous phases [50]. There are a number of approaches to controlling size and shape, including manipulating the reaction temperatures, seeding, templating, using shape-inducing agent, and fractional separation. In comparison with conventional colloids and many types of nanoparticles synthesized by reverse micelle, polymer encapsulation, and vapor-phase production, the encapsulation by alkanethiolate monolayers possesses several unique attributes, including air stability, reversible solubility in common organic solvents, and molecule-like reactivities involving both end-group functionalization and core–shell binding. The two-phase synthesis [10] of the small-sized particles, e.g., ~2 nm core sizes, is achieved by using large thiol-to-gold ratios, and larger sized nanoparticles can be formed by decreasing the ratio, which requires extensive synthetic work and a large amount of reagents. One important aspect of our synthetic approaches exploits the physical phenomenon of melting point decrease of nanoparticles for thermal processing in terms of size, shape, and composition [64,79–82]. The melting point of nanoparticles is related to that of its bulk metal by the equation [83–85],

$$\frac{T_r - T_\infty}{T_\infty} = -\frac{2}{\rho_s L r}\left[\gamma_s - \gamma_l\left(\frac{\rho_s}{\rho_l}\right)^{2/3}\right] \quad (6)$$

where T_r and T_∞ are the melting temperatures of the particle and the bulk solid, respectively, r is the radius of the particle, ρ_s and ρ_l are the density of the solid and the liquid, γ_s and γ_l are the surface energy of the solid and the liquid, and L is the heat of fusion. The implication of this relationship is that the smaller particles have a tendency of melting or partial melting at the reaction temperature. While the lowering of melting point is inversely proportional to the particle size, the surface melting temperature could be even lower (e.g., ~140 °C for 2-nm Au), and it could be influenced by the capping-dominated surface tension. Under such surface melting and surface tension effects, the driving force for the coalescence of two surface-melting particles is the reduction in free energy through a reduction in surface area, i.e., increase of size.

The theoretical model predicts a dramatic decrease of the melting point with decreasing the particle size to below a few nanometers [80]. While the temperature profile predicted by the model was higher than experimental values, improved models indeed showed a better agreement with experiments [63,85]. The key conclusion is that the melting point for the nanoparticles is much lower than that for the bulk. We consider the decrease of melting point as a function of particle size as an opportunity for us to conduct the manipulation of nanoparticle size and shape under significantly reduced heating temperatures. This means that we can use regular solvent for dissolving nanoparticles and mild temperature settings to carry out the heating process. As will be shown later, the actual temperature range, as indicated by the square in the plot, indeed allows us to induce size and shape evolution of the core–shell nanoparticles in solutions.

On the other hand, the overall temperature of a solution is limited by its boiling point (T_b). The change of T_b constitutes thus another factor for the basic consideration. As is well known for solution systems, T_b can be related to the mole fraction of the solvent (or the solute) by the relation derived under certain assumptions [86].

$$\frac{1}{T_b} = -\frac{R}{\Delta H}\left[\ln x_A - \frac{\Delta H}{R}\frac{1}{T*}\right] \tag{7}$$

where T^* is the boiling point of pure solvent, x_A is the mole fraction of the solvent, and ΔH is the associated enthalpy change. This equation predicts that the boiling point of the solution rises as a function of the concentration of the solute.

Both parameters, T_r and T_b, for a solution-containing nanoparticles as solute are therefore closely linked to the particle size and concentration. These theoretical linkages constitute the basis for the concept of temperature-manipulated size and shape evolution in solutions of thiolate-encapsulated Au NPs. Further, in an analogy to the usual high-temperature annealing, we anticipate that processes such as core coalescence and phase transitions, and reactivities such as thiolate desorption and readsorption may lead to changes in particle sizes and shapes. This type of changes can be related to the stability of different crystal facets, usually with (111) > (100) > (110) reflecting the order of surface free energy. However, positive or negative contributions to these energies also include chemisorption of thiolates, changes of solution composition, edges and corners connecting the facets, and size and shape evolution under externally supplied energies via, e.g., heating treatment.

Thermally activated core–shell reactivities serve as a strategy for inducing size and shape changes of nanoparticles from solutions of preformed thiolate-capped Au NPs [79,80], which have considerable technological applications in areas of microelectronics, optic devices, magnetic materials, catalysis, and molecular recognition [87–91]. This strategy exploits the core–shell-type reactivities of alkanethiolate-capped Au NPs under thermal activation. Larger core sizes result from an evolution of the smaller sized starting particles via sequential shell desorption, core coalescence, and shell re-encapsulation. For example, using Au NPs encapsulated with decanethiolate (DT) monolayer shells (Au_{2nm}–DT) synthesized by Brust–Schriffrin's standard two-phase method [10] (2.01 ± 0.41 nm), the solution with known concentrations of Au_{2nm}–DT particles, free DT, and TOA-Br was heated under controlled temperature ($149 \pm 1.5\,°C$). The DT-capped Au NPs (2 nm), upon the thermally activated processing in a toluene solution containing TOABr and capping molecules such as DT, were shown to undergo dramatic changes in size, shape, and homogeneity in comparison with the precursor nanoparticles [80,81]. Two morphological changes are evident. First, the size distribution is strikingly narrow, as shown in Figure 7(A) by the contrast of the size distribution histograms for both nanoparticles. The 2-nm particle core size was

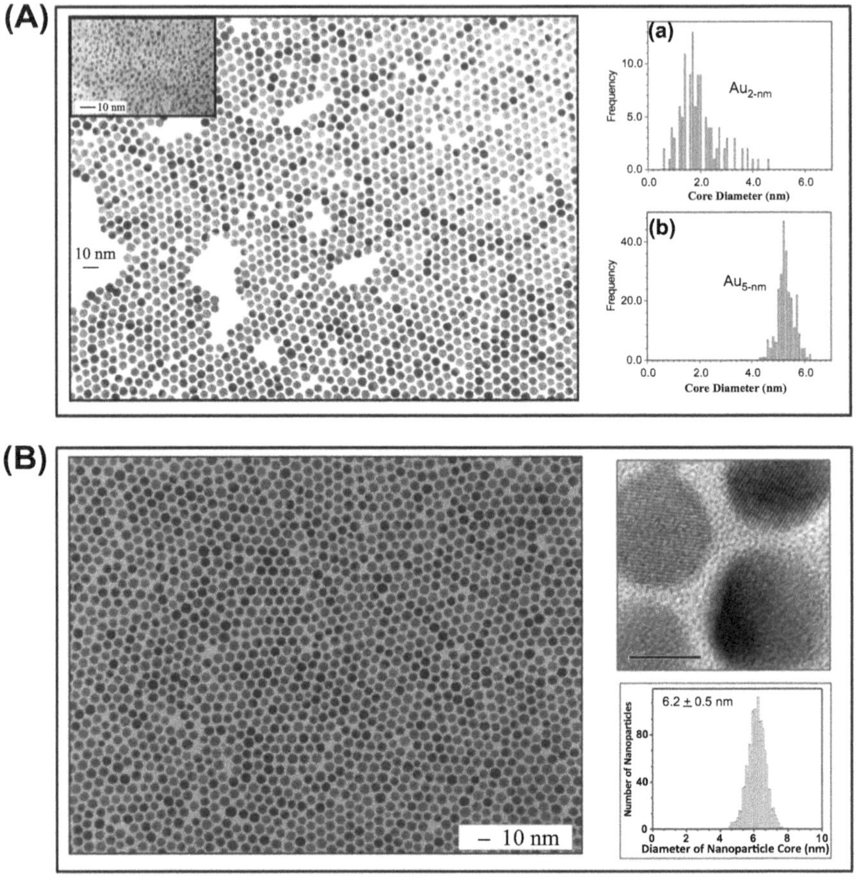

FIGURE 7 (A) TEM micrographs of DT-encapsulated Au NPs prepared by solution annealing process in the presence of excess DT following the thermally activated size evolution. The insert shows the TEM image of the precursor nanoparticles. Right: histograms of the nanocrystal core sizes comparing the precursor (a) and the evolved product (b). *Reproduced with permission from Ref. [81]. Copyright 2000 Royal Society of Chemistry.* (B) TEM, HRTEM and size distribution of DT-capped Au NPs produced by the thermally-activated processing route Refs [56], [64]. *Reproduced with permission from Ref. [56]. Copyright 2003 American Chemical Society.*

1.9 ± 0.7 nm. The particle shapes appear nonuniform polyhedrons. In sharp contrast, the data for the evolved nanoparticles display increased core size with very narrow size distribution, i.e., 5.3 ± 0.3 nm, much narrower than the 2-nm starting particles. Second, the uniform spacing between the nanoparticles with a predominant hexagonal packing array feature is remarkable. Edge-to-edge distances are estimated to range from 0.7 to 1.4 nm with an average distance of ~1.0 nm. This value corresponds closely to the distance expected for an interdigitation of alkyl chains between shells of the neighboring nanoparticles. Larger sized Au NPs can also be obtained by tuning the thermal evolution temperature, as shown by an example of Au NPs with 6.2 ± 0.5 nm diameter in Figure 7(B) [56]. The nanoparticles exhibit highly crystalline feature expected for the face-centered cubic structure.

The observed morphological features constitute clear evidence for the existence of a further resizing and reshaping effect for the evolved particles under the annealing process. Two cooperative processes are believed to be responsible for the experimental observation. The first involves thermally driven desorption of the shell components followed by coalescence of the nanoparticle cores. The second involves reshaping or resizing that minimizes the chemical potentials followed by competitive re-encapsulation of thiolate shells under the annealing condition. The latter can be analogous to thermal annealing of solid samples. The key distinction is the involvement of a series of chemical processes such as thiolate encapsulation, desorption, and re-encapsulation, during which the nanocrystal facets and corners are further developed into energetically more favorable states.

3.2 Molecularly Tuned Size Selectivity in Aggregative Growth via Capping Molecules

The combination of surface melting and surface tension effects constitutes part of the driving force for a controlled interparticle coalescence for processing size and morphology of metal, alloy, and core–shell nanoparticles in solutions (e.g., Au, Cu, AuAg, AuPt, and Fe_2O_3 (or Fe_3O_4)@Au (@ stands for "coated with")) [56,64,79–82,92] under mild elevation of temperatures (100–200 °C). In contrast to Ostwald ripening known for small particle systems, the thermally activated processing of a solution of presynthesized nanoparticles involves coupling of the molecular capping and re-encapsulation to the interparticle coalescence. For example, using presynthesized Au NPs (2 nm), the particle size growth is shown to be a function of chain length of the capping alkanethiols [64] (Figure 8). The average size of the particles is dependent on the chain length of the alkanethiols used in the thermal processing [64,80,81]. The increase in chain length of the capping alkanethiolates leads to a gain in stabilization energy due to interchain cohesive interactions, consistent with the known linear correlation of cohesive energy of the alkyl chains ($\sim$1 kcal/mol CH_2). This chain-length-dependent stabilization energy contributes to a balance of alkanethiolate desorption, core coalescence, and re-encapsulation in the size evolution [64]. The increased cohesive energy for longer chain length would favor a stronger encapsulation. Since our early demonstration of this strategy for producing highly monodispersed Au NPs [79,80], it has been reported for processing metal nanoparticles by many others [93–98]. Using this approach with slight modifications, gold-based alloy and other metal (e.g., copper) nanoparticles of different sizes have also been prepared [81,82].

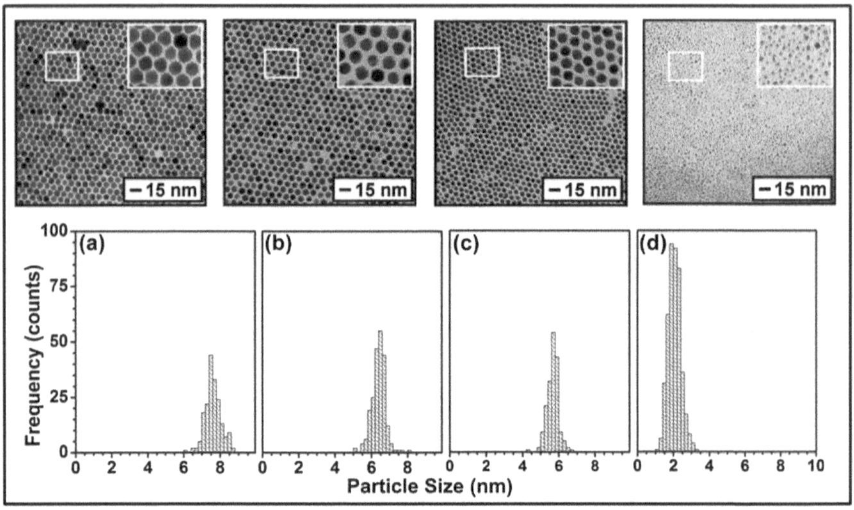

FIGURE 8 TEM images and corresponding particle size distributions of Au NPs produced by thermal processing in the presence of alkanethiols of different chain length (n): $n=5$ (a) (7.60 ± 0.43 nm), 10 (b) (6.41 ± 0.40 nm), and 15 (c) (5.86 ± 0.31 nm). TEM for the precursor Au_{2nm}-DT (d) (2.01 ± 0.41 nm) is included for comparison. *Reproduced with permission from Ref. [64]. Copyright 2006 American Chemical Society.*

The nanoparticle cores for samples evolved using shorter chain thiols were found to be larger than those evolved in the presence of longer thiols. Figure 8 shows the examples obtained from thermal processing in the presence of alkanethiols of different chain lengths under constant concentration. For example, the Au NPs obtained using alkanethiols with $n=5$ (Figure 8 (A)) show an average core diameter of 7.60 ± 0.43 nm. When the chain length is extended to $n=15$ (Figure 8 (C)), an average core diameter of 5.86 ± 0.31 nm was obtained. The thermally evolved particles show largely spherical morphology, with hexagon-like features identifiable, and are highly monodisperse with relative standard deviations typically under 5%. Note also that the size increase of the particles was also accompanied by a systematic increase of SP band.

The molecularly tuned size selectivity in thermal processing of Au NPs [64] for achieving both controlled size growth and size monodispersity from the thermal processing of small-sized monolayer-protected nanoparticles as precursors [79,80] has been demonstrated. The control of temperature provides an important means for processing nanoparticle size and monodispersity [93–98]. In view of the involvements of molecular desorption, nanocrystal core coalescence, and molecular re-encapsulation processes in the evolution of nanoparticle precursors at elevated temperatures, one fundamental question is can the size selectivity be tuned molecularly? The answer to this question would have not only significant impact toward the establishment of the thermal processing method for the precise control of size and monodispersity, but also much broader implications to many nanoparticle-based technologies. This novel tuning is accomplished in a proof-of-concept demonstration for monolayer-protected Au NPs by manipulating the chain length and the concentration of alkanethiols in the processing solution. The basic idea stems from two key promises of such molecular tuning: (1) an increase in alkanethiol chain length leads to a gain in stabilization energy due to additional interchain cohesive interactions and (2) a change in alkanethiol concentration shifts the evolution equilibrium in terms of shell desorption and re-encapsulation (Figure 9(A)) [80]. Our work demonstrated quantitative correlations between size selectivity and these two parameters.

The results from the study of chain length dependence using alkanethiols ($CH_3(CH_2)_nSH$) demonstrated for the first time that the size selectivity of nanoparticles in the thermal processing solution of presynthesized nanoparticles can be achieved by molecularly tuning the chain length and concentration of alkanethiols. As shown in Figure 9(B), there is a clear linear correlation between the size of the postheat treatment nanoparticles and the

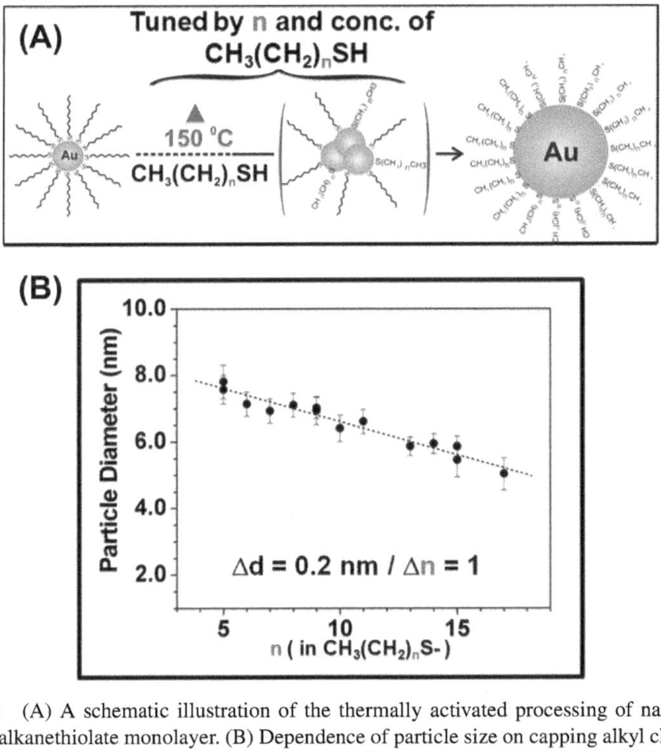

FIGURE 9 (A) A schematic illustration of the thermally activated processing of nanoparticles capped by alkanethiolate monolayer. (B) Dependence of particle size on capping alkyl chain length with 2-nm-sized Au precursor particles. *Reproduced with permission from Ref. [50]. Copyright 2010 American Chemical Society.*

chain length of the alkanethiols used in the thermal processing. Note that the concentration of the alkanethiols was constant for all chain lengths, making chain length the only independent variable. By changing the chain length of the alkanethiols from $n = 5$ to 17, nanoparticle sizes ranging from 8 to 5 nm with good monodispersity can be obtained.

The slope of the linearity, i.e., −0.2 nm/methylene, suggests that the increase of a methylene unit in the capping alkanethiol can effectively decrease the nanoparticle core size by 0.2 nm. At a higher concentration (e.g., 310 mM) of the alkanethiols, the slope was found to be smaller (~0.1 nm), which likely reflects the increased rate of re-encapsulation leading to limited growth of the nanoparticle core. In addition, the particle sizes were also shown to depend on the concentration of decanethiol added in the pre-heat-treatment solution. The observation of a reciprocal relationship with concentration is consistent with a model involving Langmuir isotherm for the adsorption—desorption equilibrium. The effect of increasing concentration on the particle size can be understood by considering that the shell desorption is favorable at low thiol concentration, and that the equilibrium shifts in favor of re-encapsulation as thiol concentration is increased. To understand the origin of the linear correlation between the particle size and the chain length, we considered the energetic aspects involved in the alkanethiolate encapsulation. Assuming a spherical model for the nanoparticle core, ~22 $Å^2$ for the average surface area per thiolate as known for a densely packed self-assembled monolayer on a planar Au surface, and ~1 kcal/mol of methylenes for the cohesive energy, the estimation showed that as particle size increases, the increasing number of thiolates in the encapsulation shells leads to an increase in cohesive energy. The longer the chain length of the capping alkanethiolate, the larger is the cohesive energy resulting from the chain—chain van der Waals interaction. The result suggests that the cohesive energy is at least partially responsible for the origin of the size—chain length correlation.

The results substantiate that shell desorption, core coalescence and growth, and shell readsorption processes in the thermally activated size evolution of Au NPs in the presence of alkanethiols of different chain lengths are dependent on the relative binding strengths of the capping thiolates and the shell composition. The desorption rate is dependent upon the energy required to break not only the gold—sulfur bond, but also the cohesive van der Waals forces between methylene groups on adjacent alkanethiolates. At thermally enhanced rates of ligand exchange, the shell composition changes and the coalescence of smaller particles into larger particles occurs. The difference in the chain length thus regulates the degree of coalescence and re-encapsulation, and ultimately results in equilibrium core size differences. This tuning is unique because it demonstrated the important role played by cohesive interactions in regulating interfacial reactivities of preformed nanoparticles in a thermal processing route, in sharp contrast to the manipulation of precursor ratios documented for a chemical synthesis route [10,99].

The same strategy can also be applied to many other nanoparticle systems involving Au NPs [50]. In contrast to Au NPs, iron oxide nanoparticles do not exhibit a coalescence under such a mild temperature elevation, which constitutes the basis for a thermally activated heterointerparticle coalescence to fabricate core@shell structured (Fe-oxide@Au) nanoparticles [92,100–102]. In this case, the surface melting temperature of the precursor 2-nm-sized Au NPs is much lower than that for the 5-nm-sized iron oxide nanoparticles. The interparticle heterocoalescence of gold-coated iron oxides leads to the formation of core–shell nanostructures. For example, the thermal processing of Fe_2O_3 (or Fe_3O_4) and Au NPs in a solution involves thermally activated heterointerparticle coalescence between gold and magnetic nanoparticles under encapsulating environment for creating the core–shell type nanoparticles, Fe_2O_3@Au (or Fe_3O_4@Au). The presence of molecularly capped iron oxide nanoparticles in the solution created a unique thermal microenvironment for the core–shell coalescence to produce monodispersed Fe-oxide@Au nanoparticles of controlled sizes. The Au-shell thickness can be manipulated depending on the relative ratio of Au and Fe-oxide precursors. The competition between growing Au and Fe-oxide@Au leading to larger-sized core@shell nanoparticles with a single or multiple Fe-oxide cores (pomegranate-like) nanoparticles. Magnetic cores coated with gold shell have also been prepared by combining bottom-up synthesis and thermal processing in a seeded growth process [92,100–103]. One example is the synthesis of highly monodispersed Fe_2O_3@Au and Fe_3O_4@Au nanoparticles [100–102]. The initial formation of Fe_3O_4 seeds in the presence of capping agents (e.g., oleylamine and/or oleic acid) [104,105] is followed by reduction of $Au(CH_3COO)_3$ in the presence of capping agents at elevated temperature (180–190 °C) which controls the thermally activated partial adsorptions of the capping layer from the core, the deposition of Au on the exposed Fe_3O_4 surface, and the subsequent re-encapsulation by the capping agent. The core–shell-type structure was supported by the Au-shell-specific assembly of the core–shell nanoparticles into thin films and the decrease of the magnetization and blocking temperature as a result of increased interparticle spacing. Such core–shell nanoparticles are useful for magnetic bioseparation and spectroscopic probes [92,103], which has been the focus of two recent reviews [51,102]. Examples of other magnetic or metal nanoparticles with gold shells and bimetallic or trimetallic alloy nanoparticles, prepared by similar methods, include MnZnFeO@Au, Fe_3O_4@Au@Pt, Pt@Au, AuAg, AuPt, PtVFe, PtNiFe, etc. [77,103,106–112], many of which were exploited for magnetic separation in detecting protein-binding activities [77,103] or for electrocatalysis in fuel cell reactions [106–112].

There has been increasing evidence supporting that aggregative growth has played a major role in the thermally activated size evolution. The importance of aggregative growth mechanism for the control of Au NPs size has been recently discussed in both experimental and theoretical studies of the kinetics and mechanism for the size evolution of Au NPs under the heating conditions described above [66,113]. By using the well-known KJMA model for aggregative growth, a straightforward analytical method was derived for analysis of the growth kinetics. The method allows extraction of separate aggregative-nucleation, aggregative-growth, and Ostwald-ripening rate constants. The established aggregative nucleation rate and size distribution, and oriented attachment of nanocrystals and nanoparticles are contrasted to classical LaMer nucleation and growth, and to Ostwald ripening. Aggregative growth is described as a nonclassical nucleation and growth process. The nucleation function is approximated with a Gaussian form, showing that the nucleation parameters height (Γ_{max}) and width (Δt_n) of the nucleation function are systematically varied by conditions that influence the colloidal stability of the small, primary nanocrystals participating in aggregative growth, and correlate with the final nanocrystal mean size and size distribution.

In addition to the size control by the thermally activated processing strategy, the control over the surface properties is also an important area of investigations in the synthesis of Au NPs [6]. In fact, organic phase synthesis was shown to exhibit advantages to enable encapsulation of special organic ligands on the Au NPs [114,115]. One example involves exploitation of two structural attributes of the dendritic arenethiol (DAT) molecules, which uses the umbrella-like structure with a single thiol as the anchorage handle and the rib with an expandable dendritic structure as a spacing-tunable cap, for tuning size and surface reactivity of the nanoparticles (Figure 10(A)) [114]. For example, in a two-phase solution containing an aqueous solution of $HAuCl_4$ and a toluene solution of TOABr in the presence of DAT, the reduction of the $AuCl_4^-$ by $NaBH_4$ was shown to produce DAT-capped Au NPs.

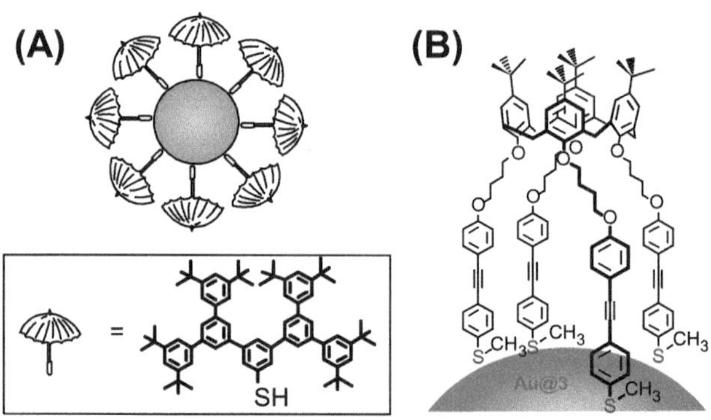

FIGURE 10 (A) Au NPs capped with dendritic arenethiols (DAT) (e.g., tert-butylbenzenethiol) where the umbrella-like DAT molecule features a single thiol as the anchorage handle and a rib with an expandable dendritic structure as a spacing-tunable cap for tuning the nanoparticle sizes and surface reactivities. (B) Au NPs capped with *t*-BCAs with four methylthio groups, where the calixarene structure features a cone cavity for host–guest recognition.

Another example involves structurally tailored *t*-butylcalixarenes (*t*-BCA) immobilized on Au NPs to tune the guest access to the calixarene cone cavity for cationic recognition (Figure 10(B)) [115]. For the synthesis of methylthio-*t*-BCA-capped Au NPs, an ethanol solution of $HAuCl_4$ was mixed with a chloroform solution of methylthio-*t*-BCA, the reduction of $AuCl_4^-$ by $NaBH_4$ was shown to produce *t*-BCAs-capped Au NPs. The immobilized *t*-BCAs structures were shown to be functional in interparticle charge-induced aggregation upon selective capture of metal cations into the nanoparticle-immobilized *t*-BCAs, which produces calorimetric changes for the detection. The value of this class of calorimetric nanoprobes will be in the area of designing advanced host–guest probes using a variety of calixarene ligands for ionic recognition in a simplistic detection format.

4. APPLICATIONS OF GOLD NANOPARTICLES IN ANALYTICAL CHEMISTRY

A few examples of recent studies of Au NPs in analytical chemistry [116–118] are highlighted here to illustrate the size- and surface-tailored functionalities, including Au NPs-assembled substrates for SERS detection of biomarkers from bacteria spores [119], SERS monitoring of DNA-mediated assembly and restriction enzyme cutting [117,118], conjugation of Au NPs with miRNAs for cell transfection [120], and detection of thiol-containing amino acids [51] or other biomarkers [121].

4.1 Enhanced SERS Detection of Bacteria Biomarker and DNA Bioactivities

Detection of bacteria biomarker [119,122–124]. Bacteria constitute different pathogens to various diseases. There is an increasing need in developing rapid and accurate detection of the bacteria. Traditionally, silver nanoparticles (Ag NPs) were used as SERS substrates for biomarker detection (e.g., *Bacillus subtilis*). Among recent examples using Au NPs as SERS substrates for detection of biological species, the work by Porter et al. demonstrated a significant SERS enhancement via a strong coupling of the surface plasmonic resonance between Au NPs and gold substrate surface [125,126]. In our recent studies, the strong SERS effect produced by the particle–particle and particle–substrate plasmonic coupling of Au NPs was demonstrated for SERS detection of biomarkers released from bacterial spores [119,122–124]. Figure 11(A) illustrates one pathway for the exploitation of plasmonic coupling of Au NPs immobilized on an Au substrate in developing SERS substrates for the detection of biomarkers that are extracted from bacterial spores.

Using gold nanoparticles synthesized by seeded growth method [4], Au NPs/PVP/Au was prepared by immobilization of Au NPs onto Au substrate via

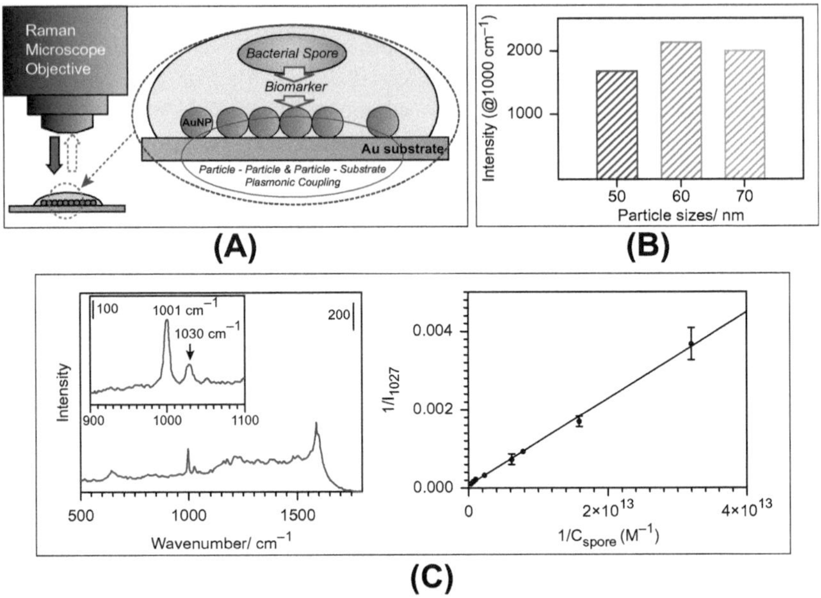

FIGURE 11 (A) Illustrations of the SERS measurement using Au NPs immobilized on an Au substrate as for the detection of biomarker molecules released from bacterial spores. (B) Peak intensity (integrated peak area ($cps \cdot cm^{-1}$)) versus the particle size for a sample of 1.0 ppm DPA. *Reproduced with permission from Ref. [122]. Copyright 2009 American Chemical Society.* (C) Left: A SERS spectrum with an AuNPs/Au substrate for a spore suspension ($C_{spore} = 7.8 \times 10^{-13}$ M) upon extraction of CaDPA using 0.02 M HNO_3. Insert: a magnified view in the 900–1100 cm^{-1} region for the SERS intensity of the ~1000 cm^{-1} peak as a function of C_{spore}: $1/I_{SERS}$ versus $1/C_{spore}$ (right). *Reproduced with permission from Ref. [124]. Copyright 2011 Elsevier Publishing.*

PVP as an adhesive layer, in which the pyrrolidone groups interact with gold surface. In addition to exploiting the particle–substrate [125,126] and the particle–particle plasmonic coupling effects for producing strong SERS effect, this type of substrates was shown to offer an effective way to improve the sensitivity of the detection down to sub-ppb level. For gold-based SERS substrates in the detection of Dipicolinic acid (DPA), the SERS intensity depends on the particle size. As shown in Figure 11 (B), the peak intensity at 1000 cm^{-1} exhibits a maximum for the 60-nm-sized particles. The enhancement factor (EF) was estimated from the integrated peak area for DPA on AuNPs (60 nm)/PVP/Au substrate and that for a powder DPA sample based on the laser spot size and its sample-penetration depth, yielding an EF of 8×10^5. This ~10^6 EF was indeed consistent with the fact that our AuNPs/PVP/Au surface functions very well as an SERS substrate. This finding is in agreement with a maximum SERS signal observed for 5,5′-dithiobis-succinimidyl-2-nitrobenzoate (reporter) labeled on Au NPs in a similar size region [125,126]. The appearance of the SP band for 60-nm Au NPs on Au substrate at a wavelength of ~650 nm slightly above the laser excitation wavelength (633 nm) was proposed to be responsible for the strong coupling between the plasmonic band and the laser excitation, leading to the maximum SERS effect [125,126].

Quantitative information can be extracted from the SERS peak intensity as a function of biomarker concentration. The intensity–concentration relationship is shown to be close to linear with a relatively large slope in the region of $C_{(DPA)} < \sim 0.01$ ppm (a), and closely linear again in the region of $C_{(DPA)} > 1$ ppm but with a much smaller slope than the low-concentration case. This relationship becomes nonlinear with a decreased slope when $C_{(DPA)}$ falls in between 0.01 and 1 ppm. These characteristics were analyzed in terms of Langmuir isotherm for the formation of monolayer or multilayer of DPA molecules. Limit of detection (LOD) of $\sim 10^{-10}$ M (or ~0.1 ppb) was obtained for the AuNPs/PVP/Au substrates. In comparison with a value of $4.9 \times 10^4\ M^{-1}$ previously reported for the equilibrium constant (K) for the adsorption of DPA on silver film-over-nanosphere (AgFON) substrates [127,128], the K value for our AuNPs (60 nm)/PVP/Au substrates in the low-concentration region ($1.7 \times 10^7\ M^{-1}$) was found to be higher (by about 3 orders of magnitude) than the reported value. Since DPA is adsorbed on the surface via carboxylate groups, the interaction of the carboxylate groups with alumina surface is likely to be stronger than that with Au particle surface.

The Au NPs-based SERS substrates were further used for detection of spores. SERS spectra were first acquired for spore suspension samples in which CaDPA were extracted using HNO_3 (0.02 M) as biomarkers (Figure 11(C)). The diagnostic peaks at 1001 and 1030 cm^{-1} are in agreement with those reported previously for CaDPA extracted similarly. As shown in Figure 11(C), the good linear relationship of $1/I_{SERS}$ vs $1/C_{spore}$ in a broad concentration range, as opposed to the limited concentration range ($>3 \times 10^{-13}$ M) for the linearity of I_{SERS} versus C_{spore}, yields a value of $4.5 \times 10^{11}\ M^{-1}$ for K_{spore}. The LOD was estimated to be 2.5×10^{-14} M or 3.0×10^3 spores per 0.2 μL sample solution, which is quite comparable with the LOD value reported for the spore detection using AgFON substrates ($1.1–2.1 \times 10^{-14}$ M or $1.4–2.6 \times 10^3$ spores per 0.2 μL) [127,128]. Au NPs have been proven viable as highly effective SERS substrates for the detection of biomarkers from bacterial spores with high sensitivity and low detection limit (~0.1 ppb), which has the potential toward establishing a highly sensitive SERS-based technique in identification of biomarkers in bacterial spores.

Detection of DNA bioactivities [117]. A gold nanoparticle-based SERS detection strategy is recently demonstrated for detection of DNA assembly and enzymatic cutting in solutions [117,129]. One of the most widely studied DNA-binding proteins, in relation to cancer, is the *p53* protein, mutations of which are found in more than 50% of all tumors. A scintillation proximity assay was recently used to analyze several known *p53* recognition sites, specifically the *mdm2*, *p21*, and *cyclin G* genes, which provided important insights for understanding the role that DNA transcription plays in diseases such as cancer. The assembly of Au NPs by complimentary strand binding of DNAs, designed using the *cyclin G* promoter sequence, has recently been demonstrated [130]. The detection of DNA assembly of Au NPs conjugated with single-stranded (ss) DNAs and the restriction enzyme cutting of the interparticle double-stranded (ds) involves protein binding/cutting processes relevant to the DNA-binding protein *p53*. SERS has rapidly emerged as a powerful technique for highly sensitive and selective detection in biomolecular recognition [51,67,77,92,131]. Due to the fact that an EF of $>10^6$ that can be obtained using silver or gold planar substrates, there have been many reports demonstrating the SERS-based detection of DNA. These include use of a fluorophore or dye-labeled DNA on gold or silver nanoparticles, and DNA assembly on a gold substrate. Our recent work [51,67,77,92] demonstrated the viability of SERS detection of Raman reporter molecules labeled on Au NPs of sizes >40 nm in an aqueous solution [67,77]. While the enhancement is smaller than with solid SERS substrates, as demonstrated for antibody−antigen binding of Au NPs on a gold thin-film substrate [77,92], the SERS detection in a solution phase exploits interparticle plasmonic coupling, or the so-called "hot-spot". This type of plasmonic coupling effect originates from the formation of small clusters of nanoparticles, such as dimers and trimers [77], and is potentially powerful for directly probing the biomolecular linkage processes in aqueous solutions containing biological samples.

Figure 12(A) illustrates a gold nanoprobe-based SERS detection strategy using a 30 base pair (bp) ds-DNA sequence derived from the regulatory region of a cell cycle gene modified to contain a site for the restriction enzyme *MspI*. One strand of the ds-DNA is attached to Au NPs of a selected size via a thiol group, and then mixed with Au NPs labeled with a Raman reporter molecule, MBA, and the complementary DNA strand. This should lead to a linkage of the two types of Au NPs via complementary binding to the DNA-conjugated Au NPs and a thiol attachment to the MBA-labeled Au NPs, producing ds-DNA−AuNPs assembly (ds-DNA/AuNPs) that generates an interparticle "hot-spot" for SERS detection. The introduction of a restriction enzyme (e.g., *MspI*) into the ds-DNA/AuNPs solution should lead to a gradual removal of the interparticle "hot-spot" due to cleavage of the ds-DNA, and consequently a decrease of the SERS signals. This strategy involves exploitation of the interparticle "hot-spot" formation and removal as a direct means for monitoring the DNA hybridization and enzyme cutting processes without use of additional reagents.

One example involves assembly of 13-nm bottom-DNA/AuNP and 62-nm MBA/AuNP by top-DNA. Upon introducing top-DNA into a mixed solution of 13-nm bottom-DNA/AuNP and 62-nm MBA/AuNP, the SERS spectra showed significant changes (Figure 12(B)). The basic characteristics of SERS spectra

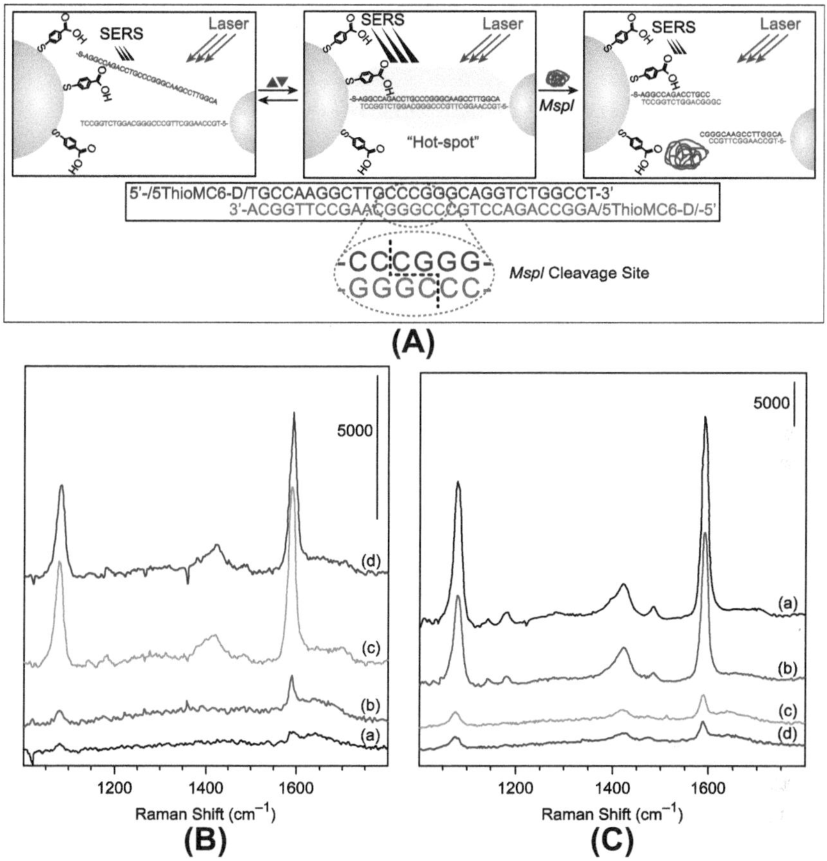

FIGURE 12 (A) Gold nanoparticle-based SERS detection strategy for a 30-bp DNA sequence containing an *MspI* restriction enzyme site (see lower panel). (B) SERS spectra for the assembly of 62-nm MBA/AuNP ($\theta \approx 0.7$) and with 13-nm bottom-DNA/AuNP by top-DNA in aqueous solution. (a) MBA/AuNP; (b) bottom-DNA/AuNP mixed with MBA/AuNPs; (c) the mixture solution of bottom-DNA/AuNP and MBA/AuNP in the presence of top-DNA (10 min); (d) ~4 h after assembly. (C) SERS spectra for *MspI* cutting of the ds-DNA/AuNPs comprised 13-nm bottom-DNA/AuNP and 62-nm MBA/AuNP with top-DNA: (a) the ds-DNA/AuNPs; (b) 1 min, (c) 5 h and (d) 8 h after the enzyme addition. (Two different oligonucleotides, (bottom-DNA: 5′-/5ThioMC6-D/AGGCCAGACCTGCCCGGGCAAGCCTTGGCA-3′and top-DNA: 5′-/5ThioMC6-D/TGCCAAGGCTTGCCCGGGCAGGTCTGGCCT-3′) were used). *Reproduced with permission from Ref. [117]. Copyright 2013 Royal Society of Chemistry.*

are very similar to those shown previously. The use of the larger sized Au NPs (62 nm) labeled with MBA (surface coverage ~0.7), in combination with the 13-nm bottom-DNA/AuNP, did not seem to change the SERS intensity significantly; however, the concentration of the 62-nm Au NPs (2×10^{-2} nM) is lower than the concentration of the 13-nm (4–5 nM) and the 39-nm Au NPs

($7-8 \times 10^{-2}$ nM). If the SERS intensity is analyzed in terms of per unit concentration of Au NPs, the 62-nm Au NPs exhibit an enhancement (3.2×10^5 counts/nM) which is nearly twice that for the 39-nm Au NPs (1.8×10^5 counts/nM). The combination of the 13- and 39-nm Au NPs displayed an average EF of 2.8×10^5, whereas that of the 13- and 62-nm Au NPs exhibited a value of 3.9×10^5. A further increase in EF would be expected by using larger sized Au NPs based on our earlier studies of particle-size effect on SERS intensity [67,77].

Another example involves *MspI* enzymatic cutting of the ds-DNA/AuNPs consisting of 13-nm bottom-DNA/AuNP and 62-nm MBA/AuNP with top-DNA. SERS spectra were obtained for the *MspI* enzymatic cutting of the ds-DNA/AuNPs composed of 13-nm bottom-DNA/AuNP and 62-nm MBA/AuNP (Figure 12(C)). A clear decrease of the intensities of the diagnostic peaks at 1078 and 1594 cm^{-1} is evident after about 5 h incubation time. This suggests that enzymatic cutting of the ds-DNA/AuNPs depends on the nanoparticle size. In comparison with data shown for combination of 13-nm bottom-DNA/AuNP and 39-nm MBA/AuNP, the use of 62-nm Au NPs as part of the assembly clearly led to a more efficient disassembly (Figure 12(C)), as evidenced by the observation of >90% reduction in the SERS signals with the larger particles compared to only 50% reduction for the smaller particles.

In summary, the formation of SERS "hot-spots" by ds-DNA linking of Au NPs and dismantling of the "hot-spots" by restriction enzyme cutting of ds-DNA in the linked Au NPs have been demonstrated for the detection of DNA linkage and restriction enzyme cutting processes. SERS signals are sensitive to a subtle change in the interparticle structure because of the EF in the order of $10^3 \sim 10^6$, and potentially provide a real-time monitoring of biomolecular reactivities occurring in the DNA assembly and restriction enzyme disassembly processes. As an expansion of this work, a novel strategy was recently demonstrated exploring bifunctional nanoparticles for spectroscopic detection and magnetic intervention of DNA assembly, disassembly, and enzyme cutting processes in a solution phase [118]. This strategy exploits magnetic MnZn ferrite nanoparticles decorated with gold or silver on the surface to retain adequate magnetization while producing sufficient plasmonic resonance feature to impart SERS function. The unique bifunctional combination of the SERS "hot-spots" and the magnetic separation capability serves as a first example of bifunctional nanoprobes for biomolecular recognition and intervention.

4.2 MicroRNA Conjugation for Cell Transfection

This section highlights an example in recent explorations of Au NPs for delivering miRNAs to multiple myeloma cells for cell transfection and gene knockdown studies [116,120]. There has been increasing interest in delivering small strands of RNA, e.g., microRNA (miRNA), into cells for genetic manipulation or cellular marking. The association of miRNA with mRNA (messenger RNA) has been identified in relation to cancer treatments, such as the resistance of multiple myeloma (MM.1S) to glucocorticoid treatment. Au NPs have been utilized for the delivery of siRNA (small interfering RNA) into cells for efficient knockdown of target genes without significant cytotoxicity. This type of delivery vehicles exploits the unique optical properties, low cytotoxicity, enhanced lifespan in the blood stream, and cancer treatment or repression of certain genes. The importance of miRNA in the treatment of cancer and for the manipulation of genetic expression has been demonstrated recently. For instance, miR-130b was found to be expressed differently in glucocorticoid-sensitive versus glucocorticoid-resistant MM.1S cell lines. The overexpression of miR-130b in the MM.1S cell line decreases the expression of a glucocorticoid receptor protein (GR-α), inhibiting glucocorticoid-induced apoptosis of cells, which causes resistance to glucocorticoids. While the delivery of miRNA into cells has used nanoparticles as carriers and other systems, the immobilization of miRNAs on the nanoparticles followed by testing cell transfection had yet to be demonstrated. Figure 13(A) shows several steps for the conjugation of miRNAs (miR-130b) to Au NPs through their use in cell transfection that was demonstrated recently in our laboratory.

For the visualization of cell transfection, the conjugation of miRNA to Au NPs involved immobilization of Cy5 (or Cy3)-labeled miR-130b on the Au NPs (miRNA−AuNPs). The conjugated dye label could also be detected using SERS [67,77,125]. The assessment of the conjugation and stability of the nanoparticles was performed using agarose gel electrophoresis. This technique takes advantage of the highly charged nature of miRNAs on Au NPs. When the Au NPs are capped with miRNAs, i.e., miRNA−AuNPs, a relatively high mobility is observed, which is consistent with the high surface negative charge of miRNA−AuNPs. When surface ligand exchange reaction is performed with miRNAs−AuNPs and BPs in the solution, no effect on the mobility is observed, indicating the absence of such surface exchange reactivity. This finding provides an important piece of evidence supporting the stability of the miRNA-conjugated Au NPs. The miRNA-conjugated Au NPs were demonstrated to be viable for cell transfection. Samples of the miRNA−AuNPs were tested in cell transfection using a multiple myeloma cell line (MM.1S). The clear uptake of the conjugated nanoparticles was evidenced by the presence of Cy5−miRNA−AuNPs (Figure 13(B)) in the MM.1S cells. The clear contrast in the images is due to fluorescence from the entrance of the dye-labeled miRNA−AuNPs. Functional luciferase assays were performed to quantitatively determine if the miRNA delivered into the cells by the Au NPs was capable of reducing luciferase expression. The reduction in gene expression scales with the concentration of miRNAs. The miRNA−AuNP solution contained ~15 miRNAs per nanoparticle in the luciferase knockdown. Based on a comparison of the knockdown efficiencies between

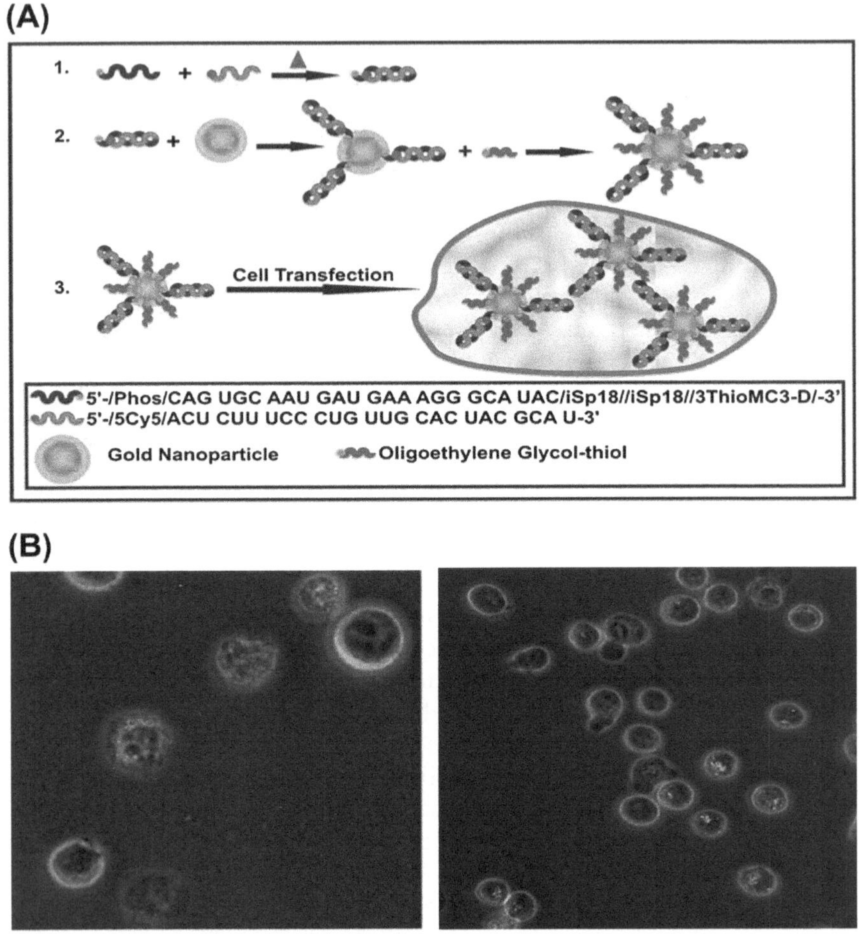

FIGURE 13 (A) Illustration (not to scale) of the preparation of miRNA–AuNP conjugates for delivering miRNAs to cells miRNA is labeled with fluorescent dye (e.g., Cy5). (B) Confocal/fluorescent composite microscopic images of the MM.1S cells 48 h after transfection were initiated with Cy5-labeled miRNA–AuNPs (B, with two different magnifications). *Reproduced with permission from Ref. [120]. Copyright 2012 American Chemical Society.*

siRNA–AuNPs (~20% at 48 h) and miRNA–AuNPs (~40% at 48 h), it appeared that the miRNA–AuNPs were somewhat more efficient at a much lower concentration. Transfection of the miRNA–AuNPs in multiple myeloma cells is clearly demonstrated.

4.3 Interparticle Assemblies for Detection of Amino Acids

The assembly of Au NPs by thiol-containing amino acids and small peptides functions as potential nanoprobes for their detection [51,78,132], especially chiral recognition in these systems [129]. Thiol-containing amino acids, such as cysteine (Cys) and homocysteine (Hcys), and small peptides, such as glutathione (GSH), are important biomolecules [51,78,131,133,134]. A level of Hcys, in both reduced and oxidized forms, above the normal range in blood (5−15 μM) is used as an indication of cardiovascular disease and other medical disorders. For Hcys- and Cys-mediated assembly of Au NPs [78,135], there is an interparticle zwitterion-type electrostatic interaction of the amino acids. In developing a colorimetric or fluorimetric nanoprobe for detecting the amino acids, one remarkable finding is the chiral recognition of cysteines driven by the nanoparticles [129]. Understanding molecular chirality of basic amino acids has important implications to medicine for specific targeting. In the nanoparticle-regulated pairwise zwitterionic interactions using enantiomeric cysteines adsorbed on Au NPs as a model system, the interparticle pairwise zwitterionic interaction of the cysteines adsorbed on Au NPs leads to the assembly of the nanoparticles via cysteine−cysteine linkages, which is characterized by the change of the SP band of Au NPs in the visible region [76,136]. Cysteines were introduced into a solution of Au NPs of 13 nm diameter (Au_{13nm}), and the SP band showed a decrease in absorbance at 520 nm and an increase in the ~630 nm region (Figure 14(A)). During this spectral evolution, the display of an isosbestic point at ~540 nm reflects the presence of two main species, the unlinked nanoparticles and the linked nanoparticles in the assembly. Based on a first-order reaction model, the apparent rate constants (*k*) obtained by curve fitting of the SP band evolution are found to show a significant difference between the interparticle homochiral and heterochiral assemblies. The *k* values for the *LL* and *DD* assemblies are found to be ~1 order of magnitude greater than that for the *DL* assembly (50% *L* and 50% *D*).

The difference between the interparticle homochiral and heterochiral interactions is further substantiated by measurements of the spectral evolution of Au NPs of different sizes in the presence of homochiral cysteines (*LL* or *DD*) or heterochiral cysteines (*DL*). Based on analysis of the three possible combinations of the enantiomeric cysteines for the interparticle pairwise dimerization using an idealized model, an excellent agreement between the experimental data and the theoretical modeling result is found for this type of interparticle chiral recognition phenomenon. If the heterochiral dimerization is less favorable than the homochiral dimerization [134], as supported by experimental data ($k_{(DL)} << k_{(LL)}$ ($\approx k_{(DD)}$)), the theoretical curve for the relative overall reaction rate is found to exhibit a minimum at an enantiomeric fraction (χ_L) of 50%. Figure 14(B) shows the correlation between the apparent rate (*r*) or rate constant (*k*) of the interparticle reactivity and the relative concentration of the enantiomer in the solution (*L*%) for Au NPs of two different sizes. A characteristic "valley" feature is revealed, with the minimum appearing at 50%*L*:50%*D*. The assembly rate for nanoparticles in the presence of heterochiral *L*- and *D*-cysteines is much slower than those in the presence of homochiral *L*- or *D*-cysteines.

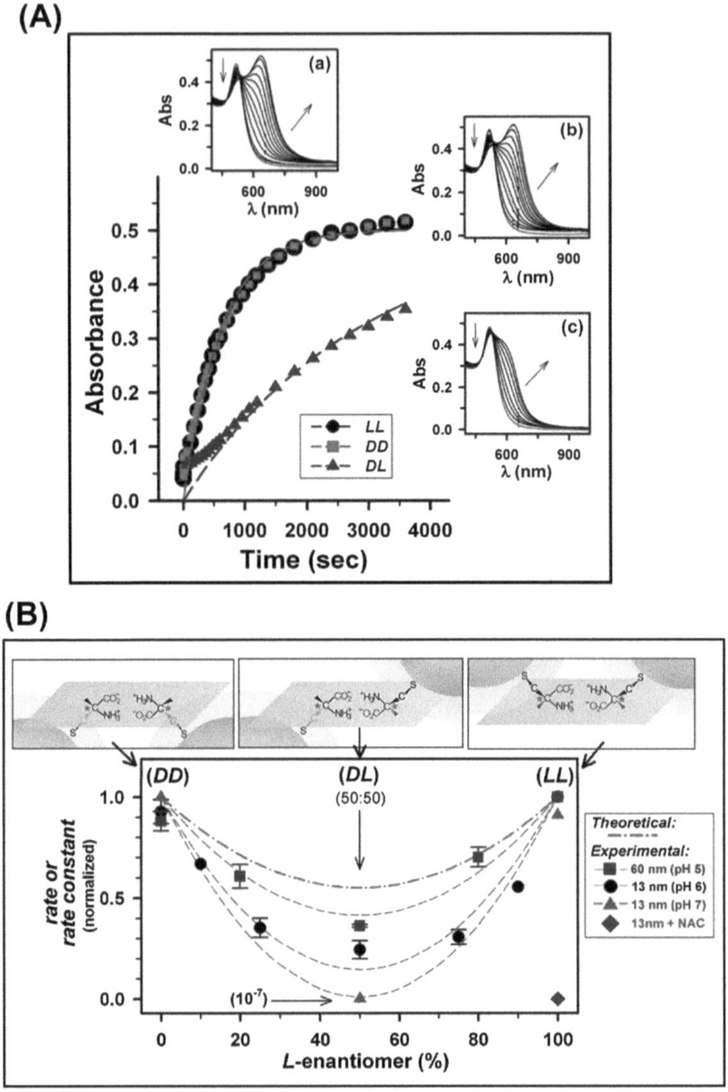

FIGURE 14 (A) Plots of the SP band absorbance (at 630 nm) as a function time for assemblies of Au_{13nm} in the presence of *LL*- (●), *DD*- (■), and *DL*- (▲) cysteines. The dotted lines represent curve fitting by a first-order reaction model ($y = a(1-e^{-kt})$) which yields $k_{(LL)} = 1.6 \times 10^{-3}\ s^{-1}$, $k_{(DD)} = 1.6 \times 10^{-3}\ s^{-1}$, and $k_{(DL)} = 3.4 \times 10^{-4}\ s^{-1}$. Insert: UV–vis spectral evolution for the Au_{13nm}-mediated assembly of *LL*- (a), *DD*- (b) and *DL*- (c) cysteines (pH = 6.2). (B) Correlation between the normalized rate or rate constant (k) and the enantiomeric percentage of cysteines (%) in the presence of Au NPs of two different sizes Au_{60nm}, Au_{13nm}, and Au_{13nm}. Top panel: A model for the interparticle pairwise zwitterionic interactions (not to scale). A hypothetical quasi-plane is drawn for illustrating the relative positions of the amino acid groups. The particles are either on the same or opposite sides of the quasi-plane. (L-, D-cysteines: $HSCH_2C^*H(NH_3^+)CO_2^-$, C*: chiral center). *Reproduced with permission from Ref. [129]. Copyright 2012 American Chemical Society.*

The inactivity observed for the assembly of Au NPs using *N*-acetyl-L-cysteine (NAC) as a mediator further confirms the important role of the interparticle pairwise zwitterionic interaction of the amino acid groups in the interparticle chiral recognition and nanoparticle assembly. The interparticle homochiral (*LL* or *DD*) vs heterochiral (*DL*) reactivity involve preferential interaction and differentiation of the different enantiomeric structures. There are important implications of the findings to the exploration of the nanoparticle-driven chiral recognition of cysteines in biomedical applications. One area stems from the linkage of elevated levels of cysteine to medical disorders (e.g., Parkinson's and Alzheimer's), where the potential use of nanoparticles to probe such elevated levels could lead to early diagnosis and identification.

Although the illustrated examples are in the areas of exploiting the ligand framework in terms of electrical and mass transfer properties, the basic principles are expected to be applicable to many other nanoparticle assemblies that are structurally defined for harnessing the collective nanoscale electrical, optical, magnetic, and spectroscopic properties. In view of these promising multifunctional properties, further studies in areas such as a detailed delineation of the interparticle parameters in terms of size, shape, composition, and surface properties and an expanded correlation of the optical, magnetic, and spectroscopic properties with different nanostructural parameters, should have a profound impact on the exploration of the molecularly mediated thin-film assembly of nanoparticles for a wide range of technological applications.

5. SUMMARY

In summary, there is a clear significance in developing the fundamental understanding of the growth mechanism in terms of sizes in the synthesis of gold nanoparticles and the properties in exploring their applications in analytical chemistry. This is in part because harnessing the nanoscale properties in most technological applications requires the abilities of controlled processing and assembly in manipulating interparticle properties. As discussed in this chapter, the molecularly engineered aggregative growth of gold nanoparticles plays an important role in manipulating size and surface properties in both aqueous and organic synthesis environments. The high monodispersity in sizes and the controllability of the surface properties for the synthesis of gold nanoparticles discussed in this chapter have enabled not only meaningful correlation between the size and the optical/spectroscopic properties but also the exploration of nanoparticles as effective nanoprobes for detection of biomarkers and biomolecules. In view of the increasing applications of gold nanoparticles, especially nanocomposite materials that combine gold and other functional nanostructures (e.g., magnetic nanoparticles), for designing sensors, biosensors, targeting, or drug delivery agents, etc., the further understanding of the molecular interactions in terms of the precursor or the capping molecule structures with multifunctional ligands, well-defined molecular rigidity and chirality, or spectroscopic signatures will provide useful information for tuning the synthesis of gold nanoparticles to achieve the desired sizes, shapes, and surface properties according to the specific applications. Indeed, as demonstrated in some of our previous and recent studies, such understanding could involve different capping or linking molecules such as thiols, thioethers, carboxylic acids, fullerenes, dyes, homocysteines, cysteines, glutathiones, proteins, and DNAs, or different core structures such as alloy, magnetic, and catalytic nanoparticles [50,51,133,137]. To further expand the applications of gold nanoparticles in analytical chemistry, in-depth understanding of factors controlling the interparticle interactions and reactivities will have a great impact for harnessing the unique electrical, optical, magnetic, and spectroscopic properties. These studies will also create new opportunities for manipulating the interparticle interaction properties and detecting the molecular signatures in terms of covalent, electrostatic, hydrogen-bonding, multidentate coordination, $\pi-\pi$ interactions, and chiral recognition. In all these aspects, a further detailed delineation of the interparticle design and control parameters should have a profound impact on the exploration of gold nanoparticles and their nanocomposite materials for a wide range of technological applications.

ACKNOWLEDGMENT

We express our appreciations to former and current members of the Zhong Research Group and collaborators who have made contributions to the work described in this article. Portions of the research are supported by the National Science Foundation (CHE 0349040, CMMI 1100736), Department of Energy – Basic Sciences (DE-SC0006877), and SUNY Research Collaboration Fund.

REFERENCES

[1] M.C. Daniel, D. Astruc, Chem. Rev. 104 (2004) 293–346.
[2] M. Faraday, Philos. Trans. R. Soc. London 147 (1857) 145–181.
[3] Y.C. Yeh, B. Creran, V.M. Rotello, Nanoscale 4 (2012) 1871–1880.
[4] K.R. Brown, D.G. Walter, M.J. Natan, Chem. Mater. 12 (2000) 306–313.
[5] D.T. Nguyen, D.J. Kim, K.S. Kim, Micron 42 (2011) 207–227.
[6] K. Saha, S.S. Agasti, C. Kim, X.N. Li, V.M. Rotello, Chem. Rev. 112 (2012) 2739–2779.
[7] E.C. Dreaden, A.M. Alkilany, X. Huang, C.J. Murphy, M.A. El-Sayed, Chem. Soc. Rev. 41 (2012) 2740–2779.
[8] J. Turkevich, P.C. Stevenson, J. Hillier, Discuss. Faraday Soc. 11 (1951) 55–75.
[9] G. Frens, Nature 241 (1973) 20–22.
[10] M. Brust, M. Walker, D. Bethell, D.J. Schiffrin, R. Whyman, Chem. Commun. (1994) 801–802.

[11] D. Jose, J.E. Matthiesen, C. Parsons, C.M. Sorensen, K.J. Klabunde, J. Phys. Chem. Lett. 3 (2012) 885–890.
[12] N.R. Jana, L. Gearheart, C.J. Murphy, Langmuir 17 (2001) 6782–6786.
[13] C.H. Kuo, T.F. Chiang, L.J. Chen, M.H. Huang, Langmuir 20 (2004) 7820–7824.
[14] J. Hu, Y. Zhang, B. Liu, J. Liu, H. Zhou, Y. Xu, Y. Jiang, J. Yang, Z.Q. Tian, J. Am, Chem. Soc. 126 (2004) 9470–9471.
[15] C. Kan, C. Wang, J. Zhu, H. Li, J. Solid State Chem. 183 (2010) 858–865.
[16] B. Nikoobakht, M.A. El-Sayed, Chem. Mater. 15 (2003) 1957–1962.
[17] T.K. Sau, C.J. Murphy, Langmuir 20 (2004) 6414–6420.
[18] A. Gole, C.J. Murphy, Chem. Mater. 16 (2004) 3633–3640.
[19] E. Oh, K. Susumu, V. Jain, M. Kim, A. Huston, J. Colloid. Interface Sci. 376 (2012) 107–111.
[20] A.C. Templeton, S.W. Chen, S.M. Gross, R.W. Murray, Langmuir 15 (1999) 66.
[21] P.J.G. Goulet, R.B. Lennox, J. Am. Chem. Soc. 132 (2010) 9582.
[22] C.K. Yee, R. Jordan, A. Ulman, H. White, A. King, M. Rafailovich, J. Sokolov, Langmuir 15 (1999) 3486.
[23] J.A. Ascencio, M. Perez, M. Jose-Yacaman, Surf. Sci. 447 (2000) 73–80.
[24] A. Sugie, T. Hatta, K. Kanie, A. Muramatsu, A. Mori, Chem. Lett. 38 (2009) 562.
[25] R. Sardar, J.S. Shumaker-Parry, Chem. Mater. 21 (2009) 1167.
[26] Y. Negishi, Y. Takasugi, S. Sato, H. Yao, K. Kimura, T. Tsukuda, J. Am. Chem. Soc. 126 (2004) 6518.
[27] G. Schmid, R. Pfeil, R. Boese, F. Bandermann, S. Meyer, G.H.M. Calis, J.W.A. Van der Velden, Chem. Ber. 114 (1981) 3634–3642.
[28] N.R. Jana, L. Gearheart, C.J. Murphy, Adv. Mater. 13 (2001) 1389.
[29] B.P. Khanal, E.R. Zubarev, J. Am. Chem. Soc. 130 (2008) 12634–12635.
[30] E. Carbó-Argibay, B. Rodráguez-González, J. Pacifico, I. Pastoriza-Santos, J. Pérez-Juste, L.M. Liz-Marzán, Angew. Chem. Int. Ed. 46 (2007) 8983–8987.
[31] J. Chen, J.M. McLellan, A. Siekkinen, Y. Xiong, Z.Y. Li, Y. Xia, J. Am. Chem. Soc. 128 (2006) 14776–14777.
[32] S.E. Skrabalak, J. Chen, Y. Sun, X. Lu, L. Au, C.M. Cobley, Y. Xia, Acc. Chem. Res. 41 (2008) 1587–1595.
[33] Z. Liang, A. Susha, F. Caruso, Chem. Mater. 15 (2003) 3176–3183.
[34] F. Kim, S. Connor, H. Song, T. Kuykendall, P. Yang, Angew. Chem. Int. Ed. 43 (2004) 3673–3677.
[35] T.K. Sau, C.J. Murphy, J. Am. Chem. Soc. 126 (2004) 8648–8649.
[36] J. Zhang, M.R. Langille, M.L. Personick, K. Zhang, S. Li, C.A. Mirkin, J. Am. Chem. Soc. 132 (2010) 14012–14014.
[37] P. Kumar, I. Pastoriza-Santos, B. Rodríguez-González, F. Abajo, L.M. Liz-Marzán, Nanotechnology 19 (2008) 015606.
[38] S.E. Skrabalak, Y. Xia, ACS Nano 3 (2009) 10.
[39] I. Hussain, M. Brust, A.J. Papworth, A.I. Cooper, Langmuir 19 (2003) 4831.
[40] P.N. Njoki, J. Luo, L. Wang, M.M. Maye, H. Quaizar, C.J. Zhong, Langmuir 21 (2005) 1623.
[41] T.K. Sau, A. Pal, N.R. Jana, Z.L. Wang, T. Pal, J. Nanopart. Res. 3 (2001) 257.
[42] D.V. Goia, E. Matijevic, New J. Chem. 22 (1998) 1203.
[43] D.V. Goia, E. Matijevic, Colloids Surf. 146 (1999) 139.
[44] A. Hengelin, J. Phys. Chem. 97 (1993) 5457.
[45] N.R. Jana, C.J. Murphy, Chem. Mater. 13 (2001) 2313.
[46] K.R. Brown, M.J. Natan, Langmuir 14 (1998) 726.

[47] R.P. Brinas, M. Hu, L. Qian, E.S. Lymar, J.F. Hainfeld, J. Am. Chem. Soc. 130 (2008) 975.
[48] K. Yu, K.L. Kelly, N. Sakai, T. Tatsuma, Langmuir 24 (2008) 5849.
[49] S.D. Perrault, W.C.W. Chan, J. Am. Chem. Soc. 131 (2009) 17042.
[50] L. Wang, J. Luo, M.J. Schadt, C.J. Zhong, Langmuir 26 (2010) 618–632.
[51] S. Lim, C.J. Zhong, Acc. Chem. Res. 42 (2009) 798–808.
[52] F.L. Leibowitz, W.X. Zheng, M.M. Maye, C.J. Zhong, Anal. Chem. 71 (1999) 5076.
[53] W.X. Zheng, M.M. Maye, F.L. Leibowitz, C.J. Zhong, Anal. Chem. 72 (2000) 2190.
[54] L. Han, M.M. Maye, F.L. Leibowitz, N.K. Ly, C.J. Zhong, J. Mater. Chem. 11 (2001) 1259.
[55] L. Han, D.R. Daniel, M.M. Maye, C.J. Zhong, Anal. Chem. 73 (2001) 4441.
[56] L. Han, J. Luo, N.N. Kariuki, M.M. Maye, V.W. Jones, C.J. Zhong, Chem. Mater. 15 (2003) 29.
[57] L. Han, W. Wu, F.L. Kirk, J. Luo, M.M. Maye, N.N. Kariuki, Y. Lin, C. Wang, C.J. Zhong, Langmuir 20 (2004) 6019.
[58] L. Wang, D. Miller, Q. Fan, J. Luo, M.J. Schadt, Q. Rendeng, G.R. Wang, J. Wang, G.R. Kowach, C.J. Zhong, J. Phys. Chem. C 112 (2008) 2448.
[59] L.Y. Wang, X. Shi, N.N. Kariuki, M.J. Schadt, G.R. Wang, Q. Rendeng, J. Choi, J. Luo, S. Lu, C.J. Zhong, J. Am. Chem. Soc. 129 (2007) 2161.
[60] G.R. Wang, L.Y. Wang, Q. Rendeng, J. Wang, J. Luo, C.J. Zhong, J. Mater. Chem. 17 (2007) 457.
[61] C.J. Zhong, M.M. Maye, Adv. Mater. 13 (2001) 1507.
[62] P.N. Njoki, J. Luo, M.M. Kamundi, S. Lim, C.J. Zhong, Langmuir 26 (2010) 13622–13629.
[63] A. Stella, P. Cheyssac, R. Kofman, in: F.C. Matacotta, G. Ottaviani (Eds.), Science and Technology of Thin Films, vol. 57, World Scientific, River Edge, NJ, 1996, p. 57.
[64] M.J. Schadt, W. Cheung, J. Luo, C.J. Zhong, Chem. Mater. 18 (2006) 5147.
[65] D. Mott, J. Yin, M.H. Engelhard, R. Loukrakpam, P. Chang, G. Miller, I.T. Bae, N.C. Das, C. Wang, J. Luo, C.J. Zhong, Chem. Mater. 22 (2010) 261.
[66] S.P. Shields, V.N. Richards, W.E. Buhro, Chem. Mater. 22 (2010) 3212.
[67] P.N. Njoki, S. Lim, D. Mott, H.Y. Park, B. Khan, S. Mishra, R. Sujakumar, J. Luo, C.J. Zhong, J. Phys. Chem. C 111 (2007) 14664.
[68] C.J. Zhong, P.N. Njoki, J. Luo, Patent No. 7524354 (2009).
[69] G. Schmid, Chem. Rev. 92 (1992) 1709.
[70] M. Avrami, J. Chem. Phys. 8 (1940) 212.
[71] G.J. Exarhos, M. Aloi, Thin Solid Films 193 (1990) 42.
[72] S.Z.D. Cheng, B. Wunderlich, Macromolecules 21 (1988) 3327.
[73] J. Luo, M.M. Maye, L. Han, N.N. Kariuki, V.W. Jones, Y. Lin, M.H. Engelhard, C.J. Zhong, Langmuir 20 (2004) 4254.
[74] S. Lim, J. Ouyang, J. Luo, L. Wang, S. Zhou, C.J. Zhong, Chem. Mater. 17 (2005) 6528.
[75] S. Lim, C. Vaiana, Z.Y. Zhang, Y.J. Zhang, D.L. An, C.J. Zhong, J. Am. Chem. Soc. 129 (2007) 5368.
[76] S. Lim, C.J. Zhong, Gold Bull. 40 (2007) 59.
[77] S. Lim, P.N. Njoki, H.Y. Park, X. Wang, L. Wang, D. Mott, C.J. Zhong, Nanotechnology 19 (2008) 11.
[78] S. Lim, W. Ip, E. Crew, P.N. Njoki, D. Mott, C.J. Zhong, Y. Pan, S. Zhou, Langmuir 23 (2007) 826–833.
[79] C.J. Zhong, W.X. Zhang, F.L. Leibowitz, H.H. Eichelberger, Chem. Commun. 13 (1999) 1211.

[80] M.M. Maye, W.X. Zheng, F.L. Leibowitz, N.K. Ly, C.J. Zhong, Langmuir 16 (2000) 490.
[81] M.M. Maye, C.J. Zhong, J. Mater. Chem. 10 (1895) 2000.
[82] D. Mott, J. Galkowski, L. Wang, J. Luo, C.J. Zhong, Langmuir 23 (2007) 5740.
[83] P.H. Buffat, J.P. Borel, Phys. Rev. A 13 (1976) 2287.
[84] J.P. Borel, Surf. Sci. 106 (1981) 1.
[85] L.L. Lewis, J. Pablo, J.L. Barrat, Phys. Rev. B 56 (1997) 2248.
[86] J. Yin, S. Shan, L. Yang, D. Mott, O. Malis, V. Petkov, F. Cai, M.S. Ng, J. Luo, B.H. Chen, M. Engelhard, C.J. Zhong, Chem. Mater. 24 (2012) 4662–4674.
[87] M.J. Hostetler, C.J. Zhong, B.K.H. Yen, J. Anderegg, S.M. Gross, N.D. Evans, M.D. Porter, R.W. Murray, J. Am. Chem. Soc. 120 (1998) 9396.
[88] M.J. Hoetetler, J.E. Wingate, C.J. Zhong, J.E. Harris, R.W. Vachet, M.R. Clark, J.D. Londono, S.J. Green, J.J. Stokes, G.D. Wignall, G.L. Glish, M.D. Porter, N.D. Evans, R.W. Murray, Langmuir 14 (1998) 17.
[89] C.J. Zhong, W.X. Zheng, F.L. Leibowitz, Electrochem. Commun. 1 (1999) 72.
[90] W.X. Zheng, M.M. Maye, F.L. Leibowitz, C.J. Zhong, Analyst 125 (2000) 17.
[91] M. Brust, J. Fink, D. Bethell, D.J. Schiffrin, C.J. Kiely, J. Chem. Soc. Chem. Commun. 1655 (1995).
[92] H.Y. Park, M.J. Schadt, L. Wang, S. Lim, P.N. Njoki, S.H. Kim, M.Y. Jang, J. Luo, C.J. Zhong, Langmuir 23 (2007) 9050.
[93] N.Z. Clarke, C. Waters, K.A. Johnson, J. Satherley, D.J. Schiffrin, Langmuir 17 (2001) 6048.
[94] T. Teranishi, S. Hasegawa, T. Shimizu, M. Miyake, Adv. Mater. 13 (2001) 1699.
[95] H. Fan, J. Gabaldon, C.J. Brinker, Y.B. Jiang, Chem. Commun. 22 (2006) 2323.
[96] F. Terzi, R. Seeber, L. Pigani, C. Zanardi, L. Pasquali, S. Nannarone, M. Fabrizio, S.J. Daolio, Phys. Chem. B 109 (2005) 19397.
[97] A.W. Shaffer, J.G. Worden, Q. Huo, Langmuir 20 (2004) 8343.
[98] S.W. Chen, J.M. Sommers, J. Phys. Chem. B 105 (2001) 8816.
[99] M.J. Hostetler, J.E. Wingate, C.J. Zhong, Y.E. Harris, R.W. Vachet, M.R. Clark, J.D. Londono, S.J. Green, J.J. Stokes, G.D. Wignall, G.L. Glish, M.D. Porter, N.D. Evans, R.W. Murray, Langmuir 14 (1998) 17–30.
[100] L. Wang, J. Luo, Q. Fan, M. Suzuki, I.S. Suzuki, M.H. Engelhard, Y. Lin, N. Kim, J.Q. Wang, C.J. Zhong, J. Phys. Chem. B 109 (2005) 21593.
[101] L. Wang, J. Luo, M.M. Maye, Q. Fan, Q. Rendeng, M.H. Engelhard, C. Wang, Y. Lin, C.J. Zhong, Mater. Chem. 15 (2005) 1821.
[102] L. Wang, H.Y. Park, S. Lim, M.J. Schadt, D. Mott, J. Luo, X. Wang, C.J. Zhong, J. Mater. Chem. 18 (2008) 2629.
[103] X. Wang, L. Wang, S. Lim, K. Bao, D. Mott, H.Y. Park, J. Luo, S. Hao, C.J. Zhong, J. Nanosci. Nanotech. 9 (2009) 3005.
[104] S.H. Sun, H. Zeng, J. Am. Chem. Soc. 124 (2002) 8204.
[105] S.H. Sun, H. Zeng, D.B. Robinson, S. Raoux, P.M. Rice, S.X. Wang, G.X. Li, J. Am. Chem. Soc. 126 (2004) 273.
[106] J. Luo, L. Wang, D. Mott, P.N. Njoki, Y. Lin, T. He, Z. Xu, B. Wanjana, S. Lim, C.J. Zhong, Adv. Mater. 20 (2008) 4342.
[107] J. Luo, H. Fan, N.N. Kariuki, L. Wang, D. Mott, C.J. Zhong, T. He, Chem. Mater. 17 (2005) 5282.
[108] J. Luo, N.N. Kariuki, L. Han, L. Wang, C.J. Zhong, T. He, Electrochim. Acta 8 (2006) 581.
[109] J. Luo, L. Wang, D. Mott, P.N. Njoki, N.N. Kariuki, C.J. Zhong, T. He, J. Mater. Chem. 16 (2006) 1665.

[110] J. Luo, P.N. Njoki, Y. Lin, D. Mott, L. Wang, C.J. Zhong, Langmuir 22 (2006) 2892.
[111] J. Luo, P.N. Njoki, Y. Lin, L. Wang, D. Mott, C.J. Zhong, Electrochem. Comm. 8 (2006) 581.
[112] N.N. Kariuki, J. Luo, M.M. Maye, A. Hassan, T. Menard, H.R. Naslund, Y. Lin, C. Wang, M.H. Engelhard, C.J. Zhong, Langmuir 20 (2004) 11240.
[113] F. Wang, V.N. Richards, S.P. Shields, W.E. Buhro, Chem. Mater. 26 (2014) 5−21.
[114] H. Fan, C. Wong, A.R. Chianese, L. Wang, J. Luo, J. Yin, C.J. Zhong, Chem. Mater. 22 (2010) 5797−6028.
[115] H. Yan, H. Xie, Q. Su, J. Luo, D. Xie, J. Yin, B.N. Wanjala, D.L. An, C.J. Zhong, Phys. Chem. Chem. Phys. 13 (2011) 5824−5830.
[116] E. Crew, S. Lim, H. Yan, S. Shan, J. Yin, L. Lin, R. Loukrakpam, L. Yang, J. Luo, C.J. Zhong, in: M. Hepel, C.J. Zhong (Eds.), Functional Nanoparticles for Bioanalysis, Nanomedicine and Bioelectronic Devices, ACS Symp. Series, ACS, 2012, pp. 119−146 (Chapter 5).
[117] E. Crew, H. Yan, L. Lin, J. Yin, Z. Skeete, T. Kotlyar, N. Tchah, J. Lee, M. Bellavia, I. Goodshaw, P. Joseph, J. Luo, S. Gal, C.J. Zhong, Analyst 138 (2013) 4941−4949.
[118] L.Q. Lin, E. Crew, H. Yan, S. Shan, Z. Skeete, D. Mott, T. Krentsel, J. Yin, N.A. Chernova, J. Luo, M.H. Engelhard, C. Wang, Q.B. Li, C.J. Zhong, J. Mater. Chem. B 1 (2013) 4320−4330.
[119] H.W. Cheng, S.Y. Huan, R.Q. Yu, Analyst 137 (2012) 3601−3608.
[120] E. Crew, M.A. Tessel, S. Rahman, A. Razzak-Jaffar, D. Mott, M. Kamundi, G. Yu, N. Tchah, J. Lee, M. Bellavia, C.J. Zhong, Anal. Chem. 34 (2012) 26−29.
[121] J. Luo, J. Luo, L. Wang, X. Shi, J. Yin, E. Crew, S. Lu, L.M. Lesperance, C.J. Zhong, Sens. Actuators B 161 (2012) 845−854.
[122] H.W. Cheng, S.Y. Huan, H.L. Wu, G.L. Shen, R.Q. Yu, Anal. Chem. 81 (2009) 9902−9912.
[123] H.W. Cheng, W.Q. Luo, G.L. Wen, S.Y. Huan, G.L. Shen, R.Q. Yu, Analyst 135 (2010) 2993−3001.
[124] H.W. Cheng, Y.Y. Chen, X.X. Lin, S.Y. Huan, H.L. Wu, G.L. Shen, R.Q. Yu, Anal. Chim. Acta 707 (2011) 155−163.
[125] J.D. Driskell, R.J. Lipert, M.D. Porter, J. Phys. Chem. B 110 (2006) 17444−17451.
[126] M.D. Porter, R.J. Lipert, L.M. Siperko, G. Wang, R. Narayanana, Chem. Soc. Rev. 37 (2008) 1001−1011.
[127] X.Y. Zhang, M.A. Young, O. Lyandres, R.P. Van Duyne, J. Am. Chem. Soc. 127 (2005) 4484−4489.
[128] X.Y. Zhang, J. Zhao, A.V. Whitney, J.W. Elam, R.P. Van Duyne, J. Am. Chem. Soc. 128 (2006) 10304−10309.
[129] S. Lim, D. Mott, M. Engelhard, Y. Pan, S. Kamodia, J. Luo, P.N. Njoki, S. Zhou, L. Wang, C.J. Zhong, Anal. Chem. 81 (2009) 689−698.
[130] S. Lim, U. Chandrachud, L. Wang, S. Gal, C.J. Zhong, Anal. Chem. 80 (2008) 6038−6044.
[131] H. Yan, S. Lim, L.C. Zhang, S.C. Gao, D. Mott, Y. Le, D.L. An, C.J. Zhong, J. Mater. Chem. 21 (1890) 2011.
[132] S. Lim, D. Mott, W. Ip, P.N. Njoki, Y. Pan, S. Zhou, C.J. Zhong, Langmuir 24 (2008) 8857−8863.
[133] L. Wang, J. Luo, S. Shan, E. Crew, Y. Lin, B. Wallek, S. Wong, C.J. Zhong, Anal. Chem. 83 (2011) 8688−8695.
[134] G. Castronuovo, V. Elia, M. Niccoli, D. Strollo, F. Velleca, Phys. Chem. Chem. Phys. 1 (1999) 5653−5657.

[135] E. Crew, S. Rahman, A. Razzak-Jaffar, D. Mott, M. Kamundi, G. Yu, N. Tchah, J. Lee, M. Bellavia, C.J. Zhong, Anal. Chem. 84 (2012) 26–29.

[136] M.M. Maye, S. Lim, J. Luo, Z. Rab, D. Rabinovich, T. Liu, C.J. Zhong, J. Am. Chem. Soc. 127 (2005) 1519–1529.

[137] J. Luo, J. Yin, R. Loukrakpam, B.N. Wanjala, B. Fang, S. Shan, L. Yang, M.S. Ng, J. Kinzler, Y.S. Kim, C.J. Zhong, J. Electroanal. Chem. 688 (2013) 196–206.

Chapter 3

Physico-Chemical Characteristics of Gold Nanoparticles

Vincenzo Amendola[1], Moreno Meneghetti[1], Mauro Stener[5], Yan Guo[2,3], Shaowei Chen[3], Patricia Crespo[4], Miguel Angel García[4], Antonio Hernando[4], Paolo Pengo[5] and Lucia Pasquato[5,*]
[1] Department of Chemical Sciences, University of Padova, Padova, Italy; [2] School of Environmental Science and Engineering, Nanjing University of Information Science and Technology, Nanjing, Jiangsu, P. R. China; [3] Department of Chemistry and Biochemistry, University of California, Santa Cruz, CA, USA; [4] Instituto de Magnetismo Aplicado and Dpto. Física de Materiales, Universidad Complutense de Madrid, Madrid, Spain; [5] Department of Chemical and Pharmaceutical Sciences, University of Trieste, Trieste, Italy
**Corresponding author: E-mail: lpasquato@units.it*

Chapter Outline

Gold Nanoparticles in Analytical Chemistry. http://dx.doi.org/10.1016/B978-0-444-63285-2.00003-1

1. INTRODUCTION

Considering their size and surface functionalization, gold nanoparticles may be divided into three main groups: gold colloids, gold nanoparticles protected by an organic monolayer (MPCs), and small gold clusters.

Gold colloids are large metal particles with a size range of 10–100 nm, usually obtained by reduction of Au(III) salts in the presence of mild reducing agents, most commonly, trisodium citrate in procedures created by Turkevich and Frens and improved by Natan [1]. The resulting particles are stabilized against aggregation by electrostatic repulsion as a result of the surface-adsorbed citrate. This stabilization, being electrostatic in nature, does not allow the use of colloids in media of elevated ionic strength; the additions of salts to a colloidal solution results in irreversible aggregation. However, gold colloids may be stabilized by adsorption of a wide variety of compounds such as proteins or polymers but also by interaction with other compounds such as thiols that form a compact, self-assembled monolayer on their surface. In describing the structure of colloids, it must be stressed that the syntheses of these compounds is very often carried out under kinetic rather than thermodynamic control. These conditions do not allow obtaining the most stable structure possible in the reaction conditions employed. As a consequence, the preparations of gold colloids frequently contain particles with a large variety of shapes and typically are polydispersed with a relatively large distribution of sizes. However, procedures for the obtainment of gold nanorods, nanocubes, nanoboxes, nanobranches, and nanobipyramids are available; generally, these procedures rely on the use of specific surface modifiers and seeds during the synthesis.

Monolayer protected clusters (MPCs) are particles with a gold core with a size in the range of 1–10 nm surrounded by a monolayer of strongly chemisorbed ligands; the most studied systems are stabilized by thiols. The resulting materials are very stable, owing to the strength of the gold–sulfur bond that is about 208 kJ/mol. MPCs became available after the synthesis developed by Brust and Schiffrin in 1994. At variance with colloids, MPCs are stable in solution and in the solid state. The presence of the passivating monolayer prevents irreversible aggregation and imparts solubility in a broad range of solvents including aqueous solutions and biological media. MPCs are, in general, prepared under kinetic control, polydispersity is typical, and metal cores with different shapes with a varying extent of twinning are common. Only recently, a theoretical thermodynamic analysis aimed at identifying the most stable morphologies of small gold nanoparticles was performed, indicating that the equilibrium shapes not only depend on the temperature but also on the size of the particles, with well-defined shape transitions occurring at certain critical sizes. In general, the most stable shapes of the particles are those with the preferential development of low energy facets. Small gold clusters represent the "lower bound" of MPCs; these are molecular entities comprising a few tens of atoms and are typically "atomically precise" or strictly monodispersed. The monodispersity of these species is the result of their inherent stability; in several cases, the preparation of small gold clusters relies on a "size focusing" or "thiol etching" process that converts polydispersed mixtures of unstable clusters to the most stable ones. The monodispersity of these systems enables their crystallization and the obtainment of several X-ray structures of gold-thiolate clusters that are so far unreported for larger MPCs.

1.1 Physical Properties, Chemical Properties, and Their Size Dependence

The physical properties of gold nanoparticles are strongly influenced by their size because the electronic structure of the gold particles changes with size and shape. Small gold clusters feature large band gaps, while in larger particles the valence electrons may be described in the framework of the *jellium* model. A clear example of this transition can be obtained by inspection of the electronic absorption spectra of these materials. The dominating optical feature of AuNPs with size of 2–100 nm is the surface plasmon resonance (SPR) which confers the characteristic bright red color to colloidal solutions of spherical AuNPs. This has attracted the interest of manufacturers since ancient times for the realization of ornamental cups, glass, and luster plates. Compared to other organic and inorganic chromophores, AuNPs with a size above about 2 nm have larger extinction cross-sections, with a possibility of reaching 100% of light-to-heat conversion efficiency, high photostability, and the ability to amplify the electromagnetic field near the metal surface. On the contrary, for smaller clusters, well-defined and specific absorption features are observed. Small thiolate-protected gold nanoparticles exhibit ferromagnetic-like behavior that is in stark contrast to the diamagnetism of bulk gold and of weakly protected nanoparticles. Another effect associated with the reduction of the nanoparticle size is the onset of quantized (single electron) charging with the nanoparticles behaving as nanoscale molecular capacitors. The chemical properties of gold nanoparticles and their catalytic activity are also strongly dependent on their size and shape. In addition, the nanoparticle size is one of the parameters governing the formation of domains with well-defined morphology on the monolayer of MPCs. In the following sections, an in-depth discussion on the properties of gold nanoparticles will be presented.

In this chapter, we discuss the relevant physical and chemical properties of gold nanoparticles that are exploited for analytical applications in a variety of different fields. This chapter was conceived to collect in a single place an overview of the properties of gold nanoparticles that are important for technological applications, with prevalent reference to the scientific literature that has appeared in the last six years. Given the extremely broad expertise needed for an in-depth discussion, the chapter is co-authored with experts in the specific field dealing with optical properties, ab initio calculations of electronic properties, magnetic, electrochemical and chemical properties.

2. OPTICAL PROPERTIES

2.1 Origin of the Surface Plasmon Absorption

Collective oscillations of conduction electrons in metals are known as *plasmons* [2,3]. According to the Fermi liquid model, plasmons can be satisfactorily described as a negatively charged electron cloud displaced from its equilibrium position around a lattice made of positively charged ions, in analogy to a real plasma [2,3]. Excitation of bulk plasmons by photons is not allowed since the energy dispersion curves of plasmons and photons never cross each other [2,3]. However, the presence of a surface in real materials allows the existence of new plasmon modes, called surface plasmons, which can be excited by electromagnetic waves [2,3]. In particular, localized surface plasmons are observed in metal nanoparticles, which are nonpropagating collective excitations of conduction electrons. Therefore, they differ from surface plasmon-polaritons, which are propagating plasmon modes allowed on the surface of bulk metals and periodic metallic patterns [2]. When AuNP size is small compared to the resonant electromagnetic wavelength, the electric field distribution along the particle is uniform, electron cloud polarization is coherent in each point of the metal, and dipolar plasmon oscillations are excited [3]. When AuNP size is comparable to the resonant electromagnetic wavelength, the electric field distribution along the particle is nonuniform, electron cloud polarization is no more coherent in each point of the metal, and multipolar plasmon oscillations are excited [3].

Surface plasmons in single-metal nanoparticles can be directly probed by electron energy loss spectroscopy (EELS), cathodoluminescence, and far field extinction microscopy, while the plasmon response of ensembles of AuNPs can be detected by optical absorption spectroscopy [2–5]. In the optical absorption spectrum of a dispersion of spherical AuNPs (Figure 1(A)), the SPR is centered around 520 nm, and it is partially overlapped with the absorption originated by interband transitions, namely by single electron optical

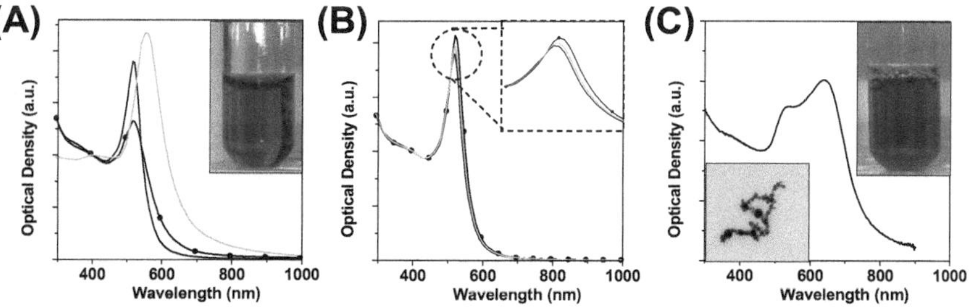

FIGURE 1 (A) Optical absorption spectra for AuNPs with size of 4 nm (black circles), 30 nm (black line), and 100 nm (grey line) in water. All spectra are scaled to 1:400 nm for clarity. Damping of SPR is observed in the 4 nm AuNPs due to intrinsic size effects, while red shift and broadening are observed in the 100 nm AuNPs due to extrinsic size effects. Inset shows the typical purple color of a AuNP colloid. (B) The SPR of a 30 nm AuNP solution in water (black line) is red shifted and slightly broadened when the medium refractive index is increased 2.5% (grey line) and 5% (black circles). Inset shows a magnification of the SPR. (C) The SPR is sensibly red shifted and broadened in aggregates of AuNPs, as those shown in the TEM image. To the naked eye, this effect is perceived as a shift to violet-blue (see inset).

excitations from the occupied 5d band to the unoccupied levels of the 6s−6p band of the metal [3,4]. The SPR intensity scales with the number of conduction electrons in the AuNP, where each gold atom contributes one conduction electron [4], and this accounts for their incredibly high extinction cross-section, which is 10^5 times larger than that of organic chromophores for a 20 nm AuNP [2,6].

The plasmon extinction band detected in the optical absorption spectra of AuNPs dispersions is originated by the sum of scattered and absorbed photons, with a proportion strongly dependent on particle size [2,3]. For AuNPs with size $d \ll \lambda$, where λ is the wavelength of the electromagnetic radiation, the absorption dominates over scattering [2,3]. However, scattering and absorption scale with the 6th and the 3rd power of a particle's size, respectively. Therefore, the two effects become comparable for 50 nm AuNPs, whereas scattering prevails above this size [2,3].

2.2 Specific Phenomena Influencing the Surface Plasmon Absorption

The SPR is primarily affected by particle size, shape, and physical−chemical environment, including the reciprocal distance between particles [2,3]. The size effects influencing the SPR are divided into extrinsic and intrinsic (Figure 1(A)) [3]. The extrinsic size effect is a retardation effect due to the excitation of multipolar plasmon modes when nanoparticle size increases toward the resonant electromagnetic wavelength [2,3]. The extrinsic size effect is a bare electromagnetic phenomenon appearing in the optical absorption spectrum as broadening and red shift of the SPR for increasing AuNP size [2,3]. The multipolar excitations are dominant for AuNPs larger than 60 nm, whereas dipolar oscillations dominate for particles of 25 nm or smaller [2,3].

Intrinsic size effects are due to the modification of the metal optical constant when AuNP size is below 30 nm [3]. The optical constant changes because of additional contributions to the free electron relaxation when the nanoparticle size is reduced [3]. In particular, the free electron scattering at particle surface is no more negligible when conduction electrons mean free path (~40 nm for Au) becomes comparable to particle size [2,3]. This scattering contribution overshadows other phenomena originated by changes in the phonons spectra, quantum size effects, changes in the phonon−electron coupling for the high surface charge during plasmon oscillation, and so on. In the optical absorption spectrum, the intrinsic size effect increases the decay rate of the SPR and, consequently, it damps for decreasing particle's size, with almost complete quenching of the plasmon band in AuNPs smaller than 2 nm [2,3].

The width, position, and number of SPRs are determined by the shape of the AuNPs [2,6]. Spherical particles have only one SPR, whereas cigar-like structures have two plasmon modes, resonating at different wavelengths and corresponding to the collective excitation of conduction electrons along the two axes of the rod [5]. By further lowering the symmetry of the Au nanostructure, new plasmon modes may arise [2,6]. A common feature of these new SPRs is that they are red shifted compared to that of spheres [3]. For instance, the red shift of the SPR corresponding to longitudinal excitation of Au rods by the electromagnetic field increases with the aspect ratio of the rod [5].

The physical-chemical environment influences the SPR in several ways. Regarding the chemical effects, the most intense mechanism is called chemical interface damping (CID) [3]. CID produces a sensible widening and red shift of the SPR when adsorbates are present on a particle's surface [3].

This holds both for chemisorption, as in the case of thiol-stabilized nanoparticles, and for physisorption, as in the case of citrate- or alkylamine-stabilized nanoparticles [3]. The SPR broadening due to CID is explained considering that adsorbates offer new relaxation pathways for both excited electrons and phonons in the metal [3]. Persson developed a model in which the empty lowest unoccupied molecular orbitals (LUMOs) of adsorbates are coupled with conduction band electrons [3]. During plasmon oscillation, the reversible electron transfer into LUMOs is possible, originating fast loss of coherence with other excited electrons and a larger bandwidth [3]. Different adsorbates have different LUMO position, then couple in a different way with the metal conduction band and originate different SPR widths [3]. Regarding the physical effects, the position, width, and intensity of the SPR are sensitive to the dielectric properties of the environment [2,3]. In AuNPs dispersed in a homogeneous transparent matrix, for example, a liquid solution or a polymeric matrix, the increase in the refractive index of the medium will correspond to an increase in intensity, red shift, and full width at half maximum (FWHM) of the SPR (Figure 1(B)) [3]. In the case of AuNPs coated with ligands, the same effect on the SPR will be observed as a function of the polarizability of the surrounding layer [2].

Moreover, plasmons on distinct nanoparticles can reciprocally influence each other, depending on the particle's distance, geometry, and relative photon polarization [2,6]. In general, multimers of spheres have new SPR because, when the nanoparticle separation decreases, the plasmons of individual nanoparticles begin to interact and hybridize with the plasmons of the other particles [2,6]. In the case of dimers, plasmon hybridization produces an SPR red shift that decays near-exponentially with the interparticle gap over a distance that is roughly 0.2 times the particle size, independent of the nanoparticle size, shape, material, and the medium dielectric constant [7].This universal decaying behavior of the SPR in nanoparticle pairs originates from the distance decay of the interparticle near-field that couples the particles together [7]. When nanoparticles are in conductive contact, the hybridization effects disappear and the SPR abruptly blue shifts to that of a particle with necklace shape [7,8].

2.3 Sensing by Surface Plasmon Absorption

The SPR finds large applications for chemical and biological sensing due to sensitive spectral response to the local surface environment, photostability, and ease of monitoring the light signal because of the large scattering and absorption cross-sections [9].The spectral sensitivity, defined as the relative shift in resonance wavelength with respect to the refractive index change of surrounding materials, is dependent on the SPR position and width, which in turn depends on metal type and on particle shape [9]. This determines a long spatial range of the decaying electromagnetic field, allowing appreciable sensitivity up to a distance of 40 nm [9]. In general, a linear relationship between the absorption or scattering band position and the surrounding refraction index exists, with a slope of about 100 nm for refraction index unit [9].

One of the most important features for a sensor is selectivity, which can be conferred to AuNPs by appropriate engineering of their surface, based on the reactivity of Au atoms with thiols and disulfides. Usually, AuNPs are functionalized with a specific molecular receptor for a target molecule capable of producing an appreciable red shift of the SPR after the change in the surrounding refraction index of metal nanoparticles upon binding of the target molecule [5]. This approach is similar to enzyme-linked immunosorbent assay (ELISA), the difference lying in that ELISA requires a second, specific fluorescent label that is able to bind the target in a sandwich structure. Whereas, these localized surface plasmon assays exploit only one binding event [8].

On the other hand, ELISA-like colorimetric sensors, based on AuNPs, have also been developed for the detection of analytes which can be coordinated on two ends [10]. In this case, the substrate and the AuNPs are both coated with a targeting molecule, for example, an antibody selective for the target protein, and the substrate is decorated with AuNPs in the presence of the analyte, with a consequent change in color from white to red-brownish [10].

Plasmon coupling in nearby AuNPs is also exploited in molecular rulers based on the SPR dependence on the distance between two metal nanoparticles [11]. Plasmon molecular rulers have been applied to the real-time sensing of DNA hybridization processes as well as to DNA folding in the presence of different saline buffers [10]. In particular, the SPR sensitivity on the coupling distance of two metal nanoparticles is so high that each step of dimer formation is measurable, included elastic recoil after the first contact [10]. AuNP of 40 nm in diameter allows monitoring distances up to 70 nm by far-field extinction microscopy of a single dimer [10].

The SPR changes due to plasmon hybridization in AuNPs aggregates can be also exploited for colorimetric assays [10]. In agglomerates of AuNPs made by tens of particles, plasmon hybridization originates two main bands, one close to the original sphere's position and another red-shifted, both sensibly broader than single-particle SPR (Figure 1(C)) [8,10,12]. When particles in the aggregates are several hundred, only one very broad red-shifted band is present [2,6,8,12]. In a typical strategy, metal nanoparticles in solution are coated with specific receptors, for instance, with single-stranded DNA polynucleotides or appositely designed peptides, which promote particle aggregation and sensible change in the color of the solution only in the presence of the target biomolecules [6,10]. This method allows analyte detection limits from femtomoles to nanomoles [6,10]. In the case of DNA detection, the target polynucleotide bridges two particles during the hybridization process with receptor oligonucleotides on the metal surface [6,10].

Since more than one receptor is present on each nanoparticle, hundreds to thousands of nanoparticles are connected together during the hybridization process [6,10]. Nanoparticle-labeled DNA also have a sharper melting hybridization—dehybridization transition than unlabeled DNA, as a consequence of the dense loading of oligonucleotides on nanoparticle surface and of the cooperative hybridization process [13]. In this way, DNA melting temperature can be optically detected and exploited for the discrimination of different polynucleotides [6,10]. SPR assays for DNA have better target specificity and sensitivity than polymerase chain reactions (PCR). SPR assays for proteins are up to 10^6 times more sensitive than conventional protein detection methods like ELISA [6,10].

2.4 Surface-Enhanced Raman Scattering and Local Field Enhancement

The electromagnetic field in the nearby AuNPs surface is enhanced by the collective oscillation of conduction electrons [2,14]. The local field enhancement scales with particle size, and it depends strongly on the assembly of AuNPs [14,15]. The local electric field at the interstices between two AuNPs is thousands of times higher than for the same isolated particles (Figure 2(A)) [14,15]. For this reason, interparticle gaps are called hot spots, and the effect of local-field amplification on nanoscale volumes is called nanolensing [14]. The amplification increases with decreasing gap up to a plateau for a separation below about 0.5 nm, when quantum tunneling effects start compensating plasmon coupling effects [16]. Other than AuNPs multimers and aggregates, linear chains of several metal nanospheres with progressively decreasing size and separation are very efficient nanolenses [14]. The maximum field amplification is achieved within a dimer when incident light is polarized parallel to the dimer axis and for a wavelength resonant with the SPR of the dimer [14].

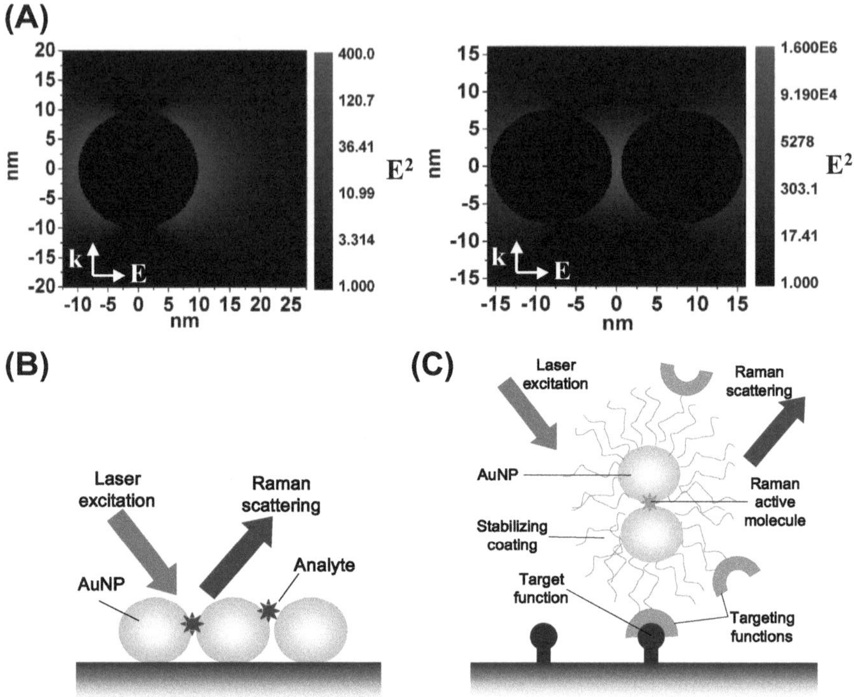

FIGURE 2 (A) The electric field is enhanced in proximity of AuNPs excited by electromagnetic wavelength. This effect is orders of magnitude higher at the junction between nanoparticles (hot spots) in particle aggregates, and this is at the base of SERS. (B–C) SERS is exploited for ultrasensitive detection of analytes by Raman spectroscopy (B) or for ultra-bright labeling of target samples (B). In (B), the analyte emits intense Raman signals when deposed on the AuNP substrate. In (C), AuNPs are loaded with a Raman-active dye and a targeting function, so that the target analyte is detected by the intense Raman signal after binding of the SERS label to the substrate.

Theoretical calculations and direct measurements on these hot spots showed that local electric field enhancement at particle interstices are useful to amplify spectroscopic signals of adsorbed organic molecules [14]. In particular, when molecules are adsorbed on metal nanoparticles or roughened metallic surfaces, a huge amplification of Raman signals is observed, called surface-enhanced Raman scattering (SERS) [14]. Enhancements up to 14 orders of magnitude for the Raman scattering cross-section have been reported [14].

When AuNPs are assembled in a way that infrared plasmon modes exist, surface-enhanced infrared absorption (SEIRA) spectroscopy is also possible [2,14].

Two distinct mechanisms can originate SERS: electromagnetic field enhancement and chemical first layer effects [2,14]. In the latter case, the electronic coupling between the molecule and the metal can produce a resonance Raman effect due to a new metal–molecule charge transfer transition [2,14]. The magnitude of chemical enhancement has been estimated to reach no more than factors of 10–100 [2,14].

The electromagnetic field enhancement is the only one due to the localized surface plasmon [2,14], and the best conditions are obtained for molecules absorbed in hot spots, as at the junctions of metal nanoparticle homodimers [15]. It is worth to point out that the hot spot is generated by exciting the SPR with the appropriate wavelength and polarization [2,14]. Therefore, molecules in the probed volume do not contribute in the same way to the SERS signal, which is generated prevalently by the molecules located in the hot spots at the junction between plasmonic nanoparticles, corresponding to about 0.01% of the total in normal conditions [14]. This also makes the exact estimation of the enhancement factors for single molecules difficult, and a careful calibration of every substrate has to be considered for analytical purposes [14].

SERS is widely exploited for ultra-sensitive detection of analytes adsorbed on AuNPs (Figure 2(B)) [10,14]. Sensitivity of SERS is so high that single-molecule Raman spectroscopy is possible [14], as well as in vivo molecular probing of cellular components [14]. A typical application of SERS is based on the ability of Raman spectroscopy to perform multiplexed analysis [15,17,18]. The Raman spectrum is like a molecular fingerprint, and in principle, the detection of each single DNA base is also possible by SERS for a single cell [19]. In the scanometric method, AuNPs are immobilized on the chip surface by complementary base pairing when the target DNA is present [19]. Then, the reduction of silver ions on the gold particles is promoted, originating larger nanoparticles which emit stronger SERS signals [19]. Combining each base with a distinct Raman-active molecule allows the identification of the target DNA sequence [19]. Advantages of Raman with respect to ordinary fluorescence assays for DNA and protein detection are various: only one excitation wavelength is necessary, there is no photobleaching, Raman spectra have higher specificity than fluorescence spectra, and a large number of Raman-active molecules are available [6,19]. AuNPs, capped with thiol-terminated oligonucleotides composed by a Raman-active molecule on the inner end and a DNA capture sequence, have been used for the multiplexed detection of femtomolar amounts of more than six different DNA sequences in the same microarray [6,19]. The advantage of SERS, namely signal intensity, multiplexing, high signal to noise ratio, time stability, and NIR excitation, have also been exploited for in vivo biolabeling of target cells, like in cancer tissues [18]. SERS labels usually consist of AuNPs that have Raman reporters anchored to their surface, are coated with a protecting/stabilizing external layer, and are equipped with a targeting function for the selective binding of molecules like DNA, proteins, or antigens on cell membranes (Figure 2(C)) [15,17]. SERS labels endowed with antibodies showed the ability to selectively bind cancer cells and to differentiate the cancer phenotype by Raman spectroscopy [20]. SERS can be also used for the quantification of the number of Raman-responsive AuNPs taken up by cells [17]. The proper choice of the Raman reporter and of AuNP size is crucial to enhance the SERS signal of several orders of magnitude, with best performances obtained for large particles with homogeneous size and Raman-active molecules which have absorption bands overlapped to the laser excitation wavelength [15]. When the latter case is satisfied, it is defined as surface-enhanced resonant Raman scattering (SERRS) [14]. The use of Raman-active molecules firmly bound to the AuNP surface is also an important requisite for use in the physiological environment [15,20].

2.5 Modeling of the Optical Properties

Optical properties of AuNPs can be successfully calculated by analytical and numerical models [2,5,8]. In the last 99 years, Mie theory has been successfully applied to the calculation of noble metal nanoparticles extinction cross-sections [2]. Gustav Mie solved the problem of light diffraction by a single sphere using classic electrodynamics [3]. Subsequent refinement of the theory by Stratton and Born preserved the main framework, consisting in the solution of Maxwell equations in spherical coordinates with proper boundary conditions and multipole expansion of the electromagnetic field diffused and inside the sphere [2,3]. The Mie model allows the accurate determination of AuNP concentration and average size by fitting the optical absorption spectra of nanoparticle dispersion, and a fitting program can be freely downloaded from the web to perform such calculations [4]. The core of the method consists in the calibration of the free electron relaxation frequency versus particle size with a few reference solutions of AuNPs with known average size [4]. The generalization of the Mie theory can be applied to the calculation of optical properties of complex ensembles of AuNPs, like fractal aggregates [12]. For isolated metal nanoparticles of arbitrary shape, the most widely used method is the discrete dipole approximation (DDA) [2,5,14,21]. In DDA, the nanostructure of interest is represented by a cubic array of N polarizable points, that is, N cubic dipoles [5,8,21]. There are no restrictions about the complexity or the shape of the target, the only limitation is the discrete structure of the target due to the finite number of dipoles utilized [5,8,21]. The core of DDA is characterized by the way in which the polarizability of each point dipole is expressed [5,8,21]. The most reliable formula has been developed by Draine and Goodman as a correction of the Clausius - Mossotti polarizability by a series expansion of the interdipole spacing [5,8,21]. It allows the calculation of target extinction spectra within 10% of the exact result, independently of size, shape, and particle composition, provided that the number of dipoles N is large enough [5,8,21]. For metal particles with nanometric size, an error smaller than 10% is achieved using N on the order of 40,000 [5,8,21]. Draine and Flatau developed a free code named DDSCAT, which adopts fast Fourier transform methods and complex conjugate gradient techniques to reduce the calculation time [5,8,21].

For modeling optical properties of periodic arrays of AuNPs, the finite difference time domain (FDTD) technique is often used, and it is an implementation of Maxwell's time-dependent equations for solving the temporal variation of electromagnetic waves within a finite space that contains a target of arbitrary shape, and it has recently become the state-of-the art method for solving Maxwell's equations for complex geometries [2,8,14,22]. In this method, the differentials in Maxwell's equations are replaced by finite differences to connect the electromagnetic fields in one time interval to the one in the next interval [2,8,14,22]. Because FDTD is a direct time and space solution, it offers the user a unique insight into all types of problems in electromagnetism and photonics [2,8,14,22]. It can also be used to obtain the frequency solution by exploiting Fourier transforms, thus a full range of useful quantities in addition to fields around particles can be calculated, such as the complex Poynting vector and the transmission/reflection of light [2,8,14,22]. In the FDTD technique, Maxwell's equations are discretized by using finite difference approximations in both time and space that are easy to program [2,8,14,22]. The unit cell of the lattice that constitutes the FDTD computational domain is called the Yee cell [2,8,14,22]. The advantage of this method is that results for a large frequency range can be obtained in a single run [2,8,14,22]. All the aforementioned methods rely on the appropriate correction of the optical constant of the AuNPs for size effects described in 2.2, and they allow the evaluation of the local electromagnetic field enhancement in the nearby AuNPs [2,3,8,14].

2.6 Surface Plasmon Radiative Dynamics

AuNPs have larger extinction cross-sections than organic dyes or other inorganic nanomaterials with the same size, but their luminescence is negligible [2,8,23]. Nonradiative energy relaxation processes, such as electron–electron scattering, are very fast in gold. Therefore the luminescence efficiency of AuNPs with size above 2 nm is only 10^{-5}–10^{-3} [2,8,23]. The band structure of particles with a size larger than 2 nm is almost that of bulk metals, and no quantum size effects are perceivable. The luminescence in AuNPs mainly occurs by the following three steps: the formation of holes in the lower lying *d* bands due to light absorption, the relaxation of excited electrons and holes in the *sp* and *d* bands, respectively, and the radiative recombination of the electron–hole couple [2,8,23]. The luminescence quantum yield of AuNPs can be, however, 10^4–10^5 times larger than bulk gold [2,8,23]. In general, surface roughness and defects can enhance the efficiency of this process [2,8,23], as well as the surface adsorption of organic impurities with luminescence efficiency larger than the metal substrate [2,8,23]. The emission enhancement is also attributed to the surface plasmon oscillation [2,8,23]. In particular, it has been suggested that the local field enhancement due to the surface plasmon absorption can favor the radiative recombination efficiency of *sp* bands electrons with *d* band holes, according to a phenomenon called the "lighting rod" effect [2,8,23]. The combination of high absorption cross-section, low luminescence yield and rapid relaxation of the SPR makes AuNPs attractive for applications requiring the efficient conversion of light into heat, like photothermal therapy or photothermal imaging [6,24]. Photothermal therapy offers a gentler alternative for cancer treatment by using optical heating for ablation of tumors [6,8]. Photothermal therapy requires that strong photo-absorbers are located selectively in the tumor region to reduce nonspecific injury to adjacent healthy tissue [6,8]. The absorption cross-section of AuNPs (σ_{abs} ~2.9 × 10^{-15} m^2 for 40 nm spheres with an absorption band around 530 nm) is five orders of magnitude larger than that of indocyanine green (σ_{abs} ~1.7 × 10^{-20} m^2 around 800 nm), a dye used in earlier demonstrations of laser photothermal tumor therapy, thus promising effective photothermal therapy at much lower irradiation energy [25]. The SPR absorption of Au nanoparticles is followed by the rapid conversion (~1 ps) of the absorbed light into heat [25]. Besides, selection of a plasmonic nanoparticle size with optimum plasmon absorption as well as scattering makes a dual imaging/therapy approach possible [25]. Therapy and imaging with visible light is suitable for in vitro studies and in vivo applications for surface-type cancers, while in vivo applications to deeper tissues require plasmon features in the near-infrared region of the spectra [25]. In general, the temperature increase due to the transformation of electromagnetic radiation into heat depends on the absorption cross-section as well as on the particle radius [24]. In particular, the temperature increase on the surface of a gold particle depends on the second power of the particle radius because of the balance between the heat generation rate, which depends on the particle volume, and the heat dissipation rate, which depends on R_{NP}^{-1} [24]. In photothermal imaging, AuNPs are heated by a laser beam, while a second weaker probe beam is exploited to locally probe the changes in refractive index due to the temperature gradient [2,8]. This technique allows monitoring very small single AuNPs (i.e., below 10 nm) inside cells [2,8].

2.7 Plasmonic Modulation of Molecular Fluorescence

Metallic nanoparticles, due to their high polarizability, show strong interactions with nearby fluorescent chromophores [26].The molecular fluorescence quantum yield can either decrease or increase depending on how nanoparticles affect the chromophore excitation rate and the radiative and nonradiative decay rates of the excited states [26]. The mechanisms of metal nanoparticle interactions with molecular chromophores can be very complicated, and several parameters govern which of the processes is predominant, like nanoparticles size and shape, molecule—nanoparticle separation and orientation, or the overlap of fluorophores excitation and emission energy with the plasmon resonance [2,8,26]. At separation distances as small as a few nanometers, quenching of fluorescence close to 100% is observed and explained with two phenomena: the increase in *nonradiative relaxation rate* due to energy transfer from the chromophore to the metal particle and the decrease in radiative relaxation rate due to the out-of-phase oscillation of molecular dipole and the dipole induced on the metal nanoparticle, when the molecule is tangentially oriented on the metal surface [2,8,26]. The Gertsen–Nitzan model used for ordinary dipole–dipole Forster resonance energy transfer (FRET) on distances of few nm is no more valid for metal–chromophore interactions, since the theoretical d^{-6} decay is faster than experimentally observed quenching effects (d is the dipole–dipole distance) [2,8,26].

Other theories afforded the specific problem of nanomaterial surface energy transfer (NSET), like the Persson–Lang model, and were successful in reproducing the d^{-4} dependence experimentally observed [2,8,26].

Though both cases concern dipole–dipole interactions, NSET has a longer range than FRET because the nanoparticle has an isotropic distribution of possible dipole vectors on its surface to accept energy from the chromophore [26]. Relaxation of the dependence on the molecular orientation implies that NSET has a higher probability of energy transfer than FRET, is less specific of the chromophore type, and can be applied to distances over 10 nm [2,8,26]. NSET can find applications in monitoring biochemical processes, such as accurately tracking DNA length scales and conformational changes in DNA following protein binding in the distance range of 30–200 Å or structure conformational change and cleavage kinetics of hammerhead ribozyme [2,8,26].

In the presence of low quantum yield fluorophores spaced more than about 10 nm from the metal surface of nanostructures having high scattering cross-sections like AuNPs larger than 100 nm, a ten- to fiftyfold emission enhancement has been measured due to a nanolensing effect [2,8,26].

The luminescence process is particularly amplified when plasmon scattering efficiency matches the molecular emission wavelength [2,8,26]. This could be an effect of near field enhancement of the optical intensity incident on the molecule or a consequence of the increase in the photon local density of states at the emission wavelength [2,8,26]. Both effects are active for a chromophore–nanoparticle distance of 5–20 nm. Therefore, fluorescence enhancement has a distance range complementary to SERS, which involves molecules closer than 5 nm to the metal surface [2,8,26].

Conversely to fluorescence quenching, the effect of radiative rate enhancement is more pronounced and easily measurable for low quantum yield molecules [2,8,26].

2.8 Nonlinear Optics of Metal Nanoparticles

AuNPs have many highly polarizable electrons, which ensure a strong nonlinear optical (NLO) response [2,8]. Indeed, the strong nonlinear properties of metal nanoparticles are usually attributed to local field enhancement effects for the plasmon oscillation because nonlinear optical susceptibility (χ) measured on nanostructures is much larger than in bulk metals [2,8].

Three main categories of nonlinear optical phenomena have been reported for AuNPs: optical Kerr effect, multiple harmonic generation, and optical limiting [2,8]. Except for optical limiting, the two other phenomena require pulses on the order of picoseconds or lower in order to reduce the contribution from thermal effects and nanoparticle degradation [2,8]. This is because AuNP has to be considered first of all as a converter of light into heat, especially at laser intensities typical of nonlinear optics [2,8].

Due to their enhanced third-order nonlinear susceptibility, especially near resonance with the plasmon frequency, AuNPs embedded in dielectric matrices have practical applications in ultrafast all-optical switching devices [2,8]. The switching phenomenon is based on the optical Kerr effect and consists in the intensity-dependent variation of the refractive index of the nanocomposites [2,8]. Z−scan and degenerate four-wave mixing (DFWM) experiments are used to estimate the optical Kerr susceptibility $\chi^{(3)}$ of such nanocomposites [2,8]. The magnitude of the nonlinear response varies with size and shape as well as the volume fraction of metal nanoclusters embedded in the dielectric matrix [2,8]. Three main mechanisms contribute to $\chi_m^{(3)}$ [2,8]: (1) a fast mechanism on a time scale of hundreds of femtoseconds due to electron−electron scattering after the plasmon absorption, driving to the electron thermalization, (2) a contribution from interband transition, which occurs on the same time scale of the former, and (3) a hot electron mechanism on a time scale of nanoseconds. Only the first two mechanisms are fast enough for optical switching applications [2,8].

Second harmonic generation (SHG) is possible in noncentrosymmetric systems when nanoparticle size is smaller than the excitation wavelength [2,8]. In metals, the SHG signal has two origins: a surface contribution arising from breaking the centrosymmetry at the metal surface and a bulk contribution arising from the electromagnetic field gradients due to the presence of the interface [2,8], which is the dominant contribution for nanometric particles [2,8]. Typically, SHG is observed in random metal nanostructures such as aggregated particles. Best SHG efficiency is achieved for excitation wavelengths resonant with particle SPR [2,8]. SHG is not allowed in isotropic dispersions of noncentrosymmetric particles, although a weak frequency doubling phenomenon is still possible, called hyper-Rayleigh scattering (HRS) [27]. HRS is a nonlinear second-order light scattering that has been used in the study of noncentrosymmetric molecules or nanoparticles in solutions [27]. The HRS relies upon the fluctuations of the density or orientation of nanoparticles, which instantaneously break the centrosymmetry of isotropic media as liquid solutions and create conditions of net frequency doubling [27]. Spherical metal nanoparticles functionalized with antibodies that conjugate a target antigen with a sandwich structure, forming a dimer of nanoparticles, have found potential application in biosensing: the symmetry break upon the target antigen coordination enables the generation of HRS signal [27]. Third harmonic generation (THG) has been also reported for metal nanoparticles on a substrate, due to third order HRS mechanism [2,8]. Local field effects due to strong localization of collective plasmon modes in fractal metal clusters have been exploited for enhancing the nonlinear properties of molecules adsorbed on metal nanoparticles [2,8]. Beyond SERS effect, previously discussed, strong enhancement of HRS signal using gold nanoparticle aggregates with adsorbed pyridine molecules have been observed [2,8]. Similarly, AuNPs have been used to enhance two photon absorption-induced fluorescence for in vivo cell imaging [2,8]. AuNPs themselves are capable of multiphoton absorption-induced luminescence, with emission yield growing with particle size and asymmetry [28]. In the case of 790 nm−100 fs laser pulse excitation, the mechanism following multiphoton absorption requires that more than one photon is emitted per laser pulse and that a three photon absorption process is dominant [28]. Observed photoluminescence spectra are broad and do not show any correlation with the plasmon absorption or scattering [28].

AuNPs are well known as excellent optical limiters with nanosecond, picosecond, and femtosecond laser pulses, exploiting a mechanism not completely understood to date (Figure 3(A)) [2,8]. For the most part, investigations on metal nanoparticles pointed out the free carriers multiphoton absorption processes, the particle fragmentation due to electron photoejection, and interparticle electronic coupling in particle aggregates as possible optical limiting (OL) mechanisms [2,8,29]. Nonlinear scattering was conjectured at the origin of the strong OL in few cases [2,8,29]. Independent of the mechanism, all the investigations on optical limiting properties of AuNPs pointed

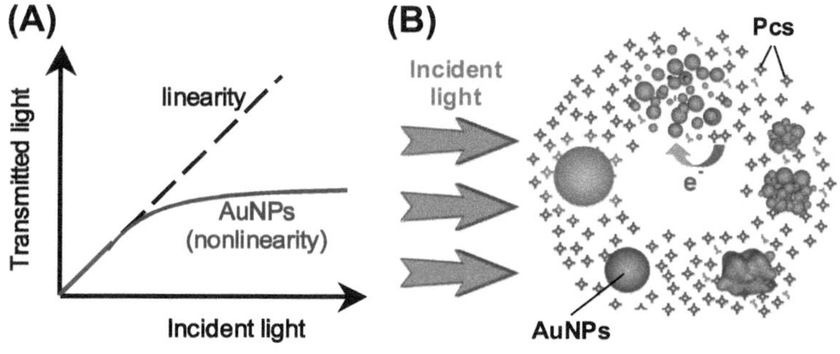

FIGURE 3 (A) AuNPs are recognized for their excellent ability to limit the intensity of transmitted light when it exceeds a critical threshold (optical limiting). (B) AuNP fragmentation, which occurs during optical limiting, can be avoided by a self-repair mechanism in the presence of phthalocyanines (Pcs), as sketched in the figure. From left in clockwise order: AuNPs absorb photons and heat up over the fragmentation threshold; electric charges are transferred to photofragmented AuNPs by the oxidation of ZnPcs, promoting particle aggregation; aggregates are photomelted into new spherical AuNPs, ready to efficiently limit light again.

out the following aspects: (1) the OL performances are amongst the best ones ever measured; (2) big nanoparticles are better optical limiters than small ones, (3) nanoparticles lose limiting efficiency after a few laser pulses due to their fragmentation, that is more rapid and has a lower threshold when particle size is big; and (4) nanoparticles have good optical limiting performances for wavelengths resonant with the surface plasmon absorption, as well as with the interband transitions [2,8,29].

In several cases, the optical limiting properties were enhanced by blending AuNPs with organic molecules, like fullerenes or β-carotene [8,29,30]. For instance, it has been reported that gold nanoparticles conjugated to a fullero-pyrrolidine by a disulfur bridge showed better optical limiting performances than the isolated metal nanoparticles, with laser pulses at 532 nm (9 ns), which was compatible with an energy transfer from the AuNPs to the fullerenes of about 60% [30]. An enhancement of optical limiting has also been found for AuNPs conjugated to fullerenes–pyridine derivatives [31]. An extremely long durability under laser irradiation was observed in the case of AuNPs blended with zinc phthalocyanines (Pc) [29]. The origin of the enhanced OL performances resides in the ability of the gold nanoparticles to self-heal during the laser pulse irradiation by interaction with the Pc (Figure 3(B)) [29]. This finding is interesting because, in general, the structural instability of nano-structured materials upon their utilization can be surmounted by reducing the interface energy or by designing a nanomaterial able to self-heal or to regenerate during the usage itself [32–34]. Therefore, the AuNP–Pc blend represents a rare example of functional nanomaterials that self-heal during use [29,32–34].

3. QUANTUM CHEMISTRY OF ELECTRONIC AND OPTICAL PROPERTIES

3.1 Introduction

"First principles" calculations of gold clusters have been performed so far at ab initio level using Hartree–Fock and post-HF wave-function-based methods or with the density functional theory (DFT) approach. For gold clusters, DFT has demonstrated itself to be an excellent compromise between computational efficiency and accuracy, so it has been extensively employed in particular when the cluster size is so large as to prevent the practicability of more accurate wave-function methods. The availability of accurate quantum chemical calculations is important both to rationalize experimental results and to describe quantitatively the electronic structures of such clusters. In this respect, cluster convergence of different properties may be investigated, and conditions can be identified to locate the cluster size in order that a given property reaches its bulk value to a reasonable accuracy. In general, different physical properties are expected to exhibit different convergence behaviors with increasing cluster size. Colloidal metals, and gold in particular, have been extensively studied with electron microscopy techniques which can find out ordered structures in nanoparticles. However, it must be considered that these colloidal metal particles are stabilized by ligand agents "coating" their surface. Although the electronic structure of stabilized nanoclusters is significantly influenced by the ligand shell, this interaction has been proven to be, in many occasions, rather confined to the top few layers at the surface of the metal cluster.

Gold exhibits strong relativistic effects, and electron correlation is known to play an important role in its compounds as well. Therefore, gold clusters are a typical field where recently developed scalar relativistic DFT methods can be successfully applied [35]. Moreover, the interest in the optical absorption of gold nanoparticles has grown in the last years, both experimentally and theoretically [36]. In fact, such systems display specific properties which make them special optical materials. This is in part due to the ability of gold to be organized in well-defined structures of nanometric size: often it is possible to tune and control the shape of such structures already at the level of a conventional chemical synthesis [37]. Therefore, it is possible to produce gold nanocluster samples with specific shape and size distributions, whose characterization is further carried on by conventional electron microscope techniques. Such samples are normally produced in solution, and gold nanoparticles are, therefore, stabilized with protective ligands: in particular, thiolates are quite popular due to their specific ability to form strong Au−S chemical bonds. Other protective agents, mainly surfactants, are also employed to stabilize nanoparticles; in this case, the gold−ligand interaction is much weaker.

The optical photoabsorption of gold nanoparticles is dominated by a strong band that, for large enough clusters, lies in the visible region [36]. This band is very sensitive to the structure of the particle and the nature of the environment: namely the size and shape of the clusters, the presence of the ligands and their nature, and finally, the solvent in which the clusters are dispersed. Such strong absorption is usually ascribed to the surface plasmon resonance (SPR) [38], and it consists of a collective excitation of the conduction electrons of the metal particle. Moreover, since SPR is a phenomenon which is localized at the surface of the nanoparticle, it is strictly connected with the finite size of the particles and, therefore, is absent in bulk material. Therefore, computational methods developed to study solid state in physics and based on periodic boundary conditions are not suitable to investigate such phenomenon. Due to the role of the SPR as a fundamental optical property of gold clusters, it has been generally considered important to develop new theoretical models with the aim of simulating this effect in order to rationalize the experimental findings and, a more ambitious goal, to perform a computational design of novel nanostructured materials with specific optical properties. In this context, the first proposed and widely used approach is the Mie theory, whose implementations are discussed in Section 2.5 of this chapter. A first improvement beyond classical Mie theory is represented by the time-dependent local density approximation (TDLDA) implemented within the *jellium* model [39]. That is, the metal nanoparticle is modelized as a background consisting of a spherical positive charge density which describes the average nuclei positions; such a positive background supports the electrons whose motion is treated in terms of quantum mechanics. Also, in this case, the *jellium* model misses the discrete atomic structure of the cluster, but the TDLDA method gives a quantum mechanical description of the electron excitations of the model system: the SPR is described as a collective excitation of the electrons that is as free electron oscillations. Moreover, the TDLDA *jellium* model offers a practical tool to consider more specific effects which can be included in the method, namely the effect of an embedding medium to simulate, for example, the solvent effects [40]. In order to improve the theoretical description, it is, therefore, appealing to consider an approach able to retain the atomic structure of the cluster, and a possible choice would be the time-dependent density functional theory (TDDFT) formalism implemented in the conventional quantum chemical linear combination of atomic orbital (LCAO) method. This choice joins the reliability of modern TDDFT approaches to describe electron excitations with the accuracy of the LCAO implementation, which avoids the drastic approximation of the effective molecular potential operated by the *jellium* approach. Optical spectra of metal clusters protected by ligands have been calculated by the TDDFT method for many years using a finite basis set of contracted Gaussian-type orbitals [41,42] and a plane-wave basis set [43]. Of course, the LCAO-TDDFT method is computationally much more demanding than *jellium* methods. However, it represents the state-of-the-art approach for the theoretical description of the optical spectra for large molecules, and therefore, it is a potential tool to understand in detail the electron excitation process and a valid instrument to help the design of new nanoparticles with specific optical properties.

Despite the great body of experimental evidence, fundamental questions still exist in this field. In particular, there is not much unambiguous information on the dependence of the shape and peak of SPR upon nanoparticle size and shape.

3.2 Electronic Structure of Gold Clusters: Quantum Chemical Methods, Relativistic Effects, and Convergence Behavior

As outlined in the introduction, the only practical quantum chemical method to study gold clusters of large size (up to hundreds of atoms) is DFT, which consists to solve the Kohn-Sham (KS) equations:

$$H_{KS}\varphi_i = \varepsilon_i\varphi_i \qquad (1)$$

The *KS* equations are Schrödinger type one-electron equations; the eigenvectors correspond to the molecular orbitals and eigenvalues to the orbital energies. The Hamiltonian of Eqn (1) has the following form:

$$H_{KS} = -\frac{1}{2}\nabla^2 - \sum_{K}^{Nucl} \frac{Z_K}{|\overline{r} - \overline{R}_K|} + \int \frac{\rho(\overline{r}')d\overline{r}'}{|\overline{r} - \overline{r}'|} + V_{XC}[\rho(\overline{r})] \tag{2}$$

In expression (2) the first term corresponds to the electron kinetic energy, the second term to the attraction between the electron and the *K*-th nucleus, the third term to the classical interelectronic repulsion, and the fourth term to the exchange-correlation potential. The electron density ρ is calculated from the occupied orbitals:

$$\rho = \sum_{i}^{occ} |\varphi_i|^2 \tag{3}$$

Since the Hamiltonian depends on the eigenvectors, Eqn (1) must be solved iteratively until self-consistency is reached. It is very important to underline the role of the exchange-correlation potential V_{XC}, a term which is approximated and, therefore, must be carefully chosen. In particular for gold clusters, it has been found that for geometry optimization, the local density approximation (LDA) is quite accurate. Whereas, for binding energies, the generalized gradient approximation is much better. In practice, the V_{XC} choice is also important according to the property of interest, in particular for photoabsorption (see Section 3.3).

From the numerical point of view, the typical quantum chemical technique of resolution of *KS* Eqn (1) consists to introduce a basis set to expand the molecular orbitals as a linear combination of atomic functions. The coefficients are then obtained by exploiting the Rayleigh-Ritz variational method. Of course, basis set size has an important impact on the accuracy of the results and on the computational effort. In the literature, various basis choices have been reported depending on the property of interest, from contracted Gaussian-type orbital (GTO) functions, for example, (19s, 15p, 10d, 6f) → (9s, 8p, 5d, 2f) to Slater-type orbitals (STO), for example, triple zeta with polarization (TZP).

A fundamental characteristic of quantum chemistry of gold compounds is the prominent role played by relativistic effects [44]. This may seem strange at first glance, since relativity should be connected with phenomena taking place around the speed of light that are far away from chemistry. Instead, relativistic effects are important for heavy atoms because the electrons of the very deep core orbitals actually move at speeds comparable to that of light. The formal treatment of relativistic quantum mechanics consists to solve the Dirac equation instead of the nonrelativistic Schrödinger equation. Although at the moment quantum chemical methods based on the Dirac equations are available, most relativistic calculations have been carried on with approximate treatment of the relativistic effects. In fact, the Dirac equation can be approximated with proper transformations in order to reduce the dimension of the problem (from four to two or one component) [45]. Such transformations, (e.g., the Douglass–Kroll (DK) and the zeroth-order regular approximation (ZORA)) lead to a Hamiltonian, which contains most of the physics introduced by relativity and can be implemented relatively easily and applied to large systems like gold nanoclusters. In particular, three new terms are introduced by relativity: the mass-velocity, the Darwin, and the spin-orbit (SO) terms. When SO interaction is neglected, the treatment is referred to as scalar relativistic (SR). SR calculations are computationally much cheaper than SO calculations, not only because new terms must be introduced in the Hamiltonian, but also because the SO coupling reduces the symmetry. So, in practice, large systems like gold nanoclusters are usually treated at the SR level. It is worth noting that to perform accurate relativistic quantum chemistry, it is necessary to employ a relativistic basis set where the atomic functions must be optimized relativistically. On the other hand, for DFT calculations, the role of relativistic effects in the exchange-correlation functional is limited. So, it is not necessary to employ functionals obtained from the relativistic electron gas.

Despite the formalism and the computational aspects, it is interesting to describe how relativistic effects have their impact on gold chemistry. Consider first what happens to gold core *1s* and *2p* electrons, because their velocities are comparable to that of light, they are stabilized and shrunk by relativity. As a consequence, the nucleus is more efficiently shielded, and the 5*d* valence orbitals are, therefore, destabilized. The extra-stabilization of the core *1s* orbital is also reflected on the valence due to orthogonality of atomic functions. So, the 6*s* atomic orbital is stabilized as well. In summary, the 5*d*-6*s* gap of gold is more than halved in a relativistic description with respect to a nonrelativistic one. It is interesting to observe that for silver, relativistic effects are negligible, and that the Ag 4*d*-5*s* gap is comparable to the Au nonrelativistic 5*d*-6*s* gap: so silver behaves like a nonrelativistic gold. The narrowing of the 5*d*-6*s* gold gap has important consequences on the gold chemistry due to the larger involvement of the 5*d* functions in the outer valence orbitals, which makes the chemical bond more "directional." This is reflected on bond lengths, binding energies, and optical properties. Besides relativity, electron correlation is also important for gold: it has been possible to treat electron correlation by coupled cluster (CC) methods on small molecules and with perturbative Møller-Plesset (MP) on larger ones. However, for large nanoclusters, only DFT methods remain practicable.

Calculations of molecular properties of gold clusters are also important to investigate the convergence behavior with increasing cluster size. This is useful to extrapolate the bulk limit but also to identify the different roles of bulk and surface atoms.

For example, the average nearest-neighbor interatomic distance has been found to correlate linearly with average coordination number of the cluster.

So, extrapolation to the bulk limit of coordination number 12 demonstrates the high accuracy of the LDA functional to optimize the geometry [35].

On the other hand, binding energies show a linear behavior with respect to $n^{-1/3}$ where n corresponds to the number of Au atoms of the cluster: such a trend is, therefore, defined by the ratio between surface/bulk atoms, which is proportional to $n^{-1/3}$ [35].

The evolution of other properties with cluster size can also be studied, provided some simplified model is employed to identify some trend rule. For example ionization potentials (IP) and electron affinities (EA) can be studied with the spherical droplet model [46], which predicts a linearity of IP and EA with respect to the reciprocal radius of an almost spherical gold cluster. In this way, it is possible to extrapolate the experimental bulk work function of Au (5.1 eV) with excellent accuracy (4.79 and 4.86 eV for IP and EA, respectively) [35].

A complete description of the electronic structure of large gold nanoclusters in terms of molecular orbitals, although possible in principle, is not practicable due to the enormous numbers of orbitals. For this reason, the most common and convenient way to describe the electronic structure of gold nanoclusters is to employ density of states (DOS) plots.

3.3 Photoabsorption: Description of the Phenomenon, Plasmons, Size, and Shape Effects

The photoabsorption of gold clusters can by described with classical electrodynamics only for sizes larger than 20 nm, when quantum size effects become negligible, and the prominent spectral feature is the surface plasmon resonance (SPR). For a cluster size smaller than 20 nm, it becomes necessary to include quantum mechanics to describe the phenomenon. However, quantum chemistry methods are too demanding to treat particles of such size. At the moment, tridimensional-protected gold clusters up to $Au_{314}(SH)_{96}$ have been treated with TDDFT [47], which correspond to the size of the metallic core around 2.0 nm. Whereas, for bare gold nanowires with a finite cross-section, a length of 3.744 nm has been reached with Au_{122} grown along the (110) direction. [48]

There exist many different implementations of the TDDFT method for quantum chemistry; however, the most popular one is the Casida approach, which consists to solve the eigenvalue equation:

$$\Omega F_i = \omega_i^2 F_i \tag{4}$$

In Eqn (4), Ω is a matrix whose dimension corresponds to all the pairs of occupied and virtual orbitals ($N_{occ} \times N_{virt}$), ω_i are the excitation energies, and F_i are the eigenvectors from which the intensity of each excitation can be extracted. So, in practice, the solution of Eqn (4) gives the calculated photoabsorption spectrum. The elements of the Ω matrix depend upon the *KS* eigenvalues and eigenfunctions, which are calculated solving the KS Eqn (1).

So, also for TDDFT calculations, a proper and careful choice of basis set and exchange-correlation functional is important. In particular, it has been pointed out that for accurate TDDFT results, it is necessary to employ V_{xc} with correct asymptotic behavior that is supporting the Coulomb tail $-1/r$ at large distance, which is a consequence of the exchange. This characteristic is not supported by standard LDA and GGA functionals, but can be introduced with the long-range correction (LC) (e.g., in the CAM-B3LYP functional) or included in the construction of a model potential (e.g., LB94 or SAOP potentials). Basis set choice is also crucial. So, for small systems, TZP basis are usually employed. However, for large systems, the smaller and more economical double zeta (DZ) has proven accurate enough. To study the emergence of the SPR, it is convenient to consider the gold nanowire Au_{122} [17] reported in Figure 4 of this chapter: the TDDFT calculation gives a very intense band at 2.06 eV, contributed by a limited number of very intense transitions. As can be seen from the figure, at higher or lower energy, only very faint intensity is calculated. In order to assess the SPR nature of the calculated band, it is useful to compare the TDDFT spectrum (solid line) with a spectrum calculated at the KS level (dashed line) that is as dipole transition moments between occupied and virtual molecular orbitals, taking the orbital eigenvalues difference as excitation energy. From Figure 4, it is evident that while in the KS spectrum the intensity is rather uniformly distributed, in the TDDFT spectrum almost all the intensity is focused in the single band. This can be explained because the TDDFT formalism is able to describe the collective electron motion, which is the typical nature of the SPR. Whereas, the KS method cannot describe collective transitions since it is a single-particle

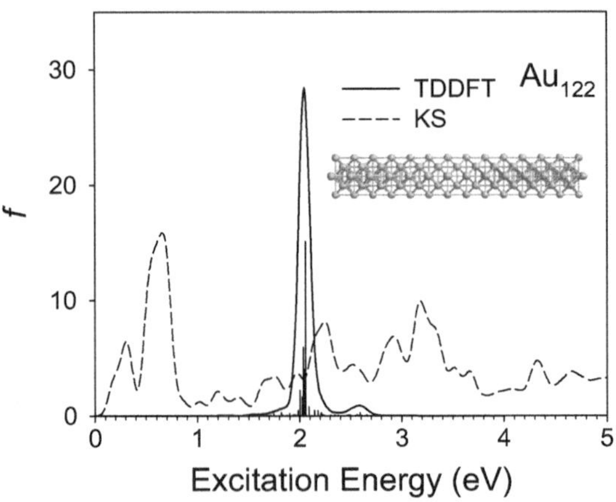

FIGURE 4 The TDDFT (full line) and KS (dashed line) photoabsorption spectra of gold nanowire Au_{122} grown along the $\langle 110 \rangle$ direction convoluted with a Gaussian function of FWHM = 0.12 eV. The vertical bars correspond to the TDDFT discrete transitions. The structure is shown in insets. *Adapted from Ref. [48], Figure 3, p. 17200.*

formalism. It is worth noting that it is very tricky to precisely identify a plasmon in a molecule. However, very recently, a systematic method based on a scalar factor has been suggested and applied successfully in a variety of different situations [49], including gold clusters. From the analysis of the effect of the longitudinal size on the optical absorption of such gold nanowires [48], it has been shown that a shortening of the nanowire induces a blue shift and an intensity reduction in the plasmon, as found in the experiment. This size effect is quite general for gold clusters and remains valid also for different shapes. Moreover, it is found that below a length of 2 nm, the plasmon disappears completely. Also, this is in good agreement with experimental findings.

In order to assess the importance of the V_{xc} choice and to identify the effect of the shape for gold clusters, in Figure 5 the analysis of a previous work of Durante is reported [50]. Four clusters are considered: (1) a cuboctahedron $[Au_{147}]^{5+}$; (2) an octahedron $[Au_{146}]^{2+}$; (3) a cube $[Au_{172}]^{4+}$; (4) an

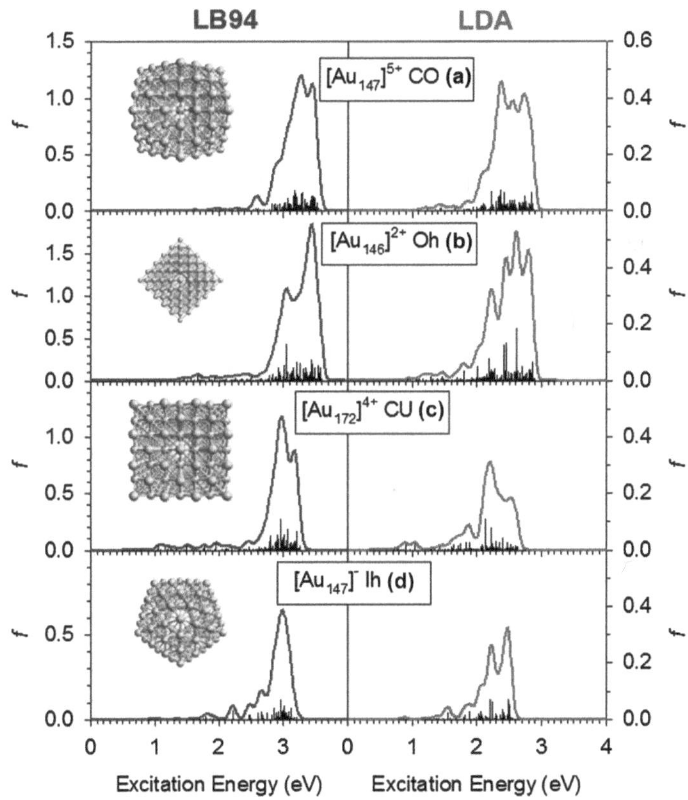

FIGURE 5 TDDFT Photoabsorption spectra calculated with LB94 and LDA potentials for Au nanoclusters. *Adapted from Ref. [50], Figure 2, p. 6280.*

icosahedron $[Au_{147}]^-$. All these structures exhibit high (O_h or I_h) point-group symmetry. (1), (2) and (3) are representative of an fcc-like (crystalline) motif, but taken with different cuts exposing only (111) facets in the case of $[Au_{146}]^{2+}$ or only (100) facets in the case of $[Au_{172}]^{4+}$, with $[Au_{147}]^{5+}$ representing an intermediate case. $[Au_{147}]^-$, instead, is representative of noncrystalline structures exhibiting fivefold symmetry axes. All these structures have been observed in actual Au nanoparticles. The charges of the clusters have been selected in order to achieve electronic shell closure. This simplification is effective in keeping symmetry high and in reducing the computational effort but hardly affects the photoabsorption spectrum.

The LB94 maximum transitions are 3.58, 3.59, 3.27, and 3.21 eV, while the LDA maximum eigenvalues are 2.88, 2.89, 2.63 and 2.55 eV for (1), (2), (3), and (4), respectively. The first and most important observation that can be drawn from an inspection of Figure 5 is that the position of the absorption peak depends weakly on the cluster shape, but it is appreciably dependent on the choice of the *xc*-functional employed in the calculations. In detail, the position of the peak maximum varies in the range from 3.0 to 3.4 eV at the LB94 level; whereas, it ranges between 2.3 and 2.6 eV at the LDA level. This provides the possibility of validating the *xc*-functional (i.e., to determine which is the most appropriate to predict the optical response of Au clusters) by comparison with available experimental data. In fact, it is known from the experiment that Au particles of roughly 2 nm diameter included in Al_2O_3 exhibit a maximum in the absorption spectrum at about 2.58 eV. This value must be corrected by the dielectric constant of Al_2O_3 and to the reduced size of the present clusters (diameter of roughly 1.7−1.8 nm). When both effects are included, one expects that the experimental peak for particles similar to those considered in Figure 5 should be around 2.9−3.0 eV. Therefore, it can be concluded that the LB94 values for the absorption maximum seem to be in better agreement with the experiment than the LDA ones. The dependence of the position of the peak on the cluster shape is instead weak. Basically, one can observe only a shift of this position to a lower binding energy by $\approx$0.2 eV at both the LDA and LB94 level for (4) and (3). The latter shift can be justified by the larger size of the (3) cluster, in agreement with the expected decrease of the position of the absorption peak with increasing cluster size. Whereas, the former shift is due to the higher symmetry of the I_h group, which narrows and reduces the number of electron−hole pairs in the excitation spectrum. The maximum of the absorption peak instead is somewhat more dependent on the cluster shape. This is due to an appreciable dependence of the *form* of the absorption onto the shape of the particle: especially at the LDA level, one can clearly see that the absorption spectrum has a roughly symmetric form for (1), and asymmetric forms for the other clusters, with a positive skewness for (2) and (4) (with the position of the maximum thus at higher energy than the centre of the peak) and negative skewness for (3) (with the position of the maximum at lower energy than the center of the peak). This is the reason for the greater sensitivity to the cluster shape of the maximum of the absorption peak with respect to its center. In Figure 5, the sudden decline of photoabsorption peak towards high energies is the consequence of the limited number (300) of eigenvalues actually extracted from Eqn (4). In fact, when very large systems are considered, it is very hard to calculate valence photoabsorption spectra over a wide excitation energy range. In fact, the Davidson algorithm employed to diagonalize very large Ω matrices is very efficient, but it is limited to extracting a relatively small number of the lowest eigenvalues and eigenvectors. It is worth mentioning that other approaches, based on the temporal evolution, do not suffer this problem and can easily calculate the spectrum over a wide energy interval [51]. However, in such methods, it is not possible to assign the spectral features to a specific electronic transition; this makes it difficult to rationalize the calculated spectrum in terms of electronic structure. On the other hand, the Casida formulation allows for a complete analysis describing each electronic transition in terms of a linear combination of one-electron excited configurations.

In conclusion, state-of-the-art quantum chemical implementations of TDDFT have proven to be competitive to describe and rationalize photoabsorption of gold nanoclusters with sizes around 2–3 nm containing some hundreds of gold atoms. Further studies to suggest new methods and algorithms to treat even much larger systems still represent a big challenge for modern quantum chemistry.

4. QUANTIZED CHARGE TRANSFER OF NANOPARTICLE MOLECULES AND ENSEMBLES

4.1 Introduction

The size dependence of the properties of metal nanoparticles has been exemplified rather extensively, in electrochemical studies of mercapto-functionalized gold nanoparticles where the nanoparticles behave as nanoscale molecular capacitors [52]. An interesting example of these size-dependent properties is the quantized capacitance charging of gold nanoparticles that is manifested as a series of rather evenly spaced voltammetric peaks, and when the diameter of the nanoparticles is reduced to the order of the metal's Fermi wavelength (~0.5 nm for Au), a semiconductor-like bandgap starts to emerge with a zero-current region observed in the voltammetric profiles [52].

Such quantized (single electron) charging of nanoparticles has been observed both with single nanoparticles by scanning tunneling microscopic/spectroscopic (STM/S) measurements as well as by solution electrochemical studies of nanoparticle ensembles [53]. In the former, the charge-transfer properties of a single nanoparticle (which can be an ill-defined component of nanogranular structures or a relatively well-defined, isolated metal nanoparticle) are probed with an STM tip in a double-junction configuration [54]; whereas, in the latter, an ensemble of nanoparticles undergo charge-transfer reactions at the electrode–electrolyte interface [53]. In further studies, nanoparticle solid films have also been found to exhibit quantized charging characteristics when the nanoparticles are reasonably monodispersed in size [55,56]. In the following, we will summarize the key results of nanoparticle charge-transfer chemistry in these three structural configurations.

4.2 STM-Based Coulomb Staircase of Single Nanoparticles

Electronic devices based on simple molecules or nanostructures represent an effective and attractive solution to the grand challenge of continuing diminishment of feature size with ever-enhanced performance, a concept pioneered by Aviram and Ratner [57]. Within such a framework, nanostructures, such as organically capped nanoparticles, have been proposed to serve as the active components to achieve the diverse functionalities of device elements in manipulating electrical outputs, analogous to silicon-based semiconductor materials, by taking advantage of the ready manipulation of their structures and properties through molecular design and nanoscale engineering [58]. One fascinating characteristic is that the electronic conduction in such nanostructures can be varied from insulating to metallic by controlling the size of the nanoparticles and the strength of coupling between them [59]. For instance, when the diameter of metal nanoparticles is reduced to the order of the metal's Fermi wavelength, the phenomenon of quantized charging will appear in the nanojunction, which is commonly known as "Coulomb staircase" in STS measurements [54]. In these studies, a double barrier tunnel junction (DBTJ, Figure 6(A) and (B), reproduced with permission from Xu and Chen [60]), which consists of an STM tip, a nanoparticle, and a conducting substrate, is generally used to interpret the charge transfer properties. One

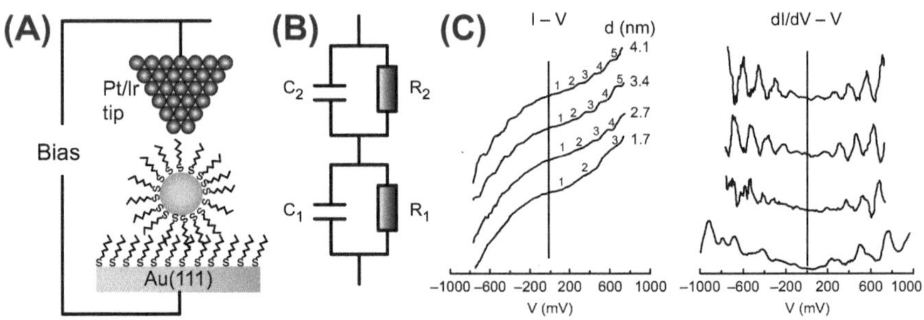

FIGURE 6 (A) Schematic of a DBTJ setup formed by an STM tip, a ligand-stabilized gold particle, and a Au substrate. (B) Corresponding equivalent circuit of the DBTJ structure. *Reproduced with permission from Ref. [60], ©2008 Elsevier B. V.* (C) *I*–*V* data of isolated Au nanocrystals of different sizes. The derivative spectra are shown alongside. *Reproduced with permission from Ref. [61], ©2000 Elsevier B. V.*

junction is formed between the nanoparticle and the STM tip, and the other between the metal nanoparticle and the conductive substrate.

Experimentally, Coulomb staircase is observed when two prerequisites are satisfied [52,62]. First, for an isolated metallic nanoparticle of capacitance (C), the energy associated with transferring one electron ($E_c = e^2/2C$) must exceed thermal kinetic energy (k_BT). At room temperature ($k_BT = 26$ meV), this dictates that the metal nanoparticle size must be below 10 nm so that the capacitance is on the order of 10^{-18} F. Second, the electrical contacts to the particles must have tunneling resistance larger than the quantum resistance ($h/4e^2 \sim 6.5$ kΩ), in order to suppress quantum fluctuations of the electron charge.

Coulomb blockade may also be observed. Coulomb blockade is an electronic phenomenon where the tunneling current is suppressed because the electron does not have sufficient energy to tunnel through the junctions. The origin of this phenomenon lies in the fundamental quantization of the electron charge, and it occurs in systems with small quantum charge fluctuations. As a result, the redistribution of junction charges necessarily associated with a current flow can be made only in quantized units of e. The correspondingly quantized electrostatic charging energy, which for small nanoscale junctions can be large compared to other relevant energies related to temperature and bias voltage, tends to block the current [63,64].

As aforementioned, one prerequisite for the appearance of Coulomb staircase is that the core size of the metal nanoparticle must be below 10 nm at room temperature. Indeed, particle size plays an important role in dictating the step width. Figure 6(C) depicts the conductance and corresponding differential conductance of Au nanoparticles of 1.7, 2.7, 3.4, and 4.1 nm in diameter [61]. In this range, the nanoparticles are small enough to have discrete charging energy but are above the nonmetallic regime [65]. The step width at higher bias in $I-V$ profile is Coulomb staircase, and the featureless region around the zero potential position is Coulomb blockade. The dI/dV plot shows uniformly spaced peaks that correspond to the steps in the $I-V$ curve. Both the step width and the number of steps do point to a size effect. Specifically, the number of steps increases, and the corresponding step width decreases with increasing size of the nanoparticles. The Coulomb blockade also increases with the decrease of the particle sizes, which is ascribed to the emergence of a HOMO-LUMO energy gap for ultrasmall particles [52]. Similar size-dependent behaviors have also been observed in organically passivated gold and ruthenium nanoparticles [66].

As the thermal kinetic energy (k_BT) is directly decided by the experimental temperature, the lower the temperature, the more likely Coulomb staircase becomes visible. For example, Au nanoparticles (mean size 3.5 nm) show well-defined Coulomb staircase curves at 20 K, even at high voltage bias [67]. But the features of Coulomb staircase become increasingly indistinguishable at 50 K. The tunneling current at 20 K is actually larger than that at 50 K because electron tunneling through the junction becomes more feasible at decreasing temperature. Similar behaviors have also been reported with Pd nanoparticles [68]. Further studies show that charge transfer through the nanoparticle junctions can also be impacted by the chemical environments, which may be exploited for the development of novel chemical sensors. The sensing mechanism is based on sorption of chemical vapors into the particle organic shells and, hence, an increase of electron transfer resistance [69]. For instance, the tunneling currents of hexanethiolate-protected gold (AuC6) nanoparticles around 6 nm exhibit a rather sensitive variation upon the exposure to varied organic vapors [60]. The $I-V$ profiles show a drastic enlargement of the Coulomb gap with increasing vapor concentration and decreasing vapor relative polarity. The differences of these experimental observations can be accounted for by the effects of vapor solvation on the nanoparticle charging energy that dictates charge transport across the tip–particle–substrate junction.

4.3 Electrochemical Quantized Charging

4.3.1 Solution Electrochemistry

In electrochemical studies, analogous quantized charging behaviors have also been observed with solutions of nanoparticles of varied size and narrow dispersity. In 1997, Murray and co-workers [52] first discovered that alkanethiolate-protected gold nanoparticles in solution exhibited quantized electrochemistry that was analogous to the STM Coulomb staircase. The main difference is that in STM measurements, a single particle is probed; whereas, in electrochemical experiments, an ensemble of nanoparticles is charged and discharged at the electrode–electrolyte interface. Additionally, in contrast to the staircase features, electrochemical responses are a series of voltammetric peaks that appear to be diffusion-controlled. These voltammetric responses are ascribed to the quantized charging of the nanoscale double-layer of the particle molecules, that is, the nanoparticles behave as diffusive nanoelectrodes.

The basic premise is that the distribution of electrons between the working electrode and the nanoparticles is determined by the potential applied to the working electrode (E_{APP}), the potential of the nanoparticles (E_P), and the capacitance of the nanoparticle:

$$E_P = E_{PZC} + \frac{ze}{C_{CLU}} \tag{5}$$

where E_{PZC} is the potential of zero charge of the nanoparticle, z is the number of electronic charges on the particle (i.e., at $z > 0$, the particle is oxidized, while at $z \ll 0$, reduced), and e is the electronic charge. From Eqn (5),

$$\Delta V = E_p - E_{pzc} = \frac{ze}{C_{CLU}} \tag{6}$$

If $z = 1$, ΔV is the potential spacing between two neighboring charging peaks, from which C_{CLU} can be evaluated, and for a spherical nanoparticle coated with a uniform monolayer (of dielectric constant ε),

$$C_{CLU} = A_{CLU}\frac{\varepsilon\varepsilon_0}{r}\frac{r+d}{r} = 4\pi\varepsilon\varepsilon_0\frac{r}{d}(r+d) \tag{7}$$

where ε_0 is the permittivity of free space, A_{CLU} the surface area of the MPC core of radius r, and d the thickness of the protecting dielectric monolayer. Thus, when the core size and thickness of the organic protecting monolayer are comparable, C_{CLU} increases with increasing core radius, and for monolayers with similar dielectric property, decreases with increasing monolayer thickness.

Figure 7 shows the differential pulse voltammogram (DPV) responses of Au nanoparticles of varied size, passivated by (A) butanethiolates or (B) hexanethiolates [52]. One can see that with particles larger than 28 kDa (diameter 1.6 nm), the voltammetric charging peaks are quite evenly separated at potentials near 0 V (with a spacing of about 0.3 V). Whereas, with smaller particles, a central gap starts to emerge between the first positive and negative charging peaks. For instance, for a particle core size of 8 kDa (Au_{38}), the central gap is about 1.2 V. This is attributed to the evolution of a HOMO–LUMO energy gap due to shrinking size of the nanoparticle core, namely, a quantum size effect. These observations clearly indicate a transition of the nanoparticle quantized charging from classical double-layer charging to molecule-like redox behaviors with decreasing nanoparticle core size.

Similar responses are also observed with Au nanoparticles protected by monolayers of alkanethiolates of varied chain lengths [70]. The average full width at half maximum of a DPV peak can be as small as 115 mV, only slightly larger than that expected for a reversible one-electron process (90 mV). The corresponding nanoparticle capacitances have also been evaluated. The calculated values (Eqn (7)) are consistent with the experimental data where the monolayer thicknesses are approximated to be the fully extended chain lengths as calculated by Hyperchem®.

These electrochemical responses, the molecule-like multivalent redox behavior, of Au nanoparticles are also sensitive to the electronic interactions between the particle core and protecting ligands [71]. For two Au clusters, $Au_{11}(PPh_3)_7C_{13}$ (triphenylphosphine-stabilized Au_{11} clusters) and $Au_{11}C12$ (dedocanethiolate-protected Au_{11} clusters), both exhibit a series of quantized charging peaks, as observed previously with larger particles. In addition, the HOMO–LUMO gaps as determined from voltammetric measurements are 1.4 V for $Au_{11}(PPh_3)_7C_{13}$ and 1.8 V for $Au_{11}C_{12}$, both larger than 1.2 V for the Au_{38} particles.

More importantly, the nanoparticle electronic energy and, hence, voltammetric properties can be manipulated by simple surface chemistry because the particle core size remains virtually unchanged, and the only difference is that

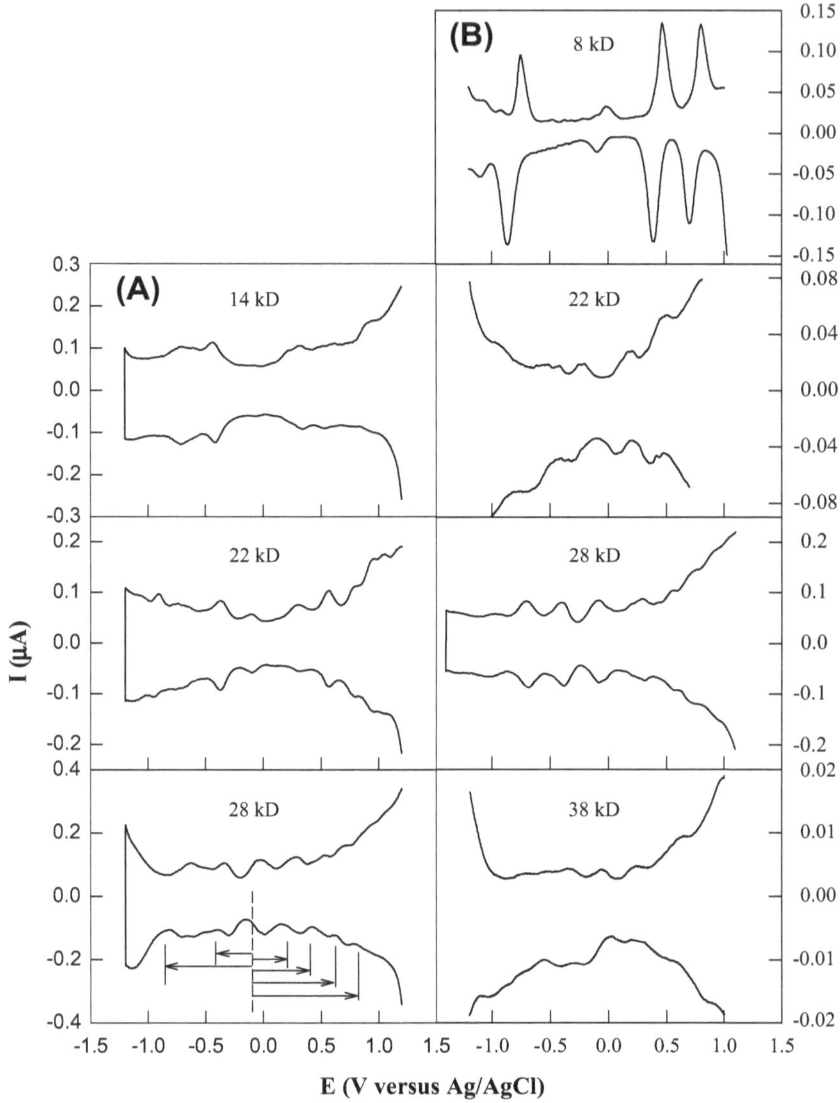

FIGURE 7 Differential pulse voltammograms for (A) butanethiolate- and (B) hexanethiolate-protected Au nanoparticles as a function of uniform core size in 0.05 M Hex_4NClO_4/toluene/acetonitrile (2/1 v:v), at 9.5 × 10-3 cm^2 Pt electrode; DC potential scan 10 mV/s, pulse amplitude 50 mV. Concentrations are: (A) 14 kD, 0.086 mM; 22 kD, 0.032 mM; 28 kD, 0.10 mM; (B) 8 kD, 0.30 mM; 22 kD, 0.10 mM; 28 kD, 0.10 mM; 38 kD, 0.10 mM. Arrows at lower left indicate ΔE potential steps. *Reproduced with permission from Ref. [52], ©1998 AAAS.*

the particle protecting monolayer is replaced from PPh_3 and Cl to dodecanethiolate (C_{12}). There are at least two areas that warrant special attention. First, the bandgap energy increases by about 0.35 V. This can be understood within the context of the stronger bonding of Au—S compared to that of Au—P (Cl), akin to the ligand-field effects on the splitting of the electronic energy of coordinated complexes [71]. Second, the Fermi level shifts positively from −0.23 V for $Au_{11}(PPh_3)_7Cl_3$ to −0.07 V for $Au_{11}C_{12}$. One possible explanation might be that the relatively weak ligand field in $Au_{11}(PPh_3)_7Cl_3$ leads to more electrons residing in the conduction band (equivalent to the high-spin state); whereas, in $Au_{11}C_{12}$, electrons are more likely to be degenerated (equivalent to the low-spin state) residing in the lower-energy valence band. Thus, the Fermi level is located at nearly halfway between the conduction band and valence band. Additionally, the electron-donating nature of P (in PPh_3) to Au is anticipated to lead to an increase of the particle core electron density and, hence, a negative (upward) shift of the Fermi level. Opposite behaviors are expected from thiol-protecting ligands, which behave as electron acceptors to gold, resulting in a decrease of the electron density of the Au cores.

4.3.2 Nanoparticle Solid Ensembles

Nanoparticle solid films are typically prepared by drop-casting/spin-casting a particle solution in a volatile organic solvent onto a flat electrode surface. The ensemble conductance has been found to be primarily driven by the thermally activated percolation effect as well as core thermal motions. Because of rampant structural inhomogeneity (disordering due to the dispersity of particle core size, shape, and chemical environments) within these particle-thick films, the energetic barriers for interparticle charge transfer most likely vary widely from site to site. Consequently, typically only featureless linear (ohmic) I—V behavior is observed, especially at a relatively high voltage bias. Such disorder is one important effect for ensemble conductivity in nanoparticle solids, and the other two are the (dipolar) exchange coupling between adjacent particles and the Coulomb repulsion of electrons (of opposite spins) on a given particle [72]. In these, the disorder within the particle ensembles will diminish the interparticle electronic coupling.

Consequently, how to prepare organized and robust assemblies of nanoparticles is one of the technological challenges. An efficient surface-immobilization of nanoparticles is developed by a two-step procedure involving place—exchange reactions and self-assembly. Alkanethiolate-protected nanoparticles first undergo surface place—exchange reactions with alkanedithiols of similar chain lengths, rendering the particle surface active with varied copies of peripheral thiol groups. Excessive dithiols and displaced thiolates are removed by phase extraction using a hexane-methanol system. Self-assembling of these surface-active nanoparticles is effected by incubating the electrode in the solution, just like monomeric alkanethiols. The electrochemistry of these surface ensembles of nanoparticles can also exhibit well-defined quantized capacitance charging features [73]. Nanoparticle assemblies fabricated on the basis of other specific interactions also show well-defined quantized charging behaviors. For instance, the cyclic voltammograms (CVs) and DPVs of nanoparticle surface ensembles by the coupling interactions between divalent metal ions and carboxylic moieties also exhibit well-defined quantized charging features. However, the peak splitting in these cases is somewhat larger than that with nanoparticles linked by alkanedithiols of similar chain lengths, implying that the nanoparticle electron transfer kinetics are impeded somewhat by the metal ion centers.

It has to be noted that the previously mentioned studies are carried out with the nanoparticle films-modified electrodes in (low-dielectric) organic media. In aqueous solutions, however, the electrochemistry of these nanoparticle assemblies is drastically different [55]. The first observation of nanoparticle electrochemical quantized charging in aqueous solutions was reported by a AuC6 self-assembled monolayer in 0.1 M NH_4PF_6 aqueous solution (Figure 8). In contrast to the behaviors in organic media, there are at least three aspects that warrant attention. First, several well-defined voltammetric peaks are found in the positive potential regime, which are ascribed to the quantized charging of nanoparticles. Second, the quantized charging features are sensitive to electrolyte composition. In the present study, the quantized charging features are only observed in the presence of PF_6^- ions but not NO_3^-. Also, the voltammetric responses do not appear to be dependent on the cationic species. Third, the voltammetric current is much larger in the positive potential regime (compared to that at the same bare electrode) but somewhat suppressed in the negative region. In essence, this behaves like a molecular diode (current rectifier). This is interpreted on the basis of the effects of the binding of hydrophobic anions like PF_6^- to the surface-confined nanoparticle molecules on interfacial double-layer capacitance. As in (low-dielectric) organic media, C_{EL} is generally smaller than C_{SAM}. Thus the double-layer charging is mainly through the surface immobilized particle molecules, which is observed at both positive and negative potentials; whereas in aqueous solutions, $C_{EL} > C_{SAM}$, the double-layer charging is generally through the naked electrode part, resulting in featureless responses only. However, in the presence of hydrophobic anions, they bind to positively charged particle molecules (at positive electrode potentials), repelling the water molecules from the interface and, hence, rendering $C_{SAM} > C_{EL}$. This ion-pairing hypothesis is further shown by the negative shift of the voltammetric profile:

$$E_f = E^o + \frac{RT}{n_aF}\ln\frac{K_2}{K_1} - (p-q)\frac{RT}{n_aF}\ln[anion] \tag{8}$$

where E_f and $E^{0'}$ are the formal potentials in the presence and absence of ion-pairing, respectively, and K_1 and q (K_2 and p) are the equilibrium constant and number of ions bound to the reduced (oxidized) form of particle molecules, respectively; other parameters have their usual significance. One can see from

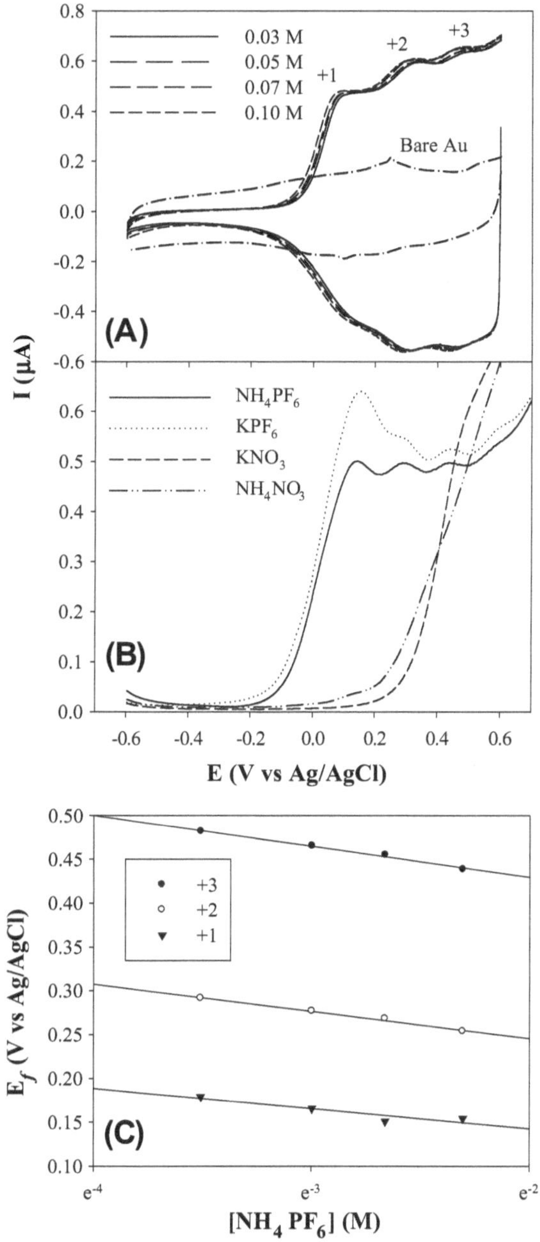

FIGURE 8 (A) CVs of a AuC6 MPC-modified Au electrode in aqueous NH_4PF_6 solutions of various concentrations. Also shown is the CV of the same bare electrode in 0.1 M NH_4PF_6. Electrode area 1.1 mm^2. Sweep rate 100 mV/s (B) DPVs of the same MPC-modified electrode in various electrolyte solutions (0.1 M). Pulse amplitude 50 mV, dc ramp 4 mV/s. (C) Variation of the formal potentials with the concentration of NH_4PF_6. Lines are linear regressions. +1, +2, and +3 refer to the three voltammetric peaks observed in (A), respectively. *Reproduced with permission from Ref. [55], ©2000 American Chemical Society.*

Figure 8(C) that all peak potentials shift negatively with increasing PF_6^- concentration, and the slopes evaluated from linear regressions are −23, −30 and −35 mV for the first, second, and third charging peaks, respectively, effectively indicating a 1:1 ratio of the number of PF_6^- ions and the MPC charge states ($n_a = 1$). It is also anticipated that the binding of anionic PF_6^- to surface MPCs increases the effective dielectric constant of the MPC-protecting monolayers, leading to the larger value of the MPC capacitance (1.05 aF).

More detailed studies involving other hydrophobic (soft) anions, such as ClO_4^- and BF_4^-, exhibit similar rectified charging features, further supporting the notion that the nanoparticle quantized charging can be manipulated by simple ion-binding chemistry. In Figure 8, one might also note that in the presence of "hard" anions (e.g., NO_3^-) the charging features are similar to those of a conventional molecular diode. At potentials more negative than the threshold value, the current is significantly suppressed; whereas at potentials more positive than the threshold potential, the current starts to increase rapidly (without the discrete charging characters). This also strongly suggests that increasing the "softness" of the electrolyte ions leads to a transition from conventional molecular diodes to single-electrode rectifiers of the nanoparticle assemblies [55].

Quantized capacitance charging has also been observed in solid-state electronic conductivity measurements of nanoparticle solid films. In these studies, Langmuir-based (LB) techniques are commonly used to fabricate monolayers and multilayers at controlled interparticle distances. The recent rapid advancement in syntheses has resulted in the ready availability of monodispersed and ultrasmall (diameter <3 nm) gold particles, rendering it possible to fabricate a clearly defined ensemble structure for the studies of their solid-state electron-transfer properties. In fact, single electron transfer can now be achieved with monolayers of these particles at controlled interparticle arrangements using the Langmuir technique. The solid-state CVs and DPVs of an LB monolayer of AuC_6 nanoparticles deposited onto an IDA electrode surface at $L = 0.72$ nm have been reported. The measurements are carried out in vacuo and at varied temperatures. When the temperature is controlled at 300 and 320 K, there are at least five pairs of well-defined and evenly spaced current peaks within the potential range of −1.0 to +1.0 V. These are, again, attributable to the single electron transfer across the nanoparticle monolayers.

4.4 Summary and Perspectives

In this section we highlighted the unique single electron transfer properties of organically capped nanoparticles. The behaviors are observed with scanning tunneling spectroscopy of single nanoparticles, electrochemical studies of nanoparticle ensembles in solution, as well as charge transfer in solid ensembles. In these studies, the nanoparticle electron transfer behaviors are found to be directly determined by the nanoparticle size, ligand length, and interparticle separation. In addition, sensitive dependence has also been observed on the chemical environments, such as temperature, vapor atmosphere, and electrolyte ions in aqueous solutions. A good understanding of the electron transfer dynamics of such nanoscale entities is of critical importance in the further manipulation of their material properties as well as applications in the fabrication of next-generation electronic circuitries and devices.

5. MAGNETISM IN GOLD NANOPARTICLES

5.1 Introduction

The observation of ferromagnetic-like response in nanoparticles (NPs) of traditionally nonmagnetic metallic systems as Pd, Pt, Ag, and Au has motivated a great interest, as well as theoretical effort, to find its possible origin. In all cases, paramagnetic and diamagnetic in bulk materials, two challenges arise: first the origin of the atomic magnetic moments and second, the origin of the ferromagnetic-like behavior. Magnetic phenomena are strongly dependent on the changes that occur at the nanoscale, arising either from the so-called size effects, associated to the geometric confinement of electrons or from the surface effects, due to the large fraction of surface atoms since both of them modify the electronic structure of the systems. The continuum that describes the band of allowed energy states for a metal becomes a discrete set of states for NPs. In addition, the density of states at the Fermi level, the parameter that governs electronic properties, is strongly affected by the nanoscale. Some of the novel properties/phenomena that appear/occur at the nanoscale are monodomain configuration instead of a multidomain configuration, superparamagnetism, surface anisotropy, and discontinuity in the short-range order at the surface, weakening of the exchange interactions, narrowing of the bands, and enhancement of the local density of states as well as of the local magnetic moments.

Surface modification of the NPs by capping with appropriate molecules could also drastically affect the magnetic behavior of the NPs with respect to the observed behavior in bulk. In bulk solids, the dependence of physical properties on the chemical environment can be defined as weak when compared to that exerted on NPs. At the nanoscale, physical properties (e.g., optical and magnetic properties) can be tuned by using a suitable coating. In other words, the electronic structure, which is extremely sensitive to the nanoscale, will also be deeply modified by the chemical bonding of the atoms.

One of the most surprising examples of such novel magnetic properties is the observation of a ferromagnetic-like behavior in materials that are not ferromagnetic in bulk. Ferromagnetism has been observed in Pd and Pt nanostructures, typical paramagnetic materials in bulk, with a particle size below 2 nm [74,75]. More surprising is the case of noble metal Au and Ag nanostructures that are diamagnetic in bulk [76,77]. The local enhancement of the density of states seems to be responsible for the observed behavior in Pd and Pt; whereas, an appropriate coating and the associated charge transfer is the origin of the magnetic behavior observed in Au and Ag nanoparticles. This work will focus on the magnetic behavior observed in gold nanostructures. It should be mentioned that the observation of ferromagnetism in Au NPs adds an important new functionality to this nanoscale system that, besides being magnetic or not, exhibits a broad potential in biology, catalysis, and nanotechnology. The main results obtained from the experimental magnetic characterization of Au NPs stabilized with different protective agents will be presented. Also, results obtained by element selective techniques, X-ray magnetic circular dichroism, and ^{197}Au Mössbauer spectroscopy will be summarized. Finally, a model for explaining several key features of the experimental results will be presented.

5.2 Experimental Studies

The first experimental work on a magnetic behavior different that a diamagnetic one in Au NPs dates back to 1999. Hori et al. [78] reported unexpected large magnetic moments in Pd and Au nanoparticles with particle size below 3 nm and embedded in poly(*N*-vinyl-2-pyrrolidone) at low temperatures.

These results were quite interesting since bulk Au behaves as a diamagnetic metal with a negative susceptibility of -1.42×10^{-7} *emu/g Oe*. The Au atomic electronic configuration is $[Xe]4f^{14}5d^{10}6s^{1}$. The 5*d* electrons are buried a few electron-volts below the Fermi level and the density of states at the Fermi level, $g(\varepsilon_F) = 0.29\ eV^{-1}atom^{-1}$, is very low. The occupancy of the 5*d* band can be obtained by using X-ray absorption near-edge structure (XANES) spectroscopy. By carrying out experiments at the Au L_2 and L_3 edges, it is possible to determine the density of unoccupied 5*d* states (i.e., the number of holes in the 5*d* band) since it is directly proportional to the intensity of the white line transitions (corresponding to $2p_{1/2} \rightarrow 5d_{3/2}$ and $2p_{3/2} \rightarrow 5d_{3/2,5/2}$ dipole transitions, respectively). For bulk gold, low intensities of the white line arising from a small population of 5*d* holes resulting from *s-p-d* hybridization are measured. Neglecting such small hybridization, the 5*d* band is nominally full with a balanced number of electrons in the spin-up and spin-down subbands.

Later, Crespo et al. [76] reported on the observation of an even more surprising magnetic behavior regarding magnetic hysteresis up to room temperature in 1.4 nm thiol-capped Au nanoparticles obtained by the Brust method [79]. These NPs were stabilized with dodecanethiol $CH_3{-}(CH_2)_{11}{-}SH$ that strongly bonds to the surface of the nanoparticle through the S atom. The unexpected magnetic behavior was compared with that obtained in Au nanoparticles that were stabilized by means of a tetraoctylammonium bromide $(C_8H_{17})_4N^+Br^-$, a weakly interacting and non-surface-binding surfactant. In this case, the experimentally observed behavior is diamagnetic (as in bulk gold) and anhysteretic.

The ferromagnetic-like behavior was also observed in thiol-capped Au glyconanoparticles (GNPs) where the Au nanoparticles were functionalized with thiol-derivatized maltose and lactose neoglycoconjugates [80]. In contrast with the thermal dependence of the ferromagnetic materials, almost no variation of the magnetization up to room temperature was observed in the thermal dependence of the magnetization; whereas, hysteresis was also observed up to room temperature. As in the previous work by Crespo et al., the neoglycoconjugates were bonded to the Au surface via an S atom.

It should be mentioned that elemental chemical analysis was carried out by means of inductively coupled plasma (ICP) analysis in order to estimate the amount of ferromagnetic impurities. It was observed that the low level of impurities could not account for the magnetization values.

Figure 9 shows an example of the magnetization curves exhibited by Au nanoparticles capped with thiol-derivatized molecules, as well as the one stabilized with the surfactant. The characteristic features of a ferromagnetic material, a hysteretic behavior with a coercive field and permanent magnetization even in a zero applied field, are clearly seen up to room temperature. Curves corresponding to nanoparticle-stabilized tetraoctylammonium bromide are also shown.

These results were analyzed on the basis of the results obtained by Zhang and Sham [81a,b]. By means of X-ray absorption near-edge structure

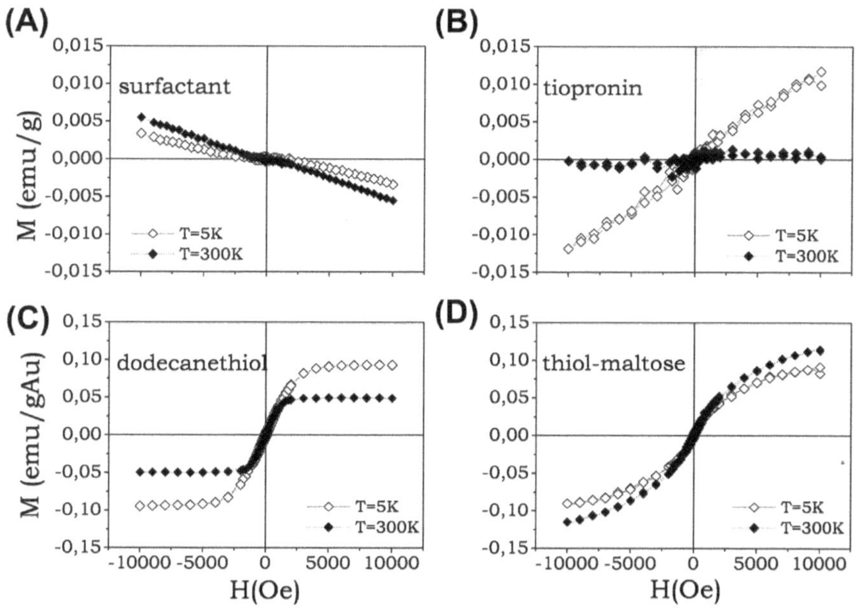

FIGURE 9 Magnetization curves at 5 and 300 K of Au nanoparticles stabilized by (A) tetraoctylammonium bromide; (B) tiopronin; (C) dodecanethiol; and (D) thiol-derivatized maltose.

(XANES) studies carried out at the *Au* $L_{3,2}$− edge, the influence of the capping system on the electronic structure of Au nanoparticles was pointed out. Au atoms in the NPs gain 5*d* electrons (relative to the bulk) when particles are capped with weakly interacting dendrimers; whereas, they lose 5*d* electrons when capped with a strongly interacting thiol molecule. This indicates that more holes are present on the gold atoms of thiol-capped particles, compared to the bulk, as a result of the Au to S charge redistribution promoted by appropriated capping. X-ray photoelectron spectroscopy corroborates this behavior. XANES studies carried out by Crespo et al. on thiol-capped Au nanoparticles showed the charge transfer from gold to sulfur atoms was extremely large in the thiol-capped Au NP; whereas, tetraalkylammonium-protected NPs exhibit a density of 5*d* holes, similar to that from bulk gold.

These changes in the electronic properties of the NPs were further studied by means of surface plasmon resonance by Crespo et al. When illuminated with light of the resonant frequency, the free electrons of the nanoparticle exhibit a collective coherent oscillation called surface plasmon resonance. The position and width of the resonance depend on the particle size, shape, as well as on the medium in which it is embedded, among other factors. The thiol-coated nanoparticles do not present surface plasmon resonance, while this feature is clearly observed for the ammonium-coated nanoparticles. This behavior was analyzed by García et al. [82] as a consequence of a decrease of the effective size of the metallic core, where conduction electrons could oscillate due to the surface localization of conduction electrons. Strong charge transfer (such as what occurs when thiols are surface-bound) localizes surface electrons, thereby broadening the surface plasmon resonance peak. For enough small particles with strongly bound thiols, this surface shell of localized electrons can become comparable or equal to the total radius of the nanoparticle, and surface plasmon resonance is suppressed due to the localized character of the electronic charges. An expression that takes into account that the region where electrons can oscillate freely could be reduced by the chemical interface damping due to the Au−S bond can be found in [82].

Thus, the ferromagnetic-like behavior of very small thiol-capped gold NPs points out that thiol bonding induces in Au NPs not only hole localization, but also permanent magnetic moments associated with the spin of extra d holes localized near the Au−S bonds. This, in combination with the strong spin-orbit coupling of Au, seems to be the origin of the observed behavior.

Following the experimental observation of magnetism in Au thin films coated with SAMs organized in ordered domains [83], the idea that orbital contribution to magnetic moment should not be ruled out, and the influence of the tendency to self-assembly of molecules at the Au surface on the magnetic behavior was investigated by Guerrero et al. [84]. Two different thiol-containing molecules with different spatial configurations were used as capping molecules of Au nanoparticles. A tiopronin-protected gold NP was synthesized, and its magnetic behavior was compared with that of dodecanethiol-capped Au NPs. Despite Au–S bonds being present in dodecanethiol- and tiopronin-capped Au NPs, the magnetic behavior was quite different. The magnetization of the tiopronin-capped NPs exhibits a paramagnetic response at five and at 300 K, which indicates that the spins induced through binding to S atoms fluctuate in orientation, and no permanent magnetic moment is observed, see Figure 9. Whereas, dodecanethiol molecules are well known to self-assemble, and this that could induce the formation of domains in which well-ordered molecules are bonded to the surface; the self-organization of tiopronin is not a straightforward issue due to the geometry of the molecule. Thus, the spatial arrangement of molecules also plays a role on magnetic behavior.

The metallic character of the nanoparticle should also be taken into consideration. As mentioned previously, the magnetism observed in the thiol-capped NPs was found to be accompanied by a partial suppression of the plasmon absorption, indicating that the itinerancy of the 5*d* electrons of Au was partially inhibited due to the strong interaction between the thiolate and the surface gold atoms. Two set of samples were prepared: a Au-thiol (polymeric) phase and the well-known dodecanethiol-capped Au NPs. XANES confirmed the formation of a –Au–S–Au–S– polymeric network without Au–Au metallic bonds for the Au-thiol (polymeric) phase. However, the number of Au–S bonds in the polymeric phase was greatly enhanced when compared with dodecanethiol-capped NP; a diamagnetic behavior was observed [85].

Thus, it was experimentally shown that the presence of a ferromagnetic-like behavior is associated with the formation of NPs with simultaneous presence of Au–Au (metallic character of the nanoparticle) and Au–S bonds, as well as with the tendency of thiol-derivatized molecules to form ordered domains. As will be discussed later, this result indicates that the contribution of orbital magnetism due to conduction electrons should be taken in consideration for explaining the magnetic behavior. The formation of domains at the surface acts as a potential barrier where quasifree electrons become trapped, leading to orbital momentum induced at the surface conduction electrons. The spin–orbit coupling at Au surfaces creates a strong anisotropy that could account for the magnetic hysteresis observed even at room temperature.

Before going into the details of the possible origin of the anisotropy in these nanoparticles, an important point should be clarified. When dealing with nanoparticles, one can argue whether the observed magnetic behavior, typically measured with a highly sensitive SQUID magnetometer, is intrinsic to the nanoparticle or arises from impurities of iron or its oxides. In most works, elemental chemical analysis is carried out by means of inductively coupled plasma (ICP) analysis in order to estimate the amount of ferromagnetic impurities. The level of ferromagnetic impurities found was, in all cases, too low to account for the obtained magnetization values.

Au(Fe) glyconanoparticles (GNPs) with particle sizes below 2 nm were prepared by introducing a controlled amount of Fe impurities by a one-step synthesis developed by Penades et al. [86] for studying the effects of impurities on the magnetic behavior of Au nanoparticles. The Fe content of the samples ranged between 0.06 wt% Fe and 2.8 wt% Fe. Surprisingly, it was observed that the spontaneous magnetization of AuFe GNPs exhibited a fast decrease with temperature that contrasted with the almost constant value of the magnetization observed in Au NPs. In addition, no ferromagnetic-like behavior was observed at room temperature. It seems reasonable not to ascribe the amazing ferromagnetism exhibited by Au NPs with the presence of magnetic impurities since the presence of Fe clearly weakened the magnetic behavior [76].

However, to evidence that the magnetism is intrinsic to the nanoparticles, element selective techniques such X-ray magnetic circular dichroism (XMCD) and Mössbauer spectroscopy should be used. XMCD is a powerful technique that consists of measuring $L_{2,3}$ absorption edges using circularly polarized X-ray photons in the presence of a magnetic field which is applied along and against the X-ray photon's momentum. The difference between the two XANES spectra yields the XMCD spectrum, which allows determining the difference between spin-up and spin-down densities of states. Furthermore, the technique allows us to separate the spin and orbital contribution to the magnetic moment.

The first experimental evidence by XMCD of a magnetic moment on gold was provided by Yamamoto et al. on PAAHC—Au nanoparticles [87]. This study confirmed that Au atoms are bearing a magnetic moment. Garitaonandia et al. [77] have combined XMCD measurements with ^{197}Au Mössbauer spectroscopy for determining the origin of the magnetization in thiol-capped Au NPs. The field dependence of the magnetization extracted from XMCD data shows a clear hysteresis, similar to that observed by SQUID magnetometry. Moreover, in the Mössbauer spectrum performed at low temperatures, 4.2 K, two contributions were observed, one arising from the core gold atoms and a second contribution that consists of a magnetically split signal attributed to the surface gold atoms. Their results demonstrate that the magnetization in thiol-capped Au NPs is localized in the 5*d* orbital of Au atoms. Although the values of the magnetic moment per Au atom, $0.3\mu_B$, were higher than those reported by others, the main conclusion of this study is that it is clearly proved that the gold atoms in the nanoparticle surface are carriers of the magnetic moment. Later, de la Venta et al. [88] conducted a similar study on thiol-capped Au nanoparticles (NPs) embedded into polyethylene. Their results also showed that show that the magnetism observed in thiol-capped Au NPS is a magnetism that is intrinsic to the nanoparticle, and it cannot be attributed to *para-* or ferro-magnetic impurities.

Ferromagnetism has not only observed in Au NPs capped with thiol-derivatized molecules. Hori et al. [89] have also reported ferromagnetic behavior in Au NPs stabilized with polyacrylonitrile and polyallyl amine hydrochloride (PAAHC). Ferromagnetic spin polarization of Au nanoparticles protected by PAAHC with a mean diameter of 1.9 nm was measured by XMCD. Au nanoparticles coated with oleic acid and oleylamine with sizes below 7 nm and a narrow size distribution were also reported to exhibited ferromagnetic behavior [90]. More recently, ferromagnetic-like behavior has been also observed in phosphine-capped Au NPs [91a,b] and thiolated azobenzenes [92].

5.3 Origin of the Magnetic Moment

Since gold NPs dispersed by a surfactant are diamagnetic, the ferromagnetic-like behavior observed for thiol-capped Au NPs must be associated with the modification introduced in the Au electronic structure by the Au—S bond. This result indicates that the origin of the magnetic behavior lies in the different degree of interaction between Au atoms and the protective molecule. Therefore, and as clearly derived from a broad spectrum of experimental reports, the unexpected characteristics of the magnetic behavior of nanoparticles of materials that are far from being magnetic in bulk, as is the case for Au, are mainly conditioned by the capping molecules. The bonding nature between these molecules and the surface atoms of the NPs determines the modification of the electronic structure of the surface that leads to the appearance of the unexpected magnetic moment.

In order to undertake a study of the intrinsic origin of the magnetization in gold nanoparticles, a more detailed understanding of the nature of the gold—alkanethiol bond is, therefore, required. In particular, understanding of the interaction between the capping molecules as alkanethiols and Au NPs is currently one of the main challenges in the study of their magnetic structure. Several articles related to the sulfur atom bonding with Au clusters have been published.

Recent ab initio calculations show that the S—Au bond is such that an electron is transferred to the Au cluster [93]. Therefore, the appearance of the magnetic moment in Au was always associated with either electrons or holes transferred from the capping molecules.

In the next paragraph, the main theoretical efforts oriented to account for the induction of the magnetic moment are outlined.

A second challenge is to account for the room-temperature hysteresis loop observed in Au NPs. In general, the presence of hysteresis in the magnetization curve of a sample indicates that the atomic magnetic moments are coupled to each other through exchange interactions. In fact, hysteresis is due to magnetic anisotropy rather than to exchange interactions. In ferromagnetic NPs as small as 1.4 nm in diameter, the expected behavior at room temperature is superparamagnetic, provided the usual values of anisotropy hold. For instance, Fe NPs are blocked if their diameters are larger than 30 nm. Then the question is:

what is the strength and origin of giant anisotropy that forces a 6 nm diameter Fe (or Au) NP to remain blocked at room temperature? Different proposals have been made in order to explain the hysteresis of the NPs. Vager et al. [94] suggested that the applied magnetic field would induce a giant paramagnetic moment on those electrons involved in the bonding to the capping molecules. The high magnetic anisotropy suggests that some kind of orbital contribution to the magnetic moment cannot be disregarded. Following this suggestion, Hernando et al. proposed a model including orbital magnetism. This model [95–97] assumes the induction of orbital magnetism by motion of surface electrons around ordered arrays of Au–S bonds.

5.4 First Principle Calculations

Theory and experiments have shown that the coverage of magic gold clusters by thiols of well-defined compositions is rather complicated [98]. Sulfur atoms are attached with a single bond to a gold atom of the cluster [99,100]. These magic or closed-shell clusters are highly ordered arrangements with full electronic sp shells, as in noble metal clusters in general [101]. Furthermore, alkanethiols with longer chains on Au surfaces have been the focus of detailed theoretical studies [102], and the presence of several different metastable states is known. In these cases, the sulfur atoms in the gold layers become arranged in a bridge configuration for the ground state.

The experimental findings in capped Au NPs are now supported by theoretical total energy calculations. Gonzalez et al. [103], based on the results of thiol-capped Au NPs, presented a simple model that addresses the induction of magnetic behavior in gold clusters on chemisorptions of one organic molecule with different chemical linkers (nitrogen and sulfur linkers). The system under consideration was a small cluster of 13 gold atoms and one chemisorbed molecule. It was found that the interaction between the S and Au orbitals was ultimately responsible for the onset of magnetism in thiol-capped gold nanoclusters, in agreement with the proposal of Crespo et al. [76]. This model showed that for a sulfur linker, energetic stabilization is accompanied by the development of spin localization preferentially in the gold atoms closest to the chemisorption site. Whereas, for a nitrogen linker, no spin symmetry-breaking leading to lower energy states is found. Later, different theoretical models have reported size-dependent magnetization and spin symmetry-breaking that would explain the experimental results found for thiol-capped gold and silver NPs [104,105].

It is now well established that the contribution of the sp levels in Au NPs is, rather than the *d*-ones, particularly significant, and it is the filling of these levels that explains the effects of the electronic shells in noble metal clusters, in general, and in gold clusters, in particular. In particular, first principle calculations point out that the S–Au bond is such that sulfur does not get a charge, and an electron is transferred to the Au cluster [93,106].

Alkanethiols with longer chains on Au surfaces have been the focus of detailed theoretical studies [102], and it was shown that there exist different metastable states in energy. The sulfur atoms in the gold layers becomes arranged in a bridge configuration for the ground state, and this must then be primarily considered in assessing their magnetic signal.

The ADF (Amsterdam Density Functional) method [107] was used for calculations involving density-functional-theory (DFT). The valence basis set was composed of the *d* gold, sulfur, carbon, and hydrogen states, which were expanded in triple zeta with two polarization functions at each atom. The relativistic effects of gold were modelled using the zeroth-order regular approximation (ZORA) [108], and its scalar relativistic version was implemented for the structural optimizations.

We carried out a study [106] with a cluster of Au_3, which allowed us to investigate a number of different coordination numbers for the sulfur atom. Au_3SCH_3 has an even number of electrons and thus has a spin moment equivalent to an even number of Bohr magnetons. The data were obtained using generalized gradient density approximations (GGA) with the parameterized exchange-correlation functional of Perdew–Burke–Ernzerhof according to [109]. The optimization of the geometry was carried out until the forces were all smaller than 0.02 eV/Å. The magnetic anisotropy (MAE) was obtained by means of self-consistent ZORA relativistic calculations, including spin-orbit. It was found that the singly or doubly occupied levels of the Au clusters are responsible for the odd-even oscillations of the magnetic moment located at the Au atom participating in the sulfur bonding.

Our calculations predict that the bonding of alkanethiols on gold clusters (i.e., Au_nSCH_3) results in a bridge arrangement with a deviated axis in a polarized ground state for even values of n. For this structure, the sulfur atom and the close pair of atoms are responsible for the chemical bonding, but it is the outer gold atoms that host the larger part of the magnetic moment. The large magnetic moment is preserved in the structure of the nanoparticles because they have lower symmetries. Thus, it should be possible for the particles of other noble metals to show similar magnetic behavior.

There are five sulfur electrons in total: the two bonds with gold atoms share two electrons, an electron is shared in the bond with the carbon atom, and two more form the lone pair in the sp^3 hybridization. Because each sulfur atom contributes a total of six electrons, a single electron is left over and is passed to the gold cluster. It is this sp electron that contributes to the spin polarization.

A related discussion that supports the back-donation of a sp electron from the sulfur to gold arises from a comparison between the electronegativity of thiols and that of their gold counterparts. We calculated the ionization potentials (IPs) and electro-affinity (EAs), thereby obtaining the Mulliken electronegativity $M = (IP + EA)/2$. This M value is the negative of the electrochemical potential. The electronegativity of the SCH_3 unit is slightly higher (5.43 eV) than the work function ($\approx$5.22 eV) for the Au surface [110] and is comparable with the M value for gold nanostructures (<6 eV) [111]. The SCH_3 unit behaves rather differently from the sulfur atom ($M = 6.21$ eV). In consequence, the transfer of charge from gold to sulfur cannot be assumed a priori. In fact, we stress that independently of the used partition scheme, the calculated transfer of charge between the thiols and the gold clusters is less than 0.1 e.

Our findings point out a gain in energy of 0.12 eV for the $-SCH_3$ adsorbate with the gold cluster in a bridge position, this being the ground-state structure with respect to the molecule in the Au_3 plane. The most important characteristic is the shift between odd-even oscillations by a single gold atom with respect to bare gold clusters. The structure is spin compensated for an even number of gold atoms in the thiol-cluster, while it is spin polarized otherwise.

It must be stressed that the sulfur atom is not negatively charged, even though XANES experiments have shown that the $5d$ orbitals of gold are slightly open. The sp contribution to the magnetic signal is larger than that of the d orbitals. The back-donated electron from sulfur to gold also sits in the sp states of the Au bond and cannot be differentiated from the previous results using the XANES method. These sp electrons are almost free in the second gold layer and could orbit, and it is these that are responsible for the magnetism. As shown in the next paragraph, these findings are consistent with experimental results in which a number of gold layers are needed in order to engender thiol-induced magnetism.

By including spin-orbit coupling, we find that thiols attached on Au_4 and Au_6 show magnetic anisotropies, MAE, of 1.43 and 0.98 meV per SCH_3 molecule, respectively, while the change of MAE in the plane parallel to the surface is less than an order of magnitude smaller. These figures are comparable with the giant magnetic anisotropy of single cobalt atoms on a Pt surface (9 meV per Co atom) [112]. Such large values of MAE demonstrate the stability of the magnetism discussed herein at room temperature and are a factor of 10 times larger than the equivalent values for tetragonal Ni (0.2 meV). The experimental results, outlined above, showed that the magnetism of gold nanoparticles covered by thiols can indeed survive almost unchanged at room temperature.

5.5 Magnetic Moment of an Electron Gas Confined on a Spherical Surface

In order to estimate the permanent magnetic moment of the electrons occupying the 4sp surface states, we must begin by calculating its energy spectrum and eigenstates. Let us initially consilder that the electrons donated by the capping organic molecules are confined by a Dirac δ potential to jump between surface atomic sites, thus forming an electron gas depicted by its effective mass. In the following paragraph, it will be shown how the unfilled surface layer gives rise to a permanent collective magnetic moment. Let us summarize the more important features of the energy spectrum of an electron gas confined on a spherical surface. The corresponding eigenstates of a noninteracting gas confined in the spherical surface of NPs [113–117] are the spherical harmonics $U_{l,m}(\theta, \varphi)$, and their associated kinetic energy eigenvalues depend on the angular quantum number l as

$$e_l = \frac{l(l+1)\hbar^2}{2mr^2} = \frac{l(l+1)}{2} e_1 \tag{9}$$

where r is the radius of the NPs, and m is the effective mass of the electron that can be written as $m = \frac{\hbar^2}{2a^2b}$, b being the hopping amplitude proportional to the surface band width ($\approx 4b \approx 4b$) and a is the average interatomic distance at the surface. Notice that $e_l = \frac{\hbar^2}{mr}$.

Let N be the number of surface atoms available to supply an electron through a bond to the capping molecule and suppose that each bonding supplies an extra electron. There are actually $N_b = \alpha N \ll$ N electrons, where the coverage factor $a \ll 1$. If we proceed by filling successively the levels of increasing energy given by Eqn (9) up to N_b, and after taking into account that the maximum number of electrons that can occupy any level is given by $2(2l + 1)$, it can be seen that the Fermi level, l_F, is implicitly defined through the following relationship [118].

$$N_b = \alpha 4\pi \left(\frac{r}{a}\right)^2 = 2(l_F + 1)^2 \tag{10}$$

From Eqn (9) and Eqn (10), the distance in energy between the Fermi level and the level immediately above, Δ, can be expressed as

$$\Delta = 2b\sqrt{2\pi\alpha}\ \frac{a}{r} \tag{11}$$

The experimentally measured width, $4b$, of surface bands varies between 0.5 and 1.0 eV [119]. Therefore, for $b = 0.25$ eV and $\alpha = 1$, Δ is given approximately by $\left(\frac{a}{r}\right)$ eV. Consequently, for $\left(\frac{a}{r}\right) > 0.023$, the distance Δ should be larger than the thermal energy at room temperature, and the transition induced by scattering to upper energy levels (upper l values) may be disregarded. For $\Delta > 0.3$ eV, the excitation of electrons to levels above the Fermi level induced by either exchange interactions or an applied magnetic field should then be negligible.

The distance Δ remains larger than the thermal energy for the sufficiently small size of the NPs. As the size of the NPs increase, the distance between adjacent levels decreases and the surface band energy spectrum approach to a continuum. It is also observed that for a given NPs size, the discreteness of the spectrum increases with the hopping term b, as is obviously expected since the increase of the hopping implies an increase of the band width keeping constant the number of levels and the coverage factor. In fact, the discreteness of the spectrum, as predicted by Eqn (11), is roughly given by the band width times the Fermi quantum number l_F.

In general, the Fermi level will not be completely filled. Let n^* be the number of electrons that occupy the Fermi level of the NP. It is useful to classify the NPs into two types, diamagnetic NPs for which $n^* = 2(2l_F + 1)$ and paramagnetic NPs for which $n^* \ll 2(2l_F + 1)$.

Associated with the n^* electrons occupying the unfilled Fermi level, there exists the magnetic moment. However, even if n^* is known, the estimation of the moment also requires knowledge of the way in which these individual moments are coupled to create the collective moment of the surface state. We need something akin to Hund's rules for the spherical two-dimensional (2DEG).

Let us consider the case of *LS* coupling. By analogy with atomic physics for an NP with total momentum quantum number *J*, the magnetic moment can be written as

$$\mu_{l_F} = \mu_B g_J J\left(l_F, n*\right) \tag{12}$$

Where g_J is the Landé factor for the particular $2J + 1$ subspace corresponding to the quantum number J that is roughly proportional to the square of l_F.

In an assembly of nanoparticles, both r/a and α will present fluctuations with respect to their average values $\langle r/a \rangle$ and $\langle \alpha \rangle$ that will give rise, according to Eqn (11), to fluctuations in l_F and n^*. For a given value of l_F and a uniform distribution of n^*, the average value of the quantum magnetic moment is

$$\langle \mu \rangle_{l_F} = \mu_B \langle g_J J(l_F, n*) \rangle \tag{13}$$

$$\mu_{at} = \frac{3\langle \mu \rangle_{l_F} a^3}{4\pi r^3} \approx \langle g_J \rangle \mu_B \frac{\alpha\, a}{r} \tag{14}$$

$$m_d = -N_b \frac{e^2 B r^2}{4m} \tag{15}$$

The order of magnitude of the average $\langle \mu \rangle_{l_F}$ when n^* varies between 1 and $2(2l_F + 1)$ can be estimated, after approximating, $g_J = 1$ and $J = \frac{1}{3} l_F^2$, as

When a magnetic field is applied along the polar or z axis, as thoroughly analyzed in [120], the spectrum of the kinetic energy splits the degenerate ground state into its $2J + 1$ components and induces a diamagnetic moment, m_d, given by

The diamagnetic component is much larger than that usually observed for atoms as a consequence of the mesoscopic diameter of the orbits [121], but it remains almost negligible for $J \neq 0$ since, for $B = 1.0$ T, it is of the order of $10^{-3}\ \mu_B N_b$.

However, the paramagnetism is expected to overweigh the diamagnetism, since with surface collective magnetic moments of the order of $g\mu_B l_F^2$, for an applied magnetic field of 1 T would induce a maximum Zeeman splitting of $2\mu_B l_F^2$ that is larger than the thermal energy at room temperature for any

$l_F > 10$. It can be concluded that the Langevin or Brillouin curve describing the paramagnetism of the NPs is close to saturation, i.e. the condition $\mu_r l_F^2 B > 3k_B T$ holds for $B = 1.5$ T and T = 300 K when l_F is larger than 10.

5.6 Final Remarks

It has been generally accepted in the literature cited in this article that the origin of the ferromagnetic-like behavior observed in NPs formed by nonmagnetic substances is originated from point defects located at the surface of the NPs, and its contribution decreases as the size of the NP increases. Its strength can be modified by tuning the interactions of the surface atoms and the coating molecules. Although the intrinsic origin is not well understood, the use of selective techniques, such as Mossbauer and XMCD, has neatly indicated that the magnetism is intrinsic to the NP atoms. The main difficulty arises in accounting for the type of exchange interaction between the localized magnetic moments.

Different experimental results confirm that the modification induced in the electronic structure of the surface by the bonding to organic molecules is associated with modifications in the magnetic properties. It has been shown that two-dimensional confinement of electrons (or holes) introduced by capping molecules (or impurities) in previously empty surface states may be the source of the magnetic moments observed in Au nanoparticles. Similar to the intrinsic ferromagnetism of atoms promoted by the Hund rule in an unfilled orbital, there exists an intrinsic ferromagnetism at the surface of NPs with an unfilled Fermi level in their surface band.

6. CHEMICAL PROPERTIES

6.1 Introduction

In this section, some of the chemical properties of gold nanoparticles will be discussed with a special focus on the nature of the gold-sulfur surface; on the chemical properties of the monolayer surrounding MPCs and on the catalytic activity of supported and unsupported gold nanoparticles. Our understanding of the chemistry of the gold-sulfur interface experienced a tremendous development after the solid-state structure of small thiolate gold clusters became available, unveiling unprecedented structural motifs that challenged the long-standing model for alkanethiols adsorption onto gold surfaces. Earlier structural models consisting of the binding of thiolates at the vertices, edges, and facets of small, idealized nanocrystals, are now superseded. The ligand shell surrounding the particles is the region where the large majority of the chemical and physical events relevant to the interactions of MPCs with other molecular entities or more complex structures take place. The monolayer of these particles inherits the curvature of the core; as a consequence, there is enough room between different chains comprising the monolayer to embed other small molecules. The monolayers are then expected to exhibit a host-guest chemistry on their own in addition to perhaps a more selective host guest chemistry imparted by a specific design of the monomers comprising the monolayer. The analysis of the structure of the monolayers containing mixtures of ligands with different properties put in evidence the existence of phase segregation phenomena with formation of domains of different sizes and shapes. This property of mixed monolayers was theoretically predicted and supported by thermodynamic arguments. The ability to govern the formation and morphology of these domains will become an essential tool for using the nanoparticles as building blocks in the preparation of self-assembled materials with predicted architecture and properties. Finally, the use of gold nanoparticles for catalytic purposes is nowadays not only a topic of mounting research interest, but also a technology that holds promise of practical applications in industrial processes. The number of reactions catalyzed by some form of heterogeneous gold nanoparticle catalyst is steadily increasing, with the gas-phase low temperature oxidation of CO remaining an archetypal process. However, gold nanoparticles protected by polymers or passivated by strongly chemisorbed ligands were also found to be catalytically active in either the supported or unsupported forms.

6.2 Surface Properties of Nanosize Gold

In discussing some details of the surface properties of nanosize gold particles, a first distinction has to be made with respect to the presence or absence of a passivating monolayer, AuNP or MPCs, respectively. Within MPCs, a second distinction must take into account the size of these materials because the present availability of structural information on atomically defined small metal clusters enables a rather precise description, while for larger passivated particles the surface properties may be inferred from our knowledge of smaller ones.

6.2.1 "Naked" Gold Nanoparticles

The reduction in the size of AuNPs leads to the onset of the size confinement effects described in other sections of this chapter; reduction in size also leads to a high surface to volume ratio, to an increase of surface atoms with reduced coordination, and to an increase of surface tension of the particle leading to bond contraction and, therefore, a contraction of the Au lattice parameters with respect to bulk gold [122], which was also detected in smaller Au clusters such as $[Au_{25}(SCH_2CH_2Ph)_{18}^- TOA^+]$. All these effects have a strong impact on the structure of the particles. The increase in surface energy, for instance, dictates their morphology but also their reactivity toward putative capping agents. The equilibrium morphology of gold nanoparticles in the size regime of 3–100 nm has been investigated by using first principle calculations and a thermodynamic model; according to the results, the predicted thermodynamic equilibrium shape of monocrystalline AuNPs of this size is a truncated octahedron. This predicted stability with respect to other shapes is the result of the balance between the surface energies of the facets and the surface to volume ratio of the particle. A further development in analyzing the thermodynamic stability of this species and its temperature dependence based on relativistic *ab initio* calculations led to the construction of a quantitative phase map for gold nanoparticles, even for particles with sizes extending below the limit of 3 nm for which the predicted equilibrium morphologies are decahedra or icosahedra. This model at the temperature of 300 K predicts an icosahedra–decahedra transition at critical size of about 10 nm and a decahedra–truncated octahedral transition at the size of 14 nm. In general, the average coordination number of the gold atoms in a particle decreases abruptly below a threshold of about 300 atoms, corresponding to a nanoparticle size of about 2 nm. Regardless of the size, the most reactive sites were considered to be those with the smallest coordination found at the edges, vertices, and terraces of the particles. These atoms are held responsible for coordination of reagents in catalytic processes, being the binding energies of relevant species such as CO, dioxygen, and atomic oxygen, monotonic functions of the coordination number of the gold atoms in the clusters [123]. However, recently, the surfaces of AuNPs with truncated octahedral morphology, a size of about 3 nm, and having composition Au_N ($N = 923 \pm 23$) were proved to contain a large number of single adatoms [124], accounting for 70% of the total surface atoms by a study employing aberration-corrected scanning transmission electron microscopy (STEM) in the high-angle annular dark-field (HAADF) imaging mode. These adatoms are located at both the {100} and {111} facets with similar probabilities, and they were found to migrate dynamically from a facet to the others under simple thermal excitation. The presence of these adatoms at the AuNP surface, even in the absence of bound ligands, will have a significant impact in improving the understanding of the surface properties of these systems and their reactivity towards putative ligands or in catalytic applications.

6.2.2 Thiolate-Protected Small Clusters and Nanoparticles

Significant advances in our understanding of the nature and properties of the Au-sulfur interface came after the elucidation of the solid-state structure of small gold nanoclusters, with the determination of the structure of Au_{102}(*p*-MBA)$_{44}$ and the subsequent theoretical investigations representing turning points [99,125]. The structural analysis of the Au_{102}(*p*-MBA)$_{44}$ cluster revealed the Au atoms are in two different states, an ensemble of 79 atoms which can be described as metallic and packed in a Marks decahedron, while the remaining gold atoms, formally Au(I), are engaged in the formation of a protecting shell of composition Au_{23}(*p*-MBA)$_{44}$ with the thiolates. The shell can also be considered as consisting of $RS(AuSR)_n$ units with $n = (1, 2)$, generally indicated as "staples." Monomeric staples if $n = 1$ (–RS–Au–SR–) or dimeric staples (–RS–Au–SR–Au–SR–) if $n = 2$ are bound to the Au_{79}

kernel by their terminal sulfur atoms [126]. Interestingly, the unit cell was found to contain an enatiomeric pair of clusters, the chirality arising from an equatorial band of 13 gold atoms bound to thiolate moieties with irregular positions. The overall symmetry of the cluster is C_2, the low symmetry resulting in the definition of a set of 22 symmetrically unique ligands. Shortly afterward, the determination of the structure of [$Au_{25}(SR)_{18}^{-}$ TOA^{+}] reinforced and corroborated the emerging staple structural model. In fact, this species may be considered as consisting of an Au_{13} core surrounded by the remaining 12 gold atoms and 18 thiolates arranged in six dimeric staples. These structural features were afterwards found to be entirely consistent with the superatom model that predicts high stability for those cluster compounds that feature a total electron count of 2, 8, 18, 34, 58, 92, 138, etc., for their inner kernel of electrons in a closed electron shell. The kernel of Au_{102}(*p*-MBA)$_{44}$ and the kernel of $Au_{25}(SR)_{18}^{-}$ feature 58 and 8 electrons, respectively.

Presently, besides the structure of the aforementioned clusters, the structures of Au_{38}(*p*-tBuC_6H_4S)$_{24}$ [127], $Au_{38}(SCH_2CH_2Ph)_{24}$, Au_{28}(*p*-tBuC6H4S)$_{20}$, and [$Au_{23}(SC_6H_{11})_{16}^{-}$ TOA^{+}] are known, and the existence of staples seems to be a sort of common structural motif in all of these species, for [$Au_{23}(SC_6H_{11})_{16}^{-}$ TOA^{+}] trimeric staple motif was also identified. These findings are also in line with recent advances on the elucidation of the nature of the gold sulfur interface for 2D-SAMs, for which the established model of an unreconstructed Au(111) surface with adsorbed alkanethiols has been challenged by the increasing evidences for the presence of AuSR monomers and, at the early stage of alkanethiol adsorption, of RS−Au−SR motifs. The robustness of these binding motifs is supported by the known stability of the 2D-SAM of alkanethiols on Au(111), to which the gold−sulfur interaction contributes with about 208 kJ/mol and the van der Waals interaction between methylene groups contribute with an average of 6.24 kJ/mol. The structural model of small gold clusters consisting in a metal core surrounded by a shell comprising gold atoms and the thiolate moieties arranged in oligomeric structures of staple-like geometry has been used to rationalize the stability of much larger nanoparticles, at least for those for which a precise stoichiometry may be defined on the basis of structural or mass analysis. This is the case for the $Au_{333}(SR)_{79}$ molecule, and that was proposed to consist of an Au_{293} kernel surrounded by a shell of composition $Au_{40}(SR)_{79}$; although, an interpretation in terms of a combination of oligomeric staples remains undefined. Besides the existence of such unitary structural motifs, there is also some evidence for a strong influence of the nature of the ligands in dictating the precise nature of the gold-sulfur interface. In fact, while for the $Au_{38}(SCH_2CH_2Ph)_{24}$ clusters, it has been reported a structure consisting of a face-fused bi-icosahedral Au_{23} core, enclosed by a second shell comprising the remaining 15 gold atoms and the 24 thiolate moieties arranged in three monomeric and six dimeric staples, the related cluster Au_{38}(*p*-tBuC_6H_4S)$_{24}$ was found to be much less stable and to transform under relatively mild conditions into a novel Au_{36}(*p*-tBuC_6H_4S)$_{24}$

cluster. In this case, where direct conjugation of the aromatic ring with sulfur is possible, the influence of such conjugation seems to extend to the electronic properties of the whole cluster, including the geometry of the inner kernel, and not only restricted to the passivating shell [127]. All of these findings suggest that the understanding of the "staples" chemistry and, in a more general way, the understanding of the interface chemistry of thiolate-protected clusters will probably be a future field of intense investigation because it is now recognized that the gold–thiolate oligomeric interfacial structures have a strong influence on the physicochemical properties of small thiolate-protected gold clusters, such as photoluminescence and chiroptical activity. An example of the rich diversity that may be encountered can be found in the recently reported solid-state structure of Au–Ag intermetallic clusters of general formula $Au_{12}Ag_{32}(SR)_{30}$ protected by different fluorinated arylthiols [128]. In this case, an otherwise unstable Au_{12} icosahedral kernel is encapsulated by an Ag_{20} cage surrounded by a metal–ligand shell consisting of six three-dimensional $Ag_2(SR)_5$ units in which Ag cations bind to three thiolate ligands in a planar $Ag(SR)_3$ configuration.

6.2.3 Reactivity at the Gold–Sulfur Interface

One of the most important chemical properties of the gold–sulfur interface of alkanethiolate-protected gold nanoparticles is the possibility to sustain exchange reactions replacing a ligand with another without affecting the structural integrity of the particles. In fact, the large majority of reports on the preparation of particles featuring mixed monolayers exploit this synthetic route. A number of evidences collected during the past years indicate that the process is associative and occurs in a 1:1 stoichiometric ratio, somehow resembling a direct displacement reaction of S_N2 character. A more detailed picture of this reaction comes after the structural elucidation of a partial exchange product on the $Au_{102}(pMBA)_{44}$ cluster [129], and by subsequent theoretical analyses of the process. The structure of the cluster $Au_{102}(pMBA)_{40}(pBBT)_4$, obtained by reacting the well-characterized cluster $Au_{102}(pMBA)_{44}$ with *p*-Bromo benzenethiol (pBBT) under kinetic control, revealed that two of the 22 symmetry unique positions on the cluster shell were quickly exchanged, with these two positions being solvent accessible, hence exposing Au atoms also accessible to the incoming ligands. The mechanism of this exchange was studied by DFT methods on a model system replacing the bound thiolates with –SH residues and the incoming ligand with methanethiol. This special reactivity at certain sites of the monolayer of small gold clusters parallels the reactivity of larger particles that were found to be amenable of functionalization at their polar position. In this case, the special reactivity was the result of a well-defined morphology of the monolayer [130]. With reference to Figure 10, the preferred pathway was found to require the development of a bond between the sulfur atom of the incoming ligand the central gold atom engaged in the

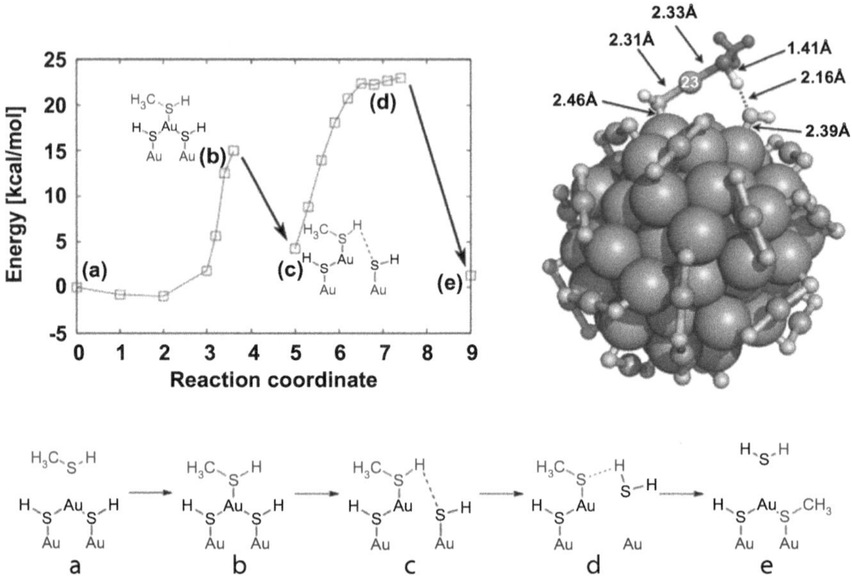

FIGURE 10 Proposed mechanism for the place exchange reaction on a $Au_{102}(SH)_{44}$ gold cluster by an incoming methanethiol as a model for the place exchange reaction of the $Au_{102}(p\mathrm{MBA})_{44}$ with *p*-bromo benzenethiol. *From Ref. [129].*

formation of a staple (b). This event is followed by a step (c) in which an intermediate is formed by dissociation of the bond connecting the gold atom to the terminal thiolate in the staple, the latter being stabilized by the formation of a hydrogen bond with the hydrogen of the incoming thiol. This hydrogen bond evolves into a full proton transfer (d) releasing the thiolate originally involved in the formation of the staple into solution as a thiol (e), leaving a kernel gold atom that connects to the sulfur of the incoming thiol, restoring the initial geometry of the system. The net balance is, therefore, the substitution of one thiolate unit with the release of a thiol into solution, the process involving the Au central center involved in the staple.

This mechanism, however, is probably a simplification and further studies are needed to understand the reactivity of nonexposed Au atoms toward the exchange reaction. Another long standing issue that will require a deeper understanding of the "staple chemistry" is the lateral mobility of the thiolates within the ligand shell. At present, evidences of such a mobility in small alkanethiolate-protected gold clusters exist [131], but the precise mechanism is unknown, being, in principle, the net result of a dissociative–associative process with the release of thiolates into solution or an "on surface" migration of the thiolates.

The presence of staples motifs on the surface of the gold core has another remarkable effect: the formation of intrinsically chiral gold clusters that do not present in their ligand shell chiral thiolates.

The chirality arises from the substitution pattern at the sulfur centers in the staple units; in fact, the sulfur atoms have three different substituents: the surface Au atom, the R residue, and the gold atom involved in the staple motif. Therefore, the sulfur is a stereocenter as it is, for instance, in trisubstituted sulfonium ions. This kind of intrinsic chirality, first identified for the $Au_{102}(p\text{-MBA})_{44}$ cluster was also found for $Au_{38}(SCH_2CH_2Ph)_{24}$ [132]. Clearly, these compounds are formed as racemic mixtures, and in the solid state, the unit cell contains two molecules of opposite handedness. In these two cases, the chiral nature of the clusters was put in evidence from the elucidation of their solid state structure by X-ray analysis. For $Au_{38}(SCH_2CH_2Ph)_{24}$, the two enantiomers were successfully separated by chromatographic methods and the enantiomeric nature of the species isolated was established by recording their CD spectra that are mirror images of one another [133]. The two enantiomers are sufficiently stable but the energy barrier for their interconversion is low, about 116 kJ/mol, leading to significant racemization at 40 °C. The compound $Au_{40}(SCH_2CH_2Ph)_{24}$, was also found to be racemic, and the two enantiomes were separated by using chiral chromatography. The intrinsic chirality of some gold clusters also has an implication on the reactivity of these clusters with chiral thiolates in the place exchange reaction. Indeed the exchange reaction of the racemic $Au_{38}(SCH_2CH_2Ph)_{24}$ cluster with *R*-1,1′-binaphthyl-2,2′-tithiol (*R*-BINAS) to give the mono- and disubstituted products $Au_{38}(SCH_2CH_2Ph)_{24\text{-}2x}(R\text{-BINAS})_x$ ($x = 1,2$) was found to be diastereoselective [134]. Another way to impart chirality to a gold cluster is to introduce chiral thiolates in their ligand shell; this often results in the induction of a cotton effect on the electronic transition of the gold core. Such behavior has been reported for the Au_{25} clusters protected by 2-phenyl-propane-1-thiol [135] or the enantiopure 1*S*, 4*R*-camphorthiol [136].

6.3 Catalytic Activity

With reference to heterogeneous catalysis, several factors contribute to the definition of the catalytic activity of these systems, primarily the nanoparticle size and the chemical nature of the support. With respect to the first aspect, nanostructures with size in the range of 0.5–5 nm seem to be preferred. An emerging strategy allows the achievement of precise control over the AuNP size consists in supporting preformed thiolate passivated nanoparticles with controlled size and composition; in this case, the protecting monolayer may be retained or removed depending on the thermal processing of the catalyst prior to use. The role of the substrate is of primary importance in modulating the properties of the supported particles, and the existence of a strong metal–support interaction (SMSI) with charge transfer between the support and the AuNP has been recognized in several cases. The substrate also provides active sites for the activation of the species involved in the reaction. In this respect, reducible metal oxides such as TiO_2, CeO_2, Fe_2O_3, and Fe_3O_4 seem to be the best substrates in addition to ZrO_2, ZnO, SiO_2, and Al_2O_3. However, other supports such as zeolites and carbonaceous materials have been successfully employed. In addition, the support provides a medium to anchor the particles, thus limiting their thermal sintering under high temperature conditions that are sometimes encountered during practical operation or catalysts pretreatment, generally resulting in catalysts deactivation. A strategy deployed to mitigate the occurrence of sintering processes exploits the preparation of encapsulated gold nanoparticles [137] or the supporting of metal nanoparticles inside the pore channels of mesoporous materials, such as mesoporous SiO_2 and TiO_2 [138]. In these cases, encapsulation also provides a mechanism of selectivity tuning of reagents and products. Alternatively, MPCs may also be used to engineer materials, as in the case of Au@CeO_2 formed by the precipitation method using MPCs comprising in their ligand shell water-soluble polyether thiolates and alkanethiolates terminating with a carboxylic group to interact with the CeO_2 precursors. After calcination, these materials were found to be active catalysts in the oxidation of CO under PROX conditions [139]. Despite the wealth of reports on AuNP catalysis, the precise nature of the catalytic sites on both the AuNPs and the support is very often unclear and is a subject of debate; either small metallic clusters of Au(0), partly oxidized ($Au^{\delta+}$), or reduced particles ($Au^{\delta-}$) formed by interaction with the substrate or in the preparation of the catalyst in addition to undercoordinated gold atoms or Au(I) and Au(III) may act as the actual active sites [140,141].

6.3.1 Catalysis in Oxidative Processes

Perhaps the most studied oxidation reaction catalyzed by supported AuNPs is the gas phase aerobic oxidation of CO, the best catalysts being Au/CeO_2 and Au/TiO_2. As mentioned above, for a given substrate, the size of the AuNPs is crucial in determining the efficiency of the catalysts, with an optimal size range of 0.5–5 nm being preferred. Concerning the low temperature CO oxidation and the somewhat related WGSR, the AuNP/CeO_2 contact interface is considered the catalytically active region [142], as could be inferred from the proportionality between the rate of CO oxidation and the length of the perimeter interface [143]. The CO adsorption sites are located on the gold particles and, in the case of AuNPs supported on nanocrystalline CeO_2, the presence of Au^{3+}, Au^{+}, and Au(0) species has been demonstrated by IR exploiting the characteristic Au–CO frequencies [144]. The reactivity of this catalyst was also correlated to the concentration of cationic Au^{3+}, but not with the concentration of Au^{+} or Au^{0} species. The ceria support is primarily involved in stabilizing dioxygen. The form of the adsorbed dioxygen being related to the precise nature of the ceria support, while nanocrystalline ceria stabilizes dioxygen as peroxide and superoxide species, precipitated ceria stabilizes dioxygen as O_{2d}^{-} or molecular O_2. Thus, simplifying, the gold particle adsorbs CO, while the ceria support activates molecular oxygen through the participation of its lattice oxygen. More recently [145], in the CO oxidation using Au/CeO_2 nanorods, a mechanistic study that included the use of isotopicaly labeled dioxygen confirmed that upon exposure of CO to the catalyst, CO_2 was immediately formed, supporting the idea that the support supplies the reactive oxygen species for the oxidation. After a pulse of $^{18}O_2$ to restore the stoichiometry of the ceria support and the subsequent introduction of CO, the production of $C^{16}O^{18}O$ was detected by mass analysis. Although in this case, the possible role of cationic Au was not explored; this picture is consistent with previous findings [144] and recent DTF calculations [146] which additionally provided substantial structural insight for the cationic nature of the species involved in the oxidation of CO. Interestingly, the typical catalysts for CO oxidation such as Au/CeO_2 and Au/TiO_2 seem to undergo morphological changes under the reaction conditions, as was recently reported by environmental transmission electron microscopy (ETEM) [147]. These findings will probably add new information in elucidating the mechanism of these processes. However, the precise mechanism of CO oxidation may be dramatically different when different supports are considered. It is now well established, for instance, that the presence of moisture in the gas stream is beneficial in increasing the catalyst efficiency [148], although the magnitude of this effect depends on the nature of the substrate and seems to be remarkable for relatively poor catalysts such as Au/SiO_2 or Au/Al_2O_3. This observation, in addition to the ability of water to regenerate inactivated catalysts, points in the direction of a direct involvement of hydroxyl groups in the catalytic cycle, at least in the case of substrates such as TiO_2 [149] and Al_2O_3 [148]. In the former case, oxygen exchange was observed between adsorbed atomic oxygen and isotopicaly labeled water. Coadsorbed water ($H_2^{18}O$) takes part in CO oxidation on oxygen adatoms precovered Au/TiO_2(110) model catalysts, leading to the formation of $C^{16}O^{18}O$ and $C^{16}O^{16}O$. For catalysts such as Au/Al_2O_3, a kinetic model comprising the formation of adsorbed hydroxyl and hydroperoxy (OOH) species was proposed, with the hydroperoxy species reacting with CO producing CO_2 and OH, since in this case, exchange between $^{18}O_2$ and $^{16}O_2$ and between $H_2^{18}O$ and $H_2^{16}O$ was not observed, while the dissociation of dioxygen was excluded as a mechanistic relevant pathway in the Au/Al_2O_3 catalyzed reaction [150]. The involvement of hydroxyl species is also considered to be relevant in the CO oxidation processes carried out in aqueous solutions [148].

As mentioned, the catalytic activity of metal nanoparticles is size dependent, with small particles being the most active. Therefore, a precise control of the particle size represents a possible strategy in improving the performance of a supported catalyst. In this respect, the methodologies developed for the synthesis of MPCs may be considered a useful tool. Indeed, materials obtained by supporting small thiolate-protected clusters were found to be active in CO oxidation; for instance, the cluster compound of well-defined composition $Au_{25}(SCH_2CH_2Ph)_{18}$ supported, by impregnation, on CeO_2 proved to be active in the CO oxidation with the onset of catalytic activity at 60 °C; whereas, a similar TiO_2 supported catalyst was found to be much less active [151]. Also in this case, the interface between the gold cluster and the support was reported to represent the catalytically active region in analogy with other supported Au/CeO_2 catalysts. The same cluster, in addition to $Au_{38}(SCH_2CH_2Ph)_{24}$, and $Au_{144}(SCH_2CH_2Ph)_{60}$ supported onto SiO_2 and hydroxyapatite without calcination treatment, thus retaining their ligand shell, displayed to be active in the oxidation of styrene by dioxygen in the presence or absence of TBHP as initiator or by TBHP as the oxidant, selectivity being toward benzaldehyde [152]. For this reaction, a tentative mechanism requiring coordination of both dioxygen and styrene on the surface of the gold clusters was proposed. Similarly, glutathione (GSH)-protected clusters of composition $Au_{10}(SG)_{10}$, $Au_{18}(SG)_{14}$, $Au_{25}(SG)_{18}$, and $Au_{39}(SG)_{24}$ supported on hydroxyapatite were found to be active in the oxidation of cyclohexane to cyclohexanol and cyclohexanone, although with limited selectivity [153].

6.3.2 Catalysis in Hydrogenation Reactions

Supported gold particles are active catalysts in the reduction of a variety of organic compounds; the majority of the studies are related to solution phase reactions carried out at high temperature and high hydrogen pressure. The reactivity in these cases is due to the heterolytic dissociation of hydrogen on the surface of nanosize gold. The heterolytic pathway requires that the formal H^- resides on gold, while the formal H^+ protonates the support as was demonstrated by inelastic neutron scattering and IR spectroscopy investigations [154,155]. The stability of the resulting Au—H is rather low when compared to the hydrides of Pd and Pt, and consequently, the concentration of Au—H entities is low. This results in a remarkable selectivity of the AuNP-catalyzed hydrogenation reactions. Typical supports for AuNP-mediated hydrogenations are SiO_2, Al_2O_3, ZrO_2, ZnO, CeO_2, and TiO_2. The spectrum of catalyzed reactions is rather wide and comprises hydrogenations of unsaturated C—C bonds, hydrogenation of carbonyl compounds, and CO_2 and hydrogenation of nitro compounds. The reactions are remarkably selective. For instance, the reduction of an aldehyde group in an α,β-unsaturated compound can be carried out safely in the presence of double bonds, and the reduction of the nitro group can be carried out in the presence of aldehyde moieties. From the mechanistic point of view, these reactions are difficult to study because of the conditions employed, and the picture is often complicated by the reactivity of ionic gold species or complexes that are themselves active catalysts under homogeneous conditions.

6.3.3 Catalysis by Unsupported Gold Nanoparticles

Although the support plays a fundamental role in dictating the reactivity of AuNPs, these species may act as catalysts even in the absence of support in reactions similar to those catalyzed by supported AuNPs. In this respect, the large majority of the cases relate to polymer-stabilized nanoparticles such as polyvinylpyrrolidone (PVP) and polyvinyl alcohol (PVA) in addition to polyamidoamine dendrimer (PAMAM)-stabilized AuNPs [156–158]. Among these catalysts, MPCs and "naked" or citrate-stabilized AuNPs are also included. Several reactions are catalyzed by these systems, comprising alcohol oxidations, epoxydations, CO oxidation, hydrogenations, and the oxidation at the benzylic position of aromatic hydrocarbons. Decreasing the size of the nanoparticles, higher catalytic activities are obtained, as it was observed for supported AuNPs. Instead, increasing the steric hindrance due to the presence of stabilizing molecules on the gold surface reduces the catalytic efficiency of the systems, even if the nanoparticle size is unchanged. Most notably, the small, thiolate-protected clusters $Au_{25}(SCH_2CH_2Ph)_{18}$ are active catalysts in the hydrogenation of benzalacetone [159] with hydrogen at atmospheric pressure, and in the reduction of *p*-nitrophenol [160]. The same system was also active in the chemoselective hydrogenation of benzalacetone without affecting the C—C double bond [161]. Other small thiolate-protected clusters such as $Au_{38}(SCH_2CH_2Ph)_{24}$ and $Au_{144}(SCH_2CH_2Ph)_{60}$ were also active in the solution phase oxidation of styrene to benzaldehyde, styrene oxide, and acetophenone, with selectivity towards benzaldehyde. When these particles retain their ligand shell during the catalytic process, a novel source of selectivity becomes available: the interaction of the reactants within well-defined "binding pockets" on the surface of the nanoclusters, as was found in the case of the stereoselective reduction of a bicyclic ketone in the presence of the $Au_{25}(SCH_2CH_2Ph)_{18}$ cluster [162].

6.4 Characteristics and Properties of Monolayers Grafted on the Gold Surface

The monolayer surrounding the nanoparticles represents the interface connecting the nanoparticles with the external environment. Looking at nanoparticles as "interfacial materials," it is in a region close to the outer surface of this monolayer that the large majority of the chemical and physical events relevant to the interactions of MPCs with other molecular entities or more complex structures take place. The monolayer of these particles inherits the curvature of the core. As a consequence, there is enough room between the different chains comprising the monolayer to embed other small molecules. The monolayers are then expected to exhibit a host—guest chemistry on their own, in addition to perhaps a more selective host—guest chemistry imparted by a specific design of the monomers comprising the monolayer. A relevant question arises when the monolayer comprises a mixture of thiolates: a mixed monolayer has to be considered as consisting of random distribution of thiolates or should phase segregation be expected in analogy with other self-assembled systems? The answer was in the positive; indeed, phase segregation with formation of domains of different sizes and shapes was experimentally evidenced, and this finding was theoretically predicted and supported by thermodynamic arguments. The evidences of domains with different properties on the surface of nanoparticles and the ability to master their preparation will become an essential tool in using the nanoparticles as building blocks for the preparation of self-assembled materials with predicted architecture and properties.

6.4.1 Monolayer Morphologies, Experiments, and Theoretical Investigations

When a mixture of thiols is assembled on the surface of gold nanoparticles, mixed monolayer gold nanoparticles (MMPCs) arise. The first direct information on the organization and microscopic structure of mixed monolayers on the surface of nanoparticles was obtained by STM in the group of Stellacci, revealing a high degree of order with the formation of well-defined stripe-like domains of unlike thiols. Further studies revealed that the formation of striped domains is favored in particles with sizes within a well-defined size regime between about 2.5 and 8.0 nm [163]. Improvements in the STM analyses allowed the achievement of quasimolecular resolution in imaging these structural features of mixed monolayer nanoparticles [164]. In particular, it was possible to highlight the presence of aligned features with a spacing of 1 nm between consecutive lines. In the same analysis, the presence of the headgroups of the thiolates within the stripes could be clearly evidenced with a spacing of about 0.5 nm. The interpretation of STM images is often cumbersome, but recently by analysis of the topographical power spectral density (PSD) of the STM data, the characteristic length scales of the features exhibited by nanoparticles in STM images could be obtained in a straightforward manner [165]. Another analytic method that proved to be valuable in addressing the structure of MMPCs is the matrix-assisted laser desorption/ionization ion mobility mass spectrometry (MALDI-IM-MS) [166]. The reason for the sensitivity of this technique in probing the monolayer structure is that the main species formed upon laser desorption/ionization, the Au_4L_4 ions, deriving either from desorption from the gold surface or from the rearrangement of staple motifs, are recognized to reflect the local composition of the monolayer. The relative abundance of fragments with a given composition is, therefore, correlated to the structure of the original monolayer. The assessment of the monolayer morphology is obtained by comparing the experimental data to theoretical abundance patterns derived from an idealized model of the monolayer, constructed assuming a specific morphology. This approach was found to be particularly useful to highlight a rather surprising feature, that is: the MMPCs obtained by direct synthesis exploiting a mixture of two thiols seem to display little phase segregation, while nanoparticles obtained by the place exchange reaction seem to display more pronounced phase segregation.

Another technique used to gain information on the structure of the monolayers is nuclear magnetic resonances, by exploiting (1) the sensitivity of the chemical shifts of the molecules in the monolayer to their chemical environment and (2) the Nuclear Overhauser Effect established between chains sufficiently close in space, Figure 11. In a random distribution of thiolates, the local and global composition of a monolayer are similar; this similarity breaks down when the monolayer presents a defined morphology. Since the chemical shift depends on the environment experienced by the probe nuclei, differences in local and global compositions and, hence, phase segregation could be evidenced by this method [167]. Notably, for nanoparticles featuring stripe-like domains, a sigmoidal dependence of the chemical shift with respect to the composition is expected (Figure 11(C)). Whereas, for Janus particles, with two well-separated macro domains, a hyperbolic dependence should arise (Figure 11(B)). If no phase segregation is present, the relationship between the chemical shift and the composition should be linear (Figure 11(A)). In the presence of monolayers with randomly distributed thiolates or with stripe-like domains, significant cross-peaks in the

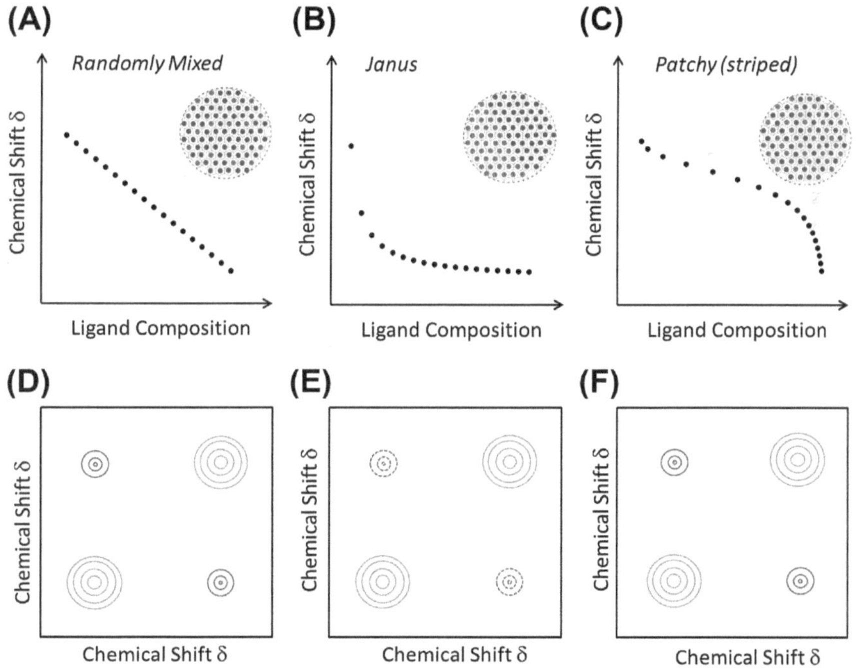

FIGURE 11 Predicted NMR patterns for nanoparticles coated with a binary mixture of ligands. Panels A, B, and C represent the expected chemical shift dependency upon monolayer composition for randomly mixed (A), Janus (B), and striped (C) nanoparticles. Panels D, E, and F represent the NOESY pattern expected for randomly mixed (A), Janus (B), and striped (C) nanoparticles. *Figure adapted from Ref. [167].*

NOESY spectra of the particles should be visible (Figure 11(F)). On the contrary, cross-peaks should be negligible if complete separation of the thiolates in two macrodomains occurs as in Janus particles (Figure 11(E)). This two experiment set enables us to characterize the monolayer morphology of MMPCs with an easily accessible technique. Indeed, nanoparticles with a core diameter of 5 nm and containing diphenylthiol and dodecanthiol with different compositions were shown by this method to present stripe-like domains in their monolayer. Smaller particles displayed the behavior expected for Janus particles. Instead, nanoparticles featuring a disordered monolayer comprising 3,7-dimethyloctanthiol and diphenylthiol displayed a linear dependency of the chemical shift toward the composition. The experimental findings on the occurrence of domains on the monolayer of nanoparticles were the subject of intense theoretical analysis, and the results fully support the experimental data. The group of Glotzer contributed to this field with a thorough investigation of the factors triggering the spontaneous formation of domains [168]. From a qualitative standpoint, it is reasonable that unlike ligands will tend to self-sort into different domains, thus minimizing their reciprocal interactions as it is observed in the case of other self-assembled systems. This phase segregation observed in MMPCs results from the interplay between thermodynamic and geometric factors, stemming from the chemical "dissimilarity" of the ligands forming the monolayer and the size of the nanoparticle core. The former comprises "true" chemical dissimilarities or a geometrical mismatch due to, for instance, differences in length of the thiolate chains. The latter is mainly related to the conformational freedom and, hence, conformational entropy of the thiolate chains located at the boundaries of different domains, which also function on the monolayer curvature. Hence, in this framework, specific morphological arrangements that allow a certain degree of conformational entropy to be achieved would be favored. In a tightly packed monolayer comprising long and short thiolates, the one that can be obtained by self-assembly on a large particle with small curvature, a reasonable gain in conformational entropy may be obtained only for those thiols located at the boundaries of the domain separating long and short ligands, with the long benefitting from an increase in conformational entropy due to the free room available for bending over the short ligands. It is, therefore, reasonable to conclude that in the case of ligands with significant differences in length, the minimization of the free energy of the system will require the maximization of the number, or the total length, of these boundary regions. For particles of size in the 2.5–8.0 nm regime, the formation of stripe-like domains is favored. However, for larger nanoparticles, the formation of worm-like domains without specific orientation is favored. For smaller particles, instead, for which a certain degree of conformational entropy is already ensured due to the free room available to the ligands because of the curvature of the particle, this picture is significantly changed, with the enthalpic term playing a major role, and the formation of Janus particles with well-separated macrodomains being favored. Hence, when a binary mixture of "dissimilar" thiolates is adsorbed on the surface of a metal nanoparticle, both experimental and theoretical evidences exist that essentially three main morphologies may arise: (1) the formation of the so-called Janus particles, with complete segregation of the different ligands on the two halves of an idealized spherical surface, (2) the formation of striated domains orthogonal to one of the axes of the idealized sphere, usually referred to as "stripes," and (3) the formation of worm-like striated domains with no orientation with respect to the symmetry elements of the sphere. This picture is the result of considering binary mixtures of thiols, when a ternary mixture or quaternary mixture of thiols is considered; a rich variety of morphologies is possible as recently highlighted by theoretical investigations by Glotzer [169]. If a ternary mixture of thiols is considered, the simplest monolayer morphology for small particles predicted is the "Cerberus" nanoparticle, with three domains displaying an interface region with one domain and another interface region with the other domain. These are the ternary equivalent of the Janus particles. For quaternary mixtures, tetrahedral particles with the four ligands that generate four domains that minimize the interface between all ligand pairs are formed, with all the longest ligands forming an interface with the shortest. This arrangement is found for small particles compared to the ligand length. If the set of ligands comprises two long and two short ligands, the predicted geometry is that of a "Brahma" particle, with four domains arranged in such a way that the two longest ligands do not share an interface. By increasing the particle size and increasing the interaction strength between the ligands, a rich variety of morphologies results. If the composition of the mixture is largely unbalanced, with one of the thiolates representing the large majority of the mixture, the formation of isolated domains or patches is possible. Patchy particles with multiple patches arranged in a well-defined geometric pattern are presently actively studied from the theoretical point of view as anisotropic building blocks to construct superstructures via the interaction of patches located on different particles [170]. This strategy, once the synthetic control over the formation of patches is achieved, would probably represent an important tool for the future development of material science.

6.4.2 Properties of Mixed Monolayers that Feature Monolayer Domains

Nanoparticles coated with striped domains have many unique structure-dependent properties that mainly derive from their interfacial interactions. MMPCs coated with binary mixtures of octanethiol and mercaptopropionic acid of different compositions display solubilities with a marked non-monotonic behavior with respect to the monolayer composition. The solubility, therefore, depends critically on the surface morphology; the same applies to the surface-wetting behavior of these systems [171]. An unprecedented property of these nanoparticles is their ability to trespass the plasma membrane of cells. Translocation across the membrane occurs by an energy-independent mechanism with no evidence of membrane poration or cell death. On the contrary, nanoparticles with the same composition but with random distributions of the thiolates on the monolayer were mostly trapped in endosomes [172]. A mechanism put forth in order to explain the observed effects relies on a two-step process which first requires the fusion of the nanoparticles with the membrane to form a transmembrane configuration, which eventually evolves toward a translocation into the cytosol. As a first step to validate this mechanism, a simplified model exploiting MMPCs contained in the monolayer 11-mercaptoundecane sulfonate (MUS) and octanethiol (OT) was tested with model lipid bilayers [173], both experimentally and theoretically. The theoretical analysis showed that nanoparticle composition and diameter, rather than morphology, were the most critical parameters in determining the free energy of interaction, the driving force being the reduction of the hydrophobic surface exposed to the solvent upon fusion with the membrane. The nanoparticle—membrane fusion is also accompanied by a significant distortion of the monolayer, forcing the charged groups out of the bilayer.

Nanoparticles with a mixed monolayer comprising 3-mercapto-1-propanesulfonic acid (MPSA) and different alkanethiols and featuring stripe-like domains were also showed to act as a catalyst in the acetylation of benzyl alcohol with acetic acid. A structure—reactivity analysis evidenced that the highest reactivity could be obtained for particles with an MPSA:octanethiol ratio of 2:1, confirming a role of the morphology on the reactivity [174]. Nanoparticles featuring stripe-like domains made of hexanethiol and polyethylene-glycol-terminated thiolates were shown to be active materials in the construction of a sensor for the determination of metal ions in solution with an unprecedented attomolar limit for the detection of methylmercury [175].

An extreme example of ligand dissimilarity driving the formation of domains on mixed monolayers is the one obtained by exploiting the known immiscibility of fluorous and hydrocarbon phases. When mixtures of PEGylated ligands that feature hydrocarbon chains and fluorocarbon chains are coassembled on the surface of gold nanoparticles, complete phase segregation takes place, even at very low loading of the fluorinated compound, as was recently inferred from ESR studies of the binding of a radical probe to the monolayer of mixed monolayer nanoparticles and by mesoscale simulations [176]. In this case, the formation of domains occurs even at loadings of the fluorinated thiolate as low as 4%.

6.4.3 Complexing Properties of Monolayers on Gold Nanoparticles

Another important feature of the monolayer surrounding the particles is its ability to complex molecules from a solution phase, thus forming a "loaded" nanoparticle. Several reports indicate that indeed monolayers have a significant ability to complex molecules, either by unspecific interaction or because of a judicious selection of the nature of the monomers. The unspecific interactions result in hydrophobic binding of hydrophobic guests to the inner regions of the monolayer. While in general, the tailored binding is due to the interaction of target molecules with specific recognition moieties on the outer surface of the monolayer; in this respect, nanoparticles may be considered exo-receptors. The outer surface of monolayers was exploited as the place for a further self-assembly, as it was demonstrated by the assembly of catalytic peptides onto the surface of cationic gold nanoparticles featuring trimethylammonium groups on their surface. The catalytic performance of the system was also enhanced by the complexing ability of the monolayer toward the substrate of choice [177]. A similar approach exploiting nanoparticles coated with a homoligand monolayer of thiolates that bear as a terminal group a triazacyclononane moiety complexed with Zn(II) [178] allowed the development of a system amenable to the interaction of oligophosphonates and oligocarboxylates in water, with binding constants in the μM region. Other such exo-receptors that exploit electrostatic interactions between ionic groups located on the surface of the nanoparticles and charge complementary guests have been developed for the interactions with proteins and for the transfection of nucleic acids.

The complexing ability of the monolayer of nanoparticles is, however, not restricted to binding events on the outer surface. Hydrophobic binding may occur at the inner region of the monolayer, as could be evidenced by early studies exploiting ESR and a hydrophobic radical probe [179]. This technique also allows the identification of the region of the monolayer involved in the binding events because of the sensitivity of the ESR spectral parameters of the probe as a function of the polarity of the environment. Extending this approach to nanoparticles featuring a mixed monolayer of hydrogenated and fluorinated thiolates bearing terminal polyether moieties allowed to highlight that the binding preferentially takes place in the fluorinated domains of the monolayer [176]. The overall binding constant depended on the composition of the monolayer and was found to be in the range of 100–180 M^{-1}; the calculation of the binding constant for the fluorinated and hydrogenated domains confirmed a preferential binding to the former, with binding constants in the range 200–762 M^{-1}. Interestingly, the binding constants increase with decreasing content of fluorinated thiolates. Mesoscale simulations aided the rationalization of this behavior by displaying that at low loading of the fluorinated thiolate, its polyether chain has sufficient conformational freedom to bend over the shorter hydrogenated ligands, allowing a better recognition of the probe molecule. The apparent overall binding constants reported for the complexation of this probe are in the same order of magnitude of those reported for the complexation of the salicylate anion [180] by using homoligand monolayer nanoparticles prepared with polyetherterminated amphiphilic thiolates. The binding constant reported was 120 ± 5/M. This study employed NOE pumping experiments that allowed the transfer of magnetization from the particle to the complexed species. Comparison of the binding constants of other anionic species allowed the conclusion that the binding was exquisitely related to the hydrophobicity of the probes. However, by introducing a zwitterionic thiolate derived from phosphorylcholine and extending the length of the hydrocarbon chain of the amphiphilic thiolate, the selectivity changed, indicating that both the size of the hydrophobic region in the coating monolayer and the specific interactions with the thiol endgroup affect the binding affinity. Another example of hydrophobic binding is the achievement of nanoparticle-based systems that are capable of transporting small molecule drugs into cancer cells [181]. These few examples serve to illustrate the potential of the monolayer surface to act as a scaffold for the assembly complex supramolecular even in the very competitive solvent water, and the potential of hydrophobic binding for the development of nanoparticle-based carrier systems.

7. CONCLUDING REMARKS

The development of novel nanoparticle-based technologies is strongly dependent on the advances in our understanding of their basic physical and chemical properties. Examples of this interconnection can be found in the ultrasensitive detection of analytes by Raman spectroscopy, achieved by exploiting the surface properties of gold nanoparticles. Other properties, such as the peculiar electron transfer dynamics of gold nanoparticles and other nanoscale entities, are of critical importance for the fabrication of next-generation electronic circuitries and devices, and the self-sorting of thiolates on MMPCs holds promise to become a tool for the bottom-up engineering of novel materials. Other examples are presented in details in this volume. The unprecedented properties of nanosize matter will likely continue to surprise and demand for better experimental and theoretical investigations and models. The understanding of the basic physical and chemical properties of gold nanoparticles is, therefore, not only important per se, but allows the development of new experimental techniques providing fundamental knowledge for the implementation of novel technological platforms.

ACKNOWLEDGMENTS

Patricia Crespo, Miguel Angel García and Antonio Hernando wish to acknowledge funding from the Spanish Ministry of Science and Innovation under project MAT2012-37109-C02-01.

Paolo Pengo and Lucia Pasquato wish to acknowledge funding from the University of Trieste (FRA 2011, 2012, 2013) and the Italian Ministry of Health, under project GR-2009-15779849.

REFERENCES

[1] K.C. Grabar, R. Griffith Freeman, M.B. Holmmer, M.J. Natan, Anal. Chem. 67 (1995) 735–743.
[2] S.A. Maier, Plasmonics: Fundamentals and Applications, Springer, 2007.
[3] U. Kreibig, M. Vollmer, Optical Properties of Metal Clusters, Springer, Berlin, 1995.
[4] V. Amendola, M. Meneghetti, J. Phys. Chem. C 113 (2009) 4277–4285.
[5] K.L. Kelly, E. Coronado, L.L. Zhao, G.C. Schatz, J. Phys. Chem. B 107 (2003) 668–677.
[6] Y. Xia, N.J. Halas, MRS Bull. 30 (2005) 338–348.
[7] P.K. Jain, M.A. El-Sayed, Nano Lett. 7 (2007) 2854–2858.
[8] V. Amendola, Synthesis of Gold and Silver Nanoparticles for Photonic Applications, Università degli Studi di Padova, Padova, 2008.
[9] K. Lee, M.A. El-Sayed, J. Phys. Chem. B 110 (2006) 19220–19225.
[10] V.M. Shalaev, S. Kawata, Nanophotonics with Surface Plasmons, Elsevier, 2006.
[11] C.A. Mirkin, R.L. Letsinger, R.C. Mucic, J.J. Storhoff, Nature 382 (1996) 607.
[12] E. Messina, E. Cavallaro, A. Cacciola, M.A. Iati, P.G. Gucciardi, F. Borghese, P. Denti, R. Saija, G. Compagnini, M. Meneghetti, V. Amendola, O.M. Marago, ACS Nano 5 (2011) 905–913.
[13] R.C. Jin, G. Wu, Z. Li, C.A. Mirkin, G.C. Schatz, J. Am. Chem. Soc. 125 (2003) 1643.
[14] K. Kneipp, M. Moskovits, H. Kneipp, Surface-enhanced Raman Scattering: Physics and Applications, Springer, 2006.
[15] V. Amendola, M. Meneghetti, Adv. Funct. Mater. 22 (2012) 353–360.
[16] R. Esteban, A.G. Borisov, P. Nordlander, J. Aizpurua, Nat. Commun. 3 (2012) 825.
[17] V. Amendola, M. Meneghetti, S. Fiameni, S. Polizzi, G. Fracasso, A. Boscaini, M. Colombatti, Anal. Methods 3 (2011) 849–856.
[18] X. Qian, X.H. Peng, D.O. Ansari, Q. Yin-Goen, G.Z. Chen, D.M. Shin, L. Yang, A.N. Young, M.D. Wang, S. Nie, Nat. Biotechnol. 26 (2008) 83–90.
[19] N.L. Rosi, C.A. Mirkin, Chem. Rev. 105 (2005) 1547–1562.
[20] M. Meneghetti, A. Scarsi, L. Litti, G. Marcolongo, V. Amendola, M. Gobbo, M. Di Chio, A. Boscaini, G. Fracasso, M. Colombatti, Small 8 (2012) 3733–3738.
[21] V. Amendola, O.M. Bakr, F. Stellacci, Plasmonics 5 (2010) 85–97.
[22] M. Futamata, Y. Maruyama, M. Ishikawa, J. Phys. Chem. B 107 (2003) 7607–7617.
[23] S. Link, M.A. El-Sayed, Intern. Rev. Phys. Chem. 19 (2000) 409–453.
[24] A.O. Govorov, H.H. Richardson, Nano Today 2 (2007) 30–38.
[25] P.K. Pain, I.H. El-Sayed, M.A. El-Sayed, Nano Today 2 (2007) 18–29.
[26] J. Lakowicz, Y. Fu, Laser Photon. Rev. 3 (2009) 221–232.
[27] C.X. Zhang, Y. Zhang, X. Wang, Z.M. Tang, Z.H. Lu, Anal. Biochem. 320 (2003) 136–140.
[28] R.A. Farrer, F.L. Butterfield, V.W. Chen, J.T. Fourkas, Nano Lett. 5 (2005) 1139–1142.
[29] V. Amendola, D. Dini, S. Polizzi, J. Shen, K.M. Kadish, M.J. Calvete, M. Hanack, M. Meneghetti, J. Phys. Chem. C 113 (2009) 8688–8695.
[30] V. Amendola, G. Mattei, C. Cusan, M. Prato, M. Meneghetti, Synth. Met. 155 (2005) 283–286.
[31] H. Fang, C. Du, S. Qu, Y. Li, Y. Song, H. Li, H. Liu, D. Zhu, Chem. Phys. Lett. 364 (2002) 290–296.
[32] V. Amendola, M. Meneghetti, Self-Healing at the Nanoscale: Mechanisms and Key Concepts of Natural and Artificial Systems, CRC Press, 2012.
[33] V. Amendola, M. Meneghetti, J. Mater. Chem. 22 (2012) 24501–24508.

[34] V. Amendola, M. Meneghetti, Nanoscale 1 (2009) 74–88.
[35] O.D. Häberlen, S.C. Chung, M. Stener, N. Rösch, J. Chem. Phys. 106 (1997) 5189–5201.
[36] C. Burda, X. Chen, R. Narayanan, M.A. El-Sayed, Chem. Rev. 105 (2005) 1025–1102.
[37] B. Nikoobakht, M.A. El-Sayed, Chem. Mater. 15 (2003) 1957–1962.
[38] C. Noguez, J. Phys. Chem. C 111 (2007) 3806–3819.
[39] B. Palpant, B. Prével, J. Lermé, E. Cottancin, M. Pellarin, M. Treilleux, A. Perez, J.L. Vialle, M. Broyer, Phys. Rev. B 57 (1998) 1963–1970.
[40] E. Prodan, P. Nordlander, N.J. Halas, Chem. Phys. Lett. 368 (2003) 94–101.
[41] K. Nobusada, J. Phys. Chem. B 108 (2004) 11904–11908.
[42] T. Iwasa, K. Nobusada, J. Phys. Chem. C 111 (2007) 45–49.
[43] H. Häkkinen, M. Walter, H. Grönbeck, J. Phys. Chem. B 110 (2006) 9927–9931.
[44] P. Pyykkö, Angew. Chem. 43 (2004) 4412–4456.
[45] K.G. Dyall, K. Fægri Jr., Introduction to Relativistic Quantum Chemistry, Oxford University Press, New York, 2007.
[46] D.M. Wood, Phys. Rev. Lett. 46 (1981) 749.
[47] S. Malola, L. Lehtovaara, J. Enkovaara, H. Häkkinen, ACS Nano 7 (2013) 10263–10270.
[48] G. Piccini, R.W.A. Havenith, R. Broer, M. Stener, J. Phys. Chem. C 117 (2013) 17196–17204.
[49] S. Bernadotte, F. Evers, C.R. Jacob, J. Phys. Chem. C 117 (2013) 1863–1878.
[50] N. Durante, A. Fortunelli, M. Broyer, M. Stener, J. Phys. Chem. C 115 (2011) 6277–6282.
[51] H-Ch Weissker, C. Mottet, Phys. Rev. B 84 (2011) 165443.
[52] S. Chen, R.S. Ingram, M.J. Hostetler, J.J. Pietron, R.W. Murray, T.G. Schaaff, J.T. Khoury, M.M. Alvarez, R.L. Whetten, Science 280 (1998) 2098–2101.
[53] S. Chen, R.W. Murray, J. Phys. Chem. B 103 (1999) 9996–10000.
[54] R.P. Andres, T. Bein, M. Dorogi, S. Feng, J.I. Henderson, C.P. Kubiak, W. Mahoney, R.G. Osifchin, R. Reifenberger, Science 272 (1996) 1323–1325.
[55] S. Chen, J. Am. Chem. Soc. 122 (2000) 7420–7421.
[56] S. Chen, R. Pei, J. Am. Chem. Soc. 123 (2001) 10607–10615.
[57] A. Arieh, R.M. Ratner, Chem. Phys. Lett. 29 (1974) 277–283.
[58] S. Chen, Z. Zhao, H. Liu, Annu. Rev. Phys. Chem. 64 (2013) 221–245.
[59] M. Brust, D. Bethell, C.J. Kiely, D.J. Schiffrin, Langmuir 14 (1998) 5425.
[60] L.-P. Xu, S. Chen, Chem. Phys. Lett. 468 (2009) 222–226.
[61] P.J. Thomas, G.U. Kulkarni, C.N.R. Rao, Chem. Phys. Lett. 321 (2000) 163–168.
[62] F.-R.F. Fan, A.J. Bard, Science 277 (1997) 1791–1793.
[63] L.Y. Gorelik, A. Isacsson, M.V. Voinova, B. Kasemo, R.I. Shekhter, M. Jonson, Phys. Rev. Lett. 80 (1998) 4526.
[64] J. Park, A.N. Pasupathy, J.I. Goldsmith, C. Chang, Y. Yaish, J.R. Petta, M. Rinkoski, J.P. Sethna, H.D. Abruña, P.L. McEuen, D.C. Ralph, Nature 417 (2001) 722–725.
[65] W.M. Lackowski, J.K. Campbell, G. Edwards, V. Chechik, R.M. Crooks, Langmuir 15 (1999) 7632–7638.
[66] Y. Guo, L.M. Chen, Y. Song, P.G. Hu, S.W. Chen, Sci. Adv. Mater. 6 (2014) 1060–1067 (accepted).
[67] K.C. Chan, P.F. Lee, J.Y. Dai, Appl. Phys. Lett. 92 (2008) 143117.
[68] E.A. Speets, B. Dordi, B.J. Ravoo, N. Oncel, A.-S. Hallbäck, H.J.W. Zandvliet, B. Poelsema, G. Rijnders, D.H.A. Blank, D.N. Reinhoudt, Small 1 (2005) 395–398.
[69] F.P. Zamborini, M.C. Leopold, J.F. Hicks, P.J. Kulesza, M.A. Malik, R.W. Murray, J. Am, Chem. Soc. 124 (2001) 8958–8964.
[70] J.F. Hicks, A.C. Templeton, S. Chen, K.M. Sheran, R. Jasti, R.W. Murray, J. Debord, T.G. Schaaff, R.L. Whetten, Anal. Chem. 71 (1999) 3703–3711.

[71] Y. Yang, S. Chen, Nano Lett. 3 (2003) 75–79.
[72] F. Remacle, R.D. Levine, J. Am. Chem. Soc. 122 (2000) 4084–4091.
[73] S. Pradhan, J. Sun, F. Deng, S. Chen, Adv. Mater. 18 (2006) 3279–3283.
[74] B. Sampedro, P. Crespo, A. Hernando, R. Litrán, J.C. Sánchez López, C. López Cartes, A. Fernández, J. Ramírez, J. González Calbet, M. Vallet, Phys. Rev. Lett. 91 (2003) 237203.
[75] M.A. Garcia, M.L. Ruiz-González, G.F. de la Fuente, P. Crespo, J.M. González, J. Llopis, J.M. González-Calbet, M. Vallet-Regí, A. Hernando, Chem. Mater. 19 (2007) 889–893.
[76] P. Crespo, R. Litran, T.C. Rojas, M. Multigner, J.M. de la Fuente, J.C. Sanchez-Lopez, M.A. Garcia, A. Hernando, S. Penades, A. Fernandez, Phys. Rev. Lett. 93 (2004) 087204.
[77] J.S. Garitaonandia, M. Insausti, E. Goikolea, M. Suzuki, J.D. Cashion, N. Kawamura, H. Ohsawa, I. Gil de Muro, K. Suzuki, F. Plazaola, T. Rojo, Nano Lett. 8 (2008) 661–667.
[78] H. Hori, T. Teranishi, Y. Nakae, Y. Seino, M. Miyake, S. Yamada, Phys. Lett. A 263 (1999) 406–410.
[79] M. Brust, M. Walker, D. Bethell, D.J. Schiffrin, R. Whyman, J. Chem. Soc. Chem. Commun. 801 (1994).
[80] P. Crespo, M.A. Garcia, E. Fernández Pinel, M. Multigner, D. Alcántara, J.M. de la Fuente, S. Penadés, A. Hernando, Phys. Rev. Lett. 97 (2006) 177203.
[81] [a] P. Zhang, T.K. Sham, Appl. Phys. Lett. 81 (2002) 736.
[b] P. Zhang, P.S. Kim, T.K. Sham, Appl. Phys. Lett. 82 (2003) 1470.
[82] M.A. Garcia, J. de la Venta, P. Crespo, J. Llopis, S. Penadés, A. Fernández, A. Hernando, Phys. Rev. B. 72 (2005) 241403.
[83] I. Carmeli, G. Leitus, R. Naaman, S. Reich, Z. Vager, J. Chem. Phys. 118 (2003) 10372.
[84] E. Guerrero, M.A. Muñoz-Marquez, E. Fernandez-Pinel, P. Crespo, A.Hernando, A. Fernandez, J. Nanopart. Res. 10 (2008) 179.
[85] E. Guerrero, T.C. Rojas, M. Multigner, P. Crespo, M.A. Muñoz-Márquez, M.A. García, A. Hernando, A. Fernández, Acta Mater. 55 (2007) 723.
[86] J.M. de la Fuente, D. Alcantara, P. Eaton, P. Crespo, T.C. Rojas, A. Fernandez, A. Hernando, S. Penades, J. Phys. Chem. B 110 (2006) 13021.
[87] Y. Yamamoto, T. Miura, M. Suzuki, N. Kawamura, H. Miyagawa, T. Nakamura, K. Kobayashi, T. Teranishi, H. Hori, Phys. Rev. Lett. 93 (2004) 116801.
[88] J. de la Venta, V. Bouzas, A. Pucci, M.A. Laguna-Marco, D. Haskel, S.G.E. te Velthuis, A. Hoffmann, J. Lal, M. Bleuel, G. Ruggeri, C. de Julián Fernández, M.A. García, J. Nanosci. Nanotechnol. 9 (2009) 6434–6438.
[89] H. Hori, Y. Yamamoto, T. Iwamoto, T. Miura, T. Teranishi, M. Miyake, Phys. Rev. B 69 (2004) 174411.
[90] P. de la Presa, M. Multigner, J. de la Venta, M.A. Garcia, M.L. Ruiz-Gonzalez, J. Appl. Phys. 100 (2006) 123915.
[91] [a] M.A. Muñoz-Márquez, E. Guerrero, A. Fernández, P. Crespo, A. Hernando, R. Lucena, J.C. Conesa, J. Nanopart Res. 12 (2010) 1307–1318.
[b] E. Guerrero, M.A. Muñoz-Márquez, A. Fernández, P. Crespo, A. Hernando, R. Lucena, J.C. Conesa, J. Appl. Phys. 107 (2010) 064303.
[92] M. Suda, N. Kameyama, M. Suzuki, N. Kawamura, Y. Einaga, Angew. Chem. Int. Ed. 47 (2008) 160–163.
[93] A. Ayuela, P. Crespo, M.A. Garcia, A. Hernando, P.M. Echenique, New J. Phys. 14 (2012).
[94] Z. Vager, R. Naaman, Phys. Rev. Lett. 92 (2004) 087205.
[95] A. Hernando, P. Crespo, M.A. García, Phys. Rev. Lett. 96 (2006) 057206.
[96] A. Hernando, P. Crespo, M.A. García, E.F. Pinel, J. de la Venta, A. Fernández, S. Penadés, Phys. Rev. B 74 (2006) 052403.

[97] A. Hernando, M.A. García, Phys. Rev. Lett. 96 (2006) 029703.
[98] O. López-Acevedo, H. Tsunoyama, T. Tsukuda, H. Häkkinen, C.M. Aikens, J. Am. Chem. Soc. 132 (2010) 8210–8218.
[99] P.D. Jadzinsky, G. Calero, C.J. Ackerson, D.A. Bushnell, R.D. Kornberg, Science 318 (2007) 430–733.
[100] M.W. Heaven, A. Dass, P.S. White, K.M. Holt, R.W. Murray, J. Am. Chem. Soc. 130 (2008) 3754–3755.
[101] J.A. Alonso, Chem. Rev. 100 (2000) 637–678.
[102] Y. Yourdshahyan, A.M. Rappe, J. Chem. Phys. 117 (2002) 825.
[103] C. González, Y. Simón-Manso, M. Marquez, V. Mujica, J. Phys. Chem. B 110 (2006) 687–691.
[104] W. Luo, S.J. Pennycook, S.T. Pantelides, Nano Lett. 7 (2007) 3134–3137.
[105] L. Puerta, H.J. Franco, J. Murgich, C. Gonzalez, Y. Simon-Manso, V. Mujica, J. Phys. Chem. A 112 (2008) 9771–9783.
[106] A. Hernando, P. Crespo, M.A. García, M. Coey, A. Ayuela, P.M. Echenique, Phys. Status Solidi B 248 (2011) 2352.
[107] G. te Velde, F.M. Bickelhaupt, E.J. Baerends, C.F. Guerra, S.J.A. van Gisbergen, J.G. Snijders, T. Ziegler, J. Comput. Chem. 22 (2001) 931–967.
[108] E. van Lenthe, A.E. Ehlers, E.J. Baerends, J. Chem. Phys. 110 (1999) 8943.
[109] J.P. Perdew, K. Burke, M. Ernzerhof, Phys. Rev. Lett. 77 (1991) 3865.
[110] B.D. Yu, M. Scheffler, Phys. Rev. B 56 (1997) 15569(R).
[111] A. Ayuela, M.J. Puska, R.M. Nieminen, J.A. Alonso, Phys. Rev. B 72 (2005) 161403(R).
[112] P. Gambardella, S. Rusponi, M. Veronese, S.S. Dhesi, C. Grazioli, A. Dallmeyer, I. Cabria, R. Zeller, P.H. Dederichs, K. Kern, C. Carbone, H. Brune, Science 300 (2003) 1130.
[113] T. Chakraborty, P. Pietiläinen, Phys. Rev. B 50 (1994) 8460–8468.
[114] A. Fuhrer, S. Lüscher, T. Ihn, T. Heinzel, K. Ensslin, W. Wegscheider, M. Bichler, Nature 413 (2001) 822–825.
[115] K. Rumpf, P. Granitzer, H. Krenn, J. Phys. Condens. Matter 20 (2008) 454221.
[116] J. Tempere, I.F. Silvera, J.T. Devreese, Phys. Rev. B 65 (2002) 195418.
[117] D. Huang, G. Gumbs, Phys. Lett. A 167 (1992) 396–400.
[118] N. Sánchez, S. Gallego, M.C. Muñoz, Phys. Rev. Lett. 101 (2008) 067206.
[119] S. LaShell, B.A. McDougall, E. Jensen, Phys. Rev. Lett. 77 (16) (1996) 3419–3422.
[120] A. Hernando, M.A. García, J. Nanopart. Res. (2011). http://dx.doi.org/10.1007/s11051-011-0257-7.
[121] A. Hernando, E. Herrero, M. Vázquez, J. Alonso, A. González, G. Rivero, J.M. Rojo, M. Vallet-Regi, J. González Calbet, Phys. Rev. B 56 (1997) 7800–7803.
[122] J.C. Azćarate, G. Corthey, E. Pensa, C. Vericat, M.H. Fonticelli, R.C. Salvarezza, P. Carro, J. Phys. Chem. Lett. 4 (2013) 3127–3138 (and references therein).
[123] B. Hvolbæk, T.V.W. Janssens, B.S. Clausen, H. Falsig, C.H. Christensen, J.K. Nørskov, Nano Today 2 (2007) 14–18.
[124] Z.W. Wang, R.E. Palmer, Nano Lett. 12 (2012) 91–95.
[125] H. Häkkinen, Nat. Chem. 4 (2012) 443–445 (and references therein).
[126] H. Häkkinen, Chem. Soc. Rev. 37 (2008) 184–1859 (and references therein).
[127] C. Zeng, H. Qian, T. Li, G. Li, N.L. Rosi, B. Yoon, R.N. Barnett, R.L. Whetten, U. Landman, R. Jin, Angew. Chem. Int. Ed. 51 (2012) 13114–13118.
[128] H. Yang, Y. Wang, H. Huang, L. Gell, L. Lehtovaara, S. Malola, H. Häkkinen, N. Zheng, Nat. Commun. 4 (2013) 2422.
[129] C.L. Heinecke, T.W. Ni, S. Malola, V. Mäkinen, O.A. Wong, H. Häkkinen, C.J. Ackerson, J. Am. Chem. Soc. 134 (2012) 13316–13322.

[130] G.A. DeVries, M. Brunnbauer, Y. Hu, A.M. Jackson, B. Long, B.T. Neltner, O. Uzun, B.H. Wunsch, F. Stellacci, Science 315 (2007) 358–361.
[131] L. Beqa, D. Deschamps, S. Perrio, A.C. Gaumont, S. Knoppe, T. Bürgi, J. Phys. Chem. C 117 (2013) 21619–21625.
[132] H. Qian, W.T. Eckenhoff, Y. Zhu, T. Pintauer, R. Jin, J. Am. Chem. Soc. 132 (2010) 8280–8281.
[133] I. Dolamic, S. Knoppe, A. Dass, T. Bürgi, Nat. Commun. 3 (2012) 798.
[134] S. Knoppe, R. Azoulay, A. Dass, T. Bürgi, J. Am. Chem. Soc. 134 (2012) 20302–20305.
[135] M. Zhu, H. Qian, X. Meng, S. Jin, Z. Wu, R. Jin, Nano Lett. 11 (2011) 3963–3969.
[136] S. Knoppe, N. Kothalawala, V.R. Jupally, A. Dass, T. Bürgi, Chem. Commun. 48 (2012) 4630–4632.
[137] Z. Ma, S. Dai, ACS Catal. 1 (2011) 805–816.
[138] R. Nafria, P. Ramírez de la Piscina, N. Homs, J.R. Morante, A. Cabot, U. Diaze, A. Corma, J. Mater. Chem. A 1 (2013) 14170–14176.
[139] M. Cargnello, C. Gentilini, T. Montini, E. Fonda, S. Mehraeen, M. Chi, M. Herrera-Collado, N.D. Browning, S. Polizzi, L. Pasquato, P. Fornasiero, Chem. Mater. 22 (2010) 4335–4345.
[140] M. Chen, D.W. Goodman, Chem. Soc. Rev. 37 (2008) 1860–1870.
[141] M.S. Chen, D.W. Goodman, Science 306 (2004) 252–255.
[142] H. Yoshida, Y. Kuwauchi, J.R. Jinschek, K. Sun, S. Tanaka, M. Kohyama, S. Shimada, M. Haruta, S. Takeda, Science 335 (2012) 317–319.
[143] M. Haruta, Angew. Chem. Int. Ed. 53 (2014) 52–56.
[144] J. Guzman, S. Carrettin, A. Corma, J. Am. Chem. Soc. 127 (2005) 3286–3287.
[145] N. Ta, J. Liu, S. Chenna, P.A. Crozier, Y. Li, A. Chen, W. Shen, J. Am. Chem. Soc. 134 (2012) 20585–20588.
[146] M. Farnesi Camellone, S. Fabris, J. Am. Chem. Soc. 131 (2009) 10473–10483.
[147] Y. Kuwauchi, H. Yoshida, T. Akita, M. Haruta, S. Takeda, Angew. Chem. Int. Ed. 51 (2012) 7729–7733.
[148] M.S. Ide, R.J. Davis, Acc. Chem. Res. 47 (2014) 825–833.
[149] T. Yan, J. Gong, D.W. Flaherty, C.B. Mullins, J. Phys. Chem. C 115 (2011) 2057–2065.
[150] M. Ojeda, B.-Z. Zhan, E. Iglesia, J. Catal. 285 (2012) 92–102.
[151] X. Nie, H. Qian, Q. Ge, H. Xu, R. Jin, ACS Nano 6 (2012) 6014–6022.
[152] Y. Zhu, H. Qian, R. Jin, Chem. Eur. J. 16 (2010) 11455–11462.
[153] Y. Liu, H. Tsunoyama, T. Akita, S. Xie, T. Tsukuda, ACS Catal. 1 (2011) 2–6.
[154] R. Juarez, S.F. Parker, P. Concepcion, A. Corma, H. Garcia, Chem. Sci. 1 (2010) 731–738.
[155] A. Lyalin, T. Taketsugu, Farad. Discuss. 152 (2011) 185–201.
[156] A. Corma, H. Garcia, Chem. Soc. Rev. 37 (2008) 2096–2126.
[157] M. Stratakis, H. Garcia, Chem. Rev. 112 (2012) 4469–4506.
[158] Y. Mikami, A. Dhakshinamoorthy, M. Alvaro, H. Garcia, Catal. Sci. Tech. 3 (2013) 58–69.
[159] Y. Zhu, H. Qian, B.A. Drake, R. Jin Angew, Chem. Int. Ed. 49 (2010) 1295–1298.
[160] A. Shivhare, S.J. Ambrose, H. Zhang, R.W. Purves, R.W.J. Scott, Chem. Commun. 49 (2013) 276–278.
[161] Y. Zhu, H. Qian, M. Zhu, R. Jin, Adv. Mater. 22 (2010) 1915–1920.
[162] Y. Zhu, Z. Wu, C. Gayathri, H. Qian, R.R. Gil, R. Jin, J. Catal. 271 (2010) 155–160.
[163] R.P. Carney, G.A. DeVries, C. Dubois, H. Kim, J.Y. Kim, C. Singh, P.K. Ghorai, J.B. Tracy, R.L. Stiles, R.W. Murray, S.C. Glotzer, F. Stellacci, J. Am. Chem. Soc. 130 (2008) 798–799.
[164] Q.K. Ong, J. Reguera, P. Jacob Silva, M. Moglianetti, K. Harkness, M. Longobardi, K.S. Mali, C. Renner, S. De Feyter, F. Stellacci, ACS Nano 7 (2013) 8529–8539.

[165] Q.K. Ong, J. Reguera, P. Jacob Silva, M. Moglianetti, K. Harkness, M. Longobardi, K.S. Mali, C. Renner, S. De Feyter, F. Stellacci, Langmuir 29 (2013) 13723–13734.
[166] K.M. Harkness, A. Balinski, J.A. McLean, D.E. Cliffel, Angew. Chem. Int. Ed. 50 (2011) 10554–10559.
[167] X. Liu, M. Yu, H. Kim, M. Mameli, F. Stellacci, Nat. Commun. 3 (2012) 1182.
[168] C. Singh, P.K. Ghorai, M.A. Horsch, A.M. Jackson, R.G. Larson, F. Stellacci, S.C. Glotzer, Phys. Rev. Lett. 99 (2007) 226106.
[169] I.C. Pons-Siepermann, S.C. Glotzer, ACS Nano 6 (2012) 3919–3924 (and reference therein).
[170] F. Romano, E. Sanz, F. Sciortino, J. Chem. Phys. 134 (2011) 174502.
[171] A. Centrone, E. Penzo, M. Sharma, J.W. Myerson, A.M. Jackson, N. Marzari, F. Stellacci, Proc. Natl. Acad. Sci. U.S.A. 105 (2008) 9886–9891.
[172] A. Verma, O. Uzun, Y. Hu, Y. Hu, H.-S. Han, N. Watson, S. Chen, D.J. Irvine, F. Stellacci, Nat. Mater. 7 (2008) 588–595.
[173] R.C. Van Lehn, P.U. Atukorale, R.P. Carney, Y.–S. Yang, F. Stellacci, D.J. Irvine, A. Alexander-Katz, Nano Lett. 13 (2013) 4060–4067.
[174] A. Ghosh, S. Basak, B.H. Wunsch, R. Kumar, F. Stellacci, Angew. Chem. Int. Ed. 50 (2011) 7900–7905.
[175] E.S. Cho, J. Kim, B. Tejerina, T.M. Hermans, H. Jiang, H. Nakanishi, M. Yu, A.Z. Patashinski, S.C. Glotzer, F. Stellacci, B.A. Grzybowski, Nat. Mater. 11 (2012) 978–985.
[176] P. Posocco, C. Gentilini, S. Bidoggia, A. Pace, P. Franchi, M. Lucarini, M. Fermeglia, S. Pricl, L. Pasquato, ACS Nano 6 (2012) 7243–7253.
[177] D. Zaramella, P. Scrimin, L.J. Prins, J. Am. Chem. Soc. 134 (2012) 8396–8399.
[178] G. Pieters, A. Cazzolaro, R. Bonomi, L.J. Prins, Chem. Commun. 48 (2012) 1916–1918.
[179] C. Gentilini, F. Evangelista, P. Rudolf, P. Franchi, M. Lucarini, L. Pasquato, J. Am. Chem. Soc. 130 (2008) 15678–15682.
[180] B. Perrone, S. Springhetti, F. Ramadori, F. Rastrelli, F. Mancin, J. Am. Chem. Soc. 135 (2013) 11768–11771.
[181] C.K. Kim, P. Ghosh, C. Pagliuca, Z.-J. Zhu, S. Menichetti, V.M. Rotello, J. Am. Chem. Soc. 131 (2009) 1360–1361.

Chapter 4

Derivatization of Colloidal Gold Nanoparticles Toward Their Application in Life Sciences[1]

Dominik Hühn* and Wolfgang J. Parak

Fachbereich Physik, Philipps Universität Marburg, Marburg, Germany

**Corresponding author: E-mail: dominik.huehn@physik.uni-marburg.de*

Chapter Outline

1. This chapter is an adopted version based on the PhD thesis of Dominik Hühn as submitted at the Philipps Universität Marburg.

Gold Nanoparticles in Analytical Chemistry. http://dx.doi.org/10.1016/B978-0-444-63285-2.00004-3

1. INTRODUCTION

In the recent decades, the number of applications involving nanotechnology for biomedical purposes has increased enormously [1,2]. Two main reasons might be considered responsible for this huge expansion: (1) the attribute that physicochemical properties change entirely when downscaling a material to the order of magnitude of a few to a few hundreds of nanometers, and (2) the opportunity to equip a single nanodevice with more than one functionality. This great opening of new fields of research not only enables advanced functions in nanomedicine [3], but also brings the duty to deal with safety aspects [4]. Herein the study of environmental as well as health effects has to go hand-in-hand with the development of new functional materials. In the introduction of this chapter, a focus will be given on gold nanoparticles (Au NPs), as those are widely used in literature and thus can be considered as one prototype of NPs. An analysis of their physicochemical properties and how those can be measured will be given. Also their potential use for biomedical purposes will be highlighted. Finally aspects like the interaction of NPs with biological materials will be discussed.

1.1 Physicochemical Properties of Gold Nanoparticles

Nowadays, Au NPs are one important material in the huge set of available materials in the rapidly growing field of nanotechnology, which involves a lot of sub areas such as applications toward biology and energy [5–9]. The great importance of this material can be attributed to the unique electronic, magnetic, and optical properties of the Au NPs. However, Au NPs have been (partly unintentionally) employed as building blocks already since ancient times, mainly as staining agent for glass or ceramics. One of the most impressive examples that directly goes back to nanometer-size related optical properties is the Lycurgus Cup (fourth century, Figure 1). In case the glass is illuminated from the front with white light (light source outside), it appears green, and in case it is illuminated from the back (light source inside the cup), it appears red. The reason is a net absorption in the green part of the electromagnetic spectrum, which causes transmittance of red light. These optical properties originate from Au and Ag NPs that are embedded in the glass. In contrast to bulk Au or Ag, these NPs (size ~50 nm) strongly absorb light of a certain wavelength, which is a direct result of their nanometer size [10–13]. This goes back to the fact that noble metals are excellent electrical conductors with the ability to show collective oscillations of the present gas of free electrons, known as plasmons. These oscillations can be excited at optical frequencies. In case an oscillation is located at a surface or an interface, the associated oscillation and electromagnetic fields are called surface plasmon polariton (SPP). These propagate alongside the surface in form of an evanescent wave while losing energy.

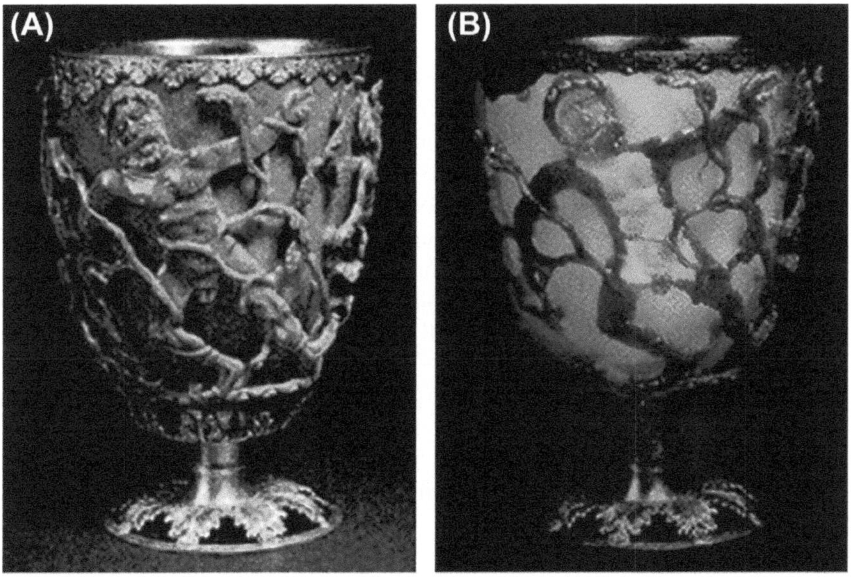

FIGURE 1 The Lycurgus cup appears (A) green in case the light source is outside the cup and (B) red in case the light source is inside. This phenomenon is directly related to the presence of gold nanoparticles embedded in the glass. *Source: http://someinterestingfacts.net/what-is-plasmons/.*

The energy loss is due to absorption by the metal on one hand, and radiation into free space on the other hand. If the frequency of the excitation light matches with the natural frequency of the electrons oscillating against the positively charged atomic nuclei, resonance occurs, called surface plasmon resonance. Upon downscaling of the material to the order of magnitude of the mean free path of electrons in the metal, this resonance is even enhanced and highly localized. Hence this phenomenon is sometimes also denoted as localized surface plasmon resonance (LSPR). For Au NPs with diameters between approximately 1 and 100 nm, this coupling occurs in the visible (vis) to near infrared (NIR) and shifts to the red with increasing diameter of the NPs. Thus the color of a colloidal Au NP suspension appears in the complementary of the LSPR: red for smaller NPs and purple/bluish for bigger NPs. As the location of the LSPR depends strongly on changes of the local dielectric environment, e.g., by adsorption of molecules onto the NP surface, LSPR spectroscopy and surface-enhanced Raman scattering (SERS) represent very sensitive techniques enabling the detection of such changes [13–15]. Moreover agglomeration of Au NPs leads to coupling of the electronic wave functions and thus to a red shift of the LSPR, which is the starting point of many colorimetric detection applications [16–19]. Additionally, the spectroscopic characteristic of the LSPR depends sensitively on the shape of the NPs.

Thus Au nanorods (NRs), i.e., NPs with different longitudinal and transverse dimension, show two LSPR peaks with their location and magnitude depending on the aspect ratio [10,20–22].

Upon excitation of Au NPs at their LSPR, one observes that parts of the energy are transferred to the atomic nuclei resulting in lattice oscillations, i.e., phonons. These oscillations usually are the source of heat, which leads to temperature increase in the Au NPs themselves, as well as of their close environment [23]. The fact that Au NPs are excellent light-to-heat converters is the starting point for the field of photothermal applications. Herein two main purposes are of great interest: (1) photothermal therapies (e.g., cancer treatment) and (2) photothermal release of materials whether bound to the NPs or accumulated in higher order structures as cargo [6,24–28]. Regarding photothermal applications in which the excitation light has to pass through the tissue to reach a certain site, i.e., the position at which absorption centers are located, one has to consider an important circumstance: visible light gets absorbed by tissue to a certain extent and the transmittance is highest in the so-called NIR window [29]. Hence, NP-based absorption centers in the NIR region are desired, so that NIR excitation of them is possible. Unfortunately small (5–50 nm) spherical Au NPs possess an LSPR around 520 nm, hence in the green part of the visible spectrum. Thus for life science applications, very often Au NPs or agglomerated Au NPs are the material of choice [21,30,31]. But also more individual constructs like nanoshells, nanocages, or nanoprisms are under investigation [32–34].

Regarding the synthesis of suspensions of colloidal Au NPs, several techniques are established. An early one is the method by Turkevich et al., which delivers citrate-capped and thus readily water-soluble Au NPs with sizes around 15 nm [35]. In contrast, the Brust method takes place in organic solvents, and the resulting NPs are hydrophobically capped with diameters around 5 nm [36]. Thus, for biomedical applications, these NPs have to be transferred to aqueous solution, for example, by overcoating them with polymers that mediate water solubility or by a ligand exchange to hydrophilic ligands. Both methods are based on the reduction of $HAuCl_4$, which finally leads to the formation of small Au seeds, which initiate the formation of NPs. There exist plenty more methods which can be seen as extensions of the mentioned ones as they make use of the same reduction principle [37,38]. What they all have in common is that they follow a bottom-up approach, hence the final NPs are the result of the fusion of smaller building blocks (Au atoms in the present case). An established method that follows a top-down approach is mediated by laser ablation [39]. Here colloidal Au NPs are seceded from a piece of bulk gold with a laser beam.

1.2 Other Inorganic Nanoparticles

Though this chapter focuses on Au NPs, many of the statements are valid also for other types of inorganic NPs. As many of them play an important role for different types of applications, in this section some of the possible opportunities of other inorganic NPs are reported.

Ag NPs as second type of noble metal NPs besides Au NPs already play an important role in daily life products, mainly due to their antimicrobial properties [40–42]. However, there is still more data warranted for a profound understanding of how and to which extend long-term effects of these NPs are hazardous when released to the environment or when getting in contact with the human body [43]. Regarding biomedical applications, the areas of usage of Ag NPs are very much similar to those of Au NPs. They can for example serve as carriers for biomolecules [44], or as sensor platform due to their plasmonic properties [45,46]. A rather novel field of application of noble metal NPs is the usage as fluorescent probes. When downscaling the material (Au or Ag) to the sub 2 nm region (then they are commonly called clusters), they show fluorescent properties [47–51]. Many synthetic routes for Ag NPs are established, but the most common ones are, similar to those of Au NPs, based on the reduction of salts of silver [52,53].

Another class of inorganic NPs made out of semiconductor materials are so-called quantum dots (QDs) [54–56]. In contrast to metals, semiconductors possess a band gap between the valence and conduction band. Thus, depending on the material, an excitation at optical frequencies can lead to the formation of an exciton, hence a bound state consisting of an electron that is shifted to the conduction band and an electron hole that remains in the valence band. Recombination thus can lead to the emission of a photon. Quantum mechanically an exciton can be described within formalism, which is similar to that of a hydrogen atom. Thus a kind of Bohr radius can be derived, although with lower binding energy and higher spatial extension compared to an H atom due to screening effects. If the spatial dimensions of the material are decreased down to the order of magnitude of the Bohr radius, the effect of quantum confinement comes into play: the exciton gets "squeezed," which results in an increased potential energy. Hence the energy gap between ground state and excited state is increased, the smaller the particle gets. The situation is comparable with the model of a "particle in a box" [57,58]. Thus, by adjusting the dimension (usually in the sub 10 nm region) of the material, the opto-electronic behavior of the QD is highly tunable. In other words, the emission wavelength of an excited QD undergoes a blue shift when decreasing the diameter. Due to their fluorescence properties, QDs are widely used as optical tracers, often in a functionalized form to address individual biomedical purposes [59–62]. Compared to conventional fluorescence markers like organic dyes, QDs possess the advantage of a rather narrow emission peak paired with a broad absorption profile, [63] which qualifies them as excellent Förster resonance energy transfer (FRET) donors [64]. Besides, QDs are much less subject to photobleaching compared to organic dyes [65]. The synthetic routes for colloidal QDs are as for the metallic NPs mainly based on wet-chemistry methods. QDs can be composed of a single material (silicon) or of compositions like chalcogenides (sulfides and selenides) of metals like Zn or Cd. A feasible method for improving the optical properties (quantum yield and brightness) of QDs is to use core–shell architectures, in which a shell of a semiconductor with higher band gap compared to the core is assembled around the latter. In this way, nonradiative exciton recombination is minimized in the core. A common combination is, e.g., CdSe/ZnS [65]. By varying the thickness ratio between core and shell and by varying the composition ratio between the components in alloyed QDs, it is possible to tune the emission wavelength under maintenance of the total diameter, a beneficial feature for size-restricted applications [66].

Also magnetic NPs obtained great attention in the recent decade due to their potential use in biomedicine [67–70], though these colloids have been known since long, for example, in the field of ferro-fluidics. Newly envisaged applications range from drug delivery systems, hyperthermia, magnetic resonance imaging (MRI) to magnetic separation, and biosensing. Applications more or less are mostly related to superparamagnetic properties of the NPs, which thus respond to an external magnetic field. Established material classes are metal oxides, like magnetite (Fe_3O_4) or maghemite (γ-Fe_2O_3), or metals (Fe, Ni, and Co) that are further passivated by a layer of a metal oxide, a noble metal, a polymer, or other surfactants. The most common synthetic routes involve wet-chemistry methods like coprecipitation (iron oxides) or thermal decomposition.

1.3 Colloidal Stabilization—Phase Transfer

Colloidal stability is an indispensible need which any high-quality NP suspension should fulfill. Otherwise the NPs would start to agglomerate and thus would no longer behave as individual objects/particles. Moreover, upon agglomeration, the physicochemical properties can change entirely [71]. As agglomeration is the result of an attractive interaction (e.g., van der Waals forces) between the surfaces of individual NPs, it can be prevented by modification of the surface [72]. This modification can act in different manner and induce repulsion by either electrostatics (i.e., repulsion between likewise charged NPs) or steric exclusion. Often the modification itself, i.e., the surfactant molecules or passivating layers, are already involved in the assembly process of the NP core and are desired for a later functionalization. This is why the topics of synthesis, stabilization, and functionalization typically go hand-in-hand and often are attained by the employment of only one type of surface modification. The attachment of such surfactants to the inorganic NP core usually is achieved via electrostatic and/or hydrophobic interaction or by covalent binding. In the case of noble metal NPs, the most common strategy is to covalently attach a molecule with a thiol-bearing end group [73–75]. The part of the surfactant pointing toward the solvent consequently determines the hydrophilic character of the NP. A possibility to achieve directly water-soluble

NPs is to choose a carboxyl-bearing end group, like given in thioglycolic acid (TGA), mercaptopropionic acid (MPA), or mercaptoundecanoic acid (MUA). In all cases, the functional end group bears a charge and ensures electrostatic repulsion between individual NPs. For electrostatically stabilized NPs, it is obvious that the suspension is sensitive to ionic strength (including pH), due to screening of the surface charge. In case the isoelectric point is reached due to charge screening, which is equivalent to neutralization of the surface charge, agglomeration occurs [76]. NPs stabilized via steric exclusion are much more resistant to such charge screening effects [77]. Polyethylene glycol (PEG) derivatives are a typical choice as surfactant for steric exclusion, which is due to the space-filling conformation of PEG. While the NPs as described so far are directly soluble in aqueous solution due to their hydrophilic surfactant molecules, in many other cases, the synthesis of the NPs involves hydrophobic surfactant molecules. To achieve water solubility, either these (hydrophobic) surfactants can be exchanged by a desired (hydrophilic) one ("ligand exchange") or amphiphilic molecules (such as polymers [78,79]) can serve as coat [72]. Such a polymer comprises in the simplest case two domains: one bearing a charge, hence providing water solubility, and another one comprising hydrophobic side chains. When the latter show a conformational similarity (at least comparable length), they can intercalate with the hydrophobic surfactants of the NP surface. If the hydrophilic domains, located at the polymer backbone, are pointing outward and the initial organic solvent is exchanged against an aqueous buffer, the polymer forms a dense shell around the NP core. Colloidal stability is maintained by electrostatic repulsion of the charged groups on the surface of this shell [80].

1.4 Functionalization Toward Biological Applications

To equip water-soluble NPs with further functionality, a feasible method is to covalently attach desired molecules to the outermost surface of the NPs. Very often an anchor point for such a functionalization is given by the stabilizing groups themselves, e.g., the carboxylic groups on the NP surface, either part of individual surfactant molecules (like in TGA, MPA, MUA, etc.) or of the backbone of an amphiphilic polymer. A good overview about established conjugation strategies between NPs and a certain substrate is given in the literature by Erathodiyil et al. [81]. (Bio-)conjugation functionalities of interest, like organic dyes for imaging and sensing, biomolecules for biological recognition/targeting, or chelators for the complexation of metal ions, can be introduced to the NP surface [78,82–85]. Another possibility to introduce functionality to an NP system is to embed additional monomers in a coating polymer, i.e., to use terpolymers [86].

In the following, a variety of such functional NP systems (typically in core + organic shell geometry, but also extended by some other examples) will be reviewed. A focus hereby will be set on sensing. Already in 2004, an immense emergence of improvements and developments according to the use of NPs in terms of applications in life sciences was predicted, concisely their use in ion and molecular sensing [87]. During 2001−2010, many of those initial ideas have been realized, such as multiplexed sensing. Hereby the NPs are often used as part of a nanostructure of higher dimension, forming a more complex device. This automatically demands for a classification of different techniques like optical/electrical/magnetic detection and their extensions to electrochemical sensing or more specific methods like SERS. Above all, one has to consider that one of the main acquisitions that nanotechnology introduced to sensing capabilities is the opportunity to combine different techniques, giving the possibility to expand the applicability of a system.

This opportunity was reflected, for example, by Cheon and Lee on the basis of multimodal imaging probes consisting of magnetic NPs decorated with further functionalities, i.e., radionuclides enabling positron emission tomography (PET), fluorescent moieties for optical tracking, or anchor molecules like antibodies/peptides/DNA/RNA giving the possibility to address specific targets [88]. The authors hereby claim that "synergistically" acting properties give the possibility for multimodal diagnosis and even for therapeutic functions (drug delivery, hyperthermia, etc.). How advantageous a magnetic NP-based multimodal system can be is also shown by Liong et al. [84]. In their study, they are dealing with multifunctional iron oxide, mesoporous silica NPs, which are detectable optically and by MRI. It is shown that the NPs are suitable for live cell imaging and therapeutic purposes by addressing them specifically to human cancer cells. Moreover, an additional important advantage of using NPs as carrier system is pointed out, which is the transfer of hydrophobic (and thus poorly water-soluble) anticancerous drugs (or other molecules) into cells, as the nature of the NP surface renders the whole system (NPs and the attached drugs) water-soluble. A comparable trimodal imaging system comprising MRI (iron oxide core), PET (chelator ligand complexing the radiotracer Cu^{64}), and fluorescence (Vivotag-680) enabling in vivo studies concerning the detection of macrophage markers was investigated by Nahrendorf et al. with regard to the detection of inflammatory atherosclerosis [89]. Due to the low required concentration of NPs and a higher target-to-background ratio, the authors claim the high relevance of these NPs for clinical perspectives. An overview and several promising applications concerning magnetic NPs are given by Jun et al. [90]. Besides the synthesis and further use in cellular MRI, the authors report in their review about the use of magnetic NPs in terms of multimodal probes comprising possibilities like hybrid "core-satellite" constructs with a fluorescent core and magnetic NPs attached, enabling a higher MRI contrast compared to individual magnetic NPs.

Regarding other core materials as starting point for the investigation of multimodal sensing devices, it is inevitable to mention gold nanostructures [91]. The excellent range of possibilities for tuning optical properties or surface chemistry and the fact of chemical inertness of the starting material makes them a versatile platform for the purpose of biomedical usage comprising a huge variety of sensing applications [92,93]. Concerning sensing of ions, one special class of materials is of particular interest, which is substances like heavy metals that are hazardous for the human health and environment. Having this in mind, Lin et al. discuss thoroughly possible routes to sense toxic metal ions like Hg^{2+}, Pb^{2+}, and Cu^{2+} [94]. In summary, they present a collection of gold NP-based probes with sensitivities that are two orders of magnitude better as compared to the current legislative limits. Two further approaches to detect Hg^{2+}, one comprising gold NP–oligonucleotide hybrids and another comprising a DNA-based machine, reached detection sensitivities of 2 and 0.2 ppb, respectively [95]. However, Li et al. admit that the measurements are made in clean aqueous solutions and potential cross-talk signals in environmental samples still have to be evaluated. A big advantage of the shown system is certainly the colorimetric signaling used as optical read-out by the eye, as it enables instrument-free and thus rapid analysis of Hg^{2+} contaminations as was also pointed out by Xue et al. [96]. That sensing of mercury is of essential importance is also due to the fact that many countries in northern Europe demand for stronger laws or even a total prohibition of this compound. Thus, it is not surprising that a lot of improvements with promising sensing capabilities are subject of recent research [97,98]. Besides hazardous materials, biological analytes are also the topic of recent sensing approaches. Especially the detection of DNA and the diagnosis of cancerous cells are an important focus of biological sensing applications based on NPs. In this purpose, Bunz and Rotello succeeded in a sensitive distinction between different proteins, bacteria, and cells (especially to discriminate between healthy and cancerous cells) based on the efficient fluorescence quenching of fluorophores, when adsorbed to gold NPs, although the concrete mechanism of quenching is not yet understood completely [99]. An approach, dealing with the indirect detection of cancer (here the detection of relevant protein biomarkers), was presented in an immunoassay by Liu et al. [100]. The assay is based on antibody-labeled Au NPs and Au NRs specific for an antigen of interest. The build-up of dimers/trimers/oligomers and the associated changes in dynamic light scattering (DLS) signals served as sensitive basis for the measurements. Another colorimetric assay based on aptamer-conjugated Au NPs was developed by Medley et al., avoiding time- and money-consuming sensing techniques for cancer detection [101]. Regarding the motivation of a multiplexed sensing device, colorimetric assays have the advantage of a simple by-eye-verification besides spectroscopic methods. The recent efforts concerning colorimetric Au NP assays for a variety of analytes was summarized by Zhao et al. [102]. Besides sensing systems within colloidal solutions, immobilized Au NPs have also been used mainly in terms of electrochemical approaches, leading to interesting applications considering enzyme or DNA detection, but also other analytes [103]. Also in the context of capillary electrophoresis–mass spectrometry (CE-MS), Au NPs can play an increasing role for molecular sensing, as it was shown for the detection of heroin and related compounds with the purpose of forensic studies or illicit drug control [104]. Although some of the presented approaches considering sensing assays based on Au NPs partly do not comprise multiple analysis techniques, they are bearing a great potential toward the same [105].

Potential for multiplexed detection has been given, for example, by Zhang and Johnson, who reported about an assay using aptamer functionalized QDs for a very sensitive detection of cocaine with possible extensions to other molecules (proteins, DNA, single ions, etc.) [106]. The study is based on FRET, a commonly exploited phenomenon for biosensing purposes, in particular considering QDs as donor or acceptor probe [64,107]. A label-free study for the detection of the cocaine metabolite benzoylecgonine was presented by Sanles-Sobrido et al. based on SERS, using Ag NP-decorated carbon nanotubes as substrate [108]. As can be seen from this example, the fact that NP-assisted SERS operates even at detection sensitivities on the single molecule level, one has to endow a particular compliance to this versatile technique [109–111]. Recent advantages and limitations of SERS in terms of biosensing are summarized by Alvarez-Puebla and Liz-Marzán [112]. However, also for the detection of DNA or in vivo tumor targeting, SERS offers great promises toward sensor applications [113,114].

1.5 Health and Environmental Aspects

Besides the potential applications which nanomaterials offer, great importance has to be addressed toward safety aspects, including health as well as environmental issues [4]. Regarding cytotoxicity, it is obvious that potentially toxic chemical substances, like cadmium (e.g., part of CdSe-based QDs), can cause cell death due to release of Cd^{2+} ions [115]. The mentioned strategy to encapsulate the NP core in a polymer shell can be a route to (1) lower the release of parts of the core material and (2) minimize the direct interaction of the core material with biological compounds ("the surface determines the chemical identity"), compared to only ligated NP cores. Also, chemically inert materials like gold can cause toxic effects, for example, induction of reactive oxygen species (ROS). There are also some particular examples like intercalation with DNA [116], which are however present under very particular conditions. Besides the size, also shape and surface chemistry certainly are the crucial parameters determining toxicity-related properties [117–119]. In general, one can say that the evaluation of toxicity of nanomaterials depends on the nanomaterial itself (including all physicochemical parameters), the cell type, but also on further external conditions such as pH, temperature and ion concentration [71,120]. As these differ enormously throughout the literature, unfortunately a unified comprehensive picture is missing. However, recently comparative studies came up, which are the starting point for such a comprehensive understanding [121].

Findings on a cellular level can be largely assigned to tissue, but aspects like distribution, degradation, or excretion of nanomaterials in organisms demand for extensive evaluation. In many in vivo studies, the liver and spleen are found to be the major accumulation sites of the NPs (and not for example the tumor where the NPs should go) [122]. Surface modifications like PEGylation promise to reduce toxic effects, e.g., due to reduced interaction with biological compounds like proteins [123]. NP-uptake pathways via the lung or the skin are also important topics of research that contribute to a greater understanding of how NPs in daily life can affect organisms [124,125]. Surface modification can also contribute indirectly to a reduced cytotoxicity, while addressing the NPs toward specific sites, like tumor cells, accompanied by drug delivery purposes [126]. Even the possibility to cross the blood–brain barrier by a transferring modification was reported [127]. Also environmental aspects can be a source of significant drawbacks regarding engineered NPs [128]. Air/water pollution and/or interaction with plants could be challenging problems, which call the use of NPs on industrial scale into question [129].

2. SYNTHESIS AND CHARACTERIZATION OF FUNCTIONAL NANOPARTICLES TOWARD BIOLOGICAL APPLICATIONS

2.1 Synthesis of Inorganic Gold Nanoparticles

Though many synthesis routes for Au NPs exist, in this chapter, only one protocol will be introduced in detail, as in particular Au NPs as synthesized with this methods were later-on used for demonstrating NP characterization and potential applications. The synthesis route for Au NPs described here as example is based on the so-called Brust method [36]. This method leads to hydrophobic NPs, which later-on are transferred to aqueous solution by overcoating them with an amphiphilic polymer. Briefly, from an aqueous solution of chloroauric acid ($HAuCl_4$, 1 eq), $AuCl_4^-$ is transferred to toluene forming an ionic pair with tetraoctylammonium bromide (TOAB, 4.5 eq), which later also acts as stabilizing agent. In the organic phase, sodium borohydride ($NaBH_4$, 10 eq) is added to reduce Au^{3+} to Au^0, which subsequently leads to the formation of colloidal Au NPs. After several washing steps, Ostwald ripening occurs during an incubation step over night, which supports the evolution of a rather monodisperse suspension, i.e., NPs with highly uniform core diameter. TOAB as surfactant molecule has two disadvantages: one is its relatively weak binding strength and the other is that the geometry of the NPs, including surfactants, is expected to be not perfectly spherical due to the potentially pyramidal appearance of TOAB at the surface. Thus a ligand exchange against 1-dodecanethiol (47.5 eq) is performed, which diminishes both shortcomings. As a result, hydrophobically capped Au NPs with a core diameter of $\sim$4 nm dissolved in toluene or chloroform are obtained, which do not show any sign of stability loss, even over several months. Figure 2 summarizes some properties of these Au NPs. In the transmission

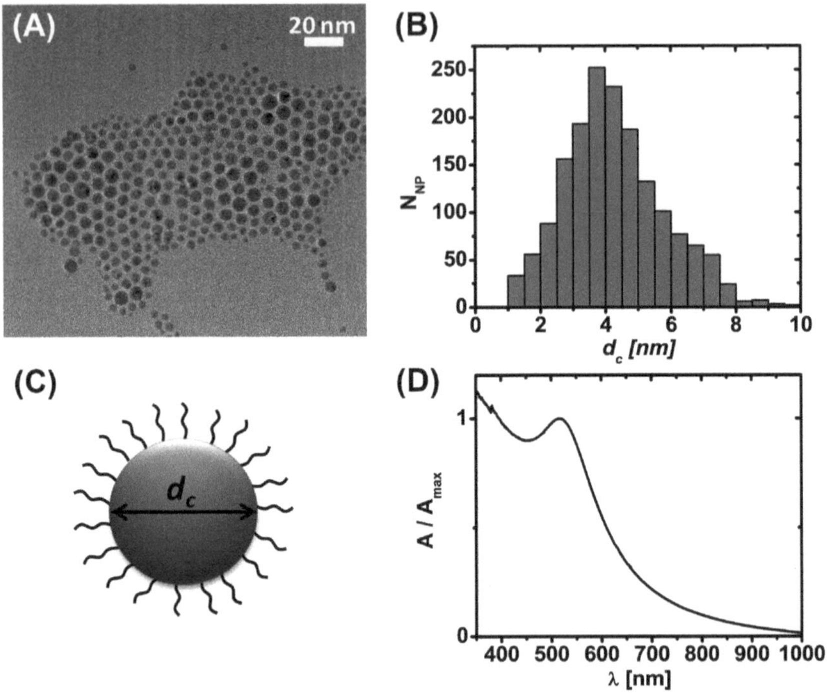

FIGURE 2 (A) Transmission electron microscopy (TEM) image of dodecanethiol stabilized gold nanoparticles (Au NPs). (B) Histogram of the size distribution of Au NPs as determined from the TEM images. In the present case, the core diameter is $d_c = (4 \pm 1.3)$ nm. (C) Schematic illustration of the Au core capped with dodecanethiol surfactants (approximately drawn to scale). (D) Extinction spectrum of dodecanethiol-stabilized Au NPs in chloroform. The graph is normalized at the localized surface plasmon resonance at approximately 520 nm. *The figure is adopted from Jimenez de Aberasturi et al. [130].*

electron microscopy (TEM) image (Figure 2(A)), the Au cores appear as gray spots with high contrast. The intermediate space between the NPs indicates the presence of dodecanethiol surfactants, which prevent the Au cores from agglomeration. The core diameter d_c is determined from the TEM images using the public domain software ImageJ in combination with a "particle size analyzer" macro freely available on the web (http://code.google.com/p/psa-macro/).

Figure 2(D) shows an extinction spectrum of the Au NPs dissolved in chloroform. A clear peak around 520 nm indicates the presence of an LSPR. For the concentration determination of Au NPs, the Lambert–Beer law (Eqn (1)) is applied for the net absorbance at the LSPR.

$$A = \varepsilon \cdot c \cdot d \tag{1}$$

Here A is the absorbance, ε is the molar decadic absorption coefficient, c is the concentration, and d is the length of the light path in the sample (cuvette length).

The relevant molar extinction coefficients are determined in the work of Liu et al., in which ε is correlated with the core size of Au NPs (Eqn (2)) [131].

$$\ln\left(\varepsilon \cdot \mathrm{M} \cdot \mathrm{cm}\right) = k \cdot \ln\left(\frac{d_c}{\mathrm{nm}}\right) + a \tag{2}$$

Here ε is the molar decadic extinction coefficient in 1/M cm, d_c is the core diameter in nm, and $k = 3.32111$ and $a = 10.80505$ are the fitting parameters which have been determined experimentally by Liu et al. [131].

2.2 Phase Transfer

For biomedical applications, it is essential that NPs possess water solubility. Especially, cell-associated experiments would suffer from hydrophobic moieties as they would lead to agglomeration and subsequently to a loss of functionality or intangible fate in a cell. Even the transfer of hydrophobic NPs to an aqueous environment would already lead to colloidal instability. In case Au NPs are synthesized directly with hydrophilic ligands, then they are in general already water soluble, though extensive purification in order to remove unreacted excess precursors and surfactant molecules still may be required. Regarding Au NPs, the method established by Turkevich et al. provides, for example, hydrophilic NPs [35], in contrast to the already mentioned Brust method [36]. In case NPs are capped with hydrophobic surfactants, as the Au NPs described as example in the previous section [36], a phase transfer has to be carried out. Synthesis in organic solvents is preferred in some cases, as higher temperatures (which are not restricted by the boiling point of water) can be used, which leads to better crystallinity due to annealing, and a larger variety of non-water-soluble organic surfactants is available. In this chapter, one universal strategy for an efficient phase transfer from organic to aqueous solvents is given as example, which utilizes a polymer-coating procedure, based on wrapping an amphiphilic polymer around the hydrophobic cores. However, many different alternative routes exist, of which each has certain advantages and disadvantages [132]. The amphiphilic polymer has to possess two crucial building blocks: (1) hydrophobic side chains, which have conformational similarity compared to the hydrophobic surfactant molecules present on the surface of the NP core, and (2) polar (e.g., typically charged) groups at the polymer backbone, that provide water solubility once the coated NP is transferred to an aqueous solvent. For the coating process, the two components (NP cores and amphiphilic polymer) are merged in an organic solvent, usually chloroform due to its low boiling point, and stirred for several minutes at $T >$ room temperature (RT $= 20$ °C). During this step, the hydrophobic side chains of the polymer are believed to intercalate with the hydrophobic surfactants of the NP core due to their conformational similarity. After evaporation of the solvent under reduced pressure, the dried NP film is redissolved in chloroform and another round of stirring is performed. This procedure is usually repeated three times. After the last evaporation step, the dried NP solid can be diluted in an aqueous buffer, as the hydrophobic parts of the polymer and the NP core surface are facing each other and thus ensure a dense packing. Moreover the hydrophilic sites at the polymer backbone, which are lying on the outer NP surface, provide colloidal stability due to electrostatic (in case of charged polymers) or steric (in case of polar end-groups such as PEG) repulsion. Figure 3 illustrates a sketch of how the coating process could be understood, though the precise geometric details have not been experimentally determined. Figure 3 is drawn in a very schematic and thus universal appearance, which is valid for the various types of polymers which are described in the following. All the used amphiphilic polymers have the possession of hydrophobic and hydrophilic parts in common, which provides them their amphiphilic nature. In addition, the polymers can comprise anchor points for the attachment of further functional molecules, indicated by the blue and red letters "F" in Figure 3.

The amount of polymer which needs to be added for the coating of the inorganic NPs can be calculated in a general way, described by the number of (hydrophobic) monomer units of the polymer that are employed per surface area regarding the effective diameter d_{eff} (Figure 3). The thickness of the layer of surfactants is usually estimated to be approximately 1 nm so that d_{eff} is calculated from the core diameter determined by TEM: $d_{\mathrm{eff}} = d_{\mathrm{c}} + 2$ nm. Equation (3) is used to calculate the required volume of polymer solution V_{P} [133]:

$$V_P = \frac{n_P}{c_P} = \frac{A \cdot R_P}{N_A \cdot c_P} = \frac{c \cdot V \cdot \pi \cdot d_{\mathit{eff}}^2 \cdot R_P}{c_P} \quad (3)$$

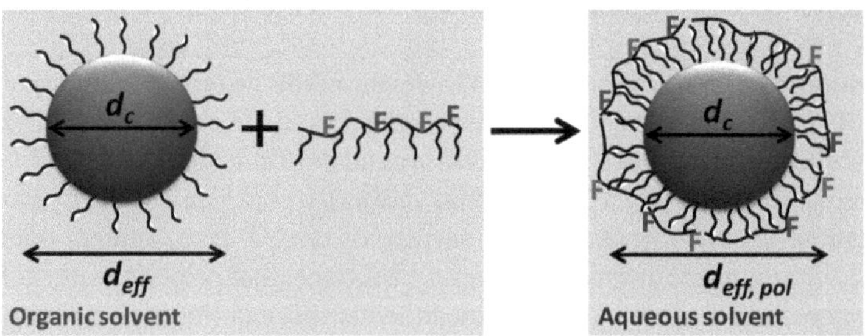

FIGURE 3 Schematic depiction of the coating process for the transfer of inorganic nanoparticles capped with hydrophobic surfactants from an organic to an aqueous solvent. An amphiphilic polymer intercalates with the hydrophobic surfactants to form a dense shell, which provides hydrophilic groups at the entire backbone. Further functionalities can be added to the polymer (blue and red letters "F"). The inner core diameter is denoted as d_{c}, the effective diameter comprising the core and the surfactant shell is denoted as d_{eff}. The resulting diameter comprising the core, surfactants, and the polymer shell is $d_{\mathrm{eff,pol}}$. The hydrodynamic diameter in solution of the water-soluble NP including a potential ionic cloud is called d_{h}. *The figure is adopted and modified from Jimenez de Aberasturi et al. [130].*

Here n_P is the molar amount of hydrophobic monomers in the polymer solution of molar concentration c_P of hydrophobic monomers. A refers to the total surface area present in the NP suspension, R_P, the number of hydrophobic monomers employed per unit area, and N_A, the Avogadro constant. A can be calculated from the known molar concentration c of the NP suspension, the employed volume of the NP suspension V, and the effective diameter d_{eff}. Usually d_{eff} is depicted in nanometers so that R_P in most of the cases is between 50 and 100/nm^2 for an efficient coating process. Successful polymer coating involves that after addition of the aqueous solvent, a transparent solution with as few agglomerates as possible is obtained. In case agglomerates are observed, an increased value of R_P can lead to an improvement, although too high values of R_P, i.e., a huge excess of polymer, should be avoided as it impedes with the subsequently following purification of the NP suspension from unbound polymer material.

Several advantages of the presented coating strategy have to be mentioned. (1) In principle, it works regardless of the composition of the inorganic core material as only the conformational similarity of the core surface and the hydrophobic part of the polymer determines the coating ability, which is driven by hydrophobic interaction. (2) The resulting physicochemical surface properties of the readily coated NPs are in first order only determined by the polymer backbone and thus intrinsic core properties can be compared under equal circumstances regarding the surface properties. This is especially advantageous with regard to the reduction of cytotoxicity. (3) Finally, the polymers can be modified with functional groups F, so that they are ideally suitable for the attachment of further functionalities.

2.3 Colloidal Characterization and Purification of Nanoparticles

Purification of NP solutions after synthesis is of utmost importance, as only in this way any interaction of the NP solution can be referred to the NPs, while otherwise it might also originate from impurities such as remaining precursors in the solution. Concerning undissolved NPs, once the NPs are coated with an amphiphilic polymer and transferred to an aqueous solvent, remaining agglomerates often can be resolved employing ultrasonification. Afterward in a first purification step, NPs can be filtered within a 0.22 μm syringe filter to get rid of bigger agglomerates.

After concentration in a centrifuge filter, a gel electrophoresis step can be applied both for purification from free polymer material and other molecules as well as for a colloidal characterization of the NP suspension [134]. For this purpose, the NP suspension is loaded onto a 2% agarose gel (UltraPure^{TM}Agarose) placed in an electrophoresis chamber (Bio-Rad Laboratories) within 0.5% tris(hydroxymethyl)aminomethane-borate–ethylenediamintetraacetic acid (Tris-borate–EDTA = TBE) buffer. A voltage of 100 V is applied (length of the gel: 10 cm → electric field ≈10 V/cm) for at least 1 h. This ensures a migration of NPs through the gel and an efficient separation from free polymer material. The free polymer material is believed to form micelle-like structures that migrate faster through the gel matrix due to a bigger charge-to-size ratio compared to the coated NPs [134]. Figure 4 shows an illustrative result as obtained with gel electrophoresis. The present system comprising Au NPs coated with a negatively charged polymer, which is modified with fluorophores (Zn^{2+}-sensitive dye and cresyl violet (CV)), will be discussed later in more detail. This example has been chosen as the empty polymer micelles can be observed by their fluorescence (due to the dye-label in the polymer), which would not have been possible in plain, unmodified polymer. After a certain migration time, the two bands regarding polymer-coated NPs and micelles are retarded enough to be extracted separately. Although the Au NPs in the example are tagged with fluorescent moieties, no fluorescence is visible in the more retarded band by eye. This can be due to higher amount of micelles than NPs, shading of dye fluorescence by the solid Au NPs which do not allow for light transmission, or quenching effects of the dye close to the Au surface. At the outer lanes of the gels, commercial 10 nm Au NPs (their original citrate shell has been exchanged by bis(*p*-sulfonatophenyl)phenylphosphine dehydrate dipotassium salt) are used as internal standard for comparison between different batches [135]. Comparison to the internal standard allows for the first estimate of the colloidal properties of the NPs. Poorly dissolved NPs, i.e., agglomerates, would lead to a significant

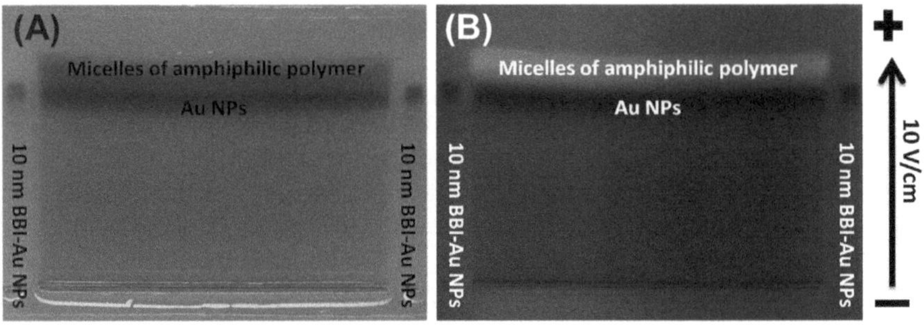

FIGURE 4 Purification of Au nanoparticles (NPs) via gel electrophoresis. The NPs are coated with a negatively charged polymer, which is premodified with two fluorescent dyes (for details *cf.* Figure 22). (A) and (B) show a view on top of the gel after having been run with gel electrophoresis, under visible and ultraviolet illumination, respectively. Before having run the gel first, the NP suspension has been loaded into the gel pockets which can be seen at the bottom of the images. Then an electric field of 10 V/cm has been applied for 1 h. After a migration toward the anode (+), two separated bands can be observed. The slow band (dark red) corresponds to the coated Au NPs and the faster band (purple in (A) and fluorescent in (B)) represents the micelle band. At the side lanes, commercial 10 nm Au NPs were used as internal control standard. *The figure is adopted from Jimenez de Aberasturi et al. [130].*

retardation. The migration distance gives information about the charge-to-size ratio and can thus verify if a certain molecule is attached to the NPs and to which extent. The width of a band represents the homogeneity, hence the monodispersity, of the sample. In case an unexpected retardation is observed, the coating procedure probably has been unfavorable and parameters like R_P have to be rescheduled. Concerning purification after having run the gel, the bands corresponding to the NPs or to the micelles can be cut out of the whole gel with a scalpel. After transfer of each gel band into a dialysis membrane (molecular weight cut-off = 50 kDa) and replenishment with TBE buffer, the different species (NPs or micelles) can be extracted from the gel matrix under the same voltage and migration direction. As neither the NPs nor the micelles are able to pass through the dialysis membrane, they can be efficiently collected in this way, followed by concentration via centrifugation with centrifuge filters [134]. After gel electrophoresis, usually another purification step with a 0.22 μm syringe filter is performed, which warrants for removal of any gel debris.

A further (alternative) level of purification and colloidal characterization is applied via size exclusion chromatography (SEC). For this purpose in the shown examples, a high-performance liquid chromatography system from Agilent Technologies was employed. For SEC-based NP purification, as described here, the stationary phase routinely consists of self-packed gel matrices, such as Sephacryl S-300 HR for negatively charged NPs and Sephadex G-25 Superfine for positively charged NPs. The mobile phase consists of a certain buffer, such as phosphate-buffered saline (PBS) or sodium borate-buffered saline (SBBS), at neutral to basic pH for negatively charged NPs, and 0.1 M NaCl (pH = 3.3 adjusted with 0.1 M HCl) for positively charged NPs. Due to interactions with the gel matrix, smaller molecules (like unbound polymer strands) remain longer in the column and can thus be separated from the NP suspension. Absorbance and emission (for fluorescent samples) profiles of the flowing sample give information about content and homogeneity. Figure 5 illustrates for two examples how the separation of NPs from free polymer material can be observed in various absorbance and/or emission channels. Collection of a narrow part of the NP suspension fraction ensures a homogeneous sample for further processing. Besides excess polymer material, also other bigger or smaller (compared to NPs) impurities can be removed with SEC. After SEC, usually an additional purification step is performed, which involved running the NP suspension through a disposable PD-10 desalting column (Sephadex G-25 Medium). Desalting (required due to salt contained in the mobile phase which forms the basis of the eluent) is of particular importance in case the NPs are equipped with ion-sensitive dyes as various ionic species in the suspension can influence the distribution of the analyte ions around NPs, and removes ionic residues originating from TBE and the SEC eluent. Afterward the suspension can be further washed or transferred to a desired buffer via centrifuge filters.

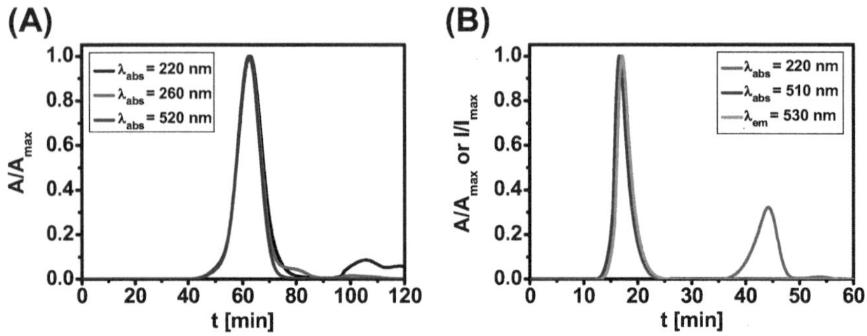

FIGURE 5 Illustrative size exclusion chromatography profiles as recorded by UV/vis absorption and fluorescence detection of the eluent: (A) Au nanoparticles (NPs) coated with a negatively charged polymer were run through a Sephacryl S-300 HR matrix with sodium borate-buffered saline (pH = 9) as mobile phase. The NP fraction passed between 50 and 70 min. In the UV/vis absorption spectra as recorded at $\lambda_{abs} = 220$ nm as well as at 260 nm, the Au cores and the polymer shell contribute to the net absorbance. At an absorption wavelength of 520 nm, only the Au cores contribute due to their localized surface plasmon resonance. Hence the detected absorbance for t = 80 min can be ascribed to excess polymer material. (B) Quantum dots (QDs) coated with a positively charged polymer are run through a Sephadex G-25 Superfine matrix with 0.1 M NaCl (pH = 3.3) as mobile phase. Here the absorbance recorded from the eluent at $\lambda_{abs} = 220$ nm is induced by the QD core and the polymer shell. The blue and green lines represent absorbance and emission at specific wavelengths for the green fluorescing QDs. Hence a fraction of polymer material (peak at 45 min) can be separated from the sample. The elution times in (A) and (B) are different due to different column sizes.

2.4 Characterization of Nanoparticles via pH Titration

When employing NPs for biomedical purposes, noble metals as core material have attracted great interest [6,9]. One reason is because of the chemical inertness of these materials, in contrast to hazardous materials which are used in many QDs of the first generation (i.e., cadmium-based ones). However, also inert materials can interact with their environment [136]. (1) "Size matters"—the size and the shape of a certain material (though they are inert) can lead to unexpected effects and thus interactions with biological systems [137]. (2) Also inert materials can possess catalytic effects and thus trigger the generation of ROS. (3) The surface chemistry of the NP shell has to be considered as crucial parameter regarding interaction with the environment, stability, and functionalization ability. These findings demand for a thorough characterization of the physicochemical surface properties of NPs in order to understand their interaction with biological environments. While a large set of available techniques, such as recoding hydrodynamic diameters with DLS, are common practice (see for example the next section), here we want to highlight a less well-known technique, pH titration, of great potential. pH titration is a feasible, time-, and cost-effective method [76]. In the following, it is shown how pH titration in combination with hydrodynamic diameter and zeta potential measurements can be applied to characterize Au NPs coated with a carboxyl-group-bearing polymer. The amphiphilic polymer used as example has significance, as it has been used in many scientific studies [78,130,133]. The here described polymer consists of a backbone of poly(isobutylene-*alt*-maleic anhydride) (PMA, M_W = 6 kDa), which is modified with dodecylamine. Usually 75% of the maleic acid anhydride rings are reacted with dodecylamine, so that after the overcoating of hydrophobic NPs with this polymer, followed by transfer to an aqueous buffer, carboxylic moieties provide a net charge at the entire polymer backbone and are anchor points for further functionality. As each anhydride ring contributes two carboxylic groups and 75% of the rings are reacted at one site, the coating terminates in $1.25 \times n$ carboxylic groups, in case n monomers are present. Figure 6(A) shows an NP after the coating has been carried out, but before the transfer into sodium borate buffer (SBB) at pH = 12, i.e., before the maleic acid anhydride rings have been opened. Thus, in Figure 6(B), a part of the polymer after coating is depicted with protonated carboxylic groups. Depending on pH, deprotonation of the carboxylic groups provides a net charge (and thus colloidal stability) to the NP surface. For this reason, alkaline pH leads to enhanced colloidal stability of the NPs as shown in Figure 6. However, this may conflict with conditions required for further functionalization. For the widely used N-(3-dimethylaminopropyl)-N′-ethylcarbodiimide (EDC)-activated coupling [138], the optimum pH is in the range 3.5−4.5 [139]. In this coupling chemistry, an amide bond is formed between a carboxylic site present in one binding partner and a primary amine present in the second binding partner. As the pK_a of aliphatic carboxylic acids is pK_a = 4−5, one might conclude that the optimal pH for EDC coupling of molecules to the surface of carboxyl-decorated NPs is 4.5, as here the NP still would be sufficiently charged to maintain colloidal stability. However, in a series of titration experiments in which a colloidal suspension of polymer-coated Au NPs was titrated against an acid, higher pK_a values were found [76]. The employed Au NPs were synthesized with a slightly different method as compared to the synthesis of polymer-coated Au NPs reported here, providing other sizes but the same surface properties [140]. For 7 nm core diameter, the pK_a was found to be 7.2, and for 2 nm core diameter, even a polybasic behavior (6.2 and 8.3) was observed. Apart from the latter, which might be related to curvature dependence, the higher pK_a values compared to plain carboxylic acids of approximately 3 pH units is an important experimental result. As the polymer-coated NPs are only negatively charged at pH values above the pK_a, these data suggest that for sufficient colloidal stability, pH should not be lower than 6−7. This however conflicts with the ideal environment for the further attachment of molecules via EDC chemistry, which requires a lower pH. A simple explanation for the higher pK_a found for the carboxyl groups present on the surface of polymer-coated NP and also for the polybasic behavior is the formation of

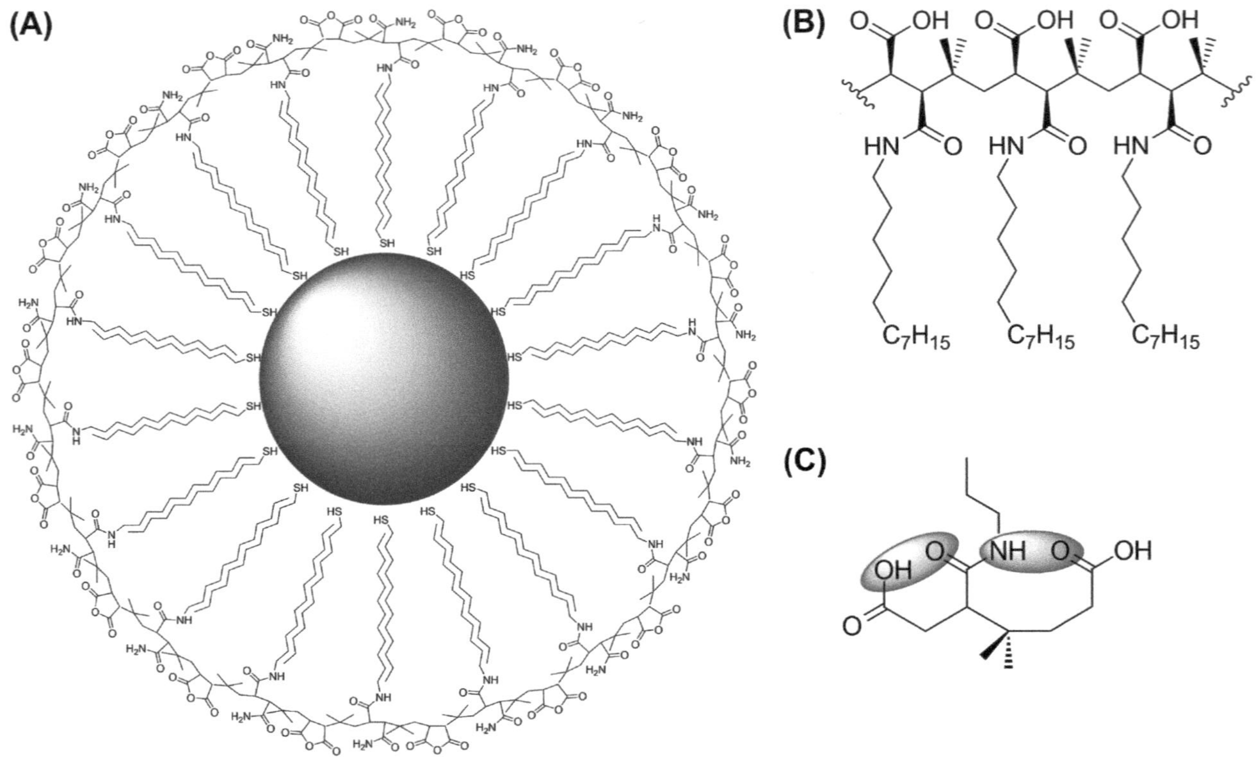

FIGURE 6 (A) Scheme of a polymer-coated nanoparticle (NP) before the transfer to aqueous buffer, as indicated by the closed maleic acid anhydride rings. The hydrophobic dodecylamine side chains of the polymer intercalate with the hydrophobic dodecanethiol surfactants present on the NP core. The scheme is not drawn to scale; the polymer is overexpressed for clarity. (B) A fraction of the polymer backbone after transfer to an aqueous buffer. The carboxylic groups ensure a negative surface charge. As not all carboxylic groups are equivalent (some have an amide bond as counterpart and some another carboxylic group), a polybasic behavior can be observed. (C) The explanation for a higher pK_a compared to plain carboxylic acids goes back to the formation of hydrogen bonds at specific sites of the polymer, which also occurs for the plain polymer without an NP core. The bonds show typical geometric parameters regarding bonding length and angles as determined by density functional theory calculations. *The figure is adopted and modified from Charron et al. [76].*

hydrogen bonds between specific sites at the polymer backbone (Figure 6(C)). At any rates, these results demonstrate why it is not straightforward to apply EDC coupling chemistry under standard conditions to link molecules to the surface of polymer-coated NPs.

Concerning a conformation regarding the higher pK_a values, measurements of the hydrodynamic diameter (d_h) and the zeta potential (ζ) of the NPs can be performed, while titrating an Au NP suspension against an acid. These measurements were performed with a Malvern Zetasizer Nano ZS with integrated MPT-2 Autotitrator via DLS for d_h and laser Doppler anemometry (LDA) for ζ. For the measurements, first the pH of the Au NP suspension was adjusted to approximately pH = 12 with an NaOH within several washing steps. The concentration of the NaOH solution was adjusted according to the number of carboxylic groups in the solution. As this number is not known, the maximum number of present carboxylic groups, regarding the added polymer amount during the coating, was used. This ensured a full deprotonation of all carboxylic groups. Incidentally the number of carboxylic groups present in the suspension could be estimated from a routine pH titration as mentioned above from the width of the buffer plateau. The NP suspension was then titrated against HCl while monitoring d_h and ζ. Concerning analysis of these data, protonation and deprotonation can be described with the Henderson–Hasselbalch equation (Eqn (4))

$$\mathrm{pH} = \mathrm{pK_a} + \log\left(\frac{f}{1-f}\right) \tag{4}$$

in which f is

$$f = \frac{c(\mathrm{COO^-})}{c(\mathrm{COO^-}) + c(\mathrm{COOH})} \tag{5}$$

and thus represents the fraction of deprotonated carboxylic groups $c(\mathrm{COO^-})$ in relation to all carboxylic groups present $c(\mathrm{COO^-}) + c(\mathrm{COOH})$ (Eqn (5)). c represents the respective molar concentrations. Thus the entire surface charge Q is then given by the maximum charge $Q_{\max}$ ($c(\mathrm{COOH}) \sim$ minimal) times the fraction f (Eqn (6)).

$$Q = Q_{max} \cdot f = Q_{max} \cdot \frac{1}{1 + 10^{\mathrm{pK_a} - \mathrm{pH}}} \tag{6}$$

Assuming a fictitious charge $Q_{\max}$ of -1 C and a pK_a of 7, the $\log(-Q/\mathrm{C})$ would show a linear behavior toward low (acidic) and high (basic) pH (Figure 7). Then the asymptotes of the acidic and basic regime would have their intersection point at the given pK_a.

Assuming now that the $\log(-\zeta/mV)$ shows a similar pH-dependent curve profile, an estimate for the pK_a can be drawn out of the titration experiment. Of course the zeta potential is not equivalent to the present surface charge, but in good approximation can be seen as a measure for it. The higher the entire

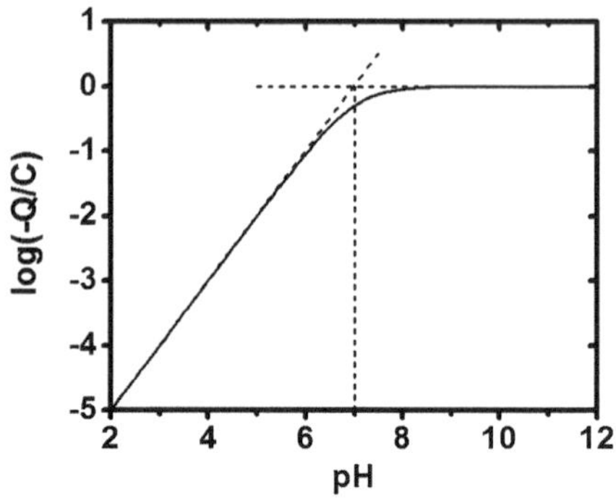

FIGURE 7 Fictitious behavior of log($-Q$/C) following the Henderson–Hasselbalch equation (Eqns (4)–(6)) for $Q_{max} = -1$ C and $pK_a = 7$. *The figure is adopted and modified from* Charron *et al. [76].*

surface charge, the higher the electrostatic potential at the slipping plane of the ambient ionic cloud present at the NP surface in an aqueous buffer. Figure 8 shows the zeta potential ζ(pH) recorded at different pH values as well as the logarithmic plot log($-\zeta/mV$) for polymer-coated Au NPs with 7 and 2 nm core diameter. In each case pK_a values between 6 and 7 are derived ($pK_{a,\zeta}$(7 nm) ≈ 6.3 and $pK_{a,\zeta}$(2 nm) ≈ 6.8). This is somewhat lower when compared to the findings of the volumetric titration, considering the obtained average pK_a values for 2 nm Au NPs ($pK_{a,V}$(7 nm) ≈ 7.2 and $pK_{a,V}$(2 nm) ≈ 7.3). However, clearly the increased pK_a compared to plain carboxylic acids can be also verified by zeta potential measurements, although the method might not have the same accuracy as the one derived with volumetric titration, resulting in a higher error in pK_a. One thus clearly can state again that standard conditions for EDC coupling reactions

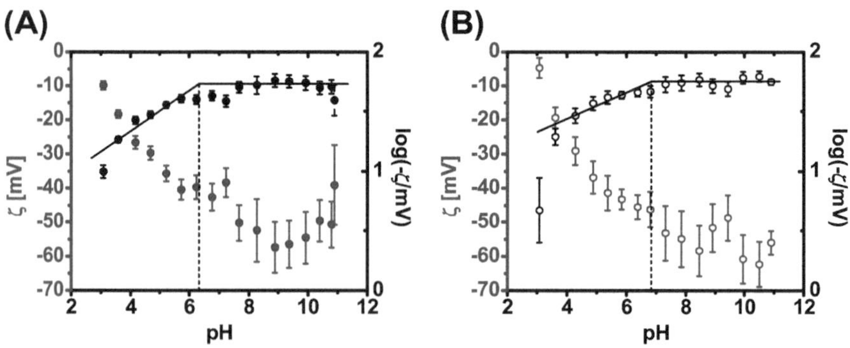

FIGURE 8 Zeta potential measurements (left/blue y-axis) for (A) 7 nm and (B) 2 nm core diameter, polymer-coated Au nanoparticles at different pH values. The pK_a of each sample is estimated from the log($-\zeta/mV$) curve at the site with the biggest curvature between the basic and acidic linear regimes (indicated by a dashed line). Each data point consists of three individual measurements. *The figure is adopted and modified from Charron et al. [76].*

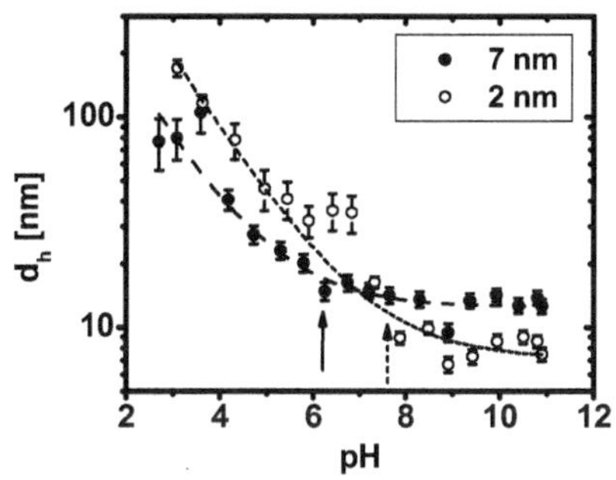

FIGURE 9 pH dependence of the hydrodynamic diameter of polymer-coated Au nanoparticles with 7 and 2 nm core diameter, as measured with dynamic light scattering. A significant increase of d_h is indicated by the arrows. The lines are guides to the eye. Each data point consists of three individual measurements. *The figure is adopted and modified from Charron et al. [76].*

need to be modified when used together with the polymer-coated Au NPs as described here.

Besides the zeta potential, which allows for a direct statement about the ionization state of the NP surface, also the hydrodynamic diameter d_h can serve as characteristic parameter. In case a sufficient surface charge is not present, as accompanied by a low absolute zeta potential, van der Waals forces might overcome the electrostatic repulsion between individual NPs and thus lead to agglomeration. This agglomeration obviously should be verifiable by a significant increase in d_h at varying pH. Figure 9 shows such a titration measurement (starting from basic pH) for polymer-coated Au NPs of 7 and 2 nm core diameter. The arrows display from which pH on a significant increase in d_h is observed, indicating agglomeration. It was assumed that the pK_a is located in the same pH region. Again this region was found to be around pH $\approx$ 7, hence significantly higher compared to the pK_a of plain carboxylic acids, what confirms the previous findings. However, the fact that the increase of d_h occurs to a greater extent for the smaller 2 nm core diameter NPs cannot be explained at this point. A possible reason could be the higher pK_a that leads to an "earlier" agglomeration. More time between the pH adjustment and the d_h measurement would be required for a more profound analysis. In summary, the here-presented DLS and LDA methods support the conventional volumetric titration measurements.

2.5 Charge-Dependent Interaction of Nanoparticles with Biological Systems

As described so far, the state of colloidal stability or functionalization ability depends on environmental parameters like pH or more general, the ionic strength present in the NP solution. However, with a look toward potential biological applications, it is also crucial to understand how the colloidal properties of NPs depend on the interaction with biological components, e.g., with protein-containing media, cell culture media, or the interaction with cells themselves. It is well established in literature that proteins adsorb to NP surfaces, and a so-called protein corona is formed [141−146]. This protein corona can have far-reaching effects as it can influence the colloidal properties of NPs, as it attributes a biological identity to NPs, which is decisive for the uptake and further processing by cells.

Here an exemplary study about the interaction of two species of NPs, namely negatively and positively charged NPs, with biological systems is presented [86]. Herein the surface charge is in first order, the only variable parameter. The employed polymers are depicted in Figure 10. Each polymer consisted of charged, hydrophobic, and potentially functional monomers [86].

For the negatively charged polymer poly((2-(methacryloyloxy)ethyl)phosphonic acid)$_x$-*stat*-poly(lauryl methacrylate)$_y$-*stat*-poly(propargyle methacrylate)$_z$ (PMAPHOS-*stat*-PLMA-*stat*-PgMA), the charge was generated by the deprotonated phosphonic acid part. For the positively charged polymer poly(*N*,*N*,*N*-trimethylammonium-2-ethyl methacrylate iodide)$_x$-*stat*-poly(lauryl methacrylate)$_y$-*stat*-poly(propargyle methacrylate)$_z$ (PTMAEMA-*stat*-PLMA-*stat*-PgMA), the charge was generated by a trimethylammonium group. In both cases, the hydrophobic part consisted of PLMA similar to the dodecylamine-modified PMA-coated NPs, which was described in Figure 6. These compounds ensured an efficient coating of NPs and transfer to aqueous solution. One has to bear in mind that a sufficient percentage of hydrophobic parts has to be present to ensure the coating of NPs, but as soon as the hydrophobic proportion rises too much, no colloidal stability in aqueous buffers can be maintained. The third monomer (*z*) was an optional one and contained the fluorescent dye perylene tetracarboxylic diimide (PDI) as functionality. For attachment, the dye was azide-modified and linked to the polymer via click chemistry. First, optimum

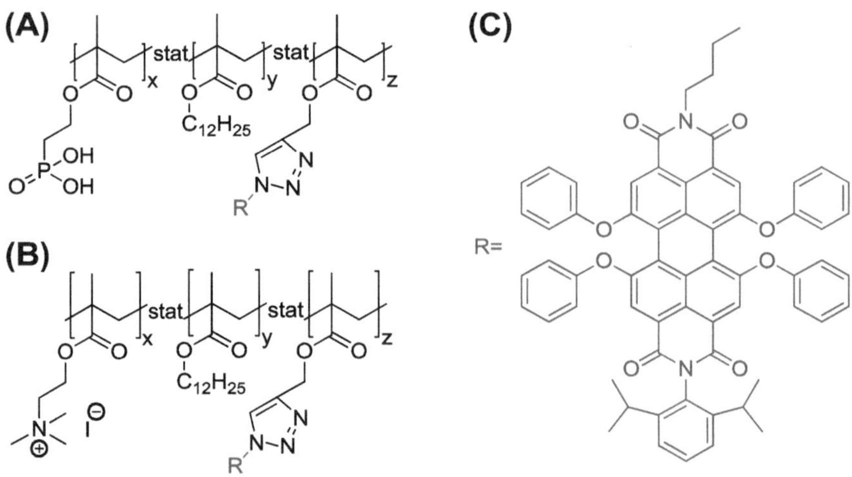

FIGURE 10 Structural formulas of the employed polymers. (A) Negatively charged terpolymer. (B) Positively charged terpolymer. (C) Fluorescent dye perylene tetracarboxylic diimide. *The figure is adopted and modified from Hühn et al. [86]. The polymer has been prepared by Christian Geidel from the group of Dr Markus Klapper.*

parameters regarding molecular weight and composition (*x*:*y*:*z*) had to be determined. Herein two criteria were of particular interest. (1) The polymer should be soluble in chloroform so that a homogeneous mixture with the hydrophobically capped NPs is possible. (2) The transfer to an aqueous buffer should lead to a colloidally stable suspension. Some experience from a previous study [79] regarding these requirements paved the way to find the final parameters as depicted in Table 1.

The employed Au NPs in this study were of the same nature as the one depicted in Figure 2. For the transfer to aqueous solvents and for the purification via SEC, 0.1 M NaCl ($pH = 3.3$, adjusted with 0.1 M HCl) was used in case of the positively charged polymer, in contrast to the already mentioned SBB at $pH = 12$ for the negatively charged polymer (as has been described for PMA). After the already described routine purification processes, the coated Au NPs were thoroughly characterized. However, it has to be noted that purification of positively charged NPs via gel electrophoresis with agarose gel was not possible, probably due to an interaction of the ammonium sites with the agarose matrix. However, by SEC and the additional desalting column and centrifugal filtration, a sufficient purification could be achieved also for positively charged NP.

The concentration determination of the Au NPs which contained the dye PDI in their shell was not as straightforward as in the case when no dye was present. This is due to the overlap of the dye absorbance with the LSPR peak (Figure 11(A) and (B)). Thus a simple approach was applied, in which a linear combination of the dye and the Au core absorbance was assumed. Herein the total absorbance of the sample A_S is given by Eqn (7):

$$A_s = x \cdot A_{ref} + y \cdot A_p \tag{7}$$

Here A_{ref} is the known absorbance of reference Au NPs coated with a copolymer without dye (Figure 11C) and A_{p} is the absorbance of the plain polymer with PDI (Figure 11D). The parameters x and y were varied to fit best to the data obtained with the terpolymers-coated Au NP sample (Figure 11E and F). Of course this method is only a rough estimate to extract the respective absorption contributions of a complex system consisting of an inorganic Au core and an organic polymer shell, but it has turned out as a viable approach. In this way, also the amount of dye in the sample can be estimated. The dye-to-NP ratio may vary, as the molar ratio of dye molecules per polymer as well as the amount of polymer strands attached per Au cores varies.

The fluorescence properties of Au NPs coated with positively and negatively charged co- and terpolymers as well as of a plain terpolymer are depicted in Figure 12.

The possibility to add fluorophores into the polymer shell offers a huge advantage. For some characterization fluorescence is required, but for other characterization fluorescence actually would impose a disadvantage. Thus NPs can be made on demand with or without fluorescence. To give some examples, fluorescent NPs are required for a characterization with techniques such as

TABLE 1 Composition of the Employed Amphiphilic Co- and Terpolymers

		Composition of the Polymer				M_n
Charge	Functionality	TMAEMA (x) (mol-%)	MAPHOS ( ) (mol-%)	LMA (y) (mol-%)	PgMA (z) (mol-%)	(g/mol)
Negative	–	–	40	60	–	9000
Positive	–	53	–	47	–	16,300
Negative	PDI	–	52	42	6	9500
Positive	PDI	48	–	48	4	17,800

PDI, perylene tetracarboxylic diimide.

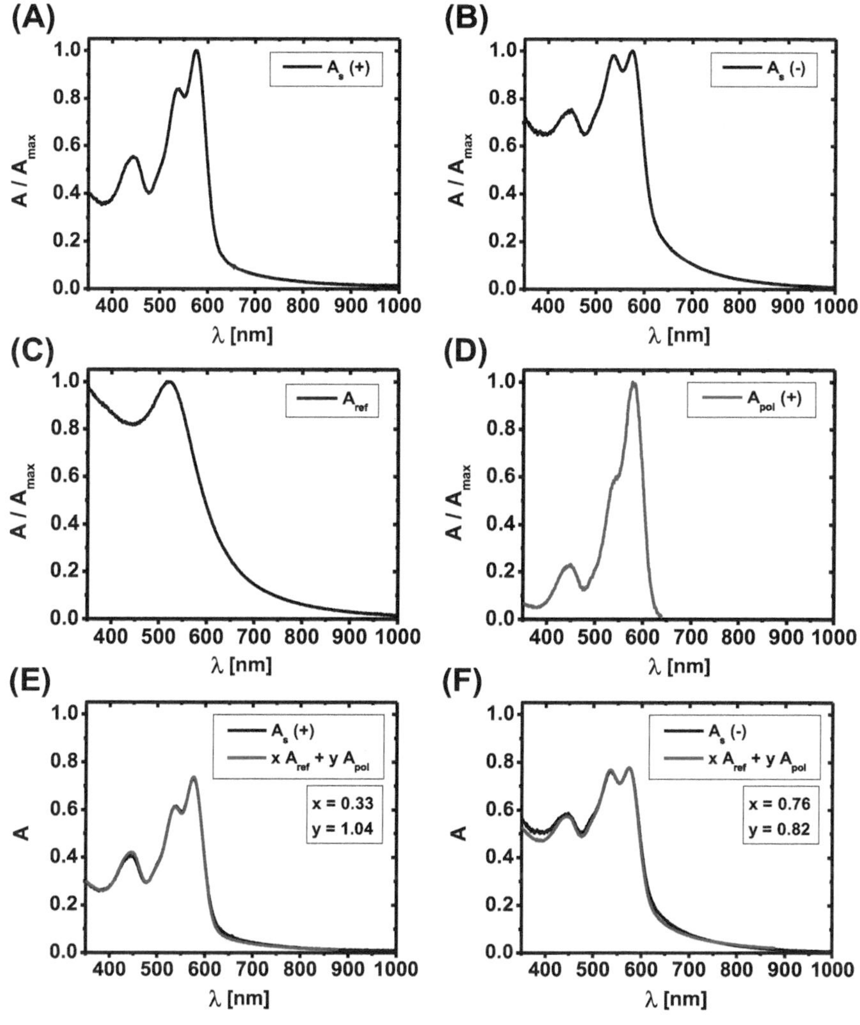

FIGURE 11 Extinction spectra of (A) positively charged and (B) negatively charged Au nanoparticles (NPs) with dye in their shell, normalized at the maximum absorbance in the region of the localized surface plasmon resonance. (C) Normalized extinction spectrum of Au NPs coated with dodecylamine-modified poly(isobutylene-*alt*-maleic anhydride as reference. (D) Normalized extinction spectrum of the plain positively charged terpolymer A_p showing the specific absorbance of perylene tetracarboxylic diimide (PDI). The absorbance of the negatively charged polymer is not shown here as it does not differ from that of the positively charged one. The absorbance by the polymers themselves is located in the UV part of the spectrum and the shown peak corresponds to PDI only. (E) and (F) show the same spectra like in (A) and (B) before normalization (black line) as well as the best fit according to the linear combination (Eqn (7)) of the dye absorbance A_p and the absorbance of a reference sample without dye in the shell A_{ref} (red line). The determined parameters x and y for each best fit are depicted in the graphs. *The figure is adopted and modified from reference Hühn et al. [86].*

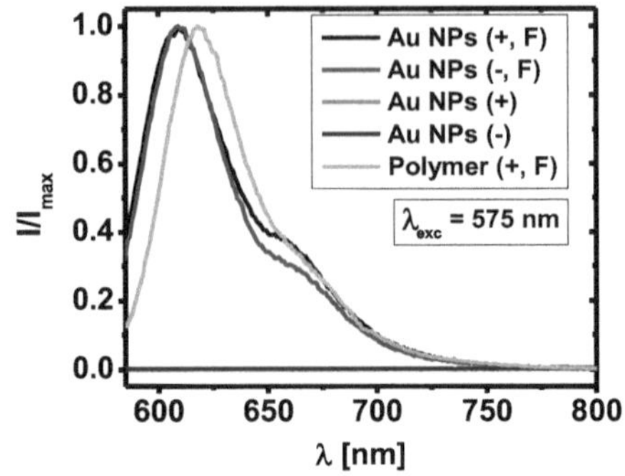

FIGURE 12 Fluorescence spectra of Au nanoparticles (NPs) coated with positively (+, F) or negatively (−, F) charged terpolymers with perylene tetracarboxylic diimide (PDI) in their shell. The spectra differ only within a small shoulder between 650 and 675 nm. The location of the maximum fluorescence is blue shifted by approximately 10 nm compared to the plain positively charged terpolymer with PDI (+, F), what might be related to changes of the microenvironment of the dye after coating. The fluorescence of the plain negatively charged terpolymer with PDI is not shown as it did not differ from that of the positively charged one. The fluorescence of Au NPs coated with positively (+) or negatively (−) charged copolymers without PDI is zero as expected. *The figure is adopted from Hühn et al. [86].*

fluorescence correlation spectroscopy (FCS) or fluorescence microscopy. Nonfluorescent NPs are preferred for a characterization via DLS and LDA, as here the fluorescence emission would interfere with the operating laser. The quantum yield of the fluorescent Au NPs was determined via the method suggested by Horiba Jobin Yvon [147]. For different concentrations of Au NPs, absorbance and emission spectra were recorded. The slope of the linear relation between the absorbance at 575 nm and the integrated emission (λ_{exc} = 575 nm) delivered the quantum yield following Eqn (8):

$$\Phi_s = \Phi_{ref}\left(\frac{Grad_s}{Grad_{ref}}\right)\left(\frac{\eta_s}{\eta_{ref}}\right)^2 \qquad (8)$$

Here Φ_s and Φ_{ref} are the quantum yields, $Grad_s$ and $Grad_{ref}$ are the slopes of the mentioned relation, and η_s and η_{ref} are the refractive indices of the solvent for each of the sample and of a reference dye. Figure 13 shows the recorded extinction and emission profiles as well as the linear relations between absorbance and integrated fluorescence intensity. The here used reference dye was cresyl violet perchlorate (CVP) diluted in methanol with a quantum yield of $\Phi_{ref} = 0.54 \pm 0.03$ [148]. The slopes as derived from the graph were $Grad_{ref} = (227.4 \pm 5.9) \times 10^8$ cps, $Grad_s(+) = (23.6 \pm 0.4) \times 10^8$ cps, and $Grad_s(-) = (22.2 \pm 0.3) \times 10^8$ cps (note that the values of ΣI in Figure 13 are divided by 10^8 for a more convenient display). For the refractive indices, $\eta_s = 1.333$ (water) and $\eta_{ref} = 1.3288$ (methanol) were used. Thus quantum yields of $\Phi_s(+) = 0.057 \pm 0.004$ and $\Phi_s(-) = 0.053 \pm 0.003$ could be determined for the positively and negatively charged Au NPs. The errors were determined via the Gaussian error propagation. The quantum yields were remarkably low, but sufficient for convenient fluorescence detection with FCS.

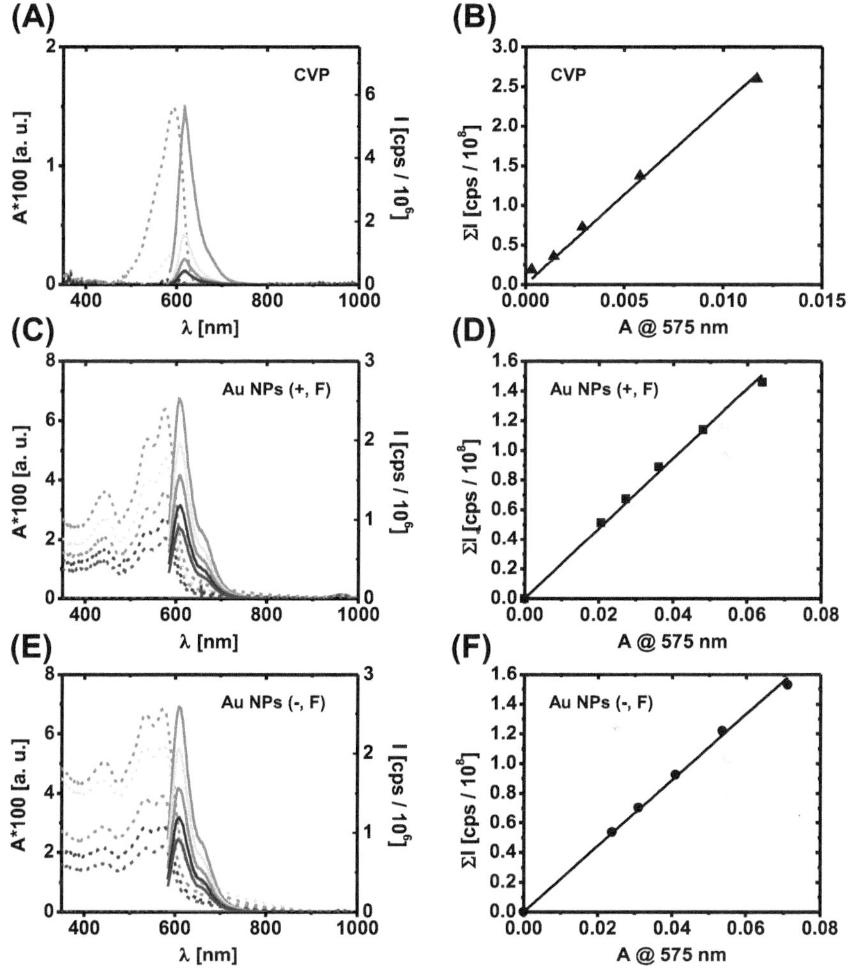

FIGURE 13 Quantum yield determination for fluorescent Au nanoparticles (NPs). Graphs in the left column (A), (C), (E) show the absorbance (left y-axis, dashed lines) and emission (right y-axis, solid lines, $\lambda_{exc} = 575$ nm) profiles of cresyl violet perchlorate, and positively and negatively charged Au NPs, respectively. Graphs in the right column (B), (D), (F) show the corresponding relation between integrated (from 588–800 nm) fluorescence intensity ΣI and absorbance at 575 nm. *The figure is adopted from Hühn et al. [86].*

For the same Au NPs, also the hydrodynamic diameter and the zeta potential were determined with DLS and LDA, respectively. Therefore, Au NPs which did not contain PDI in their shell had to be employed, as the fluorescence of the PDI would interfere with the light scattering signal. In order to demonstrate this effect, some measurements with PDI in the polymer shell were performed, and the obtained high fluctuations demonstrated the limited reliability of these data. DLS measurements were carried out in water

TABLE 2 Hydrodynamic Diameter and Zeta Potential of Positively and Negatively Charged Au NPs as Determined via DLS and LDA, Respectively

NP Charge	d_h (Water) (nm)	d_h (PBS) (nm)	ζ (Water) (mV)
+	13.0 ± 3.0	10.1 ± 0.6	9.7 ± 8.9
−	10.9 ± 3.2	14.5 ± 1.0	−39.8 ± 10.0

Each value is the average of three individual measurements. d_h was extracted from the number distribution. NP, nanoparticle; DLS, dynamic light scattering; LDA, laser Doppler anemometry; PBS, phosphate-buffered saline.

and PBS. LDA measurements in contrast were carried out only in water, as the salt of the buffer turned out not to be compatible with the used cuvettes due to undesirable deposition effects at the electrodes. In Table 2, the determined values for d_h in water and PBS and for ζ in water are given. The absolute zeta potential of the negatively charged Au NPs was higher than that of the positively charged NPs, and thus between both NP species, not only the sign of the charge but also the absolute amount of charge differs. The higher absolute zeta potential of negatively charged Au NPs is believed to be responsible for the higher hydrodynamic diameter of the NPs in PBS. Due to the ionic content in the buffer, bigger cloud of counter ions is probable, resulting in bigger hydrodynamic diameter [149]. The difference clearly is not due to limited colloidal stability as the hydrodynamic diameters of both NP species are still remarkably similar. Also the hydrodynamic diameter of the charged NPs in water showed a higher fluctuation between individual measurements (higher error) than measurements as recorded in PBS.

FCS measurements on the same NPs as performed in the group of Prof. Dr Ulrich Nienhaus by Stefan Brandholt yielded comparable results regarding the hydrodynamic diameter in PBS: $d_h(+) = 10.2 \pm 0.2$ nm and $d_h(-) = 15.8 \pm 0.4$ nm. As PDI-bearing polymers were employed for FCS measurements, the similar values indicate that the dye itself might have at best little influence on the colloidal properties of the NPs. Due to the hydrophobicity of the plain PDI, it can also be assumed that it might not sit on the outside of the NP surface, but rather in the inner hydrophobic part of the shell. This is also supported by previous results, in which the location of a polarity-sensitive dye was found to be in the hydrophobic part of the coating [150]. Figure 14 shows the result of the FCS measurements in PBS.

FCS measurements were also performed under varying concentrations of human serum albumin (HSA), again by Stefan Brandholt from the group of Prof. Dr Ulrich Nienhaus. In this way, adsorption of proteins was determined via increase of d_h (or as shown in the hydrodynamic radius $r_h = 0.5\ d_h$): The more protein is adsorbed, the bigger the NP becomes. The values given above for plain PBS correspond to the asymptote of $r_h(0)$ considering the Hill

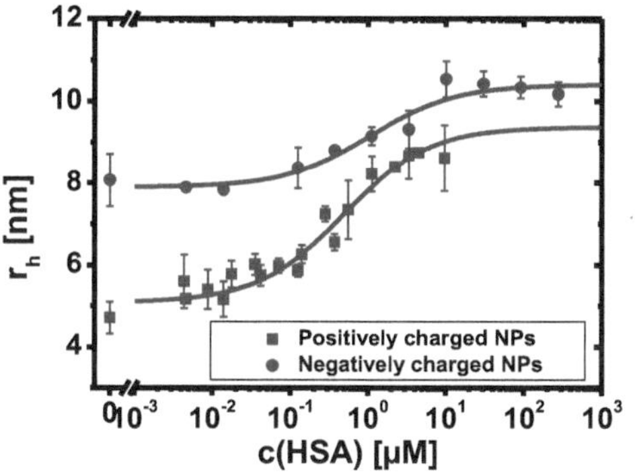

FIGURE 14 The hydrodynamic radius r_h of positively and negatively charged Au nanoparticles as recorded via fluorescence correlation spectroscopy in phosphate-buffered saline with increasing concentration of human serum albumin. Every data point consists of three independent measurements. *The figure is adopted from Hühn et al. [86] and has been performed by Stefan Brandholt from the group of Prof. Dr Ulrich Nienhaus.*

model [151]. With increasing HSA concentration, r_h increased in a sigmoidal manner until saturation was reached indicating the formation of a monolayer of proteins around the NP surface [142,143]. Data indicate that HSA adsorbed to the same amount to positively as to negatively charged NPs. This indicated that negatively and positively charged NPs after protein adsorption become more "similar," as both are coated by proteins and thus differences in their original surface chemistry are smeared out. Within the given uncertainty, one can estimate that the hydrodynamic diameters of the NPs possessing a protein corona are approximately 20 nm.

DLS measurements under the presence of proteins in the media did not deliver reliable results as their size is too similar to the size of the NPs. In case proteins were present in physiological conditions (e.g. within growth medium for cell culture), only the plain proteins were detected, overruling the scattering signal from the NP.

Besides the formation of a protein corona by proteins from biological media, these salt- and protein-containing media can also introduce colloidal instability to the NPs. In the following, it will be shown how this can be observed using UV–vis absorption spectrometry. Ions can attach to the NP surface and screen the surface charge [152,153]. This screening can lead to an attraction between NPs via van der Waals forces and thus to agglomeration. In case individual Au NPs agglomerate, their LSPR peak is no longer resolved clearly due to a coupling of the electronic wave functions, i.e., it flattens/broadens and can undergo a red shift [154]. Thus, changes in the LSPR peak can be attributed to interaction with ingredients of the medium. Some media of interest, which were used as example, are specified in Table 3. Extinction spectra of Au NPs without PDI in their shell were recorded after a blank spectrum of the entire medium had been taken. Spectra were taken every 5 min for 1 h starting 1 min after the addition of the Au NPs into the medium. The maximum net absorbance located at the LSPR peak was extracted and plotted against the time after addition of the Au NPs into the medium. Surprisingly, changes of the absorbance were observed in protein

TABLE 3 List of Media Employed for Stability Tests

Medium #							
1	2	3	4	5	6	7	8
Water[a]	PBS[b]	DMEM[c]	PBS	DMEM	PBS	DMEM	DMEM
–	–	–	1% P/S[d]	1% P/S	–	1% P/S	1% P/S
–	–	–	1% L-Glu[e]	1% L-Glu	–	1% L-Glu	1% L-Glu
–	–	–	–	–	800 μM BSA[f]	800 μM BSA	10% FBS[g]

[a]Fresh Milli-Q water.
[b]Phosphate-buffered saline.
[c]Dulbecco's modified Eagle's medium.
[d]Penicillin/streptomycin.
[e]L-glutamin.
[f]Bovine serum albumin.
[g]Fetal bovine serum.

containing media but without indications that allowed for a conclusion regarding claims of agglomeration (Figure 15).

Although the net absorbance was increasing by time for BSA containing media, no red shift was observed. Thus it can be concluded that an interaction of

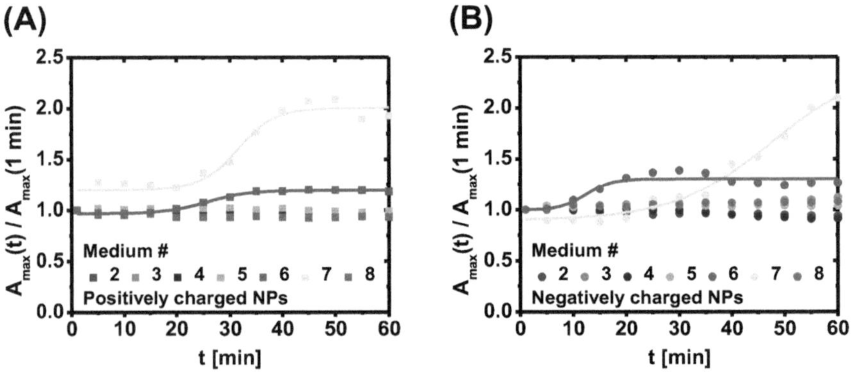

FIGURE 15 Time evolution of the maximum net absorbance of (A) positively and (B) negatively charged Au NPs at their LSPR peak in different media. The pure medium served as blank and the indicated numbers correspond to the compositions as depicted in Table 3. One minute after the addition of Au NPs into the medium, the first extinction spectrum was taken, then every 5 min for 1 h. The added volume of Au NP suspension was 10 times lower compared to the volume of present medium, so that the physicochemical properties of the medium were believed to be maintained. The absorbance was normalized to that after 1 min. A significant change of the absorbance was only observed for BSA containing media (#6 and #7). *The figure is adopted from Hühn et al. [86].*

NPs with BSA took place which influenced the absorbance and/or scattering characteristics of NPs, but did not lead to agglomeration. Indeed a sigmoidal behavior could be attributed to the evolution within media #6 and #7 in accordance to the FCS results (Figure 14), although HSA was employed in this case, indicating the formation of a protein corona. In both cases, the increase in the net absorbance was even stronger for the DMEM-based medium #7. The concentration of 800 μM BSA in media #6 and #7 ensured similar protein content in these media compared to medium #8, in which FBS provided the proteins. The other media (#2–#5) whether based on PBS or DMEM did not show any significant change in the absorbance profiles, what indicates that P/S and L-Glu did not influence the absorbance/scattering characteristics of the Au NPs.

Charge-dependent cell internalization experiments were also performed in protein-containing and protein-free media, #5, #7, and #8 (Table 3), to monitor how the presence of proteins affects NP uptake. For this purpose, 3T3 fibroblast cells were incubated with positively and negatively charged Au NPs with PDI in their shell. The amount of internalized Au NPs was quantified by verification of the cell-associated red fluorescence at various time points (Figure 16). Moreover, the fluorescence intensity was normalized considering the different amounts of fluorophores per polymer strand, the differences of the quantum yields, as well as of the brightness of a droplet of plain Au NPs under the microscope [86]. Thus, the graphs in Figure 16 allow for a quantitative comparison of internalization efficiency, in dependence of the sign of the charge of the NPs. It was found that positively charged Au NPs were taken up faster and to a greater extent than negatively charged one. This is in line with other studies dealing with charge-dependent NP internalization [155–158]. The difference in the uptake rate can be explained with the net negative surface charge of cell membranes [159–161]. Both negatively and positively charged Au NPs had in common that protein-associated uptake (media #7 and #8) was diminished compared to a protein-free incubation (medium #5). Also this finding is in line with previous studies [143,162,163].

To further verify these findings, murine C17.2 neural progenitor cells (NPCs) and human umbilical vein endothelial cells (HUVECs) were incubated by Dr Stefaan Soenen from the group of Prof. Dr Kevin Braeckmans with positively and negatively charged Au NPs and uptake affinities were analyzed by fluorescence-activated cell sorting (FACS). Figure 17 confirms that positively charged Au NPs were taken up to a greater extent whether (A) after 2 h or (B) after 24 h of incubation for both cell types. One has to note that FACS cannot distinguish between Au NPs which are internalized or adhered at the outer cell membrane [164], thus all cell-associated NPs were taken into account. Moreover, it was found that more NPs were associated with C17.2 NPCs than with HUVECs for both NP charges, although this difference was greater for the positively charged NPs. The cell-type-dependent uptake affinities may not only be attributed to the cell type itself but also reflect differences in the composition of the growth media. The NPs might have been colloidally

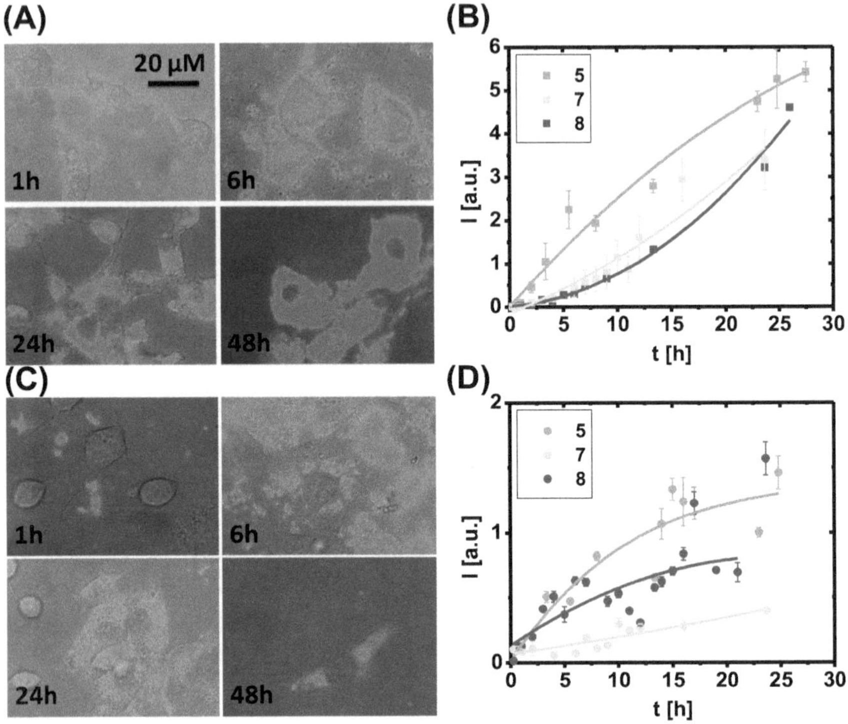

FIGURE 16 Illustrative fluorescence microscopy images of 3T3 fibroblast cells as incubated with (A) positively and (C) negatively charged Au NPs at various time points in medium #8 (c(Au nanoparticles (NPs)) = 5 nM). After normalization and background correction, the cell-associated fluorescence intensity was verified and plotted against the incubation time. (B) and (D) show the uptake evolution for positively and negatively charged Au NPs with quantitatively comparable intensity axes in different incubation media (#5, #7, and #8 according to Table 3). The uptake efficiency was higher for positively charged NPs and for protein-free medium #5 regardless of the charge. The lines in graphs (B) and (D) are guides to the eye. *The figure is adopted from Hühn et al. [86]. Uptake measurements were performed together with Karsten Kantner.*

more stable in the medium used for HUVECs with less serum content as compared to NPCs. Thus, agglomeration might have lead to the higher cell association observed for NPCs.

Results suggest that cell uptake abilities depend clearly on the NP surface charge, being in first order in the presented experiments, the only variable physicochemical parameter of the evaluated NP system. Also the growth medium composition plays a crucial role as the formation of a protein corona around NPs influences the uptake dynamics not least due to the biological identity NPs hence possess. Regarding cytotoxic effects, the positively charged NPs showed a clearer trend toward reduced cell viability as compared to negatively charged NPs, which is in line with their higher internalization [86].

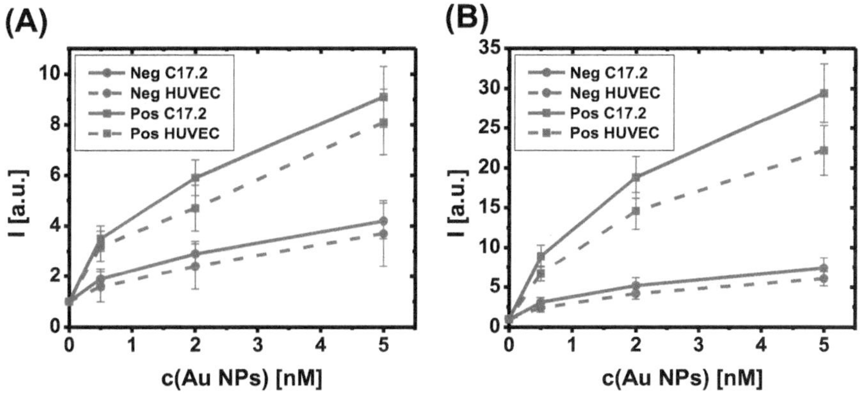

FIGURE 17 Uptake affinities as quantified with fluorescence-activated cell sorting for two different cell types (C17.2 neural progenitor cells and human umbilical vein endothelial cells) after (A) 2 h and (B) 24 h of incubation with positively and negatively charged fluorescent Au nanoparticles (NPs) at various NP concentrations. The fluorescence intensity I was normalized to that without NPs added (c(Au NPs) = 0 nM). *The figure is adopted from Hühn et al. [86]. These measurements were performed by Dr Stefaan Soenen from the group of Prof. Dr Kevin Braeckmans.*

Thus, it is a consistent finding that higher cytotoxicity for positively charged NPs correlates with a higher uptake rate.

2.6 Nanoparticles for Ion Sensing

Au NPs are an excellent platform as carrier system for biological purposes [7,165,166]. This suitability is mainly related to the chemical inertness of gold, the ease of synthesis of Au NPs of various sizes, the means of functionalization abilities, and the well-understood and tunable physicochemical properties [5]. This suggests that Au NPs could also be used as carriers of analyte-sensitive fluorophores for the purpose of intracellular analyte sensing. Regarding the detection of molecules or ions in solution, the immobilization of the analyte-sensitive fluorophore at the NP surface would offer two outstanding advantages compared to the free fluorophore: (1) the possibility to operate with intrinsically hydrophobic dyes as water solubility is mainly determined by the NP surface, and (2) the ability to immobilize several dyes or other functions at the NP surface, enabling the simultaneous detection of various specimens or introducing a normalization entity (e.g., a reference dye). Several studies have already provided insights into the NP/nanodevice-based sensing of ions as one type of important analyte within fluorescent probes [167–169]. However, experimental studies regarding ion-sensitive fluorophores attached to NP surfaces very often neglect that the NP, and especially its surface, influences the physical conditions in close proximity [82]. Herein the major factor is the charge of the NP surface as it directly influences the distribution of ions around the NP. An ion-sensitive fluorophore directly attached to the surface of an NP thus generates a fluorescence signal, which is different from the signal that the free dye would generate in the same solution [82,83]. In other words, the dye always probes its local environment regarding the distribution, hence concentration, of ions. Theoretical models include calculations based on the solution of the Poisson–Boltzmann equation [170–172] or more accurate tools based on the classical DFT [173–175]. However, there is still a lack of synergy between those theoretical models and recent experimental studies [82,83].

A possible method to partly overcome the influence of the NP surface on ion sensing is to use cross-linker molecules that serve as spacer between the NP surface and the analyte-sensitive dye molecules [82,83]. Hence the distance between NP surface and dye molecules is increased, so that the vicinity of the dyes is more similar to the bulk conditions regarding ion concentration. This strategy was effectively carried out for the case of a negatively charged NP surface and the Cl^--sensitive dye 2-[2-(6-methoxyquinolinium chloride) ethoxy]-ethanamine hydrochloride (amino-MQAE) in a previous study [83]. Herein the charge was generated by carboxylic groups which were part of the PMA-based amphiphilic polymer. The carboxylic groups also served as anchor points for efficient attachment of the dye molecules, whether directly linked to the surface or with a spacer molecule in-between via amide bond formation. In this study, it was shown that the fluorescence intensity, which was quenched with increasing Cl^- concentration, was lowered by a factor of 10 for the free dye, in case the bulk Cl^- concentration was increased from 0 to 140 μM. In contrast, the fluorescence intensity of the same dye linked directly to the polymer backbone at the Au NP surface was lowered only by a factor of approximately 1.7 for the same bulk conditions regarding Cl^- concentration. This was claimed to be due to a lower local Cl^- concentration in close proximity to the NP surface due to electrostatic repulsion. In case a spacer molecule, here PEG of varying molecular weight (i.e., length), had been inserted between the dye and the NP surface, the factor by which the fluorescence was lowered increased with increasing length of the spacer molecule.

To verify if the used sensor dye is also applicable with a positively charged NP, surface Au NPs were coated with an amphiphilic PTMAEMA-stat-PLMA-based polymer, similar to the earlier mentioned click-modified polymer (Figure 10). The anchor points (functionality z) were now represented by sulfhydril groups and the dye was attached using the cross-linker molecule sulfosuccinimidyl-4-(*N*-maleimidomethyl)cyclohexane-1-carboxylate (sulfo-SMCC) (Figure 18).

The length of the SMCC molecule was assumed to be approximately 0.83 nm, which resulted in an average distance of the dye to the NP surface of 0.5×0.83 nm $\approx$ 0.4 nm. In this case, the fluorescence intensity was lowered by a factor of approximately 1.7 for the same bulk Cl^- concentration as compared to the negatively charged Au NPs (Figure 19). As for the latter case, the same factor was gained for a distance of 0 as it was assumed that the same local Cl^- concentration is present. This fact and the finding that no saturation

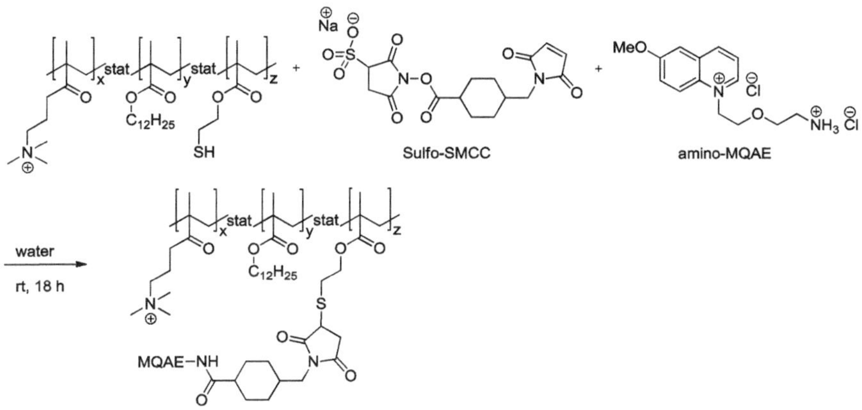

FIGURE 18 Reaction scheme for the functionalization of a positively charged amphiphilic polymer with amino-MQAE using the cross-linker molecule sulfo-succinimidyl-4-(*N*-maleimidomethyl) cyclohexane-1-carboxylate (SMCC). One equivalent of SH groups was stirred with 10 eq sulfo-SMCC and 100 eq amino-MQAE for 18 h at RT in water. After the conjugation, in which the maleimide group reacted with the SH group of the polymer and the *N*-hydroxysuccinimid residue was released under an amide bond formation with the amino function of the dye, the polymer was purified from unreacted cross-linker and dye molecules via centrifugation. After transfer of the polymer to chloroform, it was readily usable for the coating of Au nanoparticles. *The scheme is adopted from Jimenez de Aberasturi et al. [130]. The fluorophore amino-MQAE was synthesized by Fabian Dommershausen from the group of Prof. Dr Ulrich Koert. The polymers were synthesized by Christian Geidel from the group of Dr Markus Klapper.*

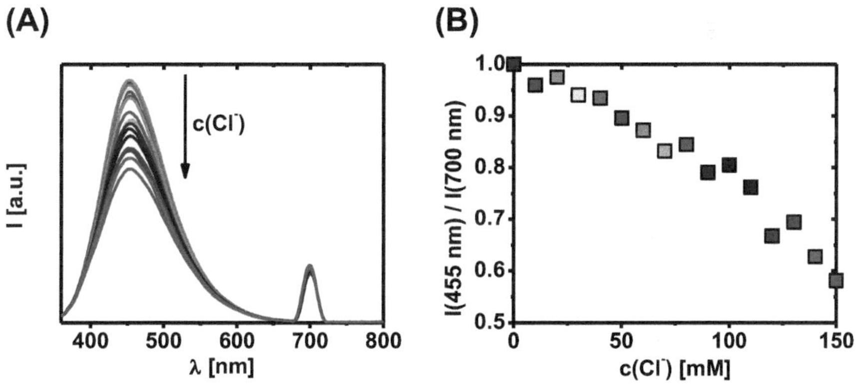

FIGURE 19 (A) Fluorescence intensity *I* of positively charged Au nanoparticles (NPs) functionalized with amino-MQAE via SMCC suspended in water for various Cl^- concentrations. The arrow indicates the direction of increasing Cl^- concentration, $\lambda_{exc} = 350$ nm, c(Au NPs) = 0.1 μM. The emission peak of amino-MQAE appeared at approximately 455 nm (15 nm red shifted compared to the free dye). The $2 \cdot \lambda_{exc}$ peak at 700 nm served for normalization. (B) The maximum intensities at approximately 455 nm (averaged from 450 to 460 nm) are plotted against the corresponding Cl^- concentration. First the intensities were divided by the corresponding $2 \cdot \lambda_{exc}$ peak intensity at 700 nm, then the data points were normalized in a way that the value 1 corresponded to 0 mM Cl^-. Thus, an intensity quenching by a factor of approximately 1.7 was observed. The sensitive region extended over the whole set of evaluated Cl^- concentrations. *The figure is adopted from Jimenez de Aberasturi et al. [130].*

was reached in the relevant bulk Cl^- concentration region lead to the conclusion that the present Cl^- ions accumulated around the surface and that the ion concentration decreased rapidly with increasing distance to the surface. To verify this assumption, positively charged Au NPs were equipped with the same dye, but with a spacer molecule of approximately the double length (1.76 nm), namely succinimidyl-([*N*-maleimidopropionamido]-2ethyleneglycol) ester (SM(PEG)$_2$) (Figure 20). The fluorescence intensity was lowered by a factor of approximately 1.3 and much higher Cl^- bulk concentrations were needed to reach a plateau (Figure 21). This finding supports the conclusion made above, as a significantly lower local Cl^- concentration needed to be present for the latter case with a longer spacer.

The presented effect that ions in solution are attracted or repulsed by colloidal NPs depending on the sign of charge does not compellingly have to be a disadvantage. It offers the possibility to tune the sensitivity of fluorescent dyes linked to NPs regarding the local concentration of ions. This has been shown in a previous study for the dye seminaphtharhodafluor (SNARF), which responds to changes in pH with a change in its emission profile [82]. Once linked to the surface of negatively charged Au NPs with spacer molecules of various lengths, the dye responds to its local pH, i.e., to the local concentration of hydronium ions. As the analyte is attracted by the NP surface, a low pH is suggested by SNARF molecules linked with a short spacer. Employment of a long spacer

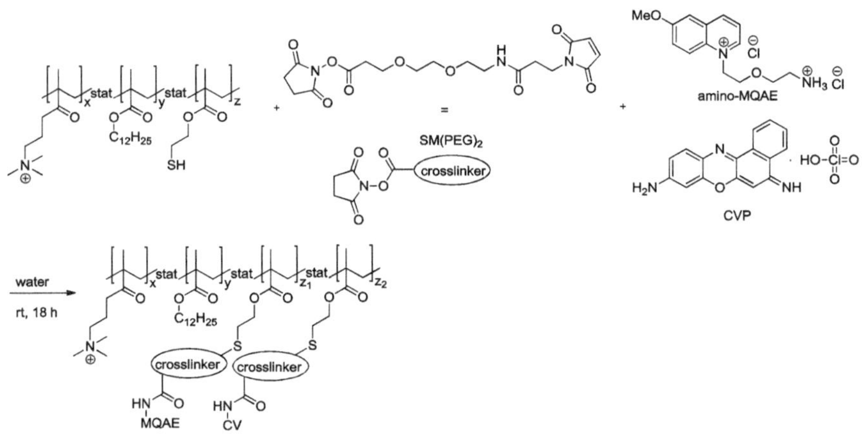

FIGURE 20 Reaction scheme for the functionalization of a positively charged amphiphilic polymer with amino-MQAE, using the cross-linker molecule SM(PEG)$_2$. One equivalent of SH groups were stirred with 10 eq SM(PEG)$_2$, 90 eq amino-MQAE, and 10 eq cresyl violet perchlorate (CVP) for 18 h at room temperature (RT) in phosphate-buffered saline/ethanol (1:2). Cresyl violet (CV) served as intrinsic fluorescent normalization probe. The remaining procedure was the same as compared to the conjugate with sulfo-SMCC (Figure 18). *The scheme is adopted from Jimenez de Aberasturi et al. [130]. The fluorophore amino-MQAE was synthesized by Fabian Dommershausen from the group of Prof. Dr Ulrich Koert. The polymers were synthesized by Christian Geidel from the group of Dr Markus Klapper.*

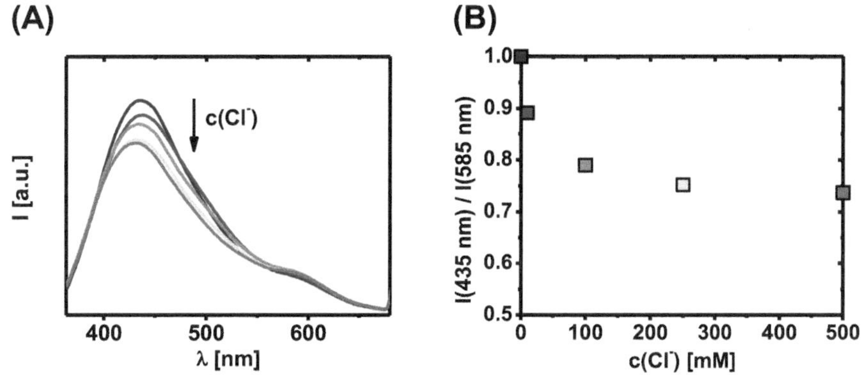

FIGURE 21 (A) Fluorescence intensity *I* of positively charged Au nanoparticle (NPs) functionalized with amino-MQAE and cresyl violet (CV) via $SM(PEG)_2$ suspended in water for various Cl^- concentrations. The arrow indicates the direction of increasing Cl^- concentration, $\lambda_{exc} = 350$ nm, c(Au NPs) = 0.1 μM. The emission peak of amino-MQAE appeared at approximately 435 nm (5 nm blue shifted compared to the free dye) and the emission peak of CV appeared as shoulder at approximately 585 nm. Although the emission of CV was partly overlapping with the emission of amino-MQAE, it served as a suitable control for normalization. (B) The maximum intensities at approximately 435 nm (averaged from 430 to 440 nm) are plotted against the corresponding Cl^- concentration. First the intensities were divided by the corresponding CV intensity at 585 nm (averaged from 580 to 590 nm), then the data points were normalized in a way that the value 1 corresponded to 0 mM Cl^-. Thus, an intensity quenching by a factor of approximately 1.3 was observed. *The figure is adopted from Jimenez de Aberasturi et al. [130].*

resulted in a read-out signal similar to that of the free dye. Thus the sensitive region of the sensor device can be tuned to be in a certain range of pH.

To verify the influence of a charged NP surface on the distribution of ions with valence 2, a Zn^{2+}-sensitive dye was attached to negatively charged NPs. Moreover, Zn is an important metabolic trace element, which motivates its detection in biological systems. The derivative **11** of reference Teolato et al. [176], in the following denoted as 4-aminomethyl-*N*-(6-methoxy-quinolin-8-yl)-benzenesulfonamide (AMQB), was chosen as it possesses a primary amino group for an effective attachment to the above-mentioned PMA-based carboxyl bearing amphiphilic polymer (Figure 22).

The fluorescence intensity of the free dye was amplified by a factor of ≥50 with increasing Zn^{2+} concentration, whereas the NP–dye conjugate showed an increased intensity by a factor of 6.5 (Figure 23), which may be related to the number of present dye molecules. Moreover, a plateau was reached for much smaller concentrations in case of the NP–dye conjugate, which goes hand-in-hand with the assumption that the Zn^{2+} ions are attracted by the NP surface, gaining a much higher local concentration at the surface compared to the bulk. Hence, a state of saturation was reached for smaller bulk concentrations.

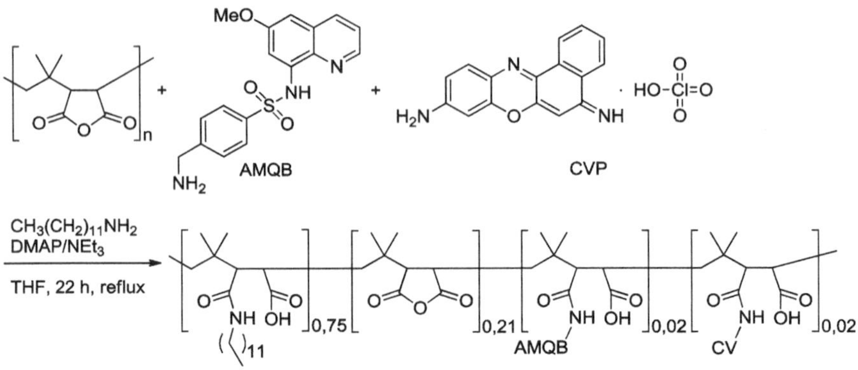

FIGURE 22 Reaction scheme for the functionalization of a negatively charged amphiphilic polymer with AMQB. One equivalent of maleic acid anhydride rings (monomers of PMA, $M_W = 6$ kDa) were stirred with 0.75 eq dodecylamine, 0.02 eq AMQB, 0.02 eq CVP, 0.1 eq 4-(dimethylamino)pyridine (DMAP), and 0.12 eq triethylamine (TEA) for 22 h at 80 °C under reflux in tetrahydrofuran (THF). After the conjugation, in which the amino groups of AMQB and CV formed amide bonds with the carboxylic groups of the polymer, the solvent was evaporated and the dry film was dissolved in chloroform. The polymer was used for coating without further purification. AMQB, 4-aminomethyl-*N*-(6-methoxy-quinolin-8-yl)-benzenesulfonamide; PMA, poly(isobutylene-*alt*-maleic anhydride); CV, cresyl violet. *The scheme is adopted from Jimenez de Aberasturi et al. [130]. The fluorophore AMQB was synthesized by Julia Schütte from the group of Prof. Dr Ulrich Koert.*

3. CONCLUSIONS AND OUTLOOK

In this report about how functional Au NPs can be synthesized and investigated regarding different fields of application is given. First, several general procedures regarding synthesis, surface modification, and purification were described, focusing in particular on one example, hydrophobically capped Au NPs. The as-synthesized Au NPs capped by hydrophobic ligands needed to be transferred into aqueous solution, which was achieved by a coating procedure, in which the inorganic Au core was enclosed within an amphiphilic polymer shell. Different scenarios regarding the functionalization ability of the polymer backbone were introduced based on the covalent attachment of molecules to one of the monomer groups. Additionally several purification steps including gel electrophoresis and SEC were introduced. Afterward three individual studies were presented ranging from the physicochemical characterization and the verification of interactions with biological systems to sensing applications.

pH titration was investigated as feasible method to characterize the physicochemical surface properties of polymer-coated colloidal Au NPs. It was found that the amphiphilic polymer, employed for the phase transfer of Au NPs, shows a polybasic behavior due to the formation of hydrogen bonds. Moreover, the pK_a values were found to be entirely higher compared to free-standing carboxylic acids. This finding is crucial as it points out the pH limits regarding colloidal stability of the NPs. Moreover, it allows for an estimation

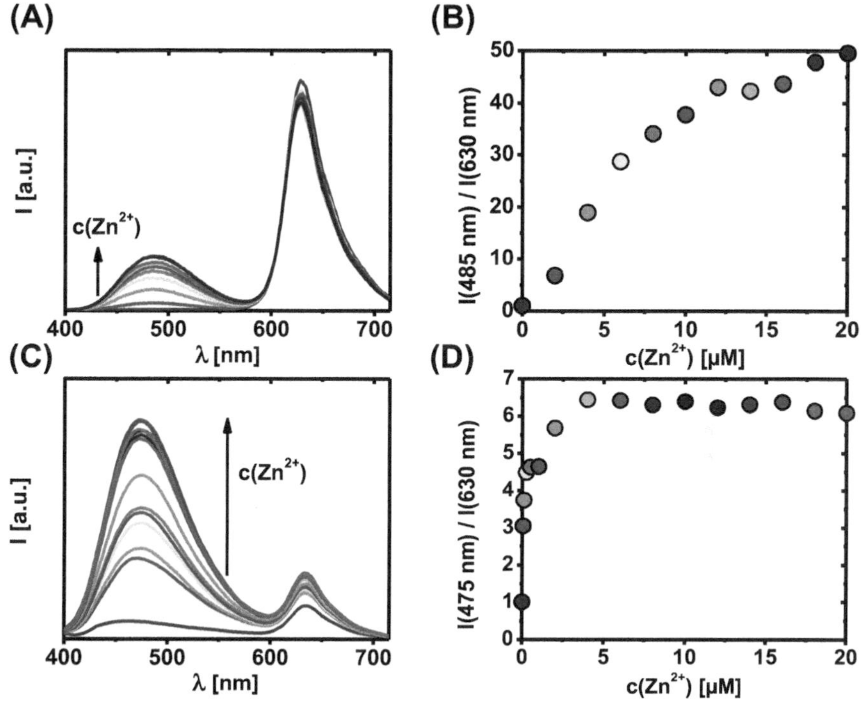

FIGURE 23 (A) Fluorescence intensity *I* of free 4-aminomethyl-*N*-(6-methoxy-quinolin-8-yl)-benzenesulfonamide (AMQB) and cresyl violet perchlorate (CVP) diluted in water/ethanol (1:1) for various Zn^{2+} concentrations. The arrow indicates the direction of increasing Zn^{2+} concentration, $\lambda_{exc} = 340$ nm. (B) The maximum intensities at approximately 485 nm (averaged from 480 to 490 nm) are plotted against the corresponding Zn^{2+} concentration. First the intensities were divided by the corresponding CVP intensity at 585 nm (averaged from 580 to 590 nm), then the data points were normalized in a way that the value 1 corresponded to 0 μM Zn^{2+}. Thus, an amplification of the intensity by a factor of ≥50 times was observed. The sensitive region extended over the whole set of evaluated Zn^{2+} concentrations. (C) Fluorescence intensity *I* of negatively charged Au NPs functionalized with AMQB and cresyl violet (CV) suspended in water for various Zn^{2+} concentrations. The arrow indicates the direction of increasing Zn^{2+} concentration, $\lambda_{exc} = 365$ nm, *c*(Au NPs) = 0.175 μM. The emission peak of AMQB appeared at approximately 475 nm (10 nm blue shifted compared to the free dye) and the emission peak of CV appeared at approximately 630 nm. Although the emission of CV was partly overlapping with the emission of AMQB, it served as a suitable control for normalization. (D) The maximum intensities at approximately 475 nm (averaged from 470 to 480 nm) are plotted against the corresponding Zn^{2+} concentration. First the intensities were divided by the corresponding CV intensity at 630 nm (averaged from 625 to 635 nm), then the data points were normalized in a way, that the value 1 corresponded to 0 μM Zn^{2+}. Thus, an amplification of the intensity by approximately 6.5 times was observed. The sensitive region extended over 0–5 μM. *The figure is adopted from Jimenez de Aberasturi et al. [130].*

of optimum conditions suitable for peptide-coupling reactions (like EDC coupling). As colloidal stability and EDC coupling are competing purposes, characterization via pH titration can be a good strategy to find a compromise between functionalization and stability.

Regarding the interaction of polymer-coated Au NPs with biological systems, a study in which two species of NPs were synthesized was explained. The fluorescence-labeled Au NPs differed in first order only in their net surface charge, which was verified by thorough purification and characterization. The interaction with proteins in biological media was quantified within several techniques including UV–vis absorption spectrometry and FCS. It was found that a protein corona forms around NPs, a phenomenon that entirely influences the behavior of NPs in biological systems due to the different physical (size and shape) and biological properties (the protein corona attributes a biological identity to the surface). Cell uptake experiments showed that positively charged NPs get internalized more efficiently compared to negatively charged NPs, probably mainly driven by the net negative charge of the cell membrane. The greater uptake efficiency was accompanied by greater cytotoxicity due to a higher intracellular NP concentration of positively charged NPs.

Additionally it was shown that Au NPs are a feasible platform for sensing applications. Positively charged NPs were equipped with a Cl^--sensitive dye and negatively charged NPs were equipped with a Zn^{2+}-sensitive dye. It can be advantageous to use NPs as template instead of the free dyes as also hydrophobic dyes can be employed and, moreover, the possibility for the implementation of further functionality is given (e.g., additional dyes as normalization). Furthermore, the NP core itself can carry certain functionality. Though a quantitative interpretation of the data seems premature, sensitivity was evidenced. These difficulties were also related to the lacking theory regarding ion distribution around colloidal NPs. Existing theories are representing scenarios, which might be too idealized to be able to render the complex NP system given.

ABBREVIATIONS

4AP 4-aminophenol
Amino-MQAE 2-[2-(6-methoxyquinolinium chloride)ethoxy]-ethanamine hydrochloride
ALP alkaline phosphatase
AMQB 4-aminomethyl-*N*-(6-methoxy-quinolin-8-yl)-benzenesulfonamide
BIIT bombardment induced ion transport
BSA bovine serum albumin
CE-MS capillary electrophoresis–mass spectrometry
CNT(s) carbon nanotube(s)
CV cresyl violet
CVP cresyl violet perchlorate
DFT density functional theory
DLS dynamic light scattering

DMAP 4-(dimethylamino)pyridine
DMEM Dulbecco's modified Eagle's medium
DNA deoxyribonucleic acid
EDC N-(3-dimethylaminopropyl)-N′-ethylcarbodiimide
EDTA ethylenediamintetraacetic acid
FACS fluorescence-activated cell sorting
FBS fetal bovine serum
FITC fluorescein 5-isothiocyanate
FRET Förster resonance energy transfer
HPLC high-performance liquid chromatography
HSA human serum albumin
HUVEC(s) human umbilical vein endothelial cell(s)
LBL layer-by-layer
LDA laser Doppler anemometry
L-Glu L-glutamin
LSPR localized surface plasmon resonance
MPA mercaptopropionic acid
MRI magnetic resonance imaging
MUA mercaptoundecanoic acid
NHS *N*-hydroxysuccinimid
NIR near infrared
NP(s) nanoparticle(s)
NPC(s) neural progenitor cell(s)
PAH poly(allylamine hydrochloride)
***p*APP** *p*-aminophenyl phosphate
PBS phosphate-buffered saline
PDI perylene tetracarboxylic diimide
PEG polyethylene glycol
PEM(s) polyelectrolyte multilayer(s)
PET positron emission tomography
PgMA poly(propargyle methacrylate)
PLMA poly(lauryl methacrylate)
PMA poly(isobutylene-*alt*-maleic anhydride)
PMAPHOS poly((2-(methacryloyloxy)ethyl)phosphonic acid)
P/S penicillin/streptomycin
PSS poly(sodium 4-styrenesulfonate)
PTMAEMA poly(*N*,*N*,*N*-trimethylammonium-2-ethyl methacrylate iodide)
QD(s) quantum dot(s)
RT room temperature
SEC size exclusion chromatography
SERS surface-enhanced Raman scattering
SMCC succinimidyl-4-(*N*-maleimidomethyl)cyclohexane-1-carboxylate
SM(PEG)$_2$ succinimidyl-([*N*-maleimidopropionamido]-2ethyleneglycol) ester
SNARF seminaphtharhodafluor
SPP surface plasmon polariton
SPR surface plasmon resonance
TBE Tris-borate-EDTA
TEA triethylamine

TEM transmission electron microscopy
TGA thioglycolic acid
THF tetrahydrofuran
TOAB tetraoctylammonium bromide
Tris tris(hydroxymethyl)aminomethane
UV ultraviolet
Vis visible

ACKNOWLEDGMENTS

In this article, figures from recently published manuscripts are replicated. The original data of some of the figures were obtained by Stefan Brandholt from the group of Ulrich Nienhaus, Christian Geidel from the group of Dr Markus Klapper, Fabian Dommershausen, and Julia Schütte from the group of Prof. Dr Ulrich Koert, and by Dr Stefaan Soenen from the group of Prof. Dr Kevin Braeckmans. The original research was supported by the German Research Foundation (SPP 1313 project PA 794/4-2) and the European Commission (project Nanognostics). The authors are also grateful to Dr Jose Maria Montenegro Martos and Jonas Hühn for proofreading.

REFERENCES

[1] H. Goesmann, C. Feldmann, Nanoparticulate functional materials, Angew. Chem. Int. Ed. 49 (2010) 1362–1395.

[2] B. Pelaz, S. Jaber, D. Jimenez de Aberasturi, V. Wulf, J.M. de la Fuente, J. Feldmann, H.E. Gaub, L. Josephson, C.R. Kagan, N.A. Kotov, L. Liz-Marzán, H. Mattoussi, P. Mulvaney, C.B. Murray, A.L. Rogach, P.S. Weiss, I. Willner, W.J. Parak, The state of nanoparticle-based nanoscience and biotechnology: progress, promises, and challenges, ACS Nano 6 (2012) 8468–8483.

[3] P. Rivera Gil, D. Hühn, L.L. del Mercato, D. Sasse, W.J. Parak, Nanopharmacy: inorganic nanoscale devices as vectors and active compounds, Pharmacol. Res. 62 (2010) 115–125.

[4] A.D. Maynard, R.J. Aitken, T. Butz, V. Colvin, K. Donaldson, G. Oberdörster, M.A. Philbert, J. Ryan, A. Seaton, V. Stone, S.S. Tinkle, L. Tran, N.J. Walker, D.B. Warheit, Safe handling of nanotechnology, Nature 444 (2006) 267–269.

[5] M.C. Daniel, D. Astruc, Gold nanoparticles: assembly, supramolecular chemistry, quantum-size-related properties, and applications toward biology, catalysis, and nanotechnology, Chem. Rev. 104 (2004) 293–346.

[6] P.K. Jain, X.H. Huang, I.H. El-Sayed, M.A. El-Sayed, Noble metals on the nanoscale: optical and photothermal properties and some applications in imaging, sensing, biology, and medicine, Acc. Chem. Res. 41 (2008) 1578–1586.

[7] R.A. Sperling, P. Rivera Gil, F. Zhang, M. Zanella, W.J. Parak, Biological applications of gold nanoparticles, Chem. Soc. Rev. 37 (2008) 1896–1908.

[8] E.C. Dreaden, A.M. Alkilany, X. Huang, C.J. Murphy, M.A. El-Sayed, The golden age: gold nanoparticles for biomedicine, Chem. Soc. Rev. 41 (2012) 2740–2779.

[9] G. Doria, J. Conde, B. Veigas, L. Giestas, C. Almeida, M. Assuncao, J. Rosa, P.V. Baptista, Noble metal nanoparticles for biosensing applications, Sensors 12 (2012) 1657–1687.

[10] S. Link, M.A. El-Sayed, Spectral properties and relaxation dynamics of surface plasmon electronic oscillations in gold and silver nanodots and nanorods, J. Phys. Chem. B 103 (1999) 8410–8426.

[11] A. Moores, F. Goettmann, The plasmon band in noble metal nanoparticles: an introduction to theory and applications, New J. Chem. 30 (2006) 1121–1132.
[12] S. Eustis, M.A. El-Sayed, Why gold nanoparticles are more precious than pretty gold: noble metal surface plasmon resonance and its enhancement of the radiative and nonradiative properties of nanocrystals of different shapes, Chem. Soc. Rev. 35 (2006) 209–217.
[13] K.A. Willets, R.P. Van Duyne, Localized surface plasmon resonance spectroscopy and sensing, Annu. Rev. Phys. Chem. 58 (2007) 267–297.
[14] S.M. Nie, S.R. Emery, Probing single molecules and single nanoparticles by surface-enhanced raman scattering, Science 275 (1997) 1102–1106.
[15] K. Kneipp, Y. Wang, H. Kneipp, L.T. Perelman, I. Itzkan, R. Dasari, M.S. Feld, Single molecule detection using surface-enhanced raman scattering (SERS), Phys. Rev. Lett. 78 (1997) 1667–1670.
[16] R. Elghanian, J.J. Storhoff, R.C. Mucic, R.L. Letsinger, C.A. Mirkin, Selective colorimetric detection of polynucleotides based on the distance-dependent optical properties of gold nanoparticles, Science 277 (1997) 1078–1081.
[17] J. Liu, Y. Lu, A colorimetric lead biosensor using DNAzyme-directed assembly of gold nanoparticles, J. Am. Chem. Soc. 125 (2003), 6642–6642.
[18] J.M. Slocik, J.S. Zabinski, D.M. Phillips, R.R. Naik, Colorimetric response of peptide-functionalized gold nanoparticles to metal ions, Small 4 (2008) 548–551.
[19] S. Chah, M.R. Hammond, R.N. Zare, Gold nanoparticles as a colorimetric sensor for protein conformational changes, Chem. Biol. 12 (2005) 323–328.
[20] T.K. Sau, C.J. Murphy, Seeded high yield synthesis of short Au nanorods in aqueous solution, Langmuir 20 (2004) 6414–6420.
[21] A.G. Skirtach, P. Karageorgiev, B.G. De Geest, N. Pazos-Perez, D. Braun, G.B. Sukhorukov, Nanorods as wavelength-selective absorption centers in the visible and near-infrared regions of the electromagnetic spectrum, Adv. Mater. 20 (2008) 506–510.
[22] W. Jo, K. Freedman, D.K. Yi, R.K. Bose, K.K.S. Lau, S.D. Solomon, M.J. Kim, Photon to thermal response of a single patterned gold nanorod cluster under near-infrared laser irradiation, Biofabrication 3 (2011) 1–7.
[23] A.O. Govorov, H.H. Richardson, Generating heat with metal nanoparticles, Nano Today 2 (2007) 30–38.
[24] X. Huang, I.H. El-Sayed, W. Qian, M.A. El-Sayed, Cancer cell imaging and photothermal therapy in the near-infrared region by using gold nanorods, J. Am. Chem. Soc. 128 (2006) 2115–2120.
[25] A.M. Alkilany, L.B. Thompson, S.P. Boulos, P.N. Sisco, C.J. Murphy, Gold nanorods: their potential for photothermal therapeutics and drug delivery, tempered by the complexity of their biological interactions, Adv. Drug Deliv. Rev. 64 (2012) 190–199.
[26] A.B.S. Bakhtiari, D. Hsiao, G.X. Jin, B.D. Gates, N.R. Branda, An efficient method based on the photothermal effect for the release of molecules from metal nanoparticle surfaces, Angew. Chem. Int. Ed. 48 (2009) 4166–4169.
[27] A.G. Skirtach, C. Dejugnat, D. Braun, A.S. Susha, W.J. Parak, H. Möhwald, G.B. Sukhorukov, The role of metal nanoparticles in remote release of encapsulated materials, Nano Lett. 5 (2005) 1371–1377.
[28] A. Muñoz Javier, P. del Pino, M.F. Bedard, A.G. Skirtach, D. Ho, G.B. Sukhorukov, C. Plank, W.J. Parak, Photoactivated release of cargo from the cavity of polyelectrolyte capsules to the cytosol of cells, Langmuir 24 (2008) 12517–12520.
[29] R. Weissleder, A clearer vision for in vivo imaging, Nat. Biotechnol. 19 (2001) 316–317.

[30] D. Hühn, A. Govorov, P. Rivera Gil, W.J. Parak, Photostimulated Au nanoheaters in polymer and biological media: characterization of mechanical destruction and boiling, Adv. Funct. Mater. 22 (2012) 294–303.

[31] R. Huschka, J. Zuloaga, M.W. Knight, L.V. Brown, P. Nordlander, N.J. Halas, Light-induced release of DNA from gold nanoparticles: nanoshells and nanorods, J. Am. Chem. Soc. 133 (2011) 12247–12255.

[32] L.R. Hirsch, R.J. Stafford, J.A. Bankson, S.R. Sershen, B. Rivera, R.E. Price, J.D. Hazle, N.J. Halas, J.L. West, Nanoshell-mediated near-infrared thermal therapy of tumors under magnetic resonance guidance, Proc. Natl. Acad. Sci. U.S.A. 100 (2003) 13549–13554.

[33] J. Chen, D. Wang, J. Xi, L. Au, A. Siekkinen, A. Warsen, Z.Y. Li, H. Zhang, Y. Xia, X. Li, Immuno gold nanocages with tailored optical properties for targeted photothermal destruction of cancer cells, Nano Lett. 7 (2007) 1318–1322.

[34] B. Pelaz, V. Grazu, A. Ibarra, C. Magen, P. del Pino, J.M. de la Fuente, Tailoring the synthesis and heating ability of gold nanoprisms for bioapplications, Langmuir 28 (2012) 8965–8970.

[35] J. Turkevich, P.C. Stevenson, J. Hillier, A study of the nucleation and growth processes in the synthesis of colloidal gold, Discuss. Faraday Soc. 11 (1951) 55–75.

[36] M. Brust, M. Walker, D. Bethell, D.J. Schiffrin, R. Whyman, Synthesis of thiol-derivatized gold nanoparticles in a 2-phase liquid-liquid system, J. Chem. Soc. Chem. Commun. (1994) 801–802.

[37] S.D. Perrault, W.C.W. Chan, Synthesis and surface modification of highly monodispersed, spherical gold nanoparticles of 50–200 nm, J. Am. Chem. Soc. 131 (2009) 17042–17043.

[38] T. Sakai, P. Alexandridis, Mechanism of gold metal ion reduction, nanoparticle growth and size control in aqueous amphiphilic block copolymer solutions at ambient conditions, J. Phys. Chem. B 109 (2005) 7766–7777.

[39] F. Mafune, J. Kohno, Y. Takeda, T. Kondow, H. Sawabe, Formation of gold nanoparticles by laser ablation in aqueous solution of surfactant, J. Phys. Chem. B 105 (2001) 5114–5120.

[40] M. Rai, A. Yadav, A. Gade, Silver nanoparticles as a new generation of antimicrobials, Biotechnol. Adv. 27 (2009) 76–83.

[41] J. Morones, J. Elechiguerra, A. Camacho, K. Holt, J. Kouri, J. Ram, M. Yacaman, The bactericidal effect of silver nanoparticles, Nanotechnology 16 (2005) 2346–2352.

[42] M.J. Hajipour, K.M. Fromm, A. Akbar Ashkarran, D. Jimenez de Aberasturi, I. R. d. Larramendi, T. Rojo, V. Serpooshan, W.J. Parak, M. Mahmoudi, Antibacterial properties of nanoparticles, Trends Biotechnol. 30 (2012) 511.

[43] S.W.P. Wijnhoven, W.J.G.M. Peijnenburg, C.A. Herberts, W.I. Hagens, A.G. Oomen, E.H.W. Heugens, B. Roszek, J. Bisschops, I. Gosens, D. Van de Meent, S. Dekkers, W.H. De Jong, M. Van Zijverden, A. Sips, R.E. Geertsma, Nano-silver – a review of available data and knowledge gaps in human and environmental risk assessment, Nanotoxicology 3 (2009) 109–138.

[44] A. Ravindran, P. Chandran, S.S. Khan, Biofunctionalized silver nanoparticles: advances and prospects, Colloids Surf. B 105 (2013) 342–352.

[45] A.J. Haes, R.P. Van Duyne, A nanoscale optical biosensor: sensitivity and selectivity of an approach based on the localized surface plasmon resonance spectroscopy of triangular silver nanoparticles, J. Am. Chem. Soc. 124 (2002) 10596–10604.

[46] A.D. McFarland, R.P. Van Duyne, Single silver nanoparticles as real-time optical sensors with zeptomole sensitivity, Nano Lett. 3 (2003) 1057–1062.

[47] Y. Bao, C. Zhong, D. Vu, J. Temirov, R. Dyer, J. Martinez, Nanoparticle-free synthesis of fluorescent gold nanoclusters at physiological temperature, J. Phys. Chem. C 111 (2007) 12194–12198.

[48] H.W. Duan, S.M. Nie, Etching colloidal gold nanocrystals with hyperbranched and multivalent polymers: a new route to fluorescent and water-soluble atomic clusters, J. Am. Chem. Soc. 129 (2007) 2412–2413.

[49] C.-A.J. Lin, T.-Y. Yang, C.-H. Lee, S.H. Huang, R.A. Sperling, M. Zanella, J.K. Li, J.-L. Shen, H.-H. Wang, H.-I. Yeh, W.J. Parak, W.H. Chang, Synthesis, characterization, and bioconjugation of fluorescent gold nanoclusters toward biological labeling applications, ACS Nano 3 (2009) 395–401.

[50] S. Huang, C. Pfeiffer, J. Hollmann, S. Friede, J.J.-C. Chen, A. Beyer, K. Volz, W. Heimbrodt, J.M. Montenegro Martos, W. Chang, W.J. Parak, Synthesis and characterization of colloidal fluorescent silver nanoclusters, Langmuir 28 (2012) 8915–8919.

[51] L. Shang, N. Azadfar, F. Stockmar, W. Send, V. Trouillet, M. Bruns, D. Gerthsen, G.U. Nienhaus, One-pot synthesis of near-infrared fluorescent gold clusters for cellular fluorescence lifetime imaging, Small 7 (2011) 2614–2620.

[52] Y. Sun, Y. Xia, Shape-Controlled synthesis of gold and silver nanoparticles, Science 298 (2002) 2176–2179.

[53] A. Panacek, L. Kvitek, R. Prucek, M. Kolar, R. Vecerova, N. Pizurova, V.K. Sharma, T. Nevecna, R. Zboril, Silver colloid nanoparticles: synthesis, characterization, and their antibacterial activity, J. Phys. Chem. B 110 (2006) 16248–16253.

[54] A.I. Ekimov, A.L. Efros, A.A. Onushchenko, Quantum size effect in semiconductor microcrystals, Solid State Commun. 88 (1993) 947–950.

[55] L. Spanhel, M. Haase, H. Weller, A. Henglein, Photochemistry of colloidal semiconductors. 20. Surface modification and stability of strong luminescing CdS particles, J. Am. Chem. Soc. 109 (1987) 5649–5655.

[56] A.P. Alivisatos, Semiconductor clusters, nanocrystals, and quantum dots, Science 271 (1996) 933–937.

[57] W.J. Parak, L. Manna, T. Nann, Fundamental principles of quantum dots, in: G. Schmid (Ed.), Nanotechnology, Principles and Fundamentals, vol. 1, WILEY-VCH, 2008, pp. 73–96.

[58] W.J. Parak, L. Manna, F.C. Simmel, D. Gerion, P. Alivisatos, Quantum dots, in: G. Schmid (Ed.), Nanoparticles: From Theory to Application, second ed., Wiley-VCH Verlag GmbH & Co. KGaA, Weinheim, 2010, pp. 3–47.

[59] M.J. Bruchez, M. Moronne, P. Gin, S. Weiss, A.P. Alivisatos, Semiconductor nanocrystals as fluorescent biological labels, Science 281 (1998) 2013–2016.

[60] W.C.W. Chan, S. Nie, Quantum dot bioconjugates for ultrasensitive nonisotopic detection, Science 281 (1998) 2016–2018.

[61] X. Michalet, F.F. Pinaud, L.A. Bentolila, J.M. Tsay, S. Doose, J.J. Li, G. Sundaresan, A.M. Wu, S.S. Gambhir, S. Weiss, Quantum dots for live cells, in vivo imaging, and diagnostics, Science 307 (2005) 538–544.

[62] I.L. Medintz, H.T. Uyeda, E.R. Goldman, H. Mattoussi, Quantum dot bioconjugates for imaging, labelling and sensing, Nat. Mater. 4 (2005) 435–446.

[63] M. Dahan, P. Alivisatos, W.J. Parak, Quantum dots: inorganic fluorescent probes for single-molecule tracking experiments in live cells, in: C. Bräuchle, D.C. Lamb, J. Michaelis (Eds.), Single Particle Tracking and Single Molecule Energy Transfer, WILEY-VCH, Weinheim, 2009.

[64] I.L. Medintz, A.R. Clapp, H. Mattoussi, E.R. Goldman, B. Fisher, J.M. Mauro, Self-assembled nanoscale biosensors based on quantum dot FRET donors, Nat. Mater. 2 (2003) 630–638.

[65] G. Gerion, F. Pinaud, S.C. Williams, W.J. Parak, D. Zanchet, S. Weiss, A.P. Alivisatos, Synthesis and properties of biocompatible water-soluble silica-coated CdSe/ZnS semiconductor quantum dots, J. Phys. Chem. B 105 (2001) 8861–8871.

[66] G. Vastola, Y.W. Zhang, V.B. Shenoy, Experiments and modeling of alloying in self-assembled quantum dots, Curr. Opin. Solid State Mater. Sci. 16 (2012) 64–70.

[67] Q.A. Pankhurst, J. Connolly, S.K. Jones, J. Dobson, Applications of magnetic nanoparticles in biomedicine, J. Phys. D: Appl. Phys. 36 (2003) R167–R181.

[68] A.K. Gupta, M. Gupta, Synthesis and surface engineering of iron oxide nanoparticles for biomedical applications, Biomaterials 26 (2005) 3995–4021.

[69] A.H. Lu, E.L. Salabas, F. Schuth, Magnetic nanoparticles: synthesis, protection, functionalization, and application, Angew. Chem. Int. Ed. 46 (2007) 1222–1244.

[70] S. Laurent, D. Forge, M. Port, A. Roch, C. Robic, L. Vander Elst, R.N. Muller, Magnetic iron oxide nanoparticles: synthesis, stabilization, vectorization, physicochemical characterizations, and biological applications, Chem. Rev. 108 (2008) 2064–2110.

[71] P. Rivera Gil, D. Jimenez de Aberasturi, V. Wulf, B. Pelaz, P. del Pino, Y. Zhao, J. de la Fuente, I. Ruiz de Larramendi, T. Rojo, X.-J. Liang, W.J. Parak, The challenge to relate the physicochemical properties of colloidal nanoparticles to their cytotoxicity, Acc. Chem. Res. 46 (2013) 743–749.

[72] T. Pellegrino, S. Kudera, T. Liedl, A.M. Javier, L. Manna, W.J. Parak, On the development of colloidal nanoparticles towards multifunctional structures and their possible use for biological applications, Small 1 (2005) 48–63.

[73] W.P. Wuelfing, S.M. Gross, D.T. Miles, R.W. Murray, Nanometer gold clusters protected by surface-bound monolayers of thiolated poly(ethylene glycol) polymer electrolyte, J. Am. Chem. Soc. 120 (1998) 12696–12697.

[74] C.S. Weisbecker, M.V. Merritt, G.M. Whitesides, Molecular self-assembly of aliphatic thiols on gold colloids, Langmuir 12 (1996) 3763–3772.

[75] J.C. Love, L.A. Estroff, J.K. Kriebel, R.G. Nuzzo, G.M. Whitesides, Self-assembled monolayers of thiolates on metals as a form of nanotechnology, Chem. Rev. 105 (2005) 1103–1169.

[76] G. Charron, D. Hühn, A. Perrier, L. Cordier, C.J. Pickett, T. Nann, W.J. Parak, On the use of pH titration to quantitatively characterize colloidal nanoparticles, Langmuir 28 (2012) 15141–15149.

[77] T. Sakura, T. Takahashi, K. Kataoka, Y. Nagasaki, One-pot preparation of mono-dispersed and physiologically stabilized gold colloid, Colloid Polym. Sci. 284 (2005) 97–101.

[78] C.-A.J. Lin, R.A. Sperling, J.K. Li, T.-Y. Yang, P.-Y. Li, M. Zanella, W.H. Chang, W.J. Parak, Design of an amphiphilic polymer for nanoparticle coating and functionalization, Small 4 (2008) 334–341.

[79] C. Geidel, S. Schmachtel, A. Riedinger, C. Pfeiffer, K. Müllen, M. Klapper, W.J. Parak, A general synthetic approach for obtaining cationic and anionic inorganic nanoparticles *via* encapsulation in amphiphilic copolymers, Small 7 (2011) 2929–2934.

[80] F. Zhang, E. Lees, F. Amin, P. Rivera_Gil, F. Yang, P. Mulvaney, W.J. Parak, Polymer-coated nanoparticles: a universal tool for biolabelling experiments, Small 7 (2011) 3113–3127.

[81] N. Erathodiyil, J.Y. Ying, Functionalization of inorganic nanoparticles for bioimaging applications, Acc. Chem. Res. 44 (2011) 925–935.

[82] F. Zhang, Z. Ali, F. Amin, A. Feltz, M. Oheim, W.J. Parak, Ion and pH sensing with colloidal nanoparticles: influence of surface charge on sensing and colloidal properties, ChemPhysChem 11 (2010) 730–735.

[83] A. Riedinger, F. Zhang, F. Dommershausen, C. Röcker, S. Brandholt, G.U. Nienhaus, U. Koert, W.J. Parak, Ratiometric optical sensing of chloride ions with organic fluorophore – gold nanoparticle hybrids: a systematic study of distance dependency and the influence of surface charge, Small 6 (2010) 2590–2597.

[84] M. Liong, J. Lu, M. Kovochich, T. Xia, S.G. Ruehm, A.E. Nel, F. Tamanoi, J.I. Zink, Multifunctional inorganic nanoparticles for imaging, targeting, and drug delivery, ACS Nano 2 (2008) 889–896.

[85] Z. Ali, A.Z. Abbasi, F. Zhang, P. Arosio, A. Lascialfari, M.F. Casula, A. Wenk, W. Kreyling, R. Plapper, M. Seidel, R. Niessner, J. Knoll, A. Seubert, W.J. Parak, Multifunctional nanoparticles for dual imaging, Anal. Chem. 83 (2011) 2877–2882.

[86] D. Hühn, K. Kantner, C. Geidel, S. Brandholt, I. De Cock, S.J.H. Soenen, P. Rivera Gil, J.-M. Montenegro, K. Braeckmans, K. Müllen, G.U. Nienhaus, M. Klapper, W.J. Parak, Polymer-coated nanoparticles interacting with proteins and cells: focusing on the sign of the net charge, ACS Nano 7 (2013) 3253–3263.

[87] P. Alivisatos, The use of nanocrystals in biological detection, Nat. Biotechnol. 22 (2004) 47–51.

[88] J. Cheon, J.-H. Lee, Synergistically integrated nanoparticles as multimodal probes for nanobiotechnology, Acc. Chem. Res. 41 (2008) 1630–1640.

[89] M. Nahrendorf, H.W. Zhang, S. Hembrador, P. Panizzi, D.E. Sosnovik, E. Aikawa, P. Libby, F.K. Swirski, R. Weissleder, Nanoparticle PET-CT imaging of macrophages in inflammatory atherosclerosis, Circulation 117 (2008) 379–387.

[90] Y.W. Jun, J.H. Lee, J. Cheon, Chemical design of nanoparticle probes for high-performance magnetic resonance imaging, Angew. Chem. Int. Ed. 47 (2008) 5122–5135.

[91] C.M. Cobley, J.Y. Chen, E.C. Cho, L.V. Wang, Y.N. Xia, Gold nanostructures: a class of multifunctional materials for biomedical applications, Chem. Soc. Rev. 40 (2011) 44–56.

[92] E. Boisselier, D. Astruc, Gold nanoparticles in nanomedicine: preparations, imaging, diagnostics, therapies and toxicity, Chem. Soc. Rev. 38 (2009) 1759–1782.

[93] Z.X. Wang, L.N. Ma, Gold nanoparticle probes, Coord. Chem. Rev. 253 (2009) 1607–1618.

[94] Y.W. Lin, C.C. Huang, H.T. Chang, Gold nanoparticle probes for the detection of mercury, lead and copper ions, Analyst 136 (2011) 863–871.

[95] D. Li, A. Wieckowska, I. Willner, Optical analysis of Hg^{2+} ions by oligonucleotide-gold-nanoparticle hybrids and DNA-based machines, Angew. Chem. Int. Ed. 47 (2008) 3927–3931.

[96] X.J. Xue, F. Wang, X.G. Liu, One-step, room temperature, colorimetric detection of mercury (Hg^{2+}) using DNA/nanoparticle conjugates, J. Am. Chem. Soc. 130 (2008) 3244–3245.

[97] H. Wang, Y.X. Wang, J.Y. Jin, R.H. Yang, Gold nanoparticle-based colorimetric and "turn-on" fluorescent probe for mercury(II) ions in aqueous solution, Anal. Chem. 80 (2008) 9021–9028.

[98] J.P. Xie, Y.G. Zheng, J.Y. Ying, Highly selective and ultrasensitive detection of Hg(2+) based on fluorescence quenching of Au nanoclusters by Hg(2+)-Au(+) interactions, Chem. Commun. 46 (2010) 961–963.

[99] U.H.F. Bunz, V.M. Rotello, Gold nanoparticle-fluorophore complexes: sensitive and discerning "noses" for biosystems sensing, Angew. Chem. Int. Ed. 49 (2010) 3268–3279.

[100] X. Liu, Q. Dai, L. Austin, J. Coutts, G. Knowles, J.H. Zou, H. Chen, Q. Huo, A one-step homogeneous immunoassay for cancer biomarker detection using gold nanoparticle probes coupled with dynamic light scattering, J. Am. Chem. Soc. 130 (2008) 2780–2782.

[101] C.D. Medley, J.E. Smith, Z. Tang, Y. Wu, S. Bamrungsap, W.H. Tan, Gold nanoparticle-based colorimetric assay for the direct detection of cancerous cells, Anal. Chem. 80 (2008) 1067–1072.

[102] W. Zhao, M.A. Brook, Y.F. Li, Design of gold nanoparticle-based colorimetric biosensing assays, ChemBioChem 9 (2008) 2363–2371.

[103] J.M. Pingarron, P. Yanez-Sedeno, A. Gonzalez-Cortes, Gold nanoparticle-based electrochemical biosensors, Electrochim. Acta 53 (2008) 5848–5866.

[104] Z.X. Zhang, B. Yan, K.L. Liu, Y.P. Liao, H.W. Liu, CE-MS analysis of heroin and its basic impurities using a charged polymer-protected gold nanoparticle-coated capillary, Electrophoresis 30 (2009) 379–387.

[105] S. Carregal-Romero, E. Caballero-Diaz, L. Beqa, A.M. Abdelmonem, M. Ochs, D. Hühn, B.S. Suau, M. Valcarcel, W.J. Parak, Multiplexed sensing and imaging with colloidal nano- and microparticles, Annu. Rev. Anal. Chem. 6 (2013) 53–81.

[106] C.Y. Zhang, L.W. Johnson, Single quantum-dot-based aptameric nanosensor for cocaine, Anal. Chem. 81 (2009) 3051–3055.

[107] E.A. Jares-Erijman, T.M. Jovin, FRET imaging, Nat. Biotechnol. 21 (2003) 1387–1395.

[108] M. Sanles-Sobrido, L. Rodriguez-Lorenzo, S. Lorenzo-Abalde, A. Gonzalez-Fernandez, M.A. Correa-Duarte, R.A. Alvarez-Puebla, L.M. Liz-Marzan, Label-free SERS detection of relevant bioanalytes on silver-coated carbon nanotubes: the case of cocaine, Nanoscale 1 (2009) 153–158.

[109] X.M. Qian, S.M. Nie, Single-molecule and single-nanoparticle SERS: from fundamental mechanisms to biomedical applications, Chem. Soc. Rev. 37 (2008) 912–920.

[110] J.F. Li, Y.F. Huang, Y. Ding, Z.L. Yang, S.B. Li, X.S. Zhou, F.R. Fan, W. Zhang, Z.Y. Zhou, D.Y. Wu, B. Ren, Z.L. Wang, Z.Q. Tian, Shell-isolated nanoparticle-enhanced Raman spectroscopy, Nature 464 (2010) 392–395.

[111] D.K. Lim, K.S. Jeon, H.M. Kim, J.M. Nam, Y.D. Suh, Nanogap-engineerable Raman-active nanodumbbells for single-molecule detection, Nat. Mater. 9 (2010) 60–67.

[112] R.A. Alvarez-Puebla, L.M. Liz-Marzan, SERS-based diagnosis and biodetection, Small 6 (2010) 604–610.

[113] Y.C. Cao, R. Jin, C.A. Mirkin, Nanoparticles with Raman spectroscopic fingerprints for DNA and RNA detection, Science 297 (2002) 1536–1540.

[114] X.M. Qian, X.H. Peng, D.O. Ansari, Q. Yin-Goen, G.Z. Chen, D.M. Shin, L. Yang, A.N. Young, M.D. Wang, S.M. Nie, In vivo tumor targeting and spectroscopic detection with surface-enhanced raman nanoparticle tags, Nat. Biotechnol. 26 (2008) 83–90.

[115] C. Kirchner, T. Liedl, S. Kudera, T. Pellegrino, A. Muñoz Javier, H.E. Gaub, S. Stölzle, N. Fertig, W.J. Parak, Cytotoxicity of colloidal CdSe and CdSe/ZnS nanoparticles, Nano Lett. 5 (2005) 331–338.

[116] M. Tsoli, H. Kuhn, W. Brandau, H. Esche, G. Schmid, Cellular uptake and toxicity of Au_{55} clusters, Small 1 (2005) 841–844.

[117] K.W. Powers, M. Palazuelos, B.M. Moudgil, S.M. Roberts, Characterization of the size, shape, and state of dispersion of nanoparticles for toxicological studies, Nanotoxicology 1 (2007) 42–51.

[118] G. Schmid, The relevance of shape and size of Au_{55} clusters, Chem. Soc. Rev. 37 (2008) 1909–1930.

[119] A. Albanese, P.S. Tang, W.C.W. Chan, The effect of nanoparticle size, shape, and surface chemistry on biological systems, in: Annual Review of Biomedical Engineering, vol. 14, vol. 14, Annual Reviews, Palo Alto, 2012, pp. 1–16.

[120] P. Rivera Gil, G. Oberdorster, A. Elder, V.F. Puntes, W.J. Parak, Correlating physico-chemical with toxicological properties of nanoparticles: the present and the future, ACS Nano 4 (2010) 5527–5531.

[121] N. Lewinski, V. Colvin, R. Drezek, Cytotoxicity of nanoparticles, Small 4 (2008) 26–49.

[122] W.H. De Jong, W.I. Hagens, P. Krystek, M.C. Burger, A.J. Sips, R.E. Geertsma, Particle size-dependent organ distribution of gold nanoparticles after intravenous administration, Biomaterials 29 (2008) 1912–1919.

[123] R. Gref, M. Lück, P. Quellec, M. Marchand, E. Dellacherie, S. Harnisch, T. Blunk, R.H. Müller, 'Stealth' corona-core nanoparticles surface modified by polyethylene glycol (PEG): influences of the corona (PEG chain length and surface density) and of the core composition on phagocytic uptake and plasma protein adsorption, Colloids Surf. B 18 (2000) 301–313.

[124] B. Baroli, M.G. Ennas, F. Loffredo, M. Isola, R. Pinna, M.A. Lopez-Quintela, Penetration of metallic nanoparticles in human full-thickness skin, J. Invest. Dermatol. 127 (2007) 1701–1712.

[125] L.K. Limbach, P. Wick, P. Manser, R.N. Grass, A. Bruinink, W.J. Stark, Exposure of engineered nanoparticles to human lung epithelial cells: influence of chemical composition and catalytic activity on oxidative stress, Environ. Sci. Technol. 41 (2007) 4158–4163.

[126] S.D. Perrault, C. Walkey, T. Jennings, H.C. Fischer, W.C.W. Chan, Mediating tumor targeting efficiency of nanoparticles through design, Nano Lett. 9 (2009) 1909–1915.

[127] K. Ulbrich, T. Hekmatara, E. Herbert, J. Kreuter, Transferrin- and transferrin-receptor-antibody-modified nanoparticles enable drug delivery across the blood–brain barrier (BBB), Eur. J. Pharm. Biopharm. 71 (2009) 251–256.

[128] B. Nowack, T.D. Bucheli, Occurrence, behavior and effects of nanoparticles in the environment, Environ. Pollut. 150 (2007) 5–22.

[129] M.N. Moore, Do nanoparticles present ecotoxicological risks for the health of the aquatic environment? Environ. Int. 32 (2006) 967–976.

[130] D. Jimenez de Aberasturi, D. Hühn, J. Hühn, L. Beqa, I. Ruiz de Larramendi, R. Pinedo, T. Rojo, J.-M. Montenegro, S. Carregal-Romero, C. Geidel, F. Dommershausen, J. Schütte, M. Klapper, U. Koert, K. Müllen, W. J. Parak, Ion selective ligands – how colloidal nano- and microparticles introduce new functionality, to be submitted for publication.

[131] X.O. Liu, M. Atwater, J.H. Wang, Q. Huo, Extinction coefficient of gold nanoparticles with different sizes and different capping ligands, Colloids Surf. B 58 (2007) 3–7.

[132] R.A. Sperling, W.J. Parak, Surface modification, functionalization and bioconjugation of colloidal inorganic nanoparticles, Phil. Trans. R. Soc. A 368 (2010) 1333–1383.

[133] T. Pellegrino, L. Manna, S. Kudera, T. Liedl, D. Koktysh, A.L. Rogach, S. Keller, J. Rädler, G. Natile, W.J. Parak, Hydrophobic nanocrystals coated with an amphiphilic polymer shell: a general route to water soluble Nanocrystals, Nano Lett. 4 (2004) 703–707.

[134] M.T. Fernández-Argüelles, A. Yakovlev, R.A. Sperling, C. Luccardini, S. Gaillard, A.S. Medel, J.-M. Mallet, J.-C. Brochon, A. Feltz, M. Oheim, W.J. Parak, Synthesis and characterization of polymer-coated quantum dots with integrated acceptor dyes as FRET-based nanoprobes, Nano Lett. 7 (2007) 2613–2617.

[135] T. Pellegrino, R.A. Sperling, A.P. Alivisatos, W.J. Parak, Gelelectrophoresis of gold-DNA nanoconjugates, J. Biomed. Biotechnol. (2007) 1–9. Article ID 26796.

[136] A.E. Nel, L. Madler, D. Velegol, T. Xia, E.M.V. Hoek, P. Somasundaran, F. Klaessig, V. Castranova, M. Thompson, Understanding biophysicochemical interactions at the nano-bio interface, Nat. Mater. 8 (2009) 543–557.

[137] Y. Pan, A. Leifert, D. Ruau, S. Neuss, J. Bornemann, G. Schmid, W. Brandau, U. Simon, W. Jahnen-Dechent, Gold nanoparticles of diameter 1.4 nm trigger necrosis by oxidative stress and mitochondrial damage, Small 5 (2009) 2067–2076.

[138] G.T. Hermanson, Bioconjugate Techniques, second ed., Academic Press, San Diego, 2008, p. 1323.

[139] N. Nakajima, Y. Ikada, Mechanism of amide formation by carbodiimide for bioconjugation in aqueous media, Bioconjug. Chem. 6 (1995) 123–130.

[140] N.R. Jana, X. Peng, Single-phase and gram-scale routes toward nearly monodisperse Au and other noble metal nanocrystals, J. Am. Chem. Soc. 125 (2003) 14280–14281.

[141] T. Cedervall, I. Lynch, S. Lindman, T. Berggård, E. Thulin, H. Nilsson, K.A. Dawson, S. Linse, Understanding the nanoparticle-protein corona using methods to quantify exchange rates and affinities of proteins for nanoparticles, Proc. Natl. Acad. Sci. U.S.A. 104 (2007) 2050–2055.

[142] C. Röcker, M. Pötzl, F. Zhang, W.J. Parak, G.U. Nienhaus, A quantitative fluorescence study of protein monolayer formation on colloidal nanoparticles, Nat. Nanotechnol. 4 (2009) 577–580.

[143] X. Jiang, S. Weise, M. Hafner, C. Röcker, F. Zhang, W.J. Parak, G.U. Nienhaus, Quantitative analysis of the protein Corona on FePt nanoparticles formed by transferrin binding, J. R. Soc. Interface 7 (2010) S5–S13.

[144] M.P. Monopoli, D. Walczyk, A. Campbell, G. Elia, I. Lynch, F.B. Bombelli, K.A. Dawson, Physical-chemical aspects of protein corona: relevance to in vitro and in vivo biological impacts of nanoparticles, J. Am. Chem. Soc. 133 (2011) 2525–2534.

[145] M. Lundqvist, J. Stigler, T. Cedervall, T. Berggard, M.B. Flanagan, I. Lynch, G. Elia, K. Dawson, The evolution of the protein corona around nanoparticles: a test study, ACS Nano 5 (2011) 7503–7509.

[146] M. Safi, J. Courtois, M. Seigneuret, H. Conjeaud, J.F. Berret, The effects of aggregation and protein corona on the cellular internalization of iron oxide nanoparticles, Biomaterials 32 (2011) 9353–9363.

[147] A.T.R. Williams, S.A. Winfield, J.N. Miller, Relative fluorescence quantum yields using a computer-controlled luminescence spectrometer, Analyst (Cambridge, U. K.) 108 (1983) 1067–1071.

[148] D. Magde, J.H. Brannon, T.L. Cremers, J. Olmsted, Absolute luminescence yield of cresyl violet — standard for the red, J. Phys. Chem. 83 (1979) 696–699.

[149] W. Brown, J.X. Zhao, Adsorption of sodium dodecyl-sulfate on polystyrene latex-particles using dynamic light-scattering and zeta-potential measurements, Macromolecules 26 (1993) 2711–2715.

[150] F. Amin, D.A. Yushchenko, J.M. Montenegro, W.J. Parak, Integration of organic fluorophores in the surface of polymer-coated colloidal nanoparticles for sensing the local polarity of the environment, ChemPhysChem 13 (2012) 1030–1035.

[151] C.B. Müller, A. Loman, V. Pacheco, F. Koberling, D. Willbold, W. Richtering, J. Enderlein, Precise measurement of diffusion by multi-color dual-focus fluorescence correlation spectroscopy, Europhys. Lett. 83 (2008) 46001.

[152] J. Loeb, The influence of electrolytes on the cataphoretic charge of colloidal particles and the stability of their suspensions. I. Experiments with collodion particles, J. Gen. Physiol. 5 (1922) 109–126.

[153] A.M. Puertas, F.J. de las Nieves, Colloidal stability of polymer colloids with variable surface charge, J. Colloid Interface Sci. 216 (1999) 221–229.

[154] E. Casals, T. Pfaller, A. Duschl, G.J. Oostingh, V.F. Puntes, Time evolution of the nanoparticle protein corona, ACS Nano 4 (2010) 3623–3632.

[155] X. Jiang, J. Dausend, M. Hafner, A. Musyanovych, C. Röcker, K. Landfester, V. Mailander, G.U. Nienhaus, Specific effects of surface amines on polystyrene nanoparticles in their interactions with mesenchymal stem cells, Biomacromolecules 11 (2010) 748–753.

[156] X. Jiang, A. Musyanovych, C. Röcker, K. Landfester, V. Mailander, G.U. Nienhaus, Specific effects of surface carboxyl groups on anionic polystyrene particles in their interactions with mesenchymal stem cells, Nanoscale 3 (2011) 2028–2035.

[157] M. Bartneck, H.A. Keul, M. Wambach, J. Bornemann, U. Gbureck, N. Chatain, S. Neuss, F. Tacke, J. Groll, G. Zwadlo-Klarwasser, Effects of nanoparticle surface coupled peptides, functional endgroups, and charge on intracellular distribution and functionality of human primary reticuloendothelial cells, Nanomedicine (New York, NY, US) 8 (2012) 1282–1292.

[158] S. Bhattacharjee, D. Ershov, J. van der Gucht, G.M. Alink, I.M.C.M. Rietjens, H. Zuilhof, A.T.M. Marcelis, Surface charge-specific cytotoxicity and cellular uptake of tri-block copolymer nanoparticles, Nanotoxicology 7 (2013) 71–84.

[159] T. Jung, W. Kamm, A. Breitenbach, E. Kaiserling, J.X. Xiao, T. Kissel, Biodegradable nanoparticles for oral delivery of peptides: is there a role for polymers to affect mucosal uptake? Eur. J. Pharm. Biopharm. 50 (2000) 147–160.

[160] L. Rabinovich-Guilatt, P. Couvreur, G. Lambert, C. Dubernet, Cationic vectors in ocular drug delivery, J. Drug Target. 12 (2004) 623–633.

[161] O. Harush-Frenkel, E. Rozentur, S. Benita, Y. Altschuler, Surface charge of nanoparticles determines their endocytic and transcytotic pathway in polarized MDCK cells, Biomacromolecules 9 (2008) 435–443.

[162] A. Muñoz Javier, O. Kreft, A. Piera Alberola, C. Kirchner, B. Zebli, A.S. Susha, E. Horn, S. Kempter, A.G. Skirtach, A.L. Rogach, J. Rädler, G.B. Sukhorukov, M. Benoit, W.J. Parak, Combined atomic force microscopy and optical microscopy measurements as a method to investigate particle uptake by cells, Small 2 (2006) 394–400.

[163] A. Lesniak, F. Fenaroli, M.R. Monopoli, C. Aberg, K.A. Dawson, A. Salvati, Effects of the presence or absence of a protein corona on silica nanoparticle uptake and impact on cells, ACS Nano 6 (2012) 5845–5857.

[164] M. Semmling, O. Kreft, A. Muñoz Javier, G.B. Sukhorukov, J. Käs, W.J. Parak, A novel flow-cytometry-based assay for cellular uptake studies of polyelectrolyte microcapsules, Small 4 (2008) 1763–1768.

[165] P. Ghosh, G. Han, M. De, C.K. Kim, V.M. Rotello, Gold nanoparticles in delivery applications, Adv. Drug Deliv. Rev. 60 (2008) 1307–1315.

[166] D.A. Giljohann, D.S. Seferos, W.L. Daniel, M.D. Massich, P.C. Patel, C.A. Mirkin, Gold nanoparticles for biology and medicine, Angew. Chem. Int. Ed. 49 (2010) 3280–3294.

[167] K.J. Wallace, Molecular dyes used for the detection of biological and environmental heavy metals: highlights from 2004 to 2008, Supramol. Chem. 21 (2009) 89–102.

[168] G. Aragay, J. Pons, A. Merkoci, Recent trends in macro-, micro-, and nanomaterial-based tools and strategies for heavy-metal detection, Chem. Rev. 111 (2011) 3433–3458.

[169] S. Sumalekshmy, C.J. Fahrni, Metal-ion-responsive fluorescent probes for two-photon excitation microscopy, Chem. Mater. 23 (2011) 483–500.

[170] H. Ohshima, T.W. Healy, L.R. White, Accurate analytic expressions for the surface-charge density surface-potential relationship and double-layer potential distribution for a spherical colloidal particle, J. Colloid Interface Sci. 90 (1982) 17–26.

[171] J.J. Lopez-Garcia, J. Horno, C. Grosse, Numerical solution of the Poisson–Boltzmann equation for suspended charged particles surrounded by a charged permeable membrane, Phys. Chem. Chem. Phys. 3 (2001) 3754–3760.

[172] H. Ohshima, Donnan potential and surface potential of a spherical soft particle in an electrolyte solution, J. Colloid Interface Sci. 323 (2008) 92–97.

[173] J.Z. Wu, Z.D. Li, Density-functional theory for complex fluids, in: Annu. Rev. Phys. Chem., vol. 58, Annual Reviews, Palo Alto, 2007, pp. 85–112.

[174] J.Z. Wu, Density functional theory for chemical engineering: from capillarity to soft materials, AIChE J. 52 (2006) 1169–1193.

[175] Y.X. Yu, J.Z. Wu, G.H. Gao, Density-functional theory of spherical electric double layers and zeta potentials of colloidal particles in restricted-primitive-model electrolyte solutions, J. Chem. Phys. 120 (2004) 7223–7233.

[176] P. Teolato, E. Rampazzo, M. Arduini, F. Mancin, P. Tecilla, U. Tonellato, Silica nanoparticles for fluorescence sensing of Zn-II: exploring the covalent strategy, Chem. Eur. J. 13 (2007) 2238–2245.

Chapter 5

Toxicity of Gold Nanoparticles

Encarnación Caballero-Díaz and Miguel Valcárcel*

Department of Analytical Chemistry, University of Córdoba, Córdoba, Spain

**Corresponding author: E-mail: qa1meobj@uco.es*

Chapter Outline

Gold Nanoparticles in Analytical Chemistry. http://dx.doi.org/10.1016/B978-0-444-63285-2.00005-5

1. INTRODUCTION

1.1 Incentives to Assess the Toxicity of Nanoparticles

The increasing use of nanomaterials in consumer and industrial products has aroused global concern about their interaction with biological systems and the possible toxicological effects resulting from this interaction [1]. The same outstanding properties that make nanoparticles so valuable for numerous applications also have the potential to increase their negative effects, and consequently a growing demand for risk assessment is required [2]. Specifically, the use of gold nanoparticles (AuNPs) in biomedical applications has led to a strong interest in investigating their potential to cause adverse effects on biological systems, and how these effects can be alleviated [3].

2. GOLD NANOPARTICLES IN THE BIOLOGICAL FIELD

2.1 Properties of Gold Nanoparticles

AuNPs have attracted enormous scientific and technological interest due to their ease of synthesis and functionalization, chemical stability, and unique optical properties [3]. AuNPs show strong absorption and scattering of light. These properties derived from the phenomenon known as *surface plasmon resonance (SPR)*, which is produced when the conduction band electrons of the nanoparticle oscillate collectively and in resonance as a result of their interaction with an incident electromagnetic radiation of an appropriate frenquency [4]. The wavelength, intensity, and shape of the SPR are dependent on the size, shape, and composition of nanoparticles, but also on alterations in the dielectric properties of the surrounding medium originated from their interactions with other molecules [4].

For spherical nanoparticles, a single plasmon band is observed within the visible region [3]. However, when nanoparticles have an anisotropic shape (e.g., a rod) there are two plasmon resonance peaks, corresponding to transverse surface plasmon resonance (TSPR) and longitudinal surface plasmon resonance (LSPR) [5]. Figure 1 shows the two absorption bands for gold nanorods (AuNRs), LSPR and TSPR, corresponding to electron oscillation along the long and short axes, respectively.

The plasmon band position can therefore be controlled by the AuNP dimensions and this raises the possibility of synthesizing nanoparticles that absorb in the near-infrared (NIR) spectrum range (800−1200 nm), which is considered "the biological water window." In this spectrum region, the light can penetrate deeper in biological tissues and the background fluorescence is low [3]. These characteristics contribute to the popularity of AuNRs and other anisotropic shapes as biomedical agents [3].

The dependence of the LSPR of AuNRs on their aspect ratio has been investigated in numerous works with the purpose of synthesizing nanoparticles with strong absorption in the "water window" [7].

However, the degree of aggregation of nanoparticles and fluctuations in the dielectric constant of the local environment also influence the plasmon band position of AuNPs. Li et al. [5] studied the influence of the concentration of ascorbic acid (AA) and cetyltrimethylammonium bromide (CTAB) during the synthesis of AuNRs on the SPR wavelength. Figure 2 shows the two plasmon resonance peaks of the synthesized AuNRs depending on CTAB and AA concentrations.

As can be seen, both surface plasmons (LSPR and TSPR) are shifted to shorter or longer wavelengths by altering the CTAB and AA concentrations used for the synthesis. In this same work, changes of the refractive index of medium affected SPR wavelength and a linear relationship between SPR wavelength and surrounding medium dielectric constant were finally obtained.

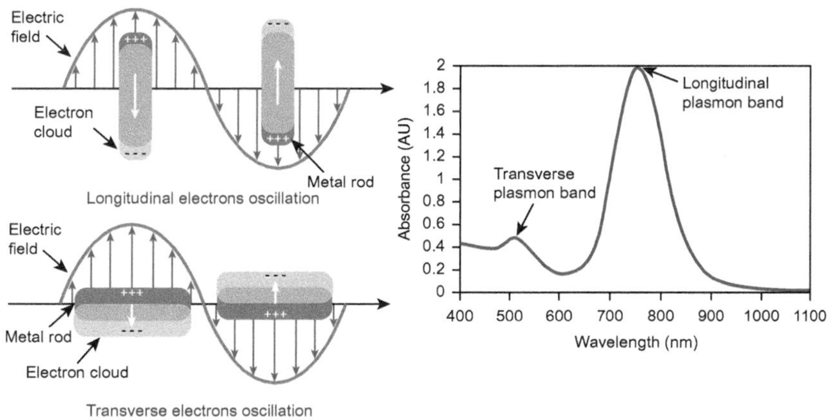

FIGURE 1 Schematic illustration of LSPR and TSPR for AuNRs. *Reproduced with permission of Ref. [6].*

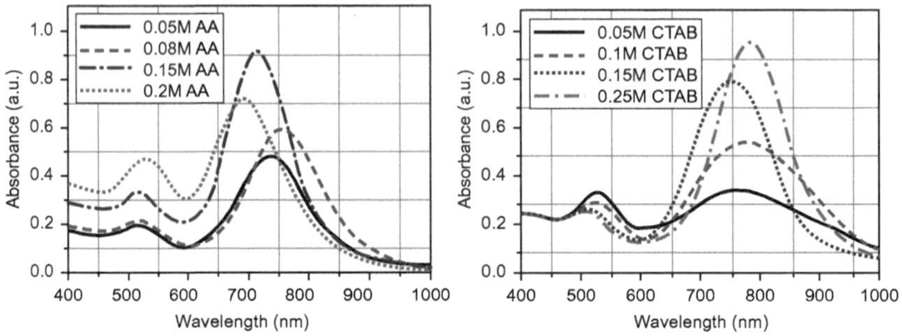

FIGURE 2 Effect of CTAB and AA concentrations on the surface plasmons of AuNRs. *Reproduced with permission of Ref. [5].*

2.2 Biofunctionalization

Nanobiocomposites that combine the recognition properties of biomolecules with the electronic, optical, and magnetic properties of nanoparticles have been considered strong candidates for biomedical applications [8].

AuNPs superficially conjugated with different agents have successfully been applied to cancer treatment. These agents include contrast agents, antibodies, peptides, ligands, drugs, and genes [9].

Polyethylene glycol (PEG) is a commonly used surface agent that confers nanoparticles an increased colloidal stability in biological environments. In addition, PEG acts as a flexible spacer for bioconjugation of targeting ligands facilitating the specific recognition of cancer cells [10].

Drugs, contrast agents, and other molecules can be conjugated with nanoparticles either by linking the ligands on the nanoparticle surface or by encapsulation. The most popular approach is *cross-linking amines and carboxylic acids following carbodiimide activation*. This strategy has some drawbacks related to the activation of side reactions, rapid hydrolysis, or low reaction yields. In addition, this conjugation method is not stable and there is a lack of control in terms of specificity, conjugation numbers, surface densities, and spatial orientation of the attached compounds [4]. A more selective strategy is the *maleimide coupling with thiols* that leads to highly specific bonds and can occur at neutral pH [4]. This approach has been described for functionalization of Au-coated magnetic glyconanoparticles with protein G [11].

Cycloaddition reactions between azides and alkynes are commonly used to link nanoparticles to biomolecules [4]. Numerous methods for the preparation of functional enzyme–AuNP conjugates using "click" chemistry via cycloaddition reactions have been developed. Nanobioprobes fabricated by conjugation of azide-AuNPs with alkyne functional horseradish peroxidase have been described in literature [12]. Gole et al. [13] demonstrated the "clicking" of azide-AuNRs to trypsin by a copper-catalyzed 1,3-dipolar cycloaddition reaction. The azide-AuNRs were first prepared by modifying the AuNRs with a copolymer containing sulfonate and maleic acid groups. Next, the maleic acid molecules participated in the linking with the amine groups of an azide linker by carbodiimide activation. The obtained azide-AuNRs were then functionalized with the enzyme to obtain the conjugate. Brennan et al. [14] described the creation of hybrids of lipase and azide-functionalized water-soluble AuNPs by copper-catalyzed cycloaddition reactions. Conjugation of Au nanoclusters to DNA and dye have been carried out by carbodiimide, succinimidyl ester, and cycloaddition chemistry, taking advantage of the carboxyl, amine, and azide groups located on the surface of nanoparticles [15].

In addition, other *noncovalent linkages* involving hydrogen bonding and hydrophobic and electrostatic interactions can be used for bioconjugation of nanoparticles [4].

Aptamer-AuNP biosensors combine the selectivity and affinity of aptamers with the optical properties of AuNPs [16]. Aptamers are single-stranded oligonucleotides, DNA or RNA, with the ability to bind to non–nucleic acid target molecules, such as peptides, proteins, drugs, organic and inorganic molecules, or even whole cells, with high affinity and specificity [17]. Colorimetric assays based on aptamer-AuNP conjugates have been utilized for detection of cancerous cells [16]. The assembly of these conjugates around diseased cells caused a shift in the absorbance spectra of AuNPs so that the cancer cells exhibited a color change while nontarget cells did not elicit any change in color.

Various methods have been used to produce protein-functionalized AuNPs involving electrostatic adsorption, chemisorption of thiol derivatives, covalent binding through bifunctional linkers, and specific affinity interactions. Jacalin (a lectin) has been conjugated with AuNPs to target human leukemia cells by in vitro experiments [18].

Conjugates combining AuNPs and antibodies have been synthesized and applied to the detection and photothermal therapy of cancer cells (L-428 Hodgkin's) [19]. The development of single glucose biosensors based on the bioconjugation of AuNPs and antibodies has also been reported [20].

2.3 Biomedical Applications

The potential of AuNPs for biomedical purposes has awakened interest owing to their unique optical properties (strong and size-tunable surface plasmon resonance, and fluorescence), easy surface functionalization, and biocompatibility. These features make them excellent candidates for disease treatment or diagnosis [21,22]. Figure 3 summarizes the main characteristics and uses of AuNPs in biomedical applications.

However, AuNPs have certain limitations in this field of application related to the toxicity of the surfactant used for their stabilization, their stability in biological environments, and the drug release kinetics [24]. For instance,

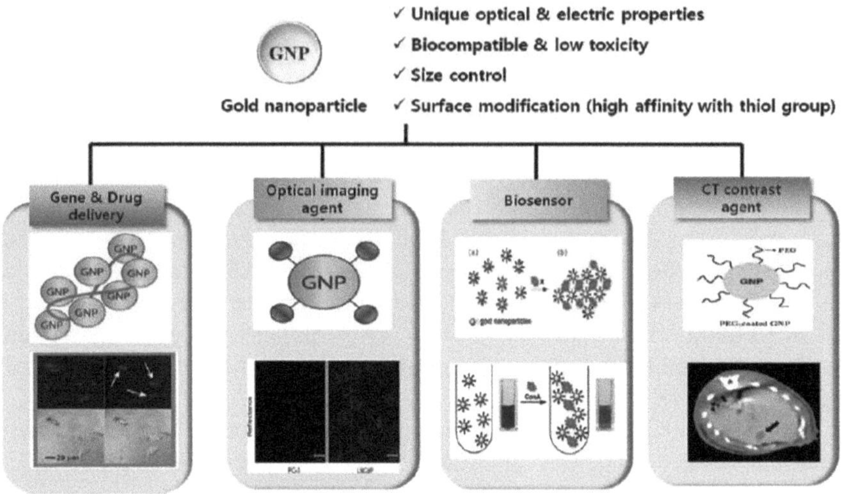

FIGURE 3 Features of AuNPs and their main applications in Biomedicine. *Reproduced with permission of Ref. [23].*

CTAB is the most widely used and convenient surfactant to synthesize and stabilize AuNPs; however, it is well known that free CTAB is toxic; consequently, other strategies involving the surface coating of AuNPs are commonly used in order to reduce the toxicity derived from this compound and to favor the stability of AuNPs in physiological media. Free CTAB molecules in solution can be originated from inadequate purification or desorption of surfactant from the AuNP surface. Some reports have confirmed that free CTAB molecules alone were responsible for the observed toxicity, and not the AuNRs themselves [3]. Recent studies state that CTAB acts by two active mechanisms: (1) direct interaction with the phospholipid bilayer that destabilizes the cell membrane, leading to cell death; and (2) catalytic action of CTAB's dissociation products, which might cause the quenching of the enzyme ATP synthase and thus lead to energy deprivation and cellular death. The second mechanism seems to be the most harmful for cells [25]. Venkatesan et al. [24] demonstrated how the toxicity of CTAB-stabilized AuNRs was reduced by further coating with a polymer (poly(sodium 4-styrenesulfonate), PSS). PSS-AuNPs were finally used as doxorubicin release vehicles in cancer therapy.

2.3.1 Imaging

Optical imaging involving photons emitted from fluorescent probes is a very useful tool to visualize and characterize early-stage disease. Also, it is a cost-effective, rapid, and easy technique compared to other imaging approaches. However, it has some handicaps related to both poor tissue penetration (due to the light absorption by proteins and water) and tissue autofluorescence [23].

AuNPs exhibit enhanced optical properties in comparison to other imaging agents. In this sense, their scattered light is stronger, they are brighter than chemical fluorophores, they are resistant to photobleaching, they can be detected at low concentrations, and they are biocompatible [23]. As noted earlier, the SPR of AuNPs, and thus their colorimetric contrast, can be easily tailored by controlling the shape, size, and surface of nanoparticles [26]. This fact explains the growing interest in the use of these nanoparticles as contrast agents in different imaging modalities for diagnosis. By tuning the SPR of AuNPs to the "phototherapeutic window" (600–1000 nm) of the spectrum, it is possible to achieve a maximum penetration power into most human tissues, which permits the deeper imaging of cancerous cells. Besides, in that spectral region, the absorption of the normal cell constituents is minimized, thus avoiding potential interferences [26]. When AuNPs are illuminated with a beam of white light, they scatter light of many colors, depending on nanoparticle shape and size. This makes it possible to carry out imaging analyses using a simple white light source [23].

Multimodal imaging studies has been developed for highly selective and sensitive detection of cancer cells by combining multi-wavelength *photoacoustic imaging* and AuNPs. For example, Mallidi et al. [27] performed imaging of cancer cells in mouse tissues by detecting shifts in the plasmon resonance frequency of AuNPs once these linked with specific receptors on cancerous cells. AuNPs became aggregated as a result of the interaction with diseased cells, and this fact caused a red shift in their plasmon band.

Recent studies are focused on utilizing AuNRs as contrast agents to enhance the image contrast in photoacoustic imaging. In this regard, the potential of AuNRs 15 nm in diameter and 45 nm long has been reported for the detection of prostate cancer. AuNRs were conjugated with a specific antibody to target prostate cancer cells and then photoacoustics was applied to image the cells targeted by conjugated nanoparticles [28]. In addition, PEG-coated AuNRs conjugated with a cancer-seeking peptide (bombesin) were used for production of a photoacoustic targeting imaging agent for detection of breast cancer [29].

The SPR of AuNPs creates an electrical field around the excited nanoparticles that amplifies the Raman scattering of nearby molecules. This phenomenon is the basis of the *surface-enhanced Raman spectroscopy (SERS)* that allows us to identify and detect single molecules [3]. Different applications of AuNPs in this field have been reported. Huang et al. [30] synthesized silica-encapsulated hollow Au nanospheres using a layer-by-layer method for detection and quantification of three breast cancer cell phenotypes expressed on cell surfaces. Au nanospheres considerably enhanced the Raman signal intensity by localizing surface electromagnetic fields through pinholes in the hollow particle structures. AuNPs loaded with pH-responsive ligands and Raman probes were engineered to aggregate in a rapid and selective way in cancerous cells. The aggregation in the presence of cancer cells provided hot spots for the SERS probe that then generated an enhancement factor above 10^4 and shifted the absorption to far-red and near-infrared regions [31]. Au nano-popcorn-based SERS has been investigated for targeted sensing and photothermal nanotherapy of LNCaP human prostate cancer cells. Multifunctional AuNPs formed several hot spots and provided a significant enhancement of the Raman signal intensity that allowed the recognition of human prostate cancer cells and also their destruction by generating localized heating under NIR irradiation [32]. Other SERS substrates involving AuNPs have also been applied to cancer diagnosis [33,34].

2.3.2 Therapy

The therapeutic applications of nanoparticles are diverse, ranging from cancer therapeutics to antimicrobial actions, vaccine delivery, gene delivery, and site-specific targeting to avoid side effects of the current therapies [35].

2.3.2.1 Photothermal and Photodynamic Therapy

Photothermal therapy is related to the use of heat as a way of action to kill pathophysiological tissue in cancer treatment [36]. It employs a nonionizing light source, usually in the visible to NIR region, which penetrates deeply into the tissues with minimal scattering [26].

Plasmonic photothermal therapy (PPTT) takes advantage of the physicochemical features of AuNPs (absorption of light and rapid conversion of excitation energy to heat) to enhance the tumor destruction in comparison to conventional phototherapy [37]. The excited electrons in the conduction band can lose their energy in the form of heat to the surrounding medium [3]. Upon absorption of light, AuNPs transfer the energy into heat, which is then transported to the surrounding area. The vapor bubbles generated in water-based media as a result of the increase of temperature cause an intense mechanical damage of the surrounding cancer tissue [38].

AuNRs have been extensively investigated as photothermal agents for cancer therapy [39]. AuNRs are better potential photothermal agents than AuNPs of other shapes due to the fact that the absorption wavelength of their surface plasmon is in the NIR region, enabling a deeper penetration in tissues [23]. Comparative studies involving Au nanocages and Au nanoshells revealed that the former were more efficient in killing breast cancer cell lines without causing any significant cytotoxic effect on normal cells [40]. Novel structures such as gold nanoflowers have also been fabricated for future applications in photothermal therapy [41].

As well as photothermal studies, AuNPs have also been employed for other tumor treatment strategies, such as *Photodynamic therapy (PDT)*, also known as *Photochemotherapy*. PDT consists in cancer cell destruction by means of reactive oxygen species (ROS) that are produced from a sequence of photochemical and photobiological processes activated by irradiation [26,37].

AuNR-based hybrid nanospheres have been synthesized and applied to tumor treatment in mice. Nanoparticles were capable of simultaneously producing sufficient hyperthermia and ROS to kill cancer cells at irradiation sites, leading to complete tumor disappearance [42]. Ai et al. [43] employed AuNPs functionalized with an aptamer and a fluorescent dye (porphyrin derivative) for both cell imaging and PDT of HeLa cells.

2.3.2.2 Drug and Gene Delivery Systems

The basis of a targeted delivery system is to concentrate the drug in the tissues of interest while reducing the concentration of medication in other surrounding tissues [22]. Nanoparticle-based therapeutic approaches have a great potential for application in this field, since nanoparticles can be delivered at the specific disease site and then the drug release is triggered for alteration of the physiological microenvironment [22]. Drugs can be loaded in nanoparticles by various strategies, such as encapsulation, surface attachment, or entrapment [35].

The principal advantages provided by nanoparticles in drug delivery applications are related to: (1) enhancing the aqueous solubility of the drug; (2) protecting it from degradation; (3) producing a more prolonged effect over time; (4) reducing the toxicity of the drug and enhancing its bioavailability; and finally (5) providing a targeted delivery [35].

Nanoparticles may release the drugs at required sites by two different strategies (Figure 4):

- *Passive targeting*. Most solid tumors exhibit enhanced vascular permeability, which will ensure a sufficient supply of nutrients and oxygen to tissues for rapid growth [35]. The *enhanced permeability and retention (EPR)*

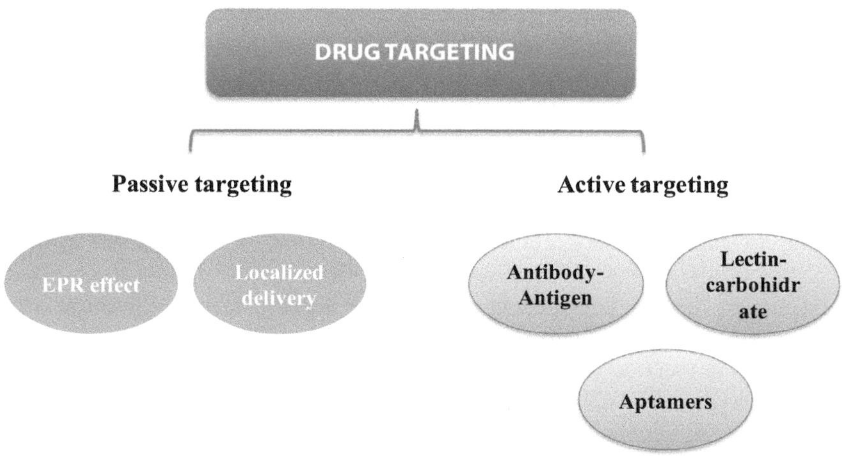

FIGURE 4 Different strategies for drug targeting.

effect considers this unique anatomical-pathophysiological nature of tumor blood vessels to facilitate the accumulation of circulating nanoparticles in the tumor tissues to a greater extent than in normal tissues [44]. Another approach in the passive targeting is the *localized delivery* of anticancer agents using nanoparticles, which can be used in the treatment of local cancers such as prostate and head and neck cancers [45]. In this case, nanoparticles enable a selective and controlled drug release at their target sites. Kennedy et al. [46] performed AuNP-mediated PPTT for the treatment of cancer in mouse studies involving the direct injection of 45 nm AuNPs and subsequent accumulation within the tumor via EPR effect. Once nanoparticles were accumulated at the target site, they were irradiated with a NIR laser and the absorbed energy by AuNPs was translated into heat, which allowed the destruction of tumor tissues.

- *Active targeting*. This approach of targeting takes advantage of the fact that cancer cells overexpress certain receptors on their surface. Nanoparticles can therefore be conjugated with specific ligands (proteins, peptides, aptamers, or antibodies) that interact selectively with those receptors located on the surface of cancer cells [29].

AuNPs show unique properties, including a nontoxic and biocompatible metal core and ease of functionalization, that make them very promising as drug nanocarriers [47].

There are many published reports about the utilization of tailor-made AuNPs for the delivery of *antitumor drugs* [26]. Photolabile plasmonic vesicles composed of AuNPs carrying poly(ethylene glycol) (PEG) and poly(2-nitrobenzyl acrylate) were used for the active delivery of a model drug (doxorubicin) in cancer cells. The release of doxorubicin was controlled by external photoirradiation [10]. Nanomedicines based on AuNPs loaded with different drugs, such as CYT-6091, are currently undergoing phase I clinical trials for the assessment of their pharmacokinetics and properties [48].

AuNPs have facilitated the targeted delivery of several important *small-molecule drugs* such as folic acid. This delivery is achieved by a combination of EPR effect—which allows the drug to be accumulated in the tumor vasculature—and receptor-mediated endocytosis promoted by the small molecules anchored to the AuNP surface [26]. AuNP–folic acid conjugates have been synthesized and applied to the specific staining of cancer cells, taking advantage of the specific interaction between folic acid and the folate receptor on the surface of diseased cells [49,50].

Gene therapy is a promising strategy for treating various diseases. Viral vectors are the most effective methods for gene delivery; however, they have some risks associated with their cytotoxicity and immunogenicity [23]. Some authors have used AuNPs to enhance the transfection efficiency of plasmid DNA into C2C12 cells [51] or as scaffolds for the development of vehicles for the RNA delivery [52].

Nonspherical AuNPs may have some advantages beyond the spherical nanoparticles as a versatile delivery system. For example, hybrid nanosystems combining carbon dots and AuNRs have been applied to the controlled release of doxorubicin into the nucleus of cancer cells upon NIR radiation (808 nm) [53].

2.3.3 Biosensors

AuNPs show remarkable characteristics for their application in the construction of optical and electrochemical biosensors, including suitable support for biomolecule immobilization, which preserves their biological activity; high surface-to-volume ratio; large surface energy; and ability to decrease the distance and favor the electron transfer between proteins and electrodes [54].

The SPR of AuNPs provides high sensitivity to short-range refractive index changes, and can be used for adsorbate concentration detection. The analyte concentration can be measured by corresponding *resonance wavelength shifts*, which depend on the material, size, shape, and distribution of AuNPs. This phenomenon is the basis of those biosensors based on ordered arrays of AuNPs on glass substrates and used for detection of breast cancer cells. In this sense, Liu et al. [55] found that the resonance peak of AuNPs was shifted in cancer cells but not in those healthy cells. The basis for the detection was the different resonance peak shift of AuNPs observed for cancer cells compared with healthy cells.

AuNR-based SPR sensors can be configured by either immobilizing AuNRs on a transparent substrate such as a glass slide or just simply leaving functionalized nanoparticles suspended in the solution in a cuvette where the detection will take place [6].

AuNPs can also be used to *amplify the detection signal*. Sensors involving AuNPs have been developed for the detection of cancer markers (carcinoembryonic antigen and epidermal growth factor receptor hEGFR). In this study, an electrode transducer was modified using AuNPs, leading to a signal enhancement [54]. Figure 5 shows the bioassay carried out for detection of cancer markers.

3. TOXICITY OF GOLD NANOPARTICLES

3.1 Cellular Uptake Processes

The uptake of extracellular material across the plasma membrane of eukaryotic cells is well described and characterized by a variety of mechanisms going from passive diffusion to active transport [56]. Endocytosis is a form of active transport in which a cell takes objects by enclosing them in vesicles or vacuoles pinched off of its cytoplasmic membrane [57]. Experimental studies have confirmed that most nanoparticles are actively incorporated into the cell via different endocytic pathways comprising phagocytosis ("cell-eating") and

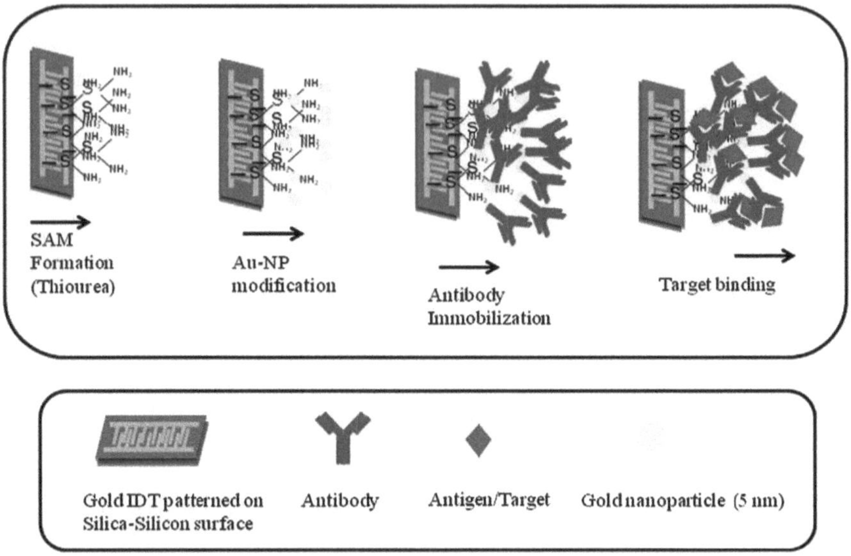

FIGURE 5 Steps of the bioassay involving an AuNP-modified electrode for detection of cancer markers: (1) thiolization of electrode surface, (2) covalent binding of AuNPs, (3) antibody immobilization, and (4) target binding with specific antigens. *Reproduced with permission of Ref. [54].*

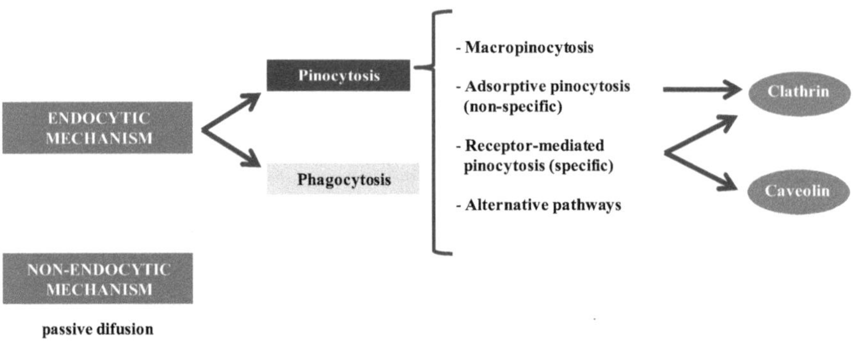

FIGURE 6 Scheme illustrating the main routes of cellular uptake of nanoparticles.

pinocytosis ("cell-drinking") [58]. Figure 6 schematizes the main internalization pathways of nanoparticles.

- *Phagocytosis* is the predominant mechanism for uptake of particles larger than 500 nm by specialized cells of the immune system, such as macrophages or monocytes. It is a process requiring activation of receptors that transmit the signal inside the cell and trigger the response (receptor-mediated process). The phagocytosis starts after opsonization of the target by opsonins in the bloodstream, in order to make the foreign substance visible to macrophages [56]. The specific binding of protein-coated nanoparticles to receptors on the macrophage surface then activates a signal cascade that provokes the formation of an extension in the cell membrane that enclosed the nanoparticles into intracellular vesicles called phagosomes. The phagosomes are transported throughout the cytoplasm and later these vesicles mature through a series of fusion and fission events with late endosomes and lysosomes, resulting in the formation of phagolysosomes [56,58].

- *Pinocytosis* is a fundamental process that occurs in most cells continuosly and independently from the needs of cell [56]. Smaller particles ranging from a few to several hundred nanometers are internalized by this pathway [58]. There are four modes of pinocytosis: macropinocytosis, adsorptive and receptor-mediated endocytosis, and alternative routes of uptake [58]. *Macropinocytosis* is one of the most important types of clathrin- and caveolae-independent endocytosis pathways, occurring in many cells [56]. Generally speaking, this process allows external fluids to be taken up in a nonspecific manner; however, whether there are nanoparticles located near the plasma membrane (with size of $>1\ \mu m$), these can accidentally penetrate together with the fluids [58]. Macropinocytosis seems to be an important mechanism for the uptake of positively charged nanosurfaces [57]. *Clathrin-mediated endocytosis (CME)* is probably the primary characterized mechanism for the cellular uptake of nanoparticles [57] and it occurs in specific regions of the cell membrane that are enriched in clathrin proteins. Nanoparticles with a size of 120–150 nm are internalized within clathrin-coated vesicles entering the endosomal/lysosomal trafficking route [58]. CME is either adsorptive or receptor-mediated [58]. In *adsorptive CME*, nanoparticles interact with complementary binding sites on the cell surface in an unspecific manner. Cationic nanoparticles may bind to the negatively charged cell surface and to be internalized by this process [58]. On the other hand, *receptor-mediated CME* is a highly selective and specific mechanism of cellular uptake [58]. In this case, a specific molecule recognizes its transmembrane receptor located on the plasma membrane, and the complex forms clathrin-coated vesicles [56]. After endocytosis, clathrin proteins dissociate from vesicles and they return to the membrane. The most prominent pathway for nanoparticles apart from CME is *caveolin-mediated endocytosis (CvME)*. This mechanism is predominant in endothelial cells [57]. Caveolae are flask-shaped invaginations of 50–80 nm in size present in the hydrophobic domains of the plasma membrane that are rich in cholesterol and glycosphingolipids [57]. These vesicles detach from the membrane using dynamin and bring their content to compartments called caveosomes [56]. Apart from the mechanisms described above, *other clathrin- and caveolin-independent alternative routes* are strongly dependent on processes involving regulation by proteins [58]. In addition, nanoparticles can also penetrate the cell membrane by *passive diffusion* mechanisms [59].

Figure 7 shows the main endocytic mechanisms described for uptake of nanoparticles by cells.

Each type of nanoparticle exhibits a preferred pathway for cellular internalization [57]. Size, shape, and surface chemistry of nanoparticles have been found to greatly determine their cellular uptake. Studies with A549 cells (human alveolar epithelial cells) and 15-nm AuNPs with different surface coatings (PEG and citrate) revealed that nanoparticle coating modulated the endocytotic uptake pathways and cellular nanoparticle trafficking. Citrate-stabilized AuNPs were preferentially internalized by macropinocytosis, whereas PEG-coated AuNPs mostly entered cells by CME and CvME [57,60]. Research focused on investigating the influence of aspect ratio and surface coating of AuNRs on their uptake by MCF-7 cells (human breast adenocarcinoma) has also been conducted [61]. The cellular uptake resulted to be greatly dependent on the geometrical aspect and surface coating of AuNRs, although the surface chemistry played a more important role. Figure 8 shows the differences observed in the internalization of AuNRs, depending on their aspect ratio and surface covering.

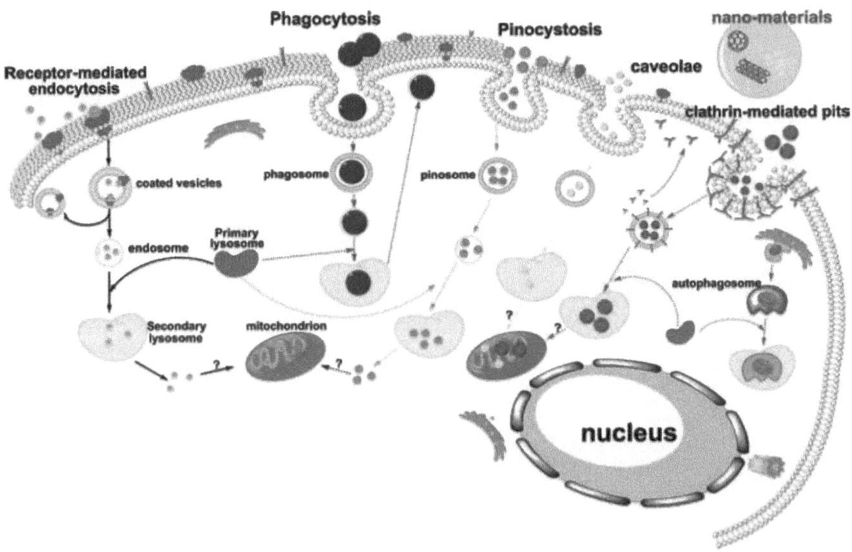

FIGURE 7 Illustration of the principal endocytic mechanisms for intracellular uptake of nanoparticles. *Reproduced with permission from Ref. [57].*

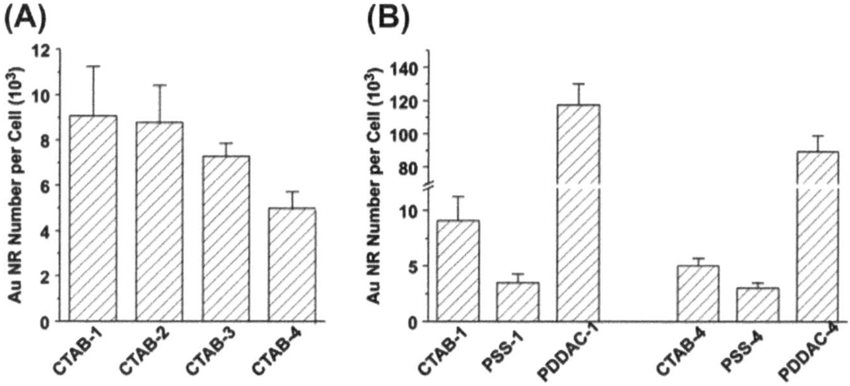

FIGURE 8 Number of AuNRs internalized per MCF-7 cell with different aspect ratio and coated with CTAB (A) or with different coatings (B). *Reproduced with permission of Ref. [61].*

3.2 Methods to Determine the Nanoparticle Uptake

The widely used methods to examine the cellular uptake of nanoparticles are flow cytometry and microscopy. *Flow cytometry* is a fast, efficient, and quantitative approach to assess the nanoparticle uptake. However, it cannot distinguish externally attached nanoparticles from those internalized. Another quantitative approach is *induced coupled plasma mass spectrometry*, which allows us to detect elements, with the exception of H, C, O, N, F, S, and inert gases. Apart from this limitation, this technique cannot differentiate between free ions and nanoparticles [58]. *Confocal scanning laser microscopy* provides qualitative information about the intracellular localization of nanoparticles [58]. These can be labeled with a fluorescent probe; the most commonly used dyes to this end are fluorescein and rhodamine [56]. When these compounds are illuminated at the wavelength of absorption, they emit fluorescence and, consequently, nanoparticles are located in samples. The main drawback of this technique lies in the prolonged exposure to laser emission that may lead to photobleaching of the samples and a decrease in the observed signal [56]. *Transmission electron microscopy (TEM)* can detect structures much smaller than those seen with visible light; it is used to confirm the subcellular nanoparticle localization with very high resolution. TEM also can be used in combination with *energy-dispersive X-ray spectroscopy* [56,58].

3.3 Intracellular Fate

After uptake of nanoparticles by cells, the nanoparticles are directed to the endosomal/lysosomal pathway in most instances [58]. The intracellular vesicles either gradually mature to late endosomes or are recycled back to the cellular surface as trafficking endosomes [58]. Nanoparticles inside late endosomes are likely to proceed to lysosomes, where they can be degradated by the action of proteases, hydrolases, and other enzymes [58]. However, some nanoparticles can escape the endosome; this phenomenon is called "proton sponge effect" [58].

3.4 Influential Factors

Here, we describe several factors that influence the toxicity of nanoparticles. Some of them correspond to the physicochemical properties of nanoparticles, and others allude to environmental conditions such as culture medium or target cell line.

3.4.1 Nanoparticle Size and Shape

The size and shape of nanoparticles determine their interactions with biological systems including their cellular uptake. Most published studies prove that the toxicity of AuNPs is greatly dependent on their size and shape [62]. Chithrani et al. [63] confirmed that the cellular uptake of AuNPs in HeLa cells was greatly dependent on the nanoparticle size. Figure 9 shows the number of AuNPs per cell depending on the AuNP size.

The same authors also investigated the uptake of transferrin-coated AuNPs of different sizes and shapes by mammalian cells [64]. Results from research indicated that nanoparticles were taken up via clathrin-mediated endocytosis pathway and the degree of cellular uptake was again determined by the AuNP size, inasmuch as smaller nanoparticles were more efficiently internalized than larger ones. Other findings reported by Pan et al. [65] showed than 1.4 nm AuNPs were found to be 100-fold more toxic, in terms of (Au), than 15 nm AuNPs, as well as particles smaller than 1 nm. Size-dependent toxicity of AuNPs was demonstrated by in vitro studies with Balb/3T3 mouse fibroblasts [66]. Results from colony-forming efficiency and trypan blue

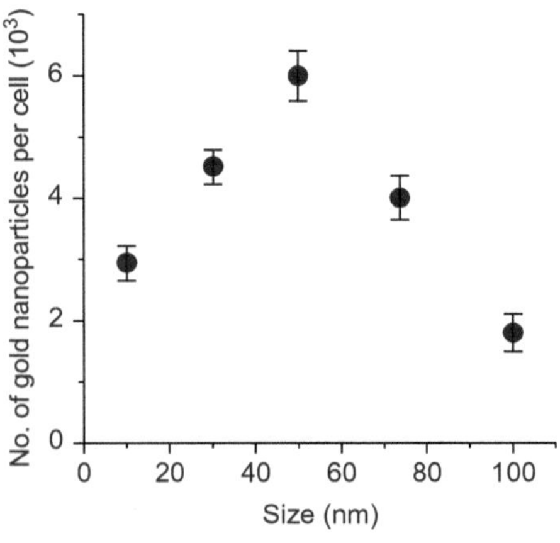

FIGURE 9 Number of AuNPs internalized per cell depending on the nanoparticle size. *Reproduced with permission of Ref. [63].*

assays revealed cytotoxicity only in cells treated with 5 nm AuNPs, whereas no cytotoxic effects were observed in cells exposed to 15 nm AuNPs.

As noted earlier, the influence of AuNP toxicity has also been shown to vary due to the different particle shapes. Schaeublin et al. [67] investigated two AuNPs with different aspect ratios (nanorods and nanospheres) on mediation of biological responses in the human keratinocyte cell line HaCaT. The cellular responses were evaluated by cell viability, ROS generation, alteration in gene and protein expression, and inflammatory response. Results revealed that gold nanospheres did not show any of the detrimental effects observed for AuNRs and, consequently, the cellular response was defined by the nanoparticle shape.

However, it should be noted that even though nanoparticles are well characterized after synthesis, their exposure to biological environments may alter their stability and provoke their agglomeration. Agglomerates with different shapes and sizes exhibit unpredictable effects that enormously complicate the results and their interpretation. Furthermore, there is a polydispersity within any batch of nanoparticles (regarding size and shape), which might generate different results in cell studies [68].

3.4.2 Nanoparticle Surface (Ligands and Charge)

Studies to assess the relationship between surface coating of AuNRs and cellular uptake have been conducted by manipulating the surface charge of these nanoparticles through layer-by-layer coating with polyelectrolytes in order to produce nanorods exhibiting a wide range of surface charges from very positive to very negative [69]. HeLa cells were exposed to these nanoparticles and the cellular uptake was quantified by inductively coupled plasma atomic emission spectroscopy. Results demonstrated that nanorods were internalized to a different degree depending on their surface charge. The greater cellular uptake for positively charged nanorods was warranted by electrostatic interactions between the surface of these nanoparticles and the cell membrane. Recent studies with AuNPs modified with three different surface agents have been carried out in order to clarify the coating influence on the uptake of these nanoparticles by primary human endothelial cells [70]. Results showed that the number of internalized nanoparticles was related to nanoparticle size and surface chemistry inasmuch as positive-charged ethanediamine-coated AuNPs were internalized to a greater extent than negative- (hydroxypropylamine-coated AuNPs) or neutral-charged AuNPs. A report by Schaeublin et al. [71] demonstrated that 1.5 nm AuNPs with different surface charges (positive, negative and neutral) led to variable effects on human keratinocyte cell line HaCaT. While the cell morphology was disrupted by all types of AuNPs, differences were observed in the mitochondrial stress and DNA damage−related gene expression for charged AuNPs compared to the neutral ones.

Alkilany et al. [72] performed toxicological assays involving AuNRs coated (in consecutive steps) with different surface agents: CTAB, polyacrylic acid (PAA), and poly(allylamine) hydrochloride (PAH). They found that PAA overcoating (PAA-CTAB-AuNRs) eliminated the observed cytotoxicity compared to CTAB-AuNRs, which may be related to the fact that polymer reduced the exposure of HT-29 cells to free CTAB molecules, which are extremely toxic. PAA-CTAB-AuNRs were further coated with PAH coating, but no significant differences were observed in comparison to PAA-CTAB-AuNRs. Figure 10 shows the differences in cell viability percentages observed for AuNRs depending on their surface coating. As can be seen, CTAB-capped AuNRs reduced the cell viability to 30% at a concentration of 0.4 nM; however, PAA- and PAH-coated AuNRs were found to be relatively nontoxic (cell viability ≈90%) at the same concentration.

Lipka et al. [73] compared the biokinetics of PEG-modified AuNPs and plain AuNPs after injection in rats. A prolonged blood circulation time was detected for PEG-AuNPs while non-PEGylated AuNPs were accumulated mostly in liver and spleen.

It should be noted that even though considering identical surface agents, their different arrangement on the nanoparticle surface plays a crucial role and can affect the interaction of nanoparticles with cell membranes [57].

3.4.3 Aggregation

The aggregation of AuNPs in cell media is a common phenomenon that can influence nanoparticle–cell interactions. A report by Albanese et al. [74] investigated this aspect by producing transferrin-coated AuNP aggregates of different sizes and characterizing their uptake and toxicity in three different cell lines. Results showed that the uptake patterns were different for single and aggregated nanoparticles. A 25% increase in cell uptake was observed for

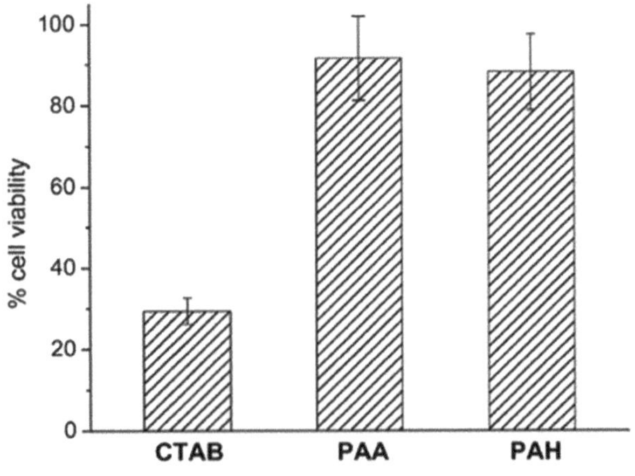

FIGURE 10 Viability of HT-29 cells exposed to AuNRs with different surface coatings for 4 days. *Reproduced with permission from Ref. [72].*

single and monodisperse nanoparticles in HeLa cells and A549 cells in comparison to aggregated AuNPs. However, there was a twofold increase in MDA-MB 435 cell uptake for the largest aggregates. Therefore, findings from this report point out the importance of studying both nanoparticle agglomeration state and cell type as determining factors to define the toxicity of nanoparticles.

Both cellular uptake and cytotoxicity are found to be dependent on the aggregation state of AuNPs. Yang et al. [75] demonstrated that nonaggregated cationic AuNPs were fourfold less toxic to HDF cells than aggregated cationic AuNPs. In addition, the uptake of nonaggregated anionic citrate AuNPs was three orders of magnitude less than that of aggregated citrate AuNPs.

3.4.4 Interaction with Biological Molecules

The ability of metal nanoparticles to interact with biological molecules, such as proteins and DNA, is of concern since this interplay can alter both the normal structure and function of biomolecules and the behavior of the nanoparticles [76].

It has been suggested that the physicochemical properties of a nanomaterial after synthesis, known as its synthetic identity, are not the same as those found in biological media. In these environments, the surface of nanoparticles becomes covered with proteins and other biomolecules in such a way that the size, aggregation state, and interfacial composition of nanoparticle are altered, providing them a biological identity [77].

Serum proteins can be adsorbed onto the outermost layer of the nanoparticles, forming a coating known as *protein corona* that favors the entry of nanoparticles into cells by receptor-mediated endocytosis [57]. The formation of protein corona is dynamic due to the continuous association and dissociation of proteins and related molecules.

Findings by Hong et al. [78] showed that the uptake of DNA-AuNPs by HepG2 cells was highly dependent on the type of end group in the DNA-AuNP complexes, since this determined different association and dissociation rates of proteins around complex.

It has been demonstrated that AuNPs strongly associate with some essential blood proteins (albumin, fibrinogen, γ-globulin, histone, and insulin) and this interaction is determined by both particle size and native protein structure. Furthermore, there is evidence that proteins may undergo conformational changes upon association with the nanoparticles and that the thickness of the adsorbed protein layer increases progressively with the AuNP size [79].

3.4.5 Cell Type

Even though the physicochemical features of AuNPs are influential factors in their cell uptake, the cell line considered in each study also plays a decisive role. Indeed, it has been described that AuNPs with positive surface charge have much higher cell internalization ability than those with negative surface charge in one type of cell line (nonphagocytic HepG2 cells) but not in another cell type (phagocytic RAW 264.7 cells) [80]. Hence, one must take into account that the cell line chosen for performing the cytotoxicity studies can determine the final results even though identical nanoparticles are investigated.

For instance, Patra et al. [81] showed that citrate-capped AuNPs, despite its chemical neutrality, caused cellular death in human carcinoma lung cell line A549 but not in other cell lines, which remained unaffected by treatment with these AuNPs. The specificity of the induction of death response in one cell line implies that these nanoparticles do not universally target all cell types.

It has been described that the mechanisms by which AuNPs induce toxicity to cells can differ depending on the cell type. For example, in vitro studies with six mammalian cell lines revealed that the action of AuNPs was mediated either by apoptosis or by cell cycle delay, depending on the considered cell line [82].

3.4.6 Cellular Media

Each cellular growth medium contains serum proteins, amino acids, vitamins, electrolytes, and other components. These interact with nanoparticles and alter their physicochemical properties, including size, aggregation state, and surface, leading to different cellular uptake and cytotoxicity [3]. Thus, different media may result in different biological responses, considering identical nanoparticles. Maiorano et al. [83] found that different cellular media (i.e., Dulbecco modified Eagle's medium (DMEM) and Roswell Park Memorial Institute medium (RPMI)) led to the formation of protein–AuNP complexes with different internalization ability into HeLa cells and, consequently, to distinct cytotoxicity degree.

3.5 Main Routes of Human Exposure

Nanomaterials are being rapidly developed and incorporated into numerous consumer and industrial products due to their unique properties arising from their nanometric size [84]. Specifically, AuNPs are widely used in cosmetics, food packaging, beverages, toothpaste, automobiles, and air handling units. Furthermore, there are a lot of biomedical products that also include these nanoparticles [85].

From a human health perspective, the most likely routes of exposure are ingestion, inhalation, and absorption through skin contact [86,87].

The *gastrointestinal tract* represents a likely route of entry for many nanomaterials, both directly through intentional ingestion and indirectly via nanoparticle dissolution from food containers or by secondary ingestion of inhaled particles [88].

One of the most common entry routes for nanoparticles is *inhalation*. Nanoparticles induce inflammatory responses and oxidative stress but may also have immune-suppressive effects, impairing macrophage function and altering epithelial barrier functions [89]. Some in vivo studies have confirmed lung inflammation as a result of exposure to nanoparticles [90]. Another report, by Sung et al. [85], found that rats exposed to AuNPs in whole-body inhalation chambers showed a minimal inflammatory response while most adverse effects were related to the increased number of macrophages and the preferential accumulation in lungs and kidneys. Since deposition of inhaled particulates is size-dependent, inhaled nanoparticles such as AgNPs and AuNPs are likely to have significant effects in the deep regions of the lung. For this reason, numerous reports have been focused on evaluating the nanoparticle-induced toxicity in human lung epithelial cells [84].

Another important route is the *skin*, from accidental exposure and use of cosmetics and other topical applications. Reports by Freese et al. [70,91] investigated the toxicity of different polymer-coated AuNPs in human dermal cell lines. Sonavane et al. [92] studied the in vitro permeation of Au particles of three different sizes (15, 102, and 198 nm) through isolated rat skin and intestine. It was observed that the permeation was time- and size-dependent. Smaller AuNPs permeated to a greater extent than larger ones and in deeper regions of skin. Larger particles (102 and 198 nm) were only located in epidermis and dermis. In addition, the permeation through intestine was higher than that through skin [76].

3.6 Mechanisms of Toxicity

AuNPs have been generally considered nontoxic, similar to bulk gold, which is inert and biocompatible [65]. However, numerous studies have confirmed that AuNPs cause adverse effects if they are taken up and stored by cells, since these nanoparticles may turn into effective catalysts.

3.6.1 Interaction with DNA (Genotoxicity)

AuNPs can interact with DNA due to two main reasons. On the one hand, AuNPs, with a size about 1.4 nm, match nearly with that of the major DNA grooves; on the other hand, gold, as the most electronegative metal, easily attracts to DNA grooves, which have a negative environment [36]. Several reports have shown that AuNPs induce DNA damage, indicative of their genotoxicity [93]. In vitro toxicological effects of AuNPs were investigated in small airway epithelial cells, which are considered the first cells of contact for inhaled nanoparticles, and results revealed a significant increase in lipid peroxidation as well as substantial DNA damage and cytotoxicity in AuNP-treated cells [94]. Di Guglielmo et al. [95] assessed the toxicity of 12 nm spherical AuNPs uncoated and coated with hyaluronic acid on BALB/c 3T3 fibroblast cells and they found that these nanoparticles induced DNA damage, probably through an indirect mechanism of oxidative stress. Predictive models about the interaction of 12 different types of nanoparticles and DNA showed that AuNPs had a high binding affinity to DNA and, consequently, their inhibitory effects on DNA replication are expected to be more pronounced than for other nanoparticles tested [96].

3.6.2 Generation of Reactive Oxygen Species

The generation of ROS represents an important mechanism in the toxicity induced by AuNPs, which may disturb the equilibrium between antioxidant and oxidant intracellular processes [93].

Under normal conditions, ROS are generated in mitochondria at low frequency and are easily neutralized by antioxidant defenses such as glutathione (GSH) and antioxidant enzymes [92].

However, when cells are exposed to environmental stress, they can generate chemically active oxygen-containing molecules in such a way that the natural antioxidant defenses are overwhelmed. Oxidative stress refers to the state in which GSH is depleted while oxidized glutathione accumulates [97]. Higher levels of ROS and over a longer time are more likely to cause cellular damage [98]. There are two different types of ROS: radical ROS (nitric oxide or hydroxide radicals) and nonradical ROS (hydrogen peroxide) [98].

Nanoparticles can generate ROS by different mechanisms: (1) direct reactivity on the metal surface; degradation of the surface coating or of the whole nanoparticle with the consequent release of free ions; (2) interaction with cellular organelles such as mitochondria following the alteration of electron transport chains; (3) interaction with redox active proteins such as NADPH oxidase; and (4) interaction with cell surface receptors and activation of intracellular signaling pathways [98]. The oxidative stress can therefore be caused by ROS produced directly by the AuNPs themselves (mechanism 1) or by secondary effects derived from AuNP endocytosis and interaction with intracellular biomolecules or organelles that lead to an oxidative response in cells (mechanism 2, 3, or 4). Generally, AuNPs generate ROS from dioxygen due to their high surface/volume ratio and the specific electronic configuration of the surface Au atoms [65,92]. Figure 11 summarizes the main mechanisms discussed above.

Efforts to clarify the ROS generation mechanism have been carried out with 1.4 nm AuNPs coated with triphenylphosphine monosulfate [65]. For this purpose, HeLa cells were incubated with AuNPs and reducing agents, alone or in combination. Findings showed that the co-incubation of AuNPs with reducing agents during the entire experiment increased cell survival considerably. Only the pretreatment of AuNPs with thiol-containing antioxidants (NAC: *N*-acetylcysteine; and GSH: glutathione) could neutralize the interaction of AuNPs with vital biological targets and reduce their toxicity associated with oxidative stress, endogenous ROS production, and depletion of the intracellular antioxidants. A report by Mateo et al. [99] investigated the effects

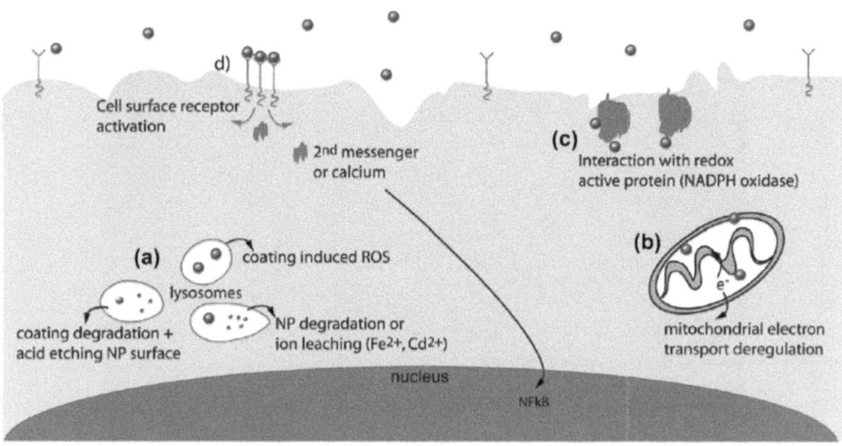

FIGURE 11 Different mechanisms of nanoparticles to induce oxidative stress in cells. *Reproduced with permission from Ref. [98].*

of three AuNPs of different sizes (30, 50, and 90 nm) on human leukemia (HL-60) and hepatoma (HepG2) cell lines. Results confirmed that cytotoxicity in both cells was mainly related to depletion of GSH content after 72 h incubation. It was found that all the tested AuNPs caused a twofold increase of ROS production in cells, which contributed to the cytotoxicity observed.

3.6.3 Mitochondrial Damage

Oxidative stress is associated with protein and polyunsaturated fatty acid oxidation, ultimately leading to a profound alteration in mitochondrial function that constitutes the main cause of cell death. A prominent indicator of mitochondrial damage is the increased permeability of the inner mitochondrial membrane to solutes <1500 Da [65]. This effect can be observed using a fluorescent dye. Studies involving 1.4 nm AuNPs and HeLa cells monitored the mitochondria leakiness by JC-1 dye in combination with fluorescence microscopy [65]. After incubation, the dye was first accumulated in the mitochondria of healthy cells, showing a red fluorescence. However, when the mitochondria became leaky, cells released the dye into the cytoplasm, exhibiting green fluorescence. As can be observed in Figure 12, AuNP-treated cells showed a progressive change in their morphology (rounding off from A to F), and only damaged cells released JC-1 into the cytoplasm, leading to green fluorescence.

The mitochondrial damage produced by spherical PEG-AuNPs on K562 cells after 24–72 h treatment was assessed by monitoring the changes in mitochondrial transmembrane potential with the help of $DiOC6_{(3)}$ dye [100]. After treatment with nanoparticles and incubation with the dye, cells were evaluated by flow cytometry in order to analyze the number of permeabilized

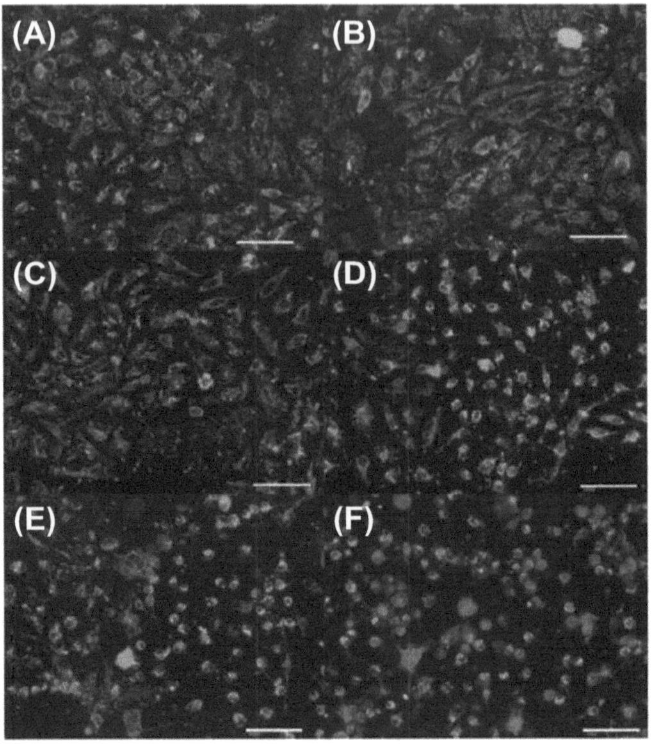

FIGURE 12 Mitochondrial damage in HeLa cells exposed to 100 μM of 1.4 nm AuNPs. (A) untreated cells; (B) 1 h exposure; (C) 6 h exposure; (D) 12 h exposure; (E) 18 h exposure; (F) 24 h exposure. *Reproduced with permission from Ref. [65].*

cells stained with the potential-sensitive dye. Morphological changes in cells exposed to PEG-AuNPs were also observed in this study by using light microscopy and Liu's stain.

As noted earlier, polyunsaturated fatty acid oxidation is a secondary effect of the oxidative stress. Studies conducted with MRC-5 human lung fibroblasts revealed that 20 nm AuNPs caused a higher lipid peroxidation after nanoparticle internalization. In addition, AuNPs also induced upregulation of antioxidants, and stress response gene and protein expression [101]. Khan et al. [102] studied the effect of 10 nm AuNPs on oxidative stress markers, including reduced GSH and malondialdehyde, in different organs of rats. The findings suggested that AuNPs produced significant lipid peroxidation in liver but not in lungs and heart.

3.6.4 Apoptosis/Necrosis

Apoptosis, autophagy, and regulated necrosis (sometimes referred to as necroptosis) are the three major forms of cell death [103]. *Apoptosis* is a form of cellular suicide than can be divided (in mammalian cells) into extrinsic (dependent on so-called death receptors expressed in the plasma membrane) and intrinsic (dependent on mitochondria-controlled signaling) [104]. *Authophagy* has an important homeostatic role, mediating the removal of dysfunctional or damaged organelles allowing the recycling of cellular building blocks [105]. However, uncontrolled autophagy can result in cell death. *Necrosis* was thought for a long time to be merely an accidental form of cell death, but recent research has shown that necrosis can be regulated and may play a role in several pathological and physiological settings [106].

When permeability transition (PT) is induced in a massive way and the mitochondrial activity rapidly reduces the ATP provision, the primary disruption of the plasma membrane, or necrosis, occurs. However, when PT is gradual, apoptogenic proteases are activated and can act on nuclear and cytoplasmic substrates to execute apoptosis [65].

Nanoparticle-induced apoptosis commonly occurs through endosomal uptake of particles and translocation to the lysosomes, followed by lysosomal destabilization and release of cathepsins (proteolytic enzymes) that active the mitochondria-controlled pathway of apoptosis (intrinsic) [103].

Cell death as a result of exposure to AuNPs is greatly dependent on several factors. One of them is nanoparticle size. Previous reports found that 1.4 nm AuNPs caused rapid cell death by necrosis, while identical AuNPs but 1.2 nm in size induced programmed cell death by apoptosis [107]. The surface charge is another factor that also comes into play to define the type of cell death induced by some nanoparticles. The surface charge of 1.5 nm AuNPs was determinant on how AuNPs induced cell death in human keratinocyte cell line HaCaT, with the charged NPs inducing apoptosis and neutral NPs leading to necrosis [71].

It is worth mentioning that a same AuNP can also induce cell death by different mechanisms, considering distinct cell lines. For example, Chueh et al. [108] found that the decrease in cell growth of three mammalian cell models (Vero, MRC-5, and NIH/3T3 cells) was associated with apoptosis in some cases and with autophagy mechanisms in others.

3.6.5 Leakage of Toxic Materials

Once internalized, nanoparticles are exposed to pH fluctuations and the action of various degradative enzymes such as cathepsin L [98]. These factors may cause the progressive degradation of the inorganic core of nanoparticles or of their stabilizing coating shell (surfactants, ligands, or polymers), with the consequent leakage of toxic components to the intracellular environment [109].

Regarding inorganic material, AuNPs have been proved to be less prone to corrosion than their Ag counterparts [109]. AuNPs, AgNPs, or Au-AgNPs were compared in terms of cytotoxicity on mammalian gametes by in vitro studies [110]. Results pointed out a greater toxicity derived from AgNPs, due mainly to their higher degree of corrosion and release of Ag+ ions, unlike other nanoparticles.

However, the organic coating can be the main cause of the toxicity observed for some nanoparticles. In this regard, there are many examples in literature that associate the toxicity derived from AuNPs with the surfactant used for their synthesis and stabilization, in particular, CTAB [72]. Similar findings were found in toxicological studies involving CTAB-stabilized AuNRs, where those nanoparticles further reduced cell proliferation of HaCaT cells due to the inherent toxicity of the cationic surfactant [111].

Several strategies have been developed to reduce the toxicity induced by the presence of CTAB as a result of the synthesis of AuNPs; these approaches are generally known as detoxification process and include the use of polymers—e.g., thiolated methoxy-poly(ethylene)glycol polymer—to displace most of the CTAB bilayer cap from the particle surface [112]. Surface modification of CTAB-stabilized AuNRs with poly(styrenesulfonate) polymer has also been performed with that same purpose [24,113].

3.6.6 Interaction with Lipids and Proteins

Several published studies have confirmed that nanoparticles interact with *lipids* of membranes, leading to harmful effects on cells. A report by Leroueil et al. [114] demonstrated by atomic force microscopy that cationic NH_2-AuNPs caused physical disruption of lipid membranes, formation of holes, and/or membrane thinning. Camesano et al. [115] found that small AuNPs (2–10 nm) caused a loss of mass in lipid bilayers, suggesting pore formation and membrane disruption. However, AuNPs larger than 20 nm did not destabilize the membrane. Therefore, the nanoparticle size was found to be a key factor that determined the cytotoxicity of AuNPs by inducing membrane lysis. The same author also carried out experiments with supported lipid bilayers and they again confirmed that depending on the AuNP size, nanoparticles promoted a more rigid or more flexible membrane [116].

Some simulation studies with lipid bilayers exposed to 2.2 nm AuNPs showed how cationic AuNPs penetrated easily into negative membranes and generated significant disruption of them. The lipids surrounding the AuNPs were highly disordered and nanoscale holes were also observed in the peripheral regions to nanoparticles [117].

Effects arising from *protein–nanoparticle* interactions on protein conformation and activity depend on the extent of protein adsorption on the nanoparticle surface. A study by Chakraborty et al. [118] about the interaction of bovine serum albumin (BSA) and different AuNPs demonstrated that BSA retained its structure and properties, except for the treatment with AuNRs. In that case, AuNRs led to a substantial loss in protein secondary and tertiary structures, thus inducing protein folding. More examples of protein–AuNP interaction have been investigated where negatively charged poly(acrylic acid)-conjugated AuNPs bound to and induced unfolding of fibrinogen, a plasma protein, finally causing an inflammatory response [119].

3.6.7 Endocrine Disruption

Endocrine disrupters or endocrine-disrupting chemicals are a highly heterogeneous group of molecules that may potentially cause disorder in the endocrine system; consequently, they suppose a severe threat to human health [120]. Evidence of endocrine-disrupting activity associated with AuNPs has been reported in literature. Stelzer et al. [121] found that 10 nm AuNPs caused an increased accumulation of estrogen in ovarian granulosa cells after incubation for 1, 3, or 5 h. In addition, some intracellular organelles involved in steroidogenesis were infiltrated and/or altered due to the presence of these nanoparticles. Increased testosterone levels in blood were found in studies with mice after treating them with PEG-modified AuNPs [122].

3.6.8 Altered Gene Expression

An extensive evaluation on the cytotoxic effects of AuNPs also should consider an analysis of possible alterations in gene expression. In this sense, Chuang et al. [82] reported an exhaustive toxicological study of AuNPs on mammalian cell lines. Results revealed that the action of AuNPs significantly modulated gene expression (436 genes in total) and protein function related to the apoptosis induction and cell cycle progression. Similar evidence was also found in a report by Ng et al. [123], who observed that AuNPs disrupted the expression of 19 genes in human fetal lung fibroblasts.

However, from a positive point of view, the potential of AuNPs to control gene expression is being used for disease treatment (gene silencing) [124].

3.6.9 Cellular Morphology Change

Various studies have found that AuNPs induce changes in cellular morphology. For example, A549 cells exposed to AuNPs for 48 h became circular due to AuNP-induced stress [81]. Figure 13 demonstrates the cellular morphology change in A549 cells. The shape was quantified by the extent of roundedness of the cell, which is an indicator of stressed cells.

4. METHODS FOR TOXICITY ASSESSMENT

4.1 In Vitro Assays

Cytotoxicity assays are based on the idea that toxic agents affect basic functions of cells, and thus their toxicity can be evaluated by measuring cellular damage [125]. The vast majority of in vitro nanotoxicity assays examine nanoparticle influence on one single cell type from the organ or tissue of interest, in order to understand the implications of the nanoparticle exposure.

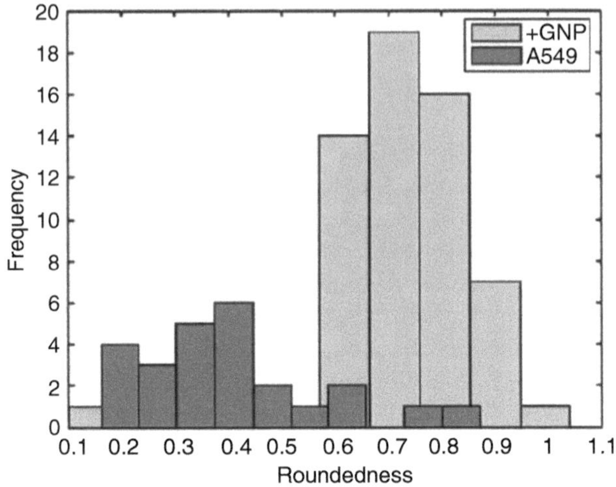

FIGURE 13 Representation by histogram of the cellular morphology change by roundedness of A549 cells after treatment for 36 h with AuNPs. *Reproduced with permission from Ref. [81].*

However, different cell lines may not be similarly affected by nanoparticle exposure, so that cell-specific effects may manifest [76]. Moreover, the observed effects also vary depending on the incubation conditions, such as exposure time and concentration range.

4.1.1 Assessment of Membrane Integrity

Assessing cell membrane integrity is one of the most widely used methods to investigate cell viability and cytotoxic effects associated with some nanoparticles [125]. Five main techniques are used to determine membrane integrity: (1) phosphatidylserine labeling with annexin V; (2) propidium iodide exclusion; (3) trypan blue exclusion; (4) neutral red staining; and (5) determination of total lactate dehydrogenase (LDH) content. The externalization of *phosphatidylserine* is one of the earliest events in apoptosis. Its translocation to the extracellular surface of apoptotic cells can be monitored by using labeled annexin V in combination with flow cytometry and fluorescence microscopy. The vital dyes *trypan blue* and *propidium iodide* can only be internalized and stain the intracellular components of cells with damaged membranes. Therefore, the live cells correspond to unstained cells [126]. The *neutral red* assay measures the ability of viable cells to incorporate and bind the supravital dye neutral red in the lysosomes. If the membrane is damaged, the cellular uptake of this dye decreases [127]. *LDH* assay measures the content of LDH in the extracellular medium, since this is an indicative factor of the damage caused on cellular membrane. When the membrane integrity is compromised, LDH is excreted out of the cell and can be quantified by an enzymatic reaction that involves the conversion of a tetrazolium salt into a colored product (formazan) that is finally detected spectrophotometrically. The more formazan product generated, the greater the cellular damage caused [125].

Within this category, the most frequently used in vitro assays to assess the cytotoxic effects of AuNPs are propidium iodide [128] and neutral red assays [129], followed by trypan blue [108] and LDH assays [130]. By contrast, we have not found any literature about the use of phosphatidylserine assay with AuNPs.

4.1.2 Assessment of Metabolic Activity

One of the traditional cytotoxicity assays to evaluate in vitro toxicity of AuNPs, in terms of reduced metabolic activity, is *3-(4,5-dimethylthiazolyl-2-yl)-2,5-diphenyltetrazolium bromide (MTT) assay*, which is based on the ability of healthy cells to reduce a tetrazolium salt (MTT) to a colorimetric product (formazan) by action of their mitochondrial reductases. The generated formazan is often dissolved by dimethyl sulfoxide and then is quantified by spectophotometry at a wavelength in the range of 500–600 nm. Changes in the metabolic activity of cells exposed to AuNPs are observed by comparing the absorbance values (from formazan) in control and treated cells [125]. Other tetrazolium salts, such as 3-(4,5-dimethylthiazol-2-yl)-5-(3-carboxymethoxyphenyl)-2-(4-sulfophenyl)-2H-tetrazolium (MTS), are also employed with this same purpose. As for MTT assay, here the reduction of MTS to formazan is monitored spectrophotometrically. For example, Chuang et al. [82] examined the cell viability of different cell lines exposed to different sizes and concentrations of AuNPs after incubation for 72 h by MTS-based colorimetric assay.

The *Alamar Blue or resazurin assay* is a colorimetric assay where the nonfluorescent alamar blue is reduced to a pink fluorescent dye by cell metabolic activity, mainly by action of enzymes such as NADP and FADH [131]. In a study by Connolly et al. [132] only the results obtained from this assay were taken as reliable for assessment of the in vitro toxicity of AuNPs, since MTT and other tested cytotoxicity assays showed strong interferences due to the absorption of AuNPs at the same wavelengths used as readout. In spite of the relevance of the results obtained by Connolly, there is no further available literature on the use of the resazurin-based assay with AuNPs.

Adenosine triphosphate (ATP)-based assays are also used to measure metabolic functions of cells. Exposure to AuNPs can inhibit the metabolic activity of cells, in particular the ATPase activity, and consequently reduce the ATP level in cells. This reduction is measured by a luciferase reaction that produces a stable luminescent signal proportional to the ATP present [130]. Studies by Vetten et al. [130] investigated the effects of AuNPs on the reaction that occurs during the ATP-based assay, consisting of the conversion of luciferin to luminescent oxyluciferin only in the presence of ATP. AuNPs resulted in a decrease of the luminescent signal, being the effect more notable at higher nanoparticle concentrations. Several reports have found that AuNPs induce a reduction in ATP levels of treated cells, meaning that they cause a clear mitochondrial dysfunction [133,134].

4.1.3 Assessment of Oxidative Stress

Reactive oxygen species are produced naturally as a result of cell oxygen metabolism. Nevertheless, when cells undergo environmental stress, their ROS levels increase significantly above normal values [125]. The assessment of oxidative stress can be conducted by *direct and indirect methods*. The direct methods allow us to quantify the amount of ROS present. On the other hand, indirect methods monitor other intracellular events derived from a prolonged oxidative stress.

The most widely used direct method to measure the intracellular production of ROS is the *dichlorodihydrofluorescein diacetate (DCFH-DA) assay*. DCFH-DA has the ability to penetrate into cells and oxidize into a stable green fluorescent product (2′,7′-dichlorofluorescein; DCF) only in the presence of ROS, including H_2O_2 and superoxide. In the general protocol, both treated and control cells are incubated with nonfluorescent DCFH-DA. This compound is rapidly oxidized to highly fluorescent DCF by ROS. Fluorescence from oxidized DCF is then detected and correlated with the intracellular ROS levels. This method has been assayed for evaluating the toxicity of AuNPs caused by oxidative stress [93,135].

With regard to indirect methods, one common strategy for analysis of oxidative stress in cells is to measure *GSH levels*. GSH is an antioxidant that plays a critical role in the above-mentioned cellular defense mechanisms. The depletion of GSH is a cytotoxicity marker associated with unusual ROS levels. Several studies have confirmed the reduction of GSH level in cells after incubation with AuNPs [128]. For instance, Gao et al. [136] investigated the role of the interaction of 8 nm AuNPs with GSH on the sequence of apoptotic signaling events in human liver cell line (HL7702 cells). After incubation with AuNPs, an early decline in cytosolic GSH of HL7702 cells was observed, followed by mitochondrial transmembrane potential depolarization and, finally, apoptosis. The depletion of GSH was related to strong Au–S bonding interactions between AuNPs and GSH. Figure 14 shows graphically the mechanism of depletion of GSH caused by interaction of AuNPs and GSH.

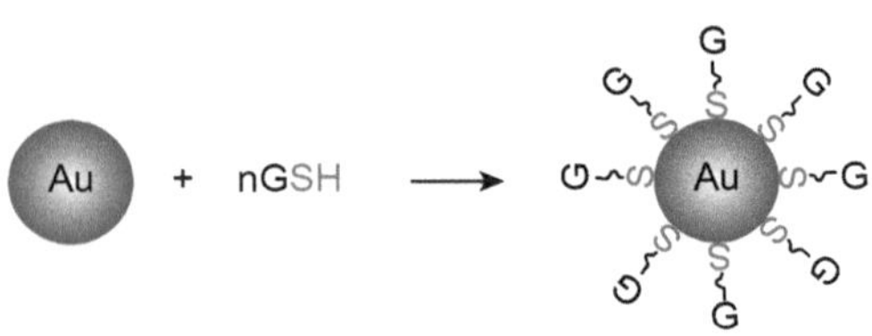

FIGURE 14 Interaction mechanism that explains the depletion of GSH induced by AuNPs. *Reproduced with permission from Ref. [136].*

Following with indirect methods, the measurement of *malondialdehyde* (MDA) formed from the breakdown of fatty acids can be used as an indicator of the extent of lipid peroxidation in cells with high ROS levels [125]. MDA is a byproduct of lipid oxidation by free radicals and ROS and this aldehyde reacts readily with protein or DNA, forming adducts that are considered to be highly genotoxic. MDA formation can be monitored by Western blotting [101].

4.1.4 Assessment of Cell Proliferation

The rate of cell growth is an important evidence of overall cell health and of the potential of nanoparticles to interfere with proliferative processes [126]. The cytotoxicity of glucose-capped AuNPs in several cell lines (HeLa, A549, Jurkat, L929, and HUVEC cells) has been evaluated by means of colony formation and MTT assays [137]. For this study, nontoxic effects were observed to the above-mentioned cell lines.

4.1.5 Assessment of Genotoxicity

Single-cell gel electrophoresis assay (comet assay) is a rapid, simple, and sensitive method for detecting DNA damage by the different migration of DNA fragments in gel electrophoresis. A study by Fraga et al. [138] evaluated the ability of AuNPs to cause DNA damage in human liver HepG2 cells. Following their experimental protocol, aliquots of cells (treated and controls) were centrifuged and the obtained pellets were resuspended in agarose. Gel electrophoresis was then performed. The extent of DNA migration depends directly on the DNA damage present in the cells. The DNA was then stained with ethidium bromide and analysis by fluorescence microscopy was carried out. The percentage of DNA in the comet tail reflected the degree of DNA damage. Finally, results revealed that both citrate and 11-mercaptoundecanoic acid AuNPs induced significant damages as compared with control cells.

4.1.6 Assessment of Apoptosis

Apoptosis can be evaluated by two common strategies. One of them is by measuring the mitochondrial membrane potential and the other one is the *annexin V assay*. This last assay is used to detect the phosphatidylserine that is externalized as a consequence of its migration from the inner surface to the outer surface of the cell membrane during the apoptosis. The presence of phosphatidylserine is therefore one of the indicators of cell apoptosis [139]. Annexin V (previously labeled with a fluorophore) can selectively bind to the exposed phosphatidylserine in the outer side of the membrane, thus allowing its detection [125]. Tsai et al. [139] reported the impact of AuNPs on the cell death of MG63 osteoblast-like cells by annexin V/propidium iodide double-labeling in combination with flow cytometry.

4.1.7 Assessment of Cellular Morphology Alteration

Morphological changes induced by exposure to nanoparticles can be examined by *scanning electron microscopy (SEM) or transmission electron microscopy (TEM)*. Unlike SEM or TEM techniques, *phase-contrast microscopy* allows examination of living cells in their natural state without need for previous fixation or dehydration steps [125]. In addition, alterations in nuclear morphology can be observed by staining the cell nucleus with a specific dye (e.g., 4-6-diamidino-2-phenylindole; DAPI) and using fluorescence microscopy as detection technique. A report by Patra et al. [81] showed how the exposure of A549 cells to AuNPs caused significant nuclear morphology changes, leading to the nuclear condensation that is one of the most important evidences of cytotoxicity.

4.1.8 Other Assays

The lysis of erythrocytes in response to nanoparticles is especially important for nanoparticles, such as AuNPs, that are intended to be directly introduced into the bloodstream [126]. Hemolysis is the damage to red blood cells, which leads to the release of the iron-containing proteins (hemoglobin) into plasma. In hemolysis assays, nanoparticles are incubated in blood, and hemoglobin is released by damaged cells and converted to red-colored cyanmethemoglobin by reagents. The absorbance of cyanmethemoglobin is measured by spectrophotometry and compared to a standard curve to determine the concentration of hemoglobin. This hemoglobin concentration is then compared to that in control samples in order to obtain the percentage of nanoparticle-induced hemolysis [140]. A report by Liu et al. [141] demonstrated by hemolysis test that polymer-modified AuNPs showed a good hemocompatibility with human red blood cells.

Table 1 summarizes the principal conclusions arising from recent in vitro assays carried out with different cell lines that were exposed to AuNPs.

4.2 In Vivo Assays

While there exists much literature about in vitro studies, there are few toxicological reports of AuNPs in animal models, which are the preferred system for toxicologic evaluation of a novel agent and should be useful to clarify the toxicity of AuNPs [21].

The in vivo toxicity is determined by different factors, including exposure dose, administration routes, metabolism, excretion, and immune response [21]. The physicochemical properties of nanomaterials, such as size, shape, and surface coating, also play an important role in this sense. It is widely conceived that AuNPs have high-level accumulation in the liver and spleen leading to further damages on the organism [160]. Therefore, it is important to consider metabolizable AuNPs for use in Biomedicine.

TABLE 1 Recent Studies (2010–2014) About In Vitro Toxicity of AuNPs

Cell Type	Nanoparticle Type	Observed Effects	References
A549 and Vero cells	AuNP–dendron conjugate	No evidence of toxicity was observed	[142]
Small airway epithelial cells	AuNPs	A significant increase in lipid peroxidation, as well as substantial DNA damage and cytotoxicity	[94]
HepG2 and PBMC cells	AuNPs capped with either sodium citrate or polyamidoamine dendrimers	In vitro genotoxicity and cytotoxicity even at very low concentrations of AuNPs	[93]
Porcine gametes	AuNPs, AgNPs, Au–AgNPs	Sperm vitality parameters were not affected by any of the tested nanoparticles	[110]
MDA-MB-231 breast cancer cells	1.9 nm AuNPs	Reduction in cell proliferation	[143]
Caco-2 cells	(<100 nm) AgNPs and AuNPs	Only AgNPs induced acute cytotoxicity	[144]
Human spermatozoa	50 nm AuNPs and 65 nm AgNPs	Both tested NPs exerted an effect on motility and viability of human spermatozoa. AgNPs seem to show a slightly elevated toxicity	[145]
MG63 cells	10 nm AuNPs	No differences in cell viability of the treated cells after 21 days. Therefore, the long-term toxicity of AuNPs is low	[139]
A549 cells	20 nm AuNPs and AgNPs	An inflammatory response was observed in response to AuNPs and AgNPs	[84]
AGS, A549, NIH3T3, PK-15, and Vero cells	39, 41, and 45 nm AuNPs	AuNPs induced a dose-dependent suppression of cell growth. Apoptosis induction or cell cycle delay were the main suppression mechanisms	[82]

Continued

TABLE 1 Recent Studies (2010–2014) About In Vitro Toxicity of AuNPs—cont'd

Cell Type	Nanoparticle Type	Observed Effects	References
HEK293 cells	1.4 nm phosphine- and thiol-stabilized AuNPs	Phosphine-stabilized AuNPs altered the gene expression, whereas thiol-stabilized AuNPs had no toxic effect	[146]
HeLa cells	Silica-coated Au nanorods	No evidence of toxicity was observed	[147]
HeLa cells	Glucose-capped AuNPs	No evidence of toxicity was observed	[148]
Balb/3T3 cells	5 and 15 nm citrate-stabilized AuNPs	Cytotoxic effects only for 5 nm AuNPs. Disruption of the actin cytoskeleton was evident	[66]
MRC-5 cells	AuNPs capped with aspartic acid (GNPA), trisodium citrate dihydrate (GNPC), or bovine serum albumin (GNPB)	All the GNPs were nontoxic to MRC-5 cells. GNPC and GNPA had slight hepatotoxic and nephrotoxic responses	[149]
Balb/3T3 cells	12 nm AuNPs uncoated or coated with hyaluronic acid	AuNPs induced DNA damage, probably through oxidative stress. The coating reduced the AuNP cytotoxicity and slowed their cell internalization	[95]
C17.2, human umbilical vein endothelial, and PC12 cells	4 nm AuNPs coated with poly(methacrylic acid)	High concentrations of AuNPs (200 nM) caused a reduced cell viability due to oxidative stress. Deformations of actin and tubulin cytoskeleton were observed at concentrations of 50 nm or higher	[150]
Vero cells	14 nm porphyran-reduced AuNPs	No evidence of toxicity was observed	[151]
Different human cell lines	Au nanorods coated with methoxypolyethylene glycol thiol	No alterations of cell viability in all the treated cells except for thyroid papillary carcinoma cells	[152]

Tumor ascites and normal peritoneal cells	Triethylene glycol-functionalized AuNPs	AuNPs did not cause cell death or morphological changes	[153]
A549 cells	AuNPs coated with serum proteins	AuNPs induced cell damage through extrinsic and intrinsic apoptotic pathways	[154]
HeLa and U937 cells	15 nm citrate-capped AuNPs	AuNPs were highly toxic to cells	[155]
HeLa and U937 cells	15 nm citrate-capped AuNPs	Cytotoxic effects were dependent on the cell culture media	[83]
MRC-5 cells	20 nm AuNPs	Oxidative stress and autophagy	[101]
PK-15, Vero, NIH3T3, and MRC5 cells	AuNPs	AuNPs induced a decrease in cell growth. This inhibitory effect was associated with apoptosis in Vero cells; DNA damage in MRC-5 cells; and autophagy in NIH3T3 cells	[108]
BEAS-2B, CHO, and HEK293 cells	14 and 20 nm citrate-stabilized AuNPs	AuNPs were relatively nontoxic; however, 20 nm AuNPs exerted higher toxicity	[130]
HeLa cells	Fe_3O_4@AuNPs	No evidence of cytotoxicity was observed	[156]
Human epithelial cells	10, 11, and 25 nm AuNPs	Increased cytotoxicity after exposure to AuNPs. The presence of sodium citrate residues was shown to affect cytotoxicity	[157]
Human cells	4.5 nm AuNPs and 7.3 nm Fe_3O_4 NPs	Cytotoxicity, immunotoxicity, and genotoxicity was low or absent	[158]
MCF-7 and PC-3 cells	20 nm AuNPs coated with citrate, starch, and gum arabic	The citrate-stabilized AuNPs showed cytotoxicity effect compared to starch- and gum arabic-stabilized AuNPs due to the acidic nature of the citrate	[129]
3T3 cells	9 nm glutathione-capped AuNPs and plain AuNPs	ROS generated by cells exposed to glutathione-capped AuNPs were lower than those of cells treated with plain AuNPs	[159]

The metabolism of AuNPs depends on their size, but to a greater extent on their surface chemistry. The surface coating may modulate the metabolism of nanoparticles; however, this coating can be degraded or modified in the physiological environment.

Regarding the effect of *surface coating* of AuNPs, Zhang et al. [160] evaluated the in vivo toxicity and renal clearance of Au nanoclusters on female mice. Nanoparticles were superficially modified with BSA or GSH and then administered to animals by intraperitoneal injection. Figure 15(A, B) represents in vivo biodistribution of two types of Au nanoclusters (GSH−AuNCs and BSA−AuNCs) after 24 h and 28 days. As can be observed, after 24 h, liver and spleen accumulated higher contents of nanoparticles than other organs. GSH-AuNCs preferred to stay in spleen, while BSA−AuNCs had high concentration in liver. However, notable differences in the biodistribution were observed after 28 days. BSA−AuNCs had very high concentrations in the liver and spleen, unlike GSH−AuNCs. Figure 15(C, D) shows the results derived from the studies on renal clearance. As can be seen, GSH−AuNCs can be

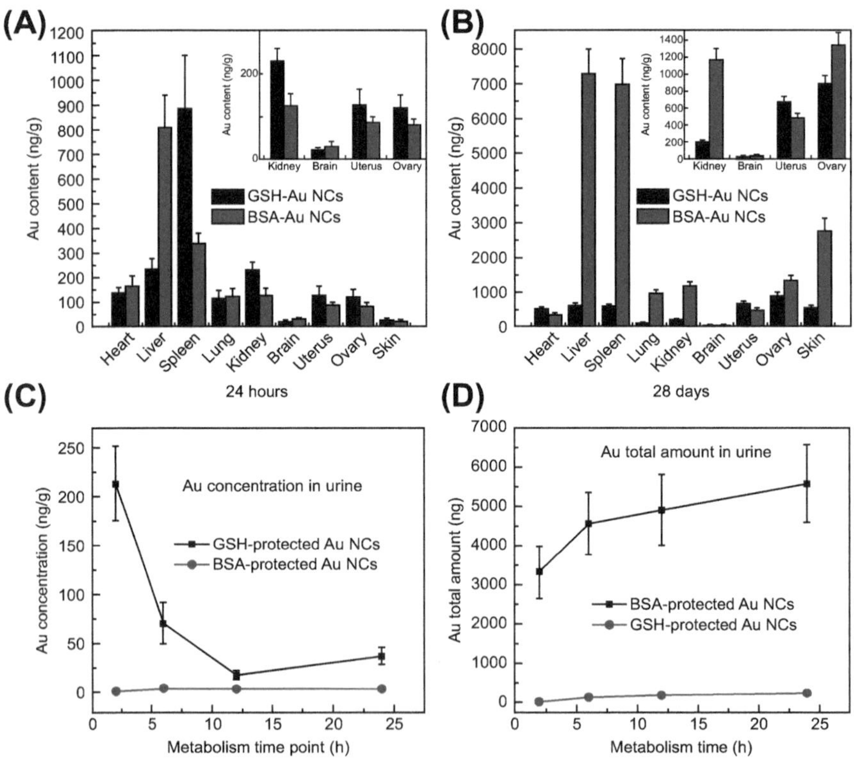

FIGURE 15 Biodistribution of GSH-AuNCs and BSA-AuNCs in treated mice after (A) 24 h and (B) 28 days. (C) Au concentration in urine (clearance rate) and (D) Au content in urine for 24 h. *Reproduced with permission from Ref. [160].*

metabolized by renal clearance to a greater extent than BSA–AuNCs, since their concentration in urine was high after 2 h injection and decreased over time.

An efficient clearance through the urinary filtration system was also observed for glutathione coated-AuNPs in studies with mice after subcutaneous administration [161]. Liver and spleen were found to be the target organs for accumulation of these nanoparticles. Figure 16 shows the average Au concentration contained in urine taken from five individual mice at different concentrations administered. As can be observed, a significant percentage of nanoparticles is cleared through renal filtration within 1 h after administration.

Although some reports have concluded that PEG-coated AuNPs are nontoxic, Cho et al. [162] found that 13 nm sized PEG-coated AuNPs caused acute inflammation and apoptosis in mouse liver. These nanoparticles were found to accumulate in the liver and spleen for up to 7 days after tail vein injection. In spite of the fact that transient inflammation might be considered as lesion induced by the host defense mechanism and negligible for determining the toxicity of PEG-coated AuNPs, the apoptosis and cumulative property of PEG-coated AuNPs is clearly considered as toxicity.

Apart from size, shape, and surface capping of nanoparticles, other factors also can induce a different degree of toxicity in mice. The toxicological effect of *different routes of administration* is still unclear. In this sense, Zhang et al. [21] also investigated the in vivo toxicity of 13.5 nm citrate-coated AuNPs administered to mice by three different routes: oral, intraperitoneal, and tail vein injection. Oral administration may cause some damage in the digestive system. Intraperitoneal injection is preferred when intravenous injection is not feasible, and it is a good and rapid route for drug absorption due to the dense blood vessels and lymph in the murine peritoneum. On the other hand, tail vein

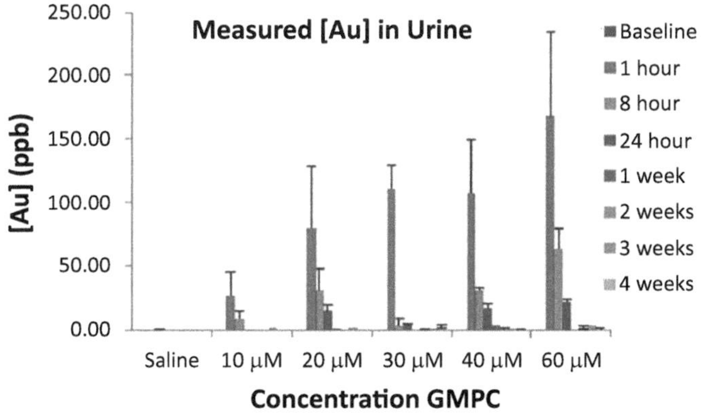

FIGURE 16 Urinalysis of glutathione-coated AuNPs as a function of time from urine taken from five mice that were administered with increasing concentrations of nanoparticles. *Reproduced with permission from Ref. [161].*

injection is widely used to study the toxicological effects of AuNPs, since it is a less invasive process. Results from this study revealed that oral and intraperitoneal injections induced some toxicity, while the tail vein injection led to a lower toxicity.

In addition to previous studies where AuNPs are administered to animals mainly by intravenous and intraperitoneal injections, other reports have evaluated the toxicity of AuNPs by *inhalation* models. This route of administration has been tested in rats [85]. The animals were exposed to these nanoparticles in a whole-body inhalation chamber. Mortality, clinical observations, body weight, food consumption, and lung function were recorded weekly. Results demonstrated that inhaled AuNPs accumulated in a dose-dependent manner in lungs and kidneys of both male and female rats. Yu et al. [163] found that AuNPs can accumulate in the olfactory bulb and translocate from the lung to other organs after inhalation exposure, causing significant effects in the exposed tissues.

Table 2 summarizes the principal conclusions arising from some recent in vivo assays.

5. LIMITATIONS IN TOXICITY STUDIES

5.1 Nanoparticle Characterization

Despite the fact that there are numerous publications concerning the toxicological effects of nanoparticles, the findings reported are disparate inasmuch as they are obtained from variable methods, testing different types of nanoparticles, on different cell lines, and under diverse incubation conditions and doses [57]. Besides, one must take into account that nanoparticles undergo stability changes in the physiological environment as a consequence of their interaction with biomolecules and other components.

Regarding toxicological assessment, there are some limitations in the current studies related to the lack of an exhaustive characterization of nanoparticles involved. This characterization should include at least size distribution, shape, surface area and chemistry, and stability of nanoparticles. In most cases, the degree of characterization performed is dependent on the available facilities in each laboratory [1]. Besides, differences in the results may have their origin in the use of various measurement techniques (e.g., TEM, SEM, or AFM techniques for nanoparticle size) or even in different operating conditions. Errors associated with the different sample preparation methods or with the fact that the samples taken for characterization are not representative of the entire sample can lead to inconsistent results.

Nevertheless, even when nanoparticles have been comprehensively characterized prior to toxicity assays, their properties may be altered as a consequence of their interaction with biological media. For example, pH fluctuations or changes in salt content modify the state of agglomeration of

TABLE 2 Recent Studies About In vivo Toxicity of AuNPs

Organism	Nanoparticle	Observed Effects	Administration Route	References
Mice	Gold nanorod core/silver shell nanostructures (Au g NRs)	The majority of the Au g NRs remained in the injection site. The released Ag ions caused oxidative injury of tissue in the injection site	Subcutaneous injection	[164]
Female and male mice	4.4, 22.5, 29.3, and 36.1 nm AuNPs	Damage in liver and kidney, although the severity of the effects was dependent on sex (female or male)	Intraperitoneal injection	[165]
Mice	Au nanoshells on silica nanorattles (GSNPs) with or without PEGylation	GSNPs without PEGylation were more toxic than those with PEGylation. Inflammatory response, damage in lung tissue, hemorrhage in the kidney, and lymphocytic infiltration in liver	Intravenous injection	[166]
Zebrafish embryos	1.3 nm AuNPs functionalized with *N,N,N*-trimethylammoniumethanethiol (TMAT–AuNPs)	TMAT–AuNPs disrupted the progression of eye development and pigmentation of zebrafish embryos	Exposure model	[167]
Rats	10 and 50 nm AuNPs	AuNPs induced significant variations on some liver enzymes	Intraperitoneal injection	[168]
Murine model	AuNPs with three different capping materials	Only AuNPs capped with trisodium citrate dihydrate caused slight hepatotoxic and nephrotoxic responses	No available information	[149]
Mice	17 and 37 nm AuNPs	Neuronal damage	Intraperitoneal injection	[169]
Rats	14 nm porphyran-reduced AuNPs	AuNPs did not produce abnormalities	Oral	[151]

Continued

TABLE 2 Recent Studies About In vivo Toxicity of AuNPs—cont'd

Organism	Nanoparticle	Observed Effects	Administration Route	References
Mice	BSA and HS–PEG–COOH-capped AuNPs	There were no effects on the normal growth of mice	Intravenous injection	[170]
Mice	GSH- and BSA-protected Au nanoclusters (AuNCs)	The kidney function was affected. However, toxicity responses of the GSH-protected AuNCs can be eliminated after 28 days	Intraperitoneal injection	[160]
Drosophila melanogaster	15 nm citrate-capped Au NPs	Mutagenic effects that may be transmitted to the descendants	Ingestion	[171]
Mice	5, 10, 30, and 60 nm PEG-coated AuNPs	10 and 60 nm PEG-coated AuNPs caused slight damage to the liver	Intraperitoneal injection	[172]
Mice	13.5 nm Citrate-coated AuNPs	Obvious effects on organ index. Oral and intraperitoneal injection showed the highest toxicity	Oral, intraperitoneal, and tail vein injection	[21]
Mice	13 nm PEG-coated AuNPs	Acute inflammation and apoptosis in the mouse liver	Tail vein injection	[162]
Mice	Glutathione-coated AuNPs	Liver and spleen were found to be the target organs for accumulation of these nanoparticles. Very effective renal clearance of nanoparticles	Subcutaneous administration	[161]

nanoparticles and this leads to a different biological response in cells exposed to nanoparticles.

5.2 Interferences with Assays

Nanoparticles often interfere with the mechanism and interpretation of results in some in vitro toxicity assays. In particular, it has been reported that uncoated AuNPs are able to traffic certain assay molecules, which contained primary amines, across the cell membrane, resulting in false-positive results [173]. For example, Sudhir et al. [174] showed that AuNPs may interfere in the annexin V/propidium iodide assay, which is used to detect necrotic cells, since these nanoparticles bind to propidium iodide and promote its internalization inside intact cells, which led to false-positive results in the detection. A strategy to prevent the enhanced permeability of the assay molecules would be to block the nanoparticle surface both sterically and chemically with molecules such as polyethylene glycol [173].

Interferences between AuNPs and the optical readout of some assays have also been reported. For instance, in a study by Connolly et al. [132] AuNP concentration-dependent interferences were detected in cytotoxicity assays carried out on human hepatocellular carcinoma cell line (HepG2) exposed to five types of AuNPs capped with peptide-biphenyl hybrid ligands. Interference studies by four types of cytotoxicity assays (MTT, neutral red or NRU, resazurin and gluthatione assays) were performed in samples containing only different concentrations of AuNPs in the absence of cells (Figure 17).

As can be seen in Figure 17(A, B), the interference of AuNPs at different concentrations on readout of MTT and NRU assays in control samples

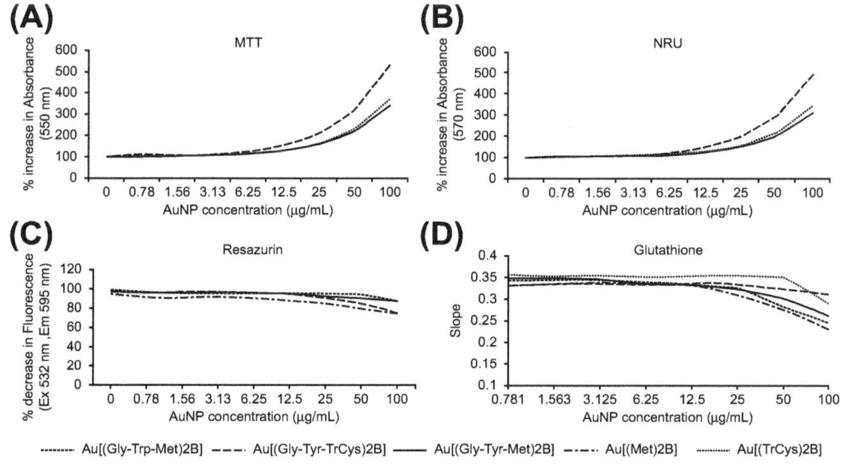

FIGURE 17 Capped AuNPs' interference with the toxicity assays: (A) MTT, (B) neutral red or NRU, (C) resaruzin, and (D) glutathione assays. *Reproduced with permission from Ref. [132].*

(without cells) is more than evident. Interferences caused by the absorbance of AuNPs at the same wavelengths that are used for detection in those assays (550 nm and 570 nm, respectively) made that signal increased about 500%. Strategies including changing the medium and washing with phosphate buffered saline (PBS) prior to incubation with MTT or neutral red did not lead to a significant reduction in interferences. For this reason, a resazurin-based cytotoxicity assay was applied (Figure 17(C)). In that case, interferences were reduced in comparison to previous assays, since the decrease in the levels of fluorescence due to the presence of different AuNP concentrations was only about 11–24%, being much lower than those observed for MTT and NRU assays (about 500%). Consequently, only data obtained from the resazurin-based assay were used. Authors also examined whether the AuNPs interfered with the glutathione assay (Figure 17(D)). They showed that the level of free glutathione decreased as the concentration of AuNPs increased. Thus, this assay was dismissed for the study of oxidative stress.

5.3 Conditions for Assessing In vivo

It is important to mention that in many in vivo studies, organisms are treated with quantities of nanoparticles much higher than those expected in the actual scenarios. High concentrations of nanoparticles favor the formation of aggregates and lead to unrealistic scenarios for toxicity assessment [1].

Taking into account the potential application of AuNPs in Biomedicine, it would be convenient to perform systematic studies that investigate a broad range of exposure conditions involving nanoparticles with different physicochemical features. This would help construct an accurate picture of the variables that influence nanostructure pharmacokinetics (absorption, distribution, metabolism, and excretion) and whether toxicity will be observed [175].

ACKNOWLEDGMENTS

The authors wish to thank Spain's Ministry of Innovation and Science for funding Project CTQ2011-23790 and Junta de Andalucía for Project FQM4801.

REFERENCES

[1] A. Dhawan, V. Sharma, Anal. Bioanal. Chem. 398 (2010) 589–605.
[2] J. Jiang, G. Oberdörster, P. Biswas, J. Nanopart. Res. 11 (2009) 77–89.
[3] A.M. Alkilany, C.J. Murphy, J. Nanopart. Res. 12 (2010) 2313–2333.
[4] C.R. Maldonado, L. Salassa, N. Gomez-Blanco, J.C. Mareque-Rivas, Coord. Chem. Rev. 257 (2013) 2668–2688.
[5] H. Li, G. Zheng, L. Xu, W. Su, Optik 125 (2014) 2044–2047.
[6] J. Cao, T. Sun, K.T.V. Grattan, Sens. Actuators B 195 (2014) 332–351.
[7] I. Gorelikov, L.M. Field, E. Kumacheva, J. Am. Chem. Soc. 126 (2004) 15938–15939.
[8] E. Katz, I. Willner, Angew. Chem. Int. Ed. 43 (2004) 6042–6108.

[9] V. Biju, Chem. Soc. Rev. 43 (2014) 744–764.
[10] J. Song, Z. Fang, C. Wang, J. Zhou, B. Duan, L. Pu, H. Duan, Nanoscale 5 (2013) 5816–5824.
[11] S. Penades Ullate, I. Garcia Martin, J. Gallo Paramo, PCT Int. Appl. EU Patent No. WO2011/036191 (2011).
[12] M.-X. Zhang, B.-H. Huang, X.-Y. Sun, D.-W. Pang, Langmuir 26 (2010) 10171–10176.
[13] A. Gole, C.J. Murphy, Langmuir 24 (2008) 266–272.
[14] J.L. Brennan, N.S. Hatzakis, T.R. Tshikhudo, N. Dirvianskyte, V. Razumas, S. Patkar, J. Vind, A. Svendsen, R.J.M. Nolte, A.E. Rowan, M. Brust, Bioconjug. Chem. 17 (2006) 1373–1375.
[15] E. Oh, F.K. Fatemi, M. Currie, J.B. Delehanty, T. Pons, A. Fragola, S. Léveque-Fort, R. Goswami, K. Susumu, A.L. Huston, I.L. Medintz, Part. Part. Syst. Charact. 30 (2013) 453–466.
[16] C.D. Medley, J.E. Smith, Z. Tang, Y. Wu, S. Bamrungsap, W. Tan, Anal. Chem. 80 (2008) 1067–1072.
[17] L. Yang, X. Zhang, M. Ye, J. Jiang, R. Yang, T. Fu, Y. Chen, K. Wang, C. Liu, W. Tan, Adv. Drug Deliv. Rev. 63 (2011) 1361–1370.
[18] V.S. Marangoni, L.M. Paino, V. Zucolotto, Colloids Surf. B 112 (2013) 380–386.
[19] X. Qu, C. Yao, J. Wang, Z. Li, Z. Zhang, Int. J. Nanomed. 7 (2012) 6095–6103.
[20] Q.-L. Liu, X.-H. Yan, X.-M. Yin, B. Situ, H.-K. Zhou, L. Lin, B. Li, N. Gan, L. Zheng, Molecules 18 (2013) 12675–12686.
[21] X.-D. Zhang, H.-Y. Wu, D. Wu, Y.-Y. Wang, J.-H. Chang, Z.-B. Zhai, A.-M. Meng, P.-X. Liu, L.-A. Zhang, F.-Y. Fan, Int. J. Nanomed. 5 (2010) 771–781.
[22] A. Kumar, X. Zhang, X.-J. Liang, Biotechnol. Adv. 31 (2013) 593–606.
[23] D. Kim, S. Jon, Inorg. Chim. Acta 393 (2012) 154–156.
[24] R. Venkatesan, A. Pichaimani, K. Hari, P.K. Balasubramanian, J. Kulandaivel, K. Premkumar, J. Mater. Chem. B 1 (2013) 1010–1018.
[25] D. Schachter, L. Fabris, The source of toxicity in CTAB and CTAB-stabilized gold nanorods (Doctoral Thesis), New Jersey, January 2013.
[26] S. Nazir, T. Hussain, A. Ayub, U. Rashid, A.J. MacRobert, Nanomed-nanotechnol. 10 (2014) 19–34.
[27] S. Mallidi, T. Larson, J. Tam, P.P. Joshi, A. Karpiouk, K. Sokolov, S. Emelianov, Nano Lett. 9 (2009) 2825–2831.
[28] A. Agarwal, S.W. Huang, K.C. Day, M. O'Donnell, M. Day, N. Kotov, S. Ashkenazi, Pro. SPIE 6437, Photons Plus Ultrasound: Imaging and Sensing 2007: The Eight Conference on Biomedical Thermoacoustics, Optoacoustics, and Acousto-optics, 64370H/1-64370H/7 (February, 2007).
[29] Z. Heidari, R. Sariri, M. Salouti, J. Photoch. Photobio. B 130 (2014) 40–46.
[30] J. Huang, K.H. Kim, N. Choi, H. Chon, S. Lee, J. Choo, Langmuir 27 (2011) 10228–10233.
[31] S. Jung, J. Nam, S. Hwang, J. Park, J. Hur, K. Im, N. Park, S. Kim, Anal. Chem. 85 (2013) 7674–7681.
[32] W. Lu, A.K. Singh, S.A. Khan, D. Senapati, H. Yu, P.C. Ray, J. Am. Chem. Soc. 132 (2010) 18103–18114.
[33] J.V. Jokerst, A.J. Cole, D. Van de Sompel, S.S. Gambhir, ACS Nano 6 (2012) 10366–10377.
[34] S. Lee, H. Chon, J. Lee, J. Ko, B.H. Chung, D.W. Lim, J. Choo, Biosens. Bioelectron. 51 (2014) 238–243.
[35] S. Parveen, R. Misra, S.K. Sahoo, Nanomed-nanotechnol. 8 (2012) 147–166.

[36] P. Rivera Gil, D. Huhn, L.L. del Mercato, D. Sasse, W.J. Parak, Pharmacol. Res. 62 (2010) 115–125.
[37] X. Huang, P. Jain, I. El-Sayed, M. El-Sayed, Laser Med. Sci. 23 (2008) 217–228.
[38] R. Letfullin, C. Joenathan, T.F. George, V.P. Zharov, Nanomedicine 1 (2006) 473–480.
[39] B. Buckway, N. Frazier, A.J. Gormley, A. Ray, H. Ghandehari, Nucl. Med. Biol. 41 (2014) 282–289.
[40] A.K. Rengan, G. Kundu, R. Banerjee, R. Srivastava, Part. Part. Syst. Charact. 31 (2014) 398–405.
[41] J. Han, J.R. Li, W.F. Jia, L.M. Yao, X.Q. Li, L. Jiang, Y. Tian, Int. J. Nanomed. 9 (2014) 517–526.
[42] R. Chen, X. Wang, X. Yao, X. Zheng, J. Wang, X. Jiang, Biomaterials 34 (2013) 8314–8322.
[43] J. Ai, Y. Xu, B. Lou, D. Li, E. Wang, Talanta 118 (2014) 54–60.
[44] J. Fang, H. Nakamura, H. Maeda, Adv. Drug Deliv. Rev. 63 (2011) 136–151.
[45] R. Shuklaa, N. Chanda, A. Zambre, A. Upendran, K. Katti, R.R. Kulkarni, S.K. Nune, S.W. Casteel, C.J. Smith, J. Vimall, E. Boote, J.D. Robertson, P. Kan, H. Engelbrecht, L.D. Watkinson, T.L. Carmack, J.R. Lever, C.S. Cutler, C. Caldwell, R. Kanna, K.V. Katti, Proc. Natl. Acad. Sci. U.S.A. 109 (2012) 12426–12431.
[46] L.C. Kennedy, A.S. Bear, J.K. Young, N.A. Lewinski, J. Kim, A.E. Foster, R.A. Drezek, Nanoscale Res. Lett. 6 (2011) 283–294.
[47] L. Yildirimer, N.T.K. Thanh, M. Loizidou, A.M. Seifalian, Nano Today 6 (2011) 585–607.
[48] S.K. Libutti, G.F. Paciotti, A.A. Byrnes, H.R. Alexander, W.E. Gannon, M. Walker, G.D. Seidel, N. Yuldasheva, L. Tamarkin, Clin. Cancer Res. 16 (2010) 6139–6149.
[49] J. Ai, Y. Xu, D. Li, Z. Liu, E. Wang, Talanta 101 (2012) 32–37.
[50] J. Qiao, X. Mu, L. Qi, J. Deng, L. Mao, Chem. Commun. 49 (2013) 8030–8032.
[51] S.M. Noh, W.-K. Kim, S.J. Kim, J.M. Kim, K.-H. Baek, Y.-K. Oh, BBA – Gen. Subj. 1770 (2007) 747–752.
[52] J.-S. Lee, J.J. Green, K.T. Love, J. Sunshine, R. Langer, D.G. Anderson, Nano Lett. 9 (2009) 2402–2406.
[53] S. Pandey, M. Thakur, A. Mewada, D. Anjarlekar, N. Mishra, M. Sharon, J. Mater. Chem. B 1 (2013) 4972–4982.
[54] Z. Altintas, S.S. Kallempudi, Y. Gurbuz, Talanta 118 (2014) 270–276.
[55] F. Liu, M.M.-K. Wong, S.-K. Chiu, H. Lin, J.C. Ho, S.W. Pang, Biosens. Bioelectron. 55 (2014) 141–148.
[56] A. Panariti, G. Miserocchi, Ilaria Rivolta, Nanotechnol. Sci. Appl. 5 (2012) 87–100.
[57] F. Zhao, Y. Zhao, Y. Liu, X. Chang, C. Chen, Y. Zhao, Small 7 (2011) 1322–1337.
[58] H. Kettiger, A. Schipanski, P. Wick, J. Huwyler, Int. J. Nanomed. 8 (2013) 3255–3269.
[59] W. Rima, L. Sancey, M.T. Aloy, E. Armandy, G.B. Alcantara, T. Epicier, A. Malchére, L. Joly-Pottuz, P. Mowat, F. Lux, O. Tillement, B. Burdin, A. Rivoire, C. Boulé, I. Anselme-Bertrand, J. Pourchez, M. Cottier, S. Roux, C. Rodriguez-Lafrasse, P. Perriat, Biomaterials 34 (2013) 181–195.
[60] C. Brandenberger, C. Mühlfeld, Z. Ali, A.G. Lenz, O. Schmid, W.J. Parak, P. Gehr, B. Rothen-Rutishauser, Small 6 (2010) 1669–1678.
[61] Y. Qiu, Y. Liu, L. Wang, L. Xu, R. Bai, Y. Ji, X. Wu, Y. Zhao, Y. Li, C. Chen, Biomaterials 31 (2010) 7606–7619.
[62] Y. Zhang, D. Xu, W. Li, J. Yu, Y. Chen, J. Nanomater. 2012 (2012) Article ID 375496, 1–7.

[63] B.D. Chithrani, A.A. Ghazani, W.C.W. Chan, Nano Lett. 6 (2006) 662–668.

[64] B.D. Chithrani, W.C.W. Chan, Nano Lett. 7 (2007) 1542–1550.

[65] Y. Pan, A. Leifert, D. Ruau, S. Neuss, J. Bornemann, G. Schmid, W. Brandau, U. Simon, W. Jahnen-Dechent, Small 5 (2009) 2067–2076.

[66] R. Coradeghini, S. Gioria, C. Pascual García, P. Nativo, F. Franchini, D. Gilliland, J. Ponti, F. Rossi, Toxicol. Lett. 217 (2013) 205–216.

[67] N.M. Schaeublin, L.K. Braydich-Stolle, E.I. Maurer, K. Park, R.I. MacCuspie, A.R.M. Nabiul-Afrooz, R.A. Vaia, N.B. Saleh, S.M. Hussain, Langmuir 28 (2012) 3248–3258.

[68] A. Verma, F. Stellaci, Small 6 (2010) 12–21.

[69] T.S. Hauck, A.A. Ghazani, W.C.W. Chan, Small 4 (2008) 153–159.

[70] C. Freese, M.I. Gibson, H-A. Klock, R.E. Unger, C.J. Kirkpatrick, Biomacromolecules 13 (2012) 1533–1543.

[71] N.M. Schaeublin, L.K. Braydich-Stolle, A.M. Schrand, J.M. Miller, J. Hutchison, J.J. Schlager, S.M. Hussain, Nanoscale 3 (2011) 410–420.

[72] M. Alkilany, P.K. Nagaria, C.R. Hexel, T.J. Shaw, C.J. Murphy, M.D. Wyatt, Small 5 (2009) 701–708.

[73] J. Lipka, M. Semmler-Behnke, R.A. Sperling, A. Wenk, S. Takenaka, C. Schleh, T. Kissel, W.J. Parak, W.G. Kreyling, Biomaterials 31 (2010) 6574–6581.

[74] A. Albanese, W.C.W. Chan, ACS Nano 5 (2011) 5478–5489.

[75] J.A. Yang, S.E. Lohse, C.J. Murphy, Small 10 (2014) 1642–1651.

[76] H.J. Johnston, G. Hutchison, F.M. Christensen, S. Peters, S. Hankin, V. Stone, Cr. Rev. Toxicol. 40 (2010) 328–346.

[77] C.D. Walkey, J.B. Olsen, H. Guo, A. Emili, W.C.W. Chan, J. Am. Chem. Soc. 134 (2012) 2139–2147.

[78] S. Hong, S. Park, J. Park, J. Yi, Colloid Surf. B. 112 (2013) 415–420.

[79] S.H. de Paoli Lacerda, J.J. Park, C. Meuse, D. Pristinski, M.L. Becker, A.Karin, J.F. Douglas, ACS Nano 4 (2010) 365–379.

[80] X. Liu, N. Huang, H. Li, Q. Jin, J. Li, Langmuir 29 (2013) 9138–9148.

[81] H.K. Patra, S. Banerjee, U. Chaudhuri, P. Lahiri, A.K. Dasgupta, Nanomed-nanotechnol. 3 (2007) 111–119.

[82] S.-M. Chuang, Y.-H. Lee, R.-Y. Liang, G.-D. Roam, Z.-M. Zeng, H.-F. Tu, S.-K. Wang, P.J. Chueh, Biochim. Biophy. Acta 1830 (2013) 4960–4973.

[83] G. Maiorano, S. Sabella, B. Sorce, V. Brunetti, M.A. Malvindi, R. Cingolani, P.P. Pompa, ACS Nano 4 (2010) 7481–7491.

[84] D. Bachand, A. Allen, M. Bachand, K.E. Achyuthan, J.C. Seagrave, S.M. Brozik, J. Nanopart. Res. 14 (2012) 1212–1222.

[85] J.H. Sung, J.H. Ji, J.D. Park, M.Y. Song, K.S. Song, H.R. Ryu, J.U. Yoon, K.S. Jeon, J. Jeong, B.S. Han, Y.H. Chung, H.K. Chang, J.H. Lee, D.W. Kim, B.J. Kelman, I.J. Yu, Part. Fibre Toxicol. 8 (2011) 16–34.

[86] G. Oberdörster, E. Oberdörster, J. Oberdörster, Environ. Health Perspect. 113 (2005) 823–839.

[87] T. Walser, S. Hellweg, R. Juraske, N.A. Luechinger, J. Wang, M. Fierz, Sci. Total Environ. 420 (2012) 119–126.

[88] I.L. Bergin, F.A. Witzmann, Int. J. Biomed. Nanosci. Nanotechnol. 3 (2013) 163–210.

[89] A. Nemmar, J.A. Holme, I. Rosas, P.E. Schwarze, E. Alfaro-Moreno, BioMed. Res. 2013 (2013) Article ID 279371, 1–22.

[90] C.S. Yah, G.S. Geoffrey, S. Iyuke, Pak. J. Pharm. Sci. 25 (2012) 477–491.

[91] C. Freese, R.E. Unger, R.C. Deller, M.I. Gibson, C. Brochhausen, H.-A. Klok, C.J. Kirkpatrick, Biomater. Sci. 1 (2013) 824–833.
[92] G. Sonavane, K. Tomoda, A. Sano, H. Ohshima, H. Terada, K. Makino, Colloid Surf. B 65 (2008) 1–10.
[93] I.M. Martinez Paino, V. Spolon Marangoni, R. de Cassia Silva de Oliveira, L.M. Greggi Antunes, V. Zucolotto, Toxicol. Lett. 215 (2012) 119–125.
[94] C.-T. Ng, J.J. Li, R.L. Gurung, M.P. Hande, C.-N. Ong, B.-H. Bay, L.-Y.L. Yung, Exp. Biol. Med. (London, U.K.) 238 (2013) 1355–1361.
[95] C. Di Guglielmo, J. De Lapuente, C. Porredon, D. Ramos-Lopez, J. Sendra, M. Borras, J. Nanosci. Nanotechnol. 12 (2012) 6185–6191.
[96] K. Li, X. Zhao, B.K. Hammer, S. Du, Y. Chen, ACS Nano 7 (2013) 9664–9674.
[97] B. Halliwell, J.M.C. Gutteridge, Free Radical. Bio. Med, Oxford Univ. Press, Oxford, 1999.
[98] S.J. Soenen, P. Rivera-Gil, J-M. Montenegro, W.J. Parak, S.C. De Smedt, K. Braeckmans, Nano Today 6 (2011) 446–465.
[99] D. Mateo, P. Morales, A. Avalos, A.I. Haza, Toxicol. Mech. Method. 24 (2014) 161–172.
[100] Y.-C. Huang, Y.-C. Yang, K.-C. Yang, H.-R. Shieh, T.-Y. Wang, Y. Hwu, Y.-J. Chen, BioMed. Res. 2014 (2014) Article ID 182353, 1–9.
[101] J.J. Li, D. Hartono, C.-N. Ong, B.-H. Bay, L.-Y. Yung, Biomaterials 31 (2010) 5996–6003.
[102] H.A. Khan, M.A.K. Abdelhalim, M.S. Al-Ayed, A.S. Alhomida, Saudi J. Biol. Sci. 19 (2012) 461–464.
[103] F. Torres Andón, B. Fadeel, Acc. Chem. Res. 46 (2013) 733–742.
[104] B. Fadeel, S. Orrenius, J. Intern. Med. 258 (2005) 479–517.
[105] B. Levine, D.J. Klionsky, Dev. Cell 6 (2004) 463–477.
[106] P. Vandenabeele, L. Galluzi, T. Vanden Berghe, G. Kroemer, Nat. Rev. Mol. Cell. Biol. 11 (2010) 700–714.
[107] Y. Pan, S. Neuss, A. Leifert, M. Fischler, F. Wen, U. Simon, G. Schmid, W. Brandau, W. Jahnen-Dechent, Small 3 (2007) 1941–1949.
[108] P.J. Chueh, R.-Y. Liang, Y.-H. Lee, Z-M. Zheng, S-M. Chuang, J. Hazard. Mater. 264 (2014) 303–312.
[109] B. Pelaz, G. Charron, C. Pfeiffer, Y. Zhao, J.M. de la Fuente, X.-J. Liang, W.J. Parak, P. del Pino, Small 9 (2013) 1573–1584.
[110] D. Tiedemann, U. Taylor, C. Rehbock, J. Jakobi, S. Klein, W.A. Kues, S. Barcikowski, D. Rath, Analyst 139 (2014) 931–942.
[111] C. Grabinski, N. Schaeublin, A. Wijaya, H. D'Couto, S.H. Baxamusa, K. Hamad-Schifferli, S.M. Hussain, ACS Nano 5 (2011) 2870–2879.
[112] S.C. Boca, S. Astilean, Nanotechnology 21 (2010) 235601/1–235601/8.
[113] H.J. Parab, H.M. Chen, T-C. Lai, J.H. Huang, P.H. Chen, R-S. Liu, M. Hsiao, C.-H. Chen, D.-P. Tsai, Y-K. Hwu, J. Phys. Chem. C 113 (2009) 7574–7578.
[114] P.R. Leroueil, S.A. Berry, K. Duthie, G. Han, V.M. Rotello, D.Q. McNerny, J.R. Baker, B.G. Orr, H. Banaszak, M. Mark, Nano Lett. 8 (2008) 420–424.
[115] T.A. Camesano, S.M. D'Angelo, 85th ACS Colloid and Surface Science Symposium, Montreal, QC, Canada, June, 2011.
[116] T.A. Camesano, E. Kamaloo, C. Bailey, K. Waterman, Abstracts of Papers, 247th ACS National Meeting & Exposition, Dallas, TX, United States, March, 2014.
[117] J.-Q. Lin, Y.-G. Zheng, H.-W. Zhang, Z. Chen, Langmuir 27 (2011) 8323–8332.
[118] S. Chakraborty, P. Joshi, V. Shanker, Z.A. Ansari, S.P. Singh, P. Chakrabarti, Langmuir 27 (2011) 7722–7731.

[119] Z.J. Deng, M. Liang, M. Monteiro, I. Toth, R.F. Minchin, Nat. Nanotechnol. 6 (2011) 39–44.

[120] X. Lu, Y. Liu, X. Kong, P.E. Lobie, C. Chen, T. Zhu, Small 9 (2013) 1654–1671.

[121] R. Stelzer, R.J. Hutz, J. Reprod. Develop. 55 (2009) 685–690.

[122] W.-Q. Li, F. Wang, Z.-M. Liu, Y.-C. Wang, J. Wang, F. Sun, Small 9 (2013) 1708–1714.

[123] C.T. Ng, B-H. Bay, L.Y.L. Yung, Abstracts of Papers, 244th ACS National Meeting & Exposition, Philadelphia, PA, United States, August, 2012.

[124] K.L. Young, A.W. Scott, L. Hao, S.E. Mirkin, G. Liu, C.A. Mirkin, Nano Lett. 12 (2012) 3867–3871.

[125] J.L. Luque-Garcia, R. Sanchez-Díaz, I. Lopez-Heras, P. Martin, C. Camara, TrAC 43 (2013) 254–268.

[126] S.A. Love, M.A. Maurer-Jones, J.W. Thompson, Y.-S. Lin, C.L. Haynes, Annu. Rev. Anal. Chem. 5 (2012) 181–205.

[127] G. Repetto, A. del Peso, J.L. Zurita, Nat. Protoc. 3 (2008) 1125–1131.

[128] M. Liu, X. Gu, K. Zhang, Y. Ding, X. Wei, X. Zhang, Y. Zhao, J. Nanopart. Res. 15 (2013) 1745–1759.

[129] S. Vijayakumar, S. Ganesan, J. Nanomater. 2012 (2012) Article ID 734398, 1–9.

[130] M.A. Vetten, N. Tloleng, D.T. Rascher, A. Skepu, F.K. Keter, K. Boodhia, L.-A. Koekemoer, C. Andraos, R. Tshikhudo, M. Gulumian, Part. Fibre Toxicol. 10 (2013) 50–65.

[131] N. Lewinski, V. Colvin, R. Drezek, Small 4 (2008) 26–49.

[132] M. Connolly, Y. Pérez, E. Mann, B. Herradón, M.L. Fernández-Cruz, J.M. Navas, Nanoscale Res. Lett. 8 (2013) 1–21.

[133] K.S. Suh, Y.S. Lee, S.H. Seo, Y.S. Kim, Biol. Trace Elem. Res. 153 (2013) 428–436.

[134] A. Jebali, B. Kazemi, Toxicol. In vitro 27 (2013) 1847–1854.

[135] Y. Cui, Y. Zhao, Y. Tianb, W. Zhang, X. Lüa, X. Jiang, Biomaterials 33 (2012) 2327–2333.

[136] W. Gao, K. Xu, L. Ji, B. Tang, Toxicol. Lett. 205 (2011) 86–95.

[137] H. Kaur, G. Pujari, A. Sarma, Y.K. Mishra, M.K. Jin, B.K. Nirala, N.K. Gohil, R. Adelung, D.K. Avasthi, Adv. Mater. Lett. 4 (2013) 888–894.

[138] S. Fraga, H. Faria, M.E. Soares, J.A. Duarte, L. Soares, E. Pereira, C. Costa-Pereira, J.P. Teixeira, M. de Lourdes Bastos, H. Carmo, J. Appl. Toxicol. 33 (2013) 1111–1119.

[139] S.-W. Tsai, J.-W. Liaw, Y.-C. Kao, M.-Y. Huang, C.-Y. Lee, L.-R. Rau, C.-Y. Huang, K.-C. Wei, T.-C. Ye, PLoS ONE 8 (2013) 1–12.

[140] B.W. Neun, M.A. Dobrovolskaia, Method. Mol. Biol. 697 (2011) 215–224.

[141] X. Liu, H. Huang, G. Liu, W. Zhou, Y. Chen, J. Ji, Nanoscale 5 (2013) 3982–3991.

[142] T.J. Cho, R.I. MacCuspie, J. Gigault, J.M. Gorham, J.T. Elliott, V.A. Hackley, Langmuir 30 (2014) 3883–3893.

[143] S. Jain, J.S. Coulter, K.T. Butterworth, A.R. Hounsell, S.J. McMahon, W.B. Hyland, M.F. Muir, G.R. Dickson, K.M. Prise, F.J. Currell, et al., Radiother. Oncol. 110 (2014) 342–347.

[144] S. Aueviriyavit, D. Phummiratch, R. Maniratanachote, Toxicol. Lett. 224 (2014) 73–83.

[145] E. Moretti, G. Terzuoli, T. Renieri, F. Iacoponi, C. Castellini, C. Giordano, G. Collodel, Andrologia 45 (2013) 392–396.

[146] A. Leifert, Y. Pan, A. Kinkeldey, F. Schiefer, J. Setzler, O. Scheel, H. Lichtenbeld, G. Schmid, W. Wenzel, W. Jahnen-Dechent, et al., P. Natl. Acad. Sci. U.S.A. 110 (2013) 8004–8009.

[147] S. Mallick, I.-C. Sun, K. Kim, D.K. Yi, J. Nanosci. Nanotechnol. 13 (2013) 3223–3229.

[148] H. Kaur, G. Pujari, M.K. Semwal, A. Sarma, D.K. Avasthi, Nucl. Instrum. Methods Phys. Res. Sect. B 301 (2013) 7–11.

[149] S. Das, N. Debnath, S. Mitra, A. Datta, A. Goswami, BioMetals 25 (2012) 1009–1022.

[150] S.J. Soenen, B. Manshian, J.M. Montenegro, F. Amin, B. Meermann, T. Thiron, M. Cornelissen, F. Vanhaecke, S. Doak, W.J. Parak, et al., ACS Nano 6 (2012) 5767–5783.

[151] V. Venkatpurwar, V. Mali, S. Bodhankar, V. Pokharkar, Toxicol. Environ. Chem. 94 (2012) 1357–1367.

[152] C.-U. Wang, Y. Arai, I. Kim, W. Jang, S. Lee, J.H. Hafner, E. Jeoung, D. Jung, Y. Kwon, J. Korean Phys. Soc. 60 (2012) 1700–1707.

[153] V. Raji, J. Kumar, C.S. Rejiya, M. Vibin, A. John, A. Abraham, J. Exp. Nanosci. 7 (2012) 174–188.

[154] S.Y. Choi, S. Jeong, S.H. Jang, J. Park, J.H. Park, K.S. Ock, S.Y. Lee, S-W. Joo, Toxicol. In vitro 26 (2012) 229–237.

[155] S. Sabella, V. Brunetti, G. Vecchio, A. Galeone, G. Maiorano, R. Cingolani, P.P. Pompa, J. Nanopart. Res. 13 (2011) 6821–6835.

[156] J. Salado, M. Insausti, L. Lezama, I. Gil de Muro, M. Moros, B. Pelaz, V. Grazu, J.M. de la Fuente, T. Rojo, Nanotechnology 23 (2012) 315102/1–315102/10.

[157] C. Freese, C. Uboldi, M.I. Gibson, R.E. Unger, B.B. Weksler, I.A. Romero, P.-O. Couraud, C.J. Kirkpatrick, Part. Fibre Toxicol. 9 (2012) 1–11.

[158] T. Pfaller, R. Colognato, I. Nelissen, F. Favilli, E. Casals, D. Ooms, H. Leppens, J. Ponti, R. Stritzinger, V. Puntes, et al., Nanotoxicology 4 (2010) 52–72.

[159] M. Pongsuchart, C. Danladkaew, T. Khomvarn, A. Sereemaspun, Int. Proc. Chem. Biol. Environ. Eng. 27 (2012) 98–102.

[160] X.-D. Zhang, D. Wu, X. Shen, P.-X. Liu, F.-Y. Fan, S.-J. Fan, Biomaterials 33 (2012) 4628–4638.

[161] C.A. Simpson, K.J. Salleng, D.E. Cliffel, D.L. Feldheim, Nanomed. Nanotechnol. Biol. Med. 9 (2013) 257–263.

[162] W.-S. Cho, M. Cho, J. Jeong, M. Choi, H.-Y. Cho, B.S. Han, S.H. Kim, H.O. Kim, Y.T. Lim, B.H. Chung, J. Jeong, Toxicol. Appl. Pharm. 236 (2009) 16–24.

[163] L.E. Yu, L.-Y.L. Yung, C.-N. Ong, Y.-L. Tan, K.S. Balasubramaniam, D. Hartono, G. Shui, M.R. Wenk, W.-Y. Ong, Nanotoxicology 1 (2007) 235–242.

[164] J. Meng, Y. Li, J. Liu, X. Cheng, H. Guo, W. Zhang, X. Wu, H. Xu, Nanotoxicology 8 (2014) 686–696.

[165] J. Chen, H. Wang, W. Long, X. Shen, Di Wu, S.-S. Song, Y.-M. Sun, P.-X. Liu, S. Fan, F. Fan, et al., Int. J. Nanomed. 8 (2012) 2409–2419.

[166] H. Liu, T. Liu, H. Wang, L. Li, L. Tan, C. Fu, G. Nie, D. Chen, F. Tang, Biomaterials 34 (2013) 6967–6975.

[167] K.-T. Kim, T. Zaikova, J.E. Hutchison, R.L. Tanguay, Toxicol. Sci. 133 (2013) 275–288.

[168] M.A.K. Abdelhalim, S.A.A. Moussa, Saudi J. Biol. Sci. 20 (2013) 177–181.

[169] Y.-S. Chen, Y.-C. Hung, M.-Y. Hong, A.A. Onischuk, J.C. Chiou, I.V. Sorokina, T. Tolstikova, G.S. Huang, J. Nanomater. 2012 (2012) Article ID 746960, 1–11.

[170] T.H.L. Nghiem, T.T. Nguyen, E. Fort, T.P. Nguyen, T.M.N. Hoang, T.Q. Nguyen, H.N. Tran, Adv. Nat. Sci. Nanosci. Nanotechnol. 3 (2012) 015002/1–015002/5.

[171] G. Vecchio, A. Galeone, V. Brunetti, G. Maiorano, L. Rizzello, S. Sabella, R. Cingolani, P.P. Pompa, Nanomedicine 8 (2012) 1–7.

[172] X.-D. Zhang, D. Wu, X. Shen, P.-X. Liu, N. Yang, B. Zhao, H. Zhang, Y.-M. Sun, L.-A. Zhang, F.-Y. Fan, Int. J. Nanomed. 6 (2011) 2071–2081.

[173] A.M. Keene, R.J. Allaway, N. Sadrieh, K.M. Tyner, Nanotoxicology 5 (2011) 469–478.

[174] Y. Sudhir, J. Pharmaceut, Biomed. Sci. 14 (2012) 1–9.

[175] H.C. Fischer, W.C. Chan, Curr. Opin. Biotechnol. 18 (2007) 565–571.

Part II

Gold Nanoparticles as Target Analytes: Characterization and Determination of Gold Nanoparticles

Chapter 6

Microscopic Techniques for the Characterization of Gold Nanoparticles

Christine Kranz* and Boris Mizaikoff
Institute of Analytical and Bioanalytical Chemistry, University of Ulm, Albert-Einstein-Allee 11, 89081 Ulm, Germany
**Corresponding author: E-mail: christine.kranz@uni-ulm.de*

Chapter Outline

1. INTRODUCTION

Among the wide variety of nanoparticles, gold nanoparticles (Au-NPs) and composite Au-NPs play a significant role in analytical and bioanalytical chemistry, and adjacent research areas derived from their unique catalytic, electrical, and optical properties [1–3]. An enormous—almost unmanageable—number of publications describing, utilizing, and characterizing Au-NPs clearly demonstrate their importance in nanoscience and nanotechnology. Analytical applications of Au-NPs range from trace analysis in environmental chemistry such as the detection of mercury in aquatic

Gold Nanoparticles in Analytical Chemistry. http://dx.doi.org/10.1016/B978-0-444-63285-2.00006-7

systems [4] to next-generation sensing schemes providing enhanced selectivity and sensitivity based on Au-NPs [5]. Last, the increased quantitative usage of surface- and tip-enhanced analytical strategies in vibrational spectroscopy benefits from significant signal-enhancement effects associated with geometrically well-controlled noble metal nanostructures. In the life sciences and bioanalytics, Au-NPs are nowadays routinely applied for immunostaining and labeling, for drug delivery, for utilizing localized heating effects, and in numerous biosensing applications [2,6], (and references therein). It should be noted that the precise cellular uptake mechanisms of nanoparticles are not yet still fully understood; hence, there is a need to advance knowledge on the parameters facilitating internalization of Au-NPs and other nanostructured materials including, e.g., the ligand shell conformation, etc., which in turn requires the application of a variety of microscopic and nanoscopic characterization strategies summarized within this chapter.

The utility of Au-NPs in a wide range of applications may also be attributed to their exquisite stability, which is associated with their largely tailorable and customizable surface chemistry. Similar to macroscopic gold surfaces, Au-NPs are readily modified via thiol attachment chemistry [7–9] or by utilizing polyelectrolytes [10]. Consequently, deliberate and specific (bio)functionalization is facilitated, which is essential, e.g., for nanoparticle-enhanced (bio)sensing concepts, and in biomedical applications. Without further detailing in this chapter, numerous examples have shown that the physical properties of Au-NPs are intimately related to the size, shape, and aspect ratio of these materials, as well as to their surface (sub)structure and morphology. Au-NPs can be synthesized in various dimensions ranging from small clusters of few nanometers in diameter to NPs at the scale of several hundreds of nanometers. A variety of shapes ranging from spherical particles to nanorods have been reported, and are summarized in detail within other chapters. However, efficient synthesis strategies for generating a desired size and shape while largely maintaining monodispersity still remain challenging. Consequently, the size- and shape-dependence of physical NP properties, the still insufficient synthesis control in terms of size, shape, etc., and the wide range of adaptable surface chemistries require a carefully selected set of characterization methods, which provides microscopic and nanoscopic information on the size, uniformity (i.e., size distribution), and—most importantly—surface morphology of Au-NPs.

A detailed characterization of Au-NPs not only plays a significant role in optimizing synthesis pathways for Au-NPs, but is also essential for the subsequent application of Au-NPs in critical applications ranging from catalysis to life sciences, where the interaction of Au-NPs with complex matrices and therein contained constituents requires fundamental knowledge of the pristine particle properties. For example, effects of the Au-NP size on their cytotoxicity were studied by Pan et al. [11] among others. Although the most common particle shape, which is obtained by simply reducing gold chloride in stabilizing solutions, corresponds to a sphere (i.e., isotropic NPs), Au nanorods and asymmetric Au-NPs (i.e., anisotropic NPs) including nano-branched or nanostar-shaped Au-NPs have also been described in literature [12] (and references herein). The formation of anisotropic structures is particularly facilitated by adding selective capping agents during the growth process. For example, a yield of 97% was reported for gold nanorods synthesized via seed-mediated growth using cetyltrimethylammonium bromide (CTAB) as the surfactant. In contrast, using commercially available gold colloids (~ 10 nm) as seed particles at the same conditions resulted in sophisticated nanostar structures.

For example, it has been shown that corners and edges of anisotropic NPs such as nanoplates and nanorods are characterized by a higher affinity for ligand binding. Moreover, such Au-NPs are suspected to exhibit more pronounced toxicological effects due to their enhanced catalytic properties, which again emphasizes the need for a thorough characterization strategy providing detailed information on structure–activity relationships.

Within the last decade, a number of routine characterization techniques have emerged for characterizing NPs, and specifically Au-NPs, which are mentioned throughout this book including optical microscopy, electron microscopy (EM), transmission electron microscopy (TEM), scanning probe techniques such as atomic force microscopy (AFM), scanning tunneling microscopy (STM), dynamic and static light scattering, X-ray photoelectron spectroscopy, X-ray diffraction spectroscopy, Fourier transform infrared spectroscopy, size exclusion chromatography, asymmetric flow-field flow fractionation, small angle neutron scattering, matrix-assisted laser desorption/ionization time-of-flight mass spectrometry, ultraviolet–visible spectroscopy, dual polarization interferometry, and nuclear magnetic resonance techniques. Hence, this chapter focuses on microscopic characterization techniques along with their specific advantages for characterizing Au-NPs, functionalized Au-NPs, and bimetallic Au–NP composites with particular emphasis on TEM/scanning transmission electron microscopy (TEM/STEM), and scanning probe microscopy (SPM) techniques including AFM, STM, and near-field scanning optical microscopy (NSOM), whereby their analytical and bioanalytical relevance rather than their fundamental physical principles will be discussed. In addition, micro- and nanoscopic electrochemical imaging techniques are attractive analytical tools as they provide information on chemical and catalytic functionality/activity beyond merely imaging Au-NPs, and will be discussed in view of their future potential for the in-situ characterization of nanoparticles. Table 1 summarizes the characterization techniques discussed herein along with their major advantage and disadvantages for analyzing Au-NPs. Finally, it should be noted that for the purpose of this chapter, a distinct division between microscopic and spectroscopic techniques is not practical, as some of the analytical devices and methods covered herein provide access to both strategies.

TABLE 1 Summary of the Main Features Pertaining the Microscopic Techniques Discussed in This Chapter for Characterizing Gold Nanoparticles

Optical Microscopy (OM)		Electron Microscopy (EM)		Scanning Probe Microscopy (SPM)	
Methods	**Abbrev.**	**Methods**	**Abbrev.**	**Methods**	**Abbrev.**
Dark-field microscopy	DFM	High-resolution scanning electron microscopy	HR-SEM	Scanning tunneling microscopy	STM
(Confocal) fluorescence microscopy	FIM	Transmission electron microscopy	TEM	Atomic force microscopy	AFM
Total internal reflection (plasmonic) microscopy	TIRM	Scanning transmission electron microscopy	STEM	Near-field optical microscopy	NSOM
				Scanning electrochemical microscopy	SECM

Features			
Resolution	>150 nm (OM); single particle (DFM, TRIM, FM); single molecule (FM)	micrometers (SEM); nanometers (HR-SEM); Ångstrom/sub-Ångstrom (TEM, STEM)	>50 nm (NSOM, SECM); atomic res. (STM, AFM)
Potential artifacts	Chromatic aberrations, spherical aberrations, astigmatism, distortions	Spherical aberrations, astigmatism, distortions, sample charging	Tip convolution
Environment	Ambient/gas, liquid	Vacuum	Ambient/gas (STM, AFM, NSOM), liquid (STM, AFM, NSOM, SECM), vacuum (STM, AFM)
Information on Nanoparticles			
Size	Yes	Yes	Yes
Aspect ratio	Yes	Yes	Yes
Height	No	No	Yes
Perimeter	Yes	Yes	Yes
Roughness	No	No	Yes
Volume	No	No	Yes
Reactivity	No	No	Yes

2. OPTICAL MICROSCOPY

Conventional optical microscopy techniques provide limited spatial resolution, which is determined by the diffraction limit, i.e., given by the wavelength of visible light (400–600 nm; resolution = 0.61 λ/NA, with λ ... wavelength, NA ... numerical aperture). For state-of-the-art instruments with numerical apertures in the range of 1.3–1.4, therefore, a resolution in the range of 200 nm can be achieved. Hence, the characterization of NPs with optical microscopy is limited to rather large particle dimensions, yet, proves useful for tests characterizing, e.g., distribution of larger particles at a surface or agglomerations/clusters of NPs. For Au-NPs dimensions usually applied in (bio)analytics, however (i.e., $<$100 nm), conventional light microscopy appears less suitable with few exceptions discussed herein.

2.1 Dark-Field Microscopy

Dark-field microscopy (DFM), a rather unconventional microscopy technique, has nowadays gained momentum due to its utility for characterizing Au-NPs. DFM experiments date back to 1902, when Siedentopf and Zsigmondy imaged diffusing Au particles with DFM. DFM utilizes an opaque circle that blocks the central light rays along the optical axis of the microscope, and therefore ensures that only scattered light is collected by the objective. Metal nanoparticles are thereby visualized due to resonant light scattering (i.e., localized surface plasmon resonances corresponding to the collective response of the conduction electrons/bands), which is predominantly dependent on the particle size and shape, but also on the dielectric properties and the surrounding temperature [13,14]. Consequently, the obtained color information enables characterizing the size of the scattering NPs. For example, spherical Au-NPs with diameters smaller than 100 nm, as well as Au nanorods and bimetallic Ag–Au nanocubes have been studied via DFM. For example, it was theoretically shown and experimentally demonstrated that elongated Au-NPs reveal a longitudinal plasmon resonance, which is proportional to their aspect [15]. As an example, Figure 1 illustrates DFM images and corresponding TEM images (insets) of Au-NPs with different shapes including spheres, hexagons, cubes, and rods [16]. Yet, from the DFM images, it may not be ruled out that individual colored spots correspond to multiple nanoparticles, which are closely spaced. Evidently, spherical Au-NPs with a diameter of 23 nm appear green, whereas 70 nm Au-NPs are yellow. Hexagonal particles are also yellow, whereas cubic Au-NPs profiles are orange-red. Nanorods with aspect ratios of 4.4 appear pale green, while those with an aspect ratio of 16 scatter dark yellow [16]. In summary, combining DFM with complementary TEM studies is an attractive strategy for characterizing Au-NPs, even though the achievable resolution is not competitive with high-resolution TEM (HR-TEM). Nonetheless, a correlation of optical properties and shape/size of Au-NPs is achievable.

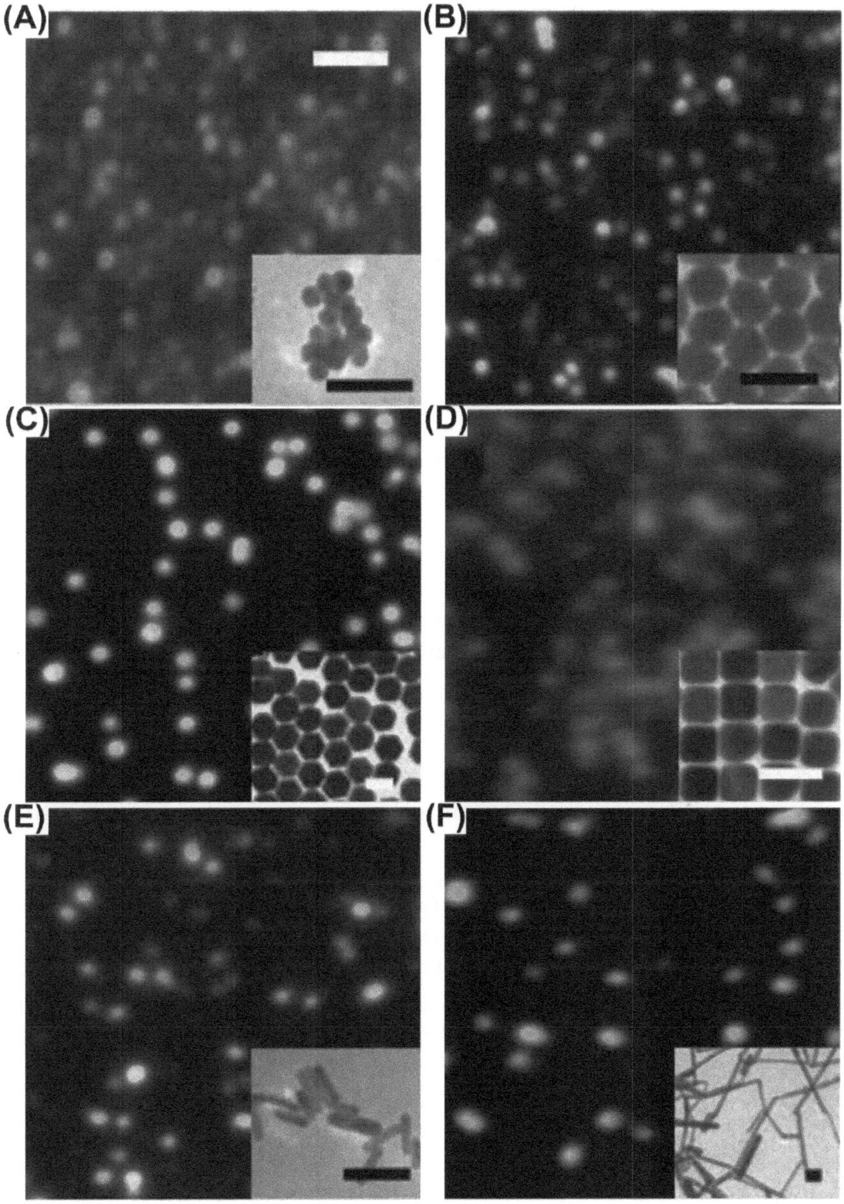

FIGURE 1 Dark-field light-scattering images of gold nanoparticles with different shapes (inset shows the corresponding TEM images): (A) 23-nm-diameter spheres, (B) 70-nm-diameter spheres, (C) hexagons, (D) cubes, (E) rods (aspect ratio ~4.4), (F) rods (aspect ratio ~16). Scale bars represent 2 μm in light-scattering images, and 100 nm in TEM images. *Reprinted from Ref. [16] with permission from Wiley-VCH.*

Advantageously, optical microscopic techniques may be performed at ambient conditions, and therefore requiring significantly less sample preparation compared to particle beam techniques operated at high-vacuum conditions.

2.2 Other Optical Techniques

As already mentioned above, besides probing the plasmonic properties of Au-NPs and in particular their sustained localized surface plasmons giving rise to exceptional optical cross-sections [17], conventional optical microscopies are limited in laterally resolving such small dimensions and features. Clearly, fluorescence microscopy (FlM) techniques along with the emerging super-resolution variants including stochastic optical reconstruction microscopy, stimulated emission depletion microscopy, and photoactivated localization microscopy provide the opportunity of analyzing even individual nanoparticles with unsurpassed sensitivity. However, FlM—and in particular, confocal fluorescence imaging techniques—is predominantly applied for localizing and characterizing fluorescent NPs in biological systems, e.g., after internalization into cells or microorganisms or when applied as fluorescent tag/label for tracking (bio)molecules such as proteins within organelles, tissues, or entire organisms, and is discussed in detail in the context of the corresponding chapters (i.e., Chapters 12 and 10). Likewise, spectroscopic techniques for characterizing NPs including mid-infrared spectroscopy/microscopy and Raman spectroscopy/microscopy along with the corresponding enhancement mechanisms (surface-enhanced infrared absorption; surface-enhanced Raman scattering; and tip-enhanced Raman scattering) are discussed in Chapters 7 and 12.

Besides FlM, confocal microscopy combined with a supercontinuum laser light source has been used for identifying individual Au-NPs with diameters <10 nm based on interferometric detection techniques. Again, access to the plasmon resonance of individual Au-NPs was facilitated via spectral analysis, thereby proving the utility of this technique not only for characterizing such nanostructures but also for future applications relevant in biophysics, e.g., tracking and/or localizing individual Au-NPs within live biological specimen [18].

Among the few conventional optical microscopy techniques facilitating the characterization of Au-NPs, total internal reflection microscopy has been demonstrated as a rather simple, yet flexible methodology for determining optical scattering spectra from individual metallic NPs. By exposing NPs to an evanescent field generated via total internal reflection at the surface of a glass prism using a simple halogen lamp as light source, Sönnichsen et al. have shown that light scattered by individual particles may be collected using a conventional optical microscope. Thereafter, the collected scattered radiation is spectrally analyzed using a conventional spectrometer, thereby providing access to the effect of the particle diameter on the dephasing time of the plasmon resonance in Au-NPs in air as well as in liquid phase [19].

3. ELECTRON MICROSCOPY

The interaction of highly energized electrons with solid matter results in secondary signals due to the interaction with the imaged sample. These secondary signals are used to obtain element-specific analytical and chemical information. Figure 2 gives an overview on the analytical signals resulting from these secondary effects. In scanning electron microscopy (SEM), secondary electrons (SEs) with energies <50 eV, Auger electrons produced by the relaxation of excited atoms, and backscattered electrons with energies close to those of the incident electrons are used for imaging or providing diffraction patterns of the specimen. In addition, spectroscopic information may be provided. Finally, characteristic X-rays and photons are generated. Dominance of secondary signals primarily depends on the energy of the incident electrons and the nature of the investigated specimen.

EM techniques are considered the gold standard for characterizing Au-NPs, as these methods provide information on the particle size and size distribution, as well as on morphology, shape, and crystal structure of the investigated NPs [20]. In SEM, an electron beam with an acceleration voltage of typically 3–30 keV is focused onto the sample, and the back-scattered

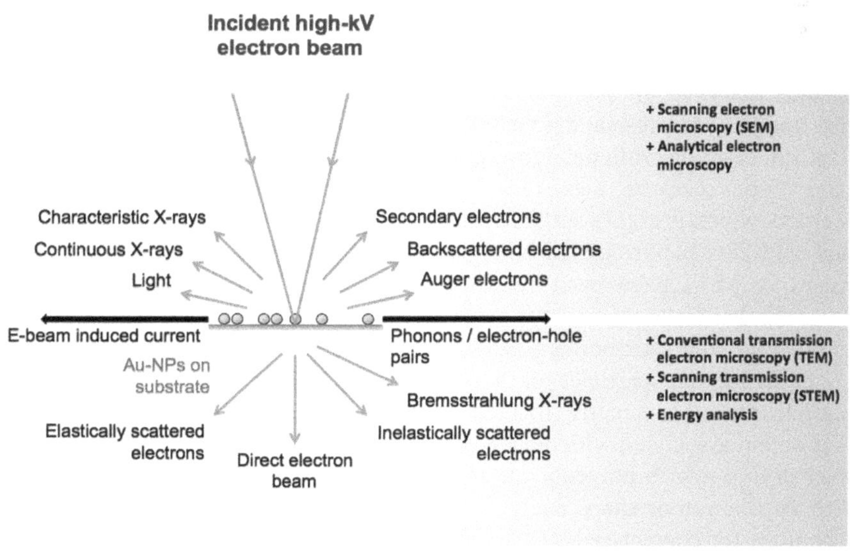

FIGURE 2 Scheme representing the interactions of an electron beam with a solid sample summarizing the generated secondary signals.

electrons or electrons ejected from the sample are used for imaging (Figure 2). SEs are generated by inelastic scattering of the electrons incident via the illuminating beam along their trajectories within the specimen, and have a lower energy. Hence, only SEs that are generated within the escape depth from the surface (i.e., a material property) are observable, which for metals amounts to ~5 nm.

3.1 High-Resolution Scanning Electron Microscopy

SEM characterization of Au-NPs down to sub-nanometer resolution may routinely be achieved with state-of-the-art scanning electron microscopes. Within the last decade, several instrumental developments have assisted in improving the achievable resolution. For example, high-end SEMs are nowadays equipped with field emission sources (i.e., cold tungsten or Schottky zirconium oxide/tungsten cathodes) instead of thermal emitters (i.e., tungsten filaments), which provide substantially increased brightness of the electron beam, yet maintaining a small energy spread (e.g., <0.35 eV for a Schottky field emitter). Such improvements combined with advanced electron optics characterized by reduced spherical and chromatic aberrations result in an achievable spatial resolution of <0.5 nm at 20 keV, and of a few nanometers at 1 keV. It should be noted that chromatic aberrations have no significant impact at high energies (i.e., >10 keV), however, pose a major limitation during low-energy operation [21]. Reducing the working distance, changing the position of the detector, or using in-lens or through-the-lens SE detectors (i.e., immersing the sample in a magnetic field) results in further improvements of the achievable resolution and contrast. In recent years, a significant effort was attributed to the development of low-voltage SEM, thereby resulting in (1) enhanced surface sensitivity, (2) reduced damage of the imaged samples, and (3) a significantly reduced electron penetration depth/interaction volume. Nowadays, a spatial resolution in the range of 0.4–1.6 nm can be achieved via low-voltage HR-SEM rendering this method, which is highly suitable for characterization of the morphology, size, and surface structure of Au-NPs. Recent developments and innovative strategies in SEM have been recently reviewed by Cazaux [22]. For more comprehensive descriptions of modern SEM, the interested reader is directed to corresponding textbooks [23,24].

In SEM, nanoparticles or a droplet of a suspension containing nanoparticles can be directly imaged after drying (i.e., solvent evaporation). A problem associated with dispersed samples is that the NPs may agglomerate after drying, which prevents a useful quantitative analysis of the nanoparticle size distribution or shape analysis. Nonetheless, SEM is considered a routine technique for characterizing Au-NPs, as also corroborated by the fact that the National Institute of Standard and Technology (NIST) has developed a standardized protocol for analyzing the size of colloidal Au-NPs via HR-SEM

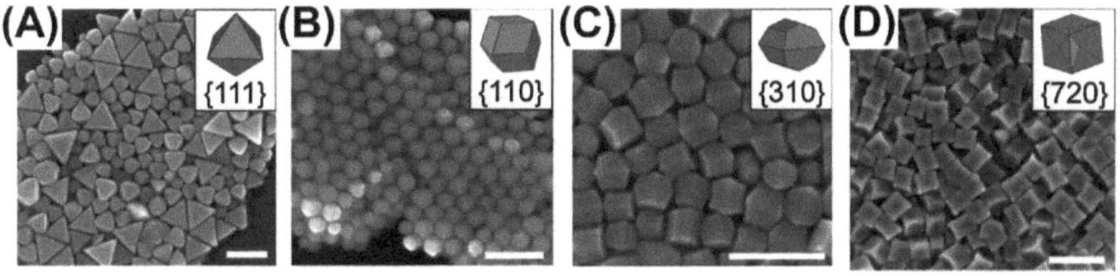

FIGURE 3 SEM images of (A) octahedra, (B) rhombic dodecahedra, (C) truncated ditetragonal prisms, and (D) concave cubes synthesized with different Ag^{+}/Au^{3+} ratios of 1:500, 1:50, 1:12.5, and 1:5, respectively. Scale bars: 200 nm. *Reprinted from Ref. [29] with permission from American Chemical Society.*

including a prescribed procedure for sample preparation [25]. Furthermore, a software package (ImageJ) is provided for image processing, analysis, and measurements. Among few others, NIST also provides Au-NP reference materials denominated RM 8011, RM 8012, and RM 8013, which refer to Au-NPs with nominal diameters of 10, 30, and 60 nm, respectively. For characterizing NPs, such reference materials are of utmost relevance as they enable validating the performance of the applied analytical or microscopic characterization techniques.

Throughout literature, it has been shown that various anisotropic gold nanostructures including nanospheres, nanoprisms, nanocubes, nanooctahedra, nanopyramids, and nanorods are predominantly synthesized by seed-mediated growth, and from reduction of gold salts, whereby the stabilizing surfactant added to the solution has a major influence on the obtained particle shape ([26,27], and references herein). For example, high-index faceted structures, such as bipyramids, rhombic dodecahedra, concave cubes, or tetrahexahedra, have been prepared using cetyltrimethylammonium halide as surfactant. Mirkin et al. have reported the seed-mediated synthesis of gold nanostructures in the presence of silver ions and halides (i.e., chloride, bromide, and iodide) resulting in the growth of 10 distinctly different classes of nanoparticles [28]. The same team has also investigated the influence of the silver ion concentration during seed-mediated growth [29]. Within these and similar studies, it has been clearly demonstrated that HR-SEM—next to TEM—is most suitable for characterizing different anisotropic forms generated in dependence of the synthesis parameters, as illustrated in Figure 3. A major advantage of SEM over TEM is the fact that the three-dimensional structure of NPs is more reliably and readily represented in the resulting SEM images.

3.2 Transmission Electron Microscopy

Advanced EM techniques such as aberration-corrected TEM and the combination of TEM and SEM, i.e., STEM allow the detailed characterization of nanoparticles revealing their structure at atomic detail, and providing compositional information at sub-Ångstrom resolution. In general, TEM and STEM along with a variety of detectors are the leading methodologies for characterizing Au-NPs, as the achievable spatial resolution ranges from the atomic scale, throughout the typical nanoparticle dimensional rage from <1 to 100 nm, and up to the micrometer level. State-of-the-art HR-TEM nowadays provides a resolving power at or close to the 1 Å (0.1 nm) level. For example, the highest point resolution realized in phase-contrast TEM is at ~0.5 Å [30]. In comparison to optical microscopy, the limitation in spatial resolution is not related to the wavelength of the incident beam since high-energy electrons are characterized by picometer wavelengths (i.e., the wavelength of an electron is given by the de Broglie equation), but is rather determined by unavoidable aberrations of the lenses, and the lack of perfectly coherent electron sources [31].

TEM imaging is based on an incident parallel electron beam, which interacts with the atoms composing the nanoparticle via elastic and inelastic scattering processes. The electron wave function is altered due to diffraction by the lattices of the crystal, thereby being reemitted from the surface of the specimen and then transmitted via an objective and subsequent magnifying lenses onto an appropriate detector establishing a corresponding image. Quantum mechanical diffraction theory is the basis for describing the interaction of electrons and matter leading to phase and amplitude changes of the electron wave function. Generally, the diffraction of electrons is the responsible contrast mechanism in bright-field (BF) (Figure 4(A)) and dark-field (DF) TEM imaging based on the variations in intensity occurring during diffraction across the sample. In BF images, the dimension of the objective aperture directly determines the area of information provided in the final image. In contrast, in DF imaging mode, such electrons are effectively excluded as shown in Figure 4(B). Consequently, the DF mode requires that the incident beam is tilted with respect to the optical axis of the objective lens; as a consequence, the electron diffraction pattern collected at the back-focal plane is shifted. This tilt is adjusted such that one Bragg reflection is in line with the optical axis of the objective aperture centered at this Bragg reflection.

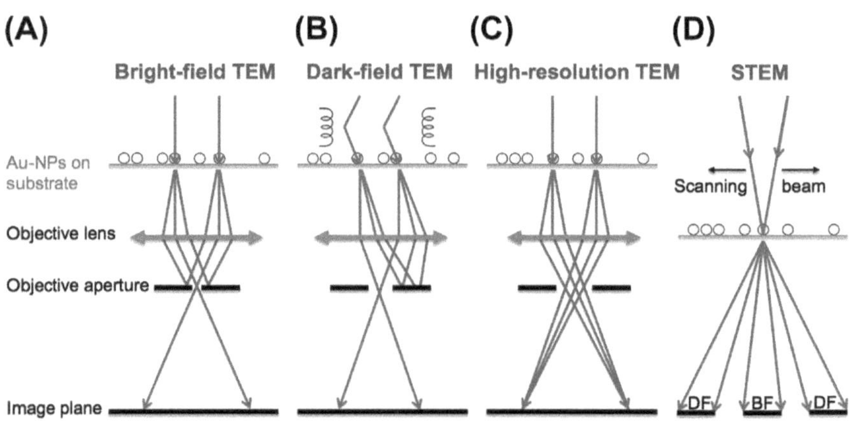

FIGURE 4 Scheme of the fundamental operation modes in electron microscopy.

All other reflections are thereby excluded via the diaphragm, and do not contribute to the image. A crystalline specimen is oriented such that a particularly diffracted beam is generated or a systematic row of reflections is apparent; however, the image remains sensitive to differences in specimen thickness, and to distortions of crystal lattices due to defects, strain, and bending. Therefore, HR-TEM imaging is usually performed in BF mode by including only few Bragg-diffracted beams within the objective aperture. "Phase contrast" and "Z-contrast" methods take advantage of the interference of Bragg-reflected beams with the central beam transmitted along the optical axis. For example, if a crystal is oriented along one of its low-index zone axes (e.g., the $\langle 110 \rangle$ or $\langle 100 \rangle$ plane), strong Bragg-reflected beams are present. For a more comprehensive physical description of the fundamental operation principles in TEM, the interested reader is referred to corresponding textbooks [31].

Inelastic scattering of electrons is the basis for obtaining spectroscopic information, which is frequently termed as the "analytical modes" in EM including methods such as such as electron-energy-loss spectroscopy (EELS), and energy-dispersive X-ray spectroscopy (EDX) providing information on phase, crystallography, and elemental composition. EELS in combination with STEM is the method of choice for characterizing small nanoparticles (i.e., NPs <10 nm), as the detection efficiency is significantly higher compared to EDX. The topic of electron beam-based spectroscopy is the focus of Chapter 7 within this book, and is therefore only mentioned for completion herein.

The sample thickness required for TEM analysis is dependent on the sample composition, the mode of analysis, and the acceleration voltage. Typically, a thickness in the range of 20–300 nm is required, which is well within the range of usual NP dimensions. Hence, elaborate sample preparations including thinning procedures, which are required for bulk samples, are therefore omitted in NP analysis. If the nanoparticles are well dispersed, TEM provides sub-Ångstrom resolution on individual NPs.

Nowadays, designated software packages are available facilitating the statistical analysis of particle dimensions and shape distributions, which is a crucial information for a wide variety of Au-NP applications requiring detailed knowledge, e.g., on their catalytic properties for studying their size-dependent toxicological effects. For example, ImageJ [32] is a highly capable software package for evaluating larger numbers of particles. Based on the size distribution, the average number (d), surface (d_s), and volume (d_v) of nanoparticles may be automatically retrieved from corresponding TEM/STEM images following,

$$d = \frac{\sum_i d_i}{n}; \quad d_s = \frac{\sum_i d_i^3}{\sum_i d_i^2}; \quad d_v = \frac{\sum_i d_i^4}{\sum_i d_i^3} \tag{1}$$

with d ... number; d_s ... surface, and d_v ... volume of the nanoparticles. n reflects the total number of investigated particles. The interested reader is also directed to an interlaboratory study on Au particle size distributions [33].

During the years, TEM methods have emerged as "the gold standard" for NP characterization. Single noble-metal nanocrystals are usually composed of the lowest index crystal planes, while ⟨100⟩ and ⟨110⟩ crystal planes of Pt and Au tend to undergo lattice rearrangements. Multiple twinned particles are frequently observed for nanoparticle and bimetallic nanoparticles, whereby typically near-fivefold symmetries could be confirmed via HR-TEM [34]. The physical and electronic properties of twinned particles are significantly different from untwinned nanoparticles, as their characteristics are strongly dependent on the surface energy and the lattice symmetry. For example, five-twinned nanoparticles with decahedral shape exhibit significantly different physical and chemical properties compared to their untwinned analogs. In general, the process of twinning results from subgrains sharing a common crystallographic plane. Multiple twinning at alternate coplanar planes results in twinned tetrahedral subunits arranged around fivefold axes. Resulting, correlating shape with reactivity requires a thorough and detailed characterization of the actual NP structure. For a comprehensive review on fivefold-twinned nanoparticles, the reader is directed to the pertinent literature [35].

Among the gold nanomaterials, gold nanorods (GNR) play a significant role in nanotechnology due to their unique properties including radiative elastic Rayleigh scattering, inelastic Raman scattering, two-photon-induced luminescence, and unique photoacoustic properties, which renders them an ideal nanomaterial for biological and biomedical applications including molecular imaging and disease diagnostics [36]. The mechanism of GNR growth was controversially debated for quite a while [37]. Finally, HR-TEM contributed significantly to the characterization of GNRs in dependence on the growth conditions. Unlike spherical NPs with an isotropic crystal structure composed of either ⟨111⟩ or ⟨100⟩ planes, the surface structure of GNRs is highly anisotropic comprising different facets. The edges (i.e., the side faces) of GNRs are composed of ⟨110⟩, and their tips consists of either ⟨111⟩ or ⟨100⟩ facets. Pure gold nanorods prepared in the absence of silver ions show a pentatetrahedral-twinned structure, again with the most stable bulk gold facets at the ends of the nanorods [38]. GNRs grown in the presence of CTAB-capped seeds and in the presence of silver ions are characterized by single-crystalline bearing ⟨110⟩ faces at the sides, and ⟨100⟩ at the ends [39].

HR-TEM was also used to investigate the reversible change in size and morphology of Au-NPs. For example, real-time in-situ HR-TEM studies demonstrated the coalescence of individual pairs of decahedral Au-NPs [40]. Changes may also be induced after NP synthesis by introducing appropriate ligands, e.g., alkanethiols (RSHs) or positively charged quaternary alkylammonium surfactants such as didodecyldimethylammonium bromide (DDAB). Stoeva et al. have demonstrated the transition of larger polyhedral nanoparticles into smaller monodispersed ligand-stabilized spherical particles, which then establish ordered nanocrystal superlattices that could be reversibly controlled by the relative ratio of alkanethiol to DDAB [41].

As another example, it has been shown by HR-TEM that moderate annealing temperatures have only minute effects on the average diameters of Au-NPs, however, their shape may change drastically including the appearance of the facets. With increasing temperature, size variations and agglomerations were also observed [34].

A limitation of TEM—as well as other electron beam-based techniques—is the fact that essentially a two-dimensional projection of three-dimensional object is provided, and hence, three-dimensional features of NPs may not effectively be captured, which may lead to erroneous interpretations of shapes and spatial relationships between NPs. Composite nanoparticles may only be investigated in terms of their composition, if the image contrast is sufficiently high (i.e., the Z-number), which is frequently the case for NPs used in catalysis. Also, the volume that may be studied by TEM is rather small, which is particularly true for high-resolution imaging.

Another issue which has to be considered is radiation damage at the sample caused by the high-energy electron beam leading to potential modifications of the imaged NPs. Radiation damage is predominantly related to the ionization of atoms, breaking of bonds, local heating, and knock-on displacements. The induced damage is dependent on the investigated material, the acceleration voltage, and the electron dose. For passivated Au-NPs, such effects have been used to generate carbon networks with embedded gold cores. Such structures have been characterized by Z-contrast STEM imaging and, complementarily via scanning probe techniques [42]. Due to local heating effects of the electron beam, the dodecanethiol-passivation layer is removed, thereby resulting in coalescence and sintering of metal cores, as shown in Figure 5 [43]. The dynamic aggregation and sintering processes are strongly dependent on the electron dose. Sintering of two particles is evident in Figure 5(B)−(D) at the center of the image forming a single larger particle after an electron dose of 33.7 $\mu C/\mu m^2$. Sintering between this new particle and an adjacent pair of nanoparticles is then observed at a dose of 73.7 $\mu C/\mu m^2$ (Figure 5(E)). After exposure to a dose of 149.8 $\mu C/\mu m^2$, an almost rectangular homogeneous island comprising sintered particles is evident (Figure 5(F)). For particles <5 nm, structural transformation and/or rotation of nanocrystals has also been observed [44].

Finally, thorough sample preparation is an important aspect in TEM analysis for omitting artifacts resulting from surface contamination, radiation damage, and structural and chemical changes during the preparation process. Typically, in TEM, the Au-NPs under study or Au colloids are dispersed at the surface of an electron-transparent substrate such as thin carbon-coated microgrids.

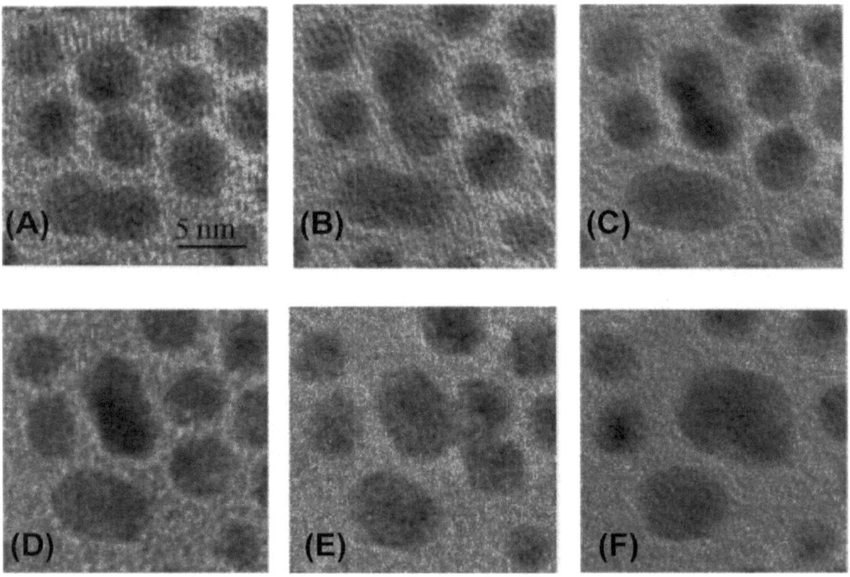

FIGURE 5 False color BF-TEM images of dodecanethiol-passivated gold nanoparticles with a core diameter of 4.8 nm: (A) before and after focused 200 keV electron beam irradiation with a dose of (B) 7.1, (C) 16.4, (D) 33.7, (E) 73.7, and (F) 149.8 $\mu C/\mu m^2$. *Reprinted from Ref. [43] with permission from the American Chemical Society.*

3.3 Scanning Transmission Electron Microscopy

In principle, STEM is operated like SEM, yet scanning a focused (i.e., vs a collimated beam in SEM) high-energy electron beam across the sample surface with electron-accelerating voltages of ~100–300 kV. In contrast to SEM, only thin specimen may be imaged, as—due to the interaction with the sample—the scattered electrons are collected in transmission mode using a series of different detectors that cover a variety of angular ranges. Dedicated high-resolution STEM (HR-STEM) instruments are available; however, also conventional SEMs may frequently be operated as STEM if an appropriate transmission detector is added. However, the acceleration voltage range is usually limited to ~30 kV. In turn, TEMs may be operated in STEM mode, if scanning coils are added. Superior resolution is provided at the sub-nanometer or sub-Ångstrom scale using field emitters and aberration corrector systems, thereby enabling the characterization of nanoparticles in terms of size, shape, defects, crystal and surface structures, and composition. In comparison to TEM, the resolution in STEM mode primarily depends on the dimension of the electron probe (i.e., the focus diameter of the electron beam), and therefore is not sensitive to chromatic aberrations of the objective lens, as indicated in Figure 4(C). Another advantage of STEM is the accurate control of the probe position at the sample surface. Standard detectors for STEM include bright-field detection systems collecting transmitted electrons at relatively low angles vs the optical axis (i.e., smaller than the incident beam convergence angle), and dark-field detectors, which are collecting transmitted electrons at relatively high angles with respect to the optical axis. Setting the inner angle of the annular dark-field detector to a large value, only elastically scattered electrons (i.e., in contrast to Bragg-diffracted electrons) are collected, which is known as "Z-contrast" or "high-angle angular dark-field" imaging (HAADF) due to the enhanced atomic number (*Z*) dependence of the imaging contrast [45,46].

The scattering behavior of a single atom is related to the scattering factor, which is given by the Mott equation as,

$$f_e(s) = e\left[Z - f_x(s)\right] / 16\pi^2 \varepsilon_0 s^2 \quad (2)$$

with f_x … X-ray scattering factor of the atom, Z … atomic number, s … a variable related to the scattering angle ($s = \sin h/k$), and k … electron wavelength.

For example, three-dimensional HAADF-STEM tomography of heterogeneous catalysts consisting of nanometer-sized noble metal particles on a heavy-metal substrate ($Ce_{0.50}Tb_{0.12}Zr_{0.38}O_{2\text{-}x}$) has been presented by J. C. Gonzalez [47]. The contrast between the atomic number (*Z*) of gold particles ($Z = 79$), and the oxide support material (average Z of ~23) is sufficient for enabling high-contrast imaging of small (i.e., ~1 nm) Au particles even at relatively thick regions of the substrate.

An important topic in heterogeneous catalysis is the fact that high-surface-area catalysts comprise a wide range of structures, which may or may not be entirely responsible for the activity, and therefore only partly contribute to the catalyzed reaction. Aberration-corrected HR-STEM was used to investigate such heterogeneous catalysts, which was convincingly demonstrated for a highly active Au/TiO_2 catalyst converting carbon monoxide. Pennycook et al. could demonstrate that a broad distribution of clusters and small nanoparticles are present in such a catalyst material of high activity [48].

Ramachandra et al. could show that spatial frequencies of 1/1.2 Å could be achieved, and that individual gold atoms comprised within Au-NPs could be resolved using 30–50-nm-thick SiN membranes as substrate. Imaging was performed via an aberration-corrected STEM at an acceleration voltage of 200 keV using HAADF detection [49]. Recently, time-resolved nanobeam electron diffraction combined with HAADF-STEM imaging has been used to determine the structure of $Au_{144}(SCH_2CH_2Ph)_{60}$ clusters with atomic resolution [50].

STEM was also applied for characterizing the factors influencing the growth and final shape of bimetallic nanostructures. The growth of core–shell bimetallic nanoparticles from a pseudospherical Au seed particle into a bimetallic Ag–Au icosahedron structure was recently demonstrated by Mirkin et al. [51]. Eight intermediate structures could be identified during the stepwise growth processes, whereby the particular ⟨111⟩-faceted particle shapes all

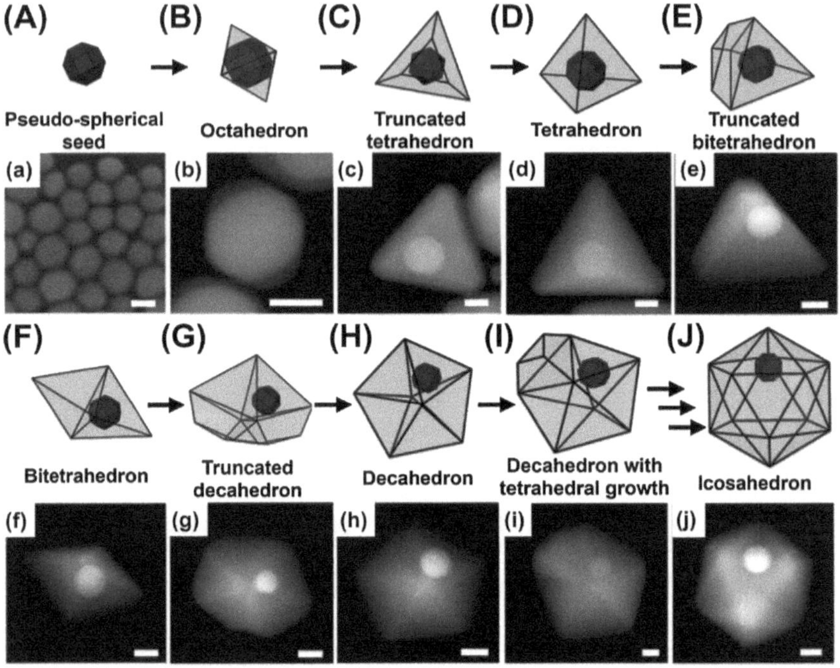

FIGURE 6 Models and representative STEM images of the growth pathway of a single-crystalline seed in plasmon-mediated reactions. (A) STEM image of pseudo-spherical single-crystalline Au seeds. (B)–(J) STEM images of bimetallic particles with (B) octahedral, (C) truncated tetrahedral, (D) tetrahedral, (E) truncated bitetrahedral, (F) bitetrahedral, (G) truncated decahedral, (H) decahedral, (I) decahedral with an additional tetrahedral growth, and (J) icosahedral morphologies, which were all obtained as products of the same reaction. Scale bars: 25 nm. *Reprinted from Ref. [51] with permission from American Association for the Advancement of Science.*

consist of 2, 5, and 20 tetrahedral subunits. For the individual structures, the spherical gold cores were differentiated from the triangular prism silver shells according to their difference in electron-scattering efficiencies, as shown in Figure 6. Different polyhedrons were produced in the same reaction from a spherical gold seed. These particular ⟨111⟩-faceted particle shapes were all related such that they consisted of 2, 5, and 20 tetrahedral subunits.

In general, HAADF-STEM is typically used for elucidating the composition of alloyed nanoparticles such as bimetallic Au/Co nanoparticles [52], Au/Ag particles [53], and Pd/Au nanoparticles [54]. Au/Co nanoparticles initially form core–shell NPs as they are synthesized via condensation in an inert gas atmosphere. After thermal treatment at ambient atmospheric conditions, thus obtained Co NPs transitioned their morphology into a fine layer forming a perfect interface with gold. The small size of the bimetallic NPs

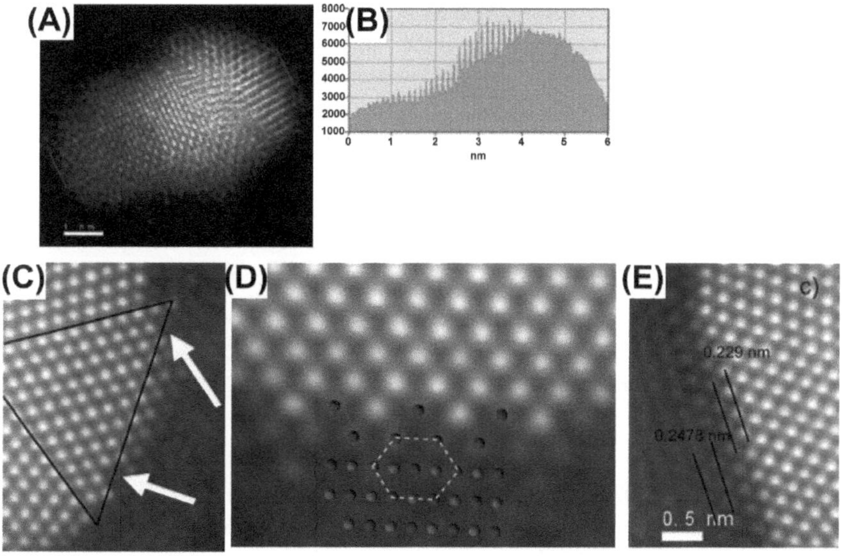

FIGURE 7 Cs-corrected STEM-HAADF images of (A) Au/Co nanoparticle with a cobalt-rich zone (left) and gold-rich zone (right). (B) Intensity profile (in arbitrary units) of the area marked in a) showing the significant difference in intensity due to the difference in the atomic number. (C) Au vacancies where the alloying is occurring. (D) Au—Co interface with a ball-and-stick model of CoO superimposed on the region rich in cobalt (blue: Co atoms; red: O atoms). (E) Interface between Au and Co and the characteristic distances between different atomic layers. *Reprinted from Ref. [52] with permission from the Royal Chemical Society.*

facilitates the oxidation of cobalt, thereby leading in case of cobalt crystals or cobalt-rich zones to the partial or complete formation of cobalt oxide. Figure 7 shows Cs-corrected STEM-HAADF images of Au/Co nanoparticle with a cobalt-rich zone (Figure 7(A)). The intensity profile (in arbitrary units) of the area marked in a) shows the difference in intensity correlated to the difference in the atomic number of Au and Co (Figure 7(B)). The atomic resolution of Au vacancies where the alloying process is occurring could be revealed, and the associated ball-and-stick model of CoO fitted particularly well with the observed Au—Co interface (Figure 7(D)).

Without further detailing herein, for investigating composite materials or bimetallic particles, analytical detectors such as EDX and EELS in combination with HR-STEM are particularly useful.

4. SCANNING PROBE MICROSCOPY

The introduction of STM [55] certainly marks a milestone in nanotechnology, as true atomic resolution and three-dimensional imaging of nanomaterials could be achieved. Derived from STM, the development of a whole series of SPM techniques was initiated, with some of them providing correlated

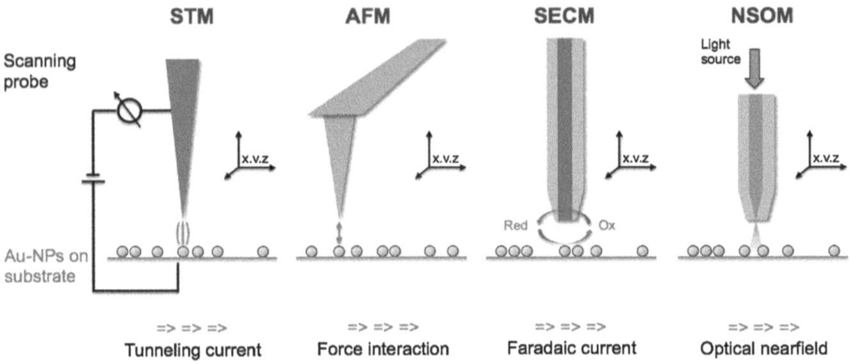

FIGURE 8 Scheme representing scanning probe techniques and the fundamental probe—sample interactions.

structure/activity information. Among those SPM techniques, AFM, NSOM, scanning electrochemical microscopy (SECM), and scanning electrochemical cell microscopy (SECCM) have been used in nanoparticle research providing important contributions for their characterization. The common theme of scanning probe techniques is the utilization of a variety of physical or chemical "near-field" interactions between a scanning tip (frequently also called a "probe") and a sample. As these interactions are recorded as a function of the tip position versus the sample, structural, morphological, physical, and chemical information associated with the sample surface is provided at microscopic to nanoscopic resolution (i.e., depending on the applied SPM technique). The respective tip—sample interaction is recorded, once the tip is scanned in close proximity across the sample surface usually positioned via piezoelectric control elements enabling highly accurate *x*,*y*,*z*-positioning [56], and a feedback-control system to maintain the probe at constant distance to the sample surface. An overview of common SPM techniques and a schematic representation of the underlying near-field interaction particularly relevant to nanomaterial research are given in Figure 8. Among the SPM techniques, AFM is certainly the most commonly used scanning probe method for nanoparticle characterization. Compared to the previously discussed vacuum-based EM techniques, the major advantages of SPMs are considered (1) that analyzing Au-NPs provides true three-dimensional information, (2) that the sample preparation is comparatively easy, and (3) that information may be obtained at ambient, vacuum, and even liquid conditions.

4.1 Scanning Tunneling Microscopy

STM essentially provides spatially resolved maps of the local density of states (LDOS) at the Fermi level of the investigated sample. The signal generation is based on mapping the tunneling current between a sharp conductive tip (e.g., W or Pt/Ir), and a conductive or semiconductive sample surface due to the overlap of the electronic wave functions between the tip and the surface, which decay exponentially within the barrier gap. Applying a small bias voltage (V_t) between sample and tip induces elastic tunneling of electrons from filled electronic states of the metallic tip into empty states of the sample or vice versa depending on the polarity of V_t at the Fermi level.

For electronic states, the electrons have to overcome the vacuum potential barrier, which is corresponding to the work function of the material. The tunneling current varies exponentially with the vacuum gap, and has a decay length of,

$$k = \frac{\sqrt{2m\Phi}}{\hbar} \tag{3}$$

with κ … decay length, m … mass of the electron, and $\hbar$ … reduced Planck constant ($= h/2\pi$). For metals, the decay length is on the order of ~ 0.4 Å. Usually, the STM tip is scanned across the sample surface at a fixed bias voltage utilizing piezoelectric positioners to control the x,y,z scan range. As the tunneling current strongly depends on the distance, changes in current reflect the position of the atoms at the sample surface. For corrugated samples, an electronic feedback circuit is used to maintain the tunneling current constant at a set value (i.e., typically between 0.5 and 10 nA). Differences between the set-point current and the measured current are converted to a voltage, which is then applied to the piezoelectric scanner and regulates the z-position of the tip to keep the tunneling current constant at the selected set-point. This procedure allows keeping the distance between tip and surface constant (i.e., also called "constant current mode"). It has to be noted that the images recorded via this procedure represent complex convolutions of surface geometry and electronic structure. Hence, such images have to be carefully interpreted as they rather represent images of the LDOS, and not simply the surface topography. Initially, STM was predominantly developed for surface investigations at ultra-high-vacuum (UHV) conditions. However, early studies have already demonstrated that this technique may also be operated at ambient conditions or in electrolyte solution [57].

STM is a noncontact technique, and hence, artifacts due to interactions of nanoparticles with the tip are usually not occurring, which is among the most significant advantages compared to electron-beam-based techniques. In EM, interactions of high-energy electrons with the sample may lead to changes and/or damage of the sample, which in turn may give rise to topological artifacts. In turn, STM is limited to imaging conductive or semiconductive substrates.

$Au_{55}(PPh_3)_{12}Cl_6$ particles have been investigated at low temperature at UHV conditions via STM, whereby the electron-rich phenyl rings of the PPh_3 ligands could be clearly resolved [58]. By changing the bias voltage, tunneling spectroscopy provided additional information on the electronic levels splitting in the Au_{55} nucleus.

STM is also commonly used to characterize monolayer-protected metal nanoparticles. Long-chain alkane thiols are widely used as surfactant molecules for stabilizing Au-NPs during synthesis via chemical reduction, thereby preventing coalescence. In addition, the nature of the surfactant molecules also determines the potential interactions between the particles and their environment. For gold, self-assembly of thiols at the surface is driven by particularly strong covalent metal–sulfur bonds (i.e., $\sim 50\ \text{kcal mol}^{-1}$). For Au-NPs, molecule–surface interactions depend on the crystallographic orientation of the NP surface. In turn, the functionality, conformity, and stability of thus established self-assembled monolayers (SAMs) remain a key factor for the application of surface-modified Au-NPs. It is known that mixed SAMs (i.e., using alkane thiols with different linker length) assemble at flat gold surfaces within phase-separated yet randomly shaped domains, which is induced by the differences in chain length [59,60], or functionalization chemistry [61]. Stellacci et al. have published a series of impressive STM investigations concerning the formation of mixed SAM layers at Au-NPs [62–65].

At Au-NP surfaces, such monolayers establish specific morphologies in dependence of the ligands and the NP core size [66]. For example, mixtures of thiol molecules co-assembled at ~5-nm-diameter Au-NPs with a ratio of the thiol of 2:1 and 1:1 resulted in stripe-like domains of ~1 nm [62,64], which could be visualized by STM measurements (Figure 9). The formation of such stripes relates to the competition between enthalpic losses and entropic gains at the phase boundaries of such mixed layers.

Ripple formation was observed at Au-NPs ranging in core diameter from 2.5 to 6 nm using different binary mixtures, thereby ensuring that the ligands contained a thiol and a significantly different linker length. These studies revealed that the surface curvature plays a significant role in the formation of SAMs at Au-NPs. An advantage of STM in contrast to TEM is that structures of the ligands could be observed next to the Au-NPs, whereas in TEM only the Au core could be visualized. Stellacci et al. have recently succeeded in the formation of threaded nanoparticles. If a curved surface is coated with ordered monolayers, at the poles of the particle, singularities must form. The molecules placed at these polar defects have a function like "chemical handles," and facilitate the formation of chains of Au-NPs [67].

An interesting instrumental development was initiated in 1988 by Spence et al. [68], which comprises the integration of an STM into a TEM. To fit in between the pole pieces of the TEM, the STM has to be implemented within a specialized mount. Almost a decade later, Takayanagi et al. have refined this concept by fitting a miniaturized STM into a standard TEM sample holder facilitating studies on the relation between electron transport and structure of quantized nanowires [69]. A commercial TEM–STM system is nowadays available from Nanofactory (Göteborg, Sweden). However, despite the

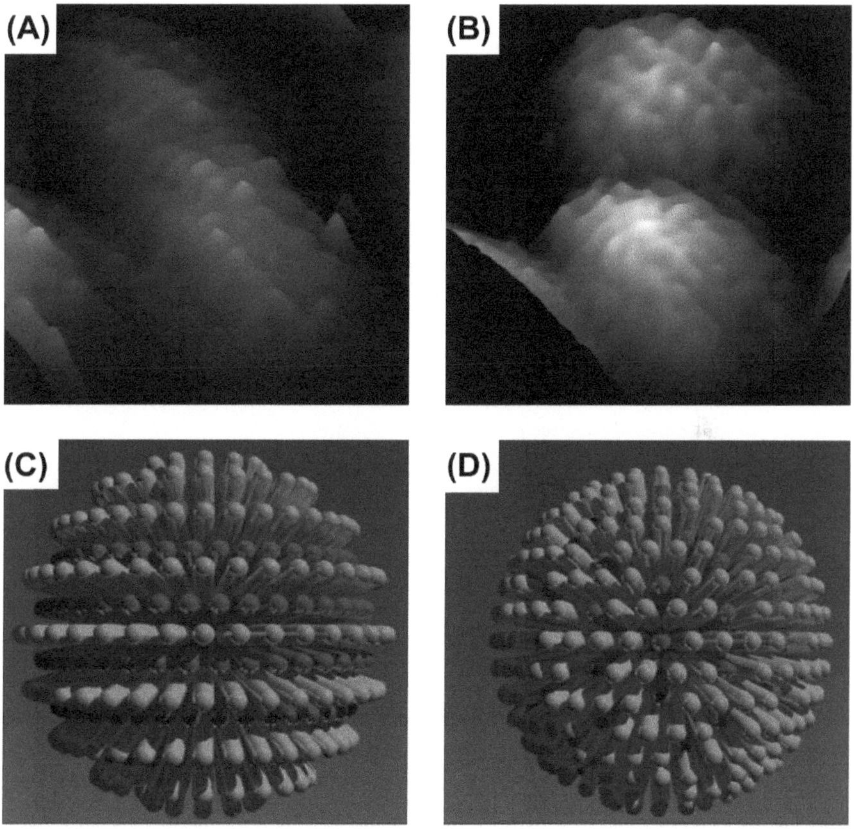

FIGURE 9 Three-dimensional rendering of STM height images of gold nanoparticles. (A) Decanethiol/mercaptopropionic acid (MPA) (2:1 M ratio) showing ripples, and (B) octanethiol (OT)/mercaptopropionic acid (MPA) (10:1 M ratio) showing packed, phase-separated domains without ripples. (C), (D) Schematics of (A) and (B), respectively. *Reprinted from Ref. [62] with permission from Nature Publishing Group.*

potential of this instrumental combination to our best knowledge, only gold nanowires [70,71] but not Au-NPs have been studied to date with in-situ STM–TEM.

4.2 Atomic Force Microscopy

AFM and its derivatives are the most commonly used scanning probe techniques in nanotechnology. AFM has the unique advantage that qualitative and quantitative information on the physical properties of NPs such as size, morphology, surface texture, and roughness can be obtained. In addition, statistical information including dimensional heterogeneity, surface area, and volume distributions may be derived at ambient conditions and requiring little

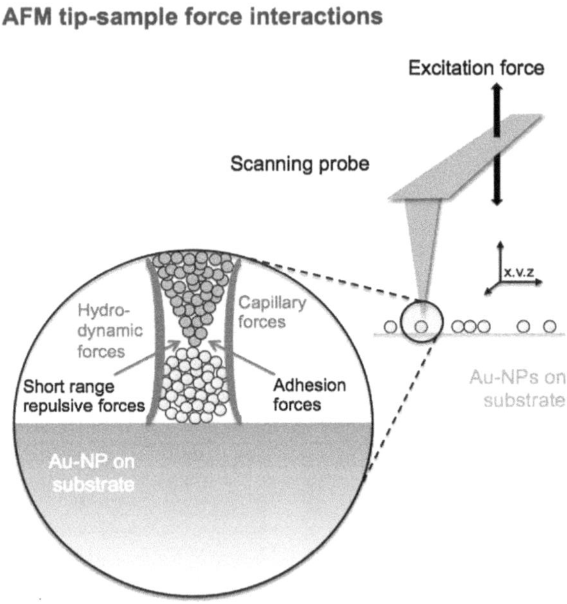

FIGURE 10 Schematic representation of the dominant forces in AFM.

sample preparation. The latter are distinct advantages compared to electron beam techniques. AFM was introduced shortly after the invention of the STM in 1986 [72]. As outlined in Figure 10, AFM is based on analyzing attractive and repulsive forces between a sharp tip located at the end of a flexible cantilever and the sample surface. Potential force interactions include long- and short-range attractive forces such as van der Waals (dipole) forces, capillary forces, electrostatic interactions, and chemical forces. Furthermore, attractive or repulsive forces such as magnetic forces, hydrophobic interactions, steric interactions, repulsive Born forces, and double-layer interactions may be used as contrast mechanism. This wide variety of force interactions provides a diverse basis for a multitude of AFM-based scanning probe techniques including lateral force microscopy (LFM), magnetic force microscopy, electrostatic force microscopy, scanning thermal microscopy, scanning capacitance microscopy, and scanning Kelvin probe microscopy. The direct relationship between the interaction force and the displacement of the cantilever is described by Hook's law. In almost all commercial AFM designs, the deflection of the cantilever is detected via a reflected laser beam and a photodetector. The measured photon current is converted into a voltage, which is translated into an error signal that is used to regulate the distance between the tip and the sample surface (i.e., constant force mode). Scanning the tip or the sample, therefore, results in a three-dimensional image of the surface topology, but may also provide information on the nanomechanical and nanotribological properties of the sample. Depending on the nature of the sample and the desired resolution, three main modes of AFM are used for imaging: (1) contact mode, (2) intermittent (i.e., dynamic) mode, and (3) noncontact mode. For relatively hard samples (e.g., for material characterization), measurements are performed in contact mode with the tip in direct physical contact with the sample surface [73]. This contact area is characterized by short-range repulsive forces resulting from overlapping orbitals of tip and sample atoms with the range and contact area of the interactions ideally at Ångstrom levels for achieving atomic resolution. However, long-range forces such as magnetic forces, electrical forces, Van der Waals forces, and capillary forces occur in addition, and increase the force imposed at the contact area. Force–distance curves adequately describe the tip–sample interaction in close proximity [74,75].

Contact-mode imaging is less suitable for the investigation of soft samples or samples with little adhesion to the substrate, since frictional forces may, e.g., "push" nanoparticles across the surface or soft samples may be damaged due to the force interaction. Dynamic operation modes can be classified into two main methods: (1) amplitude modulation (AM)-AFM [76,77], and (2) frequency modulation (FM)-AFM [78]. In AM operation, the cantilever is excited with a constant amplitude and frequency close to or at its free resonance frequency. Due to force interaction with the sample, the oscillation amplitude is damped and—as already described for the contact mode—the oscillation amplitude can be used in a feedback-back system via a set-point providing topographical information. The amplitude of the oscillation in AM-AFM is usually >10 nm (i.e., typically ~100–200 nm). The phase shift between excitation and oscillation provides additional information on the elastic properties of the sample, and hence, facilitates additional compositional contrast. The main advantage of AM-AFM, which can be used at ambient conditions [79] and in liquid [80], is the reduced tip–sample contact time resulting in minimized frictional forces, and hence, avoiding damage at soft samples. FM-AFM or non-contact-mode AFM is mainly used in UHV conditions ideally providing atomic resolution. Recently, FM-AFM with exceptional resolution has also been demonstrated in solution [81]. The cantilever is excited to oscillate at fixed amplitude, and the change in resonance frequency is recorded. The Eigen frequency is correlated to the spring constant k and effective mass m^* of the cantilever via $f_0 = 2\pi(k/m^*)^{1/2}$. The force interaction between tip and sample alternates the frequency $f = 2\pi(k^*/m^*)^{1/2}$, where $k^* = k + k_{ts}$, and which reflects the change due to the tip–sample force gradient. If k_{ts} is small compared to the force constant of the cantilever, the frequency change Δf may be simply expressed by $\Delta f = f_0 \cdot k_{ts}/2k$ provided that k_{ts} is constant over the distance range covered by the oscillating cantilever [82]. The frequency shift can again be used in a feedback circuit adjusting the sample height, thereby resulting in the topographical image. Excellent books and reviews on different operation modes of AFM are available to the interested reader [73,83,84].

For AFM studies, nanoparticles are usually only physisorbed at the substrate surface. Hence, frequently intermittent-mode AFM is preferred over contact mode for imaging NPs. However, height reductions in dynamic mode operation in air have been observed due to strong long-range attractive capillary forces resulting from water layers always present at ambient conditions. The contribution of these adhesion forces in dynamic-mode AFM may result in a significant discrepancy between the recorded heights and the real size of the imaged object, in particular for samples with varying hydrophobicity [85] (and references therein).

The lateral resolution of AFM is an important aspect, which has to be considered when imaging nanoparticles. The vertical resolution of AFM is generally determined by the force interaction between the tip and the surface, and is very accurately controlled by the z-piezo; in contrast, the lateral resolution is determined by the finite dimensions of the AFM tip. Hence, a correct sample height, yet an overestimated sample width, is usually obtained. The phenomenon that imaged objects appear wider than they are is known as "tip dilation" or "convolution," and is particularly relevant when characterizing nanoparticles [86]. If the tip moves across a spherical sample feature, the lateral dimension reflects the combination (i.e., the convolution) of the two geometries (i.e., tip and sample), as schematically illustrated in Figure 11. Several mathematical models have been proposed to deconvolute

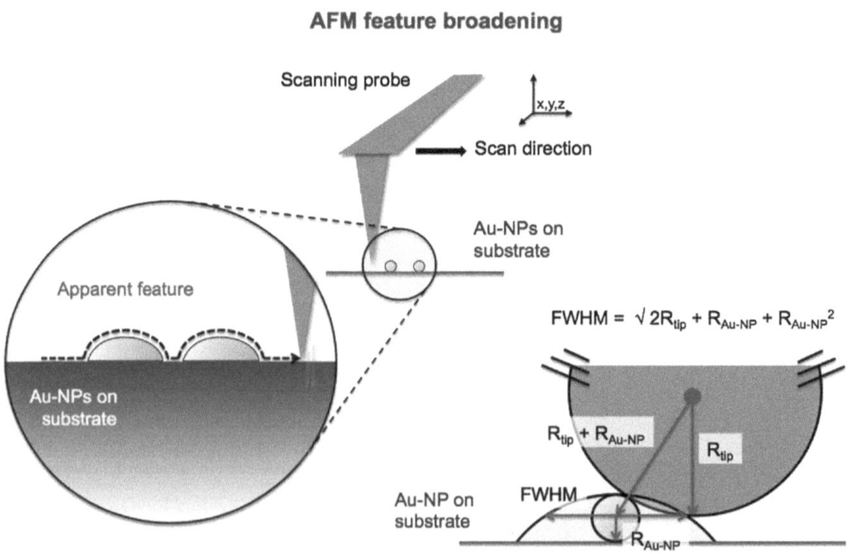

FIGURE 11 Scheme of tip convolution. The lateral dimension results from a combination of the imaged feature and the tip geometry.

the obtained data assuming a spherical, hyperbolic, or parabolic tip geometry, which is again a simplified model since the true tip geometry usually does not have a simple shape. Furthermore, these shapes may vary between different probes. Hence, an individual tip characterization by SEM or TEM is required, a tedious procedure which again does not necessarily reflect the true three-dimensional shape, if not a series of rotated images is superimposed. Hence, so-called "blind reconstruction algorithms" were developed to deconvolute AFM data, in particular, data recorded for nanoparticles. In this procedure, well-characterized nanomaterials such as nanoparticles of controlled and well-known geometry or arrayed nanostructures with known geometric parameters are frequently used to elucidate the actual AFM tip morphology. Deconvolution procedures for asymmetric and symmetric tips have been introduced during recent decades [87–91]. However, it has to be considered that the effectiveness of this procedure depends on how well these assumptions correspond to the actual tip geometry. The measurement of size distributions of spherical nanoparticles with AFM based on vertical height measurements and tip deconvolution has recently been reported by Boyd and Cuenat [92].

Alternatively, higher aspect ratio AFM probes such as carbon nanotube-modified AFM tips [93], or oxide-sharpened high-aspect-ratio probes with curvatures in the range of few nanometers may be used to minimize tip dilation artifacts. After screening the literature, it has to be noted that SEM and TEM are the predominantly used methods for characterizing size and shape of Au nanoparticles, whereas AFM is used for size distributions and preferentially for characterizing composite materials involving Au-NPs.

AFM enables analyzing three-dimensional topologies from the Ångstrom level to the scale of tens of micrometers with unprecedented resolution. Hence, AFM is frequently utilized for optimizing synthesis conditions, as the produced Au-NPs may be readily characterized with minimal sample preparation providing guidelines for adapting the synthesis strategy. For optimized imaging conditions, NPs should adhere reasonably well to a flat substrate surface. Most surfaces (i.e., substrates and nanoparticles) are negatively charged leading to poor adhesion. As an example, a method for improving the adhesion at an atomically flat substrate surface such as (freshly cleaved) mica is coating the surface with e.g., poly-L-lysine. Alternatively, hydrophobic substrates may be used for imaging in air, as the water monolayers causing strong capillary forces can be effectively avoided. The influence of several experimental parameters such as the solvent, the spin speed, pH, concentration, and nature of the substrate was investigated for spin-coated NPs by Hoo et al. [94].

For example, AFM was used to study the time-dependent growth kinetics, at different temperatures, of Au-NPs produced via Turkevich citrate synthesis [95]. Suspensions of Au-NPs prepared after different synthesis periods (1, 10, 40, and 70 min at 70 and 90 °C, respectively) have been transferred to mica substrates, and the size distribution and growth rate constants k_1 and k_2 for the different growth temperatures have been determined. In this study, it could be confirmed that based on the Finke–Watzky reaction mechanism, a higher temperature led to increased growth rates.

Lee et al. investigated the binding affinity between Au nanoparticles and chemically modified silicon substrate using quantitative nanomechanical property mapping via LFM [96]. This particular technique allows simultaneous recording of adhesion and deformation maps, thereby providing valuable information on surface modifications. Catecholamine-modified and amine-modified surfaces have been characterized in terms of their efficiency for binding Au-NPs.

The behavior of pH- or temperature-responsive Au-NP composite materials has also been studied with AFM [97–100]. pH-sensitive PVP polymer can be easily functionalized via alkylogen leading to a positively charged polymer, which is an interesting material for usage within sensors and actuators. Surface-initiated atom-transfer radical polymerization has been used to obtain pH-responsive PVP–Au-NP composite materials [97]. Thus prepared Au–PVP nanocomposites revealed a core–shell structure with gold cores and polymer shells at low pH values. It could be demonstrated via in-situ AFM measurements—as shown in Figure 12—that changing the pH results in aggregation of PVP–Au-NPs due to the collapse of the polymer layers. At pH 1.2, dispersed core–shell

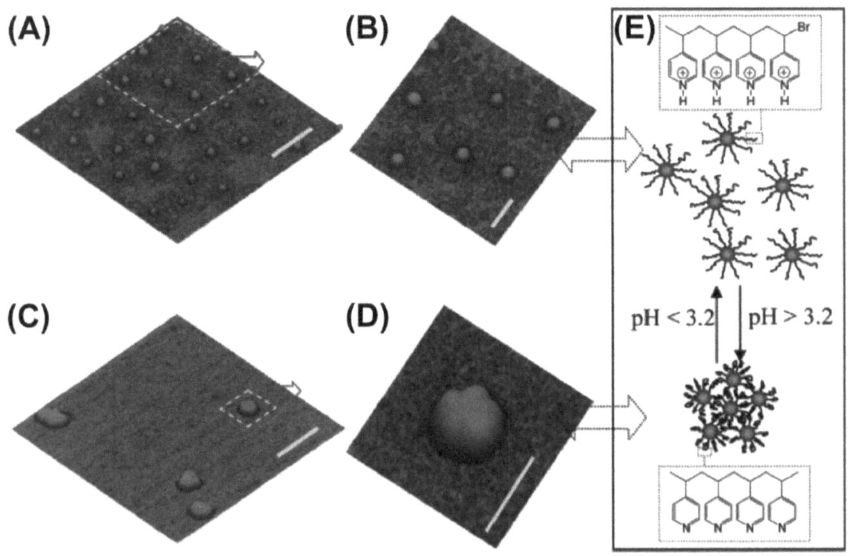

FIGURE 12 3D AFM images of gold nanoparticle and poly(4-vinylpyridine) nanoparticles (PVPAu-NPs). (A) and (B) at pH 1.2, (C) and (D) at pH 5.0 at different magnifications. (E) Schematic mechanism of the pH response of PVP–Au nanocomposites. Scale bar: 500 nm in panels (A) and (C), and 200 nm in panels (B) and (D), respectively. *Reprinted from Ref. [97] with permission from American Chemical Society.*

nanoparticle with a height of approximately 25.3 nm is evident in the AFM image (Figure 12(A)). Changing the pH to 5.0 results in aggregated PVP–Au-NPs comprising several up to tens of aggregated particles (Figure 12(C)). Figure 12(D) shows a large aggregation of nanocomposite particles with a height of 35.6 nm, which relates to six individual nanoparticles with disappeared branches. Finally, it could be shown that the pH behavior is reversible for the investigated composite nanoparticles with a translation point at pH 3.2.

Analogously to STM–TEM, also AFM was combined with TEM for in-situ friction, adhesion, and elasticity measurements at nanomaterials. Similar to STM–TEM, a standard AFM cantilever at the fixed position is placed in a TEM–STM holder, as described above [101]. However, optical detection of the cantilever deflection is not possible in this configuration, and force measurements are thus obtained via nonoptical read-out mechanisms such as capacitive or piezoresistive detection schemes. A significant redesign of the TEM to accommodate an AFM system maintaining optical read-out for studying force interactions has been demonstrated by Kizuka et al. [102]. Again, nanomaterials such as nanowires and carbon nanotubes have been characterized using such in-situ combined techniques; however, to date no reports have been published focusing on Au-NPs with this promising combined technology.

4.3 Near-Field Scanning Optical Microscopy

NSOM is a scanning probe technique that takes advantage of photonic near-field interactions between a tip providing a sub-diffraction aperture or an aperture-less near-field probe and a sample surface [103–105]. Overcoming the Abbe limit via a near-field probe enables recording optical images below the conventional resolution limits in optical microscopy, i.e., beyond ~150 nm (with green illumination). Using the near-filed aperture approach, usually a thinned optical fiber coated with a metal film at the sidewalls propagates radiation via a sub-micrometer- to nanometer-sized aperture at the apex of the fiber onto the sample surface. Alternatively, this fiber may be used for collecting light reflected off an illuminated surface through this aperture. Aperture-less approaches take advantage of light scattered at a sharp and usually metallized tip (e.g., a metal-coated AFM probe) facilitating sub-wavelength detection [106].

While NSOM may in principle be used at ambient (i.e., gas phase) conditions as well as in liquids, AFM and STM imaging operation in liquid remains challenging. For example, distance control of the NSOM probe is usually based on a shear-force detection scheme, as discussed above also for SECM operation. In this case, the fiber oscillates at its resonance frequency, which is dampened in close proximity (i.e., <15 nm) to the sample surface. This change in oscillation frequency is usually optically detected through the liquid sample, which limits the utility of this technology for liquid-phase studies. Given that the resolution remains limited to >50 nm, only certain dimensions of Au-NPs may be investigated utilizing NSOM techniques.

Recently, Chen et al. have shown studies at Au-NPs with diameters of ~10 nm via an aperture-less (i.e., scattering-type) NSOM approach at a wavelength of 633 nm. However, extensive modeling was required for understanding size effects, as well as interactions between light, tip, and sample, and for providing an accurate interpretation of the obtained NSOM images [107].

In another example, it was demonstrated that enhanced optical field distributions may be observed at one-dimensional linear array structures composed of Au-NPs via NSOM studies [108]. The characteristic distribution of the enhanced field was demonstrated to arise from interparticle interactions giving rise to propagating plasmons trapped by a localized mode at the edges of the Au-NPs, as also confirmed via finite-difference time-domain calculations.

Last but not least, of course plasmonic properties of Au-NPs can be probed via NSOM techniques. Based on near-field optical antenna effects, the line shape of the plasmon resonance obtained at individual Au-NPs was analyzed [109]. Based on Mie theory, the dephasing times of surface plasmons were extracted from the associated near-field spectra obtained at individual Au-NPs and favorably compared with corresponding calculations.

In summary, albeit providing the opportunity of performing optical microscopy beyond the classical diffraction limit, NSOM appears of limited utility for characterizing Au-NPs, as similar information may be obtained using a variety of alternative microscopic techniques, as discussed in this chapter.

4.4 Scanning Electrochemical Microscopy

In-situ characterization of nanoparticles in many applications—foremost, in catalysis—is a prerequisite for advanced applications of such materials, as it has been shown that there is a strong correlation between the structure of NPs and their catalytic activity [48,110,111]. An emerging technique, which provides information on dynamic in-situ processes at the solid/liquid interface, and which is highly suitable for determining heterogenic rate constants is SECM, which was introduced by Bard et al. [112,113]. While for the majority of applications shown, since its introduction, SECM provides limited spatial resolution for investigating nanoparticles, recent promising developments have demonstrated the potential of this microscopic analytical technique for nanoscale studies [114–119].

In conventional SECM, microelectrodes are scanned in close vicinity (i.e., at a distance of several electrode radii) across the sample surface, while local variations in electroactivity are probed by voltammetric methods revealing laterally resolved electrochemical information. Among other factors including distance and active electrode radius, the electrochemical signal is strongly dependent on the nature of the sample surface, and on occurring surfaces processes. Over the years several imaging modalities have been developed, which are discussed in several review articles [120–122]. The fabrication of nano-sized electrodes can nowadays be achieved following different (micro) fabrication routes [123], and their unique properties such as low background currents, reduced ohmic iR drop, and minimized capacitive charging render them highly useful for elucidating the relationship between particle size, structure, and electrochemical/catalytic activity.

Although Au-NPs and alloyed Au-particles are efficient catalysts, e.g., for low-temperature CO oxidation processes, selective hydrogenation of N–O bonds, and direct formation of hydrogen peroxide, most reported SECM investigations related to nanoparticles concerned Pt, Pd, and alloyed nanomaterials. Stimming et al. [124] have investigated the catalysis of single Pd nanoparticles immobilized on bulk Au ⟨111⟩ electrodes, and hydrogen evolution has been imaged at the single nanoparticle level. It was finally shown that the hydrogen evolution kinetics of Pd nanoparticles is apparently different from bulk Pd, which was related to strain at the surface of the Pd nanoparticles.

Within recent years, several novel approaches have been pursued in SECM, which succeeded to overcome some of the major limitations in conventional SECM (i.e., scanning a microelectrode in constant height across the sample surface). Scanning the microelectrode in constant height may lead to falsified electrochemical data, as topographical changes may be convoluted with electrochemical activity changes since the current signal is dependent not only on the local reactivity but also on the distance between sample surface and microelectrode. This effect is even more pronounced, if sub-microelectrodes or nanoelectrodes are used for imaging. Several developments in SECM have therefore focused on overcoming this problem by enabling strategies for independently yet synchronous recording of electrochemical and topographical information. Among these approaches, the most relevant are shear force-based constant distance modes [125–128], and combined scanning probe techniques such as atomic force scanning electrochemical microscopy (AFM-SECM) [129–133] or combined scanning ion conductance–scanning electrochemical microscopy [134–136]. Another development was shown by Unwin et al. using a pulled theta pipet as localized electrochemical cell, which was termed SECCM [137]. Both barrels were filled with electrolyte solution, and each barrel contains a quasi-reference electrode or combined reference-and-counter electrode (i.e., quasi-reference counter electrode) (Figure 13(A)). Close to contact, the meniscus at the end of the barrels forms a defined volume contacting sample and barrel, which in turn changes the measured ion current. This signal can be used for a feedback loop enabling constant distance operation.

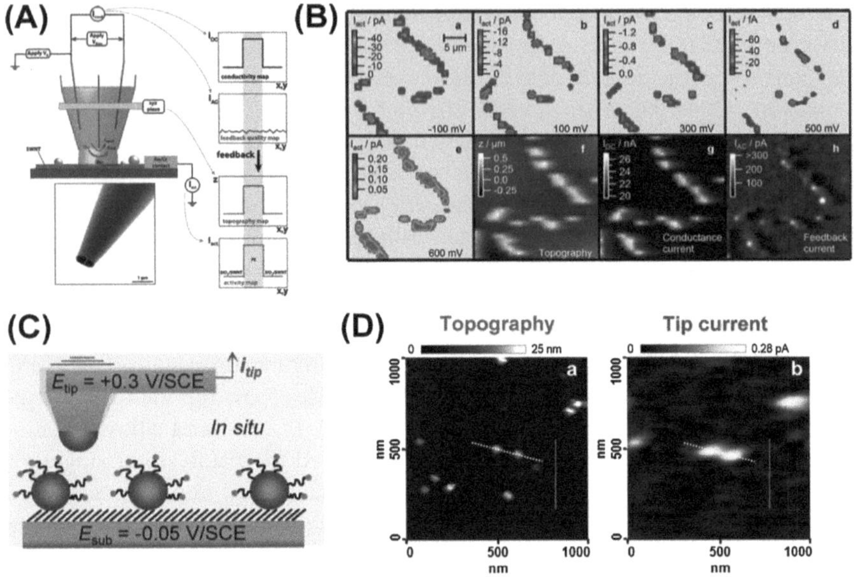

FIGURE 13 High-resolution SECM images of NPs: (A) Scheme of scanning electrochemical cell microscopy (SECCM) providing topography, conductivity, and activity maps. (B) Pt-NPs deposited at carbon nanotube, (A)–(E) electrochemical response at different potentials. (F) Surface topography. (G) DC conductance current. (H) AC component of the conductance current. (F)–(H) were recorded at a sample potential of 300 mV. *Reprinted from Ref. [152] with permission from American Chemical Society.* (C) Measurement scheme of AFM-SECM studies at Au-NPs, and (D) topography and current image of Fc-PEGylated gold nanoparticles. *Reprinted from Ref. [119] with permission from American Chemical Society.*

The electrocatalytic activity toward the oxygen reduction reaction at Au-NP spot arrays has been investigated by conventional SECM imaging using gold or platinum microelectrodes [138,139]. Due to the size of the micro-electrode, fairly large spot sizes were deposited. Sánchez-Sánchez investigated the influence of shapes of Au-NPs (cubic, sphere, and rods) on the activity toward oxygen reduction [139]. The SECM results demonstrated that cubic Au-NPs are the most active particles toward oxygen reduction in basic solution followed by spherical Au-NPs, and gold nanorods, which is related to a higher ratio of Au ⟨111⟩. A very similar study was recently presented by Wain et al. [138], investigating the activity toward oxygen reduction at spherical Au-NPs spot arrays (in generation collection mode [140], and also redox competition mode [141]). Each spot contained a narrow particle size distribution with sizes varying from 50 nm down to 5 nm within each individual spot. These studies confirmed an increase in reactivity toward oxygen reduction with decreasing particle diameter, which again can be explained by an increased Au ⟨110⟩ to Au ⟨111⟩ ratio for smaller particles. Both groups have used underpotential deposition (UPD) of Pb to determine the surface structure of Au NPs.

The voltammetric signal of Pb UPD is surface sensitive and shows contributions from the three Au basal planes.

Improvements in resolution require submicro- and nano-sized electrodes, which require significant reduced electrode-to-sample distance and hence usually a current independent positioning mode. A promising approach is the combination of AFM and SECM with an integrated electrode implemented in the AFM tip. Two different approaches have been demonstrated bearing either the electroactive area at the apex of the AFM tip (conical electrodes) [142–145] or having an electrode above the apex of an insulating sharp silicon nitride or silicon tip [133,146–149]. AFM-SECM measurements with excellent topographical and electrochemical resolution have been achieved by Demaille et al. tethering the redox mediator with a flexible polyethylene glycol (PEG) chain either to the apex of AFM-SECM probe (conical gold electrode at the tip apex) [118,150], or tethering the redox mediator to the investigated sample surface [151]. Tethering the redox mediators prevents the diffusional dispersion of the mediator. This approach was used to investigate Au-NPs, which were modified with the PEG-redox probe [119] (as shown in Figure 13(C)). As the AFM probe provides topographic information, the electrochemical current reflecting the individual Au-NPs could be recorded simultaneously. The achieved resolution was on the order of 20 nm.

The catalytic activity of individual platinum nanoparticles deposited onto carbon nanotubes was recently investigated by Unwin et al. using SECCM techniques. The catalytic properties in dependence to the morphology of the NPs were studied toward the oxygen reduction reaction [152]. By applying a constant potential between the substrate and the reference electrode within the barrel, the redox activity could be mapped. Physical characterization of particle size and morphology supported the obtained results. Furthermore, it could be shown that at the single particle level, similar particles revealed different reactivity based on small differences of their morphology. SECCM images of the reactivity of individual Pt NPs are shown in Figure 13(B).

Despite all the developmental improvements within recent years, the characterization of Au-NPs toward the structure–activity relation remains a challenging task, in particular for small particles <20 nm. A recent review highlights some of these challenges [153].

5. CONCLUSIONS

Is there a "best analytical technique" for the characterization of gold nanoparticles? This general question is frequently asked when analyzing nanoparticulate materials. Each microscopic technique discussed within this chapter has distinct advantages and disadvantages. Hence, as in most research areas, there is no single technique that provides a comprehensive characterization of Au-NPs, especially if conditions such as high sample polydispersity, sophisticated chemical or biological surface modification, and complex functionalities or material mixtures/composites have been introduced. However, emerging innovative instrumental developments such as combined scanning probe techniques, as briefly discussed herein, may establish the basis toward more comprehensive characterization tools that facilitate a multiparametric characterization of Au-NPs and other nanomaterials at in-situ and/or ex-situ conditions.

LIST OF ABBREVIATIONS

AFFFF asymmetric flow-field flow fractionation
AFM atomic force microscopy
AFM-SECM atomic force scanning electrochemical microscopy
AM-AFM amplitude modulation atomic force microscopy
Au-NPs gold nanoparticles
BF bright-field
CTAB cetyltrimethylammonium bromide
DDAB didodecyldimethylammonium bromide
DF dark-field
DFM dark-field microscopy
DLS dynamic light scattering
EDX energy-dispersive X-ray spectroscopy
EELS electron-energy-loss spectroscopy
EFM electrostatic force microscopy
EM electron microscopy
FIM fluorescence microscopy
FM-AFM frequency modulation atomic force microscopy
FTIR Fourier transform infrared spectroscopy
GNR gold nanorods
HAADF high-angle angular dark-field imaging
HR-STEM high-resolution scanning transmission electron microscopy
HR-TEM high-resolution transmission electron microscopy
LDOS local density of states
LFM lateral force microscopy
LSPR localized surface plasmon resonance
MALDI-TOF-MS matrix-assisted laser desorption/ionization time-of-flight mass spectrometry
MFM magnetic force microscopy
NBD nanobeam electron diffraction
NIST National Institute of Standard and Technology
NMR nuclear magnetic resonance spectroscopy
NSOM near-field scanning optical microscopy
PALM photoactivated localization microscopy
QRCE quasi-reference counter electrode
SAM self-assembled monolayers
SANS small angle neutron scattering
SCM scanning capacitance microscopy
SEC size exclusion chromatography

SECCM scanning electrochemical cell microscopy
SEIRA surface-enhanced infrared absorption
SERS surface-enhanced Raman scattering
SI-ATRP surface-initiated atom-transfer radical polymerization
SIC-SECM scanning ion conductance scanning electrochemical microscopy
SKPM scanning Kelvin probe microscopy
SLS static light scattering
SPM scanning probe microscopy
STED stimulated emission depletion microscopy
STEM scanning transmission electron microscopy
STEM-HAADF scanning transmission electron microscopy high-angle angular dark-field imaging
SThM scanning thermal microscopy
STM scanning tunneling microscopy
STORM stochastic optical reconstruction microscopy
TEM transmission electron microscopy
TERS tip-enhanced Raman scattering
TIRM total internal reflection microscopy
UHV ultra-high vacuum
UPD underpotential deposition
UV/Vis ultraviolet–visible spectroscopy
XPS X-ray photoelectron spectroscopy
XRD X-ray diffraction spectroscopy

REFERENCES

[1] N.M. Nanoparticles, Springer Handbook of Nanomaterials, Springer Berlin Heidelberg, Berlin, Heidelberg, 2013, pp. 303–388.

[2] D.A. Giljohann, D.S. Seferos, W.L. Daniel, M.D. Massich, P.C. Patel, C. a Mirkin, Gold nanoparticles for biology and medicine, Angew. Chem. Int. Ed. 49 (19) (2010) 3280–3294.

[3] W. Wang, N. Tao, Detection, counting, and imaging of single nanoparticles, Anal. Chem. 86 (1) (2014) 2–14.

[4] M. Rex, F.E. Hernandez, A.D. Campiglia, Pushing the limits of mercury sensors with gold nanorods, Anal. Chem. 78 (2) (2006) 445–451.

[5] K. Saha, S.S. Agasti, C. Kim, X. Li, V.M. Rotello, Gold nanoparticles in chemical and biological sensing, Chem. Rev. 112 (5) (2012) 2739–2779.

[6] R.A. Sperling, P. Rivera Gil, F. Zhang, M. Zanella, W.J. Parak, Biological applications of gold nanoparticles, Chem. Soc. Rev. 37 (9) (2008) 1896–1908.

[7] R.L. Whetten, J.T. Khoury, Z.L.W. Alvarez, S. Murthy, I. Vezmar, P.W. Stephens, C.L. Cleveland, W.D. Luedtke, U. Landman, Nanocrystal gold molecules, Adv. Mater. 8 (1996) 428–433.

[8] R. Jin, Quantum sized, thiolate-protected gold nanoclusters, Nanoscale 2 (3) (2010) 343–362.

[9] P. Zhang, A.Y.-C. Chu, T.-K. Sham, Y. Yao, S.-T. Lee, Chemical synthesis and structural studies of thiol-capped gold nanoparticles, Can. J. Chem. 87 (1) (2009) 335–340.

[10] K.S. Mayya, B. Schoeler, F. Caruso, Preparation and organization of nanoscale polyelectrolyte-coated gold nanoparticles, Adv. Funct. Mater. 13 (3) (2003) 183–188.

[11] Y. Pan, S. Neuss, A. Leifert, M. Fischler, F. Wen, U. Simon, G. Schmid, W. Brandau, W. Jahnen-Dechent, Size-dependent cytotoxicity of gold nanoparticles, Small 3 (11) (2007) 1941–1949.

[12] N. Li, P. Zhao, D. Astruc, Anisotropic gold nanoparticles: synthesis, properties, applications, and toxicity, Angew. Chem. Int. Ed. Engl. 53 (7) (2014) 1756–1789.

[13] S. Link, M.A. El-Sayed, Size and temperature dependence of the plasmon absorption of colloidal gold nanoparticles, J. Phys. Chem. B 103 (1999) 4212–4217.

[14] K.L. Kelly, E. Coronado, L.L. Zhao, G.C. Schatz, The optical properties of metal nanoparticles: the influence of size, shape, and dielectric environment, J. Phys. Chem. B 107 (2003) 668–677.

[15] C. Novo, D. Gomez, J. Perez-Juste, Z. Zhang, H. Petrova, M. Reismann, P. Mulvaney, G.V. Hartland, Contributions from radiation damping and surface scattering to the line width of the longitudinal plasmon band of gold nanorods: a single particle study, Phys. Chem. Chem. Phys. 8 (30) (2006) 3540–3546.

[16] C.J. Orendorff, T.K. Sau, C.J. Murphy, Shape-dependent plasmon-resonant gold nanoparticles, Small 2 (5) (2006) 636–639.

[17] L. Wu, B.M. Reinhard, Probing subdiffraction limit separations with plasmon coupling microscopy: concepts and applications, Chem. Soc. Rev. 43 (11) (2014) 3884–3897.

[18] K. Lindfors, T. Kalkbrenner, P. Stoller, V. Sandoghdar, Detection and spectroscopy of gold nanoparticles using supercontinuum white light confocal microscopy, Phys. Rev. Lett. Rev. Lett. 93 (3) (2004) 037401.

[19] C. Sönnichsen, S. Geier, N.E. Hecker, G. von Plessen, J. Feldmann, H. Ditlbacher, B. Lamprecht, J.R. Krenn, F.R. Aussenegg, V.Z.-H. Chan, J.P. Spatz, M. Möller, Spectroscopy of single metallic nanoparticles using total internal reflection microscopy, Appl. Phys. Lett. 77 (19) (2000) 2949–2951.

[20] H. Yao, K. Kimura, Field emission scanning electron microscopy for structural characterization of 3D gold nanoparticle superlattices, in: A. Mendez-Vilas, J. Diaz (Eds.), Modern Research and Educational Topics in Microscopy, Formatex, 2007, pp. 568–575.

[21] D.C. Joy, The theory and practice of high-resolution scanning electron microscopy, Ultramicroscopy 37 (1–4) (1991) 216–233.

[22] J. Cazaux, Recent developments and new strategies in scanning electron microscopy, J. Microsc. 217 (1) (2005) 16–35.

[23] L. Reimer, Scanning Electron Microscopy: Physics of Image Formation and Microanalysis, Springer-Verlag, New York, 1998.

[24] J.I. Goldstein, D.E. Newbury, P. Echlin, D.C. Joy, C.E. Lyman, E. Lifshin, L. Sawyer, J.R. Michael, Scanning Electron Microscopy and X-ray Microanalysis, third ed., Springer, New York, 2003.

[25] B. Ming, A.E. Vladar, Measuring the size of colloidal gold nano-particles using high-resolution scanning electron microscopy, NIST – NCL Method PCC-15 (2011) 1–20.

[26] N. Li, P. Zhao, D. Astruc, Anisotropic gold nanoparticles: synthesis, properties, applications, and toxicity, Angew. Chem. Int. Ed. 53 (7) (2014) 1756–1789.

[27] M. Grzelczak, J. Pérez-Juste, P. Mulvaney, L.M. Liz-Marzán, Shape control in gold nanoparticle synthesis, Chem. Soc. Rev. 37 (9) (2008) 1783–1791.

[28] M.R. Langille, M.L. Personick, J. Zhang, C.A. Mirkin, Defining rules for the shape evolution of gold nanoparticles, J. Am. Chem. Soc. 134 (35) (2012) 14542–14554.

[29] M.L. Personick, M.R. Langille, J. Zhang, C.A. Mirkin, Shape control of gold nanoparticles by silver underpotential deposition, Nano Lett. 11 (8) (2011) 3394–3398.

[30] C. Kisielowski, B. Freitag, M. Bischoff, H. van Lin, S. Lazar, et al., Detection of single atoms and buried defects in three dimensions by aberration-corrected electron microscope with 0.5-A information limit, Microsc. Microanal. 14 (5) (2008) 469–477.

[31] C.B. C, David B. Williams, Transmission Electron Microscopy a Textbook for Materials Science, Springer, New York, 2009.

[32] S.J. Abramoff, M.D. Magalhães, Paulo J. Ram, No title, Biophotonics Int. 11 (2004) 36–42.

[33] S.B. Rice, C. Chan, S.C. Brown, P. Eschbach, L. Han, D.S. Ensor, A.B. Stefaniak, J. Bonevich, A.E. Vladár, A.R.H. Walker, J. Zheng, C. Starnes, A. Stromberg, J. Ye, E.A. Grulke, Particle size distributions by transmission electron microscopy: an interlaboratory comparison case study, Metrologia 50 (6) (2013) 663–678.

[34] Y.Q. Wang, W.S. Liang, C.Y. Geng, Shape evolution of gold nanoparticles, J. Nanoparticle Res. 12 (2) (2009) 655–661.

[35] H. Hofmeister, Fivefold twinned nanoparticles, in: Encyclopedia of Nanoscience and Nanotechnology 3, vol. 3, American Scientific Publishers, Stevenson Ranch, 2004, pp. 431–452.

[36] X. Huang, S. Neretina, M. a. El-Sayed, Gold nanorods: from synthesis and properties to biological and biomedical applications, Adv. Mater. 21 (48) (2009) 4880–4910.

[37] H.A. Keul, M. Möller, M.R. Bockstaller, Structural evolution of gold nanorods during controlled secondary growth, Langmuir 23 (20) (2007) 10307–10315.

[38] C.J. Johnson, E. Dujardin, S.A. Davis, C.J. Murphy, S. Mann, Growth and form of gold nanorods prepared by seed-mediated, surfactant-directed synthesis, J. Mater. Chem. 12 (6) (2002) 1765–1770.

[39] M. Liu, P. Guyot-Sionnest, Mechanism of silver(I)-assisted growth of gold nanorods and bipyramids, J. Phys. Chem. B 109 (47) (2005) 22192–22200.

[40] T.H. Lim, D. Mccarthy, S.C. Hendy, K.J. Stevens, S.A. Brown, R.D. Tilley, Real-time TEM and kinetic Monte Carlo studies of the coalescence of decahedral gold nanoparticles, ACS Nano 3 (11) (2009) 3809–3813.

[41] S.I. Stoeva, V. Zaikovski, B.L. V Prasad, P.K. Stoimenov, C.M. Sorensen, K.J. Klabunde, Reversible transformations of gold nanoparticle morphology, Langmuir 21 (23) (2005) 10280–10283.

[42] J.L. Plaza, Y. Chen, S. Jacke, R.E. Palmer, Nanoparticle arrays patterned by electron-beam writing: structure, composition, and electrical properties, Langmuir 21 (4) (2005) 1556–1559.

[43] Y. Chen, R.E. Palmer, J.P. Wilcoxon, Sintering of passivated gold nanoparticles under the electron beam, Langmuir 22 (6) (2006) 2851–2855.

[44] D. Smith, A. Petford-Long, L. Wallenberg, J. Bovin, Atomic-level rearrangements in small gold particles, Science 233 (4766) (1986) 872–875.

[45] P.D. Nellist, Scanning Transmission Electron Microscopy, Springer, New York, NY, 2011, pp. 91–115.

[46] S.J. Pennycook, A.R. Lupini, M. Varela, A.Y. Borisevich, Y. Peng, M.P. Oxley, K. van Benthem, M.F. Chisholm, Scanning Transmission Electron Microscopy for Nanostructure Characterization, in: W. Zhou, Z.L. Wang (Eds.), Scanning Microscopy for Nanotechnology, Springer, New York, 2007, pp. 152–191.

[47] J.C. González, J.C. Hernández, M. López-Haro, E. del Río, J.J. Delgado, a B. Hungría, S. Trasobares, S. Bernal, P.A. Midgley, J.J. Calvino, 3 D characterization of gold nanoparticles supported on heavy metal oxide catalysts by HAADF-STEM electron tomography, Angew. Chem. Int. Ed. 48 (29) (2009) 5313–5315.

[48] S. Rashkeev, A. Lupini, S. Overbury, S. Pennycook, S. Pantelides, Role of the nanoscale in catalytic CO oxidation by supported Au and Pt nanostructures, Phys. Rev. B 76 (3) (July 2007) 035438.

[49] R. Ramachandra, H. Demers, N. de Jonge, Atomic-resolution scanning transmission electron microscopy through 50-nm-thick silicon nitride membranes, Appl. Phys. Lett. 98 (9) (2011) 93109.

[50] D. Bahena, N. Bhattarai, U. Santiago, A. Tlahuice, A. Ponce, S.B.H. Bach, B. Yoon, R.L. Whetten, U. Landman, M. Jose-Yacaman, STEM electron diffraction and high resolution images used in the determination of the crystal structure of Au144(SR)60 cluster, J. Phys. Chem. Lett. 4 (6) (2013) 975–981.

[51] M.R. Langille, J. Zhang, M.L. Personick, S. Li, C.A. Mirkin, Stepwise evolution of spherical seeds into 20-fold twinned icosahedra, Science 337 (6097) (2012) 954–957.

[52] A. Mayoral, S. Mejía-Rosales, M.M. Mariscal, E. Pérez-Tijerina, M. José-Yacamán, The Co-Au interface in bimetallic nanoparticles: a high resolution STEM study, Nanoscale 2 (12) (2010) 2647–2651.

[53] B. Rodríguez-González, A. Burrows, M. Watanabe, C.J. Kiely, L.M. Liz Marzán, Multishell bimetallic AuAg nanoparticles: synthesis, structure and optical properties, J. Mater. Chem. 15 (17) (2005) 1755–1759.

[54] F.L. Deepak, G. Casillas-Garcia, R. Esparza, H. Barron, M. Jose-Yacaman, New insights into the structure of Pd-Au nanoparticles as revealed by aberration-corrected STEM, J. Cryst. Growth 325 (1) (2011) 60–67.

[55] G. Binnig, H. Rohrer, C. Gerber, E. Weibel, Tunneling through a controllable vacuum gap, Appl. Phys. Lett. 40 (2) (1982) 178–179.

[56] G. Binnig, H. Rohrer, Scanning tunneling microscopy—from birth to adolescence, Rev. Mod. Phys. 59 (3) (1987) 615–625.

[57] R. Sonnenfeld, P.K. Hansma, Atomic-resolution microscopy in water, Science 232 (4747) (1986) 211–213.

[58] H. Zhang, G. Schmid, U. Hartmann, Reduced metallic properties of ligand-stabilized small metal clusters, Nano Lett. 3 (3) (2003) 305–307.

[59] C.D. Bain, J. Evall, G.M. Whitesides, Formation of monolayers by the coadsorption of thiols on gold: variation in the head group, tail group, and solvent, J. Am. Chem. Soc. 111 (18) (1989) 7155–7164.

[60] K. Tamada, M. Hara, H. Sasabe, W. Knoll, Surface phase behavior of n-alkanethiol self-assembled monolayers adsorbed on Au(111): an atomic force microscope study, Langmuir 13 (6) (1997) 1558–1566.

[61] R.K. Smith, S.M. Reed, P.A. Lewis, J.D. Monnell, R.S. Clegg, K.F. Kelly, L.A. Bumm, J.E. Hutchison, P.S. Weiss, Phase separation within a binary self-assembled monolayer on Au {111} driven by an amide-containing alkanethiol, J. Phys. Chem. B 105 (6) (2001) 1119–1122.

[62] A.M. Jackson, J.W. Myerson, F. Stellacci, Spontaneous assembly of subnanometre-ordered domains in the ligand shell of monolayer-protected nanoparticles, Nat. Mater. 3 (5) (2004) 330–336.

[63] Q.K. Ong, J. Reguera, P.J. Silva, M. Moglianetti, K. Harkness, M. Longobardi, K.S. Mali, C. Renner, S. De Feyter, F. Stellacci, High-resolution scanning tunneling microscopy characterization of mixed monolayer protected gold nanoparticles, ACS Nano 7 (10) (2013) 8529–8539.

[64] A.M. Jackson, Y. Hu, P.J. Silva, F. Stellacci, From homoligand- to mixed-ligand- monolayer-protected metal nanoparticles: a scanning tunneling microscopy investigation, J. Am. Chem. Soc. 128 (34) (2006) 11135–11149.

[65] A. Centrone, Y. Hu, A.M. Jackson, G. Zerbi, F. Stellacci, Phase separation on mixed-monolayer-protected metal nanoparticles: a study by infrared spectroscopy and scanning tunneling microscopy, Small 3 (5) (2007) 814–817.
[66] R.P. Carney, G.A. DeVries, C. Dubois, H. Kim, J.Y. Kim, C. Singh, P.K. Ghorai, J.B. Tracy, R.L. Stiles, R.W. Murray, S.C. Glotzer, F. Stellacci, Size limitations for the formation of ordered striped nanoparticles, J. Am. Chem. Soc. 130 (3) (2008) 798–799.
[67] G.A. Devries, M. Brunnbauer, Y. Hu, A.M. Jackson, B. Long, B.T. Neltner, O. Uzun, B.H. Wunsch, F. Stellacci, Divalent metal nanoparticles, Science 315 (5810) (2007) 358–361.
[68] J.C.H. Spence, W. Lo, M. Kuwabara, Observation of the graphite surface by reflection electron microscopy during STM operation, Ultramicroscopy 33 (2) (1990) 69–82.
[69] H. Ohnishi, Y. Kondo, K. Takayanagi, Quantized conductance through individual rows of suspended gold atoms, Nature 395 (1998) 780–783.
[70] H. Ohnishi, Y. Kondo, K. Takayanagi, Quantized conductance through individual rows of suspended gold atoms, Nature 395 (1998) 2–5.
[71] D. Erts, H. Olin, L. Ryen, E. Olsson, A. Thölén, Maxwell and Sharvin conductance in gold point contacts investigated using TEM-STM, Phys. Rev. B 61 (19) (2000) 12725–12727.
[72] G. Binnig, C.F. Quate, C. Gerber, Atomic force microscope, Phys. Rev. Lett. 56 (9) (1986) 930–933.
[73] P. West, P. Eaton, Atomic Force Microscopy, Oxford University Press, 2010.
[74] Y. Martin, C.C. Williams, H.K. Wickramasinghe, Atomic force microscope—force mapping and profiling on a sub 100-A scale, J. Appl. Phys. 61 (10) (1987) 4723–4729.
[75] H.-J. Butt, B. Cappella, M. Kappl, Force measurements with the atomic force microscope: technique, interpretation and applications, Surf. Sci. Rep. 59 (1–6) (2005) 1–152.
[76] T.R. Rodrıguez, R. García, Tip motion in amplitude modulation (tapping-mode) atomic-force microscopy: comparison between continuous and point-mass models, Appl. Phys. Lett. 80 (9) (2002) 1646–1648.
[77] R. García, Amplitude Modulation Atomic Force Microscopy, Wiley-VCH, 2010.
[78] F.J. Giessibl, Advances in atomic force microscopy, Rev. Mod. Phys. 75 (3) (2003) 949–983.
[79] Q. Zhong, D. Inniss, K. Kjoller, V.B. Elings, Fractured polymer/silica fiber surface studied by tapping mode atomic force microscopy, Surf. Sci. Lett. 290 (1–2) (1993) L688–L692.
[80] P.K. Hansma, J.P. Cleveland, M. Radmacher, D.A. Walters, P.E. Hillner, M. Bezanilla, M. Fritz, D. Vie, H.G. Hansma, C.B. Prater, J. Massie, L. Fukunaga, J. Gurley, V. Elings, Tapping mode atomic force microscopy in liquids, Appl. Phys. Lett. 64 (13) (1994) 1738–1740.
[81] S. Rode, N. Oyabu, K. Kobayashi, H. Yamada, A. Ku, True atomic-resolution imaging of (1014) calcite in aqueous solution by frequency modulation atomic force microscopy, Langmuir 25 (2009) 2850–2853.
[82] F.J. Giessibl, AFM's path to atomic resolution, Mater. Today 8 (5) (2005) 32–41.
[83] R. Wiesendanger, Scanning Probe Microscopy and Spectroscopy, Methods and Applications, Cambridge University Press, 1994.
[84] Y. Seo, W. Jhe, Atomic force microscopy and spectroscopy, Rep. Progr. Phys. 71 (1) (2008) 016101.
[85] B. Bonanni, S. Cannistraro, Gold nanoparticles on modified glass surface as height calibration standard for atomic force microscopy operating in contact and tapping mode, AZoNano — Online J. Nanotechnol. 1 (2005) 1–14.

[86] C. Wong, P.E. West, K.S. Olson, M.L. Mecartney, N. Starostina, Tip dilation and AFM capabilities in the characterization of nanoparticles, JOM 59 (1) (2007) 12–16.
[87] F. U, M.F. Tabet, Deconvolution of tip affected atomic force microscope images and comparison to Rutherford backscattering spectrometry, J. Vac. Sci. Technol. B 15 (4) (1997) 800–804.
[88] P. Markiewicz, M.C. Goh, Atomic force microscopy probe tip visualization and improvement of images using a simple deconvolution procedure, Langmuir 10 (1) (1994) 5–7.
[89] P. Markiewicz, M.C. Goha, Atomic force microscope tip deconvolution using calibration array, Rev. Sci. Instrum. 66 (5) (1995) 3186–3190.
[90] S. Yuan, F. Luan, X. Song, L. Liu, J. Liu, Reconstruction of an AFM image based on estimation of the tip shape, Meas. Sci. Technol. 24 (10) (2013) 105404.
[91] E.E. Flater, G.E. Zacharakis-Jutz, B.G. Dumba, I.A. White, C.A. Clifford, Towards easy and reliable AFM tip shape determination using blind tip reconstruction, Ultramicroscopy 146 (2014) 130–143.
[92] R.D. Boyd, A. Cuenat, New analysis procedure for fast and reliable size measurement of nanoparticles from atomic force microscopy images, J. Nanoparticle Res. 13 (1) (2010) 105–113.
[93] C.L. Cheung, J.H. Hafner, C.M. Lieber, Carbon nanotube atomic force microscopy tips: direct growth by chemical vapor deposition and application to high-resolution imaging, Proc. Natl. Acad. Sci. USA 97 (8) (2000) 3809–3813.
[94] C.M. Hoo, T. Doan, N. Starostin, P.E. West, M.L. Mecartney, Optimal sample preparation for nanoparticle metrology (statistical size measurements) using atomic force microscopy, J. Nanoparticle Res. 12 (3) (2009) 939–949.
[95] P. Georgiev, A. Bojinova, B. Kostova, D. Momekova, T. Bjornholm, K. Balashev, Implementing atomic force microscopy (AFM) for studying kinetics of gold nanoparticle's growth, Coll. Surf. A 434 (2013) 154–163.
[96] G. Lee, H. Lee, K. Nam, J.-H. Han, J. Yang, S.W. Lee, D.S. Yoon, K. Eom, T. Kwon, Nanomechanical characterization of chemical interaction between gold nanoparticles and chemical functional groups, Nanoscale Res. Lett. 7 (1) (2012) 608–619.
[97] D. Li, Q. He, Y. Cui, J. Li, G. Potsdam, Fabrication of pH-responsive nanocomposites of gold Nanoparticles/Poly (4-vinylpyridine), Chem. Mater. 19 (12) (2007) 412–417.
[98] S. Gupta, M. Agrawal, P. Uhlmann, F. Simon, M. Stamm, Poly(N -isopropyl acrylamide)–gold nanoassemblies on macroscopic surfaces: fabrication, characterization, and application, Chem. Mater. 22 (2) (2010) 504–509.
[99] V. Kozlovskaya, E. Kharlampieva, S. Chang, R. Muhlbauer, V.V. Tsukruk, Ph-responsive layered hydrogel microcapsules as gold nanoreactors, Chem. Mater. 21 (10) (2009) 2158–2167.
[100] N. Higashi, T. Takagi, T. Koga, Layer-by-layer fabrication of well-packed gold nanoparticle assemblies guided by a β-sheet peptide network, Polym. J. 42 (1) (2010) 95–99.
[101] D. Erts, A. Lohmus, R. Lohmus, H. Olin, Instrumentation of STM and AFM combined with transmission electron microscope, Appl. Phys. A Mater. Sci. Process. 72 (2001) S71–S74.
[102] T. Kizuka, H. Ohmi, T. Sumi, K. Kumazawa, S. Deguchi, M. Naruse, S. Fujusawa, A. Yabe, Y. Enomoto, Simultaneous observation of millisecond dynamics in atomistic structure, force and conductance on the basis of transmission electron microscopy, Jpn. J. Appl. Phys. 40 (2001) 170–173.
[103] E.H. Synge, A suggested method for extending the microscopic resolution into the ultramicroscopic region, Phil. Mag. 6 (1928) 356–362.

[104] A. Lewis, A. Harootunian, M. Isaacson, E. Kratschmer, E. Betzig, Near field scanning microscopy, Biophys. J. 49 (1986) 269–279.

[105] D.W. Pohl, W. Denk, M. Lanz, Optical stethoscopy: image recording with resolution λ/20, Appl. Phys. Lett. 44 (1984) 651–653.

[106] W.F. Zenhausern, M.P. O'Boyle, H.K. Wickramasinghe, Apertureless near- field optical microscope, Appl. Phys. Lett. 65 (1994) 1623–1625.

[107] W. Chen, A. Kimel, A. Kirilyuk, T. Rasing, Apertureless SNOM study on gold nanoparticles: experiments and simulations, Phys. Status Solidi (B) 247 (8) (2010) 2047–2050.

[108] T. Shimada, K. Imura, H. Okamoto, M. Kitajima, Spatial distribution of enhanced optical fields in one-dimensional linear arrays of gold nanoparticles studied by scanning near-field optical microscopy, Phys. Chem. Chem. Phys. 15 (12) (2013) 4265–4269.

[109] T. Klar, M. Perner, S. Grosse, G. von Plessen, W. Spirkl, J. Feldmann, Surface-plasmon resonances in single metallic nanoparticles, Phys. Rev. Lett. 80 (19) (1998) 4249–4252.

[110] T. V, W. Janssens, B.S. Clausen, H. Falsig, C.H. Christensen, Catalytic activity of Au nanoparticles, Nanotoday 2 (4) (2007) 14–18.

[111] Y. Mikami, A. Dhakshinamoorthy, M. Alvaro, H. García, Catalytic activity of unsupported gold nanoparticles, Cat. Sci. Technol. 3 (1) (2013) 58–69.

[112] H.-Y. Liu, F.R.F. Fan, C.W. Lin, A.J. Bard, Scanning electrochemical and tunneling ultramicroelectrode microscope for high-resolution examination of electrode surfaces in solution, J. Am. Chem. Soc. 108 (1986) 3838–3839.

[113] A.J. Bard, F.R.F. Fan, D.T. Pierce, P.R. Unwin, D.O. Wipf, F. Zhou, Chemical imaging of surfaces with the scanning electrochemical microscope, Science 254 (5028) (1991) 68–74.

[114] F.O. Laforge, J. Velmurugan, Y. Wang, M. V Mirkin, Nanoscale imaging of surface topography and reactivity with the scanning electrochemical microscope, Anal. Chem. 81 (8) (2009) 3143–3150.

[115] R.A. Lazenby, K. McKelvey, M. Peruffo, M. Baghdadi, P.R. Unwin, Nanoscale intermittent contact-scanning electrochemical microscopy, J. Solid State Electrochem. 17 (12) (2013) 2979–2987.

[116] N. Ebejer, A.G. Güell, S.C.S. Lai, K. McKelvey, M.E. Snowden, P.R. Unwin, Scanning electrochemical cell microscopy: a versatile technique for nanoscale electrochemistry and functional imaging, Annu. Rev. Anal. Chem. 6 (March 2013) 329–351.

[117] S.E.F. Kleijn, S.C.S. Lai, T.S. Miller, A.I. Yanson, M.T.M. Koper, P.R. Unwin, Landing and catalytic characterization of individual nanoparticles on electrode surfaces, J. Am. Chem. Soc. 134 (45) (2012) 18558–18561.

[118] E. Anne, A. Cambril, A. Chovin, C. Demaille, C. Goyer, Electrochemical atomic force microscopy using a tip-attached redox mediator for topographic and functional imaging of nanosystems, ACS Nano 3 (10) (2009) 2927–2940.

[119] K. Huang, A. Anne, M.A. Bahri, C. Demaille, Probing individual redox PEGylated gold nanoparticles by electrochemical-atomic force microscopy, ACS Nano 7 (5) (2013) 4151–4163.

[120] P. Bertoncello, Advances on scanning electrochemical microscopy (SECM) for energy, Energy Environ. Sci. 3 (11) (2010) 1620–1633.

[121] S. Amemiya, A.J. Bard, F.-R.F. Fan, M. V Mirkin, P.R. Unwin, Scanning electrochemical microscopy, Annu. Rev. Anal. Chem. 1 (2008) 95–131.

[122] A. Schulte, M. Nebel, W. Schuhmann, Scanning electrochemical microscopy in neuroscience, Annu. Rev. Anal. Chem. 3 (2010) 299–318.

[123] C. Kranz, Recent advancements in nanoelectrodes and nanopipettes used in combined scanning electrochemical microscopy techniques, Analyst 139 (2) (2014) 336–352.

[124] J. Meier, K.A. Friedrich, U. Stimming, Novel method for the investigation of single nanoparticle reactivity, Faraday Discuss. 121 (2002) 365–372.

[125] B. Ballesteros Katemann, A. Schulte, W. Schuhmann, Constant-distance mode scanning electrochemical microscopy (SECM)—Part I: adaptation of a non-optical shear-force-based positioning mode for SECM tips, Chem. Eur. J. 9 (9) (2003) 2025–2033.

[126] P.I. James, L.F. Garfias-Mesias, P.J. Moyer, W.H. Smyrl, Scanning electrochemical microscopy with simultaneous independent topography, J. Electrochem. Soc. 145 (1998) L64–L66.

[127] M. Ludwig, C. Kranz, W. Schuhmann, H.E. Gaub, Topography feedback mechanism for the scanning electrochemical microscope based on hydrodynamic forces between tip and sample, Rev. Sci. Instrum. 66 (4) (1995) 2857–2860.

[128] M. Etienne, A. Schulte, W. Schuhmann, High resolution constant-distance mode alternating current scanning electrochemical microscopy (AC-SECM), Electrochem. Commun. 6 (3) (2004) 288–293.

[129] C. Kranz, G. Friedbacher, B. Mizaikoff, A. Lugstein, J. Smoliner, E. Bertagnolli, Integrating an ultramicroelectrode in an AFM Cantilever: combined technology for enhanced information, Anal. Chem. 73 (2001) 2491–2500.

[130] J.V. Macpherson, P.R. Unwin, Combined scanning electrochemical – atomic force microscopy, Anal. Chem. 72 (2) (2000) 276–285.

[131] L. Patrick, T.M. Frederix, M.R. Gullo, T. Akiyama, A. Tonin, N.F. De Rooij, U. Staufer, A. Engel, Assessment of insulated conductive cantilevers for biology and electrochemistry, Nanotechnology 16 (8) (2005) 997–1005.

[132] A. Davoodi, J. Pan, C. Leygraf, S. Norgren, Integrated AFM and SECM for in situ studies of localized corrosion of Al alloys, Electrochim. Acta 52 (27) (2007) 7697–7705.

[133] S.E. Pust, M. Salomo, E. Oesterschulze, G. Wittstock, Influence of electrode size and geometry on electrochemical experiments with combined SECM-SFM probes, Nanotechnology 21 (10) (2010) 105709, 1–105709/12.

[134] B. Paulose Nadappuram, K. McKelvey, R. Al Botros, A.W. Colburn, P.R. Unwin, Fabrication and characterization of dual function nanoscale pH-scanning ion conductance microscopy (SICM) probes for high resolution pH mapping, Anal. Chem. 85 (2013) 8070–8074.

[135] D.J. Comstock, J.W. Elam, M.J. Pellin, M.C. Hersam, Integrated ultramicroelectrode – nanopipet probe for concurrent scanning electrochemical microscopy and scanning ion conductance microscopy, Anal. Chem. 82 (2010) 1270–1276.

[136] Y. Takahashi, A.I. Shevchuk, P. Novak, Y. Murakami, H. Shiku, Y.E. Korchev, T. Matsue, Simultaneous noncontact topography and electrochemical imaging by SECM/SICM featuring ion current feedback regulation, J. Am. Chem. Soc. 132 (2010) 10118–10126.

[137] N. Ebejer, M. Schnippering, A.W. Colburn, M.A. Edwards, P.R. Unwin, Localized high resolution electrochemistry and multifunctional imaging: scanning electrochemical cell microscopy, Anal. Chem. 82 (22) (2010) 9141–9145.

[138] A.J. Wain, Imaging size effects on the electrocatalytic activity of gold nanoparticles using scanning electrochemical microscopy, Electrochim. Acta 92 (2013) 383–391.

[139] C.M. Sánchez-Sánchez, F.J. Vidal-Iglesias, J. Solla-Gullón, V. Montiel, A. Aldaz, J.M. Feliu, E. Herrero, Scanning electrochemical microscopy for studying electrocatalysis on shape-controlled gold nanoparticles and nanorods, Electrochim. Acta 55 (27) (2010) 8252–8257.

[140] J.L. Fernandez, A.J. Bard, Scanning electrochemical microscopy 50. Kinetic study of electrode reactions by the tip generation-substrate collection mode, Anal. Chem. 76 (8) (2004) 2281–2289.

[141] K. Eckhard, X. Chen, F. Turcu, W. Schuhmann, Redox competition mode of scanning electrochemical microscopy (RC-SECM) for visualisation of local catalytic activity, Phys Chem Chem Phys 8 (45) (2006) 5359–5365.

[142] C.E. Gardner, J.V. Macpherson, Atomic force microscopy probes go electrochemical, Anal. Chem. 74 (2002) 576–584.

[143] J. Abbou, C. Demaille, M. Druet, J. Moiroux, Fabrication of submicrometer-sized gold electrodes of controlled geometry for scanning electrochemical-atomic force microscopy, Anal. Chem. 74 (24) (2002) 6355–6363.

[144] M.R. Gullo, P.L.T.M. Frederix, T. Akiyama, A. Engel, N.F. DeRooij, U. Staufer, Characterization of microfabricated probes for combined atomic force and high-resolution scanning electrochemical microscopy, Anal. Chem. 78 (15) (2006) 5436–5442.

[145] A.J. Wain, D. Cox, S. Zhou, A. Turnbull, High-aspect ratio needle probes for combined scanning electrochemical microscopy — atomic force microscopy, Electrochem. Commun. 13 (1) (2011) 78–81.

[146] A. Kueng, C. Kranz, B. Mizaikoff, A. Lugstein, E. Bertagnolli, Combined scanning electrochemical atomic force microscopy for tapping mode imaging, Appl. Phys. Lett. 82 (10) (2003) 1592–1594.

[147] C. Kranz, G. Friedbacher, B. Mizaikoff, A. Lugstein, J. Smolliner, E. Bertagnolli, Integrating an ultramicroelectrode in an AFM cantilever : combined technology for enhanced information, Anal. Chem. 73 (2001) 1–13.

[148] A. Eifert, W. Smirnov, S. Frittmann, C. Nebel, B. Mizaikoff, C. Kranz, Atomic force microscopy probes with integrated boron doped diamond electrodes: fabrication and application, Electrochem. Commun. 25 (2012) 30–34.

[149] A. Davoodi, J. Pan, C. Leygraf, S. Norgren, In situ investigation of localized corrosion of aluminum alloys in chloride solution using integrated EC-AFM/SECM techniques, Electrochem. Solid-State Lett. 8 (6) (2005) B21–B24.

[150] A. Anne, C. Demaille, C. Goyer, Electrochemical atomic-force microscopy using a tip-attached redox mediator. Proof-of concept and Perspectives for functional probing of Nanosystems, ACS Nano 3 (4) (2009) 819–827.

[151] J. Abbou, A. Anne, C. Demaille, Probing the structure and dynamics of end-grafted flexible polymer chain layers by combined atomic force-electrochemical microscopy. Cyclic voltammetry within nanometer-thick macromolecular poly(ethylene glycol) layers, J. Am. Chem. Soc. 126 (32) (2004) 10095–10108.

[152] S.C.S. Lai, P.V. Dudin, J.V. Macpherson, P.R. Unwin, Visualizing Zeptomole (Electro) catalysis at single nanoparticles within an Ensemble, J. Am. Chem. Soc. 133 (28) (2011) 10744–10747.

[153] S.E.F. Kleijn, S.C.S. Lai, M.T.M. Koper, P.R. Unwin, Electrochemistry of nanoparticles, Angew. Chem. Int. Ed. 53 (14) (2014) 3558–3586.

Chapter 7

Spectroscopic Techniques for Characterization of Gold Nanoparticles

Georg Ramer and Bernhard Lendl*
Institute of Chemical Technologies and Analytics, Vienna University of Technology, Vienna, Austria
**Corresponding author: E-mail: bernhard.lendl@tuwien.ac.at*

Chapter Outline

1. ELECTROMAGNETIC MODES OF GOLD NANOPARTICLES

The interaction between electromagnetic fields and metal particles much larger than the wavelength can adequately be described using geometrical optics. For particles much smaller than the wavelength, simple approximations are also

Gold Nanoparticles in Analytical Chemistry. http://dx.doi.org/10.1016/B978-0-444-63285-2.00007-9

available in the form of Rayleigh scattering and Gans theory. However in the range approximately equal to the wavelength of the light, the scattered field is somewhat harder to describe. Since the spectroscopic analysis of gold nanoparticles often includes a step of comparing the measurement data with model-based reference data, the most common theoretical models for the interactions of electromagnetic radiation with these particles will be succinctly introduced below.

1.1 Mie Theory

In 1908, Gustav Mie published a thorough examination of the scattering of light by spherical particles [1]. Using Mie's theory, the scattering cross-section, the absorption cross-section, and the intensity distribution of the scattered light around a sphere can be calculated to any needed degree of accuracy. The general idea of Mie's work was to rewrite Maxwell's equations in spherical coordinates and then to solve them for an incident plane wave expanded in the same coordinate system. A sketch of the problem is depicted in Figure 1.

In the following only Mie's main results as pertinent to the topic of gold nanoparticles will be shown as laid out by Bohren and Huffman [2].

According to Mie theory, the extinction cross-section of a spherical particle is

$$C_{ext} = \frac{2\pi}{k^2}\sum_{n=1}^{\infty}(2n+1)\mathrm{Re}(a_n + b_n) \tag{1}$$

and the scattering cross-section is

$$C_{sca} = \frac{2\pi}{k^2}\sum_{n=1}^{\infty}(2n+1)\left(|a_n|^2 + |b_n|^2\right) \tag{2}$$

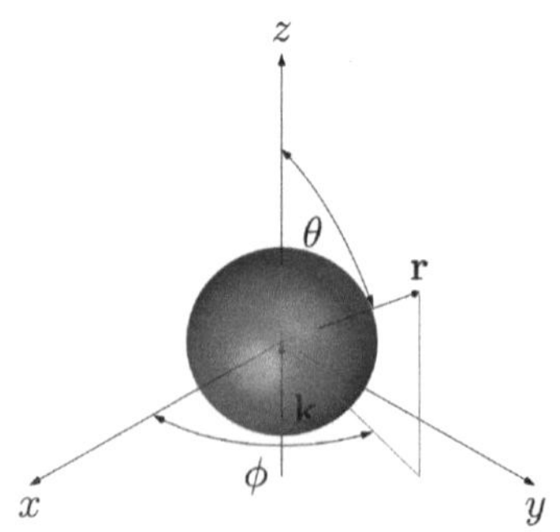

FIGURE 1 Sketch showing the treated in Mie's 1908 paper. A plane wave (wave vector **k**) is incident on a gold sphere (gray) placed at the origin of a spherical coordinate system.

Here k is the wave vector $\left(k = \frac{2\pi}{\lambda}\right)$, and a_n and b_n are the coefficients that have to be calculated numerically.

$$a_n = \frac{m\psi_n(mx)\psi_n'(x) - \psi_n(x)\psi_n'(mx)}{m\psi_n(mx)\xi_n'(x) - \xi_n(x)\psi_n'(mx)} \tag{3}$$

$$b_n = \frac{\psi_n(mx)\psi_n'(x) - m\psi_n(x)\psi_n'(mx)}{m\psi_n(mx)\xi_n'(x) - m\xi_n(x)\psi_n'(mx)} \tag{4}$$

ψ and ξ are the functions that depend on the spherical Bessel function. x is the size parameter that relates the radius R of the particle and the wave vector of the incident light, $x = kR$. m is the relative refractive index, $m = \frac{N_1}{N}$, where N_1 is the (complex) refractive index of the particle and N is that of the surrounding medium.

C_{ext} pertains to the total intensity of light removed from the plane wave, C_{sca} only to that part which is removed by scattering alone. Their difference is the part removed by absorption in the particle, C_{abs}. While C_{ext} and C_{sca} contain infinite series, it can be shown that with decreasing size of the sphere—or, more specifically, decreasing size parameters—higher summands become small relative to the lower ones and that $b_{n+1} < a_n$ [1]. For particles in the range of $2R < \lambda/10$, only the first coefficient contributes significantly to the extinction [3]. The edge case of such a small particle results in Rayleigh scattering [1,2] through simplifications of Mie's formulas for small values of x.

For particles too large for the approximation above either more summands of the analytical solution or different, purely numerical methods have to be used. Some headway has been made by researchers who were able to show analytical solutions for particles of several other shapes, such as infinite cylinders, spheroids [2], coated spheres [4], and two spheres [5]. Work to improve the stability and accuracy of numerical methods to calculate a_n and b_n is also ongoing.

1.2 Gans Theory

If the particle examined is very small relative to the wavelength, all electrons inside the particle experience the same electric field and therefore act as one dipole [6]. This is often called the nonretarded limit as the delay of the electrons (i.e., their retardation) due to their finite speed is neglected. The nonretarded limit allows calculating the absorption and scattering of a wide range of different particles, especially those of ellipsoidal shape by replacing the particle by three orthogonal dipoles of different polarizability placed along the principal axis of the particle. The extinction cross-section then is (in the notation of, e.g., Myroshnychenko et al. or Bohren and Huffman)

$$C_{ext} = \frac{2\pi}{\lambda N}\mathrm{Im}\left(\frac{\alpha_1 + \alpha_2 + \alpha_3}{3}\right) \tag{5}$$

where α is the polarizability of the particle. α depends on the volume of the particle (V), the refractive indices of the medium, and the particle and depolarization factors L_j [6,7].

$$\alpha_j = V N^2 \frac{N_1^2 - N^2}{N^2 + L_j\left(N_1^2 - N^2\right)} \tag{6}$$

$$L_j = \frac{R_1 R_2 R_3}{2} \int_0^\infty \frac{ds}{\left(s + R_j^2\right)\sqrt{\left(s + R_1^2\right)\left(s + R_2^2\right)\left(s + R_3^2\right)}} \tag{7}$$

R_1, R_2, and R_3 are the semidiameters of the ellipsoid. Equation (6) will have a maximum at $N_1 \approx N\left(1 - \frac{1}{L_j}\right)$ meaning that C_{ext} (5) will have up to three absorption peaks that solely depend on the shape and size of the particle. The frequencies at which the electron plasma exhibits resonant oscillations are so-called plasmon resonances. The influence of the semidiameter on the positions of the resonances shows that the extinction spectra of gold nanoparticles are defined by their shape. A striking example of the influence of the shape of the particles on the absorption spectrum is depicted in Figure 2.

FIGURE 2 Color of suspensions of gold nanoparticle of different aspect ratios and their corresponding micrographs. *Reproduced with permission from Jorge Peréz-Juste et al. [8]. Copyright 2005 Elsevier.*

The nonretarded limit also allows calculating the scattering of small coated ellipsoidal particles [2]:

$$\alpha_j = V \frac{(\varepsilon_2 - \varepsilon_m)\left(\varepsilon_2 + (\varepsilon_1 - \varepsilon_2)\left(L_j^{(1)} - fL_j^{(2)}\right)\right) + f\varepsilon_2(\varepsilon_1 - \varepsilon_2)}{\left(\varepsilon_2 + (\varepsilon_1 - \varepsilon_2)\left(L_j^{(1)} - fL_j^{(2)}\right)\right)\left(\varepsilon_m + (\varepsilon_2 - \varepsilon_m)L_j^{(2)}\right) + fL_j^{(2)}\varepsilon_2(\varepsilon_1 - \varepsilon_2)} \tag{8}$$

As for the uncoated particle, the volume V is $4\pi a_1 b_1 c_1$. f is the ratio of the volumes of the inner ellipsoid (i.e., the particle) and the outer ellipsoid (i.e., particle and coating). The geometrical factors for the particle $L_j^{(1)}$ and the coating $L_j^{(2)}$ are calculated as in Eqn (7).

1.3 Numerical Methods

A multitude of numerical algorithms exists, which can be used to determine the scattering on particles of arbitrary shape. Following are the most widespread methods to calculate the EM fields in gold nanoparticles [2,7]. The best method for a given problem depends on the geometry of the system and the information needed as well as technological constraints (computational power and time).

- *Boundary element method* considers only the field along the boundaries of homogeneous areas [7] and thus reduces the number elements to include in the calculation in comparison to volume-based methods. Further reductions in the number of elements can be achieved by considering symmetries of the system. Calculations can be performed for retarded and nonretarded systems [9].
- For *Discrete-dipole approximation* (DDA), a particle is divided into an array of polarizable points [10]. These dipolar elements have to be much smaller than the skin depth of the material for the calculations to be accurate; anisotropy and inhomogeneities can be modeled by adjusting the polarizability of the dipoles [7]. To model the shape of particles correctly, the elements can be increased as necessary [10]. DDSCAT (currently in version 7.2) is a freely available implementation of the DDA algorithm [11].
- *Finite difference time domain* calculates the electric field with a spatial and temporal discretization. Not only the system of interest but also a space around it has to be included in the calculations. Calculations for multiple wavelengths can be performed at once when a wave packet is used for the excitation [7].

1.4 Optical Constants of Gold Nanoparticles

The refractive index N describes the relation between the speed of light in the medium (v) and in vacuum (c), $N = \frac{v}{c_0}$. For absorbing materials, the refractive index will be a complex number $N = n' + in''$, where the additional imaginary part describes the attenuation of the wave in the medium. The real and the imaginary parts both depend on the wavelength (dispersion).

The imaginary part of the refractive index is proportional to the product of absorption cross-section and concentration in Eqn (17) as intensity of a plane wave traveling for length l in an absorbing medium decreases by $\ln(I/I_0) = -\frac{4\pi n''}{\lambda_0} l = -\rho c l$.

$$\frac{4\pi}{\lambda_0} n'' = \rho c \tag{9}$$

A material constant closely related to the refractive index is the relative permittivity ε_r,

$$N = \sqrt{\varepsilon_r \mu_r} \tag{10}$$

The relative permeability μ_r of almost all common optical materials, including gold, is very close to 1 therefore the following approximation is commonly used:

$$N \approx \sqrt{\varepsilon_r} \tag{11}$$

When N is complex, the relative permittivity will also become complex: $\varepsilon_r = \varepsilon' + i\varepsilon''$.

One major caveat when it comes to optical constants used for nanoparticles is that tabulated optical constants are determined for bulk materials and may not be valid for particles below a certain diameter; for gold nanoparticles the deviation of optical constants is stated to begin in the range of several tens of nanometers [7]. In particles with diameters in this range diffuse collisions of electrons with the particle boundary start to dominate instead of collisions with lattice vibrations as the particle length is in the range of the mean free path of electrons in bulk metal [2,12]. A common solution to this problem is to model the influence of the particle boundary by correcting the bulk optical properties ε_{bulk} with terms based on the Drude model [2,12,13]:

$$\varepsilon = \varepsilon_{bulk} - \varepsilon_{bulk}^{Drude} + \varepsilon_{R}^{Drude} \tag{12}$$

The ε^{Drude} terms are the permittivity of bulk gold and nanoparticles of radius R, respectively, calculated according to the Drude model:

$$\varepsilon_{bulk}^{Drude} = 1 - \frac{\omega_p^2}{\omega^2 + i\left(\frac{1}{\tau_s}\right)\omega} \tag{13}$$

and

$$\varepsilon_{R}^{Drude} = 1 - \frac{\omega_p^2}{\omega^2 + i\left(\frac{1}{\tau_s} + \frac{v_F}{L}\right)\omega} \tag{14}$$

TABLE 1 Constants Used in the Drude Model of Gold (12)

Constant	Quantity	Source
ω_p	1.44×10^{-16}/s	[13]
	9.0 eV	[14]
$\frac{1}{T_s}$	1.075×10^{14}/s	[15]
$\varepsilon(\lambda)$	Tabulated: 188–1937 nm	[15]
	Tabulated: 0.1250–9919 nm	[16]

ω is the circular frequency of the electromagnetic wave, τ_s is the average time between collisions, v_F is the Fermi velocity (speed of electrons at the Fermi level), ω_p is the bulk plasma frequency, and L is a size-dependent factor given by Bohren and Huffman as $\frac{4}{3}R$. The Drude model describes the conduction electrons in a metal as freely moving particles that only interact with the metal atoms through elastic collisions. The speed of the values and sources for ω_p and τ_s can be found in Table 1.

Although Drude correction works well for small gold nanoparticles, for gold clusters of a diameter below 3 nm, Alvarez et al. [17] found absorption features that could not be accurately predicted. The authors suggested a better fit by adding an increased electron density to the model. For size below 2 nm, a fine structure was observed in the absorption spectra that stemmed from intraband transitions due to the small size of the clusters [18].

2. PHOTON-BASED TECHNIQUES

The ultraviolet–visible–near-infrared (UV–VIS–NIR) range (300–900 nm) is rich in information about the properties of gold nanoparticles. As outlined in the last chapter, shape and size of nanoparticles influence its spectrum in this spectral range. Furthermore, the fact that this spectral range is mostly encompassed in the range of the human eye makes it easy to perceive general tendencies of the measurement without the need for any instruments at all.

2.1 UV–VIS Transmission Spectroscopy

In transmission (extinction) spectroscopy, the gold nanoparticles are usually suspended in a solvent and placed in a cuvette of known thickness. The transmission through the sample is recorded as a function of the wavelength.

For suspensions or solutions of particles that are small compared to a light beam traversing the liquid, a simple relation can be found between the attenuation of light and the concentration of the particles contained in it. We start with a volume of length dl and that is traversed by a light beam with a

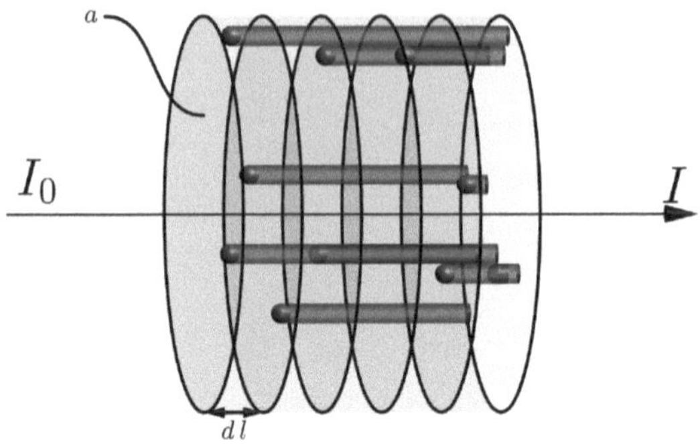

FIGURE 3 Sketch showing a light beam traversing a series of volumes containing particles. The space shaded by the black particles is drawn in dark gray.

cross-section ρ (Figure 3). In this solution, spherical particles with a cross-section ρ are suspended. ρ describes the probability for an interaction between light and particle that decreases the light intensity in the beam path (e.g., C_{ext} from Eqn (1) or (5)) and does not have to be equal to the geometrical cross-section of the particle at all, it can be smaller as well as larger.

Light that traverses the volume in the direction of dl will be blocked if it falls within a particle's cross-section. At a concentration of c, we will find ca dl particles in each volume, leading to a cumulative area of ρca dl or as fraction of the beam cross-section $\frac{\rho ca \, \mathrm{d}l}{a} = \rho c \mathrm{d}l$. That means after passing one volume, the beams intensity I_0 will be reduced by $I_0(\rho c \mathrm{d}l)$ from its original intensity. Since a part of it has been blocked, the light beam will only touch $(a - \rho ca \, \mathrm{d}l) \, \mathrm{d}l$ of the volume.

More generally, if the coverage until the current volume was α, the light beam can only interact with the volume $(a - \alpha)$ dl. The change of coverage in one volume, therefore, can be written as $d\alpha = \rho c(a - \alpha) \, \mathrm{d}l$. We can find the coverage after an arbitrary length l by integrating

$$\int_0^\alpha \frac{d\alpha}{a - \alpha} = \int_0^l \rho c \mathrm{d}l \tag{15}$$

and end up with

$$\mathrm{In}\left(1 - \frac{\alpha}{a}\right) = -\rho cl \tag{16}$$

To determine $1 - \frac{\alpha}{a}$, we can measure the transmission $T = \frac{I}{I_0}$, the ratio of the light intensity before I_0 and after I passing through the volume as $T = 1 - \frac{\alpha}{a}$. We finally have what is known as Beer's law.

$$\ln(T) = \ln\left(\frac{I}{I_0}\right) = -\rho cl \tag{17}$$

Sample extinction is calculated by using the transmission through a cuvette containing either air or pure solvent as I_0. Depending on the instrument design, the reference is either measured at the same time as the sample (double beam) or in a second measurement (single beam).

Spectral resolution in a UV–VIS spectrometer is achieved via a dispersive element, usually a diffraction grating. Light that has passed through the sample is first directed through a slit or aperture that restricts the beam dimensions and then onto the grating. The achieved spectral resolution is inherently linked with the optical throughput of the instrument through the slit width. A larger slit will increase the amount of light transmitted to the detector and thereby improve the signal-to-noise ratio; it will, however, also decrease the resolution by allowing a broader wavelength range to pass onto the detector.

The cuvettes in which liquid samples are usually measured have several important parameters:

- Path length
- Material
- Volume
- Shape.

The most common path length for UV–VIS is 1 cm, where strongly absorbing samples will necessitate shorter cells, while weakly absorbing ones will need long path cells. Standard cuvettes have a prism shape with a 1 cm × 1 cm base.

Common materials are crown glass, quartz, poly methyl methacrylate (PMMA), and polyethylene (PE). The main argument for the use of quartz cuvettes is their superior transmission range in comparison to the other materials, which ranges from 150 to 4000 nm. Glass cuvettes have a slightly reduced transmission range of 340–2500 nm but are usually less expensive than quartz cuvettes. PE and PMMA are used mainly because their low cost allows constructing single-use cuvettes and thereby eliminates cleaning time and possible cross-contamination of samples.

To reduce the amount of sample needed to fill the flow cell, the cross-section of the cell should be adjusted to the diameter of the beam. Sample volume that is never touched by the light beam does not contribute to the measurement and can therefore be reduced by reducing the width of the flow cell. It is however important that the whole beam travels through the same path length of the cell, as partial coverage of the beam cross-section will lead to deviations from the linear dependence of the absorption on the concentration given by Beer's law.

2.1.1 Particle Shape and Size

The influence of the shape of an elliptical particle on its extinction spectrum is described by Eqns (5)–(7). While the formulae suggest that the number of peaks show the asymmetry of this particle type, broad and overlapping

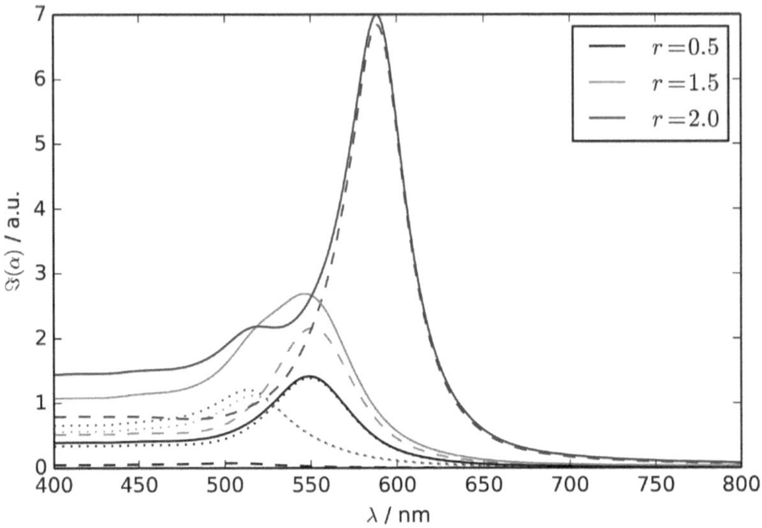

FIGURE 4 Simulated values for the polarizability of spheroidal gold nanoparticles in aqueous solution for shapes of $R_1 = R_2 = R_3/r = 25$ nm. Dotted lines describe the $\frac{\alpha_1+\alpha_2}{3}$ term, the dashed line describes $\frac{\alpha 3}{3}$.

absorption bands may obfuscate this information (Figure 4). An empirical formula to determine the diameter of spherical particles in water is given in the following equation [19]:

$$d = \frac{\ln\left(\frac{\lambda_{spr} - 512\,\mathrm{nm}}{6.53\,\mathrm{nm}}\right)}{0.0216/\mathrm{nm}} \tag{18}$$

Amendola and Meneghetti [20] developed a fitting model that allows determining the distribution of the excentricity, the particle size, and the concentration from UV–VIS absorptions spectra with an average error of 6%.

2.1.2 Concentration

To determine the concentration of an analyte in a solution, its extinction cross-section has to be known. The cross-section can be either calculated from theoretical consideration taken from literature or determined experimentally from a series of samples with known concentrations.

For aqueous suspensions of spherical gold particles of known diameter, Haiss et al. [19] gave an empirical formula to calculate the number density of particles N from the diameter d (in nanometer) and the absorption at 450 nm A_{450}:

$$N = \frac{A_{450} \times 10^{14}}{d^2\left[-0.295 + 1.36\exp\left(-\left(\frac{d-96.8}{78.2}\right)\right)\right]} \tag{19}$$

2.1.3 Medium Refractive Index

The medium refractive index influences the extinction, scattering, and absorption cross-section of nanoparticles. This can be seen from Eqns (3) and (4), where the medium refractive index appears in the form of the relative refractive index *m*.

Following Gans Theory (Section 1.2), the location of the plasmon resonance peak is given by $N^2 + L_j(\mathrm{Re}(N_1^2) - N^2) \cong 0$. For a small change ΔN in the medium refractive index, the wavelength of the resonance will shift by [21]

$$\Delta\lambda = -\frac{2N\Delta N\frac{1-L_j}{L}}{\frac{d\mathrm{Re}\left(N_1^2\right)}{d\lambda}} \tag{20}$$

Since $\frac{d\mathrm{Re}(N_1^2)}{d\lambda}$ is negative for the major part of the visible range, the plasmon resonances will show an increase in the wave length with increasing refractive index.

Underwood and Mulvaney have found a good match between theoretical calculations using Mie theory and measurement value for the peak position of spherical gold nanoparticles for the solvent refractive indices of hexane, dodecane, decaline, and range of mixtures *o*-butyl acetate and CS_2 [22].

2.2 Spatial Modulation Spectroscopy

In spatial modulation spectroscopy (SMS), single-particle spectroscopy is achieved moving a single particle quickly in and out of the tight focal spot of the light source [23]. Particles are deposited onto a transparent carrier, dilute enough that their distance to each other is larger than the diameter of the focal spot to assure the observation of a single particle. The carrier is then placed in the focal plane of the light source and actuated in a sine wave motion in the plane (Figure 5). This will move a single particle through the focal spot. By filtering the transmission signal with a lock-in amplifier, the extinction caused by the particle can be sensitively detected. Billaud et al. [24] report the use of tunable lasers as well as white-light sources and detection of single gold nanoparticles of sizes down to 5 nm [23] with SMS.

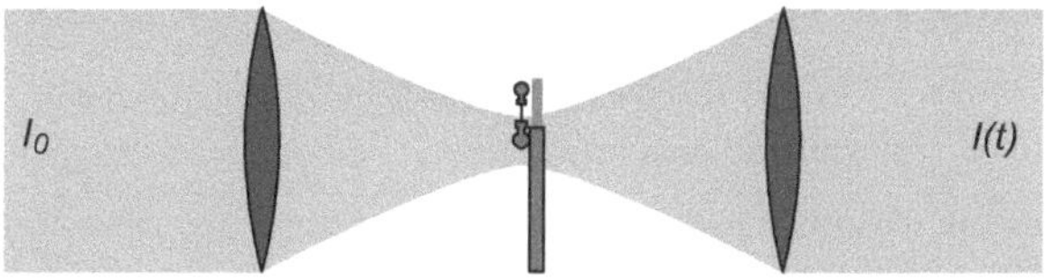

FIGURE 5 Experimental layout in SMS. The particle (black) is placed in the focal spot of the laser and then translated vertically with an amplitude larger than the focal spot size. Thus the transmitted intensity *I*(*t*) is modulated.

2.3 Dark-Field Microscopy

Dark-field microscopy is an optical microscopic technique that allows to specifically view the light that is scattered off the sample. Like in a standard bright-field microscopic setup, the sample is placed on a sample carrier and is then illuminated through a condensing optic and observed through a collimating optic and an ocular. In contrast to bright-field microscopy, in the dark-field microscopy, light rays cannot pass straight through the sample (Figure 6), instead only the scattered light reaches the ocular. This is done by blocking central part of the light cone so that only oblique rays can reach the sample. After the sample rays outside, the central cones are blocked. Therefore only light scattered by the sample into this central cone reaches the ocular or the detector.

Since scatterers show up as light spots against a dark background, very small amounts of gold nanoparticles down to single particles can be detected in dark-field microscopy. In the microscopic image, similar particles will show up as having similar colors.

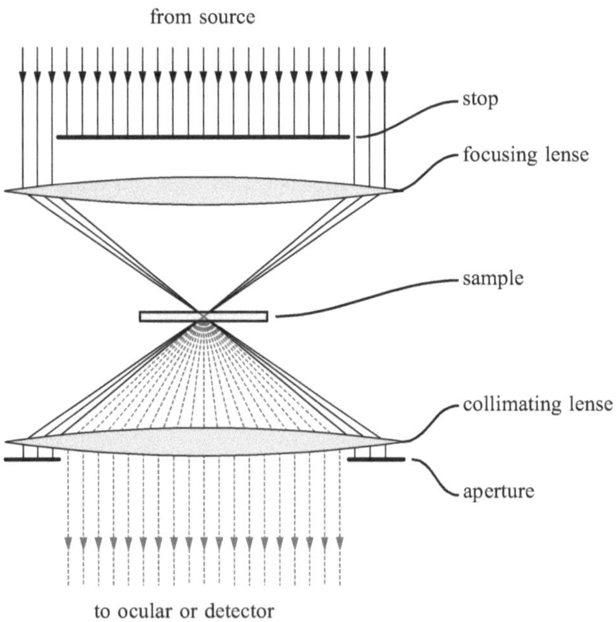

FIGURE 6 Sketch of the central part of a dark-field microscope. The incident rays are drawn in solid lines, and the scattered rays are drawn in dashed lines. Only light that has been scattered at the sample can pass the aperture to the ocular.

Dark-field microscopy introduces several effects that are not seen in bulk spectroscopic techniques:

- The orientation of the particles is fixed, meaning that instead of the average spectra around all angles of incidence, each particle will only give a scattering spectrum that corresponds to the range of angles defined by the angle of the illuminating light cone, the collection optics, and its fixed orientation on the substrate [25].
- When gold nanoparticles are placed directly on the sample carrier, their surroundings do not have homogeneous refractive index and Mie theory is not completely applicable [7].
- While the technique allows detection of nanoparticles far below the spatial resolution achievable with the illuminating light, it is difficult to distinguish between single particles and aggregates as both show up as "dots" in the dark-field image.

2.4 Scanning Near-Field Optical Microscopy

Spatial resolution in optical microscopy has a lower limit in the order of magnitude of the wavelength. A common value for the lowest distance d that can be resolved in dependence of the wavelength and the numerical aperture (NA) of the objective is given by the Abbe diffraction limit as

$$d = \frac{\lambda}{2\mathrm{NA}} \tag{21}$$

This limit can be overcome if instead of imaging the sample in the far field, a near-field probe that generates a highly localized UV—VIS signal is scanned across the sample. Probes that generate such localized signals can be tapered waveguide fibers (aperture-based scanning near-field optical microscopy (SNOM)) or metalized tips (scattering SNOM) [26]. Positioning of the probe is done in a similar way as in atomic force microscopes by using the change in amplitude of the oscillations of the tip above the sample as a measure for the sample—probe distance (intermittent contact mode) [26] or by measuring the shear force by mounting a tuning fork on the fiber.

SNOM techniques map the photonic local density of states (LDOS) at a given wavelength and thereby give direct access to the spatial distribution of photonic wave functions of the sample [27]. The signal in aperture-based techniques is related to the field component in the sample plane while scattering instruments give signals depending on the field component normal to the sample plane [28]. The spatial resolution in aperture-based techniques depends on the size of the aperture, while in scattering-type measurements, depends on the curvature of the tip; however, both techniques can achieve resolutions in the range of 50—10 nm [27].

Aperture-based techniques as applied to gold nanoparticles can be performed in

- *transmission*, where the sample is placed on a transparent carrier and the light transmitted from the aperture through the sample is measured;
- *two photon photoluminescence (TPP)*, where modes of the gold nanoparticle are excited by illumination with a femtosecond laser and the induced photoluminescence is detected in the far field [28].

An advantage of TPP is that radiated photons have higher energy than incident photons, which allows almost background-free measurements [27]. For examples of transmission-type SNOM and TPP SNOM images, see Figures 7 and 8, respectively.

Spectra collected in the near field cannot be directly compared to those collected in far-field transmission spectroscopy. In fiber probe SNOM setups, the evanescent waves in the fiber can couple with surface plasmon modes of the nanoparticle, thereby converting evanescent fiber modes into propagating modes [29]. This causes the transmission through the setup to increase on top

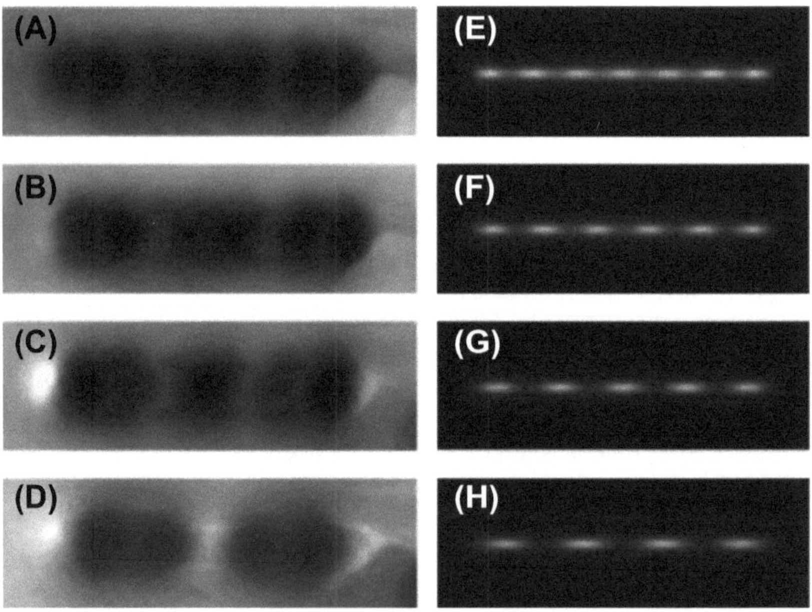

FIGURE 7 Transmission SNOM image of a gold nanorod (diameter: 20 nm, length: 520 nm) at different wavelengths of (A) 647–666 nm, (B) 666–-686 nm, (C) 705–725 nm, (D) 822–842 nm. On the right-hand side (E–F), simulated photonic LDOS are depicted for (E) 636 nm, (F) 679 nm, (G) 722 nm, and (H) 829 nm. *Reproduced with permission from Kohei Imura and Hiromi Okamoto [27].*

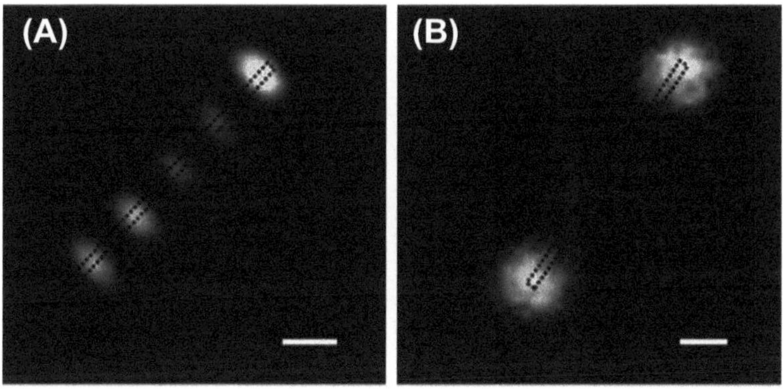

FIGURE 8 TPP SNOM image of gold nanorods (A) diameter 20 nm, length: 540 nm; (B) diameter: 21 nm, length: 565 nm. *Reproduced with permission from Kohei Imura and Hiromi Okamoto [27].*

of gold nanpparticles to values above the transmission achieved on the bare substrate.

A modification of SNOM where the scattering instead of the transmission is measured was introduced by Eah et al. [21]. The authors used an inverted microscope to select a single gold nanoparticle, which was then picked up with an SNOM fiber with a microfabricated axicon lens at the tip as a collection optic. Through the fiber, the light scattered off the particle could be collected. The connection to the fiber allowed placing the particle in different solvents to analyze the influence of the solvent refractive index on the plasmon peak of said single particle.

2.5 Dynamic Light Scattering

Dynamic light scattering (DLS) is a photon-based technique for analyzing nanoparticles that looks at the movements of the particles in a solution to deduce their shape and size. A thorough explanation of DLS can be found in *Dynamic Light Scattering with applications to chemistry, biology and physics* by Berne and Pecora, concise introductions are presented in Refs [30,31].

A DLS instrument consists of a monochromatic (laser) light source, a liquid cell, and a photodetector placed at an angle θ to the beam path. In the most common implementation of DLS, photon correlation spectroscopy, the detector is connected to an autocorrelator circuit which calculates a discrete approximation of the correlation function [32],

$$\langle I(0)I(\tau)\rangle = \lim_{T\to\infty} \frac{1}{T}\int_0^T I(t)I(t+\tau)dt \tag{22}$$

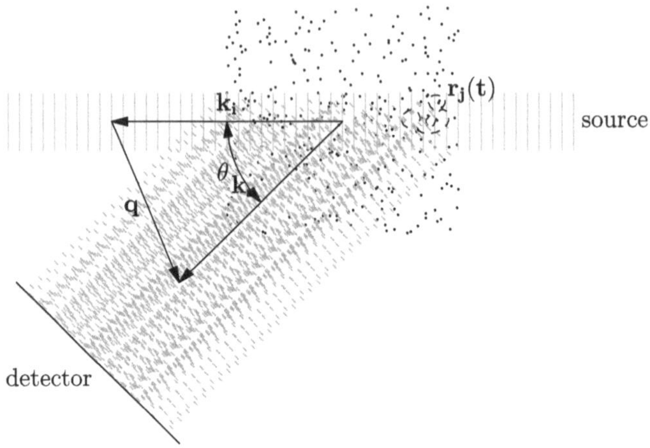

FIGURE 9 Measurement setup for a DLS experiment. The detector is placed at an angle 9 from the incident beam. Instruments using multiple detector positions are common. The image is not to scale; in actual setups, the beam diameter will be far larger than the particles and the detector will be placed significantly further from the scatterers.

from the intensity $I(t)$ output by the detector. In comparison, the signal detected in standard transmission absorption or scattering experiments is the time average of $I(t)$ (Figure 9):

$$\langle I \rangle = \lim_{T\to\infty} \frac{1}{T} \int_0^T I(t)dt \quad (23)$$

In the following paragraphs, we will give a short derivation for the shape of the DLS signal; for more thorough explanations, the reader is directed to Refs [30,32]. We begin the derivation of the signal in DLS by considering similar particles suspended in a transparent solvent. The particles are moving randomly due to Brownian motion. For a single scattering particle, the phase shift of the electric field E_s on the detector depends approximately on the scattering vector $\mathbf{q}$ and the position of the particle j $\mathbf{r(t)}$ [30]:

$$E_j(t) = E_0 e^{-i\mathbf{q}\cdot\mathbf{r(t)}+i\omega\mathbf{t}} \quad (24)$$

$\mathbf{q}$ is defined as the difference between the wave vectors of the scattered and the incident wave $\mathbf{q} = \mathbf{k_s} - \mathbf{k_i}$. For multiple similar scattering particles, the electric field will be the sum of all scattered fields

$$E(t) = \sum_{j=1}^{N} E_0 e^{-i\mathbf{q}\cdot\mathbf{r_j(t)}\,+\,i\omega\mathbf{t}} \quad (25)$$

The intensity seen by the detector is (with * indicating complex conjugation)

$$I(t) \propto E^*(t)E(t) \tag{26}$$

$$\propto \sum_{i,j} E_i(t)E_j^*(t) \tag{27}$$

which allows us to write the autocorrelation of I as

$$\langle I(t)I(t+\tau)\rangle \propto \langle E(t)E^*(t)E(t+\tau)E^*(t+\tau)\rangle \tag{28}$$

$$\propto \left\langle \sum_{i,j,k,l} E_i(t)E_j^*(t)E_k(t+\tau)E_l^*(t+\tau)\right\rangle \tag{29}$$

The position-independent factors in the electric fields cancel as we are only looking at monochromatic light. Equation (29) can be simplified by only keeping terms where $i = l$ and $j = k$. When $i = l$ and $j = k$, the term is equal to E_0^4. Since the positions of different particles are uncorrelated terms, where either $i \neq l$ or $j \neq k$ are zero or lead to high-frequency modulations that are not picked up by the detector [30].

$$\langle I(t)\mathrm{I}(t+\tau)\rangle \propto N^2E_0^4 + \sum_{i,j}\left\langle E_i(t)E_j^*(t)E_j(t+\tau)E_i^*(t+\tau)\right\rangle \tag{30}$$

$$\propto N^2E_0^4 + \sum_{i,j}\left\langle E_i(t)E_j^*(t)E_j(t+\tau)E_i^*(t+\tau)\right\rangle \tag{31}$$

$$\propto N^2E_0^4 + \sum_{i,j}\left\langle E_i(t)E_j^*(t+\tau)\right\rangle\left\langle E_j^*(t)E_i^*(t+\tau)\right\rangle \tag{32}$$

and since all particles are similar, this can be further simplified to [30]

$$\propto N^2E_0^4 + \left(N^2 - N\right)E_0^4\left|\left\langle e^{i\mathbf{q}\left(\mathbf{r_j}(\mathbf{t}+\tau)-\mathbf{r_j}(\mathbf{t})\right)}\right\rangle\right|^2 \tag{33}$$

The expression $\langle E_i(t)E_i^*(t+\tau)\rangle$ only depends on the autocorrelation function for the field scattered by a single particle which can be shown to be proportional to $\exp(-q^2Dt)$ [32] for a Brownian particle $\left(q = |\mathbf{q}| = \frac{4\pi n}{\lambda}\sin\left(\frac{\theta}{2}\right)\right)$. Finally, after normalizing Eqn (33) to the average intensity on the detector, we end up with the following expression to describe the DLS signal [30]:

$$g(\tau) = \frac{\langle I(t)I(t+\tau)\rangle}{\langle I\rangle^2} = 1 + e^{-2Dq^2\tau} \tag{34}$$

Hence, we can determine the diffusion coefficient of the particles from the detector read out.

For spherical particles, the diffusion coefficient is indirectly proportional to the particle radius R (*Einstein relation*):

$$D = \frac{k_B T}{6\pi\eta R} \tag{35}$$

where k_B is the Boltzmann constant, T is the temperature, and η is the viscosity of the solvent.

In contrast to other UV–VIS techniques, DLS allows determining the distribution of the radii from bulk measurements. When particles of different radii are suspended in the solvent, the detector signal will be the weighted sum of several functions $g(t)$ with different diffusion coefficients. Next to least squares fits, several other methods can be used to analyze the autocorrelation signal. The method of cumulants is used for unimodal distributions. Here the expression

$$B + \beta \exp(-2q^2 D\tau)\left(1 + \frac{\mu_2}{2!}\tau^2 - \frac{\mu_2}{3!}\tau^2 + \ldots\right) \tag{36}$$

is fit to $\frac{\langle I(t)I(t+\tau)\rangle}{\langle I\rangle^2}$ [33]. The cumulants μ_2 and μ_3 give information about the shape of the distribution of the particle sizes: corresponds to the $\mu_2$variance, μ_3 describes the skewness [33], and B and β are there to fit deviations from theory (i.e., baseline $B = 1$). The CONTIN algorithm [34] uses inverse Laplace transform to determine the distribution function $A(\Gamma)$ for $\Gamma = q^2 D$ in the expression $\int_0^\infty A(\Gamma)\exp(\Gamma t)d\Gamma$ [31].

Nonspherical particles also show rotational diffusion, defined by the rotational diffusion coefficient D_R in addition to the translational diffusion (D_T) seen in spherical particles [32,35]. The rotational and the translational contribution can be picked apart by using a polarized light source and an analysator in front of the detector. When the polarizer is oriented vertically (relative to the plane of source and detector) and the analysator is oriented horizontally, the autocorrelation signal behaves as [36]

$$g_2^{VH}(q,t) - 1 = \beta' \exp\left(-\left(2D_T q^2 + D_R\right)\right) \tag{37}$$

In a plot of $g_2^{VH}(q,t)$ as a function of q^2, the intercept will be $12D_R$ and the slope will be $2D_T$ [36]. Rodríguez-Fernandez et al. [36] also recommend a different approach whereby Eqn (37) and

$$\begin{aligned} g_2^{VV} - 1 = \beta\big(&A^2 \exp(-2q^2 D_T t) + 2AB \exp\left(-\left(2q^2 D_T + 6D_R\right)t\right) \\ &+ B^2 \exp\left(-\left(2q^2 D_T + 12D_R\right)t\right)\big) \end{aligned} \tag{38}$$

are fitted to measurement data by using a nonlinear least-squares method.

2.6 Other Photon-Based Techniques

In addition to the techniques above, several other photon-based spectroscopic techniques have been used for the analysis of gold nanoparticles. Some of these are introduced in the following paragraphs.

Kalkbrenner, Håkanson, and Sandoghdar [37] demonstrated the optical tomography on a single gold nanoparticle of a diameter around 5 nm. This was done by mounting the particle on a sharp glass tip and then using this tip to place the particle in the beam of a xenon lamp (Figure 10). Light scattered off the particle was collected by a microscope objective and analyzed using a spectrometer. Scattering spectra were collected in dependence of the polarization of the incident light. Furthermore, by rotating the glass fiber, the particle itself could be rotated, allowing to collect the scattering spectra for different orientations of the particle. The researchers found that the glass tip introduced only a slight uncertainty in their measurement results.

Recently Lopez-Lorente et al. [38] used mid-infrared attenuated total reflection spectroscopy to analyze the formation of gold nanoparticles. Instead of detecting the particles themselves, the researchers observed the influence of the gold nanoparticles on the infrared absorption of the surrounding medium, specifically the resonant enhancement of the absorption of the H_2O bending mode around 1650 cm^{-1} by the gold nanoparticles.

3. ELECTRON-BASED TECHNIQUES

Additionally to photons treated in the last chapter, electrons traveling on a path through or past a nanoparticle will also interact with the electrons of said particle. Electron-based techniques allow resolutions <1 Å due to the small de

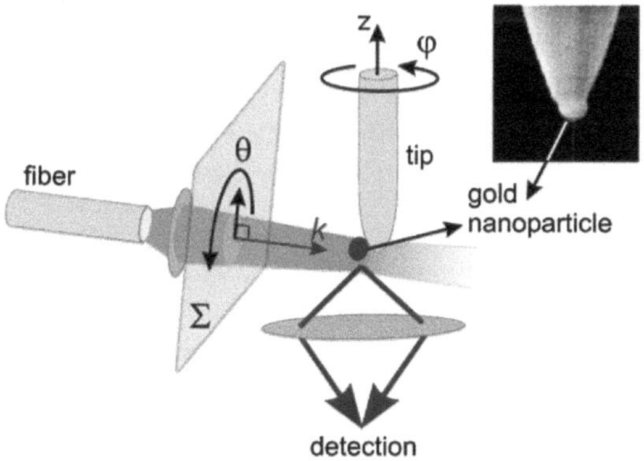

FIGURE 10 Schematics of the measurement setup used for optical tomography of a single gold nanoparticle by Kalkbrenner, Hakanson, and Sandoghdar. A xenon lamp coupled to a multimode fiber is used to illuminate the particle. θ is the angle of polarization in plane $\sum$, which is orthogonal to the direction of the light beam k. ϕ is the angle of the particle (relative to its starting angle). *Reprinted with permission from Thomas Kalkbrenner, Ulf Hakanson, and Vahid Sandoghdar [37]. Copyright 2004 American Chemical Society.*

Broglie wavelength of electrons at sufficiently high kinetic energies (>20 keV) [39].

3.1 Interaction of Nanoparticles and Electron Beams

The change in kinetic energy of the electron before and after the interaction with a particle in part depends on the dielectric function of the particle's plasmons [40]. In contrast to the techniques used for optical analysis of nanoparticles where signals are recorded as a function of the incident wavelength, electron-based techniques use a "monochromatic" source, i.e., electrons of similar kinetic energy. Frequency-specific information is gained from the change in the kinetic energy of the electron (see EELS in Section 3.2) or from the wavelength of the light emitted by the particle after excitation by electrons (see cathodoluminescence in Section 3.3).

The differential probability of energy loss $\Gamma_{sph}^{NR}(\omega)$ of a given circular frequency ω for an electron traveling at a distance b and velocity v from a spherical particle of radius a is in the nonretarded approximation given by [39,41]:

$$\Gamma_{sph}^{NR}(\omega) = \frac{4e^2}{\pi \hbar \upsilon^2} \sum_{l=1}^{\infty} \sum_{m=-l}^{l} \frac{(\omega/\upsilon)^{2l}}{(l+m)(l-m)} K_m^2\left(\frac{\omega b}{\upsilon}\right) \mathrm{Im}\left(\alpha_l^{N\,R}(\omega)\right) \quad (39)$$

Where,

$$\alpha_l^{N\,R}(\omega) = a^{2l+1} \frac{l\varepsilon(\omega) - l}{l\varepsilon(\omega) + l + 1} \quad (40)$$

e is the charge of an electron and K is the reduced Planck constant. A formula that does not neglect retardation can be found in Ref. [42].

3.2 Scanning Transmission Electron Microscopy Electron Energy-Loss Spectroscopy and Energy Filtered Transmission Electron Microscopy Electron Energy-Loss Spectroscopy

Scanning transmission electron microscopy electron energy-loss spectroscopy (STEM EELS) and energy filtered transmission electron microscopy electron energy-loss spectroscopy (EFTEM EELS) are closely related techniques that are performed by accelerating electrons to energies in the 100 keV range and directing them toward the sample [43]. An analysator below the sample analyzes the kinetic energies of the electrons. The energy loss of the electrons contains information about the interactions between the electrons and the sample. Mainly, these interactions are the following [43]:

- Removal of inner shell electrons
- Excitation of inner shell electrons to unoccupied levels
- Excitation of plasmons

Excitation of plasmons are so-called low-loss excitations in the 5−50 eV range, while the energies necessary for the excitation of inner shell electrons will be found above 50 eV [43]. Around a loss of 0 eV in EEL spectra, the zero-loss peak (ZLP) is situated. This very intensive feature can overlap the plasmon peaks and thus hinder the application of STEM EELS and EFTEM EELS to the analysis of nanoparticles. Only recent improvements in measurement technique have made it possible to measure the plasmon peaks close to the ZLP [44].

An in-depth introduction to the theoretical background and implementation details of STEM EELS and EFTEM EELS would be out of the scope of this chapter. Such an introduction can be found, e.g., in EELS and EFTEM [43]. We will instead focus on a rough functional description of these methods as pertaining to the analysis of gold nanoparticles.

In STEM EELS, a spectrum of electron energy loss is collected for a single measurement position at a time. An electron optic placed below the instrument redirects the electrons toward an energy filter system that deflects the electrons according to their kinetic energy. Energy information is then generated either by detecting the electrons that pass through a slit placed after the filter or by using an array detector to detect electrons with different deflections simultaneously. In contrast, EFTEM EELS illuminates a larger area of the sample at once. After interacting with the sample, these electrons are band pass filtered for a selected range of kinetic energies. They are then directed either to a screen for viewing or toward an array detector. This way the energy-loss probability can be imaged across a larger area at once. Spectra can be collected in EFTEM EELS by changing the transmission range of the band pass filter. A direct comparison of STEM EELS and EFTEM EELS can be found in Ref. [45].

EELS of plasmons can also be performed in the so-called *aloof configuration*, where electrons do not pass through the particles but closely by them. This allows collecting EEL plasmon spectra with reduced sample damage [46] as no ionization takes place.

To perform any one of the TEM techniques on a sample, it has to be electron transparent, i.e., thinner than 100 nm [47]. This can be achieved for bulk sample by etching or milling them down. Small particle can also be mounted on a mesh grid [40].

The peak positions in the electron energy loss probability measured in EEL spectra (39) are similar to those in the imaginary part of the inverse of the dielectric constant of the plasmon modes $\mathrm{Im}(1/\varepsilon_p(\omega))$ [40]. In contrast, the peak position in optical spectra are similar to those in $\mathrm{Im}(\varepsilon_p(\omega))$. A comparison of the electron energy loss probability, $\mathrm{Im}(1/\varepsilon_p(\omega))$ and $\mathrm{Im}(\varepsilon_p(\omega))$ for gold spheroids can be seen in Figure 11. A closer look at the connections between LDOS for photons, the electron energy loss probability in TEM, and the intensity of the cathodoluminescence spectra (see next section) is taken, for example, in Refs [48] and [49]. Examples for the signals in each of these three

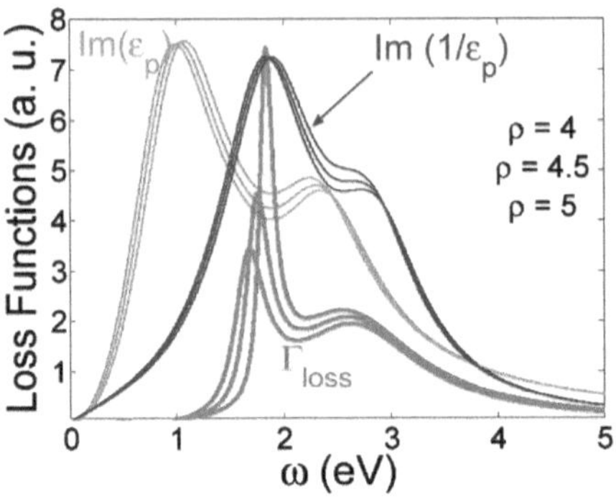

FIGURE 11 Comparison between analytical representations of the energy loss peak for three Au prolate spheroids. All calculations were performed for particles in vacuum. *p* is the aspect ration of the particles. *Reprinted with permission from Moussa N'Gom et al. [40]. Copyright 2008 American Chemical Society.*

techniques are reprinted in Figure 12. Asenjo-Garcia and García de Abajo have shown theoretically that additional information about nanoparticles can be gained when EELS is performed while the particle is illuminated with a laser light source. This electron energy-gain spectroscopy [50] gives a higher energy resolution than EELS and allows to include the effects of the light polarization into the experiments [51].

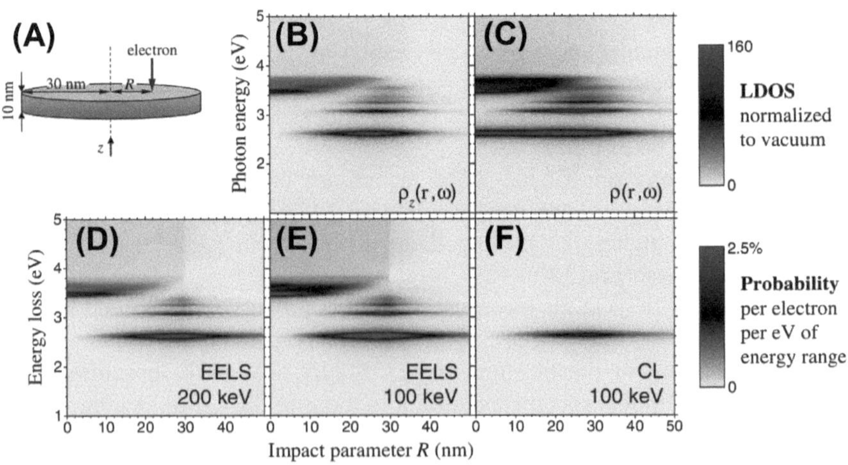

FIGURE 12 Comparison of EELS, CL, and photon local density of states for an Ag disk of 10 nm height and 30 nm diameter. (B) and (C) show the local density of states, (D) and (E) show the EELS energy loss probability for 200 keV and 100 keV, respectively. (F) shows the CL emission. Note that only a subset of the modes shows up in CL. *Reproduced with permission from F. J. Garcia de Abajo and M. Kociak [48].*

3.3 Cathodoluminescence (CL)

Cathodoluminescence is the radiative decay of modes that have been excited by electrons. Secondary electron microscopes (SEM) or TEMs outfitted with a photon detector—usually a photomultiplier tube—can be used to record the spatially resolved cathodoluminescence of nanoparticles.

In contrast to EELS, CL analysis does not need electron-transparent samples. The spatial resolution depends on the instrument used to excite the CL and is given as <1 Å for TEM excitation [39] and more than 2 nm [39] or around 10 nm [52] for SEM excitation. The spatial resolution is, however, not only defined by the exciting electron beam alone, but also influenced by the shape of the excited mode which can either be locally confined or stretch over the whole particle [47].

CL instruments can be employed with a monochromator to gain wavelength-specific information or in panchromatic (panCL) mode, where all photons are directed toward the detector [47]. Additional information can be gained by using a polarization filter in front of the photon detector.

Nanoparticles can exhibit so-called dark modes, which show now optical emission [39] (Figures 12 and 13). These can be detected by EELS but not by CL spectroscopy.

4. SUMMARY

4.1 Comparison of Methods

The spectroscopic analysis of gold nanoparticles spans a wide range of different methods, ranging from very cheap and quick methods such as

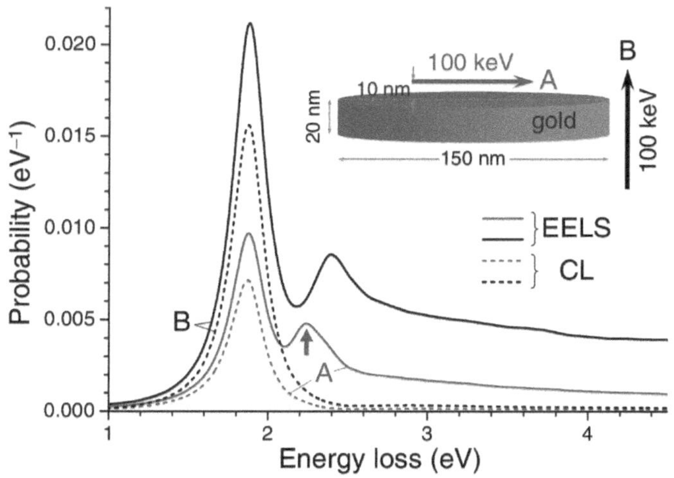

FIGURE 13 Comparison of EELS and CL intensities in a gold nanodisk of the geometry depicted in the inset. A (orange) and B (black) show different trajectories of the electron beam. *Reproduced with permission from Garcia de Abajo [39].*

UV–VIS–NIR transmission spectroscopy, to more complicated tools such as SMS and DLS and finally to expensive and complicated room-sized equipment such as TEM. However, it is of course not possible to generally rank the methods, instead they should be seen as complementary: While EELS TEM is able to spatially resolve single particles, it is very tedious to use this method to get information about the totality of all particles in the sample. Even more, the particles might appear differently in the high vacuum (e.g., aggregation, hydration of the cover layer) needed for TEM than in its intended application at ambient conditions. Transmission spectroscopy, on the other hand, will be able to analyze particles at ambient conditions but will not readily reveal the finer details about the shapes of individual particles.

Listed in Table 2 are all methods from the preceding chapter in a quick overview.

TABLE 2 Information Content of the Techniques Treated in this Book Section

		Information Content		
Method	Resolution	Concentration	Size and Shape	Near Field
Photon Based				
Transmission	Bulk	Calibration	Calculation	No
Dark-field microscopy	Single particle	No	Calculation	No
SNOM	Subparticle	No	Yes	Yes
Optical tomography	Single particle	No	Calculation	No
SMS	Single particle	No	Calculation	No
MIR-ATR	Bulk	No	No	No
DLS	Bulk	Calibration	Calculation	No
e^- Based				
EELS TEM	Subparticle	No	via TEM	Yes
EELS EFTEM	Subparticle	No	via TEM	Yes
SEM CL	Single particle	No	via SEM	Yes

REFERENCES

[1] Gustav Mie, Beitrage zur Optik triiber Medien, speziell kolloidaler Metalllosungen, Ann. Phys. 4(25) (1908) 389–445.

[2] Craig F. Bohren, Donald R. Huffman, Absorption and Scattering of Light by Small Particles, John Wiley & Sons, 1983.

[3] Stephan Link, Mostafa A. El-Sayed, Shape and size dependence of radiative, non- radiative and photothermal properties of gold nanocrystals, Int. Rev. Phys. Chem. 19 (3) (July 2000) 409–453. http://dx.doi.org/10.1080/01442350050034180. ISSN: 0144-235X, http://dx.doi.org/10.1080/01442350050034180.

[4] Arthur L. Aden, Milton Kerker, Scattering of electromagnetic waves from two concentric spheres, J. Appl. Phys. 22 (10) (1951) 1242–1246, http://dx.doi.org/10.1063/1.1699834. ISSN: 00218979. http://link.aip.org/link/JAPIAU/v22/i10/p1242/s1%5C&Agg=doi.

[5] Raphael Ruppin, Optical absorption of two spheres, J. Phys. Soc. Japan 58 (4) (1989) 1446–1451.

[6] R. Gans, Über die Form ultramikroskopischer Goldteilchen, Ann. Phys. (Leipzig) 342 (5) (1912) 881–900.

[7] Viktor Myroshnychenko, et al., Modelling the optical response of gold nanoparticles, Chem. Soc. Rev. 37 (9) (September 2008) 1792–1805. http://dx.doi.org/10.1039/b711486a. ISSN: 0306-0012, http://www.ncbi.nlm.nih.gov/pubmed/18762829.

[8] Jorge Peréz-Juste, et al., Gold nanorods: synthesis, characterization and applications, Coord. Chem. Rev. 249 (September 2005) 1870–1901. http://dx.doi.org/10.1016/j.ccr.2005.01.030. ISSN: 00108545. http://linkinghub.elsevier.com/retrieve/pii/S0010854505000287.

[9] F. García de Abajo, a Howie, Retarded field calculation of electron energy loss in inhomogeneous dielectrics, Phys. Rev. B 65 (11) (March 2002) 115418. http://dx.doi.org/10.1103/PhysRevB.65.115418. ISSN: 0163-1829. http://link.aps.org/doi/10.1103/PhysRevB.65.115418.

[10] Bruce T. Draine, Piotr J. Flatau, Discrete-dipole approximation for scattering calculations, J. Opt. Soc. Am. A 11 (4) (April 1994) 1491. http://dx.doi.org/10.1364/JOSAA.11.001491. ISSN: 1084-7529, http://www.opticsinfobase.org/abstract.cfm?URI=josaa-11-4-1491.

[11] Bruce T Draine and Piotr J Flatau. User guide for the discrete dipole approximation code DDSCAT 7.2. In: c (Feb. 2012), p. 95. arXiv: 1202.3424. http://arxiv.org/abs/1202.3424.

[12] Claes-Göran Granqvist, O. Hunderi, Optical properties of ultrafine gold particles, Phys. Rev. B 16 (8) (1977).

[13] Paul Mulvaney, Surface plasmon spectroscopy of nanosized metal particles, Langmuir 12 (3) (January 1996) 788–800. http://dx.doi.org/10.1021/la9502711. ISSN: 0743-7463, http://pubs.acs.org/doi/abs/10.1021/la9502711.

[14] A. Lambrecht, S. Reynaud, Casimir force between metallic mirrors, Eur. Phys. J. D 8 (2000) 309–318.

[15] P.B. Johnson, R.W. Christy, Optical constants of the noble metals, Phys. Rev. B 1318 (1970) (1972) 4370–4379.

[16] Edward Palik (Ed.), Handbook of Optical Constants of Solids, Elsevier, 1998, ISBN 978-0-12544420-0.

[17] Marcos M. Alvarez, et al., Optical absorption spectra of nanocrystal gold molecules, J. Phys. Chem. B 5647 (96) (1997) 3706–3712.

[18] T.G. Schaaff, et al., Isolation of smaller nanocrystal Au molecules: robust quantum effects in optical spectra, J. Phys. Chem. B 5647 (97) (1997) 7885–7891. http://pubs.acs.org/doi/abs/10.1021/jp971438x.

[19] Wolfgang Haiss, et al., Determination of size and concentration of gold nanoparticles from UV-vis spectra, Anal. Chem. 79 (11) (June 2007) 4215–4221. http://dx.doi.org/10.1021/ac0702084. ISSN: 0003-2700, http://www.ncbi.nlm.nih.gov/pubmed/17458937.

[20] V. Amendola, M. Meneghetti, Size evaluation of gold nanoparticles by UV – vis spectroscopy, J. Phys. Chem. C 113 (11) (2009) 4277–4285.

[21] Sang-Kee Eah, et al., Plasmon scattering from a single gold nanoparticle collected through an optical fiber, Appl. Phys. Lett. 86 (3) (2005) 031902. http://dx.doi.org/10.1063/1.1851011. ISSN: 00036951, http://link.aip.org/link/APPLAB/v86/i3/p031902/s1%5C&Agg=doi.

[22] Sylvia Underwood, Paul Mulvaney, Effect of the solution refractive index on the color of gold colloids, Langmuir 10 (10) (October 1994) 3427–3430. http://dx.doi.org/10.1021/la00022a011. ISSN: 0743-7463, http://pubs.acs.org/doi/abs/10.1021/la00022a011.

[23] A. Arbouet, et al., Direct measurement of the single-metal-cluster optical absorption, Phys. Rev. Lett. 93 (12) (September 2004) 127401. http://dx.doi.org/10.1103/PhysRevLett.93.127401. ISSN: 0031-9007, http://link.aps.org/doi/10.1103/PhysRevLett.93.127401.

[24] P. Billaud, et al., Optical extinction spectroscopy of single silver nanoparticles, Eur. Phys. J. D 43 (1–3) (May 2007) 271–274. http://dx.doi.org/10.1140/epjd/e2007-00112-y. ISSN: 1434-6060, http://www.springerlink.com/index/10.1140/epjd/e2007-00112-y.

[25] Viktor Myroshnychenko, et al., Plasmon spectroscopy and imaging of individual gold nanodecahedra: a combined optical microscopy, cathodoluminescence, and electron energy-loss spectroscopy study, Nano Lett. 12 (8) (August 2012) 4172–4180. http://dx.doi.org/10.1021/nl301742h. ISSN: 1530-6992, http://www.ncbi.nlm.nih.gov/pubmed/22746278.

[26] Marc Richter, Volker Deckert, Scanning near-field optical microscopy (SNOM), in: Gernot Friedbacher, Henning Bubert (Eds.), Surface and Thin Film Analysis, 2011, pp. 481–497. Chapter 31.

[27] Kohei Imura, Hiromi Okamoto, Development of novel near-field microspectroscopy and imaging of local excitations and wave functions of nanomaterials, Bull. Chem. Soc. Japan 81 (6) (2008) 659–675. http://dx.doi.org/10.1246/bcsj.81.659. ISSN: 0009-2673, http://joi.jlc.jst.go.jp/JST.JSTAGE/bcsj/81.659?from=CrossRef.

[28] Hiromi Okamoto, Kohei Imura, Visualizing the optical field structures in metal nanostructures, J. Phys. Chem. Lett. 4 (13) (July 2013) 2230–2241. http://dx.doi.org/10.1021/jz401023d. ISSN: 1948-7185. http://pubs.acs.org/doi/abs/10.1021/jz401023d.

[29] Kohei Imura, Tetsuhiko Nagahara, Hiromi Okamoto, Characteristic near-field spectra of single gold nanoparticles, Chem. Phys. Lett. 400 (4–6) (December 2004) 500–505. http://dx.doi.org/10.1016/j.cplett.2004.10.152. ISSN: 00092614. http://linkinghub.elsevier.com/retrieve/pii/S0009261404017269.

[30] W.I. Goldburg, Dynamic light scattering, Am. J. Phys. 67 (12) (December 1999) 1152. http://dx.doi.org/10.1119/1.19101. ISSN: 00029505, http://link.aip.org/link/?AJP/67/1152/1%5C&Agg=doi.

[31] R. Pecora, Dynamic light scattering measurement of nanometer particles in liquids, J. Nanopart. Res. 2 (2000) 123–131.

[32] Bruce J. Berne, Robert Pecora, Dynamic Light Scattering with Applications to Chemistry, Biology and Physics, John Wiley & Sons, Inc, 1976.

[33] Barbara J. Frisken, Revisiting the method of cumulants for the analysis of dynamic light-scattering data, Appl. Optics 40 (24) (2001) 4087–4091. http://www.opticsinfobase.org/abstract.cfm?uri=A0-40-24-4087.

[34] Stephen W. Provencher, Contin: a general purpose constrained regularization program for inverting noisy linear algebraic and integral equations, Comput. Phys. Commun. 27 (1982) 229–242. http://www.sciencedirect.com/science/article/pii/0010465582901746.

[35] B.N. Khlebtsov, N.G. Khlebtsov, On the measurement of gold nanoparticle sizes by the dynamic light scattering method, Colloid J. 73 (1) (Febuary 2011) 118–127. http://dx.doi.org/10.1134/S1061933X11010078. ISSN: 1061-933X, http://link.springer.com/10.1134/S1061933X11010078.

[36] J. Rodriguez-Fernandez, et al., Dynamic light scattering of short Au rods with low aspect ratios, J. Phys. Chem. C 111 (13) (April 2007) 5020–5025. http://dx.doi.org/10.1021/jp067049x. ISSN: 1932-7447. http://pubs.acs.org/cgi-bin/doilookup/?10.1021/jp067049x.

[37] Thomas Kalkbrenner, Ulf Håkanson, Vahid Sandoghdar, Tomographic plasmon spectroscopy of a single gold nanoparticle, Nano Lett. 4 (12) (December 2004) 2309–2314. http://dx.doi.org/10.1021/nl048694n. ISSN: 1530-6984, http://pubs.acs.org/doi/abs/10.1021/nl048694n.

[38] Ángela Inmaculada López-Lorente, et al., Infrared attenuated total reflection spectroscopy for the characterization of gold nanoparticles in solution, Anal. Chem. 86 (2014) 783–789.

[39] F.J. García de Abajo, Optical excitations in electron microscopy, Rev. Mod. Phys. 82 (1) (February 2010) 209–275. http://dx.doi.org/10.1103/RevModPhys.82.209. ISSN: 0034-6861, http://link.aps.org/doi/10.1103/RevModPhys.82.209.

[40] Moussa N'Gom, et al., Single particle plasmon spectroscopy of silver nanowires and gold nanorods, Nano Lett. 8 (10) (October 2008) 3200–3204. http://dx.doi.org/10.1021/nl801504v. ISSN: 1530-6984. http://pubs.acs.org/doi/abs/10.1021/nl801504v.

[41] T.L. Ferrell, P.M. Echenique, Generation of surface excitations on dielectric spheres by an external electron beam, Phys. Rev. Lett. 55 (14) (1985) 1526–1529. http://dipc.ehu.es/Echenique/admin/documentos/archivos/publicaciones/037PRL1985.pdf.

[42] F.J. Garcia de Abaja, Relativistic energy loss and induced photon emission in the interaction of a dielectric sphere with an external electron beam, Phys. Rev. B 59 (4) (1999) 3095–3107.

[43] Reinhard Schneider, Electron energy-loss spectroscopy (EELS) and energy-filtering transmission electron microscopy (EFTEM), in: Gernot Friedbacher, Henning Bubert (Eds.), Surface and Thin Film Analysis, Wiley-VCH, 2011, pp. 67–91. Chapter 4.

[44] Jaysen Nelayah, et al., Mapping surface plasmons on a single metallic nanoparticle, Nat. Phys. 3 (5) (April 2007) 348–353. http://dx.doi.org/10.1038/nphys575. ISSN: 1745-2473, http://www.nature.com/doifinder/10.1038/nphys575.

[45] Bernhard Schaffer, et al., Comparison of EFTEM and STEM EELS plasmon imaging of gold nanoparticles in a monochromated TEM, Ultramicroscopy 110 (8) (July 2010) 1087–1093. http://dx.doi.org/10.1016/j.ultramic.2009.12.012. ISSN: 03043991, http://linkinghub.elsevier.com/retrieve/pii/S0304399109002903.

[46] J.M. Pitarke, et al., Theory of surface plasmons and surface-plasmon polaritons, Rep. Prog. Phys. 70 (1) (January 2007) 1–87. http://dx.doi.org/10.1088/0034-4885/70/1/R01. ISSN: 0034-4885, http://stacks.iop.org/0034-4885/70/i=1/a=R01?key=crossref.5b1052b473009d0de2f21c63e367f570.

[47] Pratik Chaturvedi, et al., Imaging of plasmonic modes of silver nanoparticles using high-resolution cathodoluminescence spectroscopy, ACS Nano 3 (10) (October 2009) 2965–2974. http://dx.doi.org/10.1021/nn900571z. ISSN: 1936-0851, http://pubs.acs.org/doi/abs/10.1021/nn900571z.

[48] F.J. García de Abajo, M. Kociak, Probing the photonic local density of states with electron energy loss spectroscopy, Phys. Rev. Lett. 100 (10) (March 2008) 106804. http://dx.doi.org/10.1103/PhysRevLett.100.106804. ISSN: 0031-9007, http://link.aps.org/doi/10.1103/PhysRevLett.100.106804.

[49] Ulrich Hohenester, Harald Ditlbacher, Joachim Krenn, Electron-energy-loss spectra of plasmonic nanoparticles, Phys. Rev. Lett. 103 (10) (August 2009) 106801. http://dx.doi.org/10.1103/PhysRevLett.103.106801. ISSN: 0031-9007, http://link.aps.org/doi/10.1103/PhysRevLett.103.106801.

[50] A. Asenjo-Garcia, F.J. García de Abajo, Plasmon electron energy-gain spectroscopy, New J. Phys. 15 (10) (October 2013) 103021. http://dx.doi.org/10.1088/13672630/15/10/103021. ISSN: 1367-2630, http://stacks.iop.org/1367-2630/15/i=10/a=103021?key=crossref.fab53d6d6335655edc339c7b89c41f37.

[51] F.J. García de Abajo, M. Kociak, Electron energy-gain spectroscopy, New J. Phys. 10 (7) (July 2008) 073035. http://dx.doi.org/10.1088/1367-2630/10/7/073035. ISSN: 1367-2630, http://stacks.iop.org/1367-2630/10/i=7/a=073035?key=crossref.2dec4824a26fb28465428f671607d875.

[52] Vighter Iberi, Nasrin Mirsaleh-Kohan, Jon P. Camden, Understanding plasmonic properties in metallic nanostructures by correlating photonic and electronic excitations, J. Phys. Chem. Lett. 4 (7) (April 2013) 1070–1078. http://dx.doi.org/10.1021/jz302140h. ISSN: 1948-7185, http://pubs.acs.org/doi/abs/10.1021/jz302140h.

Chapter 8

Mass Spectrometry for the Characterization of Gold Nanoparticles

Laura Trapiella-Alfonso, José M. Costa-Fernández, Jorge Ruiz Encinar, Rosario Pereiro and Alfredo Sanz-Medel*

Department of Physical and Analytical Chemistry, University of Oviedo, Oviedo, Spain

**Corresponding author: E-mail: asm@uniovi.es*

Chapter Outline

Gold Nanoparticles in Analytical Chemistry. http://dx.doi.org/10.1016/B978-0-444-63285-2.00008-0

1. MASS SPECTROMETRY IN NANOTECHNOLOGY

Nanotechnology is a very active research area, providing a great number of new nanomaterials with novel exciting applications more and more appreciated and valued in the pharmaceutics, electronics, food, cosmetics, or materials industries. Nanoparticles (NPs), in particular, display rich physical and chemical phenomena due to the high size restriction (quantum confinement), explaining their fascinating and unusual properties. That is, NP properties differ from those of bulk material of the same chemical composition and may be tuned by varying their size, shape, and chemical environment. Therefore, to understand and exploit the high potential of NPs in novel applications, a deep knowledge of their synthesis, morphological and chemical properties, and the final quality of the nanoproducts (e.g., homogeneity and stability) is needed. Moreover, the increasing number of NP-based products is bringing about an unavoidable growing exposure of people and their environment to NPs. As consequences that they may cause are still unknown, the risk to health is not predictable so far. Eventual toxic effects of NPs will strongly depend on their composition, size, and shape, as well as on their surface chemistry, including the ligands that protect the nuclei [1–3]. Therefore, a detailed characterization of the whole nanoparticle (i.e., core and ligands) is important not only for the improvement of the synthesis and final properties of the NPs but also for a better understanding of their behavior, final fate, and health safety studies. So far several analytical techniques have been successfully applied for the characterization of nanomaterials in general and gold nanoparticles (AuNPs) in particular [1,4,5]. Table 1 summarizes the tools most utilized for such characterizations.

Very often, conventional analytical methodologies typically used for bulk material analysis cannot be directly applied for the characterization of NPs of the same composition. In fact, new instruments and new strategies are required to assess the adequate characterization of NPs. Of course, NP analysis requires methods capable of detecting NPs at relevant concentrations (sometimes at trace levels) and avoiding sample interferences. Thus, the combination of existing detection methods with new separations to ensure selective and sensitive detection is a current research trend. In particular, mass spectrometry (MS) is a good example of how known techniques and methodologies can be adapted and evolve to become one of the more reliable, versatile, and sensitive analytical tools for tackling a holistic characterization of NPs. Figure 1 shows a general scheme summarizing the types of information that can be obtained with the different mass spectrometric techniques in the search for a characterization of NPs as complete as possible.

This chapter describes recent advances and the evolution of mass spectrometric techniques to control the synthesis and to achieve a complete characterization of AuNPs through analytical measurements. First, the application of these MS techniques, which allow elemental information to be obtained, will be

TABLE 1 Overview of Common Analytical Techniques Used for Nanoparticle Characterization

Analytical Technique	Information Obtained	Advantages	Limitations
UV–visible spectroscopy	Bulk concentration, optical features, and some structural information (size and shape)	• Noninvasive, cost-effective, simple, and fast • Changes in the spectra can be associated with aggregated states	• Particles must be in suspension • Possible interferences from turbidity and dissolved organic matter
ICP-OES	Elemental chemical composition of bulk NPs, concentration (number, mass, content)	• Simultaneous elemental composition of tens of elements • Very sensitive (μg/L)	• Requires an acid digestion of the NP or NM isolated from the sample • Destructive
Scanning electron microscopy (SEM)	Size, size distribution, and shape of the NPs. Aggregation state (morphology)	• High-resolution image • Suitable for elemental analysis by complementary technique (EDS)	• Destructive • Incompatible with wet or liquid samples. Extensive sample preparation Sample drying may change NM structure or aggregation
Transmission electron microscopy (TEM)	Size, size distribution, shape, and structural information of the NPs. Aggregation state (morphology)	• Very high-resolution images • Suitable for elemental analysis by complementary technique (EDS, EELS, SAED)	• Requires extensive sample preparation and operates under vacuum • Destructive
Near-field scanning optical microscopy (NSOM)	Size and shape of NPs	• Nondestructive • High-resolution surface analysis at ambient conditions	• Small surface area analyzed • Currently a "specialist" technique

Continued

TABLE 1 Overview of Common Analytical Techniques Used for Nanoparticle Characterization—cont'd

Analytical Technique	Information Obtained	Advantages	Limitations
Atomic force microscopy (AFM)	Size, size distribution, and shape of the NPs	• Dry, moist, or liquid samples under ambient conditions • Capable of 3D mapping of the sample surface	• Sample under study must be immobilized onto a suitable surface and sample preparation can be complex • Technique requires extensive optimization and interpretation
Dynamic light scattering (DLS)	Hydrodynamic size, size distribution	• Nondestructive • Rapid, simple, and relatively cheap • Measurements can be performed in any liquid media	• Typically assumes spherical shape • Resolution issues • Sample preparation is extremely important
X-ray diffraction (XRD)	Structure/crystallinity, elemental composition of single NP	• Proven method for the analysis of minerals and crystal structure	• Needs a dry powder or a thin film placed on a target • Requires a large amount of sample 1–3% wt limit
ζ potential	Surface charge		
Fluorescence spectroscopy	Optical features, surface modifications	• Nondestructive, sensitive	• Limited to fluorescent species
Mass spectrometry (MS): ICP–MS, ESI, MALDI, hyphenated techniques (HPLC, DMA, AF4), Sp-ICP–MS	Elemental and molecular information on the composition, structure, and chemical state	• Highly versatile, able to couple with different separation techniques and different ways of sample introduction • Complementary information	• Requires expensive equipment and expertise • Destructive

Modified from Refs. [2,4,5].

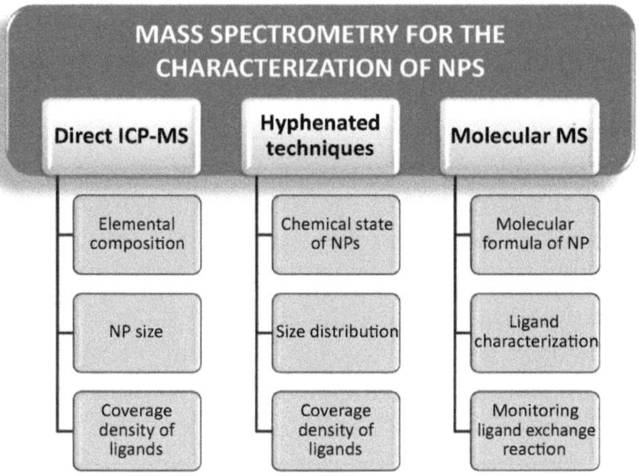

FIGURE 1 Mass spectrometric-based approaches used for NP characterization showing the different types of information attainable.

described, in particular, inductively coupled plasma (ICP)—MS techniques; NP chemical composition, core/surface ligand stoichiometries, and even NP size can be measured with such techniques. Next, the use of hyphenated approaches resorting to a previous separation step, which allow the characterization of polydispersed samples, will be discussed. Finally, the chapter focuses on molecular mass spectrometry, a discipline with amazing capabilities for NP characterization; today, molecular mass spectrometry techniques are indispensable tools for obtaining information about the NPs and also to evaluate their dynamics. The chapter ends with a brief overview of the perspectives expected from the use of MS in this exciting field of nanoscience.

2. ELEMENTAL MASS SPECTROMETRY

2.1 Inductively Coupled Plasma—Mass Spectrometry for Elemental Characterization of Engineering Gold Nanoparticles

In general, synthesis control and analytical characterization of colloidal NPs have been mainly carried out using molecular spectrophotometry, photoluminescence, X-ray photoelectron spectroscopy (XPS), X-ray diffraction, and electron microscopy techniques, among others. Surprisingly, considering their metallic core, only a few papers have paid attention so far to the elemental content of NPs and their influence on the physicochemical properties. In this context, the inherent quantitative character of the ICP—MS response signals could be of great value for achieving a more precise elemental characterization of the corresponding nanomaterial. This, in turn, could shed new light on NP properties and enable optimization of their synthesis.

ICP—MS is currently one of the more versatile and sensitive analytical tools for elemental analysis in a broad range of fields, including environmental, geological, clinical, and industrial. The very low detection limits achievable, in the range of picograms per gram, along with the virtual independence of ICP–MS signals with the sample matrix (facilitating ultratrace metal determinations in complex matrices) are among their more salient features. Moreover, multielement analysis (most elements of the periodic table can be easily ionized simultaneously in the ICP source) and a wide dynamic range (of up to 9 orders of magnitude) make ICP–MS the preferred technique for the elemental analysis of both major constituents and trace components of the same sample. In addition, quantification is comparatively easy and usually requires only standard solutions of the inorganic elements sought. Furthermore, ICP–MS allows the straightforward applications of isotope ratio measurements (e.g., isotope dilution-based techniques) [3,6,7].

Taking into account the elemental composition of most metal NPs existing today and the exceptional analytical features of ICP–MS for multielemental analysis at ultratrace concentration levels, new approaches for the characterization of nanomaterials (even nonmetal and metalloid based) by ICP–MS have gained increasing importance in recent years. ICP–MS is one among the few characterization techniques offering rich information about size, size distribution, composition, presence of impurities, and chemical state of the NPs, via direct ICP–MS (to be discussed in this section) and ICP–MS hyphenated techniques (that will be discussed in the following section).

In direct ICP–MS elemental analysis both colloidal and ionic solutions may be analyzed. The ICP can easily vaporize, atomize, and ionize NPs with radii typically up to 100 nm in solution. Thus, ICP–MS is being exploited as a highly sensitive and selective analytical tool for the elemental characterization of colloidal NPs in different applications, as it has already demonstrated its potential for quantification of NPs in terms of solution in mass and number concentration. Additionally, ICP–MS determinations of total elemental concentrations of such colloidal suspensions have also been proposed to determine, with high accuracy and precision, elemental stoichiometries all along the synthesis process [3].

Also, ICP–MS detection of metal NPs can be used in biomolecule assays. In fact, functional groups present in biomolecules can be labeled with appropriate metal NPs which would then be determined quantitatively, via ICP–MS, with high sensitivity. Many bioanalytical applications of ICP–MS based on the use of NPs as biomolecule labels have been reported during the past few years and illustrative examples will be described later on.

2.1.1 Elemental Characterization of Engineering Gold NPs

The direct quantification of the bulk NP elemental content and the NP number concentration in solution usually require conventional acid digestion for sample preparation (e.g., by use of a microwave oven to minimize the risk of losses). However, to overcome labor-intensive microwave digestion procedures, direct introduction of particle slurries has also been tested, using matrix- and slurry-tolerant nebulizers or electrothermal vaporization. In this sense, in a recent publication [8] three different methods of sample preparation for the quantification by ICP–MS of total concentration of different metal-based NPs, including AuNPs, were evaluated. Microwave-assisted digestion, acidification of suspensions, and the direct analysis of colloidal NPs without further preparation by ICP–MS were critically compared. In the particular case of AuNPs two different sizes (10 and 200 nm) were analyzed. The results obtained showed a clear influence of both the NP size and the sample pretreatment eventually used, revealing the actual importance of the kind of sample introduced into the ICP–MS on the sample treatment to be selected for analysis. Thus, whereas the three methods of scrutinizing sample pretreatments produced the same analytical results for 10 nm AuNPs, strong differences were observed for 200 nm AuNPs, depending on the particular pretreatment used. A comparison of the relative total concentration of 200 nm AuNPs obtained by MW-assisted digestion (100%), with acidification of the sample (96.5%), and by direct analysis (61.1%) shows the influence of sample preparation.

In the same work [8], the influence of the size of the NP in the ICP–MS direct determination of the total NP concentration by analyzing 10, 30, 80, and 200 nm AuNPs was also evaluated. The investigation showed that NPs up to 10 nm have a behavior similar to that of the ICP–single element standard while higher sizes provoke instabilities and signal decreases. Therefore, the total elemental concentration can be determined if an appropriate sample preparation method is ensured for each type and size of NP.

2.1.2 Single Particle Analysis Strategy

When characterizing colloidal NPs, such as AuNPs, for bioanalytical applications there are some other additional key NP parameters to consider, including aggregation state, dissolved metal fraction, and solubility which could affect the ICP–MS measurements. Particularly, nowadays it is well recognized that information on just the "elemental total concentration" of a nanoparticle sample is not enough to achieve a complete chemical or toxicological characterization of these NPs. Actually, many published reports use the total mass–concentration (e.g., mass of gold/volume) to indicate the amount of nanomaterial in a particular assay. However, depending on the particle size, the particle number and the total surface area can differ by several orders of magnitude. Considering that NP chemical and toxicological effects are mainly related to the surface area, this can drastically change the intrinsic behavior of the NPs. Therefore, expressing the concentration of nanoparticles in a sample in terms of the total number of particles present in a given volume would be much more relevant. In this sense, the development of analytical strategies enabling information about the "particle number concentration" in a specific solution is highly demanded. The hyphenated techniques, enabling analytical approaches to obtain information about the "particle concentration," also called "number concentration," will be discussed in the next section.

Recent research in instrumental development concerning sample introduction in the ICP–MS has allowed the development of novel methodologies for the determination of particle number concentration in a colloidal sample, based on a chemometric study of the MS spectra. This method provides a new tool for the investigation of the size distribution of AuNPs in colloidal suspensions [9] and is based on special ICP–MS detection: the ICP–MS "single-particle operation mode" (SP–ICP–MS). It consists of a nebulization of very dilute solutions, so that statistically only one NP reaches the plasma within a certain time window. Such event will produce an intensity signal at the MS detector which will be proportional to the size of that given particle. Using a colloid suspension injection coupled to the SP–ICP–MS system, AuNPs of 80–250 nm were analyzed in relatively very short analysis times (less than 10 min) and the method was shown to provide a size detection limit of just 25 nm [9].

Considering that many AuNPs typically used in bioapplications may have smaller diameters, the overall ion transfer efficiency in the mass spectrometer should be improved to detect lower sized NPs. In this sense, a remarkable advance in SP–ICP–MS was the application of a microdroplet generator (MDG) as a sample introduction system [10], improving the sample transport efficiency nearly to 100%. With this system detection size-related limits of quantification for AuNPs were around 20 nm. More recently, the same research group [11] has improved the MGD system, evaluating vertical and horizontal configurations, to find the more appropriate for engineered NP characterization (quantitative mass and size determinations). The horizontal experimental setup provided better results due to a better transport efficiency of droplets in the range below 40 μm. Using such SP–ICP–MS configurations the quantification of AuNPs of different commercial brands (including BAM and NanoComposix) and different NP diameters (25–125 nm) was carried out. In this SP–ICP–MS approach, the number of metal ions is proportional to the total NP mass whose volume is calculated by assuming a perfect sphere for the NP shape.

It should be stressed that SP–(MDG)–ICP–MS provides quantitative analysis without the need for NP standard reference materials, which is an important feature considering the wide variety of NPs that currently need to be characterized. However, conventional internal standardization was required for appropriate quantification in order to minimize strong matrix effects. AuNP sizes obtained by this technique are in good agreement with those provided by the commercial suppliers and by other complementary techniques, including transmission electron microscopy (TEM) and asymmetric flow field-flow fractionation system (AF4) with dynamic light scattering detection (DLS). In addition, SP–(MDG)–ICP–MS may provide valuable information about mass distributions by counting individual particles, along with information on the statistically significant number of particles/agglomerates. Finally, the high transport efficiency (>95%) offered by this technique allows the quantification of the total number of particles in a defined volume of the original sample, a difficult task using other techniques (e.g., AF4–DLS and TEM). In any case, the main drawback of this promising NP characterization technique is the reduced sensitivity (only particles >20–30 nm can be detected). Thus, in order to exploit the full potential of the SP–ICP–MS technique, further improvements in the sensitivity of the final ICP–MS signals are needed.

2.1.3 Determination of NP Core/Surface Ligand Stoichiometries

The surface of colloidal engineering NPs, such as AuNPs, must be covered by appropriate ligands in order to ensure their solubility in aqueous media and allow their further (bio)applications. The determination of the ligand packing density, also known as the degree of functionalization or the "coverage density," that allows the eventual estimation of core/ligand stoichiometry, is critically important and a current chemical challenge. X-ray crystallography, XPS, and surface plasmon resonance, among others, are common techniques for performing the determination of the coverage density. Of course, molecular mass spectrometry (able to give a molecular formula) is the more powerful and well developed analytical tool for such a goal, but ICP–MS can also provide valuable information for tackling this challenge [12,13].

Most of the ligands used to protect, solubilize, and functionalize AuNPs contain sulfur in their structure; thus, it is possible to establish a relationship of Au/S that identifies the core/ligand stoichiometry. In a first attempt [12], ICP–MS was assayed as a robust and accurate technique for the simultaneous measurement of mass concentrations of Au and S in digested solutions of L-cysteine-modified AuNPs (25 and 50 nm). A Dynamic reaction cell was used for the accurate measurement of sulfur concentration in ICP–MS; oxygen was employed as reaction gas to convert the highly interfered $^{32}S^+$ ion into the less interfered $^{32}S^{16}O^+$ ion (the final molecular ion detected). NP coverage densities of cysteines observed were of 0.81 ± 0.17 and 0.99 ± 0.19 nmol/cm^2 for AuNPs of 25 and 50 nm, respectively (assuming a perfect sphere shape), which are in agreement with previous results obtained by XPS for a self-assembled monolayer of aliphatic thiols on planar Au (111) films (0.77 nmol/cm^2). The mole ratios of gold to sulfur observed for these samples were 51 ± 10 (25 nm AuNPs) and 83 ± 14 (50 nm AuNPs) with quantitative recoveries (96% Au, 102% S) and relatively low uncertainty in the measurements (Au, 6.1%, and S, 6.9%).

Recently, Hinterwirth et al. [13] extended this ICP–MS methodology, based on the measure of Au/S ratios, not only to quantify ligand densities on the surface of AuNPs but also to evaluate the influence of ligand chain length in that coverage density. They found that the ligand density linearly dropped with increasing ligand chain length. The surface densities ranged between 6.3 molecules/nm^2 for the short ligand MPA (spacer length = 0.68 nm) and

4.3 molecules/nm^2 for the longer PEG7 ligand (spacer length = 3.52 nm). This strategy not only allowed the estimation of the average number of ligands attached to the surface for each AuNP but also the actual available ligands per volume of AuNP solution (the practically relevant figure in many applications).

2.1.4 Element-Based Direct Quantification of Colloidal NPs in Biotoxicology Studies

Over the past few years, many studies have been conducted to establish the biodistribution patterns of NPs in living species. When metal-based nanomaterials are used for biological applications, their biological safety must be evaluated. Therefore, there is an urgent need to evaluate the biological distributions (absorption, deposition, metabolism, and elimination) of these new nanomaterials before developing new bioscience applications. That information is of great value in understanding the possible toxic impact after a possible exposure of a given organism to the NPs studied. Moreover, to optimize the use of NPs in pharmacology and diagnosis, the biodistribution profiles of NP conjugates (e.g., used as drug carriers) should be established. Such evaluation can be performed via tracking the distribution of the metallic constituents of the nanoparticles in various tissues and organs after exposure.

Thus, the exceptional elemental sensitivity of the ICP–MS is of great value in monitoring metal NPs even at very low concentration levels. Thus, ICP–MS has already been used to study the biodistribution of AuNPs after intravenous administration in mice [14,15]. As an example, AuNPs of different sizes (15, 50, 100, and 200 nm) suspended in sodium alginate solution were administered to mice, and then blood and various mice tissues/organs were collected, 24 h after administration, to obtain the biodistribution of the NPs in the animal. The determination of Au in the samples was carried out quantitatively by ICP–MS and the results revealed that the tissue distribution of AuNPs was size dependent, with the smallest NPs (15 nm) showing the most widespread organ distribution. Moreover, the majority of the metal was demonstrated to be present in liver and spleen for all AuNP sizes assayed.

2.1.5 Quantification of Biomolecules by ICP–MS Using Functionalized AuNPs as Elemental Labels

During the past decade ICP–MS has been shown to be of great value for the sensitive detection of biomolecules [16]. However, despite the reported elemental advantages of ICP–MS for metals, the quantification of key elements of great importance in biomolecules (e.g., P and S) is hampered due to their high ionization potentials and their very abundant polyatomic interferences. Thus, whereas metals can be readily detected at nanogram per liter levels, the detection limits for P and S are usually much worse. The use of high-resolution mass analyzers or triple quadrupole instruments can overcome such interferences [17]. Additionally, ICP−MS instruments equipped with collision or reaction cells can alleviate such problems, at least to some degree [18].

Alternatively, the above-noted problems can be avoided if the scrutinized biomolecule is labeled with a metallic element or a metal NP containing elements easily detected by ICP−MS [19] (e.g., a nanogold cluster), which is measured instead of P or S. The idea behind such a strategy is to make any biomolecule detectable by ICP−MS and thus benefit from the very high sensitivity, isotopic ability, linearity, and matrix robustness characterizing this technique. This would allow the detection and determination of nearly all biomolecules, provided that the stoichiometry of the label and the efficiency of the bioconjugation reaction are known. Such an approach has been used to develop different ICP−MS-linked immunoassays that employ metal NP-labeled antibodies as secondary antibodies in surface-linked immunoreactions.

Within the field of atomic spectrometry, the concept of elemental labeling with elemental nanostructures was pioneered some years ago by Baranov et al. [19] and by Zhang et al. [20] and still offers a high potential for future applications and continuous development. Baranov et al. [19] developed a set of novel immunoassays for the determination of different proteins using antibodies covalently labeled with 1.4-nm nanogold clusters. The immunoassays were directly coupled with ICP−MS detection to quantify the metal (Au) of the NP-labeled antibodies. Experiments with four different immunoassay formats (centrifugal filtration, protein A affinity, size-exclusion gel filtration, and ELISA) with ICP−MS as elemental detector system demonstrated the achievable high sensitivity and precision for determinations of proteins even in complex biological samples. Also Zhang et al. [20] successfully demonstrated the determination of rabbit-anti-human immunoglobulin G (IgG) using antibodies labeled with colloidal AuNPs and a sandwich-type immunoreaction. The antibody was first captured by immobilized human IgG, and then incubated with Au-NP-labeled goat-anti-rabbit IgG. After release with 2% HNO_3, gold quantification was carried out by ICP−MS.

Also, metal NP-labeled antibodies were used for a multiplex determination of two different proteins in a complex mixture [21]: labeling of two different antibodies with AuNPs and an Eu cluster, respectively, allowed the simultaneous determination of two different antigens in a maleic anhydride plate immunoreaction, after appropriate digestion of labeling NPs with 10% HCl. The antigen contents, determined in cultured cell lysates, agreed well with results obtained using a conventional Western blot. However, a significantly higher precision was achieved using the ICP−MS detection of the labeling NPs.

Later, AuNPs were also used as elemental labels for the determination of DNA in water samples [22]: in this case, AuNPs modified with anti-mouse IgG were used to trace oligonucleotides carrying a c-myc peptide (c-myc is a regulator gene that codes a transcription factor). Oligonucleotide–peptide conjugates were first immobilized by UV radiation into a nitrocellulose membrane. They were then incubated with an anti-c-myc monoclonal antibody, followed by incubation with the conjugated AuNP secondary antibody (anti-mouse IgG), and finally determined by measuring Au by ICP–MS, after its appropriate dissolution with aqua regia. The limit of detection for peptide-modified DNA was as low as 0.2 pmol for the oligonucleotide and the reported sensitivity turned out to be approximately 40 times better than that obtained by the conventional dot–blot reference method. In this way, ICP–MS-based approaches might have a significant potential for nonradioactive DNA detection and simultaneous determinations of various DNA sequences (by multiple labeling with different kinds of inorganic NPs).

From all of the examples above, one could conclude that the use of metal NPs as labels, AuNPs in particular, combined with elemental MS detection opens new avenues for extremely sensitive detection of biomolecules and so offers considerable potential for future novel immunoassays to carry out targeted analysis of biomolecules (e.g., quantification for targeted proteomics) at very low concentration levels in real samples.

3. HYPHENATED TECHNIQUES WITH ELEMENTAL MASS SPECTROMETRY

As noted before, an important goal in modern nanotechnology is the synthesis and production of NPs with controlled and well-characterized sizes, as NP properties are strongly dependent on such a parameter. In addition, size distribution should also be controlled when evaluating the NPs quality and stability in time (the presence of free NPs, aggregates, ligands, degraded products, or their mixtures may coexist in a given NP colloidal solution). Thus, isolation of the desired nanomaterial in its native dispersed state is a critical step for useful NP production. Moreover, even initially monodisperse NPs tend to aggregate (or to partially decompose) when introduced in complex environmental or biological media. This latter fact poses a further challenge to the so far conventional bulk characterization techniques, stressing the current need for size separation methods to enable such further characterizations of produced NPs.

Most established characterization techniques (e.g., TEM, XPS, UV–Vis spectroscopy, and Raman spectroscopy) are detection tools demanding special sample preparation or selective previous isolation from other species that might interfere in the final analysis. So far, the coupling of a powerful separation technique with one of these sensitive detectors is considered one of the more reliable approaches for real-time size distribution assessment of mixtures of NPs (and/or the elucidation of a given NP chemical nature).

Hybrid techniques for size distribution analysis of NPs in polydispersed samples include first a separation technique (e.g., liquid chromatography, gel electrophoresis, field flow fractionation (FFF), or differential mobility analysis (DMA)) coupled on-line with different possible spectroscopic detectors (including UV/Vis spectroscopy, light scattering, and ICP–MS). In the next sections, some selected applications for AuNP characterization are illustrated. Table 2 summarizes the main separation techniques that have already been coupled to elemental MS for the analysis of colloidal AuNPs.

3.1 Liquid Chromatography

Chromatographic techniques provide powerful separation methods for shape and size nanocrystal distributions. Size-exclusion chromatography (SEC) is most commonly used for this purpose. Its low running costs, fast operation, and reliability make SEC an attractive option for standardizing nanomaterials directly in the solution state [23,25,27,28]. SEC–ICP–MS, in particular, has been successfully applied to assess size distributions and elemental characterizations of AuNPs [29]. Additionally, other chromatographic separations (e.g., reverse phase with a C_{18} column) have been coupled to ICP–MS: gold colloids (5–20 nm) were successfully size characterized and separated on a C_{18} column coupled on-line to ICP–MS (retention time increased proportionally to the decrease of AuNP size) [26].

Another emerging alternative, with high potential for NP size characterization, is hydrodynamic chromatography (HDC), which allows NP separation within a range from 5 to 300 nm. The technique exhibited short analysis time (less than 10 min) requiring minimum sample pretreatment. In this sense, the coupling of HDC separation to an ICP–MS detector (working with Collision cell) allowed a tool to define a range of AuNPs [24] (employing external size calibration standards as well as internal retention time markers). The technique was evaluated to test different types of NPs in environmental samples and to study their final fate. HDC has also been coupled to the SP–ICP–MS configuration mode enabling the determination of NP size, NP number concentration, and NP metal content [30]. Most recently, a useful comparison of HDC and asymmetric flow field-flow fractionation (AF4) separation techniques has been carried out for the analysis of AuNPs [31]. The main strengths and limitations of each approach have been discussed: AF4 offers a higher separation resolution, wheras HDC presents higher recoveries and lower analysis time. A comparison of both separation techniques for the analysis of given AuNP samples is presented in Figure 2.

Alternatively, the gel electrophoresis technique has also been reported for size analysis of AuNPs, offering a reasonable NP size resolution [26,32]. Conversely, despite the potential appropriateness of capillary electrophoresis for size separation of water-soluble NPs, no reports on applications of on-line coupling of capillary electrophoresis systems to ICP–MS detectors have been published to date.

TABLE 2 Hyphenated Techniques for the Characterization of NPs Published so Far

Separation Technique	Size Range	Resolution	Recovery	Time of Analysis	References
GE–ICP–MS	1 nm–10 μm	10–15 nm		30–45 min	[1,23,24]
SEC–ICP–MS	3–100 nm	10 nm		≥30 min	[22,25]
HPLC–ICP–MS	5–100 nm	10–15 nm		10–40 min	[1,23,24]
HDC–ICP–MS	5–300 nm	≥50 nm	75–100%	10 min	[26]
FFF–ICP–MS	1–100 nm	10–15 nm	4–90%	≥30 min	[26]
DMA–ICP–MS	3 nm–2 μm	0.3 nm		5 min	[1]

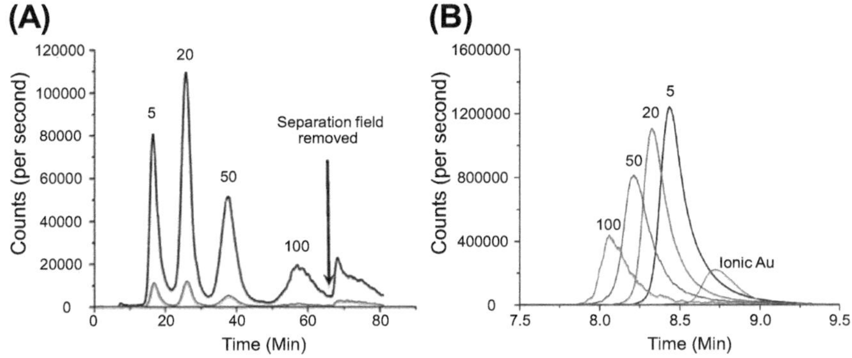

FIGURE 2 (A) Separation of mixtures containing 5, 20, 50, and 100 nm AuNPs by AF4–ICP–MS at two levels of concentration: 50 (line red) and 500 (line black) µg/L. (B) Overlap of the chromatograms obtained for the analysis of 5, 20, 50, and 100 nm AuNPs compared with the injection of gold ionic solution using HDC–ICP–MS. *Figure adapted from Ref. [31] with permission of the Royal Society of Chemistry.*

Unfortunately, the chromatographic techniques reported so far have some important limitations for fractionations of low diameter NPs. For instance, SEC suffers from unspecific surface adsorptions in the stationary phase and the limited size separation range. This hinders SEC application to sizing primary NPs and their (bio)conjugates or aggregated products. Also, the capabilities of HDC and gel electrophoresis for NP size fractionation seem to be lower than those of SEC and further improvements of separation resolution are still required. Finally, at present, elemental recoveries reported are far from quantitative.

3.2 Differential Mobility Analysis

DMA is just one of several ion mobility existing techniques. In general, ion mobility techniques measure how fast an ion moves in a gaseous viscous medium under the influence of an electrical field. Its interest here is that depending on the design, particle sizes from subnanometer to several hundred micrometers can be separated by DMA. This technique typically addresses the size distribution of NPs dispersed in a gas and its separation principle is the differential electrical mobility of aerosol NPs that were previously charged. Thereby NP ions of different mobilities will undergo different trajectories and so separate each other (see Figure 3).

One of the most studied configurations for the characterization of bionanoparticles (defined here as nano-sized particles), either biologically or chemically synthesized, or even functionalized to integrate into a bioconjugate, is electrospray ionization (ESI) coupled to DMA. This technique is similar to ESI–MS but measures the effective particle size, rather than molecular mass. The main applications of this technique in nanotechnology are

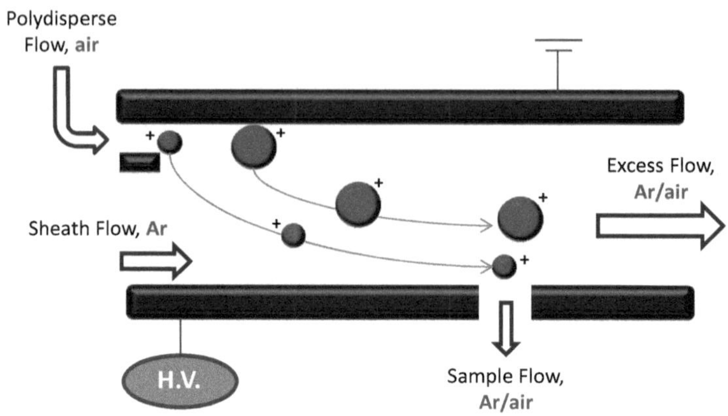

FIGURE 3 Scheme of the size discrimination process in DMA systems. *Figure reprinted with permission of Springer from Ref. [33].*

determinations of size distributions of NPs (3 nm up to 2 μm), studies on NP aggregation kinetics, quantification of ligand adsorption onto the NP surface, and calculation of the absolute number concentration of NPs in solution [34]. In this sense, several applications of ESI–DMA for the characterization of AuNPs have been described. As an example, the characterization and quantification of surface coverage of single-strand DNA molecules on 10 nm AuNPs were one of the first studies performed by ESI–DMA in the nanotechnology field [35]. Similar studies reported for self-assembled alkane thiol monolayers (SAMs) on 10 and 60 nm AuNPs found that the coverage of SAMs was independent of the size of the AuNPs [36]. Research on AuNP–protein interactions has been also carried out, using BSA as a model protein, to study this interaction as a function of protein concentration and pH [37].

Although ESI–DMA is today a well-established technique for NP characterization, its hyphenation to other detectors such as ICP–MS presents some difficulties, including the incompatibility of air (sheath gas flow typically used in ESI–DMA analysis) with the ICP. This could explain why very few reports have been published until now on such couplings even if much effort has been made trying to overcome such important limitations [33,38,39]. Nevertheless, the coupling of ESI–DMA to ICP–MS, as shown in these references, offers important benefits in the characterization of NPs, including: (1) determination of number- and mass-based distributions; (2) elemental compositions of nanoparticles classified by size, where the size classification and elemental analysis are performed within a single experiment; (3) particle concentrations in both solution (before size discrimination) and aerosol (after size discrimination) phases, and finally (4) the number of atoms per NP or the NP density can be obtained.

The future challenges of ESI–DMA/ICP–MS techniques include reduction of the air (particularly O_2) content, which interferes with the determination of S and other key elements in the ICP−MS, validation of calibration methods for a broader range of nanomaterials, and optimization of the methodology for the analysis of more complex NPs (such as core/shell particles and NP bioconjugates).

3.3 Flow-Field Fractionation

FFF is considered a highly promising and versatile separation tool for NPs in real-life environmental matrices. This flow-assisted technique is suitable for separating NPs by their hydrodynamic diameter in the absence of a stationary phase (in this way, irreversible and unspecific interactions are reduced and so the morphology and the size of NPs during operation are preserved). FFF offers an analytical continuous size fractionation range from 1 to 100 nm, which is very adequate for NP separations. In function of the applied field to induce the separation of the particles (e.g., flow, thermal, sedimentation, gravitational) different modes of FFF can be observed. When a secondary flow is perpendicularly applied to the conductor flow the modality is referred to us as flow field-flow fractionation (F4) and the separation is based on the diffusion coefficient of the NPs. In particular, F4 techniques (symmetrical or asymmetrical) have been used to monitor the synthesis and growth of different NPs (such as water-soluble AuNPs or TiO_2NPs) with different detectors [40,41]. Moreover, characterization of NPs by the on-line coupling of F4 with ICP−MS has already been described, allowing the development of rather flexible methods to characterize the size, polydispersity, and metal concentrations as a function of diameter [42−45].

A complete study of different AF4 parameters that could have a strong influence in the separation of different size AuNPs (e.g., type of membrane and focus time) has been recently performed [46]. The results obtained for size and particle concentration in the analysis of certified materials of AuNPs are in good agreement with the certified values given by NIST for such material.

AF4 coupled on-line to diverse detectors (MALS, DLS, and ICP−MS) has been extensively used for studies related to the evaluation of the behavior of NPs in food and environmental samples. Very recently the presence and physicochemical properties of AuNPs in rat livers after alkaline (TMAH) extraction were studied by employing the hyphenated technique AF4−ICP−MS. Au content in the different fractions eluting from the AF4 was monitored by ICP−MS. However, in this matrix the separation of different sizes of AuNPs was not possible due to the association of these NPs to the liver tissue that remains in the extract [47].

To conclude it should be noted that FFF is in its earliest stages, so important challenges still remain, including poor recoveries that can strongly fluctuate between 4 and 90% owing to NPs sticking to the membrane, and the lack of strategies for external and internal accurate ICP−MS calibrations.

4. MOLECULAR MASS SPECTROMETRY

NPs are generally coated by ligands of different nature to confer them higher colloidal stability or to provide the surface with a specific functionality. Such surface-coating ligands may determine the NP surface chemistry, which governs any further functionalization processes in subsequent (bio)applications. Therefore, the characterization of such protecting ligands layers is a fundamental aspect for designing further bioapplications of NPs. On the other hand, assigning formulas to coated NPs is far more informative than simply giving a diameter dimension. In fact, one of the most challenging analytical steps in characterizing stable NPs for bioanalysis is determining their molecular formula. In this sense, molecular MS techniques (e.g., ESI and matrix-assisted laser desorption/ionization (MALDI)) can provide such key information, correlating the entire clusters and/or aggregates with the actual chemical composition even of complex NP-based bioconjugates. In this vein, they can also provide information about the evolution of the NP composition itself along with the synthesis, showing preferential association of atoms to conform "stable" structures for the NP (the so-called "magic numbers").

Some of the more recent and relevant examples of using molecular MS for AuNP characterization will be discussed below, followed by a look forward to future work on this research line. At the same time, it is important to note that nanotechnology developments are useful in turn to improve already established MS techniques (e.g., MALDI–MS ionization conditions).

4.1 Matrix-Assisted Laser Desorption/Ionization

MALDI–MS is a very well-known technique routinely used today for characterization of biomolecules. However, adaptation of this technique for the analysis of NPs was not straightforward. The first challenge was finding an appropriate matrix to minimize the formation of fragments from the NP itself that may limit the possibility to assign unequivocally the correct molecular formula. In this sense, the research conducted by Dass et al. [48] showed that *trans*-2-[3-(4-*tert*-butylphenyl)-2-methyl-2-propenylidene]malononitrile (DCTB) is preferred over conventional matrixes for MALDI (e.g., sinapinic acid, 4′-hydroxyazobenzene-2-carboxylic acid, α-cyano-4-hydroxycinnamic acid, or the Universal MALDI matrix which is a 1:1 mixture of dihydroxybenzoic acid) because it produces less fragmentation of the analyzed cluster ion (see Figure 4(A)). The key to DCTB action is its method of production of charged ions from the NPs. Instead of the transfer of a proton, as typical MALDI matrices, DCTB assists ionization of the sample through electron transfer properties, inducing less fragmentation. In addition, DCTB allows working in both modes of MALDI, positive and negative, although using the positive mode guarantees less analyte fragmentation (see Figure 4(B)). After these pioneering studies, extensive research on NP characterization by MALDI using DCTB as matrix has been reported [49–53].

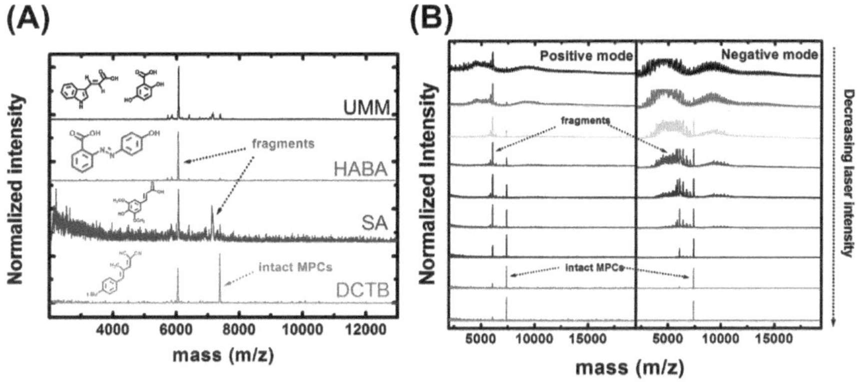

FIGURE 4 MALDI−TOF−MS spectra (Applied Biosystems Voyager-DE Pro) of $Au_{25}(SCH_2CH_2Ph)_{18}$ (A) in different MALDI matrices, showing the superiority of DCTB matrix to obtain the molecular ion (over fragment ions); (B) in DCTB matrix with varying laser intensity, delineating the molecular ions from fragment ions in positive and negative linear mode. *Reproduced with permission of J. Am. Chem. Soc. from Ref. [48].*

Such MALDI−MS investigations demonstrated good capability for the characterization of a broad range of AuNPs (with cores ranging between 16 and 144 atoms of gold), and providing enough resolution to be able to discriminate between Au_{38} and Au_{40} clusters. However, there are still some limitations, including the robust determination of unbroken higher mass NPs and NPs protected with hydrophilic ligands. These two different points represent the next challenges of the field for the near future.

The work of Dass [48,54] also reveals that the evolution of ligand exchange reaction in NP solubilization processes can be followed using MALDI−MS, and mixtures of NPs having the same core atom count but with different mixed ligand shell compositions can be perfectly analyzed. Figure 5 shows a ligand exchange experiment, where $Au_{25}(SCH_2CH_2Ph)_{18}$ is reacted with hexanethiol (HSC_6) for varying times. The ligand mass difference in this case is only 20 Da per ligand, and the average mass steadily moves to lower values as exchange proceeds. The putative limit of complete exchange is represented by separately synthesized $Au_{25}(SC6)_{18}$ monolayer-protected clusters, which gives a spectrum at the expected mass. The exchange spectra show that a wide distribution of mixed monolayer compositions is produced. The 12 h reaction product consists of nanoparticles containing from 3 to 12 new ($-SC_6$) ligands.

The determination of NP stoichiometry and(or) the quantification of ligand mixtures that protect the metal core are currently important aspects of current research in this field and more sophisticated MS-based strategies have been recently investigated with this purpose. Harkness et al. [55] developed an approach based on MALDI−ion mobility−MS to measure the relative quantity of ligands on an AuNP surface: three samples of mixed-ligand AuNPs (tiopronin:glutathione of average diameter 2.5 nm, octanethiol:decanethiol of

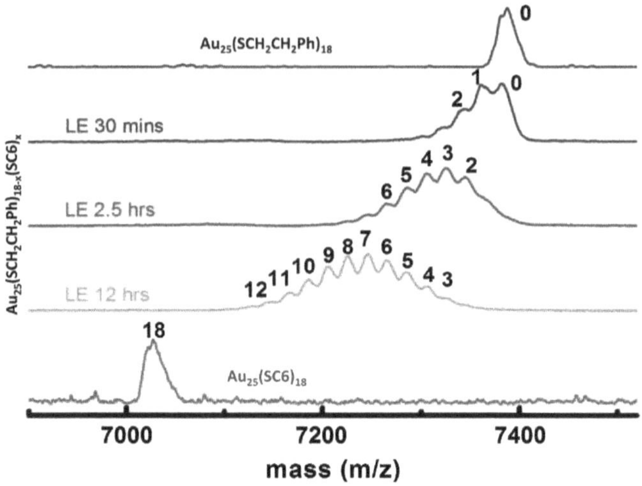

FIGURE 5 Example of the ligand exchange reaction evolution monitored by MALDI—TOF—MS using DCTB as matrix, working in positive linear mode. The initial $Au_{25}(SCH_2CH_2Ph)_{18}$ NP (top blue spectrum) visualization of the intermediates at different reaction times $Au_{25}(SCH_2CH_2Ph)_{18-x}(SC6)_x$ and the final reaction product $Au_{25}(SC6)_{18}$ NP (bottom red spectrum) can be clearly observed. *Figure modified from Ref. [48].*

average diameter 3.6 nm, and tiopronin:11-mercaptoundecyl(polyethylene glycol) of average diameter 2.5 nm) were studied. The results were successfully validated by nuclear magnetic resonance (NMR) spectroscopy and MS without ion mobility separation (in fact, average differences of less than 1% relative abundance were reported).

To conclude this section, it is worth noting a recent complementary trend: the actual use of NPs to enhance the performance of common MS tools. As an example, NP-based substrates (e.g., carbon nanotubes, quantum dots, and Au, SiO_2, and TiO_2 NPs) are being proposed to improve the sensitivity of laser desorption/ionization MS techniques. Interestingly, NPs seem to offer important advantages over more conventional organic matrices, due to the fact that they are excellent laser-energy-absorbing materials. This, of course, facilitates analyte desorption and ionization processes, so enabling improvements of spectra acquisition, particularly in the low-mass range (e.g., amino acid, peptide, polymer, and even metal analysis) where a lot of ions derived from the organic matrices are typically detected [56,57]. For example, Huang et al. used a matrix-free strategy where nitrocellulose membranes modified with AuNPs were used for detecting lead ions in biofluids using pulsed-laser desorption ionization MS [58]. A nanocomposite film was formed by trapping bovine serum albumin-modified AuNPs on a nitrocellulose membrane due to strong hydrophobic interactions. The AuNPs can interact with thiosulfate in solution forming Au—thiosulfate complexes which facilitates Pb atom deposition in the presence of lead ions as Pb–Au alloys. Soft and enhanced ionization of Pb–Au alloys from the AuNP surface was observed with the prepared nanocomposite film.

4.2 Electrospray Ionization

A complementary technique to MALDI for NP characterization is the ESI–MS. ESI is a soft ionization technique that overcomes the propensity of molecules to fragment when they are ionized from a solid matrix (e.g., MALDI), and so it constitutes a popular ionization source for liquid samples, where the molecular ion of the analyte can be easily observed. Another advantage of ESI over MALDI is the production of multiply charged ions, effectively extending the mass range of the analyzer to accommodate the kDa–MDa orders of magnitude (extremely important in protein analysis). The corresponding ions observed by mass spectrometry are created by the addition of a hydrogen cation (denoted $[M+H]^+$), or of another cation such as sodium ion, $[M+Na]^+$, or the removal of a hydrogen nucleus, $[M-H]^-$. Multiple charged ions such as $[M+nH]^{n+}$ are often observed in ESI sources. In this context, information about the charge states of NP-based compounds can be also obtained by analyzing samples by ESI–MS [59,60]. In Figure 6 all features noted above for ESI can be observed and a comparison made of MALDI–MS and ESI–MS for analysis AuNPs. Of course, both can be complementary characterization techniques: as an example, both techniques

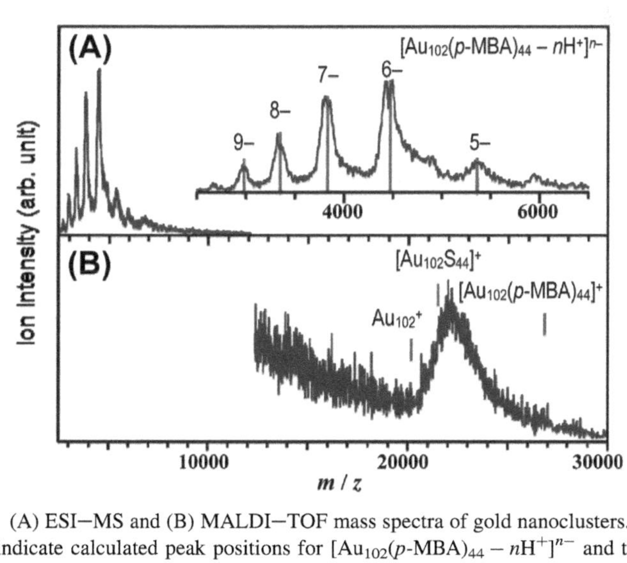

FIGURE 6 (A) ESI–MS and (B) MALDI–TOF mass spectra of gold nanoclusters. The red bars in panel A indicate calculated peak positions for $[Au_{102}(p\text{-MBA})_{44} - nH^+]^{n-}$ and the breadth of the peaks could reflect slight contamination by particles of different compositions, such as $Au_{101}(p\text{-MBA})_{43}$ and $Au_{102}(p\text{-MBA})_{42}$. *Reprinted with permission of J. Am. Chem. Soc. (Ref. [53]).*

have been used in order to characterize $Au_{102}(p\text{-MBA})_{44}$NPs (*p*-MBA: *para*-mercaptobenzoic acid) [53]. Information about molecular weight (22 kDa) is given by MALDI−MS whereas ESI−MS is used to determine the chemical formulas and charge states of thiolate-protected gold clusters. Both types of results are in agreement and allow the elucidation of the final molecular formula of the coated clusters of gold: $Au_{102}(p\text{-MBA})_{44}$.

ESI−MS has also demonstrated its great potential and capabilities in the analysis of more complex samples containing NPs. Particularly, analysis of samples containing bimetallic protected NPs (e.g., Au−Ag, Pd−Ag, Au−Pd, and Cu−Au core−shell NPs) has been carried out and the products from ligand exchange reactions (e.g., $Au_{25-x}Pd_x(SC_2Ph)_{18-y}(SPEG)_y$), i.e., $SPEG = -S(CH_2CH_2O)_5CH_3$, could be monitored [61]. For the case of Au/Pd bimetallic NPs a more rapid ligand exchange during PEGylation was observed. In addition, ESI−MS has also been used as a reference technique to validate results obtained by alternative new analytical tools. As an example, in a pioneer work about the use of diffusion ordered NMR spectroscopy for size determination of thiol-stabilized gold nanoparticles, ESI−MS was used to confirm the results obtained by the NMR method [62].

To conclude, it should be noted that this soft ionization source has been very recently employed to develop a novel strategy for hyphenated techniques (ESI−DMA−ICP−MS) with the aim of achieving complementary molecular and elemental information in a single run (see previous section) [33].

5. A FORWARD LOOK INTO THE FUTURE OF MASS SPECTROMETRY IN THE NANOWORLD

In the past decade great effort has been made to integrate the available MS techniques to obtain both elemental and molecular information, to design new MS instrumentation, and to investigate novel operation modes and processes to facilitate a more reliable and complete characterization of metal-based nanomaterials. Actually, the use of MS-based approaches (both elemental and molecular) is now commonplace for the characterization of colloidal gold nanoparticles and lately of their bioconjugate architectures, a still pending critical challenge. The ongoing MS developments will facilitate the controlled synthesis and production of NP-based elemental labels and definitely will provide invaluable information about possibly synthesized novel NP bioconjugates, to tackle absolute and reliable quantitative analysis of biomolecules (e.g., proteins) using Au−NPs as elemental labels.

Elemental MS today does not only allow for a reliable determination of the elemental NP content, but also provides information which could be of great value for estimation of the NP mass and number concentration, for evaluation of core/ligand ratios, and for determination of elemental stoichiometry within the NP.

Particularly, the development of an SP–ICP–MS operation mode has been an important breakthrough in the direct NP analysis. However, much effort still should be directed to enable the analysis of smaller NPs (the calculated size LD, up to now, around 25 nm that is far from the typical NP sizes used in bioanalytical applications, around 1–20 nm). In addition, another important limitation is the time resolution of the mass spectrometer used for the analyses. To solve this problem the development of new novel mass spectrometers able to monitor fast transient signals (as those originated normally by atomization of a single particle in a plasma) represents a current challenge. A magnetic sector analyzer equipped with an array detector (multicollector) and a fast time-of-flight analyzer are the mass spectrometers better positioned (due to enabling simultaneous or quasi-simultaneous, respectively, full mass spectra acquisition) [63].

As we have discussed throughout this chapter, MS techniques (elemental and molecular) are good alternatives to other more conventional characterization techniques of NPs. MS-based techniques enable valuable multiparametric information to be attained; this includes NP size, composition, stoichiometry, size distribution, number and mass elemental concentration, coverage density of ligands, synthesis and reaction processes monitoring, and elucidation of molecular formulas. Nevertheless, the improvement of both instrumentation and methodologies is still necessary to achieve more accurate results, with better resolution and more reproducibility.

In this context, one of the critical challenges still pending is "nanoparticle handling and sample preparation." Of course, sample preparation and the suitability of an analytical procedure can strongly depend on the nanoparticle type, on the coating, and on the nature of the sample (e.g., dispersion, powder and presence of surfactant). Thus, the preparation of stable dispersions is probably a key challenge to be addressed especially for size-fractionated NP analysis.

Regarding coatings characterization, actual bonding mechanisms of coatings to nanoparticles (e.g., covalent or electrostatic bonding, steric or charge stabilized nanoparticles) are not well known yet and so specific tests are needed along with the development of reliable standardized sample preparation procedures.

Also, since the characterization of NPs by MS techniques is becoming more and more a routine task there is an urgent need for standards and stable certified reference materials in order to validate new characterization methodologies. Standard materials are required mainly in two directions: (1) size-specific reference materials (fortunately there are already available some standards of gold colloids and silica particles) and (2) certified reference materials, in size and composition, in the matrices of interest (for instance, AuNPs in tissues or body fluids are not yet available).

Last but not least, further advances of molecular MS (MALDI and ESI) applications in the characterization of NPs in general should be directed to: (1) deeper analysis of NPs coated with hydrophilic ligands; (2) novel approaches to minimize the fragmentation of the ligands/NPs association, and (3) increase in the certainty of molecular formula assignments, particularly via sample quality, resolving power, and mass measurement accuracy improvements to minimize nearly isobaric permutations currently observed.

To conclude, one of the more widespread and promising fields of applications of AuNPs in bioanalysis is their use as elemental labels after bioconjugation to an appropriate recognition element. Unfortunately, the analytical potential of such approach still relies on eventual knowledge/determination of NPs—biomolecules stoichiometries as well as on the purity and stability of such bioconjugates. Again, MS (elemental and molecular) tools could play a pivotal role here to achieve a more complete physicochemical characterization of the engineered biofunctionalized AuNPs to help reliable synthesis and control of the produced nanomaterial and, in this way, to boost the number of their applications, particularly in bioscience.

ACKNOWLEDGMENTS

The authors would like to express their gratitude for the financial support from "Plan Nacional de I+D+I" (Spanish Ministry of Science and Innovationand FEDER Program) through CTQ2010-16636.

REFERENCES

[1] B.M. Simonet, M. Valcárcel, Monitoring nanoparticles in the environment, Anal. Bioanal. Chem. 393 (2009) 17—21.

[2] P. Krystek, A. Ulrich, C.C. Garcia, S. Manohard, R. Ritsema, Application of plasma spectrometry for the analysis of engineered nanoparticles in suspensions and products, J. Anal. At. Spectrom. 26 (2011) 1701—1721.

[3] A.R. Montoro Bustos, J. Ruiz Encinar, A. Sanz-Medel, Mass spectrometry for the characterisation of nanoparticles, Anal. Bioanal. Chem. 405 (2013) 5637—5643.

[4] K.E. Sapsford, K.M. Tyner, B.J. Dair, J.R. Deschamps, I.L. Medintz, Analyzing nanomaterial bioconjugates: a review of current and emerging purification and characterization techniques, Anal. Chem. 83 (2011) 4453—4488.

[5] B. Ferreira da Silva, S. Pérez, P. Gardinalli, R.K. Singhal, A.A. Mozeto, D. Barcelo, Analytical chemistry of metallic nanoparticles in natural environments, Trends Anal. Chem. 30 (2011) 528—540.

[6] B. Fernández, J.M. Costa, R. Pereiro, A. Sanz-Medel, Inorganic mass spectrometry as a tool for characterisation at the nanoscale, Anal. Bioanal. Chem. 396 (2010) 15—29.

[7] P. Rodríguez-González, J.M. Marchante-Gayón, J.I. García Alonso, A. Alfredo Sanz-Medel, Isotope dilution analysis for elemental speciation: a tutorial review, Spectrochim. Acta Part B At. Spectrosc. 60 (2005) 151—207.

[8] A.L. Fabricius, L. Duester, B. Meermann, T.A. Ternes, ICP-MS-based characterization of inorganic nanoparticles—sample preparation and off-line fractionation strategies, Anal. Bioanal. Chem. 406 (2) (2014) 467—479.

[9] C. Degueldre, P.-Y. Favarger, S. Wold, Gold colloid analysis by inductively coupled plasma-mass spectrometry in a single particle mode, Anal. Chim. Acta 555 (2006) 263–268.
[10] S. Gschwind, L. Flamigni, J. Koch, O. Borovinskaya, S. Groh, K. Niemax, D. Günther, Capabilities of inductively coupled plasma mass spectrometry for the detection of nanoparticles carried by monodisperse microdroplets, J. Anal. At. Spectrom. 26 (2011) 1166–1174.
[11] S. Gschwind, H. Hagendorfer, D.A. Frick, D. Günther, Mass quantification of nanoparticles by single droplet calibration using inductively coupled plasma mass spectrometry, Anal. Chem. 85 (2013) 5875–5883.
[12] W. Jiang, D.B. Hibbert, G. Moran, R. Akter, Measurement of gold and sulfur mass fractions in L-cysteine-modified gold nanoparticles by ICP-DRC-MS after acid digestion: validation and uncertainty of results, J. Anal. At. Spectrom. 27 (2012) 1465–1473.
[13] H. Hinterwirth, S. Kappel, T. Waitz, T. Prohaska, W. Lindner, M. Lämmerhofer, Quantifying thiol ligand density of self-assembled monolayers on gold nanoparticles by inductively coupled plasma-mass spectrometry, ACS Nano 7 (2013) 1129–1136.
[14] G. Sonavanea, K. Tomoda, K. Makinoa, Biodistribution of colloidal gold nanoparticles after intravenous administration: effect of particle size, Colloids Surf. B Biointerfaces 66 (2008) 274–280.
[15] W.H. De Jong, W.I. Hagens, P. Krystek, M.C. Burger, A.J.A.M. Sips, R.E. Geertsma, Particle size-dependent organ distribution of gold nanoparticles after intravenous administration, Biomaterials 29 (2008) 1912–1919.
[16] A. Sanz-Medel, M. Montes-Bayón, J. Bettmer, M.L. Fernández-Sanchez, J. Ruiz Encinar, ICP-MS for absolute quantification of proteins for heteroatom-tagged, targeted proteomics, Trends Anal. Chem. 40 (2012) 52–63.
[17] S. Diez-Fernández, N. Sugishama, J. Ruiz-Encinar, A. Sanz-Medel, Triple quad ICPMS (ICPQQQ) as a new tool for absolute quantitative proteomics and phosphoproteomics, Anal. Chem. 84 (2012) 5851–5857.
[18] S.D. Tanner, V.I. Baranov, D.R. Bandura, Reaction cells and collision cells for ICP-MS: a tutorial review, Spectrochim. Acta Part B 57 (2002) 1361–1452.
[19] V.I. Baranov, Z.A. Quinn, D.R. Bandura, S.D. Tanner, A sensitive and quantitative element-tagged immunoassay with ICPMS detection, Anal. Chem. 74 (2002) 1629–1636.
[20] C. Zhang, Z.Y. Zhang, B.B. Yu, J.J. Shi, X.R. Zhang, Application of the biological conjugate between antibody and colloid Au nanoparticles as analyte to inductively coupled plasma mass spectrometry, Anal. Chem. 74 (2002) 96–99.
[21] Z.A. Quinn, V.I. Baranov, S.D. Tanner, J.L. Wrana, Simultaneous determination of proteins using an element-tagged immunoassay coupled with ICP-MS detection, J. Anal. At. Spectrom. 17 (2002) 892–896.
[22] A. Merkoci, M. Aldavert, G. Tarrason, R. Eritja, S. Alegret, Toward an ICPMS-linked DNA assay based on gold nanoparticles immunoconnected through peptide sequences, Anal. Chem. 77 (2005) 6500–6503.
[23] A.M. Al-Somali, K.M. Krueger, J.C. Falkner, V.L. Colvin, Recycling size exclusion chromatography for the analysis and separation of nanocrystalline gold, Anal. Chem. 76 (2004) 5903–5910.
[24] K. Tiede, A.B.A. Boxall, D. Tiede, S.P. Tear, H. Davide, J. Lewis, A robust size-characterisation methodology for studying nanoparticle behaviour in 'real' environmental samples, using hydrodynamic chromatography coupled to ICP-MS, J. Anal. At. Spectrom. 24 (2009) 964–972.
[25] J.P. Wilcoxon, J.E. Martin, P. Provencio, Size distributions of gold nanoclusters studied by liquid chromatography, Langmuir 16 (2000) 9912–9920.

[26] A. Helfrich, W. Brüchert, J. Bettmer, Size characterisation of Au nanoparticles by ICP-MS coupling techniques, J. Anal. At. Spectrom. 21 (2006) 431–434.

[27] J.P. Wilcoxon, P. Provencio, Etching and aging effects in nanosize Au clusters investigated using high-resolution size-exclusion chromatography, J. Phys. Chem. B 107 (2003) 12949–12957.

[28] S. Knoppe, J. Boudon, I. Dolamic, A. Dass, T. Bürgi, Size exclusion chromatography for semipreparative scale separation of $Au_{38}(SR)_{24}$ and $Au_{40}(SR)_{24}$ and larger clusters, Anal. Chem. 83 (2011) 5056–5061.

[29] H. Weinberg, A. Galyean, M. Leopold, Evaluating engineered nanoparticles in natural waters, Trends Anal. Chem. 30 (2011) 72–83.

[30] S.A. Pergantis, T.L. Jones-Lepp, E.M. Heithmar, Hydrodynamic chromatography online with single particle-inductively coupled plasma mass spectrometry for ultratrace detection of metal-containing nanoparticles, Anal. Chem. 84 (2012) 6454–6462.

[31] E.P. Gray, T.A. Bruton, C.P. Higgins, R.U. Halden, P. Westerhoff, J.F. Ranville, Analysis of gold nanoparticle mixtures: a comparison of hydrodynamic chromatography (HDC) and asymmetrical flow field-flow fractionation (AF4) coupled to ICP-MS, J. Anal. At. Spectrom. 27 (2012) 1532–1539.

[32] A. Helfrich, J. Bettmer, Analysis of gold nanoparticles using ICP-MS-based hyphenated and complementary ESI-MS techniques, Int. J. Mass Spectrom. 307 (2011) 92–98.

[33] S. Elzey, D.H. Tsai, L.L. Yu, M.R. Winchester, M.E. Kelley, V.A. Hackley, Real-time size discrimination and elemental analysis of gold nanoparticles using ES-DMA coupled to ICP-MS, Anal. Bioanal. Chem. 405 (2013) 2279–2288.

[34] S. Guha, M. Li, M.J. Tarlov, M.R. Zachariah, Electrospray–differential mobility analysis of bionanoparticles, Trends Biotechnol. 30 (2012) 291–300.

[35] L.F. Pease, D.H. Tsai, R.A. Zangmeister, M.R. Zachariah, M.J. Tarlov, Quantifying the surface coverage of conjugate molecules on functionalized nanoparticles, J. Phys. Chem. C 111 (2007) 17155–17157.

[36] D.H. Tsai, R.A. Zangmeister, L.F. Pease, M.J. Tarlov, M.R. Zachariah, Gas-phase ion-mobility characterization of SAM-functionalized Au nanoparticles, Langmuir 24 (2008) 8483–8490.

[37] D.H. Tsai, F.W. Del Rio, A.M. Keene, K.M. Tyner, R.I. MacCuspie, T.J. Cho, M.R. Zachariah, V.A. Hackley, Adsorption and conformation of serum albumin protein on gold nanoparticles investigated using dimensional measurements and in situ spectroscopic methods, Langmuir 27 (2011) 2464–2477.

[38] C. Carazzone, R. Raml, S.A. Pergantis, Nanoelectrospray ion mobility spectrometry online with inductively coupled plasma-mass spectrometry for sizing large proteins, DNA, and nanoparticles, Anal. Chem. 80 (2008) 5812–5818.

[39] E.A. Kapellios, S.A. Pergantis, Size and elemental composition of nanoparticles using ion mobility spectrometry with inductively coupled plasma mass spectrometry, J. Anal. At. Spectrom. 27 (2012) 21–24.

[40] T. Rameshwar, S. Samal, S. Lee, S. Kim, J. Cho, I.S. Kim, Determination of the size of water-soluble nanoparticles and quantum dots by field-flow fractionation, J. Nanosci. Nanotechnol. 6 (2006) 2461–2467.

[41] C. Contado, A. Pagnoni, TiO_2 in commercial sunscreen lotion: flow field-flow fractionation and ICP-AES together for size analysis, Anal. Chem. 80 (2008) 7594–7608.

[42] A. Siripinyanond, R.M. Barnes, Flow field-flow fractionation-inductively coupled plasma mass spectrometry of chemical mechanical polishing slurries, Spectrochim. Acta Part B 57 (2002) 1885–1896.

[43] M. Bouby, H. Geckeis, F.W. Geyer, Application of asymmetric flow field-flow fractionation coupled to inductively coupled plasma mass spectrometry (ICPMS) to the quantitative characterization of natural colloids and synthetic nanoparticles, Anal. Bioanal. Chem. 392 (2008) 1447–1457.

[44] A.J. Bednar, A.R. Poda, D.M. Mitrano, A.J. Kennedy, E.P. Gray, J.F. Ranville, C.A. Hayes, F.H. Crocker, J.A. Steevens, Comparison of on-line detectors for field flow fractionation analysis of nanomaterials, Talanta 104 (2013) 140–148.

[45] J.M. Pettibone, J. Gigault, V.A. Hackley, Discriminating the states of matter in metallic nanoparticle transformations: What are we missing? ACS Nano 7 (2013) 2491–2499.

[46] H. Hagendorfer, R. Kaegi, J. Traber, S.F.L. Mertens, R. Scherrers, C. Ludwig, A. Ulrich, Application of an asymmetric flow field flow fractionation multi-detector approach for metallic engineered nanoparticle characterization – prospects and limitations demonstrated on Au nanoparticles, Anal. Chim. Acta 706 (2011) 367–378.

[47] B. Schmidt, K. Loeschner, N. Hadrup, A. Mortensen, J.J. Sloth, C.B. Koch, E.H. Larsen, Quantitative characterization of gold nanoparticles by field-flow fractionation coupled on-line with light scattering detection and inductively coupled plasma mass spectrometry, Anal. Chem. 83 (2011) 2461–2468.

[48] A. Dass, A. Stevenson, G.R. Dubay, J.B. Tracy, R.W. Murray, Nanoparticle MALDI-TOF mass spectrometry without fragmentation: $Au_{25}(SCH_2CH_2Ph)_{18}$ and mixed monolayer $Au_{25}(SCH_2CH_2Ph)_{18-x}(L)_x$, J. Am. Chem. Soc. 130 (2008) 5940–5946.

[49] H. Qian, R. Jin, Controlling nanoparticles with atomic precision: the case of $Au_{144}(SCH_2CH_2Ph)_{60}$, Nano Lett. 9 (2009) 4083–4087.

[50] S.M. Reilly, T. Krick, A. Dass, Surfactant-free synthesis of ultrasmall gold nanoclusters, J. Phys. Chem. C 114 (2010) 741–745.

[51] H. Qian, Y. Zhu, R. Jin, Isolation of ubiquitous $Au_{40}(SR)_{24}$ clusters from the 8 kDa gold clusters, J. Am. Chem. Soc. 132 (2010) 4583–4585.

[52] P.R. Nimmala, A. Dass, $Au_{36}(SPh)_{23}$ nanomolecules, J. Am. Chem. Soc. 133 (2011) 9175–9177.

[53] Y. Levi-Kalisman, P.D. Jadzinsky, N. Kalisman, H. Tsunoyama, T. Tsukuda, D.A. Bushnell, R.D. Kornberg, Synthesis and characterization of Au_{102}(*p*-MBA)$_{44}$ nanoparticles, J. Am. Chem. Soc. 133 (2011) 2976–2982.

[54] A. Dass, K. Holt, J.F. Parker, S.W. Feldberg, R.W. Murray, Mass spectrometrically detected statistical aspects of ligand populations in mixed monolayer $Au_{25}L_{18}$ nanoparticles, J. Phys. Chem. C 112 (2008) 20276–20283.

[55] K.M. Harkness, B.C. Hixson, L.S. Fenn, B.N. Turner, A.C. Rape, C.A. Simpson, B.J. Huffman, T.C. Okoli, J.A. McLean, D.E. Cliffel, A structural mass spectrometry strategy for the relative quantitation of ligands on mixed monolayer-protected gold nanoparticles, Anal. Chem. 82 (2010) 9268–9274.

[56] M. Rainer, M.N. Qureshi, G. Bonn, Matrix-free and material-enhanced laser desorption/ionization mass spectrometry for the analysis of low molecular weight compounds, Anal. Bioanal. Chem. 400 (2011) 2281–2288.

[57] R. Pilolli, N. Ditaranto, C. Di Franco, F. Palmisano, N. Cioffi, Thermally annealed gold nanoparticles for surface-assisted laser desorption ionisation–mass spectrometry of low molecular weight analytes, Anal. Bioanal. Chem. 404 (2012) 1703–1711.

[58] Y.-C. Liu, C.–K. Chiang, H.–T. Chang, Y.F. Lee, C.C. Huang, Using a functional nanogold membrane coupled with laser desorption/ionization mass spectrometry to detect lead ions in biofluids, Adv. Funct. Mater. 21 (2011) 4448–4455.

[59] C.S. Ho, M.H.M. Chan, R.C.K. Cheung, L.K. Law, L.C.W. Lit, K.F. Ng, M.W.M. Suen, H.L. Tai, Electrospray ionisation mass spectrometry: principles and clinical applications, Clin. Biochem. Rev. 24 (2003) 3–12.

[60] J.J. Pitt, Principles and applications of liquid chromatography-mass spectrometry in clinical biochemistry, Clin. Biochem. Rev. 30 (2009) 19–34.

[61] C.A. Fields-Zinna, M.C. Crowe, A. Dass, J.E.F. Weaver, R.W. Murray, Mass spectrometry of small bimetal monolayer-protected clusters, Langmuir 25 (2009) 7704–7710.

[62] K. Salorinne, T. Lahtinen, J. Koivisto, E. Kalenius, M. Nissinen, M. Pettersson, H. Häkkinen, Nondestructive size determination of thiol-stabilized gold nanoclusters in solution by diffusion ordered NMR spectroscopy, Anal. Chem. 85 (2013) 3489–3492.

[63] G.D. Schilling, F.J. Andrade, J.H. Barnes, R.P. Sperline, M.B. Denton, C.J. Barinaga, D.W. Koppenaal, G.M. Hieftje, Continuous simultaneous detection in mass spectrometry, Anal. Chem. 79 (2007) 7662–7668.

Chapter 9

Separation Techniques of Gold Nanoparticles

Ángela Inmaculada López-Lorente and Miguel Valcárcel*

Department of Analytical Chemistry, University of Córdoba, Córdoba, Spain

**Corresponding author: E-mail: qa1meobj@uco.es*

Chapter Outline

Gold Nanoparticles in Analytical Chemistry. http://dx.doi.org/10.1016/B978-0-444-63285-2.00009-2

1. INTRODUCTION

1.1 Separation of Nanoparticles in Analytical Nanoscience and Nanotechnology

Separation within the Analytical Nanoscience and Nanotechnology scope can be considered from two points of view. One is the separation of nanoparticles (NPs) on the basis of their differences in size and shape (e.g., by capillary electrophoresis (CE), size-exclusion chromatography (SEC), hydrodynamic chromatography (HDC), or field-flow fractionation (FFF)), which is related to the characterization of NPs. On the other hand, NPs can be separated from their sample matrix (e.g., by ligand-assisted extraction [1], cloud-point extraction [2], and ionic liquid microliquid–liquid extraction [3]), which is a crucial step when determining NPs, for example, in complex environmental or biological matrices.

Separation of NPs is achieved by using different properties of NPs, namely, (1) ζ potential of the NPs, which is employed, for example, for electrophoretic separations; (2) charge-to-size ratio of the NPs, which can be influenced by the functionalization of the surface of the NPs with different ligands; (3) hydrodynamic volume, which is the basis of, for example, HDC; (4) chemical characteristics of the NPs; and finally (5) solubility of the NPs in different solvents.

Separation systems can be designed so that a detector is coupled online; thus it is possible to perform qualitative or quantitative analysis of the NPs at the same time that they are separated. For example, CE is commonly coupled to ultraviolet (UV)–vis detectors, which provide information about the optical characteristics of the AuNPs. Most of the separation techniques have been hybridized with inductively coupled mass spectrometry (ICP–MS) [4], which enables the determination of the concentration of the separated metallic species.

A challenge of analytical nanoscience and nanotechnology is the development of effective procedures for industrial-scale separation and purification of nanomaterials. Many synthetic methods of NPs lead to mixtures of NPs with different sizes or shapes. Classification and purification of the desired nanomaterials are mandatory to use these NPs in their prospective applications.

1.2 Most Usual Separation Techniques

Various methods of separation of NPs are available nowadays, based on the different properties of NPs. Technologies for the fractionation and collection of NP fractions in liquid mediums include SEC. It has the advantages of being fast, as well as the capacity to be coupled to optical spectroscopy. However, irreversible adsorption of stationary phase toward NPs is a problem for NP separation, as a consequence of the high surface area of the stationary phase combined with the high surface activity of NPs.

CE could reduce this problem by decreasing the surface effect of separation systems. It has the advantage of low consumption of sample and reagents. Nevertheless, sample collection to obtain monodispersed NPs by CE is more challenging [5].

Other chromatographic techniques, such as high-performance liquid chromatography (HPLC) or HDC, have enabled the separation of AuNPs.

Flow techniques, such as flow FFF, have been also used for the separation and quantification of AuNPs, usually in combination with ICP–MS. It allows separation of the NPs in a liquid matrix of a desired composition and provides a size distribution instead of an average value. However, this method requires complex and expensive instrumentation and may suffer from incomplete recovery and analyte loss [6].

Common methods for NP separation are centrifugation and ultracentrifugation, although they have the drawback of possible irreversible agglomeration or aggregation, which distorts the original morphology of NPs. In the case of smaller NPs, centrifugation provides lower yields. Other methods are filtration, which relies on the use of controlled pore-size materials, which can suffer from fouling [7].

Compared to other conventional approaches, cloud point extraction (CPE) presents an attractive method for extraction and has advantages over the traditional liquid–liquid approach. These involve nonstructured molecular solvents due to high extraction efficiencies and preconcentration factors achieved and the fact that CPE is water based and avoids the use of expensive and toxic solvents. Moreover, the presence of surfactants can reduce extractant losses [2].

2. ELECTROPHORESIS

2.1 Introduction

Electrophoresis has emerged as a useful tool for the characterization of nanomaterials, such as gold NPs (AuNPs). It is potentially faster than electron microscopies, with fewer surface effects than in other techniques, such as SEC, which, as will be seen later, is prone to irreversible adsorption of the NPs onto the stationary phase, which hampers their use. Another advantage of electrophoresis, especially in the case of CE modality, is the low consumption of sample and reagents [8].

Among the different modalities of electrophoresis, CE has been the most widely employed for the characterization and separation of AuNPs. Both capillary zone electrophoresis (CZE) and the so-called micellar electrokinetic chromatography have enabled the separation and size characterization of AuNPs with different functionalizations. Moreover, gel electrophoresis (GE) has gained exposure for the separation of nanomaterials, for example, preparative electrophoresis with agarose gels has been employed for the size and shape separation of AuNPs. Other electrophoretic techniques, such as isoelectric focusing (IEF), commonly employed to determine the isoelectric point of proteins and enzymes, has broadened its field of application to include the separation of colloidal metal and semiconductor NPs.

Several overviews have given theoretical descriptions of the physical mechanism involved in the electrophoretic transport of NPs and colloids [9–12] at the same time that present representative examples of the potential of electrophoretic techniques for the separation of nanomaterials [8,13].

2.2 Capillary Electrophoresis

2.2.1 General Considerations

Capillary electromigration methods allow the separation of dispersions according to differences in their particle size or ζ potential. In the electrolyte solution, a dispersion of charged NPs will be surrounded by an "ionic atmosphere," and the thickness of the ion cloud is of the same order of magnitude as the radius of the particle. Thus, the ζ potential can be calculated from the electrophoretic mobility by using Hückel or Smoluchowski equations [9].

Several forces are involved during an electrophoretic event. These include the driving force exerted on the NPs by the external electric field, Stokes friction as a consequence of the viscosity of the medium the NP is immersed in, the so-called electrophoretic retardation force due the movement of the medium in an opposite direction, and, finally, the relaxation effect caused by the distortion of the ion cloud as a function of the mobility of the ions in the cloud [13].

In the case of charged colloidal particles, the dominant forces among the particles are said to be the repulsive forces due to electrostatic interaction and London–van der Waals interactions resulting in attractive forces [9]. In order to be separated, NPs must remain stable in the capillary, so the buffer electrolyte plays a crucial role in achieving this condition. A stable solution is obtained when the repulsive energy is enough to avoid aggregation of NPs. When the concentration of the buffer ions is increased, they comprise the ion cloud with the consequent decrease of repulsion energy and, thus, stability of the colloidal solution. The valence of the counterions has also a very strong impact on the stability of the system; those with a lower charge number are more suitable [13].

As has been mentioned, in order to be separated by electrophoresis, NPs must present a stable charge in the electrophoretic system. The most common methods of production of AuNPs (e.g., citrate reduction) lead to AuNPs that are negatively charged as a consequence of the binding of citrate ions onto their surface. Charged ligands adsorbed on the surface of the NPs form an electric double layer, which stabilizes NPs and prevents them from agglomeration and avoids adsorptive interactions with the capillary wall when it is negatively charged [8]. Different strategies of functionalization/derivatization devoted to increase (or in the case of uncharged NPs to provide) the number of charges in the NPs, which improve their electrophoretic separation, have been attempted. These comprise a covalent bond to a hydrophobic interaction, for example, with surfactant molecules. Proposed functionalizations include the use of thiol derivatives [14–17], carboxylic groups [18], as well as coating the NPs with charged polymers [16] or surfactant molecules [16,19–29].

The use of CE for the separation of charged species is based on the charge-to-size ratio of the analyte. Therefore, AuNPs with different diameters have different charge-to-size ratios and will migrate with different velocities in the capillary. This is a function of both the size and surface charges of AuNPs [19].

2.2.2 CE Separation of AuNPs

AuNPs have been separated in CE according to their size, shape, and charge. For example, Hwang et al. [30] separated a mixture of AuNPs with sizes of a 5 and 20 nm diameter by CE and proved the suitability of the technique for the separation of NPs of different natures, namely, polystyrene and AuNPs. Schnabel et al. [31] characterized colloidal AuNPs according to their size by CZE using acetate buffer at pH 5 without any sieving additive. A marker was introduced to measure the electroosmotic flow mobility and, thus, reduce possible bias of the colloid particle mobility data. CE has been also employed for the characterization of gold nanoclusters fractionated by the addition of an electrolyte [32].

As previously mentioned, different strategies have been used to stabilize AuNPs inside the capillary by the introduction of charges on their surfaces. The use of surfactant molecules was first introduced by Wei and Liu [19]. They employed CZE for the separation of AuNPs according to their size. For this task, they employed a buffer composed of 7.5 mM tetradodecyltrimethylammonium bromide (TTAB) and 12.5 mM of tetraborate at pH 9. The cationic surfactant TTAB is above the critical micelle concentration to form a protecting layer around the AuNPs, which stabilizes the NPs by preventing them from aggregation and providing charges on the outer surface; thus, NPs will behave as molecular charged species. They established a correlation between NP size and migration time. Under optimal conditions, AuNPs with a mean diameter as small as 1 nm can be separated by CZE.

The same authors reported an enhancement in the CE separation of AuNPs when sodium dodecyl sulfate (SDS) was added to the running electrolyte [20]. SDS molecules associate onto the surface of AuNPs and cause a change in the charge-to-size ratio, which is a function of the surface area of NPs and the surfactant concentration of running electrolyte. At high concentrations of the surfactant, the surface of AuNPs will be fully occupied with SDS molecules, and a linear relationship was observed between electrophoretic mobility and diameter of NPs. It should be pointed out that there is no kinetic equilibrium between micelles and NPs.

Moreover, they employed such methodology for characterizing the sizes of AuNPs synthesized by a microwave heating method, since a linear relationship between electrophoretic mobilities and sizes of NPs was found [21]. The accuracy of the method was confirmed by scanning electron microscopy (SEM), and an excellent correlation was obtained. Liu et al. [22] also characterized by CE the sizes of Au/Ag core/shell NPs in a similar manner to that previously reported.

CE electrophoresis of AuNPs stabilized by SDS surfactant was employed for the characterization of AuNPs synthesized with pyrene as additive during the synthesis in an SDS micellar solution [23]. Chang et al. [24] employed this technique for the rapid characterization of AuNPs prepared through seed-assisted synthesis. Using CE, they found that the rate of addition of the reducing agent affected the size distribution of NPs while the presence of NaOH in the synthesis medium minimized the sizes of AuNPs.

A drawback of CE is the low sensibility of UV detection systems. Because of the small dimensions of CE capillaries, only very small sample volumes can be loaded, and the path length is reduced. To preconcentrate and increase the amount of sample loaded without damaging the separation performance, online sample enhancement techniques, such as reversed electrode polarity stacking mode (REPSM), have been developed [33,34]. Liu employed REPSM for the online enhancement of the detection and separation of AuNPs, and observed that the sensitivity toward the AuNPs produced through seed-assisted synthesis improved up to 260-fold [25]. The procedure was also applied to the enhanced separation of Au/Ag NPs [26]. He also further optimized the REPSM conditions and investigated the influence of the addition of salt additive (NaH_2PO_4) to the AuNPs sample solutions [27]. An increase in the amount of sample additive leads to a decrease in the migration times of AuNPs as a consequence of relaxation effect on their migration. An enhancement up to 380-fold as compared with no REPSM conditions was observed. The influence of the nature of added salts, pH of the sample matrix, as well as the introduction of a buffer plug after injecting the NPs samples were also investigated [28]. Finally, two parameters were varied to further enhance the concentration and separation of AuNPs, namely, the rate of polarity switching (from −20 to +20 kV) during REPSM and the length of the capillary column [29]. The sensitivity enhancement factors for 5.3- and 40.1-nm-diameter AuNPs were improved by approximately 20- and 500-fold, respectively.

Another strategy is functionalization of AuNPs with thiol derivatives. Lo et al. [15] reported a CE technique for separating samples of negatively charged gold monolayer-protected cluster (Au MPC) protected by monolayers of *N*-acetyl-L-cysteine (NAC). Larger-core-sized Au MPCs emerged first from the capillary. The addition of aliphatic alcohols to the run buffer improved the separation by reducing the EOF and changing the selectivity between the Au MPCs.

Moreover, as can be seen in Figure 1, they also employed CZE to characterize nanometer-sized thiolated α-cyclodextrin-capped AuNPs (α-CD-S-AuNPs) [35]. The addition of tetrabutylammonium (Bu_4N^+) ions to

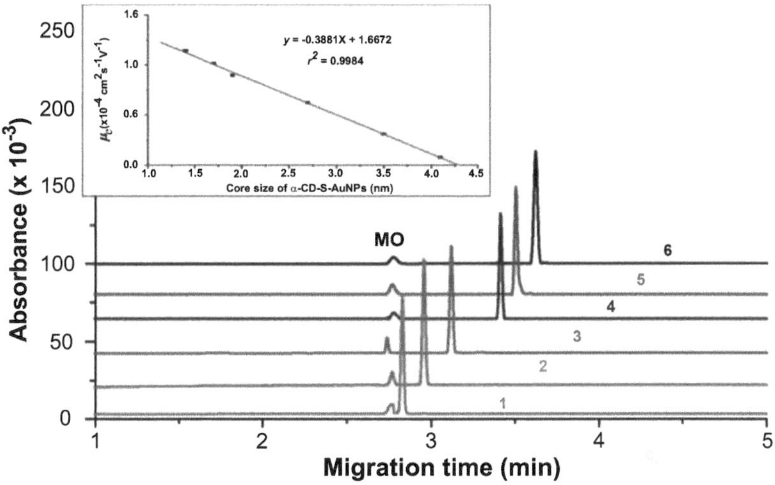

FIGURE 1 Electropherograms of various core sizes of α-CD-S-AuNPs: (1) 4.1 nm, (2) 3.5 nm, (3) 2.7 nm, (4) 1.9 nm, (5) 1.7 nm, and (6) 1.4 nm. The first peaks in the electropherograms are of mesityl oxide. The inset displays the inverse linear relationship between the electrophoretic mobility and core size of α-CD-S-AuNPs. *Reproduced with permission from Ref. [35].*

the run buffer greatly narrows the migration peak of the NPs. The migration mechanism is based on the formation of inclusion complexes between Bu_4N^+ and α-CD-S-AuNPs, which induces changes in their charge-to-size ratio.

López-Lorente et al. [17] reported the capillary electrophoretic behavior of citrate-capped gold and silver NPs in an aqueous medium when applying a ligand-exchange surface reaction with thiols. They investigated how ligand packing density influences the migration time of NPs of a similar size, shape, and shell in CE due to the formation of self-assembled monolayers of thioctic acid (TA) and thiomalic acid (TMA) on AuNPs regarding their strong affinity.

CE has been employed for other purposes related to AuNPs, for example, the study of the layer-by-layer assembly of poly-diallyldimethylammonium chloride and AuNPs on capillary walls, which causes surface charge neutralization, producing changes in the EOF [36]. Li et al. [37] resolved stable AuNP–protein complexes (bovine serum albumin, BSA) from the free NP and the proteins by CZE. This technique, in conjunction with laser-induced fluorescence (LIF) detection, has been used for investigating the specificity of biocatalysis by trypsin when it is conjugated to AuNPs [38].

Recently, a system for interfacing dark field microscopy with CE has been developed and applied to investigate the movement of single approximately 60-nm cetyl trimethyl ammonium bromide (CTAB) and SDS-capped AuNPs visually and "in real time" for the first time [39] (Table 1).

TABLE 1 Representative Examples of Capillary Electrophoretic Methods for the Separation of AuNPs

Type of NP	NP Modification	Buffer	Separation Conditions	Detection	References
Au	–	50 Mm Tris solution, pH 9.2	28 kV	UV–vis: 520 nm	[30]
Au	–	Acetic acid/sodium acetate (0.3–6 mM), pH 5.0	20 kV	UV–vis: 214, 440 and 520 nm	[31]
Au	–	Sodium phosphate (30 mM), pH 10, 20% v/v ethanol		UV–vis: 214 nm and LIF: 488/550 nm	[32]
Au	Stabilized by cationic surfactant	TTAB (7.5 mM), 12.5 mM of tetraborate at pH 9	20 kV	UV–vis: 546 nm	[19]
Au	Stabilized by anionic surfactant	SDS (70 mM) + CAPS (10 mM), pH 10	20 kV	UV–vis: 546 nm	[20]
Au	Stabilized by anionic surfactant	SDS (70 mM) + CAPS (10 mM), pH 11	18 kV	UV–vis: 400–800 nm	[21]
Au/Ag	Stabilized by anionic surfactant	SDS (40 mM) + CAPS (10 mM), pH 9.7	20 kV	UV–vis: 400 nm	[22]
Au	Stabilized by anionic surfactant	SDS (20 mM) + $Na_2B_4O_7$ (10 mM) and NaH_2PO_4 (10 mM), pH 8	300 V/cm	UV–vis: 254 nm	[23]

Au	Stabilized by anionic surfactant	SDS (70 mM) + CAPS (10 mM), pH 10	20 kV	Vis: 520 nm	[24]
Au	Stabilized by anionic surfactant	SDS (70 mM) + CAPS (10 mM), pH 10	20 kV (REPSM)	UV–vis: 520 nm	[25]
Au, Au/Ag	Stabilized by anionic surfactant	SDS (40 mM) + CAPS (10 mM), pH 10	20 kV (REPSM)	Vis: 520 nm	[26]
Au	Stabilized by anionic surfactant	SDS (70 mM) + CAPS (10 mM), pH 10 [sample additive: NaH_2PO_4 (10 mM)]	20 kV (REPSM)	UV–vis: 520 nm	[27]
Au	Stabilized by anionic surfactant	SDS (70 mM) + CAPS (10 mM), pH 10 [sample additive: NaH_2PO_4 (5 mM)]	20 kV (REPSM)	UV–vis: 520 nm	[28]
Au	Stabilized by anionic surfactant	SDS (70 mM) + CAPS (10 mM), pH 10	20 kV (REPSM)	UV–vis: 520 nm	[29]
Au MPCs	–	Phosphate buffer (40 mM) + 20% v/v ethanol, pH 10	−20 kV	UV–vis: 214 nm	[15]
α-CD-S-AuNPs	–	30 mM Phosphate and 10 mM Bu_4N^+ ion, pH 12	15 kV	UV–vis: 254 nm	[35]
Au	TA and TMA functionalized inside capillary	40 mM SDS, 10 mM CAPS, 0.1% v/v methanol, pH 9.7 (buffer additives: 0.1% TA/TMA)	20 kV	UV–vis: 539 nm	[17]
Au/ BSA–Au	–	10 mM Borate, pH 8.3	25 kV	UV–vis	[37]

2.3 Gel Electrophoresis

In GE, charged particles migrate in a porous matrix (gel) under an electric field, and particle mobility depends on their charge and size [40].

GE hyphenated to inductively coupled plasma–mass spectrometry (GE–ICP–MS) has been used for the characterization of AuNPs stabilized with citrate ligands [41]. During electrophoretic separation, it can be assumed that the complex stability decreases and that free citrate is released, which is in agreement with the loss of charges on the AuNPs. Consequently, the electrophoretic mobility is lower than that of the original AuNP, and results in significant peak tailing. When mercaptosuccinic acid (MSA) modified AuNPs were analyzed, the formerly observed peak tailing was reduced to a nearly symmetrical peak.

GE–ICP–MS has been also employed for the determination of Au+/s+ ratios in AuNPs covered by MSA by simultaneous detection of the sought elements [4].

AuNPs coated with different ligands such as glycine, lysine, cysteine, arginine, polyethylene glycol (PEG), polyvinylpyrrolidone (PVP), SDS, Triton X-100, or CTAB have been studied by GE in 0.8% agarose gel prepared in Tris–acetate–ethylenediaminetetraacetic acid (EDTA) buffer [16].

Conjugates of NPs with one, two, and three PEG molecules per NP have been separated using agarose gel electrophoresis [42]. In addition, amine-terminated poly(amidoamine) (PAMAM) dendrimer-stabilized AuNPs have been analyzed by polyacrylamide gel electrophoresis, which indicated that the conjugates are stable and do not separate from each other during the electrophoretic process [43]. Hanauer et al. [44] separated AuNPs according to their size and shape by agarose gel electrophoresis after coating them with a charged polymer layer (SH-PEG-COOH). The separation was monitored optically using the size- and shape-dependent plasmon resonance of AuNPs and confirmed by transmission electron microscopy (TEM). Doane et al. [45] described GE experiments and a theoretical model that quantified the role of the PEG-ligand molecular weight on NP transport in agarose gels.

Hasenoehrl et al. [46] described a novel sensitive method for detecting bis-sulfonato (phenyl)phenylphosphine (BSPP)-capped AuNPs in agarose gels, which involves staining the gel with the common organic fluorophore flourescein to produce AuNP band intensities that are linear with NP concentration and almost an order of magnitude larger than those obtained without staining the gel.

Arnaud et al. [47] employed IEF gel electrophoresis to analyze the size distribution of AuNPs synthesized by a chemical route with MSA as ligand. In a typical IEF gel electrophoresis, the charged molecules migrate in an immobilized pH gradient polyacrylamide gel. The isoelectric point of the NPs was shown to be size dependent, and allowed fractionation by electrophoresis.

AuNPs modified by 11-mercaptoundecanoic acid (MUA) have been separated and characterized by agarose gel electrophoresis [48]. The electrophoretic mobility depends on the number of MUA molecules per AuNP.

Further, Xu et al. [49] separated AuNPs according to their size, shape, and charge by agarose gel preparative electrophoresis. Size separation was demonstrated with 5, 15, and 20 nm spherical gold nanoclusters, while the ability to separate AuNPs by shape was also shown by the purification of a mixture of gold spheres, plates, and long rods. To introduce charges to the NPs, they were capped with 11-MUA. The introduction of carboxylic groups provide negative charges and NPs migrate to the anodic electrode.

GE has been also employed to analyze AuNP bioderivatives. Colloidal nanocrystal/DNA conjugates hold the promise of their use as probes in biological diagnostics, and are important to control the number of oligonucleotides attached to the nanocrystal. GE was employed to isolate 5- and 10-nm nanocrystals bearing discrete numbers of single-stranded DNA. It can be also employed to determine the average number of strands per particle [40]. They also employed GE to study the mobility of Au/DNA hybrid materials to form small groupings of particles, such as dimers and trimers [50].

Moreover, Parak et al. [51] determined the electrophoretic mobility on 2% agarose gel of thiol-modified single-stranded oligonucleotides of different lengths (8−135 bases) to the surface of 10-nm Au nanocrystals with different DNA/Au ratios. The effective diameter of the DNA/Au conjugates was also determined. GE was also employed to confirm the binding of AuNPs with isolated DNA fragments [52] as well as to isolate AuNP/oligo-DNA conjugates having a defined number of oligo-DNAs [53]. In addition, Qin and Yung [54] prepared AuNP−DNA conjugates bearing a specific number of long double-stranded DNA strands by GE. NP conjugates bearing a specific number of strands of short DNA with <20-base were achieved.

Agarose gel electrophoresis was used to confirm the relative mobility of triangular and spherical NPs and determine their surface charge [55]. A study of the ability of agglomerates and aggregates of NPs to travel through an agarose matrix, which was significantly hindered compared to the primary particles in 5% dextrose or water, was performed by GE [56]. In addition, it was used for determining the capacity and association constant for the binding of chymotrypsin (ChT) to polyacrylate-coated AuNPs (Figure 2) [57]. Agarose gel electrophoresis has been also employed to purify Au/Ag alloys [58] and separate a mixture of gold nanospheres and nanoplates with a wide standard deviation after density gradient centrifugation [59]. Finally, GE enabled the evaluation of the adsorption of binding proteins to negatively charged 5-, 15-, and 80-nm AuNPs [60].

Recently, Beskorovaynyy et al. [61] presented a high throughput two-dimensional-grid GE cell-based method, capable of optimizing the surface modification of as many as 96 samples of AuNPs in approximately 1 h. The method was tested for optimizing pH and concentration of the

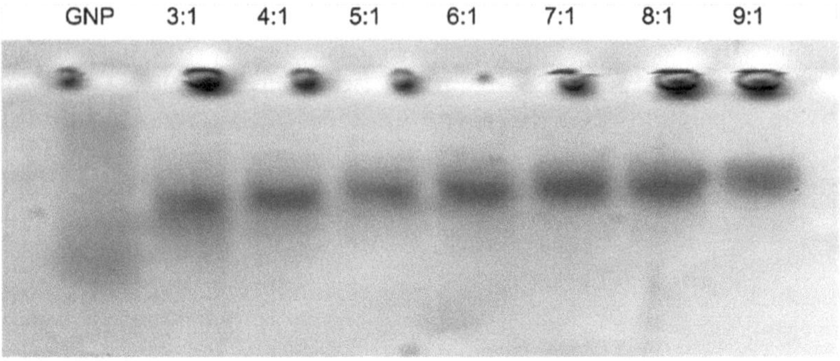

FIGURE 2 Agarose gel electrophoresis of the AuNPs as a function of ChT concentration. The ChT:AuNP ratio is varied from 0 to 9:1. The mobility of the AuNP is decreased by added protein (resulting from binding), but levels off between approximately 7–9:1, reflective of the binding stoichiometry. *Reprinted with permission from J. Hou, D.M. Szaflarski, J.D. Simon, J. Phys. Chem. B 117 (2013) 4587–4593. Copyright 2013 American Chemical Society.*

modifiers for functionalizing gold nanorod thiol-containing acidic agents (Table 2).

2.4 Dielectrophoresis

Dielectrophoresis (DEP) is defined as the motion of neutral particles due to the application of an external nonuniform electric field. The motion occurs due to the polarization of natural particles, which then experience a force called the dielectrophoretic force [62]. Such a force can be positive (attractive) or negative (repulsive) depending on whether the particle is more or less polarizable than the medium.

Strobel et al. [63] reported on the successful dielectrophoretic trapping and electrical characterization of DNA-coated AuNPs on vertical nanogap devices. Moreover, Chuang et al. [64] developed a microfluidic chip for programmable manipulations of polymethylmethacrylate nanoparticles (PMMA-NPs) based on DEP.

In addition, DEP has been employed to fabricate AuNP sensors [65–67] as well as AuNP single-electron transistors [68]. DEP is an effective method for capturing NPs and assembling them into nanostructures [69–72]. The frequency of the dielectrophoretic alternating current (AC) electric field greatly influences the morphology of resultant NP assemblies [73]. They have been assembled, for example, in nanogap electrodes [74,75] to fabricate continuous nanobridges [76]. AuNPs of different diameters (5, 10, and 20 nm) have been assembled between 20 nm gap electrodes using the AC (DEP) process [77,78], while citrate-stabilized colloid AuNPs with a diameter of 40 nm have been assembled in chains between microelectrodes [79,80].

Moreover, nanowires from AuNPs have been fabricated by AC DEP and ink-jet delivery [81]. Arrays of AuNPs and Au nanorods were fabricated in the

TABLE 2 Examples of Separation of AuNPs by Gel Electrophoresis (GE)

Type of NP	NP Modification	Gel Composition	Separation Conditions	Detection	References
Au	Citrate and MSA ligands	2–2.4% (w/w) SeaKem® LE agarose	400 V	ICP–MS	[41]
Au	Citrate and MSA ligands	2.0% and 2.4% (w/w) SeaKam® LE agarose	400 V	ICP–MS	[4]
Au	Au-PEG	2% Agarose gel, TBE buffer	10 V/cm		[42]
Au	Au-PAMAM	Polyacrylamide, Tris–glycine native buffer	200 V		[43]
Au and Ag	Au-SH-PEG-COOH	Agarose gel (0.2%), Tris (44.5 mM) + borate (44.5 mM) + EDTA (1 mM), pH 9	150 V	UV–vis	[44]
Au	Au-PEG	Agarose buffer, TAE buffer[b]			[45]
Au	BSPP-Au	0.2% Agarose gel, TBE buffer[c]	3.1 V/cm		[46]
Au	MSA ligands	Polyacrylamide pH gradient gel	Ramped 0–1000 V (2 h)	UV–vis	[47]
Au	MUA-Au	0.5% Agarose gel, TBE buffer[c]	100 V	UV–vis	[48]
Au	11-Mercaptoundecanoic Acid	Agarose gel (2%), Tris (89 mM) + borate (89 mM) + EDTA (2 mM)	100 V	UV/Vis	[49]
Au/DNA	DNA Au conjugates	2–3% Agarose gel			[40]
Au/DNA	DNA Au conjugates	3% Agarose gel, TBE buffer[c]	5 V/cm		[50]
Au/DNA	DNA Au conjugates	2% Agarose gel, TBE buffer[c]			[51]
Au	Au-ODA[a]	1% Agarose gel, TAE buffer,[b] pH 7.4	50 V		[55]
Au	ChT-Au	1% Agarose gel, 5 mM sodium phosphate buffer, pH 7.4	100 V		[57]
Au	Au nanospheres and nanoplates	0.9% Agarose gel, TBE buffer[c]	100 V	UV–vis	[59]

[a] *ODA = octadecylamine.*
[b] *TAE = tris-acetate ethylene diamine tetraacetic acid.*
[c] *TBE = tris-borate ethylene diamine tetraacetic acid.*

spatially localized pores of a porous alumina membrane using positive DEP [82,83]. A fast, single step, room temperature deposition of AuNPs on polyethylene terephthalate (PET) and biaxially oriented PET (BoPET) substrate has been performed by employing DEP [84].

2.5 Other Electrophoretic Techniques

Electrophoresis studies have been performed to calculate the charge of NPs or their ζ potential. For example, van der Zande et al. [85] characterized PVP-stabilized gold rods by electrophoresis. The electrophoretic mobility was measured with an alternating electric field so that electroosmosis was suppressed. The measured electrophoretic mobility was converted to ζ potential by using the Hückel equation. Electrophoresis revealed that Au rods were negatively charged with an ζ potential of -47 mV, sufficient to stabilize AuNPs. The electrophoretic mobility of AuNPs with different sizes has been also calculated by using platinum-coated alumina membranes as a separation system [86]. It has been also calculated for colloidal AuNPs of different sizes in aqueous potassium chloride and sodium phosphate electrolyte solutions [87]. Electrophoretic mobility measurements allowed the characterization of the surface of AuNPs functionalized with human serum albumin and creatine [88]. Makino and Ohshima developed a systematic method to determine the surface charge density of a spherical colloidal particle on the basis of the particle electrophoretic mobility data [89].

In addition, electrophoresis has been also employed to characterize the sign of the surface charges of AuNPs synthesized using sucrose by seeding formation [90]. For this purpose, changes in absorption spectrum were measured during electrophoresis performed in a glass U tube with two Pt electrodes inserted at the end of the tube.

An electrophoresis device composed of two glass vessels with two platinum electrodes connected by a glass tube was used to separate hydrophobic AuNPs with a particle size <10 nm dispersed in an organic solvent [91]. TEM and UV–vis spectra showed that the NPs in the vicinity of the cathode have a mean diameter smaller than those near the anode.

The technique of isotachophoresis (ITP) permits one to generate plugs in which the coions and counterions of the original colloidal solution are removed and replaced by the buffering counterions of the leading electrolyte [92]. ITP was employed for the calibration-free determination of the molar (particle) concentration of dispersed colloidal AuNPs (either coated with mercaptoethanol (ME) or BSA) [92].

Bücking and Nann [14] performed a two-dimensional, online coupled ITP CZE system to study the size, electrophoretic mobility, and coagulation kinetics of AuNPs. Citrate-coated AuNPs were not suitable as the ligand was desorbed during analysis and NP solutions degraded. Thus, AuNPs were functionalized with mercaptocarboxylic acids due to the chemisorptions of thiol compounds on Au surfaces.

IEF is a technique that is commonly employed to determine the isoelectric point of proteins and enzymes. Gole at al [18] separated colloidal AuNPs of different sizes derivatized with carboxylic acids groups from their mixtures in a miniscale IEF unit developed in house. In addition to separating differently sized NPs, this technique may also be used to improve the polydispersity of colloids.

As previously described in GE, IEF in a polyacrylamide pH gradient gel enabled the size characterization of AuNPs [47].

Continuous free-flow electrophoresis is a fractionation technique that isolates monodisperse particle sizes into several different collection vials on the tens of milligram scale. It has been employed for the size fractionation of *N*-(2-mercaptopropionyl)glycine (tiopronin) monolayer-protected gold clusters into monodisperse NP samples on the basis of their electrophoretic mobilities [93] (Table 3).

3. FIELD-FLOW FRACTIONATION

3.1 Introduction

The use of FFF for the separation and characterization of NPs has increased in the last few decades. FFF belongs to a family of chromatography-like separation techniques that provide a continuous and high-resolution separation of NPs as a function of their diffusion coefficient.

The FFF theory was first proposed by Gidding [94]. FFF is a flow-assisted hydrodynamic separation technique that enables the physical separation of small quantities of particles [95]. The principle of separation consists of injecting the analytes into a flat channel with a laminar tangential flow (having a parabolic velocity profile), and then applying a transverse field perpendicular to the laminar flow to retain (and thereby fractionate) particles according to different physicochemical properties. The term "retain" means that particles are situated in a zone retained by their confinement to flow lines having velocities that are lower than the average velocity present in the FFF channel [96]. Particle retention is a function of diffusion coefficients, and those particles with different values of this parameter are separated [97].

The external field drives the particles toward the so-called accumulation wall from where they diffuse back into the channel. The particles form diffusional clouds in the channel, where the thickness of the clouds depends on the diffusion coefficient of the particles and the counteracting external force applied perpendicular to the carrier-driven flow trajectories of the particles in the channel (Figure 3) [95].

Depending on the relationship between the NP size and the thickness of the diffusional cloud formed by the NPs against the applied field force, two different elution modes are described, namely, normal or Brownian mode and steric mode separation. The first one takes place when particles are small enough to undergo significant Brownian motion and their size is negligible as

TABLE 3 Characterization and Separation of NNPs with isotachophoresis (ITP) and isoelectric focusing (IEF) Techniques

Type of NP	Electrophoretic Modality	NP Modification	Buffer	Detection	References
Au	ITP	BSA and ME	Leading electrolyte: HCl (1 mM) titrated with BTP to pH 7.88. Terminating electrolyte: 10 mM MES titrated with Tris to pH 6/ HEPPS titrated BTP to pH 8 (polyacrylamide-coated capillary).	UV–vis: 230 nm and conductivity	[92]
Au	ITP-CZE	Mercaptocarboxylic acid functionalization	Leading electrolyte: HCl (10 mM) + histidine (20 mM) pH 6. Terminating electrolyte: MES or glutamic acid (5 mM).	Vis: 520 nm	[14]
Au	IEF technique	Carboxylic groups using self-assemble bifunctional surfactants	Electrolyte solutions: (1) 0.1 M ortho-phosphoric acid anode solution, (2) 0.1 NaoH cathode. Separation solution: 1.5 mL of glycerol with 1 mL water. Gradient solutions: 3 mL glycerol, 2 mL hydrosols, 0.2 mL carrier electrolyte.	Vis spectra	[18]

BTP = 1,3-bis(tris(hydroxymethyl)-methylamino)propane.
HEPPS = 3-[4-(2-Hydroxyethyl)-1-piperazinyl]propanesulfonic acid.
MES = 2-morpholinoethanesulfonic acid monohydrate.

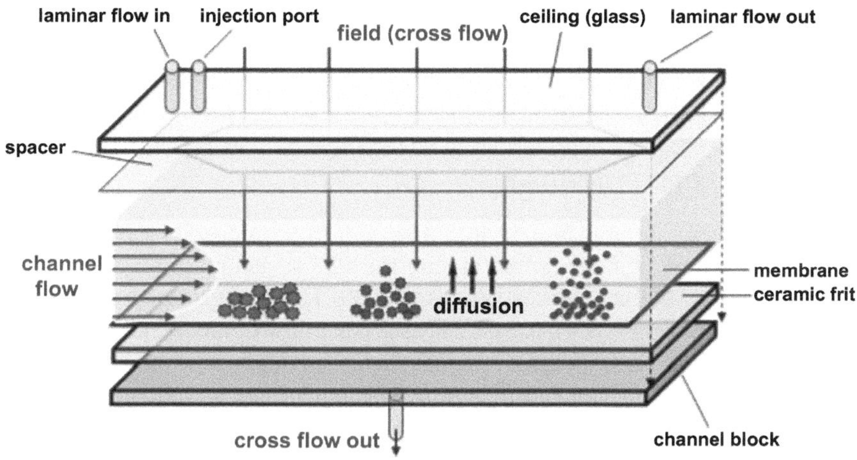

FIGURE 3 Schematic diagram of an asymmetric flow field flow fractionation (AFFF) fractionation channel, which illustrates the modes of flow and principle of hydrodynamic separation. *Reproduced with permission from Ref. [98].*

compared to the thickness of the diffusional cloud [95]. In this modality, smaller particles elute before larger ones do, since the higher the cloud is, the larger is the mean travel velocity of the particle comprised on the cloud. The steric mode however occurs when particles are comparable to the diffusional cloud. Large particles elute earlier since the travel velocity along the channel is determined by the extension of particles into the channel instead of diffusional clouds [95]. In the case of NPs, as their sizes are on the nanometer scale, the normal mode is predominant.

FFF possesses several advantages over other separation techniques. For example, it can separate NPs continuously, nondestructively, and at a high resolution, in the size range of 1 nm−100 μm. In addition, it is possible to couple both online and offline a wide range of detectors, such as UV, ICP−MS, multiangle light scattering (MALS), and microscopic techniques such as TEM or atomic force microscopy (AFM). Particle separation occurs in an open channel, which does not contain a stationary phase, thus, minimizing particle adsorption into the stationary phase or size exclusion [97].

However, this technique also presents some drawbacks, such as material losses, particle−membrane interaction, sample dilution, washing of sample components, and overloading [97]. Material losses originate from the permeation of NPs through the accumulation wall membrane and adsorption onto the surface of the membrane, as a result of attractive forces (e.g., van der Waals, hydrophobic interactions). Another limitation of this technique is the lack of established protocols and standards for its application [96].

When coupled with detectors such as MALS, dynamic light scattering (DLS), absorbance, or luminescence spectrophotometry, FFF allows one to perform multidimensional characterization of NPs and provides information about size distribution of the NPs, aggregation of NPs in their native conditions, separation of the unbound constituents of the functional NPs, optical features of the NPs together with the correlation of spectroscopic properties with NP size [99].

3.2 Types of FFF Techniques

FFF techniques can be classified into different groups depending on the field applied. The type of field differs among the FFF techniques and determines which property of the particles is responsible for retention and separation (Figure 4). In flow FFF (F4), the field is as second flow of mobile phases that is applied across the longitudinal, laminar flow. Retention is inversely proportional to the hydrodynamic diffusion coefficient of the analyte and, consequently, to its hydrodynamic size [100]. Sedimentation FFF (SedFFF) uses a centrifugal force, which results in a separation according to particle volume and particle buoyant mass. Other field types include electric fields (electrical-FFF), magnetic fields (magnetic-FFF), and thermal gradients (thermal-FFF) [95]. Among these techniques, F4 has been the most widely employed for the separation and characterization of AuNPs.

Regarding F4, two different modalities have been described: symmetric (sF4) and asymmetric (AF4), depending on how the perpendicular flow-field is produced [95]. In sF4, the top wall of the channel is permeable and the crossflow is independently generated through the top frit. However, in AF4, the top wall of the channel is impermeable and only the accumulation wall is composed of a permeable ceramic frit. Carrier flow enters the channel through the inlet at the tip of the channel and exits the channel through both the accumulation wall membrane and the channel outlet. Thus, the inlet flow generated both the crossflow and the longitudinal flow in the channel [97]. AF4 is achieving increased popularity nowadays and is the modality employed

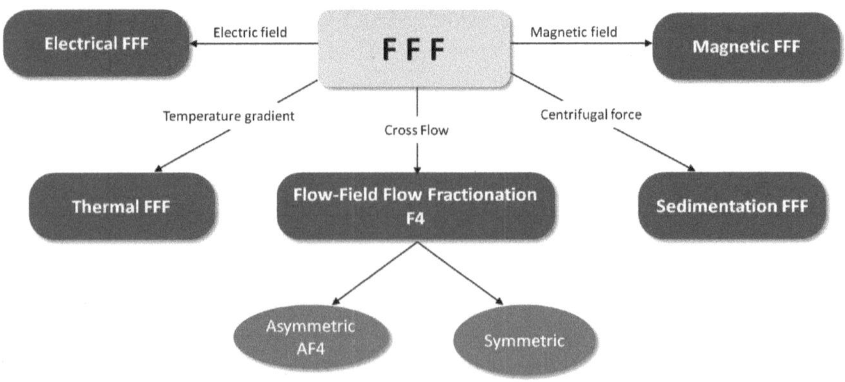

FIGURE 4 Examples of the different modalities of field-flow fractionation (FFF) depending on the force or field applied.

in most of the examples explained below. Some authors also described a third variant, which is the so-called hollow-fiber F4 (HF5) [99,101].

3.3 Application to the Separation of AuNPs

As previously mentioned, AF4 is the most developed and applied FFF variant. A hydrodynamic field is generated by splitting the longitudinal flow into a second flow stream (crossflow) across the only permeable wall that constitutes the fractionation channel. AF4 thus allows the focusing of the sample in a narrow band prior to elution, increasing fractionation efficiency [102]. The crossflow, force opposed by the Brownian diffusion of the NPs, drives constituents toward the accumulation wall (permeable membrane). Larger NPs equilibrate at a relatively low mean elevation above the accumulation wall, where they migrate through the channel more slowly. Contrarily, smaller NPs form a more diffuse layer that extends upward to increase their migration rate along the channel. Particles are separated into fractions according to their hydrodynamic size [98].

Retention of the sample particle in F4 is affected by the property of the accumulation wall, which is composed of a polymer membrane. Song et al. [103] studied the retention behavior of 5-, 10-, 20-nm sized gold colloid standards on two different types of membranes; a hydrophilic regenerated cellulose membrane and a hydrophobic polysulfone membrane, both with the same molecular weight cutoff of 10 kDa, in the presence of a nonionic surfactant in the mobile phase.

Mercaptosuccinic acid-capped AuNPs have been analyzed by AF4, and such measurements were compared with TEM and DLS off-line measurements [104].

AF4 with MALS detection has been employed to separate and characterize polymer-coated AuNPs in dispersion [102]. NPs were size fractionated, and their hydrodynamic radius was calculated. In addition, separation from the unbound components of the NP structures that are present in the solution has been carried out. To provide additional information on the "bound" and "unbound" components, online photodiode-array UV/vis and refractive index measurements were also performed.

AF4 coupled with MALS and ICP–MS was also employed by Schmidt et al. [105] to separate and quantify AuNPs. The three differently sized separated AuNPs were quantified by ICP–MS and recovered at 50–95% of the injected masses. AuNPs adhering to the membrane in the separation channel were found to be a major cause for incomplete recoveries.

F4 has also been employed for the size characterization of AuNPs as well as for examination of the capability of α-tocopherol to induce enlargement of AuNPs [106]. Measured particle sizes obtained from F4 for commercial 10-nm AuNPs and those synthesized in the laboratory were in good agreement with sizes determined by TEM.

Cho and Hackley [98] reported the analysis of individual AuNP components in a size range from 10 to 60 nm in an aqueous mobile phase and their fractionation from multicomponent mixtures. The effects of flow rates, membrane types, mobile phases, and surfactants were investigated. Measurements were performed with AF4 coupled in-line with MALS, DLS, and UV−Vis detectors.

Moreover, AF4 combined with UV/vis, light scattering, and ICP-MS detection was employed for the characterization of well-defined AuNP dispersions [107]. Investigation of different membrane materials revealed that interaction of NPs with the channel membrane leads to an unspecific recovery from the AF4 channel. For NPs with a size <10 nm the use of carriers with already 1 mM NaCl showed to destabilize the NP system during the AF4 separation.

Calzolai et al. [108] also employed AF4 coupled to a UV−vis detector for the efficient separation, size determination, chemical characterization, and quantification of the relative number of particles of different sizes in the range 5−60 nm in complex AuNP mixtures.

Tsai et al. [109] performed simultaneous size classification and characterization of functional AuNP clusters (AuNCs) based on AF4, which enabled well-resolved separation of purified AuNCs fractions, comparable in purity to density gradient centrifugation. Separation was validated using UV−vis spectroscopy and ex situ field emission SEM and AFM. They also used AF4 to study the adsorption and conformation of BSA on AuNPs [110].

Further, Gray et al. [111] compared the performance of HDC and AF4 coupled to ICP−MS in the separation of mixtures of 5-, 20-, 50-, and 100-nm AuNPs. AF4 showed a greater resolution than HDC did, and differences were more pronounced in the smaller size range. However, recoveries for AF4 ranged from 4% to 89%. Low recoveries are related to the largest NPs at the lowest concentrations examined. The study was, in fact, applied to dilute concentrations that would mimic expected environmental NP concentrations.

Gigault et al. [112] applied AF4 to the separation of a low aspect ratio gold nanorods. This work introduced two advancements: one is the fractionation of positively charged NPs by AF4, despite the interaction that exists between the negatively charged native membrane and the NPs, which can be overcome by controlling the composition of the mobile phase (mixture of ammonium nitrate and CTAB). Another novelty is the demonstration of shape-based separation of gold nanorods in AF4 normal mode elution (Brownian mode). These authors also demonstrated a clear influence of the core NP material nature on the retention process in the AF4 channel using several complementary detectors [113]. The study was accomplished with NPs of different natures; citrate-stabilized nominally 100-nm-diameter AuNPs are among them.

However, sedimentation FFF (SedFFF) has also been applied for the separation of AuNPs (Figure 5). Oppenheimer and Smith reported for the first time the characterization of gold and silver sols using this technique [114].

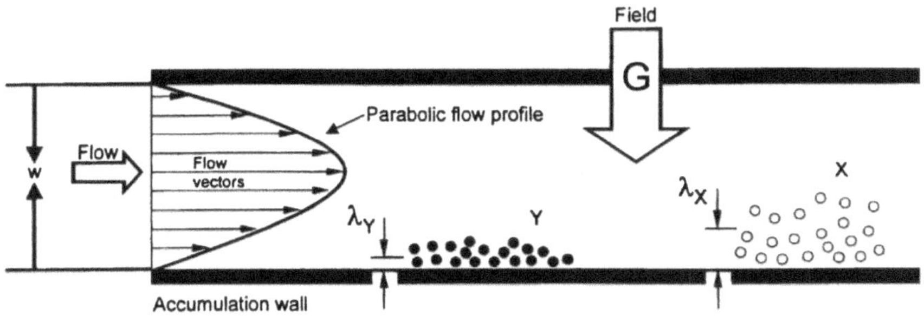

FIGURE 5 Principle of Sedimentation FFF (SedFFF). Two different particle populations (X and Y) form zones with different layer thicknesses under the influence of the sedimentation field. *Reproduced with permission from Ref. [115].*

Anger et al. [115] also employed this technique for the high-resolution size determination of 20-nm colloidal AuNPs. A surfactant (FL-70) was used in the carrier for dispersing the AuNPs, which allowed the determination of the diameter of single AuNPs. The polydispersity of the NP population can be calculated on the basis of the plate height.

SedFFF has been also employed by Contado and Argazzi [116] for the size characterization of AuNPs synthesized in aqueous medium down to diameters of the order of 12 nm. An advantage of the technique is that the sample can be analyzed directly without sample pretreatments that may alter their nature.

The potential of electrical field-flow fractionation (ElFFF) for characterization of metal NPs was investigated by Somchue et al. [117] by using NPs with different types of stabilizers, including citrate and tannic acid. Citrate-stabilized AuNPs eluted earlier than tannic acid-stabilized AuNPs of the same size. ElFFF can be used for characterization of AuNPs with different particle sizes including 10, 20, and 40 nm with a fixed stabilizing agent.

4. CHROMATOGRAPHIC TECHNIQUES

4.1 Introduction

Chromatographic techniques have also proved their suitability to achieve separation of AuNPs. High-pressure liquid chromatography (HPLC) approaches to the separation of NPs are of two basic types [118]. The first relies on the different chemical affinities of molecules for different surface chemical groups attached to solid media, typically porous SiO_2, which depends on the different affinities for molecules in the solid or stationary phase of the column as compared to the mobile phase.

A second approach, commonly known as SEC, uses porous hydrophobic microgels (typically crosslinked polystyrene) to separate molecules by size. In this modality, large molecules cannot penetrate the pores of the column media as effectively as small ones can and result in different times of elution. The key to the application of SEC to the analysis of inorganic nanoclusters is effective surface passivation of the clusters using nonpolar organic ligands. SEC columns were designed to have little or no specific chemical affinity for nonpolar organic molecules while they can irreversibly adsorb inorganic ions or NPs [118].

In HDC, the separation mechanism relies on the size of the NPs and is independent of the coating or the surface charge of the NP. HDC separates small particles based on flow through a packed bed of nonporous beads.

4.2 Size-Exclusion Chromatography

The so-called steric exclusion chromatography was firstly employed for measuring the size distribution of colloidal gold sols [119]. Separation of 3- to 20-nm sized AuNPs was achieved using columns containing 25 μm silica particles (pore size 50–100 nm) using aqueous trisodium citrate as eluent.

SEC separations involve the selection of a suitable stationary phase with a proper pore size and the use of the adequate eluent. However, irreversible adsorption of the NPs onto the stationary phase may occur due to the high surface area of the stationary phase and the high surface activity of the NPs. An approach to overcome this limitation was proposed by Wei and Liu [5], which is the addition of surfactants into the mobile phase to reduce surface adsorption of the NPs (Figure 6). They employed SEC for the separation of AuNPs in the range from 5.3 to 38.3 nm with a polymer-based column of a 100-nm pore size. The association of surfactants with the NPs stabilizes them by preventing their agglomeration.

Moreover, they compared the effects of three different mobile phase additives, sodium citrate (SC), NaCl, and SDS on the separation of AuNPs with diameters of 5.3 and 38 nm [120]. When SC and NaCl were used as mobile phase additives adsorption problems occurs, however, the addition of SDS overcomes such problem. SEC is a good tool not only to separate NPs but also to find suitable protection reagents that stabilize NPs. The mechanism by which NPs associate with protection reagents may be studied by the use of an SEC system.

Liu also reported the use of SEC for the rapid determination of the sizes and size distributions of AuNPs obtained by seed-assisted synthesis [121]. The separation was accomplished by using a polymer-based column of a 400-nm pore size, using 10 mM of SDS as the mobile phase. The addition of SDS to the synthesis medium led to the production of AuNPs that could be stored for longer durations. In addition to observing the elution times of AuNPs, their size-dependent optical properties were also evaluated by diode array detector (DAD) and the stability of AuNPs products was monitored. In a later study, Liu also demonstrated that the presence of NaOH in the synthesis medium dramatically decreased the sizes of the AuNPs [122]. AuNPs with sizes ranging from 12.1 to 79.1 nm eluted in a linear manner with respect to elution time.

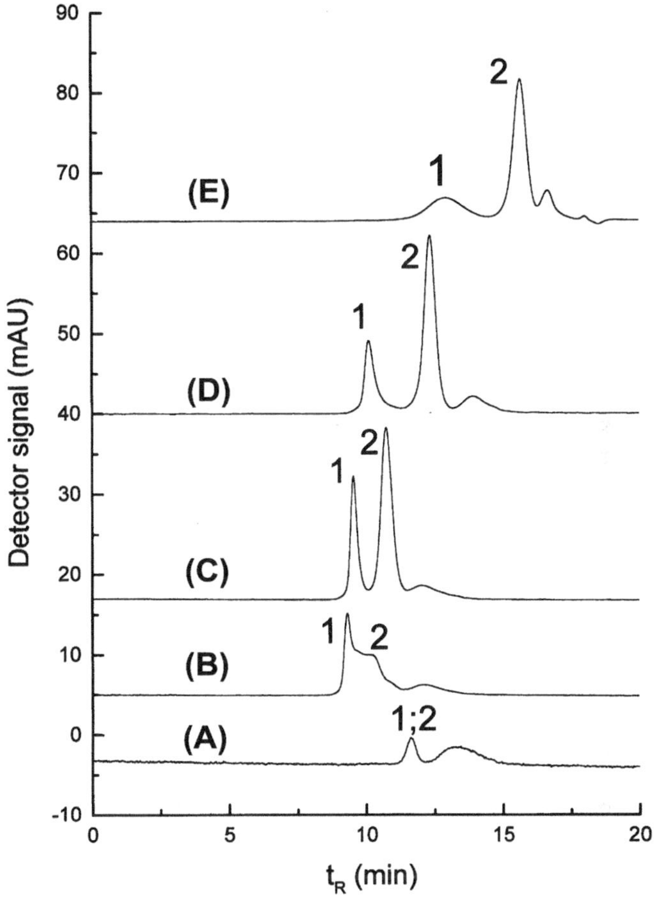

FIGURE 6 Sodium dodecyl sulfate (SDS) concentration effect on the separation of 5.3- and 38.3-nm AuNPs by SEC. (A) 0, (B) 0.1, (C) 1, (D) 5, and (E) 80 mM SDS; 1:38.3-nm AuNPs, 2:5.3-nm AuNPs. *Reproduced with permission from Ref. [5].*

Further, SEC has been used to characterize the sizes of fabricated Au/Pt [123] and Au/Pd [124] core/shell NPs by using a polymer-based SEC column. When using SEC to analyze synthetic products, the sizes of the core/shell NPs can be controlled via modification of the volumes of the seed, SDS, and sodium hydroxide solutions. A good correlation was observed between the sizes determined through SEC and TEM.

SEC has been also proposed for the separation of nanometer-sized AuNPs with different shapes [125]. Shape separation of the NPs was achieved by adding a mixed surfactant system containing SDS and poloxyethylene (23) dodecanol (Brij-35) into the eluent, which affects differently from the adsorption behavior of rodlike and spherical AuNPs onto the column packing materials. Separation was identified by DAD and confirmed by TEM images of the fractional collections.

This technique has been also employed for isolating enriched fractions of phenylethynyl-bridged AuNP dimers and trimers from the monomeric particle starting material [126]. This method enabled the isolation of enriched fractions without causing significant sample aggregation or replacement of the phenylethynyl bridge. SDS was again added to the mobile phase to prevent irreversible binding to the stationary phase.

Al-Somali et al. [127] demonstrated the application of both closed-loop and alternate-recycling size-exclusion chromatography to the analysis and separation of thiol-passivated gold nanocrystals. RSEC is a practical method for increasing the column length in an SEC system, and it overcomes the increase in back pressure associated with long columns by recycling the column output back into the column. The resolution increases with each pass through the column, and helps achieve baseline separation of populations of particles that differ by 6 Å in diameter in a polydisperse AuNP sample.

Moreover, recycling gel permeation chromatography was employed by Tsunoyama et al. [128] to fractionate Au:SC_x clusters into Au_{38}:SC_x and Au_{55}:SC_x. These were further identified by laser desorption ionization mass spectrometry.

The sizes of Au and Au/Pd core/shell NPs subjected to thermal treatment were evaluated by SEC to analyze the surface plasmon absorption of NPs after elution [129]. In the absence of an adequate stabilizer, thermal treatment resulted in longer retention times for AuNPs and shorter times for the Au/Pd core/shell NPs in the chromatograms. The suitability of 3A-amino-3A-deoxy-(2AS,3AS)-β-cyclodextrin (H_2N-β-CD) as stabilizer was proved, as retention times and absorbance of eluted AuNPs and Au/Pd NPs remained constant.

SEC has been also employed for assessing the size stabilization of AuNPs in the presence of salt and organic solvents (e.g., NaCl and methanol) [130]. In the absence of an adequate stabilizer, their presence leads to agglomeration of the AuNPs with the consequent disappearance of the signal of the AuNPs in the elution spectra after SEC separation. However, when capped again with an adequate stabilizer, such as H_2N-β-CD, the elution time of the signal of the AuNPs and their elution spectra were barely affected.

In addition, SEC has been also used to evaluate variations in size of AuNPs after microwave irradiation by monitoring the changes through diode array detection [131]. The sizes of citrate-capped AuNPs decreased upon increasing the microwave irradiation temperature, while AuNPs capped with SDS increased in size. When the AuNPs were capped with H_2N-β-CD, their dimensions were barely affected by the irradiation.

Wilcoxon and Provencio employed high-resolution SEC to investigate the effects of aging of Au nanoclusters in the presence of surfactant ligands [132]. They showed that a narrowing of the size dispersion and change in average size can occur with time under ambient conditions. In the presence of various chain length alkanethiols, an etching and size decrease usually occur; in the case of weakly bound alkylated poly(ethylene oxide) surfactants, an increase in size with time is observed. SEC has been also employed to study the growth of a series of discrete sized generations of AuNPs by the heterogeneous deposition of atoms onto monodisperse seed nanocrystals [133]. The size dispersion of each generation was determined from the SEC elution line widths and the spectral homogeneity of the elution peaks.

Recently, SEC has been also employed to characterize the green synthesis of AuNP–nanocluster composite nanostructures directly using trypsin as the linking and reducing agent [134].

4.3 Liquid Chromatography

LC is a rapid, economic technique for separating NPs. Reversed phase liquid chromatography (RPLC) has been applied to the separation of cationic metal cluster compounds [135]. It has been also used to rapidly characterize Au/Pt core/shell NPs produced through seed-assisted synthesis [136] with diameters in the range of 5.3–40.1 nm. It was employed to evaluate the effect of the rate of addition of the reducing agent and the volumes of the added seeds, shell precursor metal ion, and reducing agent solution have on the size of the NPs. A good correlation was found between the sizes of the NPs determined by RPLC and those determined by TEM, which confirms the suitability of this technique for the rapid size evaluation of NPs.

High-pressure liquid chromatography (HPLC) has been also employed for studying the size distributions of nanosize gold clusters dispersed in organic solvents [118,137]. HPLC is sensitive enough to discern changes in hydrodynamic volume corresponding to only two carbon atoms of the passivating agent or metal core size changes of <4 Å.

Helfrich et al. [41] coupled online LC and ICP–MS for separating AuNP standards. Separations by LC showed size-dependent retention behavior with good reproducibility. The results were compared with those of complementary methods, such as DLS and TEM, and showed a good agreement. Their first experiments for separating AuNPs with reversed phase chromatography showed that the analytes behave in a comparable way to that with SEC. Thus, with increasing retention time, NPs that are smaller in size elute first. SDS was added to the buffer to avoid the irreversible adsorption of analytes on the packing material.

Helfrich and Bettmer [4] also employed LC coupled to ICP–MS for the characterization of AuNPs. Such an LC–ICP–MS approach was optimized for monitoring the formation of citrate-stabilized AuNPs. For HPLC separation, different reversed phase columns were connected to the ICP–MS. LC separation required the use of SDS as a modifier in the mobile phase.

Polydisperse samples of AuNPs protected with monolayers of hexanethiolate ligands (C6 MPCs) and with mixed monolayers of hexanethiolate and MUA (C6/MUA MPCs) have been chromatographically separated using C8 120-Å columns and acetone/toluene mobile phase [138]. The smallest MPCs were eluted first in the case of C6 MPCs, whereas the smallest C6/MUA MPCs were eluted last, due to the intermolecular interactions with the stationary phase, dominant for the C6 MPC, and suppressed by the heightened polarity of the monolayer surface of the C6/MUA MPCs, leading to a size-exclusion mechanism.

Samples of polydisperse gold nanoclusters protected with monolayers of NAC have been chromatographically separated by a C18 column (4.6 × 250 mm) using a gradient elution program with a mobile phase of methanol (MeOH)/water containing tetrabutylammonium fluoride ($Bu_4N^+F^-$) and sodium chloride [139]. The separation coupled with absorption, fluorescence, electrochemical detection, and MS proved to be an extremely powerful method to identify core size and spectral feature of the separated AuNCs. The elution order followed with their core sizes from the smallest to the largest. However, this proposed HPLC method is not suitable for separating AuNCs protected by hydrophobic ligands.

Reversed phase HPLC (RP-HPLC)chromatographic separations of Au MPCs in the size range 1−2 nm (diameter) have been detected electrochemically on the basis of double layer charging of the metal-like NPs by the detector electrode [140]. In the case of alkanethiolate MPCs, the elution order from the HPLC column was small clusters first. Significant differences in reversed phase chromatographic retention times for clusters with apparently very similar core sizes over the range of 1- to 2-nm diameter were observed. They also employed fast-scan cyclic voltammetric for the detection of small (<2-nm core diameter) organothiolate monolayer-protected gold clusters (MPCs) separated HPLC [141].

Reverse-phase HPLC with absorption and fluorescence detection and ion chromatography (IC) have been also used for the separation and analysis of polydisperse water-soluble gold nanoclusters (AuNCs) stabilized with *N,N′*-dimethylformamide (DMF) to obtain more monodisperse AuNC fractions [142]. This RP-HPLC methodology was applied to separate AuNC of various Au atoms as well as DMF-stabilized ligands. The composition of separated DMF−AuNCs was confirmed by matrix-assisted laser desorption/ionization time-of-flight mass spectrometry (MALDI-TOF-MS) and electrospray ionization mass spectrometry (ESI-MS).

Moreover, RP-HPLC has also enabled the fractionation of gold nanoclusters stabilized with histidine (His-AuNCs) [143] with core diameters <1 nm by using a binary mixture of methanol and ammonium acetate in water with an optimal solvent elution program. At least nine separated AuNCs species were observed in a single chromatogram, which proves the presence of various AuNCs species in the His-AuNC product.

A separation system for AuNPs has been developed using monolithic silica capillary columns prepared via an in situ sol−gel process [144].

Agglomeration of AuNPs was prevented by the introduction of SDS in the mobile phase, and the NPs were retained by adsorption on the silica surface. Separation results obtained with nonporous silica packed column and monolithic silica columns with and without growth were compared, and it was found that better resolution was achieved with the mesopore-less monolithic column.

4.4 Hydrodynamic Chromatography

Another powerful technique for the characterization of AuNPs is HDC. The separation mechanism relies on size being independent of the coating or the surface charge of the NP. HDC is suitable for sizing NPs within the range of 5–300 nm. HDC separates small particles based on flow through a packed bed of nonporous beads. Particle separation takes place on the basis of the velocity gradients developed within the capillaries between beads. Larger NPs spend less time near the edges of the capillaries, so they are subjected to a higher linear velocity and elute faster than do smaller NPs [111].

A hyphenated methodology has been developed, which utilizes the extensive size separation range of HDC combined with the multielement selectivity of ICP–MS [145]. This has been applied to the analysis of metal-based NPs in environmental samples. AuNPs were employed as calibration standards and internal retention time markers. HDC–ICP–MS shows advantages such as rapid analysis time (<10 min), extended sizing range (5–>300 nm), and limited sample pretreatment requirement.

Pergantis et al. [146] coupled HDC online to inductively coupled plasma mass spectrometry (ICP-MS), and provided an extremely selective and sensitive analytical tool for the detection of AuNPs. Single particle ICP–MS coupled online to HDC proved to be suitable for simultaneously determining NP size, NP number concentration, and NP metal content. Retention times of NPs on the HDC column were correlated to NP size.

HDC coupled with single particle ICP–MS has been also applied to the analysis of AuNPs agglomerated under controlled conditions [147]. When released into the environment, it is expected that AuNPs will form homoagglomerates and heteroagglomerates, which is thus necessary to differentiate them from primary particles. ICP–MS was employed to detect single particles eluted from an HDC column, and to give information about mass and effective diameter.

Finally, Gray et al. [111] employed HDC coupled to ICP–MS for the separation of a mixture of AuNPs with sizes of 5, 20, 50, and 100 nm. Recoveries ranged from 77–96%, and the limit of detection was approximately 5 $\mu g\ L^{-1}$. HDC proved to be capable of separating a dissolved signal from an NP sample.

4.5 Other Chromatographic Techniques

AuNPs have been also characterized by gas chromatography coupled to mass spectrometry [148]. It has been employed to evaluate the presence of residual

SMe_2 after the synthesis of AuNPs by thermolysis of $AuCl(SMe_2)$ in the presence of akylcarboxylic acids under controlled conditions.

Ion-pair chromatography has also proved its efficiency for the separation of samples of charged, polydisperse, water-soluble AuNPs protected by monolayers of NAC and of tiopronin ligands [149]. The NP mixtures were resolved; the smallest NPs eluted first, on an octadecylsilyl (C18) column using isocratic elution with a methanol/water mobile phase containing tetrabutylammonium fluoride ($Bu_4N^+F^-$) and phosphate buffer. The retention mechanism is dominated by ion pairing in either the mobile phase or at the stationary/mobile-phase interface, while size-exclusion effects are not significant.

5. OTHER SEPARATION TECHNIQUES

5.1 Centrifugation

Sedimentation by centrifugation in the solvent used for the reaction is a direct approach to collect the reaction products generated in the synthesis of gold nanorods. Rate-zonal centrifugation employs the differences in hydrodynamic behavior to separate objects. Akbulut et al. [150] developed an aqueous multiphase system (MuPS) as a medium for the rate-zonal centrifugation to separate NPs of different shapes and sizes (Figure 7). They demonstrated the

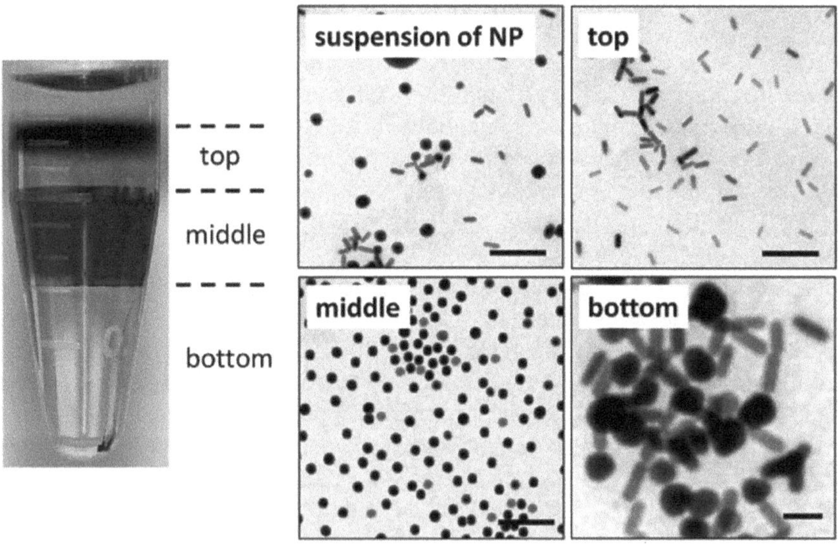

FIGURE 7 Left: Image of an aqueous three-phase system composed of Brij-35 (8.7% v/v), poly(2-ethyl 2-oxazoline) (10% w/v), and Ficoll (11.7% w/v) after sedimenting the reaction products of a AuNP synthesis for 8 min at 16,000 g. Right: TEM images of a suspension of NPs and samples collected from the layers as shown in (A). The scale bar in each image corresponds to 200 nm. *Reprinted with permission from O. Akbulut, C.R. Mace, R.V. Martinez, A.A. Kumar, Z. Nie, M.R. Patton and G.M. Whitesides, Nano Lett., 12:4060–4064, 2012. Copyright 2012 American Chemical Society.*

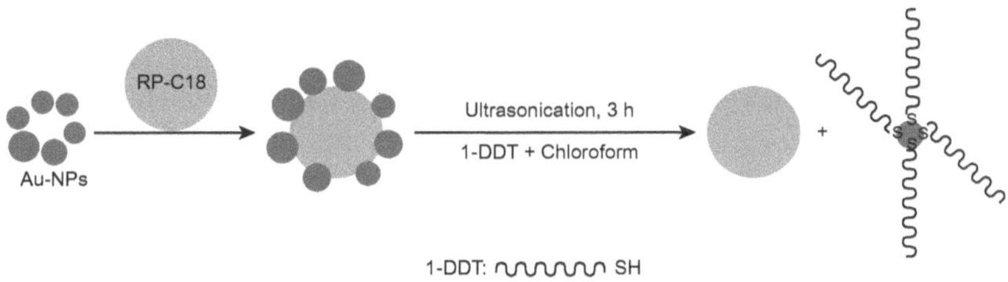

FIGURE 8 Schematic illustration of the strategy for separation and concentration of AuNPs from waters by a two-step extraction. *Reprinted with permission from L. Li, K. Leopold, Anal. Chem., 84 (2012) 4340–4349. Copyright 2012 American Chemical Society.*

separation of the reaction products (nanorods, nanospheres, and large particles). Aqueous MuPSs are phase-separated mixtures of water-soluble polymers and/or surfactants [151]. When submitted to centrifugation, the phases are ordered according to their densities (Figure 8).

The purification of citrate-capped AuNPs by centrifugation is complicated because of NP aggregation upon sample concentration and instability when combined with high ionic strength or nonaqueous solvents [126]. To prevent aggregation of the NPs during centrifugation, Novak et al. [126] exchanged the citrate layer of the NPs by bis(*p*-sulfonatophenyl)phenylphosphine (BSPP), whose adsorption energy is intermediate between citrate and thiol ligands. Density gradient centrifugation was used to separate particles with different densities.

Sharma et al. [151] demonstrated the use of centrifugation for the efficient separation of colloidal gold nanorods from a mixture of nanorods and nanospheres. By analyzing the hydrodynamics of rods and spheres, they were able to elucidate the conditions under which centrifugation leads to the shape separation of NPs. The understanding of the parameters and physics that drives separation is a necessary step in the widespread use of centrifugation to shape-separate particles.

Different particle sizes of AuNC fractions were centrifuged down as a function of the electrolyte concentration or lipophilicity of the solution by adding an electrolyte and an organic solvent in the aqueous medium [32]. Small AuNCs were separated from the polydispersed product under various phosphate/ethanol conditions.

Gradient ultracentrifugation has been employed to separate primary particles, aggregates, and agglomerates from a mixture using a discontinuous glycerol gradient [56]. Density gradient centrifugation has been also employed to separate a mixture of Au nanospheres and nanoplates synthesized using a *Cacumen platycladi* leaf extract [59]. A four-layer step gradient (from 30% to 60% sucrose) was used.

5.2 Extraction

A two-step extraction procedure was proposed by Li et al. [1] for the separation and preconcentration of AuNPs in a size range from 10 to 80 nm from aqueous samples by means of solid phase extraction followed by ligand-assisted liquid extraction. First, AuNPs are loaded onto a reversed phase C-18 (RP-C18) column, and then ligand-assisted extraction into chloroform is performed by using 1-dodecanethiol as a selective ligand for quantitative extraction under ultrasonic conditions. The optimized procedure enabled the separation and preconcentration of AuNPs with an enrichment factor of up to 250, ensuring no changes in size and/or shape of the NPs. Effects resulting from different coatings of the original AuNPs were observed with Tween-20, a nonionic surfactant although nonquantitative adsorption onto the RP-C18 was observed.

The same authors also reported a method capable of extracting AuNPs from real environmental waters based on noncovalent reversible adsorption onto an ionic exchange resin, Amberlite IRN-78, which contains positively charged ammonium groups [7]. MSA was used to bind the surface of the AuNPs through its thiol group, with its two carboxylic acid groups exposed to the medium. The MSA-modified AuNPs were loaded onto the resin due to electrostatic interaction of positively charged ammonium groups and negatively charged carboxylic groups from MSA ligands. Adsorbed NPs were extracted using an 8% (v/v) formic acid solution in methanol by shaking at room temperature. Preservation of size and shape of the NPs was proved by TEM analysis. The electrostatic adsorption onto the resin is reversible so that the resin can be reused for the extraction of NPs after conditioning with 5% (v/v) hydrochloric acid and 5% (v/v) sodium hydroxide solution.

López-Lorente et al. [3] demonstrated the possibility to perform microliquid—liquid extraction of AuNPs by the combination of imidazolium-based ionic liquid with a cationic surfactant. Thanks to the high affinity of the imidazolium group for the AuNPs, it is possible to preconcentrate them into a small volume of ionic liquid. Aqueous samples containing AuNPs were treated with cetyl trimethyl-ammonium chloride and ionic liquid 1-butyl-3-methyl-imidazolium hexafluorophosphate. The system was manually shaken for 30 s and allowed to stand to achieve ionic liquid from water phase separation. Both absorbance and Raman measurements were employed for quantify the samples.

5.3 Cloud-Point Extraction

A promising approach for the separation of AuNPs is CPE. This separation technique is based on the aggregation behavior of surfactant in aqueous solutions as a consequence of variations in temperature. If a solution containing a surfactant at a concentration above the so-called critical micellar concentration is heated until cloud point, the surfactant molecules will form micelles that generate an additional phase. Depending on the nature of the surfactant employed, the surfactant-rich phase may be lighter or heavier than the aqueous phase and can be separated from it [6].

Liu et al. [152] reported the thermoreversible recovery/concentration/separation and dispersion of various nanomaterials in the aqueous phase with CPE by using a low cost and commercially available nonionic surfactant, Triton X-114. 0.2% of Triton X-114 was added into the NPs aqueous suspensions and heated to a temperature above the lower critical solution temperature. NPs were extracted from 10 mL of aqueous suspensions into 0.1 mL of a surfactant-rich phase after the addition of NaCl and centrifugation. The size and shape of NPs were preserved during the phase transfer and storage in the surfactant-rich phase, as the switch over was based on the formation of noncovalently bonded NM micelles.

Nazar et al. [2] reported a viable cost-effective approach employing mixtures of nonionic surfactants Triton X-114/Triton X-100 and subsequent CPE to concentrate and recycle AuNPs in aqueous media. AuNPs have been presynthesized in aqueous phases and stabilized by sodium 2-MES ligands, then dispersed in aqueous nonionic surfactant mixtures. As depicted in Figure 9, heating the NP-micellar systems induced cloud-point phase separations, resulting in concentration of the NPs in lower phases after the transition. UV/vis absorption was employed to quantify the recovery and recycle efficiency after five repeated CPE cycles.

Hartmann and Schuster [6] also developed a simple and cost-effective separation method for the preconcentration and extraction of AuNPs by CPE using the surfactant Triton X-114. Effective separation of ionic gold species and AuNPs with an enrichment factor of 80 was achieved by using sodium thiosulfate as a complexing agent. The proposed method does not affect the size of the particles that undergo extraction.

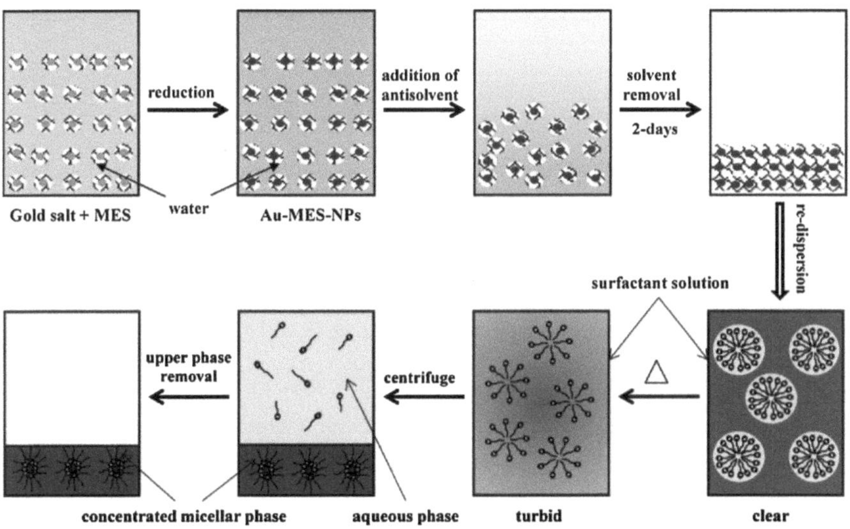

FIGURE 9 Synthesis and CPE of AuNPs. *Reproduced with permission from Ref. [2].*

Park et al. [153] demonstrated a robust and efficient procedure for the shape and size separation of AuNPs through the formation of reversible flocculates by surfactant micelle induced depletion interaction. AuNP flocculates form at a critical surfactant micelle molar concentration, where the number of surfactant micelles is sufficient to induce an attractive potential energy between the AuNPs. Since the magnitude of this potential depends on the interparticle contact area of AuNPs, separation is achieved even for the NPs of the same mass with different shapes by tuning the surfactant concentration and extracting flocculates from the sediment by centrifugation or gravitational sedimentation. Those separated NPs were redispersed by decreasing the surfactant concentration to reduce the effective attractive potential.

5.4 Filtration

Filtration is finding more application in recent years as alternative to other separation techniques such as chromatography, electrophoresis, or centrifugation. Ultrafiltration is a pressure-driven separation process in which porous membranes retain particles larger than the membrane cutoff (ranging from 2−100 nm) [154]. Filtration enables rapid separation, requires small solvent volumes, and is suitable for separating and purifying various NPs [155].

Krieg et al. [156]. reported the development of supramolecular membranes prepared by the self-assembly of compound PP2b (5,5′-bis(1-ethynyl-7-polyethylene glycol-*N,N′-bis*(ethylpropyl) perylene-3,4,9,10-tetracarboxylic diimide)-2,2′-bipyridine) in water/tetrahydrofuran mixtures (Figure 10). The membranes are robust, which enables their application to the size-selective separation of NPs. AuNPs with different sizes and functionalizations (e.g., mercaptopropionic acid, PEG, citrate, and CTAB-stabilized AuNPs) were separated.

5.5 Precipitation

Precipitation of nanoclusters from solution by adding a nonsolvent has been employed by Wilcoxon et al. [118]. This is difficult in the case of nanoclusters in alkane media as the polar, organic nonsolvents (e.g., methanol and ethanol) used to precipitate the alkane thiol-stabilized nanoclusters are immiscible. They found that cold acetone or 2-propanol allows nanocluster precipitation with low recoveries in comparison with that in liquid-phase extraction.

Guo et al. [157] achieved highly effective separation of colloidal gold nanorods/nanoplates from coproduced isotropic NPs by selective precipitation triggered by salt. They found that isotropic NPs hold much better solution stability on adding NaCl solution to the mixtures and remain in the supernatant, while nanorods/nanoplates are subject to aggregation followed by precipitation. This enables the high throughput separation of the differently shaped AuNPs.

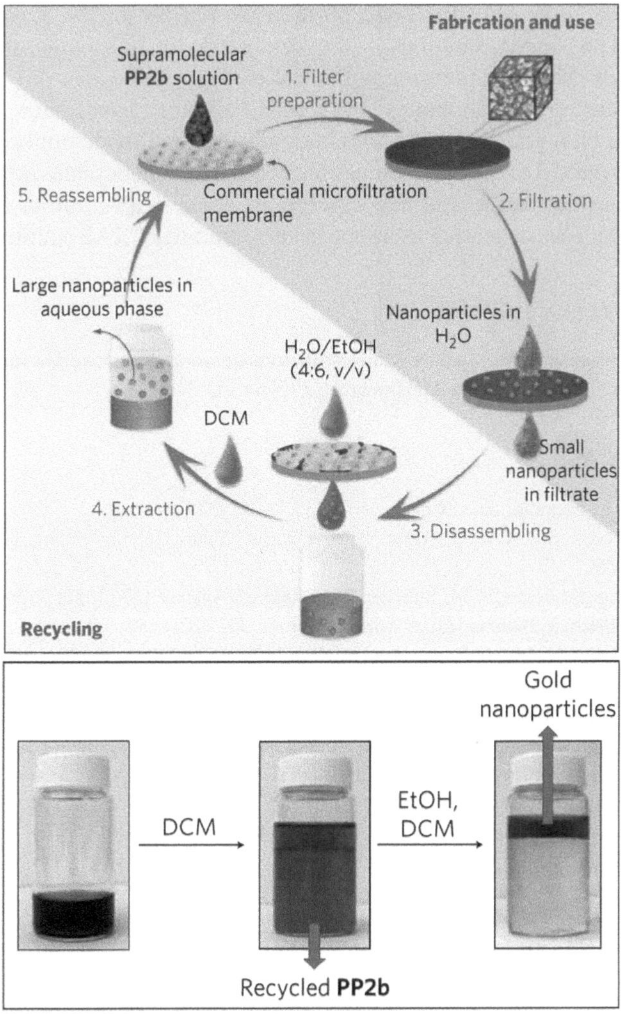

FIGURE 10 Top: Scheme of fabrication, use, and recycling of the supramolecular membrane. Bottom: Photographs showing the retrieval of PP2b and AuNPs from a water/ethanol mixture. *Adapted with permission from Ref. [156].*

Sequential size-selective precipitation (SSSP) can be employed for simple and large-scale preliminary separation of polydisperse water-soluble AuNCs into different fractions with a relatively narrower size distribution. Water-soluble gold nanocluster stabilized with penicillamine ligands (AuNC-SR) have been separated and characterized in aqueous medium by SSSP and MS [158]. The highly polydisperse crude AuNC-SR product with an average core diameter of 2.1 nm was synthesized by a one-phase solution method and subsequently precipitated and separated from larger to smaller ones by progressively increasing the concentration of acetone in the aqueous AuNC solution. The SSSP fractions were analyzed by UV–vis spectroscopy, MALDI-TOF-MS, and thermogravimetric analysis.

Triangular gold nanoplates in CTAB solution have been selectively precipitated on a glass wall in ambient conditions and the nanoplates could be easily recovered by a brief sonication [159]. The nanoplates appear to be isolated from the rest of the nanocrystals of other shapes due to their planar and fixed micelle structures thereon, in the saturated CTAB solution.

ACKNOWLEDGMENTS

The authors wish to thank Spain's Ministry of Innovation and Science for funding Project CTQ2011-23790 and Junta de Andalucía for Project FQM4801.

REFERENCES

[1] L. Li, K. Leopold, Anal. Chem. 84 (2012) 4340–4349.
[2] M.F. Nazar, S.S. Shah, J. Eastoe, A.M. Khan, A. Shah, J. Colloid Interf. Sci. 363 (2011) 490–496.
[3] A.I. López-Lorente, B.M. Simonet, M. Valcárcel, Analyst 137 (2012) 3528–3534.
[4] A. Helfrich, J. Bettmer, Int. J. Mass Spectrom. 307 (2011) 92–98.
[5] G.T. Wei, F.K. Liu, J. Chromatogr. A 836 (1999) 253–260.
[6] G. Hartmann, M. Schuster, Anal. Chim. Acta 761 (2013) 27–33.
[7] L. Li, K. Leopold, M. Schuster, Chem. Commun. 48 (2012) 9165–9167.
[8] A.I. López-Lorente, B.M. Simonet, M. Valcárcel, Trends Anal. Chem. 30 (2011) 58–71.
[9] U. Pyell, Electrophoresis 31 (2010) 814–831.
[10] N. Surugau, P.L. Urban, J. Sep. Sci. 32 (2009) 1889–1906.
[11] M.A. Rodriguez, D.W. Armstrong, J. Chromatogr. B 800 (2004) 7–25.
[12] S.P. Radko, A. Chrambach, Electrophoresis 23 (2002) 1957–1972.
[13] A.I. López-Lorente, B.M. Simonet, M. Valcárcel, Compr. Anal. Chem. 59 (2012) 33–89.
[14] W. Bücking, T. Nann, IEE Proc. Nanobiotechnol. 153 (2006) 47–53.
[15] C.K. Lo, M.C. Paau, D. Xiao, M.M.F. Choi, Electrophoresis 29 (2008) 2330–2339.
[16] H.K. Patra, D. GuhaSarkar, A.Kr Dasgupta, Anal. Chim. Acta 649 (2009) 128–134.
[17] A.I. López-Lorente, M.L. Soriano, M. Valcárcel, Microchim. Acta (2014). http://dx.doi.org/10.1007/s00604-014-1218-5.
[18] A.M. Gole, C. Sathivel, A. Lachke, M. Sastry, J. Chromatogr. A 848 (1999) 485–490.
[19] G.T. Wei, F.K. Liu, J. Chin. Chem. Soc. 45 (1998) 47–52.
[20] F.K. Liu, G.T. Wei, Anal. Chim. Acta 510 (2004) 77–83.
[21] F.K. Liu, Y.Y. Lin, C.H. Wu, Anal. Chim. Acta 528 (2005) 249–254.
[22] F.K. Liu, M.H. Tsai, Y.C. Hsu, T.C. Chu, J. Chromatogr. A 1133 (2006) 340–346.
[23] J.P. Deng, C. Wu, C.H. Yang, C.Y. Mou, Langmuir 21 (2005) 8947–8951.
[24] T.H. Chang, F.K. Liu, Y.C. Chang, T.C. Chu, Chromatographia 67 (2008) 723–730.
[25] F.K. Liu, J. Chromatogr. A 1167 (2007) 231–235.
[26] K.H. Lin, T.C. Chu, F.K. Liu, J. Chromatogr. A 1161 (2007) 314–321.
[27] F.K. Liu, J. Chromatogr. A 1216 (2009) 2554–2559.
[28] F.K. Liu, Anal. Sci. 26 (2010) 1145–1150.
[29] F.K. Liu, Anal. Chim. Acta 694 (2011) 167–173.

[30] W.M. Hwang, C.Y. Lee, D.W. Boo, J.G. Choi, Bull. Korean Chem. Soc. 24 (2003) 684–686.
[31] U. Schnabel, C.H. Fischer, E. Kenndler, J. Microcolumn Sep. 9 (1997) 529–534.
[32] C.K. Lo, M.C. Paau, D. Xiao, M.M.F. Choi, Anal. Chem. 80 (2008) 2439–2446.
[33] J.P. Quirino, S. Terabe, J. Chromatogr. A 791 (1997) 255–267.
[34] S. Almeda, L. Arce, M. Valcárcel, Curr. Anal. Chem. 6 (2010) 126–143.
[35] M.C. Paau, C.K. Lo, X. Yang, M.M.F. Choi, J. Chromatogr. A 1216 (2009) 8557–8562.
[36] Q. Liu, L. Yao, Q. Shen, Z. Nie, M. Guo, S. Yao, Chem. Eur. J. 15 (2009) 12828–12836.
[37] N. Li, S. Zeng, L. He, W. Zhong, Anal. Chem. 82 (2010) 7460–7466.
[38] Y.F. Huang, C.C. Huang, H.T. Chang, Langmuir 19 (2003) 7498–7502.
[39] L. Li, H. Yu, D. Liu, T. You, Analyst 138 (2013) 3705–3710.
[40] D. Zanchet, C.M. Micheel, W.J. Parak, D. Gerion, A.P. Alivisatos, Nano Lett. 1 (2001) 32–35.
[41] A. Helfrich, W. Brüchert, J. Bettmer, J. Anal. Spectrom. 21 (2006) 431–434.
[42] R.A. Sperling, T. Pellegrino, J.K. Li, W.H. Chang, W.J. Parak, Adv. Funct. Mater. 16 (2006) 943–948.
[43] X. Shi, T.R. Ganser, K. Sun, L.P. Balogh, J.R. Baker Jr., Nanotechnology 17 (2006) 1072–1078.
[44] M. Hanauer, S. Pierrat, I. Zins, A. Lotz, C. Sönnichsen, Nano Lett. 7 (2007) 2881–2885.
[45] T.L. Doane, Y. Cheng, A. Babar, R.J. Hill, C. Burda, J. Am. Chem. Soc. 132 (2010) 15624–15631.
[46] C. Hasenoehrl, C.M. Alexander, N.N. Azzarelli, J.C. Dabrowiak, Electrophoresis 33 (2012) 1251–1254.
[47] I. Arnaud, J.P. Abid, C. Roussel, H.H. Girault, Chem. Commun. (2005) 787–788.
[48] J.Y. Kim, H.B. Kim, D.J. Jang, Electrophoresis 34 (2013) 911–916.
[49] X. Xu, K.K. Caswell, E. Tucker, S. Kabisatpathy, K.L. Brodhacker, W.A. Scrivens, J. Chromatogr. A 1167 (2007) 35–41.
[50] D. Zanchet, C.M. Micheel, W.J. Parak, D. Gerion, S.C. Williams, P.A. Alivisatos, J. Phys. Chem. B 106 (2002) 11758–11763.
[51] W.J. Parak, T. Pellegrino, C.M. Micheel, D. Gerion, S.C. Williams, A.P. Alivisatos, Nano Lett. 3 (2003) 33–36.
[52] A. Mishra, S.K. Tripathy, S.I. Yun, Process. Biochem. 47 (2012) 701–711.
[53] T. Tamaki, N. Miyoshi, T. Uehara, Y. Ohya, Chem. Lett. 39 (2010) 1084–1085.
[54] W.J. Qin, Y.L. Yung, Langmuir 21 (2005) 11330–11334.
[55] R. Pasricha, A. Singh, M. Sastry, J. Colloid Interface Sci. 333 (2009) 380–388.
[56] A.M. Keene, K.M. Tyner, J. Nanopart. Res. 13 (2011) 3465–3481.
[57] J. Hou, D.M. Szaflarski, J.D. Simon, J. Phys. Chem. B 117 (2013) 4587–4593.
[58] N.M. Nori, K. Abdi, M.R. Khoshayand, S.H. Ahmadi, N. Lamei, A.R. Shahverdi, J. Exp. Nanosci. 8 (2013) 442–450.
[59] W. Wu, J. Huang, L. Wu, D. Sun, L. Lin, Y. Zhou, H. Wang, Sep. Purif. Technol. 106 (2013) 117–122.
[60] M. Schäffler, M. Semmler-Behnke, H. Sarioglu, S. Takenaka, A. Wenk, C. Schleh, S.M. Hauck, B.D. Johnston, W.G. Kreyling, Nanotechnology 24 (2013) 265103.
[61] A.V. Beskorovaynyy, D.S. Kopitsyn, A.A. Novikov, M. Ziangirova, G.S. Skorikova, M.S. Kotelev, P.A. Gushchin, E.V. Ivanov, M.D. Getmansky, I. Itzkan, A.V. Muradov, V.A. Vinokurov, L.T. Perelman, ACS Nano 8 (2014) 1449–1456.
[62] A.T.J. Kadaksham, P. Singh, N. Aubry, Electrophoresis 25 (2004) 3625–3632.

[63] S. Strobel, R.A. Sperling, B. Fenk, W.J. Parak, M. Tornow, Phys. Chem. Chem. Phys. 13 (2011) 9973–9977.
[64] C.H. Chuang, Y.W. Huang, H.P. Wu, S.Y. Lin, IEEE Int. Conf. Nano/Micro Engineered Mol. Syst. 2011 (2011) 1012–1015.
[65] S. Wang, Y. Qu, Z. Dong, Y. Yang, L. Zhou, Weinadianzi Jishu 48 (2011) 809–815.
[66] C. Leiterer, N. Jahr, S. Berg, G. Bronstrup, A. Csaki, S. Christiansen, W. Fritzsche, GMM-Fachbericht 68 (2011) 42–45.
[67] S.L. Leung, M. Li, W.J. Li, J.D. Mai, Sens. Act. A 178 (2012) 32–39.
[68] S.I. Khondaker, K. Luo, Z. Yao, Nanotechnology 21 (2010) 095204/1–095204/4.
[69] X. Xiong, A. Busnaina, S. Selvarasah, S. Somu, M. Wei, J. Mead, C.L. Chen, J. Aceros, P. Makaram, M.R. Dokmeci, Appl. Phys. Lett. 91 (2007) 063101/1–063101/3.
[70] R.J. Barsotti Jr, M.D. Vahey, R. Wartena, Y.M. Chiang, J. Voldman, F. Stellacci, Small 3 (2007) 488–499.
[71] N. Khanduja, S. Selvarasah, C.L. Chen, M.R. Dokmeci, X. Xiong, P. Makaram, A. Busnaina, Appl. Phys. Lett. 90 (2007) 083105/1–083105/3.
[72] H. Ding, W. Liu, J. Shao, Y. Ding, L. Zhang, J. Niu, Langmuir 29 (2013) 12093–12103.
[73] B.C. Gierhart, D.G. Howitt, S.J. Chen, R.L. Smith, S.D. Collins, Langmuir 23 (2007) 12450–12456.
[74] R. Kretschmer, W. Fritzsche, Langmuir 20 (2004) 11797–11801.
[75] C.T. Harrower, D.R. Oliver, J. Mater. Sci. 41 (2006) 8166–8172.
[76] D.U. Cheon, S. Kumar, G.H. Kim, IEEE Conf. Nanotechnol. 2009 (2009) 261–263.
[77] S.H. Yoon, S. Kumar, G.H. Kim, Y.S. Choi, T.W. Kim, S.I. Khondaker, J. Nanosci, Nanotechnol 8 (2008) 3427–3433.
[78] D.G. Cheon, S. Kumar, G.H. Kim, Appl. Phys. Lett. 96 (2010) 013101/1–013101/3.
[79] A. Zoy, A.G. Nassiopoulou, Phys. Stat. Solidi. A 205 (2008) 2621–2624.
[80] A. Zoy, A.A. Nassiopoulos, A.G. Nassiopoulou, Nanotechnology 18 (2007) 345608/1–345608/7.
[81] J. Tang, D. Tsoukalas, J. Nano Res. 6 (2009) 67–74.
[82] H.J. Lee, T. Yasukawa, M. Suzuki, S.H. Lee, T. Yao, Y. Taki, A. Tanaka, M. Kameyama, H. Shiku, T. Matsue, Sens. Actuat. B 136 (2009) 320–325.
[83] H.J. Lee, T. Yasukawa, M. Suzuki, Y. Taki, A. Tanaka, M. Kameyama, H. Shiku, T. Matsue, Sens. Actuat. B 131 (2008) 424–431.
[84] S.P. Koiry, M.E. Celestin, R. Ratnadurai, P. Veerender, S. Krishnan, D.K. Aswal, S. Bhansali, Electrophoresis 34 (2013) 1182–1188.
[85] B.M.I. van der Zande, J.K.G. Dhont, M.R. Böhmer, A.P. Philipse, Langmuir 16 (2000) 459–464.
[86] H. Yuan, P.S. Cheow, J. Ong, C.S. Toh, Sens. Actuat. B 134 (2008) 127–132.
[87] S.M. Agnihotri, H. Ohshima, H. Terada, K. Tomoda, K. Makino, Langmuir 25 (2009) 4804–4807.
[88] J. López-Viota, S. Mandal, A.V. Delgado, J.L. Toca-Herrera, M. Möller, F. Zanuttin, M. Balestrino, S. Krol, J. Colloid Interface Sci. 332 (2009) 215–223.
[89] K. Makino, H. Ohshima, Langmuir 26 (2010) 18016–18019.
[90] Z.M. Qi, H.S. Zhou, N. Matsuda, I. Honma, K. Shimada, A. Takatsu, K. Kato, J. Phys. Chem. B 108 (2004) 7006–7011.
[91] Y. Yun, M. Jin, J. Long, Chin. Sci. Bull. 50 (2005) 1591–1594.
[92] U. Pyell, W. Bücking, C. Huhn, B. Herrmann, A. Merkoulov, J. Mannhardt, H. Jungclas, T. Nann, Anal. Bioanal. Chem. 395 (2009) 1681–1691.
[93] R.R. Peterson, D.E. Cliffel, Anal. Chem. 77 (2005) 4348–4353.

[94] C. Giddings, Sep. Sci. 1 (1966) 123–125.
[95] F. von der Kammer, S. Legros, E.H. Larsen, K. Loeschner, T. Hofmann, Trends Anal. Chem. 3 (2011) 425–436.
[96] J. Gigault, J.M. Pettibone, C. Schmitt, V.A. Hackley, Anal. Chim. Acta 809 (2014) 9–24.
[97] M. Baalousha, B. Stolpe, J.R. Lead, J. Chromatogr. A 1218 (2011) 4078–4103.
[98] T.J. Cho, V.A. Hackley, Anal. Bioanal. Chem. 398 (2010) 2003–2018.
[99] A. Zattoni, B. Roda, F. Borghi, V. Marassi, P. Reschiglian, J. Pharm. Biomed. Anal. 87 (2014) 53–61.
[100] P. Reschiglian, D.C. Rambaldi, A. Zattoni, Anal. Bioanal. Chem. 399 (2011) 197–203.
[101] K.G. Wahlund, J. Chromatogr. A 1287 (2013) 97–112.
[102] A. Zattoni, D.C. Rambaldi, P. Reschiglian, M. Melucci, S. Krol, A.M. Coto-Garcia, A. Sanz-Medel, D. Roessner, C. Johann, J. Chromatogr. A 1216 (2009) 9106–9112.
[103] J.H. Song, W.S. Kim, D.W. Lee, J. Liq. Chromatogr. Relat. Technol. 26 (2003) 3003–3035.
[104] T. Rameshwar, S. Samal, S. Lee, S. Kim, J. Cho, I.S. Kim, J. Nanosci. Nanotechnol. 6 (2006) 2461–2467.
[105] B. Schmidt, K. Loeschner, N. Hadrup, A. Mortensen, J.J. Sloth, C.B. Koch, E.H. Larsen, Anal. Chem. 83 (2011) 2461–2468.
[106] W. Sermsri, P. Jarujamrus, J. Shiowatana, A. Siripinyanond, Anal. Bioanal. Chem. 396 (2010) 3079–3085.
[107] H. Hagendorfer, R. Kaegi, J. Traber, S.F.L. Mertens, R. Scherrers, C. Ludwig, A. Ulrich, Anal. Chim. Acta 706 (2011) 367–378.
[108] L. Calzolai, D. Gilliand, C.P. Garcìa, F. Rossi, J. Chromatogr. A 1218 (2011) 4234–4239.
[109] D.H. Tsai, T.J. Cho, F.W. DelRio, J. Taurozzi, M.R. Zachariah, V.A. Hackley, J. Am. Chem. Soc. 133 (2011) 8884–8887.
[110] D.H. Tsai, F.W. DelRio, A.M. Keene, K.M. Tyner, R.I. MacCuspie, T.J. Cho, M.R. Zachariah, V.A. Hackley, Langmuir 27 (2011) 2464–2477.
[111] E.P. Gray, T.A. Bruton, C.P. Higgins, R.U. Halden, P. Westerhoff, J.F. Ranville, J. Anal. Spectrom. 27 (2012) 1532–1539.
[112] J. Gigault, T.J. Cho, R.I. MacCuspie, V.A. Hackley, Anal. Bioanal. Chem. 405 (2013) 1191–1202.
[113] J. Gigault, V.A. Hackley, Anal. Bioanal. Chem. 405 (2013) 6251–6258.
[114] L.E. Oppenheimer, G.A. Smith, Langmuir 4 (1988) 144–147.
[115] S. Anger, K. Caldwell, H. Niehus, R.H. Muller, Pharm. Res. 16 (1999) 1743–1747.
[116] C. Contado, R. Argazzi, J. Chromatogr. A 1216 (2009) 9088–9098.
[117] W. Somchue, A. Siripinyanond, B.K. Gale, Anal. Chem. 84 (2012) 4993–4998.
[118] J.P. Wilcoxon, J.E. Martin, P. Provencio, Langmuir 16 (2000) 9912–9920.
[119] T. Siebrands, M. Giersig, P. Mulvaney, C.H. Fischer, Langmuir 9 (1993) 2297–2300.
[120] F.K. Liu, G.T. Wei, Chromatographia 59 (2004) 115–119.
[121] F.K. Liu, Chromatographia 66 (2007) 791–796.
[122] F.K. Liu, Chromatographia 68 (2008) 81–87.
[123] F.K. Liu, Chromatographia 72 (2010) 473–480.
[124] F.K. Liu, Chromatographia 70 (2009) 7–13.
[125] G.T. Wei, F.K. Liu, C.R.C. Wang, Anal. Chem. 71 (1999) 2085–2091.
[126] J.P. Novak, C. Nickerson, S. Franzen, D.L. Feldheim, Anal. Chem. 73 (2001) 5758–5761.
[127] A.M. Al-Somali, K.M. Krueger, J.C. Falkner, V.L. Colvin, Anal. Chem. 76 (2004) 5903–5910.
[128] H. Tsunoyama, Y. Negishi, T. Tsukuda, J. Am. Chem. Soc. 128 (2006) 6036–6037.

[129] F.K. Liu, Y.C. Chang, Chromatographia 74 (2011) 767–775.
[130] F.K. Liu, Chromatographia 75 (2012) 1099–1105.
[131] F.K. Liu, ISRN Chromatogr. 2012 (2012) 970685.
[132] J.P. Wilcoxon, P. Provencio, J. Phys. Chem. B. 107 (2003) 12949–12957.
[133] J.P. Wilcoxon, P. Provencio, J. Am. Chem. Soc. 126 (2004) 6402–6408.
[134] L. Zou, W. Qi, R. Huang, R. Su, M. Wang, Z. He, ACS Sustainable Chem. Eng. 1 (2013) 1398–1404.
[135] W. Bos, J.J. Steggerda, S. Yan, J.A. Casalnuovo, A.M. Mueting, L.H. Pignolet, Inorg. Chem. 27 (1988) 948–951.
[136] F.K. Liu, Y.C. Chang, J. Chromatogr. A 1217 (2010) 1647–1653.
[137] J.P. Wilcoxon, J.E. Martin, P. Provencio, J. Chem. Phys. 115 (2001) 998–1008.
[138] V.L. Jimenez, M.C. Leopold, C. Mazzitelli, J.W. Jorgenson, R.W. Murray, Anal. Chem. 75 (2003) 199–206.
[139] Y. Zhang, S. Shuang, C. Dong, C.K. Lo, M.C. Paau, M.M.F. Choi, Anal. Chem. 81 (2009) 1676–1685.
[140] Y. Song, V. Jimenez, C. McKinney, R. Donkers, R.W. Murray, Anal. Chem. 75 (2003) 5088–5096.
[141] Y. Song, M.L.A.V. Heien, V. Jimenez, R.M. Wightman, R.W. Murray, Anal. Chem. 76 (2004) 4911–4919.
[142] S. Xie, M.C. Paau, Y. Zhang, S. Shuang, W. Chan, M.M.F. Choi, Nanoscale 4 (2012) 5325–5332.
[143] Y. Zhang, Q. Hu, M.C. Paau, S. Xie, P. Gao, W. Chan, M.M.F. Choi, J. Phys. Chem. C 117 (2013) 18697–18708.
[144] Siswoyo, L.W. Lim, T. Takeuchi, Anal. Sci. 28 (2012) 107–113.
[145] K. Tiede, A.B.A. Boxall, D. Tiede, S.P. Tear, H. David, J. Lewis, J. Anal. Spectrom. 24 (2009) 964–972.
[146] S.A. Pergantis, T.L. Jones-Lepp, E.M. Heithmar, Anal. Chem. 84 (2012) 6454–6462.
[147] D. Rakcheev, A. Philippe, G.E. Schaumann, Anal. Chem. 85 (2013) 10643–10647.
[148] M. Yamamoto, Y. Kashiwagi, M. Nakamoto, Z. Naturforsch. 64b (2009) 1305–1311.
[149] M.M.F. Choi, A.D. Douglas, R.W. Murray, Anal. Chem. 78 (2006) 2779–2785.
[150] O. Akbulut, C.R. Mace, R.V. Martinez, A.A. Kumar, Z. Nie, M.R. Patton, G.M. Whitesides, Nano Lett. 12 (2012) 4060–4064.
[151] V. Sharma, K. Park, M. Srinivasarao, Proc. Natl. Acad. Sci. U.S.A. 106 (2009) 4981–4985.
[152] J.F. Liu, Y.G. Yin, G.B. Jiang, Chem. Commun. (2009) 1514–1516.
[153] K. Park, H. Koerner, R.A. Vaia, Nano Lett. 10 (2010) 1433–1439.
[154] W.J. Koros, Y.H. Ma, T. Shimidzu, J. Membr. Sci. 120 (1996) 149–159.
[155] A. Akthakul, A.I. Hochbaum, F. Stellacci, A.M. Mayes, Adv. Mater. 17 (2005) 532–535.
[156] E. Kreig, H. Weissman, E. Shirman, E. Shimoni, B. Rybtchinski, Nat. Nanotechnol. 6 (2011) 141–146.
[157] Z. Guo, X. Fan, L. Xu, X. Lu, C. Gu, Z. Bian, N. Gu, J. Zhang, D. Yang, Chem. Commun. 47 (2011) 4180–4182.
[158] X. Yang, Y. Su, M.C. Paau, M.M.F. Choi, Anal. Chem. 84 (2012) 1765–1771.
[159] T.H. Ha, Y.J. Kim, S.H. Park, Chem. Commun. 46 (2010) 3164–3166.

Chapter 10

Determination of Gold Nanoparticles in Biological, Environmental, and Agrifood Samples

Ángela Inmaculada López-Lorente and Miguel Valcárcel*
Department of Analytical Chemistry, University of Córdoba, Córdoba, Spain
Corresponding author: E-mail: qa1meobj@uco.es

Chapter Outline

1. INTRODUCTION

Nanotechnology is raising growing interest in the analytical community. Gold nanoparticles (AuNPs) have aroused great attention and offer a huge potential for their application in many fields, such as nanomedicine and nanotechnologies, as they provide several opportunities in imaging, diagnostics, and therapies [1].

Gold Nanoparticles in Analytical Chemistry. http://dx.doi.org/10.1016/B978-0-444-63285-2.00010-9

AuNPs show great potential as they can be easily tailored to a desirable size from 0.8 to 200 nm. Moreover, they are easy to prepare and can be functionalized with proteins or molecular species without altering the biological activity of the conjugated species [2]. They show unique optical properties, as a consequence of the collective oscillation of electrons "surface plasmon oscillation," which occurs at a frequency usually in the visible region and results in strong surface plasmon resonance (SPR) absorption. This property enables the use of AuNPs in applications such as Raman sensors [3] or photoelectrochemical materials [4]. They are widely used in biomedical imaging and diagnostic tests by virtue of the high chemical stability of Au(0) against, for example, cancer. They have also been used as catalysts on the grounds of their ability to lower the activation energy barriers or increase the rates of some reactions—and the yields of the target products as a result [5]. AuNPs are suitable for many biological applications [6], such as chromophores for immunoreactions and nucleic acid hybridization, gene delivery into cells, or photothermal agents in hyperthermia.

Engineered nanomaterials are being manufactured in increasing quantities and are finding applications in a wide range of products. The increasing use of engineered nanoparticles (NPs) in industrial and household applications will probably lead to the release of such materials into the environment as well as their accumulation in biological systems. Millions of tons of NPs are likely to be produced annually from both manufacturing and as by-products of human activity through unintended emissions [7]. However, little is known about the occurrence and fate of NPs, partly because there are still a few analytical methods for their quantification in complex matrices, that is, water, soil, food, or biological matrices.

The proliferation of nanotechnology has prompted concerns about the safety of these nanomaterials on human health and the environment; the exposure of workers and consumers and the release into the environment is unavoidable. Humans will be exposed to NPs, for example, as a result of migration of NPs from food packaging into food. Moreover, both unintended—from waste, wastewaster, or sludge—and intended, in the case of remediation, release of NPs into the environment will expose humans to NPs, for example, via drinking water or via entering the food chain [8]. Therefore, it is mandatory to understand the behavior of engineered NPs in food materials, consumer products, and environmental matrices, as well as their potential toxicity to humans and the environment.

To ensure the safety of NPs and reduce environmental and (eco)toxicological impact, the development of regulatory frameworks and an appropriate risk assessment is urgent; this comprises the development of validated and easily applicable analytical methods [9]. Until now, the detection, characterization, and quantification of engineered NPs, especially in complex matrices, are still a challenge, and the analytical methods required are most often laborios, expensive, and/or demand specialized and trained operators [10].

Those protocols applied to the determination of chemicals are usually not applicable to NPs since they do not provide complementary information about physicochemical parameters (e.g., size, shape, agglomeration, aggregation state, and surface area) that are important for NP characterization, while many techniques are only suitable for specific samples or sample matrices or limited to a certain range of concentration or size [10].

2. ANALYTICAL REQUIREMENTS

A number of challenges have to be addressed when analyzing environmental, biological, or food materials, including extraction and sample preparation; the presence of analytical artifacts; lack of reference materials; and problems of distinction between natural and engineered NPs.

Effectively monitoring NPs in environmental samples entails the use of analytical methods capable of detecting environmentally relevant concentrations, which fall in the nanogram per liter or the picogram per liter range, as well as avoiding the potential interference of natural NPs frequently present in environmental samples [7]. This could require a number of separation and enrichment steps prior to their analytical determination. Thus, analytical procedures for the determination of NPs in different matrices usually include sampling, sample treatment, separation of the NPs from the matrix, and finally, their detection.

Sampling is a critical step in the characterization/determination of NPs, since it can alter the dispersion state of NPs. This possibly leads to either further aggregation or partial disruption of the aggregation already present. Chemical or physical changes may affect NP stability and dispersion, since they are usually present in the environment in dynamic nonequilibrium. The fate of NPs in the environment (water column and sediments) and their separation from environmental matrices are greatly affected by their aggregation or stability, as well as by pH or ionic strength. The presence of natural NPs and organic matter complicates analytical procedures. Because of these difficulties, adequate and undisturbed separation of NPs from environmental matrices is necessary [11].

Further more, adsorption of inorganic NPs onto sample bottles has to be taken into account. For example, it has been observed that AuNPs show higher adsorption behavior in sample introduction and preparation systems if they are made of materials other than quartz or borate glass, while porous plastic walls lead to a high loss of NP concentration [12]. Thus, methods causing minimal perturbation from sampling to analysis are required.

In situ determination of the NPs would reduce (or overcome) those drawbacks; nevertheless, few methods enable this due to the lack of sensitive instrumentation for this purpose. Depending on the technique, different sample preparation or digestion is needed. Results obtained from pretreated or digested samples can differ from the state in which NPs are characterized in situ [13]. Moreover, NPs can aggregate or dissolve over time, so an effort must be made to preserve the original state of the sample.

When measuring NPs in different matrices, information about size distribution and properties of NPs is necessary in addition to data of the concentration present in the sample. Thus, a range of analytical techniques will be required for the complete characterization of the sample since there is no single technique that can provide such information [8]. Many of the techniques usually employed for the characterization of NPs, such as microscopies and dynamic light scattering (DLS), are not applicable in the case of complex matrices, since the requirements imposed by such kinds of samples substantially differ from their analysis in pure or neutral media. In complex matrices, such as environmental, biological, or food related, the elemental compositions will vary from one to another, while the sample contains more than one type of NPs. Many of them are heterogeneous in terms of dispersion and size distribution of NPs.

Biological measurements require the use of analytical methods that enable size measurements, since toxicity of NPs will depend on the size and shape of NPs. Different techniques, such as electron microscopy (EM), DLS, or ultracentrifugation, may provide different values of size and size distribution. The determination of the absolute size of NPs is still a challenge. Correct size measurements are difficult, and lead to artifacts, depending on the technique and the medium (e.g., coating not visible in EM). Moreover, one of the biggest challenges in the environmental analysis of NPs is that all their dimensions are important; environmental effects are related to NP chemical composition, dimensional cross-section, and volume [11].

The lack of reference nanomaterials and standards aggravate the problem [14]. Only a few size- but not mass- and number concentration-certified NP reference materials are available (e.g., National Institute of Standards and Technology reference materials 8011–8013 [15], which consist of colloidal suspensions of spherical AuNPs having primary particle diameters of nominally 10, 30, and 60 nm, respectively). Thus, the validation of analytical methods for the determination/characterization of NPs is challenging, and often multimethod approaches are needed to provide reliable data to assess the limitations of the different techniques [16].

It is also difficult to discriminate between engineered NPs and those naturally present in such samples. There is an increasing interest in determining concentrations of NPs in both consumer products and the environment since the number of engineered NPs reaching the environment is unknown [17]. The proposal to solve this problem comprise nanomaterial labeling, for example, isotopic labeling and tracking of metal NPs. Typically, NPs have been detected in cells and tissues labeled with radioactive, magnetic, or fluorescent markers [18]. However, common inorganic NPs do not have a convenient label that enables their detection. In addition, discrimination between NPs, submicron particles, and soluble species with the same nominal elemental composition is usually a difficult task [19].

In addition, it has to be borne in mind that the properties of NPs may differ among each other and from their (chemically identical) bulk material and may vary over time or in dependence on the surrounding matrix [10]. It is crucial to distinguish between the particulate and the dissolved fraction, since they may have a strong impact on toxicity, in order to gain information on the fate and transport of the particles and ionic forms. The dissolved fraction can reach a stable status, and complete dissolution could be possible [20].

In the case of food samples, the form of the nanomaterials present in the sample may differ from the nanomaterial that was originally added to food. Therefore, when analyzing NPs in different food matrices, it is necessary to determine the physical and chemical properties of the NPs in addition to the chemical composition and concentration [21].

3. ANALYTICAL TOOLS AND PROCEDURES FOR THE ANALYSIS OF AUNPs

3.1 Microscopic Techniques

Microscopic techniques have been intensively used for AuNP characterization. They include scanning and transmission electron microscopy (SEM and TEM) and atomic force microscopy (AFM) and scanning tunneling microscopy (STM) among the scanning probe microscopes.

Electron microscopies (EMs) enable visualization and characterization of nanosized objects, and provide a great resolving power, as a consequence of the application of an electron beam with a wavelength below the nanometer range. Two main EMs are employed for the characterization of nanomaterials, namely, SEM and TEM. Such techniques are usually employed for the physical characterization of colloidal AuNPs [22], since metals scatter light efficiently. However, many biological compounds (e.g., proteins) are invisible to TEM, without heavy metal staining procedures, as they do not sufficiently deflect an electron beam.

In SEM, the image of the sample is created by scanning the sample surface with a low-energy beam of electrons (1–30 keV), and those scattered off the sample are detected. This technique allows acquiring images with a great depth of field, with the whole micrograph in focus [23]. In the case of focused ion beam SEM (FIB-SEM), the ion beam can cut into the material, and a solid specimen can be sectioned for three-dimensional (3D) imaging [24].

In contrast, in TEM, a high-energy electron beam (80–300 keV) is transmitted through a very thin slice of the sample. Part of the beam is scattered from the sample, while some electrons can be transmitted with little change in energy or direction.

These techniques provide information about the type and characteristics (shape, size, etc.) of the NPs. However, one of their most important shortcomings is the low representativity of the results due to the small sample volume that is analyzed; this has an impact on the limit of detection and quantification (e.g., particle-number concentration). Thus, ideally, a suitable number of replicates have to be prepared for each sample and, if the sample is not concentrated, separation and preconcentration steps prior to microscopic analysis are advisable. In this sense, a challenge is the analysis of samples with very low concentrations of NPs, such as foodstuffs, in which nanomaterials can arise from migration from packing materials. Autometallographic silver enhancement has been proposed as a histochemical tool for tracing a variety of metal containing nanocrystals (e.g., pure gold and silver nanoclusters), which consists of the enlargement of NPs to dimensions that can be observed in the microscope [25]. For biological samples (e.g., cells and tissues), methods for sampling and quantitative analysis by TEM have been developed [26].

The most common pitfall of microscopy is that visual observation is often biased by subjective evaluation of image content. A number of studies reported the use of automatic tools for quantitatively analyzing nanostructures in food substrates [27].

In SEM, organic samples (e.g., biological or foodstuffs) need to be covered with a layer of a conductive substance, and the energy of the electron beam should be adjusted to avoid damaging the specimen. Bulk samples can be introduced, while in TEM, they need to be cut into thin sections; sample preparation in SEM is easier than in TEM.

Both conventional SEM and TEM methods usually need complex processing of the specimen before introducing the sample into the vacuum chamber of the electron microscope for imaging. Sample preparation involves drying, which might result in aggregation of the sample and thus a false representation of the sample size in tissue [19]. This shortcoming can be amended partially by the use of environmental SEM (ESEM) and environmental TEM (ETEM), being possible to maintain a higher vapor pressure around the sample, which can be imaged while hydrated. It is also possible to encapsulate hydrated samples behind thin electron-transparent membranes or to image under cryogenic conditions [23].

As previously mentioned, common EM equipments operate under vacuum conditions; thus, sample preparation in the case of complex matrices, such as biological tissues and food samples, is a key step. Different procedures have been proposed for dealing with hydrated samples, such as chemical fixation followed by dehydration or drying for conventional SEM or TEM; employment of ESEM or liquid capsules; and freezing followed by cryo-TEM or cryo-SEM [23].

Fixation of the specimen (i.e., with glutaraldehyde or osmium tetroxide [28], as shown, e.g., in Figure 1) followed by chemical dehydration and drying for SEM is useful, for example, for analyzing food with cellular structures to minimize changes in the matrix and observe NPs within cells and tissues. Dehydration of the sample with ethanol can be also applied for food matrices since it helps to preserve the sample, although it cannot be used for products

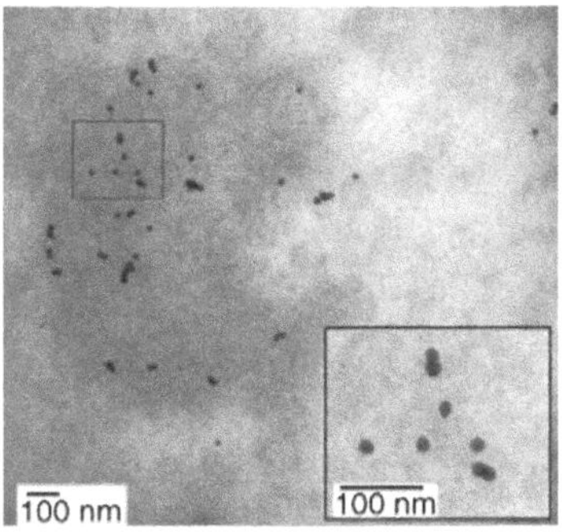

FIGURE 1 Electron micrograph of a cell containing nanoparticles (NPs). Cells were exposed to NPs for 24 h, fixed with osmium tetroxide, sectioned, and visualized with an electron microscope. *Adapted with permission from Ref. [28].*

with a high content of saturated fats or water-soluble carbohydrates [23]. Drying of the sample prevents loss of components; it can be accelerated by partially absorbing the liquid, which reduces agglomeration artifacts in dried suspensions [29]. It can be applied to food samples with a liquid matrix whose density is low so that a thin film is formed. To measure in TEM, the sample needs to be very thin so that electrons can pass through it. Complex samples can be embedded in a resin and cut into thin sections by ultramicrotomy [30] or FIB milling [23].

When using ESEM with the liquid cell technique, the specimen can be imaged without any sample treatment, although sometimes centrifugation or other treatments to improve specimen adhesion can be used [31]. Freezing of the sample avoids sample alteration, for example, in the case of biological samples or food matrices, in which drying will change the structure of the sample since water forms a part of it. There are different freezing methods available, depending on the dimension of the sample (e.g., plunge freezing, contact freezing, spray freezing, and high-pressure freezing). By cryo-TEM, it has been possible to image with high-resolution 15-nm AuNPs with a 0.2-nm lateral resolution [32].

EMs as analytical techniques are not sufficient for the complete characterization of nanomaterials, since they do not provide chemical information. Apart from providing information on particle aggregation, dispersion, size, structure, and shape, additional elemental information can be obtained by coupling EMs to energy-dispersive spectrometry [33] or to electron energy loss spectrometry [34].

AFM has been also employed for the characterization of AuNPs [35]. It relies on tapping a particle (either in solution or dried to a surface) and so has a limited resolution of flexible compounds (e.g., proteins), which may move under the force applied by the instrument tip. With this technique, it is possible to analyze samples under moist conditions or even in liquids, which affords minimum perturbation. However, under liquid conditions, particles can eventually stick to the cantilever, which leads to imaging artifacts. This smearing effect could be minimized by using a noncontact scanning mode where the tip is not touching the NPs but only feels its forces.

AFM offers the capability of tridimensional visualization and provides information on many physical properties including size, morphology, surface texture, and roughness. Statistical information, such as size, surface area, or volume distributions, can be also obtained. AFM can measure high and lateral distances; however, in the latter case, when the size of the NPs is similar or smaller than the radius of curvature of the tip, a systematic bias is introduced such that height measurements are more accurate [36].

STM imaging has been one of the key approaches to determine nanoscale domains of AuNPs protected by a binary mixture of ligand molecules [37]. Imaging of NP surface faces is difficult due to steep surface curvature, complex molecular structures, and the possibility of imaging artifacts in the same size range.

3.2 Spectroscopic Techniques

Optical analysis techniques have proved to be useful for the characterization of NPs, in particular when taking advantage of the pronounced SPRs attributed to AuNPs. AuNPs have a characteristic red color as a consequence of the collective oscillation of the electrons in the conduction band, known as surface plasmon oscillation [5]. When the wavelength of light is much larger than the NP size, it can set up standing resonance conditions that causes the free electrons in the metal to oscillate. As the wave front of the light passes, the electron density in the particle is polarized to one surface and oscillates in resonance with the light's frequency, which causes a standing oscillation [5]. In such NPs having an anisotropic shape, such as a rod, two plasmon bands occur as a result of electron oscillation along the two axes [38] (Figure 2).

Haiss et al. [39] reported a method to calculate the particle diameter from UV−Vis spectra and established analytical relationships between the extinction efficiency and diameter; this allows the determination of particle concentration.

In addition, a method for the evaluation of the average size of AuNPs based on the fitting of their UV−Vis spectra by the Mie model for spheres has been proposed [40]. The method gives good results and uses a calibration of the dumping frequency of the SPR and accounts for the presence of nonspherical AuNP in solution by the Gans model for spheroids. It has been successfully applied to free and functionalized AuNPs in various solvents with diameters in the 4- to 25-nm range. The method can be applied to determine not only the size but also the concentration and aggregation level of AuNPs, which are key points for their applications.

Optical biosensors, based on SPR phenomenon, provide a real-time and automated analysis with a relatively high capacity. For example, a

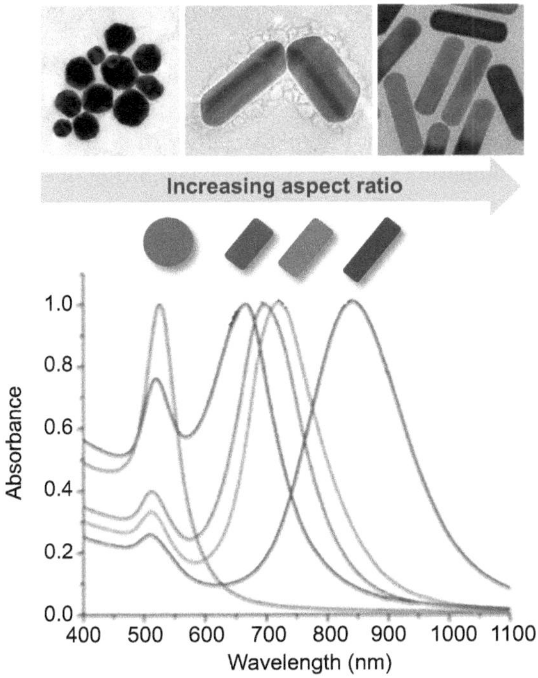

FIGURE 2 UV–Vis visible absorption of gold nanoparticles of different sizes and shapes.

metallothionein-based SPR sensor for the detection of silver NPs has been developed [41]. Similar SPR sensors could be applied to AuNPs.

AuNPs also possess fluorescence properties [42]; however, due to their low quantum yields, such properties have not attracted great attention, despite showing excellent behaviors as well as nonphotobleaching. The fluorescent features of AuNPs were investigated in solution by Xu et al. [43] by using single molecule spectroscopy–fluorescence correlation spectroscopy. AuNPs showed a high photosaturation feature and stronger brightness per particle when illuminated with stronger light.

In general, AuNPs do not show midinfrared (2.5–25 μm) absorptions, unless they are functionalized with organic ligands attached to the particle surface, which gives rise to characteristic infrared (IR) spectra. For example, AuNPs with cetyltrimethylammonium bromide (CTAB) molecules adsorbed in ionic liquid media have been characterized by Fourier transformation IR spectroscopy [44].

IR-attenuated total reflection (ATR) spectroscopy has been recently proposed as an innovative analytical tool that is applicable for the characterization of AuNPs. It provides a straightforward and nondestructive characterization strategy [45]. The adsorption of water molecules at AuNPs has been utilized for in situ monitoring of the synthesis of AuNPs via surface-enhanced

IR absorption-ATR (SEIRA-ATR) spectroscopy. As NPs were formed and deposited at the surface of the ATR waveguide, an enhancement of the water absorption features was observed; this correlated with the amount of AuNPs present at the surface (Figure 3). Moreover, the generic suitability of SEIRA-ATR spectroscopy for the investigation of AuNP solutions has been studied via the evaluation of effects such as the addition of salts or variation of the pH on the aggregation state of AuNPs in solution, as well as monitoring the involved sedimentation processes.

In addition, Raman spectroscopy has been applied for the characterization and quantification of AuNPs previously extracted by ionic liquids [46]. Duval [47] reported that Raman spectra of Au nanomaterials showed a band in the low-frequency range due to the excitation of the acoustic vibrations of the roughness features of the NPs. Among the two types of vibrational modes—torsional and spheroidal—only, the second one can scatter light. Such

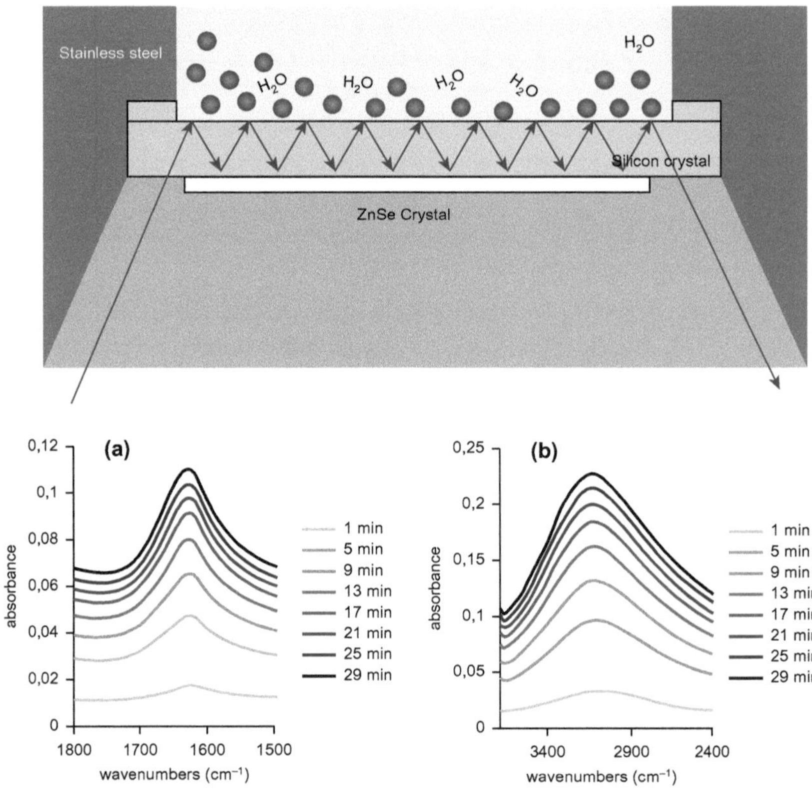

FIGURE 3 Top: schematic of the attenuated total reflection (ATR) configuration used for monitoring the synthesis of gold nanoparticles (AuNPs). Bottom: in situ surface-enhanced infrared absorption-ATR spectra obtained during the synthesis of AuNPs inside the ATR unit in an aqueous environment: (a) H_2O bending mode; (b) OH stretching mode. *Reprinted with permission from Ref. [45]. Copyright 2014 American Chemical Society.*

spheroidal vibrations can be detected by Raman scattering owing to their strong interaction with the dipolar plasmon. The fundamental radial mode (breathing mode) is described to be the more dominant one in time-resolved experiments with AuNPs [48].

Particle-induced X-ray emission has been applied to the determination of NPs [49] (AgNPs and SiO_2 NPs), although its use could be extended also to AuNPs. It is a technique that can be used to detect trace metallic contaminants and/or NPs in liquids without sample preparation. In this sense, NPs have been determined using this technique from food matrices such as mineralized water, coffee, or milk. Moreover, it has been also applied to the trace detection of NPs in rat lungs [50].

Dynamic light scattering (DLS) is a very commonly used method to determine particle size and their aggregation in suspensions [51]. The advantages of DLS are the rapid and simple operation, readily available equipment, and minimum perturbation of the sample. A drawback of this technique is that it cannot be appropriate for in situ measurements or in a digested biological sample, because of multiple scattering issues [52]. DLS is sensitive to flexible biological molecules, such as polymers, proteins, and antibodies, causing frictional drag [53], which can influence the rate of the particle's motion under Brownian diffusion. Moreover, DLS may be influenced by NP aggregation, although further analysis by TEM can ensure that the measured sizes represent single NPs and not an ensemble of them.

In addition, other light scattering techniques, such as static light scattering [54] or resonance light scattering, have been employed for investigating the aggregation of biotinylated bovine serum albumin AuNPs [55].

Mass spectroscopy (MS) has gained much importance in the last five years as a powerful tool for reliable analytical characterization of NPs [56]. It is useful for the identification and quantization of AuNPs at the subnanometer level. For example, *N*-acetyl-L-cysteine monolayer-protected gold clusters have been detected by matrix-assisted laser desorption and ionization time-of-flight mass spectrometry (MALDI-TOF-MS) [57].

Inductively coupled plasma (ICP) spectrometry is commonly used to determine the total element concentration of colloidal solutions as well as for the size characterization of NPs [58]. The single-particle mode allows the determination of discrete ion clouds generated by the atomization of a single NP in the plasma; the signal is proportional to the size of the particle. A very dilute solution is introduced so that statistically one particle reaches the plasma within a certain time regime.

ICP–MS is a sensitive and accurate technique for the determination of AuNP solutions. It has become one of the most employed tools for the determination of the total concentration of inorganic NP suspensions [59]. There is a controversy regarding the determination of mass concentration by direct application of suspensions via the nebulizer and the spray chamber. For daily routine analysis, fast and simple sample preparation protocols are needed, and in the case of fractionation techniques coupled online to ICP–MS (e.g., field-flow fractionation (FFF), hydrodynamic chromatography (HDC), or size-exclusion chromatography (SEC)), a digestion of the sample prior to quantification is impossible.

Fabricius et al. [10] investigated the suitability of three sample preparation approaches, namely, microwave-assisted digestion, acidification of the suspensions, and direct measurement via ICP–MS without further preparation. Using a microwave system, suspensions were digested in a two-step program of a temperature ramp of 60 min to 240 °C, followed by an irradiation of 30 min at 240 °C. AuNPs were digested in a mixture of HNO_3 and HCl. Digested suspensions were compared with acidified samples (same mixture but not digested) and directly measured NP suspensions, which were diluted in ultrapure water. For AuNPs, differences between the microwave-assisted digestion and a gravimetric approach were found, so concentrations were additionally determined by graphite furnace-atomic absorption spectrometry. The authors concluded that without any further sample preparation, it is very likely that imprecise results and/or instabilities of the measurements occur; microwave-assisted digestion is suggested to be a reliable method.

Single-particle ICP–MS is an increasingly applied technique, although it demands trained operators since several special considerations (e.g., instrumental parameters, sample characteristics, handling, and data interpretation) have to be taken into consideration [60]. The single-particle mode ICP–MS method has been applied to the determination of antibody-conjugated AuNPs [61].

Direct determination of AuNPs in colloid solutions without previous digestion/dissolution, avoiding the use of hazardous reagents and reducing the time of the analysis, has been proposed by Allabashi et al. [62] (Figure 4). ICP–MS measurements are strongly dependent on the matrix of the sample, while in the case of suspended NPs, the stability of the suspension plays an important role. They investigated different approaches for dilution of samples with deionized water, 1% hydrochloric acid, or 1% trisodium citrate, and showed that there is no necessary preliminary digestion of AuNP colloid solutions with aqua regia. This leads to better sensitivity and lower analysis cost.

Immunoglobulin-conjugated AuNPs have been also analyzed by ICP–MS after digestion in aqua regia [63]; oligonucleotide-linked AuNPs were also treated similarly [64]. ICP–MS combined with other separation techniques, such as liquid chromatography (LC) or gel electrophoresis, has been also employed for AuNP size characterization [65]. However, although being an excellent quantitative tool, ICP–MS is a destructive technique, and cannot differentiate between NPs adsorbed onto the surface of the cell and internalized into them [38].

ICP–optical emission spectrometry (OES) has been also employed for the quantification of the AuNPs in solution [35] as well as graphite furnace-atomic absorption spectroscopy (GFAAS), which has been also used for evaluating NP dimensions based on the difference in atomization kinetics for true and

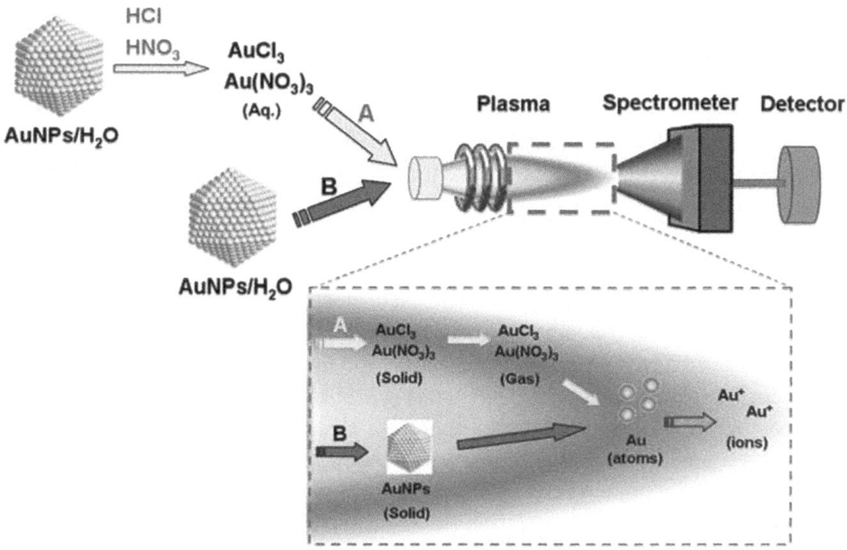

FIGURE 4 Scheme of the processes involved in the inductively coupled plasma–mass spectrometry analysis of gold nanoparticles (AuNPs) with (A) and without (B) previous gold dissolution. *Reproduced with permission from Ref. [62].*

colloidal solutions [66]. The method makes it possible not only to analyze the elemental composition of colloidal NPs but also to determine their dimensions.

3.3 Separation Techniques

Different chromatographic techniques have been employed for the separation of AuNPs, such as SEC, high-performance liquid chromatography (HPLC), or HDC. SEC has been employed for the analysis of AuNPs in aqueous media [67]. It can be used to characterize nanometer-sized spherical AuNPs ranging from 3 to 20 nm [68]. A drawback of this technique is the irreversible adsorption of NPs onto the stationary phase due to the high surface activity of NPs. The use of surfactants, such as sodium dodecyl sulfate as additives to the mobile phase, prevents aggregation and adsorption of the AuNPs [69].

Reversed-phase HPLC has been also employed for the separation of AuNPs protected by a nonpolar surface-coating layer (alkanethiolate-AuNPs) by using a diode array detector [70]. In addition, HPLC in combination with cyclic voltammetric detection has been employed to separate AuNPs with a core diameter of <2 nm [71]. Moreover, HDC in combination with other techniques, such as ICP–MS, has been also employed for the characterization of AuNPs [72].

Capillary electrophoresis (CE) has emerged as a useful technique, and is potentially faster than EM. It has fewer surface effects than in other techniques (e.g., SEC), with a low consumption of sample and reagents [73]. AuNPs have been characterized by capillary zone electrophoresis (CZE) according to their size [74].

To be separated, metallic NPs must present a stable charge in the electrophoretic system. The charge can be the result of a colloidal stabilization with a ligand or can result from the introduction of charged organic groups (e.g., through formation of thiol derivatives) [75]. Surfactants have been widely employed as stabilizers for the size-selective preparation of AuNPs [76], since they prevent agglomeration by electrostatic repulsion. Another strategy is functionalization of AuNPs with thiol derivatives [77].

A drawback of CE is the low sensitivity of UV detection systems because of the small dimensions of CE capillary. Thus, a method for enhancing detection sensitivity in CE, called reversed electrode polarity stacking mode, has been applied to the characterization of AuNPs [78].

Gel electrophoresis has also been used for the characterization of AuNPs, for example, using an isoelectric focusing technique in a polyacrylamide pH gradient gel [79] or agarose gel preparative electrophoresis [80].

FFF is a technique that is being widely used for separation of AuNPs [81]. For example, an analytical platform coupling asymmetric flow field-flow fractionation (AF^4) with multiangle light scattering (MALS), DLS, and ICP–MS has been used for the separation and quantitative determination of the size and mass concentration of AuNPs between a 10- and 60-nm diameter in aqueous suspensions [82].

Common methods for the separation of NPs include centrifugation and ultracentrifugation [83]. However, these methods are prone to irreversible agglomeration/aggregation, which hampers the effective evaluation of the original morphology of the NPs in the environmental or biological system. In addition, their efficiency depends on the characteristics of the NPs, with those smaller showing lower yields. Other methods such as diafiltration [84] and ultrafiltration [85] rely on controlled-pore-size materials and can suffer from fouling.

Cloud-point extraction (CPE) has been also employed for the preconcentration and extraction of AuNPs. CPE presents advantages over the traditional liquid–liquid approach involving nonstructured molecular solvents [86]. High extraction efficiencies and preconcentration factors can be achieved, and it avoids the use of expensive and toxic solvents. Liu et al. [87] introduced an interesting approach to recover various NPs by exploiting CPE. Triton X-114 has been employed for this purpose as well as mixtures of Triton X-114 and Triton X-100.

3.4 Other Techniques

AuNPs have been also determined by means of electrochemical methods. For example, populations of gold clusters with a core of Au_{147} and Au_{38} have been measured by the transition time of chronoamperometry [88]. A direct amperometric method has been used as one of the detectors for AuNP separation by LC [89]. Cathodic linear sweep voltammetry can be also employed for the analysis of AuNPs. Once adsorbed on a graphite electrode, the AuNPs can be oxidized to $AuCl_4^-$, and those anions are then reduced and deposited on the surface of the electrode by a linear sweep of the electrode potential. A cathodic peak appears as a result of the reduction of $AuCl_4^-$ to Au metal [90].

Stripping analysis is an extremely sensitive electrochemical technique. When metallic gold in the AuNPs is chemically oxidized to Au(III), the Au(III) ions can be deposited on the electrode by electrochemical reduction, and, then, the thin layer of Au metal can be stripped by anodic sweeping the electrode potential from 0V to 1.0V [90]. Moreover, a method to determine the concentration of AuNPs in colloidal solutions based on silver electrodeposition on the particles adsorbed on a carbon paste electrode has been proposed by Hernández-Santos et al. [91]. A pretreated carbon paste electrode was immersed in colloidal AuNP solution, and then activated to form oxides on the surface of the AuNPs. Silver electrodeposition followed by anodic stripping voltammetry was performed; a linear relationship was obtained between the area of anodic stripping peak and the number of colloidal AuNPs.

Pumera et al. [92] developed a voltammetric method for the direct determination of AuNPs, based on the adsorption of AuNPs on the rough surface of a graphite–epoxy composite electrode followed by their electrochemical oxidation, which results in $AuCl_4^-$ ions that are detected by differential pulse voltammetry (DPV) (Figure 5).

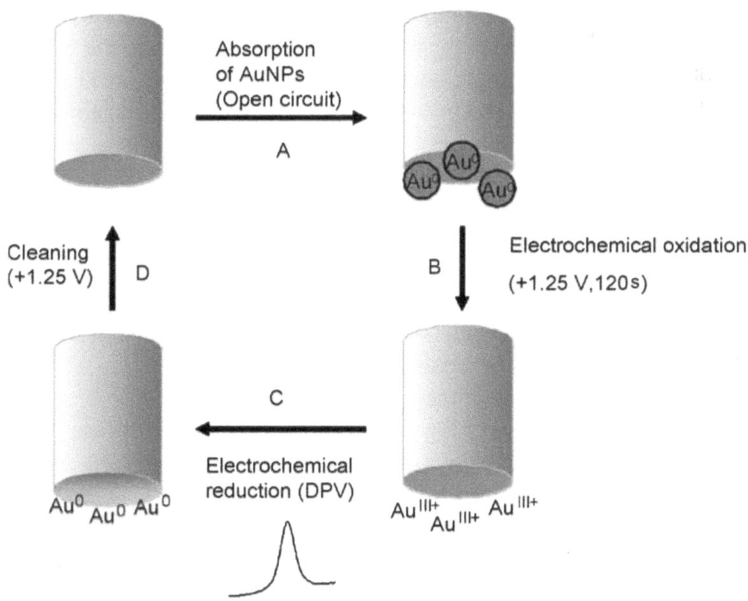

FIGURE 5 Schematic diagram of the electroanalytical protocol for direct voltammetric determination of gold nanoparticles (AuNPs) using a graphite–epoxy composite electrode. DPV, differential pulse voltammetry. *Reproduced with permission from Ref. [92].*

4. DETERMINATION OF AUNPs IN ENVIRONMENTAL SAMPLES

4.1 Occurrence and Fate of NPs in the Environment

The growing production and use of NPs will inevitably lead to their release and accumulation in environmental compartments, and have implications for human and environmental health. The US Environmental Protection Agency and the European Union are paying attention to the fate, transport, and health effects of nanomaterials in the environment, being already included in list of emerging pollutants [93]. Despite this fact, there are few reports on the presence and behavior of NPs in environmental matrices; only generalized estimations of potential environmental concentrations have been carried out through modeling and estimates of production and usage [94–96].

The fate and stability of NPs in the environment play important roles in assessing their environmental distributions and control human exposure. Fate depends on their physical and chemical properties (e.g., size, surface and properties) and other chemical species in water [97]. The physicochemical properties of NPs, such as surface area, will affect their transport and bioavailability. NPs present in the environment may interact with natural organic matter (NOM), larger particles, and other NPs; thus, determining the stability of such NPs requires studying their ability to aggregate and interact with other particles [98]. They can suffer from transformations by action of light, oxidants, or microorganisms [99]. Despite environmental transformations and changes in their surrounding environment, metallic NPs tend to exist as stable colloidal aggregated or dispersions [11]. Surface coating, aggregation, and disaggregation largely determine the bioavailability and behavior of nanomaterials.

The aggregation of NPs plays an important role in their fate in the aqueous environment, depending on parameters such as pH, ionic strength, and ionic composition. Such parameters affect the stability of surface charge and controls particle interactions (Figure 6). Aggregated NPs can be trapped or eliminated through sedimentation; they are less mobile, but can undergo uptake by filter feeders and sediment-dwelling animals leading to a biomagnifications in the food chain [100].

NP stability may be also influenced by the presence of dissolved NOM; it imparts them with a negative charge and thus increases their surface potentials. The presence of electrolytes (e.g., Ca^{2+}) induces aggregations since they neutralize the negative charge of the natural organic matter. Transport of aggregates will differ from primary constituent particles, controlling fate and speciation [101]. Stabilized NP assemblages may be more easily transported through environmental compartments than larger particulate phases [102].

Moreover, NPs can exert toxicity by direct interaction with a living cell or through interaction among themselves (self-interaction NP–NP), with

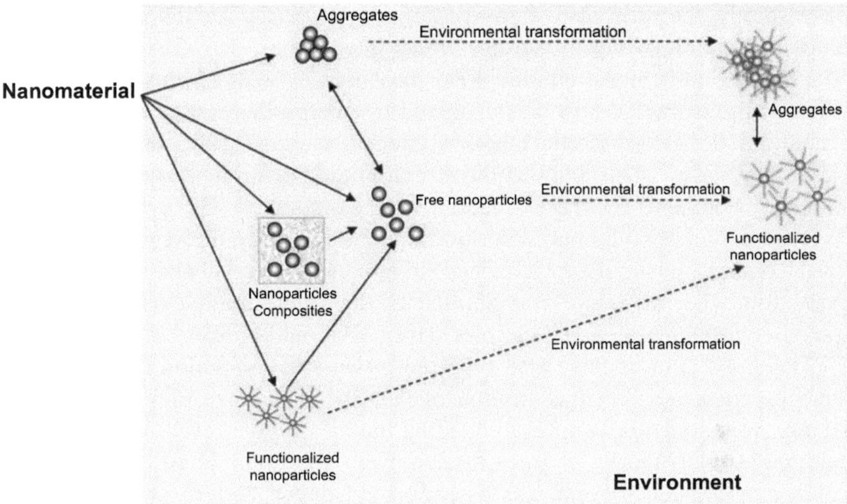

FIGURE 6 Modification of nanoparticles in the environment. *Reproduced with permission from Ref. [100].*

colloids, or dissolved organic matter (DOM), through the process of stabilization and aggregation [11]. NPs can also act as carriers, and transport toxins or nutrients by themselves or as encapsulated layers. This affects the distribution of secondary pollutants.

Various models predict that the level of AuNPs in environmental waters will reach 140 ng L^{-1} by 2018 [95]. Thus, tracking NPs and monitoring their presence in environmental systems are important. Until now, characterization methods have focused mainly on synthetic products; these are usually not suitable for complex mixtures, such as environmental samples, since they cannot detect low concentrations [103]. Recent results of laboratory exposure studies suggest that AuNPs can be biomagnified within the food chain, partially due to their uptake by cells [104].

4.2 Representative Examples

Li et al. [103] reported the effective and selective extraction of AuNPs based on noncovalent reversible adsorption onto an ionic exchange resin (Amberlite IRN-78), which contains positively charged ammonium groups. The bifunctional ligand mercaptosuccinic acid (MSA) was employed to bind to the surface of the NPs through its thiol group, while its two carboxylic groups are exposed to the medium. MSA-modified NPs were loaded onto the resin via the electrostatic interaction of negatively charged carboxylic groups with positive amino groups of the resin. Those adsorbed NPs were subsequently eluted with a solution of 8% (v/v) formic acid in methanol at room temperature.

Interestingly, the resin also showed a capacity to retain AuNPs without MSA functionalization owing to van der Waals interactions. The concentrations of NPs in the runoff of the column were measured by GFAAS, and the composition of the extracted NPs was analyzed by energy-dispersive X-ray fluorescence and IR spectroscopy. Possible interferences of DOM and metal ions were investigated. This showed the effect on the performance of the method, which was suitable for the extraction of NPs stabilized with citrate, PVP10, PVP40, and tetraoctylammonium bromide. Additionally, TEM measurements revealed that no significant changes in particle size or shape occur during extraction. The procedure was applied to the determination of NPs in real spiked environmental water samples (river, lake, and brook).

The same authors have also reported a two-step extraction procedure for the separation and preconcentration of AuNPs, in the size range from 10 to 80 nm, from aqueous samples by loading the NPs onto a reversed phase C-18 column [105] (Figure 7). Reextraction from the solid phase into a small volume of organic solvent was assisted by 1-dodecanethiol (1-DDT) under ultrasonic conditions. An enrichment factor of 250 was obtained, and the suitability of the method was assessed with real water samples. Selectivity toward ionic gold species and other NPs was investigated as well as that due to the presence of DOM, which can be overcome by pretreatment with hydrogen peroxide. Preservation of the size and shape of NPs was proven by UV–Vis and TEM.

It has been demonstrated that ionic liquid with a cationic surfactant is a good combination to extract AuNPs from river water [46]. Thanks to the high affinity of the imidazolium group for the AuNPs, it is possible to preconcentrate them into a small volume of ionic liquid. River water samples containing AuNPs were treated with cetyltrimethylammonium chloride (CTAC) and ionic liquid 1-butyl-3-methyl-imidazolium hexafluorophosphate (BMIM PF_6). The system was manually shaken for 30 s and allowed to stand to achieve ionic liquid from water phase separation. Both absorbance and Raman measurements were employed to quantify the samples.

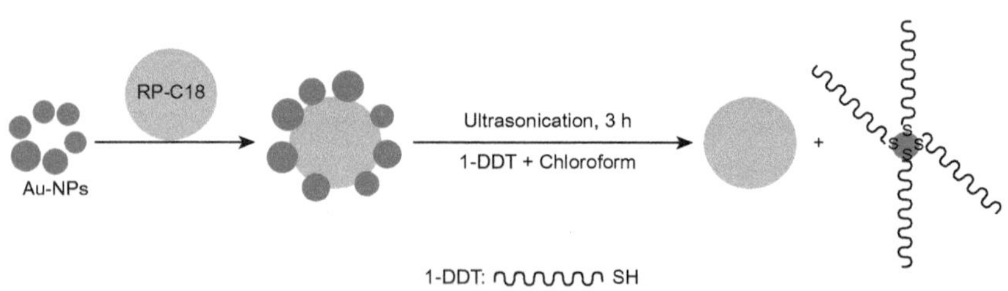

FIGURE 7 Scheme of the strategy for the separation and concentration of gold nanoparticles (AuNPs) from waters by two-step extraction. 1-DDT, 1-dodecanethiol. *Reprinted with permission from Ref. [105]. Copyright 2012 American Chemical Society.*

The determination of AuNPs in environmental samples requires sample pretreatment that ideally combines preconcentration and selectivity. Hartmann and Schuster [106] have developed a preconcentration method using CPE using Triton X-114, which enables effective separation of ionic gold species and AuNPs by using sodium thiosulfate as a complexing agent. Quantification was accomplished by electrothermal atomic absorption spectrometry, to obtain an enrichment factor of 80 with a limit of detection of 5 ng L^{-1}. The CPE process did not lead to particle size changes, as confirmed by TEM. NOM present in the media up to a concentration of 10 mg L^{-1} showed no interference. This procedure was applied to the analysis of samples of wastewater treatment plant influent and effluent as well as river water samples.

The use of FFF for the separation and characterization of NPs has increased in the last two decades. FFF belongs to a family of chromatography-like separation techniques that provides a continuous and high-resolution separation of NPs as a function of their diffusion coefficient. This technique has been employed for the separation and purification of AuNPs and the analysis and determination of natural colloids and manufactured NPs in environmental analysis; thus, it demonstrates its potential to determine also AuNPs from such matrices [107,108].

5. DETERMINATION OF AUNPs IN BIOLOGICAL MATRICES

5.1 Effects of NPs in Biological Systems

The specific physicochemical properties of NPs are expected to result in an increased reactivity with biological systems. Thus, in spite of their beneficial effects, engineered NPs may also represent a potential hazard to human health. Several studies indeed suggest that these NPs have a different toxicity profile compared with that of larger particles [109].

Owing to the wide range of application of AuNPs, it is important to assess the potential for these nanomaterials to accumulate in the body and produce undesired side effects. Despite the huge potential benefit of AuNPs in the field of biomedical and industrial applications, very little is known about their toxicity and tissue bioavailability of AuNPs in animals. Once AuNPs have been released into the environment and humans have been exposed to them, the effects of those NPs in the body may be diverse depending on the characteristics of the NPs (e.g., size, shape, surface properties such as surface modifications or coatings, aggregation state.). Since their toxicity depends on such factors, not only the composition and concentration but also the physical and chemical properties within the sample and the chemical characteristics of any capping/functional layer on the particle surface have to be determined [8].

The toxicity of several types and sizes of gold nanoconjugates has been investigated by several research groups. The most salient conclusions are that the toxicity of gold nanoconjugates is dependent on the chemical composition of the surface ligands, which is usually what induces toxicity [110]. For example, CTAB-functionalized AuNPs do not cause cytotoxicity when excess of ligand is removed [28]. Goodman et al. [111] found that amine-functionalized particles were only mildly toxic while those functionalized with carboxylic acids were nontoxic. One study investigating human dermal fibroblasts reported that the rate of cell proliferation, spreading, and adhesion was slower in the presence of citrate-capped NPs [112]. It has been also reported that the size of the conjugate determines its toxicity, for example, 1.4-nm-diameter NPs were toxic, while 15-nm NPs showed no toxicity even at higher concentrations [113].

Many investigations about adverse effects of NPs have been carried out in mammalian systems, to study the respiratory system, gastrointestinal tract, and skin routes of exposure. Effective physiological barriers, such as pulmonary and gastrointestinal mucosa, may not prevent NPs from entering the organism. Exposure to silver and gold NPs can lead to kidney damage and to a blue–gray discoloration of the eyes, nose, and nasal septum, throat, and skin [114].

In vivo toxicity of 13-nm-sized AuNPs coated with PEG was studied by Cho et al. [115]. They found that acute inflammation and apoptosis in the liver were induced. These NPs were found to accumulate in the liver and spleen for up to 7 days after injection and to have long blood circulation times. NPs could also be identified as foreign by immune cells, and promote their reaction against surface or core components to mount an inflammatory response [116].

It has been reported that the gold level in blood did not increase when AuNPs were administered in several doses in mice, whereas in all the organs investigated (brain, liver, kidney, spleen, and lungs), there was a proportional increase in gold, indicating tissue uptake [117]. Although the brain was the organ in which less content of gold was found, it is suggested that the AuNPs are capable of crossing the blood–brain barrier and accumulating in neural tissue. The major percentage of AuNPs was accumulated in the liver, followed by the kidney and the spleen.

Similar studies have revealed that AuNPs of different sizes are accumulated in organs such as the liver, the lung, and the spleen, depending on particle size [118]. Smaller NPs (15 nm) showed the most widespread organ distribution, and were accumulated in higher amounts in the blood, liver, lung, spleen, kidney, brain, heart, and stomach. This confirmed, as reported by other authors that AuNPs are able to cross the blood–brain barrier. Two-hundred-nanometer-sized AuNPs were shown to be present in the blood, brain, stomach, and pancreas.

Hillyer and Albrecht [119] reported that after oral administration of AuNPs of decreasing sizes (58, 28, 10, and 4 nm) to mice, an increased distribution to other organs was observed. The smallest particle (4 nm) administered orally resulted in an increased presence of gold particles in the kidney, liver, spleen, lungs, and even the brain. The biggest particle (58 nm) tested was detected almost solely inside the gastrointestinal tract.

In addition, Jong et al. [120] performed a study to determine the influence of particle size on the in vivo tissue distribution of spherical-shaped AuNPs in rats. They were intravenously injected into the tail vein with AuNPs with a diameter of 10, 50, 100, and 250 nm, respectively. For all AuNP sizes, the majority of the gold was demonstrated to be present in the liver and spleen. The 10-nm particles were present in various organ systems including the blood, liver, spleen, kidney, testis, thymus, heart, lung, and brain, whereas the larger particles were only detected in the blood, liver, and spleen. Thus, results demonstrate that tissue distribution of AuNPs is size dependent with the smallest 10-nm NPs showing the most widespread organ distribution.

A systematic study of the cytotoxicity of water-soluble AuNPs stabilized by triphenylphosphine derivatives ranging in size from 0.8 to 15 nm in four cell lines representing major functional cell types with barrier and phagocyte function has been performed [121]. Connective tissue fibroblasts, epithelial cells, macrophages, and melanoma cells proved to be most sensitive to 1.4-nm AuNPs while 15-nm AuNPs were nontoxic at up to 60-fold higher concentrations.

CTAB-functionalized gold nanorods were rapidly cleared from the blood after injection into the tail vein of a mouse model [122]. Another study on similar nanorods found that they were accumulated in the liver after 72 h; however, when functionalized with PEG, very few particles remained in the liver after 72 h [123].

A series of AuNPs have been examined for the uptake and acute toxicity in human leukemia cells by Connor et al. [28]. The results indicate that although some precursors of NPs may be toxic, the NPs themselves are not necessarily detrimental to cellular function.

Chithrani et al. [124] studied the effect of NP size, shape, concentration, and incubation time on the cellular uptake using colloidal AuNPs with sizes varying between 14 and 100 nm. A large difference in the uptake of the different sized and shaped AuNPs was observed. In addition, a quantitative comparison of citrate-stabilized AuNPs versus transferrin-coated NPs showed a greater uptake of citrate-stabilized ones.

Internalization and subcellular localization in HeLa cells of AuNPs modified with the SV40 large T antigen were studied as a function of incubation time, temperature, NP diameter, and large T surface coverage [125]. Increasing the large T coverage was found to afford higher cell internalization and nuclear targeting. In addition, quantitative evaluation of cytotoxicity revealed that higher large T coverages also resulted in greater cytotoxicity.

The uptake and release of AuNPs in Daphnia magna have been studied by Lovern et al. [126]. Particle clearance was investigated by placing the organism in fresh water and observing retention of NPs in the gut tissue over time.

Finally, Cho et al. [127] pointed out the importance of sedimentation and diffusion velocities of the AuNPs on their cellular uptake. Their results suggest that sedimentation needs to be considered when performing in vitro studies for large NPs.

There have been several other studies performed about toxicity and cellular uptake of AuNPs, which would be beyond the scope of this chapter. Further information can be found in reviews such as in Ref. [38].

5.2 Representative Examples

Lasagna-Reeves et al. [117] evaluated the bioaccumulation and toxic effects of different doses of AuNPs upon intraperitoneal administration in mice every day for 8 days. To determine the distribution of AuNPs in tissues, they employed two different methods to measure gold in various tissues, including the brain, liver, kidney, spleen, and lungs. With both ICP–MS and GFAAS, the amounts of gold detected in each tissue were similar. The analytical procedure used comprised drying of tissue samples before digestion to white ash using 30% hydrogen peroxide at 50–60 °C followed by digestion with concentrated ultrapure trace-metal free nitric acid. The toxicological assessment was performed by serum biochemical analysis, as well as by hematological and histopathological analyses.

Sonavane et al. [118] evaluated the biological distribution of differently sized AuNPs after intravenous administration in mice. Synthesized AuNPs were characterized by SEM to determinine their size distribution by means of a particle size analyzer. The determination of gold in tissue samples was performed by ICP–MS. In order to extract gold and prepare tissue for ICP–MS analysis, isolated tissues were cut into small parts to obtain fine pieces, homogenized, and completely lysed in aqua regia. After evaporating at 126 °C in an oil bath, the residual aqua regia, the residue was redissolved in 0.05 M HCl, ultrasonicated for 20 min, and the sample was submitted to ICP–MS.

ICP–MS was also employed to determine the presence of AuNPs with a diameter between 10 and 250 nm intravenously injected in rats [120]. Gold suspensions were diluted by adding phosphate buffer saline for injection. After 24 h, the rats were sacrificed, and blood and various organs were collected for gold determination. The organs were weighed, and tissue samples were homogenized and frozen for determination of gold content.

ICP coupled with atomic emission spectroscopy (ICP–AES) has been employed to quantify the uptake of Au nanorods coated with polyelectrolytes by HeLa cervical cancer cells [128]. TEM was also employed for imaging the presence of the Au nanorods inside the cells. AuNPs incubated in HeLa cells have been determined by detaching of the cells from the Petri dish surface using the enzyme trypsin, homogenized, and the concentration of Au was measured by ICP–AES [124]. All results were further verified by UV–Vis spectrophotometry.

Ryan et al. [125] demonstrated the efficacy of ICP–OES as an analytical tool for use in biological assays involving AuNPs. ICP–OES allows for the efficient quantification of many delivery parameters (e.g., differences in internalization over time and degree of nuclear localization), which would be difficult to examine using only standard microscopic imaging techniques such as TEM, confocal microscopy, or video-enhanced color differential interference contrast microscopy. Internalization and subcellular localization in HeLa cells of AuNPs modified with the SV40 large T antigen were quantified by using this technique.

AF^4 coupled with MALS, DLS, and ICP–MS has been applied to the characterization of AuNPs (10–60 nm) in livers of rats by intravenous injection [82]. The homogenized livers were solubilized in tetramethylammonium hydroxide, and the recovery of AuNPs from the livers amounted to 86–123% of their total Au content. In spite of successful stabilization with bovine serum albumin even in alkaline medium, separation of the AuNPs by AF^4 was not possible due to association with undissolved remains of the alkali-treated liver tissues as demonstrated by EM images.

Amendola et al. [129] demonstrated that it is possible to correlate the surface enhanced Raman scattering (SERS) signal with the concentration of SERS labels based on AuNP by fitting their UV–Vis absorption spectra with a Mie-Gans model. This method was applied to the determination of the number of AuNPs taken up by U937 macrophage cells and a good correlation between the results obtained by ICP–MS was observed.

A protocol for the quantitative analysis of gold in solution and gold uptake by macrophages by detecting the decrease in the fluorescence of a dye upon forming a complex with gold and a ligand has been proposed [130]. It provides a preliminary first step for the determination of AuNPs, which must be first dissolved in aqua regia.

Xie et al. [131] have developed a nondestructive optical assay for the calculation of gold nanoshell concentrations in blood and estimated the delivered dose of intravenously injected NPs using DLS. Triton X-100 was applied to the system as both a blood lysing agent and a quantitative comparative standard for the colloidal nanoshells. The concentration of nanoshells was determined to be linearly proportional to the relative light scattering attributed to nanoshells via DLS as compared with the Triton X-100 micelles in calibration samples. This relationship was found to remain valid in blood samples taken from a murine tumor model as confirmed by neutron activation analysis.

DLS in combination with microscopic techniques, such as TEM or AFM (Figure 8), have been used to examine the size and charge of colloidal particles incubated with plasma [132], and provided an insight into the molecular identities of the plasma proteins that bind the colloids. Gold colloids with 30- and 50-nm nominal diameters were examined before and after incubation in plasma. Because DLS measures a particle's hydrodynamic diameter, this measurement can also be influenced by particle agglomeration. To determine if the observed increase in particle hydrodynamic diameter upon plasma

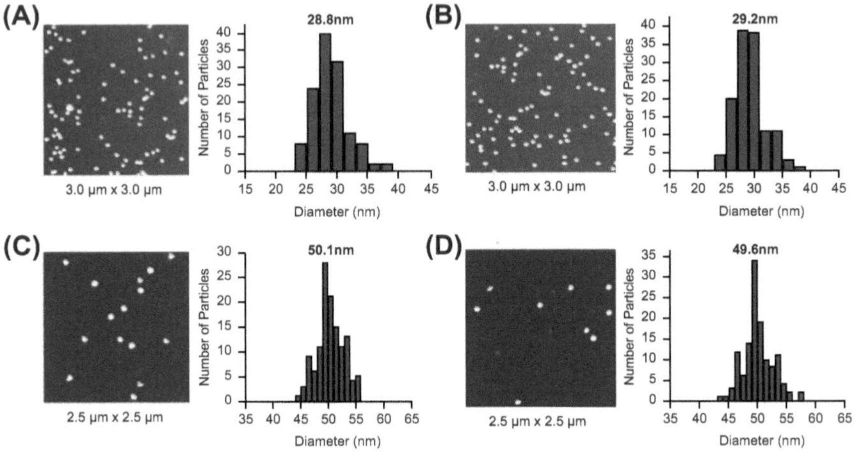

FIGURE 8 Analysis of nanoparticle size by atomic force microscopy (AFM). Representative AFM micrograph of 30-nm gold colloids before (A) and after (B) plasma incubation, and that of 50-nm gold colloids before (C) and after (D) plasma incubation. All AFM measurements were performed by the tapping-in-liquid mode to minimize lateral interaction between the AFM tip and the gold colloid and plasma proteins. *Reproduced with permission from Ref. [132].*

incubation could be attributed to a change in agglomeration state, the same samples measured by DLS were also studied by TEM and AFM. The zeta potential (the electrostatic potential generated by the accumulation of ions at the surface of the colloidal particles) was monitored before and after plasma incubation. Two-dimensional polyacrylamide gel electrophoresis and mass spectrometry were used to identify bound proteins.

ESEM and ETEM have been used for studying collagen crosslinking with AuNPs for application in drug delivery. SEM micrograph allowed the authors to evaluate the influence of NPs on the porosity of collagen gel, while TEM revealed the organization of the NPs within the gel [133]. EM has been also used to image AuNPs with a 4-nm resolution within a cell by using high-angle annular dark-field detector placed in a circle around the electron beam of TEM [134].

Chithrani and Chan [135] investigated and captured the uptake process of transferrin-coated AuNPs entering cells by EM (TEM), established the mechanism of uptake, and quantitated the exocytosis process. Confocal fluorescence microscopy was also employed to determine their uptake by labeling the transferring protein with an organic fluorophore.

Dark-field optical microscopy can be performed on living cells to visualize the location of AuNPs (~200 nm which is the diffraction limit of the instrument), by taking advantage of the elastic light scattering properties of the AuNPs [136].

Adsorption of the most abundantly binding proteins onto AuNPs was studied after 24 h in vitro incubation of monodisperse, negatively charged 5-, 15-, and 80-nm gold spheres (AuNP) in mouse serum by a two-step analysis: proteomic protein identification and quantitative protein biochemistry [137]. The adsorbed proteins were separated from nonadsorbed proteins by centrifugation and gel electrophoresis and identified using an MALDI-TOF-MS-Proteomics Analyzer.

6. DETERMINATION OF AUNPs IN AGRIFOOD SAMPLES

6.1 Nanotechnology Applications in Agriculture and Food

Engineered nanomaterials are increasingly being used in the food industry. The use of nanotechnology-derived food ingredients, additives, supplements, and contact materials is attracting great attention. Over 200 companies are conducting R&D on the use of nanotechnology in agriculture, engineering, processing, packaging, or delivery of food and nutritional supplements [138].

This increased use is raising concerns over the safety of applying them in food and food-related products [139]. However, the lack of available analytical techniques to detect and characterize nanomaterials in food and beverage matrices is preventing the industry from rapidly developing "nanofood" [23]. The high surface-to-volume ratio of nanomaterials enables their interaction with proteins, carbohydrates, and fats in food, in addition to nucleic acid, ions, minerals, and water. It is thought that the type of proteins surrounding the NP can determine the fate and impact of nanomaterials in the organism [140].

Applications of nanotechnology for the food sector are of the following main categories [138]: (1) food ingredients processed or formulated to form nanostructures; (2) nanosized, nanoencapsulated, or engineered NP additives; (3) nanomaterials incorporated to develop improved, "active," or "intelligent" materials for food packaging; and, finally, (4) nanotechnology-based devices and materials used, for example, for nanofiltration, water treatment, nanosensors for food safety, and traceability.

NPs of different natures are used in the food industry, such as AgNPs in food packaging, owing to their antibacterial properties or silica NPs usually employed as food additive [49]. Currently, some food additives with nanoingredients are being commercialized, which implies that NPs are present in the food.

Potential sources of unintentional nanomaterial contaminants to food include the environment, nanomaterials used during plant and animal production, unintentional release from NP-containing food packaging materials, and residues in food from nanomaterials used as food-processing aids or surface coating on food equipments [21].

Many food substances or ingredients have nanostructures in nature, with a size of micrometers or nanometers; some examples are food proteins, linear polysaccharides with 1D nanostructures, and starch polysaccharides that have small 3D crystalline nanostructures. These natural nanomaterials in food are not included in the scope of this chapter; however, they contribute to the complexity of the analysis since they should be distinguished from engineered or contaminating nanomaterials and, on the other hand, because of their specific physicochemical properties NPs could interact with proteins; lipids; carbohydrates; nucleic acids; ions; minerals and water in food, feed, and biological tissues [141].

6.2 Representative Examples

AuNPs can be generated through the interaction of Au(III) ions with oat and wheat biomasses; thus, procedures for their recovery need to be developed. Armendariz et al. [142] developed a method for the extraction of AuNPs from oat and wheat biomasses using sodium citrate and CTAB. After extraction by sonication, samples were centrifuged, and supernatants were analyzed using UV−Vis spectroscopy and TEM to corroborate the NP extraction. Biomass samples were also analyzed using X-ray absorption spectroscopy to observe changes in the NPs on the biomass after extraction.

López-Lorente et al. [46] described an approach for the determination of AuNPs in chicken liver samples based on the stabilization of AuNPs with a cationic surfactant followed by a microliquid−liquid extraction in ionic liquids, and further determination by UV−Vis and Raman spectroscopy (Figure 9). Chicken liver samples were dried with acetone and pulverized until a uniform blend is formed. Analysis was performed by adding trichloroacetic acid to denature proteins that precipitate and may interfere with measurements and ethylenediaminetetraacetic acid as complexing agent of possible interfering ions to powdered chicken liver in water. The solution was spiked with AuNPs and stirred for 1 h to promote interactions with the biological compounds. The solid phase was then separated by centrifugation, and AuNPs were extracted into an ionic liquid BMIM PF_6 in the presence of CTAC.

AuNPs have been detected by ICP-optical emission spectroscopy mainly in the digestive gland of marine organisms (*Mytilus edulis*) exposed to AuNP

FIGURE 9 Photograph of the water solution containing gold nanoparticles before and after extraction and preconcentration treatment with 0.3 g of water-insoluble ionic liquid (left). Scheme of the extraction process that implies the use of a cationic surfactant (hexadecyltrimethylammonium chloride) and the ionic liquid with an imidazolium group (right). *From Ref. [46]. Reproduced by permission of the Royal Society of Chemistry.*

[143]. Two-dimensional electrophoresis of digestive gland extracts, from AuNP and control groups, showed decreased levels of thiol proteins in response to AuNP, which can be attributed to oxidation.

Gatti et al. [144] investigated the presence of inorganic microsized and nanosized contaminants in wheat bread and in biscuits from 14 different countries by means of ESEM. The X-ray microprobe of an energy dispersion spectroscope was employed to identify their chemical composition. The results indicate that 40% of the samples analyzed contained foreign bodies as ceramic and metallic debris of probable environmental or industrial origin.

FFF is a promising technique for the study of NPs in complex food samples. It has been applied for the separation of various types of NPs (e.g., organic macromolecules, inorganic NPs) in food samples [107]. FFF can be coupled to different detectors and offers additional information and specificity. It is expected that determination of AuNPs from food samples can be achieved by using this technique.

7. TRENDS

Ideally, techniques for the determination of NPs must fulfill several requirements: (1) They must be capable of dealing with heterogeneous samples, (2) they must avoid artifacts, and (3) they must provide the major possible amount of information not only of concentration but also of their physicochemical properties. Although promising advances have been achieved in NP analysis, further developments are needed, such as in situ analysis, which will decrease sample disturbance from their original state in the sample, as well as reliable techniques for improving size determination.

Such ideal methods are not still developed. Until now, the determination of AuNPs in complex samples usually involves, on the one hand, the combination of microscopic and spectroscopic techniques to identify and characterize the nanomaterial as well as to detect their chemical composition. On the other hand, other procedures are focused only on determining the chemical composition or quantifying the NPs present in the sample. As no one technique can address all the information needed for the complete characterization of NPs in complex samples, the combination of complementary techniques is required. In this context, new instrumentation combining in situ spectroscopic and microscopic techniques (e.g., AFM-IR and AFM-Raman) is being developed, which will be very useful to provide information about both size and shape of the NPs as well as chemical information simultaneously.

Efforts in the area of the determination of AuNPs in complex samples should be directed to the development of standard testing nanomaterials and testing guidelines. New protocols of sampling and calibration need to be developed specifically for AuNPs, which may vary depending on the average size of AuNPs or their size distribution.

ACKNOWLEDGMENTS

The authors wish to thank Spain's Ministry of Innovation and Science for funding Project CTQ2011−23,790 and Junta de Andalucía for Project FQM4801.

REFERENCES

[1] R.A. Rippel, A.M. Seifalian, J. Nanosci. Nanotechnol. 11 (2011) 3740−3748.
[2] O. Olorundare, S. Simmons, R. Albrecht, Blood 79 (1992) 99−109.
[3] A.I. López-Lorente, B.M. Simonet, M. Valcárcel, B. Mizaikoff, Anal. Chim. Acta 788 (2013) 122−128.
[4] Y. Kuwahara, T. Akiyama, S. Yamada, Langmuir 17 (2001) 5714−5716.
[5] S. Eustis, M.A. El-Sayed, Chem. Soc. Rev. 35 (2006) 209−217.
[6] R.A. Sperling, P. Rivera-Gil, F. Zhang, M. Zanella, W.J. Parak, Chem. Soc. Rev. 37 (2008) 1896−1908.
[7] B.M. Simonet, M. Valcárcel, Anal. Bioanal. Chem. 393 (2009) 17−21.
[8] K. Tiede, A.B.A. Boxall, S.P. Tear, J. Lewis, H. David, M. Hassellöv, Food Addit. Contam. 25 (2008) 795−821.
[9] OECD, Guidance on Sample Preparation and Dosimetry for the Safety Testing of Manufactured Nanomaterials, OECD Environment, Health and Safety Publications, 2012 (No. 36).
[10] A.L. Fabricius, L. Duester, B. Meermann, T.A. Ternes, Anal. Bioanal. Chem. 406 (2014) 467−479.
[11] B. Ferreira da Silva, S. Pérez, P. Gardinalli, R.K. Singhal, A.A. Mozeto, D. Barceló, Trends Anal. Chem. 30 (2011) 528−540.
[12] A. Scheffer, C. Engelhard, M. Sperling, W. Buscher, Anal. Bioanal. Chem. 390 (2008) 249−252.
[13] D.J. Burleson, M.D. Driessen, R.L. Penn, J. Environ. Sci. Health A 39 (2004) 2707−2753.
[14] J.R. Lead, K.J. Wilkinson, Environ. Chem. 3 (2006) 159−171.
[15] National Institute of Standards & Technology (NIST). http://www.nist.gov/mml/csd/inorganic/goldnanorms.cfm (accessed 10.01.14).
[16] H. Weinberg, A. Galyean, M. Leopold, Trends Anal. Chem. 30 (2011) 72−83.
[17] K. Tiede, A.B.A. Boxall, D. Tiede, S.P. Tear, H. David, J. Lewis, J. Anal. At. Spectrom. 24 (2009) 964−972.
[18] J.T. Kwon, S.K. Hwang, H. Jin, D.S. Kim, A. Minai-Tehrani, H.J. Yoon, M. Choi, T.J. Yoon, D.Y. Han, Y.W. Kang, B.I. Yoon, J.K. Lee, M.H. Cho, J. Occup. Health 50 (2008) 1−6.
[19] A.I. López-Lorente, B.M. Simonet, M. Valcárcel, Front. Biosci. E4 (2012) 1024−1042.
[20] J.Y. Liu, R.H. Hurt, Environ. Sci. Technol. 44 (2010) 2169−2175.
[21] C. Blasco, Y. Picó, Encyclopedia of Analytical Chemistry, 2013. http://dx.doi.org/10.1002/9780470027318.a9285.
[22] B. Pergolese, A. Bigotto, M. Muniz-Miranda, G. Sbrana, Appl. Spectrosc. 59 (2005) 194−199.
[23] A. Dudkiewicz, K. Tiede, K. Loeschner, L. Helene, S. Jensen, E. Jensen, R. Wierzbicki, A.B.A. Boxall, K. Molhave, Trends Anal. Chem. 30 (2011) 28−43.
[24] W. Denk, H. Horstmann, PLoS Biol. 2 (2004) 1900−1909.
[25] G. Danscher, M. Stoltenberg, Prog. Histochem. Cytochem. 41 (2006) 57−139.
[26] T.M. Mayhew, C. Mühlfeld, D. Vanhecke, M. Ochs, Ann. Anat. − Anat. Anz. 191 (2009) 153−170.

[27] G. Impoco, N. Fucà, C. Pasta, M. Caccamo, G. Licitra, Comput. Electron. Agric. 84 (2012) 26–35.
[28] E.E. Connor, J. Mwamuka, A. Gole, C.J. Murphy, M.D. Wyatt, Small 1 (2005) 325–327.
[29] J.P. Novak, C. Nickerson, S. Franzen, D.L. Feldheim, Anal. Chem. 73 (2001) 5758–5761.
[30] M. Kaláb, P. Allan-Wojtas, S.S. Miller, Trends Food Sci. Technol. 6 (1995) 177–186.
[31] C. Lorenz, K. Tiede, S. Tear, A. Boxall, N. von Götz, K. Hungerbühler, Int. J. Occup. Environ. Health 16 (2010) 406–428.
[32] O. Balmes, J.O. Malm, N. Pettersson, G. Karlsson, J.O. Bovin, Microsc. Microanal. 12 (2006) 145–150.
[33] P.V. Asharani, Y.L. Wu, Z. Gong, S. Valiyaveettil, Nanotechnology 19 (2008) 255102.
[34] A.E. Porter, M. Gass, K. Muller, J.N. Skepper, P.A. Midgley, M. Welland, Nat. Nanotechnol. 2 (2007) 713–717.
[35] A.I. López-Lorente, B.M. Simonet, M. Valcárcel, S. Eppler, R. Schindl, C. Kranz, B. Mizaikoff, Talanta 118 (2014) 321–327.
[36] R.F. Domingos, M.A. Baalousha, Y. Ju-Nam, M.M. Reid, N. Tufenkji, J.R. Lead, G.G. Leppard, K.J. Wilkinson, Environ. Sci. Technol. 43 (2009) 7277–7284.
[37] Q.K. Ong, J. Reguera, P.J. Silva, M. Moglianetti, K. Harkness, M. Longobardi, K.S. Mali, C. Renner, S. De Feyter, F. Stellacci, ACS Nano 7 (2013) 8529–8539.
[38] A.M. Alkilany, C.J. Murphy, J. Nanopart. Res. 12 (2010) 2313–2333.
[39] W. Haiss, N.T.K. Thanh, J. Aveyard, D.G. Fernig, Anal. Chem. 79 (2007) 4215–4221.
[40] V. Amendola, M. Meneghetti, J. Phys. Chem. C 113 (2009) 4277–4285.
[41] S.R. Raz, M. Leontaridou, M.G.E.G. Bremer, R. Peters, S. Weigel, Anal. Bioanal. Chem. 403 (2012) 2843–2850.
[42] E.M. Goldys, M.A. Sobhan, Adv. Funct. Mater. 22 (2012) 1906–1913.
[43] Z. Xu, X. Huang, C. Dong, J. Ren, Microchim. Acta 181 (2014) 723–730.
[44] B. Wang, X. Wang, W. Lou, J. Hao, J. Coll. Interf. Sci. 362 (2011) 5–14.
[45] A.I. López-Lorente, M. Sieger, M. Valcárcel, B. Mizaikoff, Anal. Chem. 86 (2014) 783–789.
[46] A.I. López-Lorente, B.M. Simonet, M. Valcárcel, Analyst 137 (2012) 3528–3534.
[47] E. Duval, Phys. Rev. B Condens. Matter. 46 (1992) 5795–5797.
[48] N. Del Fatti, C. Voisin, F. Chevy, F. Vallée, C. Flytzanis, J. Chem. Phys. 110 (1999) 11484–11487.
[49] O. Lozano, J. Mejia, T. Tabarrant, B. Masereel, J.M. Gogné, O. Toussaint, S. Lucas, Anal. Bioanal. Chem. 403 (2012) 2835–2841.
[50] O. Lozano, J. Mejia, B. Masereel, O. Toussaint, D. Lison, S. Lucas, Nanotoxicology 6 (2012) 263–271.
[51] A. Ledin, S. Karlsson, A. Ker, B. Allard, Water Res. 28 (1994) 1539–1545.
[52] C.Y.M. Hsu, H. Uludag, BMC Biotechnol. 8 (2008) 23.
[53] K.W. Powers, S.C. Brown, V.B. Krishna, S.C. Wasdo, B.M. Moudgil, S.M. Roberts, Toxicol. Sci. 90 (2006) 296–303.
[54] V.A. Bogatyrev, L.A. Dykman, B.N. Khlebtsov, A.V. Alekseyeva, A.A. Polyakov, N.G. Khlebtsov, Proc. SPIE 5475 (2004) 38–47.
[55] I. Delfino, Biophys. Chem. 177–178 (2013) 7–13.
[56] A.R. Montoro-Bustos, J. Ruiz-Encinar, A. Sanz-Medel, Anal. Bioanal. Chem. 405 (2013) 5637–5643.
[57] Y. Zhang, S.M. Shuang, C. Dong, C.K. Lo, M.C. Paau, M.M.F. Choi, Anal. Chem. 81 (2009) 1676–1685.
[58] C. Degueldre, P.Y. Favarger, S. Wold, Anal. Chim. Acta 555 (2006) 263–268.

[59] B. Fernández, J.M. Costa, R. Pereiro, A. Sanz-Medel, Anal. Bioanal. Chem. 396 (2010) 15–29.

[60] J.W. Olesik, P.J. Gray, J. Anal. At. Spectrom. 27 (2012) 1143–1155.

[61] S.H. Hu, R. Liu, S.C. Zhang, Z. Huang, Z. Xing, X.R. Zhang, J. Am. Chem. Soc. Mass Spectrom. 20 (2009) 1096–1103.

[62] R. Allabashi, W. Stach, A. de la Escosura-Muñiz, L. Liste-Calleja, A. Merkoçi, J. Nanopart. Res. 11 (2009) 2003–2011.

[63] A. Merkoçi, G. Tarrasón, R. Eritjia, S. Alegret, Anal. Chem. 77 (2005) 6500–6503.

[64] S.L. Kerr, B. Sharp, Chem. Commun. (2007) 4537–4539.

[65] A. Helfrich, W. Brüchert, J. Bettmer, J. Anal. At. Spectrom. 21 (2006) 431–434.

[66] A.D. Levin, Y.M. Sadagov, L.L. Koroli, Meas. Tech. 54 (2012) 1186–1190.

[67] G.T. Wei, F.K. Liu, J. Chromatogr. A 836 (1999) 253–260.

[68] T. Siebrands, M. Giersig, P. Mulvaney, C.H. Fischer, Langmuir 9 (1993) 2297–2300.

[69] G.T. Wei, F.K. Liu, C.R.C. Wang, Anal. Chem. 71 (1999) 2085–2091.

[70] R.L. Wolfe, R.W. Murray, Anal. Chem. 78 (2006) 1167–1173.

[71] Y. Song, M.L.A.V. Heien, V. Jimenez, R.M. Wightman, R.W. Murray, Anal. Chem. 76 (2004) 4911–4919.

[72] D. Rakcheev, A. Philippe, G.E. Schaumann, Anal. Chem. 85 (2013) 10643–10647.

[73] A.I. López-Lorente, B.M. Simonet, M. Valcárcel, Compr. Anal. Chem. 59 (2012) 33–89.

[74] U. Schnabel, C.H. Fischer, E. Kenndler, J. Microcolumn Sep. 9 (1997) 529–534.

[75] A.I. López-Lorente, B.M. Simonet, M. Valcárcel, Trends Anal. Chem. 30 (2011) 58–71.

[76] F.K. Liu, G.T. Wei, Anal. Chim. Acta 510 (2004) 77–83.

[77] A.I. López-Lorente, M.L. Soriano, M. Valcárcel, Microchim. Acta (2014). http://dx.doi.org/10.1007/s00604-014-1218-5.

[78] F.K. Liu, J. Chromatogr. A 1167 (2007) 231–235.

[79] I. Arnaud, J.P. Abid, C. Roussel, H.H. Girault, Chem. Commun. (2005) 787–788.

[80] M. Hanauer, S. Pierrat, I. Zins, A. Lotz, C. Sönnichsen, Nano Lett. 7 (2007) 2881–2885.

[81] E.P. Gray, T.A. Bruton, C.P. Higgins, R.U. Halden, P. Westerhoff, J.F. Ranville, J. Anal. At. Spectrom. 27 (2012) 1532–1539.

[82] B. Schmidt, K. Loeschner, N. Hadrup, A. Mortensen, J.J. Sloth, C.B. Koch, E.H. Larsen, Anal. Chem. 83 (2011) 2461–2468.

[83] X.J. Ma, Y. Kuang, L. Bai, Z. Chang, F. Wang, X.M. Sun, D.G. Evans, ACS Nano 5 (2010) 3242–3249.

[84] S.F. Sweeney, G.H. Woehrle, J.E. Hutchison, J. Am. Chem. Soc. 128 (2006) 3190–3197.

[85] J.C. Trefry, J.L. Monahan, K.M. Weaver, A.J. Meyerhoefer, M.M. Merkopolous, Z.S. Arnold, D.P. Wooley, I.E. Pavel, J. Am. Chem. Soc. 132 (2010) 10970–10972.

[86] A. Favre-Réguillon, M. Draye, G. Lebuzit, S. Thomas, J. Foos, G. Cote, A. Guy, Talanta 63 (2004) 803–806.

[87] J. Liu, R. Liu, Y. Yin, G. Jiang, Chem. Commun. (2009) 1514–1516.

[88] B.M. Quinn, P. Liljeroth, V. Ruiz, T. Laaksonen, K. Kontturi, J. Am. Chem. Soc. 125 (2003) 6644–6645.

[89] Y. Song, M.L. Heien, V. Jimenez, R.M. Wightman, R.W. Murray, Anal. Chem. 76 (2004) 4911–4919.

[90] L. Yu, A. Andriola, Talanta 82 (2010) 869–875.

[91] D. Hernández-Santos, M.B. González-García, A. Costa-García, Electrochim. Acta 46 (2000) 607–615.

[92] M. Pumera, M. Aldavert, C. Mills, A. Merkoçi, S. Alegret, Electrochim. Acta 50 (2005) 3702–3707.

[93] S.D. Richardson, Anal. Chem. 81 (2009) 4645–4677.
[94] A.B.A. Boxall, K. Tiede, Q. Chaudhry, Nanomedicine 2 (2007) 919–927.
[95] N.C. Mueller, B. Nowack, Environ. Sci. Technol. 42 (2008) 4447–4453.
[96] B. Nowack, T.D. Bucheli, Environ. Pollut. 150 (2007) 5–22.
[97] C. Tso, C. Zhung, Y. Shih, Y.M. Tseng, S. Wu, R. Doong, Water Sci. Technol. 61 (2010) 127–133.
[98] Q. Zhao, Y. Li, X. Cai, W. Li, J. Nanosci. Nanotechnol. 6 (2006) 1357–1364.
[99] H. Hyung, J.D. Fortner, J.B. Hughes, J.H. Kim, Environ. Sci. Technol. 41 (2007) 179–184.
[100] M. Farré, K. Gajda-Schrantz, L. Kantiani, D. Barceló, Anal. Bioanal. Chem. 393 (2009) 81–95.
[101] B. Mukherjee, J.W. Weaver, Environ. Sci. Technol. 44 (2010) 3332–3338.
[102] M.E. Mackay, A. Tuteja, P.M. Duxbury, C.J. Hawker, B. Van Horn, Z. Guan, G. Chen, R.S. Krishnan, Science 311 (2006) 1740–1743.
[103] L. Li, K. Leopold, M. Schuster, Chem. Commun. 48 (2012) 9165–9167.
[104] H.F. Krug, P. Wick, Angew. Chem. Int. Ed. 50 (2011) 1260–1278.
[105] L. Li, K. Leopold, Anal. Chem. 84 (2012) 4340–4349.
[106] G. Hartmann, M. Schuster, Anal. Chim. Acta 761 (2013) 27–33.
[107] F. von der Kammer, S. Legros, E.H. Larsen, K. Loeschner, T. Hofmann, Trends Anal. Chem. 3 (2011) 425–436.
[108] M. Baalousha, B. Stolpe, J.R. Lead, J. Chromatogr. A 1218 (2011) 4078–4103.
[109] G. Oberdörster, E. Oberdörster, J. Oberdörster, Environ. Health Perspect. 113 (2005) 823–839.
[110] D.A. Giljohann, D.S. Seferos, W.L. Daniel, M.D. Massich, P.C. Patel, C.A. Mirkin, Angew. Chem. Int. Ed. 49 (2010) 3280–3294.
[111] C.M. Goodman, C.D. McCusker, T. Yilmaz, V.M. Rotello, Bioconjug. Chem. 15 (2004) 897–900.
[112] N. Pernodet, X.H. Fang, Y. Sun, A. Bakhtina, A. Ramakrishnan, J. Sokolov, A. Ulman, M. Rafailovich, Small 2 (2006) 766–773.
[113] Y. Pan, S. Neuss, A. Leifert, M. Fischler, F. Wen, U. Simon, G. Schmid, W. Brandau, W. Jahnen-Dechent, Small 3 (2007) 1941–1949.
[114] E. Casals, S. Vázquez-Campos, N.G. Bastús, V. Puntes, Trends Anal. Chem. 27 (2008) 672–683.
[115] W.S. Cho, M. Cho, J. Jeong, M. Choi, H.Y. Cho, B.S. Han, S.H. Kim, H.O. Kim, Y.T. Lim, B.H. Chung, J. Jeong, Toxicol. Appl. Pharmacol. 236 (2009) 16–24.
[116] M.A. Dobrovolskaia, S.E. McNeil, Nat. Nanotechnol. 2 (2007) 469–478.
[117] C. Lasagna-Reeves, D. Gonzalez-Romero, M.A. Barria, I. Olmedo, A. Clos, V.M. Sadagopa Ramanujam, A. Urayama, L. Vergara, M.J. Kogan, C. Soto, Biochem. Biophys. Res. Commun. 393 (2010) 649–655.
[118] G. Sonavane, K. Tomoda, K. Makino, Colloids Surf. B 66 (2008) 274–280.
[119] J.F. Hillyer, R.M. Albrecht, J. Pharm. Sci. 90 (2001) 1927–1936.
[120] W.H.D. Jong, W.I. Hagens, P. Krystek, M.C. Burger, A.J.A.M. Sips, R.E. Geertsma, Biomaterials 29 (2008) 1912–1919.
[121] Y. Pan, S. Neuss, A. Leifert, M. Fischler, F. Wen, U. Simon, G. Schmid, W. Brandau, W. Jahnen-Dechent, Small 3 (2007) 1941–1949.
[122] H.F. Wang, T.B. Huff, D.A. Zweifel, W. He, P.S. Low, A. Wei, J.X. Cheng, Proc. Natl. Acad. Sci. U.S.A. 102 (2005) 15752–15756.
[123] T. Niidome, M. Yamagata, Y. Okamoto, Y. Akiyama, H. Takahashi, T. Kawano, Y. Katayama, Y. Niidome, J. Control. Release 114 (2006) 343–347.

[124] B.D. Chithrani, A.A. Chazani, W.C.W. Chan, Nano Lett. 6 (2006) 662–668.
[125] J.A. Ryan, K.W. Overton, M.E. Speight, C.N. Oldenburg, L. Loo, W. Robarge, S. Franzen, D.L. Feldheim, Anal. Chem. 79 (2007) 9150–9159.
[126] S.B. Lovern, H.A. Owen, R. Klaper, Nanotoxicology 2 (2008) 43–48.
[127] E.C. Cho, Q. Zhang, Y. Xia, Nat. Nanotechnol. 6 (2011) 385–391.
[128] T.S. Hauck, A.A. Ghazani, W.C.W. Chan, Small 4 (2008) 153–159.
[129] V. Amendola, M. Meneghetti, S. Fiameni, S. Polizzi, G. Fracasso, A. Boscaini, M. Colombatti, Anal. Methods 3 (2011) 849–856.
[130] P. Aggarwal, M.A. Dobrovolskaia, in: S.E. McNeil (Ed.), Characterization of Nanoparticles Intended for Drug Delivery, Methods in Molecular Biology, vol. 697, 2011, pp. 137–143.
[131] H. Xie, K.L. Gill-Sharp, D.P. O'Neal, Nanomed. Nanotechnol. Biol. Med. 3 (2007) 89–94.
[132] M.A. Dobrovolskaia, A. Patri, J. Zheng, J.D. Clogston, N. Ayub, P. Aggarwal, B.W. Neun, J.B. Hall, S.E. McNeil, Nanomed. Nanotechnol. Biol. Med. 5 (2009) 106–117.
[133] L. Castaneda, J. Valle, N. Yang, S. Pluskat, K. Slowinska, Biomacromolecules 9 (2008) 3383–3388.
[134] R.S. Niels de Jonge, Nanotechnology, in: T. Vo-Dinh (Ed.), Biology and Medicine–Methods, Devices and Applications, CRC Press, Boca Raton, FL, USA, 2007, p. 13.
[135] B.D. Chithrani, W.C.W. Chan, Nano Lett. 7 (2007) 1542–1550.
[136] J.W. Stone, P.N. Sisco, E.C. Goldsmith, S.C. Baxter, C.J. Murphy, Nano Lett. 7 (2007) 116–119.
[137] M. Schäffler, M. Semmler-Behnke, H. Sarioglu, S. Takenaka, A. Wenk, C. Schleh, S.M. Hauck, B.D. Johnston, W.G. Kreyling, Nanotechnology 24 (2013) 265103.
[138] Q. Chaudhry, R. Aitken, R. Scotter, J. Blackburn, B. Ross, A. Boxall, L. Castle, R. Watkins, Food Addit. Contam. 25 (2008) 241–258.
[139] Q. Chaudhry, K. Groves, in: Q. Chaudhry, L. Castle, R. Watkins (Eds.), Nanotechnologies in Food, RSC Publishing, Cambridge, UK, 2010, p. 69.
[140] I. Lynch, K.A. Dawson, Nano Today 3 (2008) 40–47.
[141] C. Blasco, Y. Picó, Trends Anal. Chem. 30 (2011) 84–99.
[142] V. Armendariz, J.G. Parsons, M.L. Lopez, J.R. Peralta-Videa, M. Jose-Yacaman, J.L. Gardea-Torresdey, Nanotechnology 20 (2009) 105607.
[143] S. Tedesco, H. Doyle, J. Blasco, G. Redmond, D. Sheehan, Comp. Biochem. Physiol. C 151 (2010) 167–174.
[144] A.M. Gatti, D. Tossini, A. Gambarelli, S. Montanari, F. Capitani, Crit. Rev. Food Sci. Nut. 49 (2009) 275–282.

Part III

Gold Nanoparticles as Analytical Tools

Chapter 11

Applications of Gold Nanoparticles in Electroanalysis

Marek Trojanowicz
Department of Chemistry, University of Warsaw, Poland and Laboratory of Nuclear Analytical Methods, Institute of Nuclear Chemistry and Technology, Warsaw, Poland
E-mail: trojan@chem.uw.edu.pl

Chapter Outline

1. INTRODUCTION

In contemporary science and technology, it might be difficult to find a field more intensively developed in the past two decades than nanotechnology and the synthesis, properties, and applications of nanomaterials. This has been the subject of thousands of research papers and patents and many books and

Gold Nanoparticles in Analytical Chemistry. http://dx.doi.org/10.1016/B978-0-444-63285-2.00011-0

specialized journals. The extraordinary properties of numerous nanomaterials, after that short period of intense investigation and development, have found surprisingly wide applications. Since the 1990s, nanomaterials have also been increasingly employed in different fields of analytical chemistry [1–3]. Their main areas of application, resulting from the unique physicochemical properties of nanomaterials and nanostructures, include use as sorbents in inorganic [4] and organic [5] analysis, in high-performance separation methods for modification of resolution in chromatography [6,7] and electromigration methods [8], and improvement of ionization in mass spectrometry [9]. Much attention has also been focused on their applications in spectroscopy, especially in spectroscopic methods of surface analysis [10], and in the design of optical sensors and biosensors [11].

One can find an especially broad scope of applications of nanomaterials and nanostructures in electrochemical methods of analysis [12,13] and the construction of electrochemical sensors and biosensors [14–16]. Based on nanomaterials, electrochemical sensors are being developed for various purposes, including safety and security applications [17] and medical diagnoses [18].

Electrochemical applications of nanomaterials are predominated by carbon and metallic nanoparticles, which are employed in modifying active surfaces of conventional macro- and microelectrodes. Such applications of gold nanoparticles (AuNPs) are the subject of this chapter. Another field of application of nanotechnology in electrochemistry and electroanalysis is the manufacture of nanodisk electrodes. Assuming that their diameters are smaller than 100 nm, they allow higher current densities, reduced capacitance of double layer, increased ratios of signal to noise, and short analysis times [19]. They can be manufactured by different methods, e.g., an etched microwire or pulled glass capillaries containing microwires. However, they also have several disadvantages, such as difficult and laborious fabrication, leaking of electrolyte through seals, and extremely low measurable currents; hence, much more attention is focused on modifying surfaces of conventional electrodes with nanomaterials. When metallic nanoparticles are used for this purpose, one of the most common methods employed is electrochemical fabrication, which may result in the formation of a wide variety of three-dimensional nanostructures [20]. Chemical deposition is also widely employed, in which nanoparticles are formed by means of chemical reduction with different reagents [21].

2. GOLD NANOPARTICLES IN ELECTROCHEMISTRY: BEGINNINGS

Besides carbon nanotubes [22], AuNPs are the most commonly used nanomaterials in electrochemistry; this has been presented in several review papers [23–25]. As especially valuable reference articles on various aspects of the synthesis, properties, and applications of AuNP, two are noteworthy, published by Daniel and Astruc [21] and Saha et al. [24].

Long before their introduction for common use in material science with terms such as "nanotechnology," "nanomaterials," and "nanoparticles," AuNPs found several applications in electroanalysis as colloidal gold. The review mentioned previously [21] indicated that the chemical formation of the colloidal gold was reported already in 1857 by Faraday [26]. Since then, colloidal gold has been the subject of scientific research and physicochemical, biological, and medical applications [27]. Modern applications in sensor technology appeared in early 1990 and dealt with the use of colloidal gold in the design of electrochemical enzymatic biosensors, in which enzymes were immobilized by adsorption to colloid gold sols [28,29]. There were also developed electrochemical methods to evaluate AuNP numbers in colloidal solutions and the number of AuNPs in which molecules of protein were adsorbed [30].

The first colloid chemical approach was published regarding the preparation of self-assembled spherical ultra-microelectrodes using AuNPs and nonanethiol molecules [31]. A three-dimensional conductive structure on an indium-tin oxide (ITO) electrode was built by alternating treatment with 12 nm AuNPs and cyclo-bis (paraquat-*p*-phenylene) tetrachloride as sensitive matrix for dopamine, adrenaline, and several other analytes [32]. A few years later, "nanowiring" of redox enzymes by AuNPs was documented, showing that incorporation of AuNPs in the enzyme greatly increases its maximum turnover rate [33]. Those works initiated further quick progress in the application of AuNPs in the design of electrochemical sensors and biosensors.

3. KINDS OF GOLD NANOSTRUCTURES EMPLOYED IN ELECTROCHEMICAL SENSORS

The physical properties of AuNPs significantly depend on their size and shape, the interparticle distance, and their application in protecting/stabilizing the organic shell. With the application of a well-resolved differential pulse voltammetry (DPV) and Pt working microelectrode, for an AuNP solution stabilized with hexane thiol, 15 evenly spaced peaks were recorded corresponding to 15 oxidation states of AuNP [34]. This observation is a good illustration of how complex the electrochemical properties of AuNPs can be, which will affect the characteristics of developed sensors in measurements with different detections.

There are two main methods for fabricating AuNPs for different applications, including the construction of electrodes for electroanalytical measurements. The one most commonly used is based on the reduction of Au(III) compounds in aqueous solutions using citrate, pioneered by Turkevitch et al. [35] in 1951. Spherical AuNP about 20 nm in diameter can be obtained. In another method, based on a two-phase system described by Faraday [26] and commonly reported as the Brust–Schiffrin method [36], Au(III) is reduced by applying thiol ligands that strongly bind reduced gold and stabilize nanoparticles. In the pioneering work $AuCl_4^-$ was transferred to toluene using tetra-octylammonium bromide and reduced by $NaBH_4$ in the presence of dodecanethiol. Depending on the value of the ratio of thiol to gold, AuNPs of different sizes from 1 to 5 nm can be obtained.

Those methods are mainly employed for fabricating AuNPs for application in sensors, where they are deposited directly on the electrode surface, or on the layer of polymer or sol–gel, which covers the electrode surface. Usually it is not discussed, but it is commonly assumed that nanoparticles fabricated in such a way are spherical; however, in numerous cases the reduction of Au(III) is carried out in modified conditions. For instance, octanethiol stabilizes AuNPs when manufactured with a modified Brust method [37], whereas AuNPs 5 ± 1 nm in diameter have been prepared by reduction with $NaBH_4$ in aqueous solution [38]. The latter method was also employed in the preparation of AuNPs directly on cellulose triacetate membrane, with immersion in reacting Au(III) solution and then the addition of $NaBH_4$ [39].

The reductive deposition of AuNPs has also been carried out in the presence of other reagents, which are introduced to stabilize them in solutions or to obtain a more complex composites and conjugates. For application in immunoassay, for example, glutathione-protected 5-nm AuNPs were obtained by $NaBH_4$ reduction [40]. They were deposited on the electrode on an underlayer of polycation and served to immobilize the antibody, which allowed two different strategies of immunoassay to be performed. In a similar way, cystamine-capped AuNP was produced for a deoxyribonucleic acid (DNA) biosensor [41]. Analogous procedures were used to create a (3-aminopropyl)triethoxysilane-derived silicate sol–gel three-dimensional AuNP network and for the simultaneous voltammetric determination of several reducing analytes [42] and poly(L-lactide) stabilized AuNPs [43]. *N,N'*(4-dimethylamino) pyridine-protected AuNPs were fabricated by adding the modifying reagent to nanoparticles obtained in a two-phase system with tetraoctylammonium bromide in toluene [44]. The reduction of Au(III) with citrate in fabricating AuNPs for their application as composite with carbon nanotubes (CNTs) was carried out in the presence of β-cyclodextrin in reaction solution [45]. Gold nanoparticles capped in polyvinyl alcohol molecules were prepared by adding alcohol to aqueous $HAuCl_4$ solution, which was heated for 1 h at 100°C under reflux [46].

Similar procedures were reported for carrying out more complex modifications of AuNPs or composites directly during the chemical reduction of Au(III). Dendrimer-encapsulated AuNPs were produced by adding COOH group-functionalized dendrimer to $HAuCl_4$ solution, and then $NaBH_4$ [47]. Bimetallic AuPd nanoparticles functionalized with β-cyclodextrin were obtained from mixing $HAuCl_4$ and K_2PdCl_4 with mercapto-β-cyclodextrin by reduction with $NaBH_4$ [48]. Several other nanoparticles were combined with

AuNPs directly during the process of Au(III) reduction. Two such composites were obtained with graphene. To design a hemoglobin-based sensor for nitrite, graphene–AuNP hybrids were created by adding graphene with biocompatible chitosan to $HAuCl_4$ solution containing hydrazine, and then adding citrate [49]. To obtain a tracing tag for immunosensing, acrylic microbeads were added to $HAuCl_4$ with citrate, and $NaBH_4$ was also introduced to the reaction mixture [50]. Antibody-functionalized magnetoparticles, starting from the core–shell $CoFe_2O_4$ – nanogold composite nanoparticles, were prepared by adding citrate to a boiling solution of $CoFe_2O_4$ and $HAuCl_4$ [51]. Polypyrrole-Au–coated NiO nanocomposites were obtained by adding a pyrrole monomer solution to NiO nanoparticles dispersed in ethanol and $HAuCl_4$ solution [52]. Similarly, AuNP–β-cyclodextrin–graphene composite was produced [53], whereas to prepare AuNP encapsulated-tubular TiO_2, 12-phosphotungstate-modified TiO_2 nanotubes were first ultraviolet irradiated and then added to $HAuCl_4$ solution without another reductant [54].

As already mentioned, one can assume that in these procedures the AuNPs obtained are spherical; spherical ones were also produced on graphene [49,53], polymer microparticles [50], magnetic nanoparticles [51], and NiO [52] or TiO_2 nanotubes [54]. Gold nanoclusters can be obtained by adding Au(III) to bovine serum albumin (BSA) solution, based on the reduction ability of BSA solution at pH ~12 [55]. Stabilized BSA–AuNP nanoclusters, which exhibit strong fluorescence, can be employed to obtain electrochemiluminescence sensors based on Au nanocluster–silica nanoparticle composites [56]. Gold atomic clusters were also electrogenerated on polycrystalline gold electrode by cycling in cetyltrimethylammonium bromide solution, and in the voltammetric nonenzymatic detection of pesticide methyl parathion they showed better sensitivity than that obtained with bulk gold and AuNP-modified electrodes [57].

Several other gold nanostructures were employed in developing electrochemical sensors. The synthesis of nanogold with complex three-dimensional structures such as flower structures has been reported in the literature [58]. For the application in the design of voltammetric immunosensors, gold nanoflowers with an average size of 50 nm in diameter were prepared by reduction of $HAuCl_4$ by ascorbic acid in the presence of chitosan [59]. Gold nanowire electrodes can be produced using a hybrid electron beam-photolithography process on silicon wafers, and their electrochemical characteristics show good promise for a variety of sensor applications [60]. Rod-shaped AuNPs were prepared from $HAuCl_4$ solution containing hexadecyltrimethylammonium bromide and $AgNO_3$ by reduction with ascorbic acid [61]. Their self-assembling on Pt electrode resulted in a nonenzymatic potentiometric glucose sensor.

Two other different applications of gold nanostructures in the construction of electrochemical sensors should be also mentioned. Solid-state potentiometric sensors were developed by immobilizing dithiolated-ionophores in

AuNPs [62]. Nanopore arrays with nominal diameters between 15 and 80 nm were formed by electro-less deposition of gold onto the surface of polycarbonate membranes. In a model system with Ag^+-selective thiacalixarene, a Nernstian response to Ag(I) ions was observed. Voltammetric trace mercury(II) determination was reported with a screen-printed gold nanofilm electrode, in which a 70- to 80-nm gold layer was fabricated by sputtering gold onto a silver conductive layer in a vacuum atmosphere [63].

4. DEPOSITION OF AuNPs ON THE SURFACE AND IN THE BULK MATERIAL OF ELECTRODES

Technically, the simplest way to incorporate the appropriately fabricated AuNPs into the bulk material of the electrode is to prepare them containing carbon paste. Several examples of such electrodes were reported in the design of ion-selective electrodes (ISEs) for potentiometric determination. In the case of Al(III) ISE, sol–gel-AuNP was prepared from AuNP and 3-(mercaptopropyl)-trimethoxysilane and was incorporated with ionophore DNMA (2,2′-dihydroxy-1-napththylidine-1′-naphthylmethylamine) in the paste of graphite and silicon oil [64]. To design a graphite paste ISE sensitive to Cu(II), ionophore *bis*(*o*-[*N*-methylidenamino-2-thiol-1,3,4-thiadiazole-5-*yl*]phenoxy) ethane was self-assembled on AuNPs fabricated by the citrate method and incorporated into the graphite paste in that form [65]. In an Ag(I) paste ISE, as an ionophore novel graphene oxide nanosheets covalently grafted with 2-thiophenecarboxylic acid were used, which, together with AuNPs, were added to graphite paste with paraffin oil [66].

There are also examples of the construction of amperometric paste sensors. A graphite paste amperometric sensor for L-cysteine was prepared by incorporating into the paste a composite of Prussian blue (PB) and AuNP [67]. The composite was prepared by mixing AuNP with PB solution and graphite powder. After ultrasonication and evaporation, the obtained powder was used to prepare a paste with Nujol. It was shown that the most sensitive response was obtained when nanostructured palladium was added to the active carbon powder.

In the design of a potentiometric biosensor, in the final step of biosensor preparation, antibody-functionalized magnetic nanoparticles with adsorbed AuNPs via an Au–S bond for antibody attachment were in final step of biosensor preparation attached to the surface of a carbon paste electrode with a magnet placed on the other side of the carbon paste [51].

To deposition AuNPs or their composites and conjugates on the surface of working or indicating electrodes, there are three main strategies for preparation (besides electrodeposition, which is separately discussed in the next part). First is the physical adsorption of nanoparticles on different materials of electrodes and the incorporation of nanoparticles in polymer layers formed on the electrode surface, or the strongest chemical binding to the surface using different interactions.

The most commonly used method is surface adsorption. For various carbon-based electrode materials, AuNPs were directly deposited, e.g., on the surface of a screen-printed carbon electrode [68], glassy carbon (GC) [69,70], or GC modified with carbon nanotubes [71], or with graphene together with carbon nanotubes [72]. In a similar way, AuNPs were deposited onto the surface of carbon fiber coated with aligned carbon nanotubes [73] and by sputtering deposition of AuNPs onto vertically aligned carbon nanotubes [74]. In all of these methods except the last one, the process is usually carried out by evaporation of AuNPs solutions in different solvents. The layer containing AuNPs can also be covered, e.g., with a film of chitosan [70]. A three-dimensional structure of such a layer can be obtained by multiple formations of CNT and AuNP layers in the design of a voltammetric cytosensor [71] (Figure 1(A)). Deposition on nonmodified AuNPs was also employed on the surface of a polycrystalline gold disk electrode [77] and with a complex preparation on three-dimensional Pt-coated AuNP arrays [78]. In the latter method nanoparticle trapping was employed at the interface of immiscible solvents and subsequent transfer on the substrate for electrochemical ultrathin layer coating. Gold nanoparticles were also deposited onto the surface of indium-tin oxide (ITO) electrodes derivatized with (3-aminopropyl) trimethoxysilane [79]. In another report, inkjet printing of AuNPs onto the cellulose ester membrane was described, which was further used for seed growth after immersing the membrane into $HAuCl_4$ solution with NH_2OH [80]. A similar procedure was used to deposit AuNPs on paper or cotton gauze [81]. Such produced materials have been placed onto the surface of a GC electrode and fixed tightly with a nylon mesh membrane. In the construction of a potentiometric biosensor for Hg(II), based on the inhibition of urease, AuNPs were deposited directly onto the surface of a PVC membrane [82].

Many differently modified AuNPs were directly adsorbed onto surfaces of electrodes. On GC electrodes several composites of AuNPs with other nanomaterials were deposited, such as CNTs obtained with microwaves and dispersed in dimethylformamide [83], or with graphene [49]. In a similar way AuNP–β-cyclodextrin–graphene composite [53] and NiO/polypyrrole–Au nanocomposite [52] were deposited. On GC electrodes AuNPs embedded in (3-aminopropyl) triethoxysilane-derived silicate sol–gel [42] were deposited and thiolated or modified with dithizone [37]. Glutathione-decorated AuNPs have been deposited onto pyrolytic graphite disks [84], whereas AuNPs embedded in the polyvinyl alcohol-tetraethyl orthosilicate matrix were deposited onto a graphite electrode [46]. For the same purpose, carbon SPE electrodes were used with immobilization of glutathione-modified AuNPs [40], and poly(L-lactide) AuNPs [43].

For more resistance, immobilization of nanoparticles in polymeric layers formed on the surfaces of solid working or indicating electrodes are considered. Such a method was used both in the construction of potentiometric sensors, as well as voltammetric ones. Modified AuNPs or their composites are

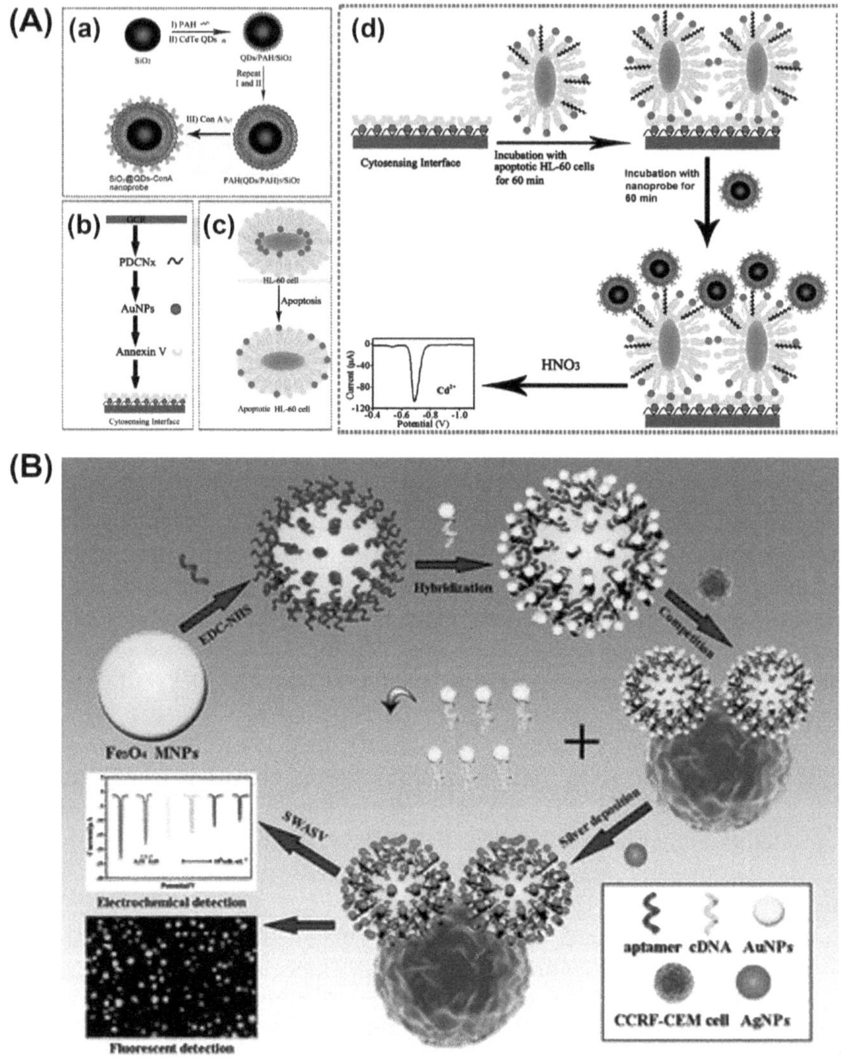

FIGURE 1 Examples of the design of AuNP-based voltammetric cytosensors. (A) [71]: (a) Scheme of the preparation of an SiO_2−QDs-ConA nanoprobe via layer-by-layer assembly; (b) Illustration of the fabrication process of the electrochemical cytosensing interface; (c) Schematic presentation of the exposure of phosphatidylserine residues on the outer surface of the cell membrane during apoptosis; (d) sandwich electrochemical strategy for the detection of apoptotic cells. (B) [75]: Schematic illustration of the fabrication process of a competitive hybridization-based sensing platform. The amino-modified aptamers were assembled on the surface of Fe_3O_4 MNPs, and AuNP-conjugated cDNA were then mixed with MNPs for hybridization with target cells. (C) [76]: Scheme of nanoprobe assembly and electrochemical strategy for in situ detection of mannose groups on living cells. Con A, concanavalin A, a mannose group specific lectin; SWNHs, single-walled carbon nanohorns; RGDS, arginine-glycine-aspartic acid-serine tetrapeptide; *o*-PD, *o*-phenylenediamine; DAB, 2,2′-diaminoazobenzene.

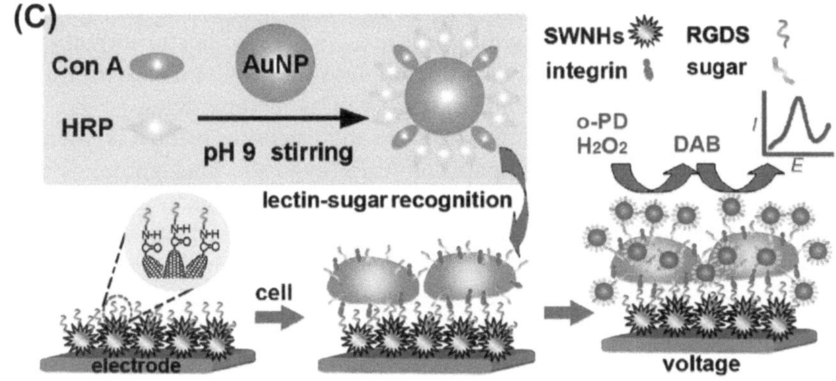

FIGURE 1 cont'd

most commonly immobilized in the layer of cation-exchange polymer Nafion, which is formed by evaporating its solutions in organic solvents. This method has been employed for deposition of AuNPs with anti-paraoxon antibody on a GC electrode [85]; as another example, a Pt disk electrode was first covered with a Nafion layer and then AuNPs with a hepatitis B surface antibody were entrapped on top of it by a gelatin matrix [86]. In Nafion layers on GC electrode a composite of AuNPs was immobilized [45], and AuNP-encapsulated TiO_2 nanotubes, modified with 12-phosphotungstic acid, together with thiolated horseradish peroxidase (HRP) [54], and to design a electrochemiluminescence sensor, an AuNP−BSA−silica nanocomposite [56]. To construct amperometric H_2O_2 enzymatic biosensor, modified electrodes were prepared by directly electrospinning AuNP--composite Nafion nanofibers on the stainless-steel working electrode, and then the positively charged HRP was immobilized on the nanofibrous electrode via electrostatic adsorption [44]. On GC electrode AuNPs were also entrapped into the sol−gel network by the hydrolysis of ethylmethoxysilane for amperometric detection of some biologically relevant analytes [87]. The xerogel-based biosensors reported recently were composed of a Pt electrode modified with a composite film of (3-mercaptopropyl) trimethoxysilane with glucose oxidase (GOD), hexanethiolate-stabilized AuNPs ($Au_{223}(C6)_{75}$) monolayer protected clusters, and coated with an outer polyurethane layer [88]. The potentiometric immunosensor of diphtheria was fabricated by means of self-assembly to immobilize diphtheria antibody on a Pt electrode modified with a layer of polyvinyl butyral containing Au, Ag, or silica nanoparticles [89]. An amperometric enzymatic biosensor employing a polyphenol oxidase was developed by modifying CNTs and AuNPs within a poly(allylamine hydrochloride) film, which then was treated with cysteamine and glutaraldehyde to immobilize the enzyme [90]. In addition, the in situ formation of stable AuNPs in polymer inclusion membranes [39] offers the possibility of tailoring the properties of membranes as well as nanoparticles.

Another strategy for binding AuNPs to the electrode surface is chemical binding, in which the chemisorption of AuNP onto the thiol groups is most commonly used. They are introduced onto the surface of electrodes by the use of specific reagents or by the deposition of polymers. For instance, by using cystamine as a covalent attachment cross-linker, a multilayer enzymatic biosensor was developed [91]. The surface of a gold electrode was covered with cysteamine, and then oxidized glucose oxidase was covalently attached, again covered with a cysteamine layer to which AuNPs were attached. In the optimum design, six such layer were used. Similarly, in the construction of a voltammetric immunosensor for proteins, a gold electrode was modified with cysteamine and then with mercaptoethanol, and then AuNP-encapsulated dendrimer was bound through the formation of amide bonds between the NH_2 groups of cysteamine and COOH groups of dendrimer [47] (Figure 2(A)). Finally, on such a modified surface, antibody was co-immobilized with thionine as an electron mediator for the voltammetric detection of specific proteins. In the design of an amperometric immunosensor for α-fetoprotein, the gold working electrode was first modified with a mixture of two thiols (cysteamine and 1,6-hexanedithiol); on such an self-assembled monolayer (SAM)-coated electrode AuNPs were attached together with antibody [92]. Frequently employed in similar systems, thionine has a double role: It acts as reagent forming an Au−S bond and it is also an electron mediator. In another example of amperometric sensors for α-fetoprotein, a gold electrode first was covered with a Nafion layer and then was soaked in a thionine solution [93]. In the next step, AuNPs were adsorbed with antibody. Immobilization of thionine was assisted electrostatically owing to the positive charge of thionine and the negative sulfonic groups of Nafion. Thionine was also employed to immobilize CNT−AuNP composites onto a GC electrode in the design of an enzymatic biosensor for H_2O_2 [94]. In this case, it was used as a cross-linker to construct a CNT−thionine−AuNP composite, and suspension of composite in ethanol was drop-coated onto the GC electrode. Furthermore, the enzyme solution was dropped to obtain a final biosensor.

Voltammetric and potentiometric sensitivities to Cu(II) ions were observed in the thin film assembly of AuNPs, which was linked by 11-mercaptoundecanoic acid on the surface of a GC electrode [95]. Gold nanoparticles encapsulated with decanethiol monolayer were obtained by the standard two-phase Brust method. To produce a more stable attachment of immobilized species, reagent bearing −SH groups are polymerized on substrate surfaces or polythiophene derivatives are used. This was employed, e.g., in the layer-by-layer deposition of polythiophene−AuNPs onto Au and ITO electrodes [96]. In the case of a gold electrode, its surface was first covered by

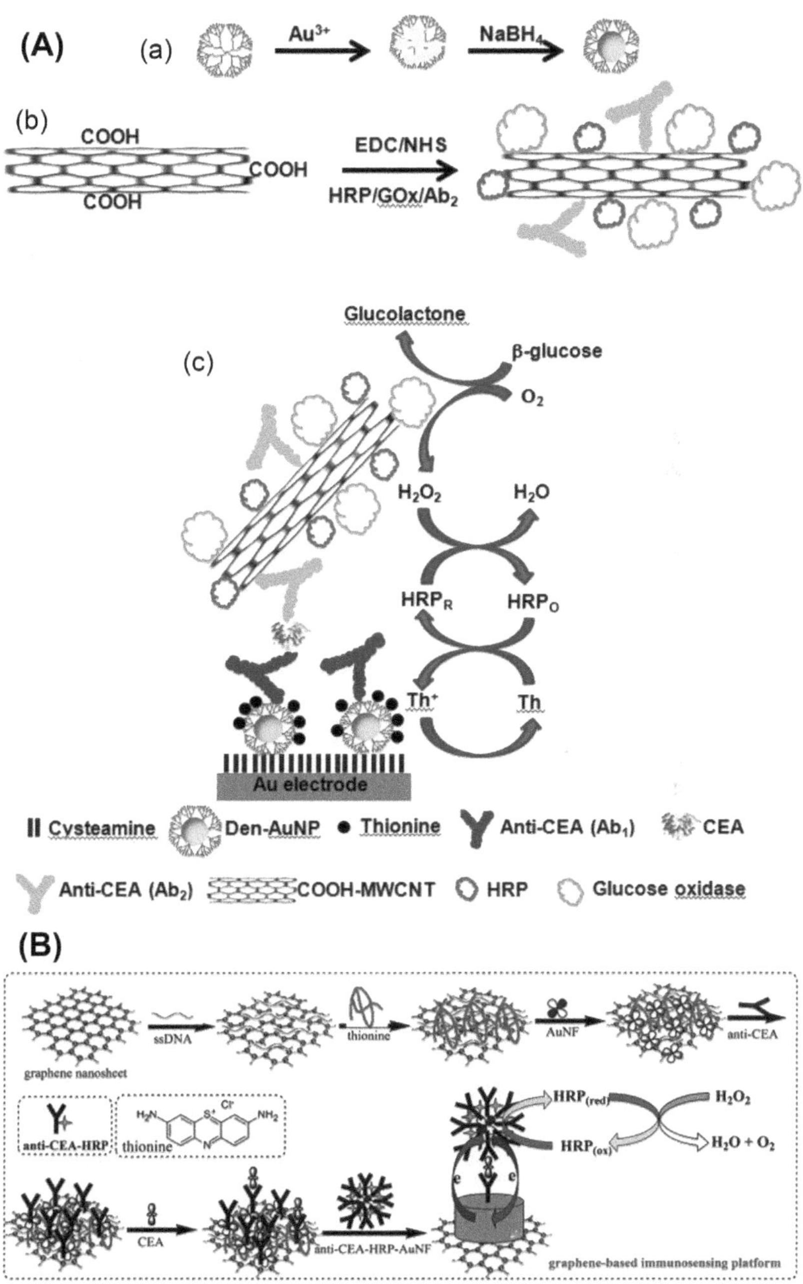

FIGURE 2 Examples of the design of electrochemical immunosensors involving AuNPs and HRP labeling for amperometric detection. (A) Schematic illustrations of the fabrication of voltammetric carcinoembryonic antigen (CEA) immunosensors and detection principle [47]. (B) Schematic illustration of the graphene-based electrochemical platform and sandwich-type protocol for CEA immunoassay [59].

2-mercaptoethanesulfonate and then multilayer films were built by alternatively dipping in aqueous solutions of poly(1-methyl-3-[3-[3-thienyloxy]-propyl]-1H-imidazolium) fluoride and AuNPs obtained by citrate reduction. The electrocatalytic properties of those systems were tested for ascorbic acid.

Polythiolated systems were also reported for the immobilization of AuNPs and biomolecules on substrates other than gold. In the design of an impidimetric immunosensor, a GC electrode was covered by electropolymerization of thionine and then immersed in AuNP solution, and finally in antibody solution [97]. The electropolymerization of 3-octylthiophene on a Pt electrode was also a first step in the preparation of a potentiometric nonenzymatic glucose sensor [61]. Rod-shaped AuNPs, before their deposition onto a poly(3-octylthiophene)–covered Pt disk, were dispersed in an ethanol solution of 1-decanethiol and 11-mercaptoundecanoic acid to prepare Au-SAMs. Another approach can be self-assembling gold AuNPs to a thiol-containing sol–gel network [98,99]. In preparing an HRP amperometric biosensor, a gold electrode was immersed in a hydrolyzed (3-mercaptopropyl) trimethoxysilane sol–gel solution and then AuNPs were chemisorbed onto the thiol groups; finally, HRP was adsorbed onto the surface of AuNPs [98]. In the fabrication of a diphtheria potentiometric immunosensor, after immersing the Au electrode into the solution of the same silane, a polymerization step was carried out before depositing a mixture of Au and Ag nanoparticles and antibody [99]. The use of AgNPs in optimized proportion to AuNPs resulted in a certain increase in sensitivity of the potentiometric response. The same sol–gel was also applied for silanization of an Si/SiO_2-sensing surface of a capacitive chip in the preparation of a glucose sensor [100] (Figure 3(A)). On such a modified surface, tetraoctylammonium bromide AuNPs were deposited, and then the binding of GOD to the AuNPs modified surface was performed through a bifunctional coupling reagent dithio-(succinimidyl proprionate). Co-immobilization of ferrocene in this system resulted in a twofold increase in response sensitivity. Another application of the same silanizing compound was reported in the fabrication of a gold nano-gapped electrode system developed for the amperometric detection of polychlorinated biphenyl [101] (Figure 3(B)). In this case, deposited AuNPs were further modified thiolated HS-β-cyclodextrin employed for capturing polychlorinated biphenyls.

Two other examples of attachment of AuNPs to the surface of solid electrodes involve application of poly(*p*-aminobenzene sulfonic acid) with L-cysteine [102] and sulfonated graphene [103]. The amperometric H_2O_2 biosensor based on the use of hemoglobin was designed by covering a Pt electrode by electropolymerizing aminobenzene sulfonic acid, immersing in a solution of L-cysteine, and adsorbing AuNPs. Finally, on such modified support, hemoglobin was adsorbed from aqueous solution [102]. A selective amperometric nitrite sensor was obtained by covering a GC electrode with sulfonated graphene, and then by self-assembling positively charged AuNPs onto the surface of negatively charged sulfonated graphene [103].

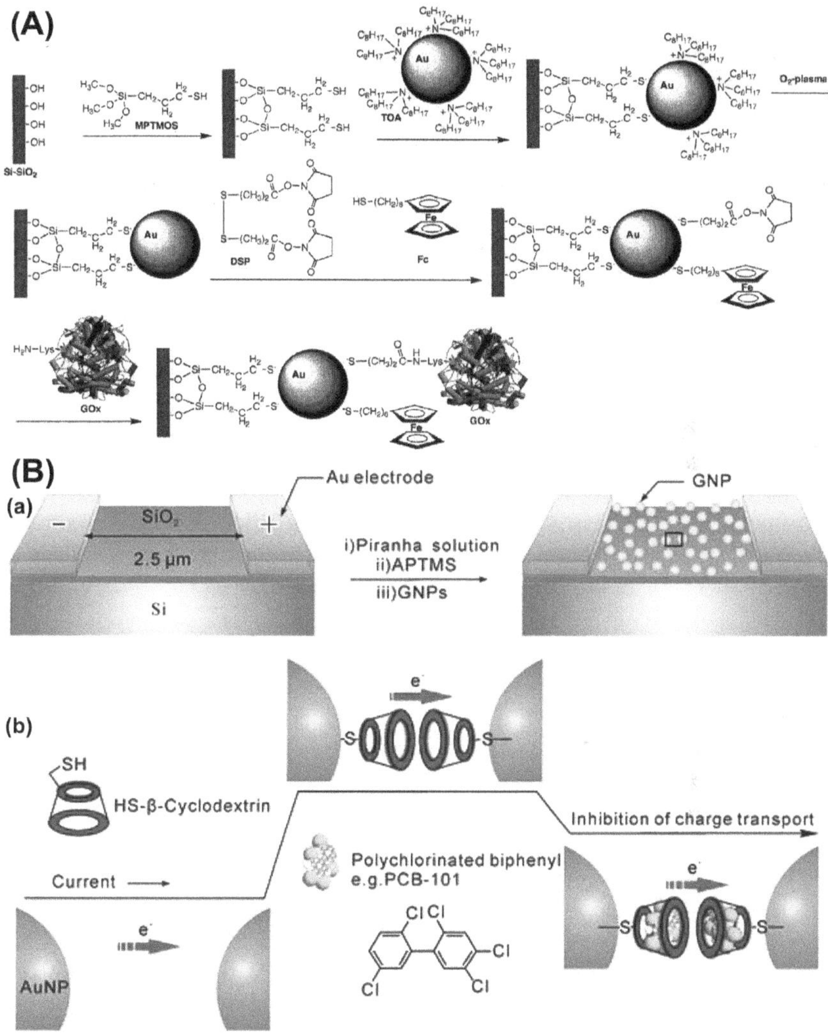

FIGURE 3 Examples of the application of polythiolated self-assembly in construction of AuNP-based electrochemical sensors. (A) Modification steps in the preparation of a ferrocene/GOD-functionalized Si-based EIS field-effect structure for AuNP-based constant capacitance mode biosensor for glucose [100]. (B) Schematic of an electrical nanogap device modified with β-cyclodextrin (CD) for polychlorinated biphenyl detection [101]. (a) Electrical nanogap is constructed by a nanogapped gold particle film on an interdigitated microelectrode with a 2.5-μm gap. (b) Conceptual illustration of specific inhibition of charge transport. Functionalized point traps made through the modification of HS-β-CD between the gaps in an electrical nanogap device. Cavities of HS-b-CD act as specific grabbers to analytes. Charge transport can be inhibited by electrical barriers (analyte molecules) in the transport channels.

5. ELECTRODEPOSITION OF AuNPs FOR ELECTROCHEMICAL APPLICATIONS

Preparation of AuNPs by the electrochemical reduction of Au(III) is one of the most efficient approaches to obtaining structured gold nanomaterials [104]. It is also employed in modifying electrodes in the construction of electrochemical sensors and biosensors. Electrodeposition is carried out on different substrates and under various electrochemical conditions. In the case of the deposition of AuNPs onto the surfaces of electrodes that were previously modified with other nanomaterials, it leads to the formation of different composites.

The most common method of electrodeposition of AuNPs on various substrates is carried out under potentiostatic conditions from acidified $HAuCl_4$ solutions, with applied potentials from 0.18 [105] to −1.2 V [106]. In the latter case, deposition was carried out on a gold electrode and flower-like gold nanostructures were formed [106]; such prepared electrodes were used in the simultaneous voltammetric determination of dopamine and ascorbic acid using differential pulse voltammetry (DPV). The electrodeposition of AuNPs was also carried out under galvanostatic conditions [107,108] using cyclic voltammetry [109−111] or linear sweep voltammetry [112]. In electrodeposition onto an ITO substrate with cyclic voltammetry, the reduction of Au(III) was carried out with a solution of $KAu(CN)_2$ complex in phosphate buffer [109,110].

Among work on the electrodeposition of AuNPs, resulting from the reduction of Au(III), this method was most commonly exploited in modifying carbon screen-printed electrodes (SPE) [111−118]. Some were employed directly in the voltammetric determination of particular analytes [113,116], but in many cases they were further modified with biomolecules to produce amperometric biosensors. By cross-linking with glutaraldehyde, the immobilization of urease to obtain an Hg(II) inhibitive-based biosensor was reported [114], or by directly covering solution the biosensor with aptamer solution for the detection of MUC1 tumor marker [111]. To immobilize protein in the design of a sensor to detect specific immunoglobulin G (IgG) antibodies, AuNP-modified SPE was covered with a layer of 8-marcaptopropionic acid, which after activation with carbodiimide covalently conjugated antigen [118]. In another design of an SPE biosensor to detect antibodies, before the electrodeposition of AuNPs, an electrode was modified with a thin film formed by dropping Al_2O_3, but then the AuNP layer was directly covered with antigen protein [115]. For amperometric detection of glycated hemoglobin, the AuNP-modified SPE was covered with the conducting polymer, followed by the covalent attachment of amino-phenyl boronic acid as a host to capture the analyte [112]. Similarly, in the design of a DNA-based amperometric biosensor for the detection of anticancer drugs, the same polymer layer electrodeposited onto AuNP-SPE was used to immobilize ssDNA [117].

As substrates for the electrodeposition of AuNPs also GC electrodes [105,119], mentioned already ITO [109,110], gold [106], and boron-doped diamond [107] were reported. Glassy carbon-AuNP electrodes were used for direct ASV determination of arsenite [105] or served as substrates for the electrochemical deposition of molecularly imprinted polymers for the voltammetric determination of organophosphorus pesticides [119] (Figure 4). On the surface of ITO-AuNP electrodes the adsorptive immobilization of enzyme superoxide dismutase was reported [109], and they were also used to examine the effect of Fenton's reagent on reducing the catalytic activity of AuNPs [110]. The electrodeposition from $HAuCl_4$ solution, containing ethylenediaminetetraacetic acid and Na_2SO_3 was reported for the preparation of an Au nanowire array as a result of deposition onto porous aluminum oxide templates [108]. Those electrodes were also modified by the electrodeposition of platinum nanoparticles and adsorption of glucose oxidase. In the case of a boron-doped diamond electrode, the layer of electrodeposited AuNPs was modified by immersion into cysteamine solution, and then glutaraldehyde tyrosinase was immobilized for the voltammetric determination of phenol [107].

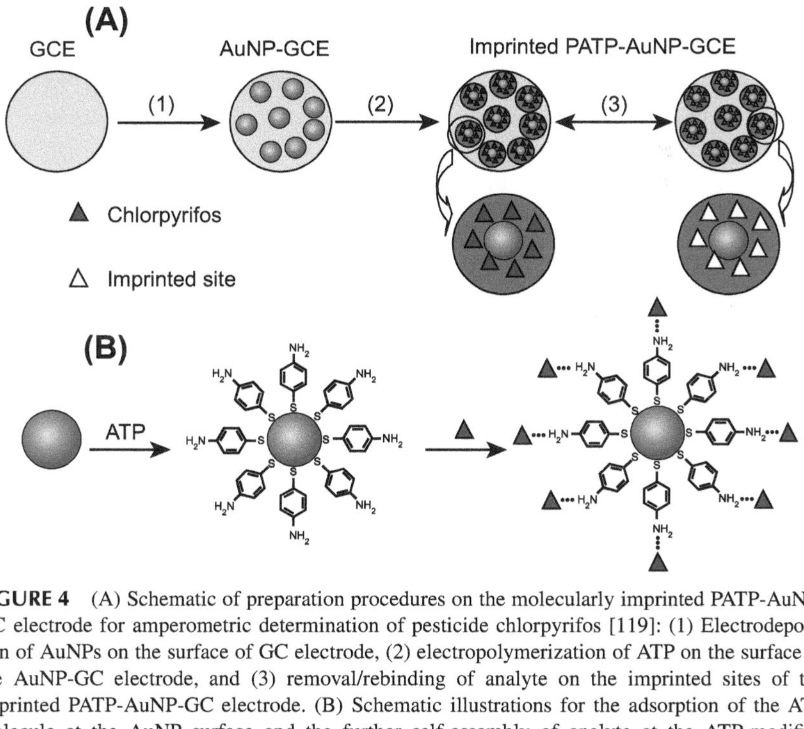

FIGURE 4 (A) Schematic of preparation procedures on the molecularly imprinted PATP-AuNP-GC electrode for amperometric determination of pesticide chlorpyrifos [119]: (1) Electrodeposition of AuNPs on the surface of GC electrode, (2) electropolymerization of ATP on the surface of the AuNP-GC electrode, and (3) removal/rebinding of analyte on the imprinted sites of the imprinted PATP-AuNP-GC electrode. (B) Schematic illustrations for the adsorption of the ATP molecule at the AuNP surface and the further self-assembly of analyte at the ATP-modified AuNP-GC electrode.

The electrodeposition of AuNPs onto different substrates was also widely employed in the preparation of nanocomposites with AuNPs. Using cyclic voltammetry or chronocoulometry, the arrays of AuNPs were fabricated on single-walled carbon nanotubes films deposited onto a flexible poly(ethylene terephthalate) substrate for the amperometric detection of hydroxylamine [120]. To create a tramadol drug the GC electrode covered with CNT suspension and then with electrodeposited AuNPs was employed [121]. Graphene SPE before the electrodeposition of AuNPs was covered with MWCNT or graphene oxide films and employed in the ASV determination of Hg(II) [122]. For the voltammetric determination of nitrite, GC and ITO electrodes were used with ultrasonic electrodeposition of AuPt alloy nanoparticles on MWCNT-ionic liquid composite film [123]. Other similar systems were reported for the use of graphene. A GC electrode used for the voltammetric detection of 4-nitrophenol was first modified by the electrochemical deposition of graphene oxide and then AuNP from Au(III) solution [124]. The aptamer-linked biosensor for the voltammetric determination of thrombin in blood serum was more complex [125]. First, a GC electrode was modified by evaporating graphene oxide dispersion containing thionine. Then, AuNPs were deposited by cyclic voltammetry and aptamer was adsorbed onto the modified electrode surface.

Other AuNP modifications of electrode surfaces may involve the use of, e.g., chitosan or conducting polymers. The one-step electrodeposition of AuNP-chitosan on a GC electrode was conducted from an $HAuCl_4$ solution containing chitosan under strongly reductive conditions at −2.5 V [126]. Onto such a modified electrode surface, the GOD solution containing cytochrome c was applied to obtain biosensors for the voltammetric determination of glucose. The potentiometric immunosensor for hemoglobin was based on an electrochemically synthesized polypyrrole−AuNP composite [127]. The deposition was carried onto the surface of gold thin electrodes of an ion-sensitive field-effect transistors integrated chip from $HAuCl_4$ solution containing pyrrole by cyclic voltammetry. The surface of the PPy-AuNP film was covered by adsorbed antibody.

Composites of polyaniline and AuNP have been used in the design of a potentiometric DNA biosensor for the detection of nucleic acids based on hybridization with an immobilized ssDNA probe [41]. On the surface of a gold electrode, composite polymeric multilayer films were formed, first by chemical modification with 3-marcapropropionic acid and the deposition of cysteamine-capped AuNPs from $HAuCl_4$ solution containing cysteamine. This was followed by immersion into sulfonated polyaniline solution; finally, the amino groups of cysteamine-AuNP formed an electrostatic affinity with the phosphate skeleton of DNA molecule for immobilization.

The electrochemical deposition of gold inside the silica nanopore allows the fabrication of Au disk nanoelectrodes with diameters down to ~8 nm [128]. They can be also be modified with chemically fabricated AuNPs or 6-(ferrocenyl) hexanethiol.

In the modification of a GC electrode, it was also demonstrated that the electrodeposition of AuNPs can be carried out with internal electrolysis [129].

6. GOLD NANOPARTICLES AS PROBES IN ELECTROCHEMICAL BIOSENSING

In most of these applications of AuNP in electroanalysis, using them to modify an electrode surface or as a component of electrode bulk material resulted in catalytic enhancement of the analytical signal and improvement in the sensitivity of detection. They can also act as convenient support for the immobilization of biomolecules in the construction of enzymatic biosensors, immunosensors, or DNA genosensors. In discussing the construction of electrochemical biosensors with AuNPs, we should not neglect their application as labels (tags or probes) in multistage determinations carried out with electrochemical detection. Numerous such applications can be found, especially recently, for the development of novel immunosensors and biosensors based on interactions of analytes with aptamers or DNA.

Many immunochemical assays with electrochemical detection and using AuNPs were developed for the determination of proteins. In sandwich immunoassay IgG employing conjugate of AuNPs with antibody (Ab), potentiometric detection with a silver ion-selective electrode was reported [130]. Gold nanoparticles were covered with silver and functioned as a tag for bound antigen (Figure 5(A)); after oxidative dissolution with hydrogen peroxide, they were determined potentiometrically. In another approach to IgG immunochemical determination, the voltammetric detection with an ITO electrode modified with ferrocenyl-tethered dendrimers was used [132]. Using an external magnet, the conjugate of magnetic beads with Ab is bound to its surface, and in the presence of antigen IgG a sandwich is formed with AuNP–Ab conjugate (Figure 6(A)). Quantitative determination is based on measurement of the amount of AuNPs bound, which catalyzes the reduction of *p*-nitrophenol to *p*-aminophenol in the presence of $NaBH_4$, and which then is reoxidized at the electrode. In the voltammetric immunoassay of carcinoembryonic antigen, a polystyrene microbead was covered with chemically deposited AuNPs, which were also employed as a tracing tag to label signal antibody [50] (Figure 7(A)). Gold nanoparticles induced the deposition of silver nanoparticles, and finally silver was determined using anodic stripping voltammetry (ASV) with linear sweep, using a GC electrode modified with graphene oxide and chitosan. As a signal tag in the immunoassay of chloramphenicol AuPd bimetallic nanoparticles functionalized with β-cyclodextrin (CD) were employed [48]. Adamentine, used for labeling antibody, strongly bound β-CD–functionalized nanoparticles, and voltammetric detection, was based on the electrocatalysis of AuPd nanoparticles toward $NaBH_4$ oxidation on GC electrodes modified with CNTs. Also, in an electrochemical immunoassay for the detection phosphorylation, AuNPs modified with HRP-loaded

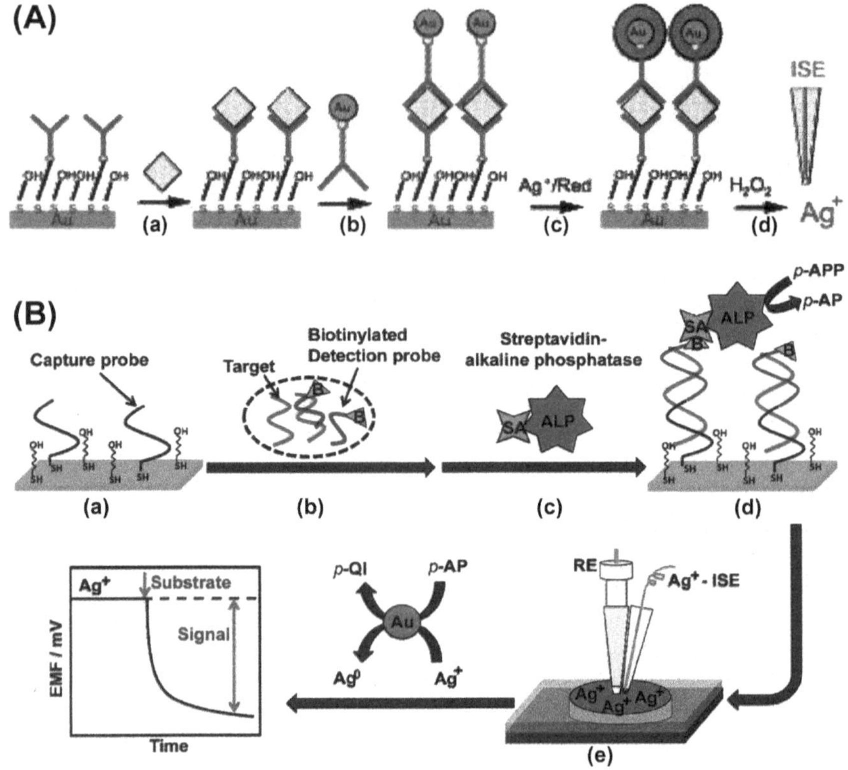

FIGURE 5 Application of potentiometric detection using Ag^+ ISE in AuNP-based sensing. (A) Schematic illustration of potentiometric detection in a sandwich immunoassay of IgG protein [130]: (a) antigen addition, (b) capture of the AuNPs-labeled anti-mouse IgG antibody, (c) catalytic deposition of silver ions on the conjugated AuNPs, (d) silver dissolution and potentiometric detection using Ag^+-selective electrode. (B) Scheme of the application of potentiometric detection of DNA hybridization [131]: (a) formation of the mixed thiol monolayer (thiolated DNA capture probe and MCH), (b) hybridization of the target DNA—biotinylated detection-probe mixture with the surface capture probe, (c) binding of the SA-ALP enzyme, (d) addition of the ALP substrate to initiate the enzymatic reaction, (e) potentiometric detection of changes in the level of the silver ion upon adding an aliquot of the enzymatic reaction mixture to the Ag^+-ISE cell.

streptavidin were employed [134]. For detection purposes, the oxidation of tetramethylbenzidine catalyzed by HRP is used on a gold electrode.

The most common application of biosensors with immobilized oligonucleotides DNA is in the detection of hybridization, leading to the detection and quantitative determination of specific nucleic acid sequences. Their functioning can based on the measurement of electrical signals resulting from the change in concentration of redox label associated with the formation of the DNA duplex. The signal changes can be also induced by DNA hybridization itself. The magnitude of such signals can be significantly enhanced by applying appropriate labels providing large changes in electrochemical

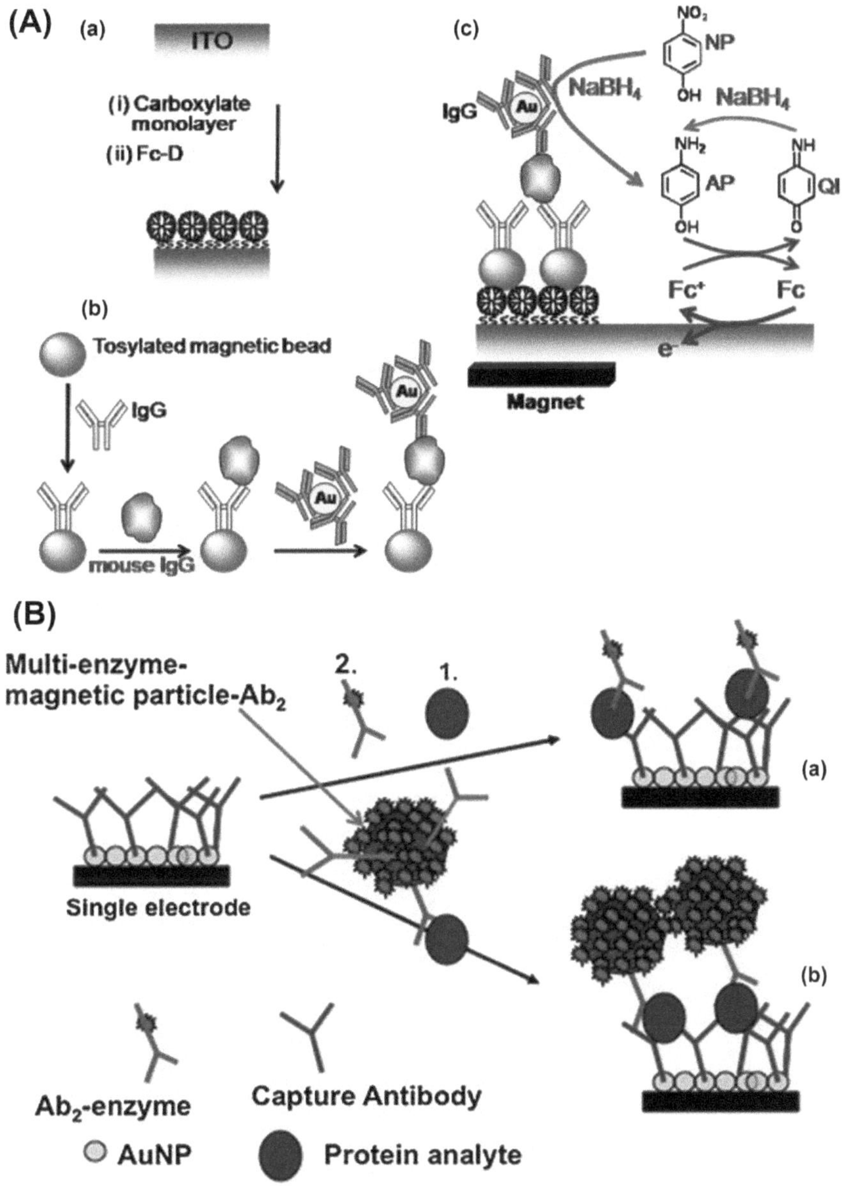

FIGURE 6 Schemes of construction of sample electrochemical AuNP-based immunosensors involving magnetic microparticles (MP). (A) Voltammetric immunosensor for the detection of protein IgG [132]: (a) preparation of an Fc-D—modified electrode, (b) scheme of the preparation of an immunosensing complex, (c) scheme of the attraction of the immunosensing complex to an Fc-D—modified ITO electrode and the electrochemical detection of IgG. (B) Alternative strategies of amperometric protein immunosensor featuring AuNP surfaces with capture antibodies attached [40]. (a) Immunosensor after incubating with protein analyte (1) and flowing in a conventional single-enzyme (HRP)-labeled antibody (2,Ab_2); (b) Immunosensor after analyte protein captured offline with MP-Ab_2—HRP—analyte bioconjugates are trapped by antibodies on the electrodes, and amperometric signals are then developed by injecting a solution of mediator and H_2O_2.

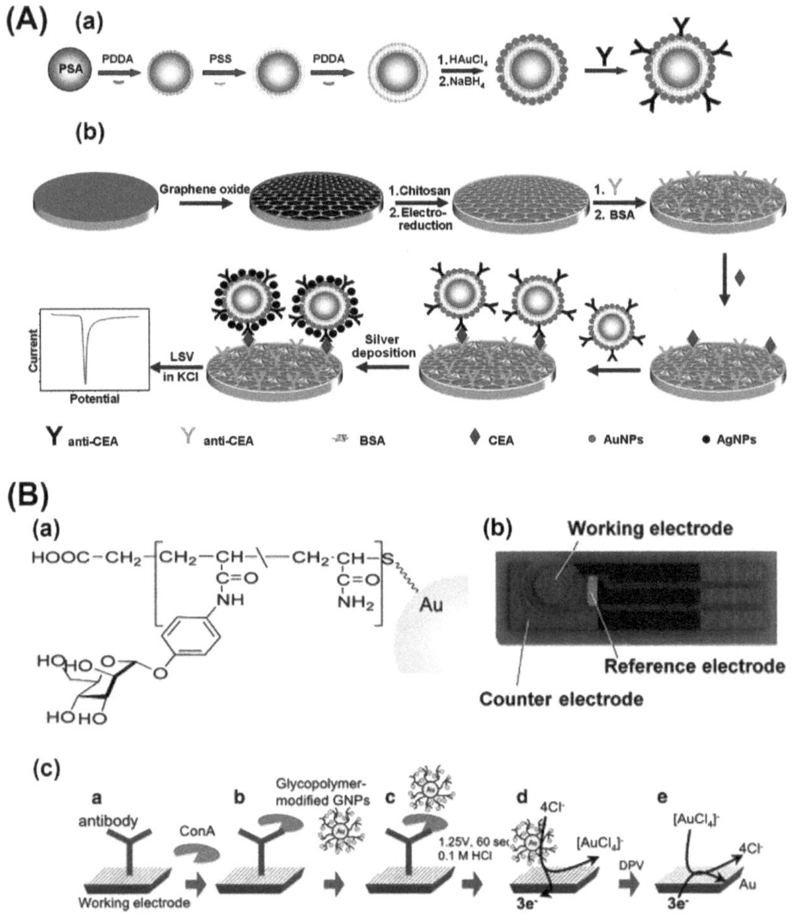

FIGURE 7 Voltammetric immunosensors employing AuNPs as tracing tags. Upper: Schemes of the preparation of tracing tag and labeled Ab_2 (A), and immunosensor fabrication and sandwich-type immunoassay procedure of anticarcinoembryonic antigen (CEA) (B) [50]. Lower: Schemes concerning a voltammetric immunoassay employing glycopolymer-modified AuNPs [133]: (A) Structure of a glycopolymer consisting of copolymer of *p*-acrylamidophenyl-α-D-mannoside. (B) Structure of SPE. (C) Schematic illustration of glycopolymer-AuNP–based electrochemical protein assay for the detection of Con A. The Con A antibody was immobilized directly on the working electrode surface of the SPE (a). A protein-containing sample was applied and Con A was captured (b). Formation of a protein-glycopolymer–modified AuNP complex (c). A high potential of 1.25 V was applied for the oxidation of AuNPs (d) and then voltammetric measurement was performed (e).

signals. For this purpose, DNA is used with enzyme labels (e.g., HRP, alkaline phosphatase [ALP]) or electroactive tags (ferrocene or anthraquinone). Because of their particular properties, for the same purpose nanoparticle-based tags can be employed. For instance, AuNP tags were employed in determinations with potentiometric stripping detection [135]. Determinations were carried out using immobilized DNA probe onto magnetic beads and biotinylated target with captured AuNPs. The amount of AuNP bound as result of hybridization is measured by its chemical dissolution and by anodic potential stripping at carbon SPE.

Potentiometric detection employing silver ISEs was also reported in DNA hybridization monitoring with application of AuNPs [131] (Figure 5(B)). The detection of hybridization is based on the depletion of silver ions, induced by the biocatalytic reaction of the ALP enzyme tag. After the hybridization process was carried out on an array of thiolated Au electrodes, the amount of bound enzyme tag was determined by adding substrate *p*-aminophenyl phosphate. The product of the enzyme reaction, *p*-aminophenol, reduces silver ions, which is monitored by ISE. The process of Ag(I) reduction is significantly faster in the presence of AuNPs in reaction solution owing to the presence of the AuNP nucleation sites. In another electrochemical DNA biosensor, the progress of hybridization was monitored by the voltammetric detection of thionine, which was capped on DNA–AuNP conjugates [136] (Figure 8(A)). In a DNA-based assay of nuclear factor-κB, AuNPs were functionalized with triplex-forming oligonucleotide (TFO) [137]. In the developed procedure, TFO acts with dsDNA self-assembled on a gold electrode (Figure 8(B)). The measured chronocoulometry signal is a function of the presence of hexamineruthenium(III) chloride, which can be bound to TFO-modified AuNPs. The presence of the analyte decreases the binding of AuNP-TFO to dsDNA and results in a small electrical signal.

Gold nanoparticles are also applied in the development of aptamer-based electrochemical sensors. The application of aptamers is based on their ability to incorporate different labels, linkers, or other moieties, or to undergo analyte-dependent conformational changes. In the development of microfluidic paper-based assays of IgE protein, the AuNP–aptamer conjugate was employed and AuNPs were used to catalyze the deposition of silver nanoparticles, whereas detection was based on the potentiometric determination of Ag(I) after dissolving silver nanoparticles in H_2O_2 [138]. In a voltammetric aptamer biosensor developed for to determine carcinoembryonic antigen, AuNPs were employed as conjugate with aptamer and 6-ferrocenyl hexanethiol [139]. In the presence of analyte, a two-aptamer sandwich complex is formed (Figure 9(A)) and detection of ferrocene is carried out by DPV. Gold nanoparticles have yet another role in the determination of thrombin with application of aptamer and electrochemiluminescence (ECL) detection based on the use of CdS quantum dots (QDs) [38]. In this determination a GC electrode was employed with immobilized QDs as the source of ECL, and on a modified electrode surface thrombin aptamer was attached (Figure 9(B)). Then the aptamer was hybridized to the AuNP–ssDNA conjugate and as a result, AuNPs were placed about 12 nm from the QDs, which resulted in significant enhancement of the ECL signal. In the presence of thrombin protein, a part of

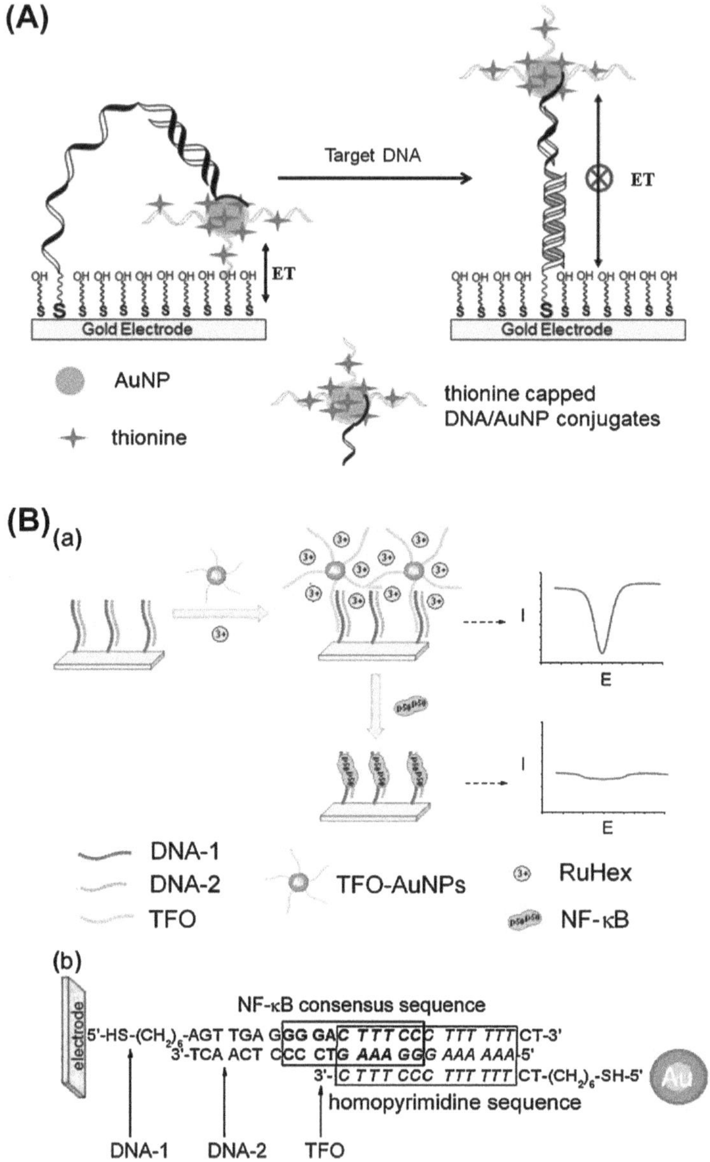

FIGURE 8 Examples of DNA sensors employing AuNPs. (A) Fabrication and application of the DNA voltammetric sensor using thionine-capped DiDNA/AuNP conjugates as signal tags [136]. (B) (a) A schematic illustration for the chronocoulometry detection of nuclear factor-κB detection with the help of triplex DNA and AuNPs [137]. (b) Sequences of the three oligonucleotides, in which the analyte consensus sequence and the homopyrimidine sequence are outlines in the frames.

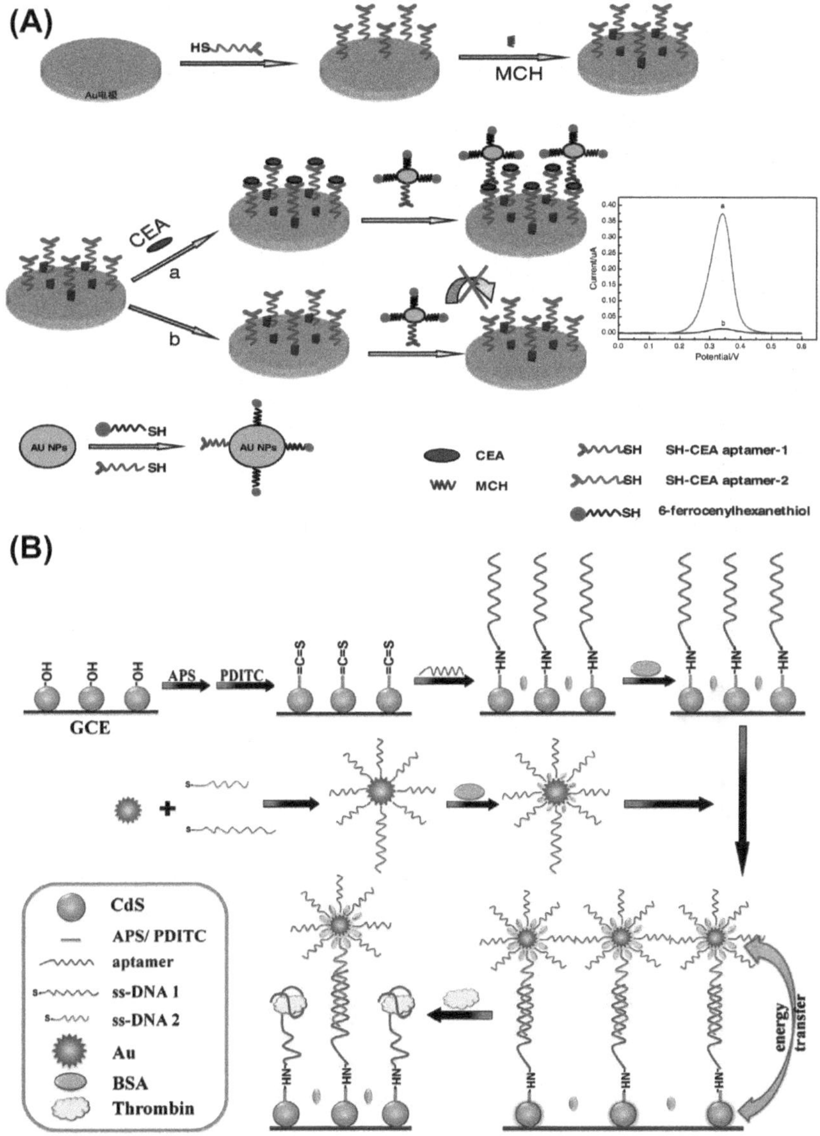

FIGURE 9 Examples of construction of response signal of electrochemical aptamer-based sensors employing AuNPs. (A) Schematic diagram of the voltammetric aptamer biosensor based on AuNP amplification for the detection of CEA [139]. (B) Scheme of the electrochemiluminescence aptamer sensing platform based on energy transfer between CdS nanocrystals and AuNPs developed for ultrasensitive thrombin detection [38]. (C) left, (A) cyclic voltammograms of the sensor after each immobilization step a – with base electrode, b – after self-immobilization of aptamer, and c – after backfilling with 2-mercaptoethanol); (B) Nyquist plot of impedance spectra obtained with aptamer-based impedimetric sensor for bacterial typing of *Salmonella enteritidis* in the range of 1×10^3 (curve a) to 1×10^5 (curve e) CFU/mL [68]. The right inset represents a modified Randles circuit applied to fit to the measured data. The left inset represents a calibration plot of the resistance to the charge transfer R_{CT} versus log concentration of *S. enteritidis*.

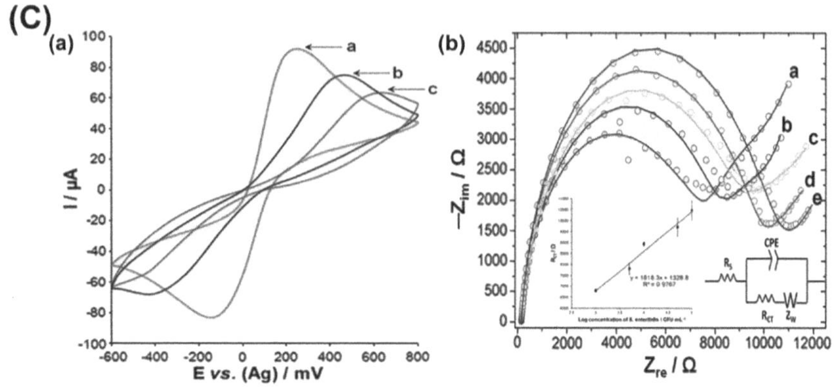

FIGURE 9 cont'd

the AuNP–ssDDA conjugate was released, which resulted in a decrease in ECL signal magnitude, measured as an analytical signal.

Gold nanoparticles were also employed in the construction of electrochemical biosensor-based saccharide–protein interactions [133]. The detection of concanavalin A (Con A) protein was reported with the use of graphite SPE, with surface-immobilized Con A antibody, which binds the antigen from the sample solution (Figure 7(B)). As a result of the strong binding of protein with saccharide, the glycopolymer-modified AuNPs are bound to antigen. The detection of bound AuNP conjugate is based on oxidation at a constant potential, and DPV measurement is performed. During the whole procedure AuNPs are employed as a label of the progress of that process in the presence of an analyte.

Cytosensing can be considered the most advanced type of biochemical sensing, in which the detection of a particular type of cells is carried out. It is usually based on exploiting known intermolecular interactions, in which a given target molecule can be identified as a label of a particular cell. Recent literature provides some examples of electrochemical cytosensors employing AuNPs. For instance, a lectin-based biosensor functionalized with enzyme was developed for the determination of carbohydrate expression on living cells [76]. The complexity of the electrochemical system, which is used for in situ detection of mannose groups on living cells, is illustrated by Figure 1(C). In this case, AuNPs were used to enhance the analytical signal acting as a substrate of HRP immobilization. Selective recognition, fundamental for such detection, was achieved by applying natural protein lectins, which specifically recognize sugar epitopes. To cells bound on the electrode surface, as a nanoprobe, AuNPs with ConA and HRP were bound. Under optimized conditions, in DPV measurements the limit of detection (LOD) was 15 cells in the determination of the K562 cell line, a human chronic myelogeneous leukemia line. In the construction of another electrochemical cytosensor, competitive binding of cell-specific aptamers was used [75]. A satisfactory level of LOD (10 cells) was obtained as a result of signal amplification by using Fe_3O_4 magnetic nanoparticles as a carrier to load a large amount of AuNPs and AuNP-catalyzed silver deposition in ASV determinations. The principle of molecular recognition in this case was competition with aptamer between determined cells and the AuNP–cDNA conjugate, which can form a duplex with aptamer (Figure 1(B)).

7. ELECTROCHEMICAL SENSING EMPLOYING AuNPs WITH DIFFERENT DETECTION METHODS

A vast body of original research publications in the current chemical literature indicates that AuNPs have been used to improve electroanalytical measurements with almost all commonly employed detection methods. Their function is most commonly to improve the detectability and sensitivity of detection, although also often they allow modification of the selectivity of response. They can also contribute to more convenient construction of electrochemical sensors and biosensors, e.g., by efficient immobilization of the molecular recognition elements. In each detection method discussed below, AuNPs are used to determine ions and small molecules in the construction of enzymatic and immunochemical biosensors and DNA genosensors, as well as aptamer-based biosensors.

7.1 Potentiometry

Potentiometry based on measurements of the potential difference of electrodes at the open circuit of an electrochemical cell belongs instrumentally to the simplest electrochemical detections. Currently, its predominated applications are based on the use of membrane ion-selective electrodes (ISE). In such systems AuNPs do not have obvious applications except, for instance, their use in building a solid contact for ISE with a poly(vinyl chloride) (PVC)-based membrane, which was reported for potassium ISE [69]. The presence of AuNPs, stabilized with aliphatic thiols at the backside of the membrane resulted in highly stable potentiometric responses.

Gold nanoparticles were used to design non-membrane potentiometric sensors for different inorganic ions. A GC electrode coated with thin films of thiolate, encapsulated AuNPs linked by 11-mercaptoundecanoic acid, was sensitive to a Cu(II) concentration below 1 mg/L [95]. The potentiometric response to Cu(II) ions was also demonstrated for a GC electrode covered with dithizone-modified AuNPs with higher potentiometric selectivity than typical polymeric ISEs with ionophores in the membrane [37]. Sensitivity to nanomolar concentrations of Cu(II) ions was also shown for a carbon paste electrode modified by a self-assembled mercapto compound on AuNPs [65]. When carbon paste contained sol–gel AuNPs in the presence of 2,2′-dihydroxy-1-naphthidine-1′-naphthyl methyl amine as ionophore, a sensitive Nernstian response was observed for Al(III) ions with an impressive LOD of 0.054 μg/L [64]. A graphene paste potentiometric sensor for Ag(I) ions was then constructed using as ionophore grapheme oxide nanosheets covalently grafted with 2-thiophene carboxylic and decorated with AuNPs (Figure 10) [66].

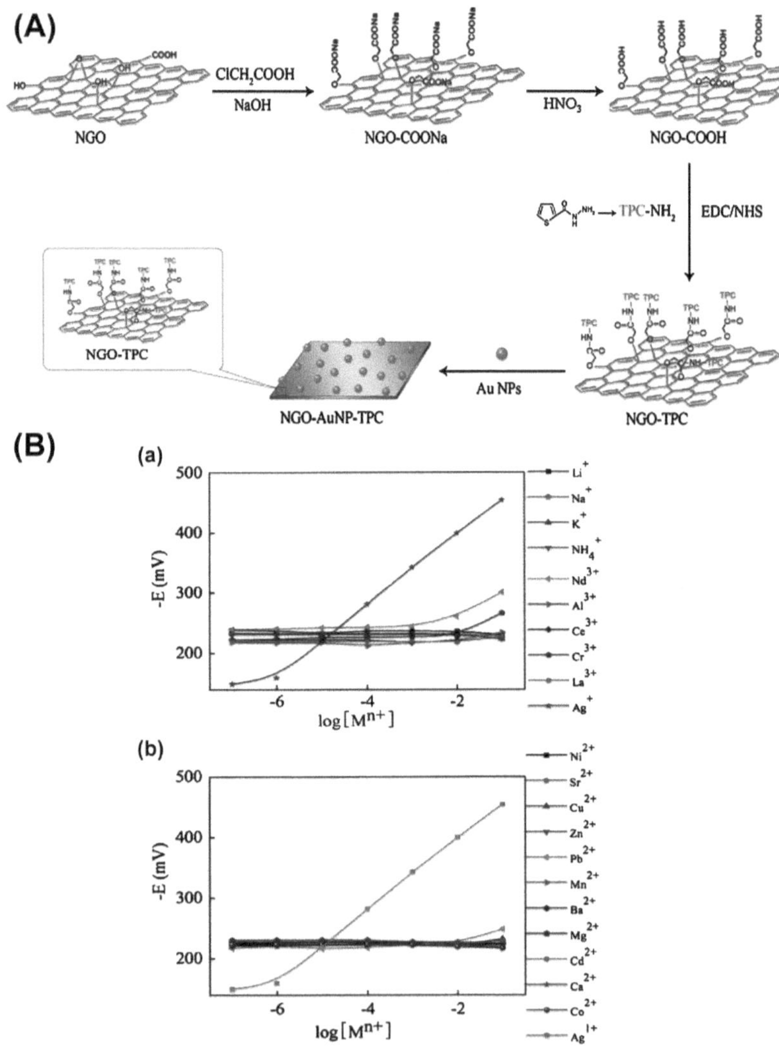

FIGURE 10 Scheme of construction and potentiometric response of silver ion graphene paste electrode employing AuNPs [66]. (A) Schematic representation of the preparation procedure for graphene oxide nanosheets (NGO)—AuNP—2-thiophenecarboxylic (TPC) hybrid material. (B) Potentiometric responses of NGO—AuNP—TPC paste electrode for different metal ions.

A different concept of potentiometric sensor with the use of nanostructured gold was based on ionophore-modified nanopore arrays, which was described as solid-state ion channels [62]. Onto the walls of Au nanopores, using Au−S bonds, a synthetic Ag^+-selective thiacalixarene ionophore was immobilized providing a Nernstian potentiometric response down to 10 nM Ag(I).

Some examples of potentiometric sensors for non-ionized small molecules can be also found in the literature. A nonenzymatic potentiometric glucose sensor was developed electropolymerizing 3-octylthiophene on a Pt disk electrode and then covering it with gold nanorods functionalized by self-assembling 4-mercaptophenyl boronic acid and 11-mercaptoundecanoic acid [61]. A linear potentiometric response was observed in the concentration range of 5−30 mM glucose. A semilogarithmic linear potentiometric response was also showed for glucose for a Pt nanoparticle-modified Au nanowire array on a copper substrate with glucose oxidase immobilized on the surface by adsorption [108] (Figure 11). A linear response with sensitivity of 33.4 mV/decade was observed in the concentration range of 0.1−7 mM glucose. Another example of potentiometric enzymatic biosensor employing AuNPs was reported for Hg^{2+} and based on the inhibition of urease [82]. Gold

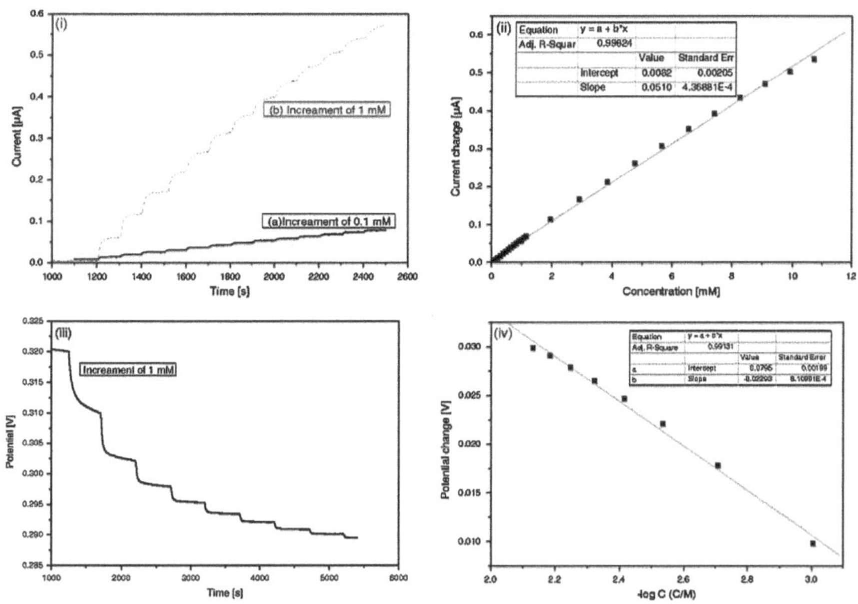

FIGURE 11 Electrochemical responses of Pt nanoparticle-modified Au nanowire array in detection of glucose [108]. Amperometric and potentiometric detection of glucose. (i) Chrono-amperograms for increasing glucose concentrations. (ii) Typical calibration curve for amperometric response. (iii) Chronopotentiogram for increasing glucose concentration. (iv) Nernstian plot of the potentiometric response for measurements in 50 mM phosphate buffer, pH 7.0.

nanoparticles were adsorbed onto the aminated PVC membrane pH electrode surface and served as a support for urease immobilization.

Gold nanoparticles were also employed in the construction of numerous potentiometric immunosensors. One immunosensor developed for the detection of hepatitis B surface antigen was based on a Pt electrode covered with Nafion layer, in which antibody and AuNPs were entrapped by a gelatin matrix [86]. In the development of an immunosensor for the determination of diphtheria antigen, a gold electrode was covered with a silica sol—gel monolayer formed from 3-mercaptopropyl trimethoxysilane, and then AuNPs and silver nanoparticles were chemisorbed onto the thiol groups [99]. Finally, the antibody was adsorbed onto the surface of nanoparticles. Both of those immunosensors exhibit similar LOD values: 1.3 and 3.7 ng/L for hepatitis B surface antigen [86], and diphtheria antigen [99], respectively. In other attempts to design a potentiometric immunosensor to detect diphtheria, the application of silica, silver and gold nanoparticles, and polyvinyl butyral as matrices was compared for the immobilization of diphtheria antibody [89]. The SiO_2-modified immunosensor was considered optimum.

The potentiometric immunosensor to detect hemoglobin (Hb) employed an electropolymerized polypyrrole (PPy)—AuNP composite, which was deposited on a gold micro-electrode surface of an ISFET sensor and acted as the support for the adsorptive immobilization of antigen [127]. Another original design for an immunosensor was reported for the detection of α1-fetoprotein [51]. It was based on antibody-functionalized core—shell—shell magnetic nanocomposite particles, which were attached to the surface of carbon paste electrode by the external magnet placed on the other site of the carbon paste electrode. The response was based on the modification of potential owing to the interfacial change of charge resulting from the immunochemical reaction.

The potentiometric immunoassay can employ also conventional membrane ISE and AuNPs as labels. This was demonstrated in a sandwich format immunoassay of proteins, in which Ab was conjugated to AuNP tags onto which silver ions were deposited [130]. After completion of the sandwich assay, the precipitated silver is oxidatively dissolved with H_2O_2 and detected by Ag-ISE (Figure 5(A)).

Several different approaches were also reported for the application of AuNPs in the potentiometric detection of DNA hybridization. Hybridization of target DNA to the probe can be followed by binding the streptavidin-coated AuNPs to the captured DNA and then dissolution of a gold tag and potentiometric stripping measurements of the dissolved metal tag [135]. In another protocol, in which the ALP enzyme tag was employed (Figure 5(B)), detection is based on the application of Ag-ISE [131], similar to the immunoassay mentioned above. The product of enzymatic reaction can reduce silver ions to metallic silver; this can be monitored by Ag-ISE. However, this process is too slow without AuNP nucleation sites; hence, they are employed in the optimized procedure in a measured solution. This procedure was developed to monitor 16SrRNA of *Escherichia coli* pathogenic bacteria. The cysteamine-capped AuNPs were used to immobilize ssDNA probe for the chronocoulometric detection of hybridization using $[Fe(CN)_6]^{3-/4-}$ in solution as a mediator [41]. The target DNA with 20 bases could be detected down to 2.13×10^{-13} M.

Protein assays with potentiometric detection can be carried out using protein aptamer–AuNP conjugates [138]. Similar to the applications mentioned earlier, AuNPs catalyzed the deposition of silver nanoparticles that by oxidative dissolution generated silver ions for the detection with Ag-ISE. The developed procedure was demonstrated in a microfluidic paper-based format for the determination of IgE.

7.2 Voltammetry

Voltammetry is the most commonly used detection method for diagnostics of electrochemical properties of electrode systems modified in different ways with gold nanostructures. Such measurements often precede further detailed optimization of analytical procedures, e.g., with potentiometric or amperometric detection. The diagnosis of electrochemical properties is efficiently carried out first by cyclic voltammetry, whereas the most suitable modes for obtaining improved selectivity and detectability are differential pulse or stripping voltammetric modes.

In this case, methods employing AuNPs are developed for the determination of inorganic ions and small organic molecules. Those methods are widely used for detection with enzymatic and immunochemical biosensors, DNA and aptamer-based biosensors, and cytosensors. This chapter presents an arbitrary selection of developments derived mostly from articles published in the past few years.

Voltammetry is routinely employed method in the determination of heavy metal ions, and numerous articles describe important novel aspects for using AuNPs with such methods. In the determination of Hg(II) using square-wave anodic stripping voltammetry (SW-ASV), the deposition of AuNPs on disposable gold SPE allows for better sensitivity than with a gold disk electrode, with an LOD of 0.8 μg/L in under 3 min accumulation time [63]. For trace Hg(II) ASV determination, a GC electrode was used after surface modification by a CNT–AuNP nanocomposite, synthesized by microwave radiation [83]. In this case, an even lower LOD was reported: 0.06 μg/L at 2 min accumulation. For the same purpose carbon SPEs were first covered with CNTs or graphene oxide and then with AuNP layers [122]. In SW-ASV measurements, a better LOD was obtained with a nanohybrid surface of CNT and AuNP.

Voltammetric determinations of arsenite can also be improved using AuNPs. The electrochemical deposition of AuNPs on the GC electrode surface was employed in the square-wave voltammetric determination of As(III) in a flow system in which for 2 min accumulation time an LOD of 0.2 μg/L was reported [105]. For the same purpose, internal electrolysis–deposited AuNPs on a GC electrode was used with an LOD of 0.07 μg/L in ASV with a linear sweep [129]. To adhere the AuNPs firmly onto a GC electrode, the layer of AuNPs can be also be covered by a chitosan layer, which was reported, e.g., in SW-ASV determination of As(III) [70].

Several attempts at similar approaches were reported for the voltammetric determination of chromium. In the determination of Cr(VI) by DPV with a carbon SPE modified with electrodeposited AuNPs, the LOD was reported to be 0.4 μM, even in the presence of a high concentration of Cr(III) [113]. Voltammetric procedures were also developed for the simultaneous determination of Cr(III) and Cr(VI). Two carbon SPEs were used, connected in array mode, covered with mercury film for Cr(III), and with AuNPs for Cr(VI) determination. Differential pulse voltammetry measurements in wastewater were reported [116]. For the LSV determination of Cr(III) and Cr(VI), a carbon SPE was modified with a gold polymer paste with dendric gold, with particle sizes in the range of low micrometers, providing an LOD of 38.8 and 4.4 μM for Cr(III) and Cr(VI), respectively.

In cyclic voltammetry of Cu^{2+} at a thin film of AuNPs linked by 11-mercaptoundecanoic acid deposited on a GC electrode, a linear dependence of peak current versus Cu(II) concentration was shown with an LOD below 1 mg/L [95].

Gold nanoparticle–modified working electrodes were also successfully employed in the determination of several inorganic anionic species such as sulfite, sulfide, and nitrite. The DPV determination of nitrite was reported with GC or ITO electrodes, modified by the co-deposition of gold–platinum nanoparticles on a CNT–ionic liquid composite layer [123]. The reported LOD was down to 1.0 nM of nitrite. The determination of hydrogen sulfide can be carried out by monitoring As(III) at a poly(L-lactide)–stabilized, AuNP-modified electrode [43]. The measured differential pulse–ASV response is based on inhibition of the oxidation current of As(III), and an LOD of 40 nM was found. When AuNPs are embedded in a silicate sol–gel three-dimensional network on the surface of a GC electrode, the simultaneous voltammetric detection of sulfite, nitrite, and hydrazine is based on well-separated oxidation peaks [42] (Figure 12(A)). A linear plot of LSV signals was obtained in the sub-millimolar range of analyte concentrations.

In voltammetric determinations of low-molecular organic compounds, working electrodes are most commonly modified with AuNP composites or multilayers. In voltammetric enzyme-less glucose sensing, carbon fiber working microelectrodes were first coated with aligned CNTs and then with an AuNP layer [73]. Gold nanoparticles were deposited on GC electrodes covered with CNTs for the DPV of pharmaceutical tramadol hydrochloride in the nanomolar concentration range [121]. The cyclic voltammetry of ascorbic acid was demonstrated on Au and ITO electrodes modified by the layer-by-layer

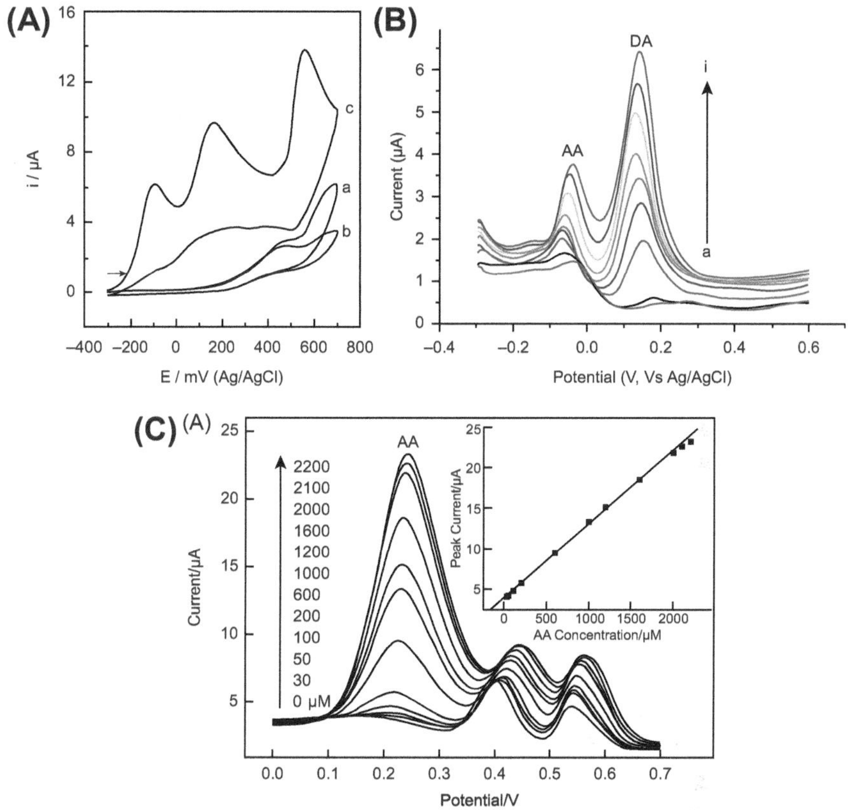

FIGURE 12 Examples of recorded voltammetric signals recorded in multianalyte detection with AuNP-based working electrodes. (A) Comparison of cyclic voltammograms recorded with different electrodes for mixture of 0.25 mM hydrazine, sulfite, and nitrite at GC electrode (a), (3-aminopropyl) triethoxysilane sol–gel modified GC electrode (b), and GC electrode modified with sol–gel containing AuNP at a scan rate of 20 mV/s [42]. (B) Differential pulse voltammograms recorded using a gold working electrode with a flower-like gold nanostructure for a mixture of different concentrations of ascorbic acid AA (60–500 µM) and dopamine DA (1–150 µM) [106]. (C) Square-wave voltammograms recorded using an AuNP–β-cyclodextrin–graphene–modified GC electrode for mixtures containing different concentrations of ascorbic acid (AA) up to 2.2 mM, and 5.0 µM of dopamine (DA) and uric acid (UA) [53]. The inset shows the plot of peak current versus AA concentration.

deposition of a polythiophene–AuNP multilayer [96]. The AuNP modification of a working electrode was helpful in optimizing electrochemical conditions for simultaneous voltammetric determinations of ascorbic acid with other electroactive analytes. It was shown for the simultaneous DPV determination of dopamine and ascorbic acid with a working electrode modified with a flower-like Au nanostructure [106] (Figure 12(B)). The simultaneous square-wave voltammetric determination of ascorbic acid, dopamine, and uric acid was reported with an AuNP–β-cyclodextrin–graphene–modified GC electrode [53] (Figure 12(C)). In this case, graphene covered with AuNPs was synthesized by the in situ thermal reduction of graphene oxide and $HAuCl_4$ with β-cyclodextrin under alkaline conditions, and linear responses for measured analytes were observed in the micromolar concentration range.

Voltammetric detection was also employed in the determination of nitroaromatic compounds and organophosphorus pesticides using AuNP-modified working electrodes. A GC electrode covered with Nafion film containing AuNPs and CNTs was used as working electrode in ASV to detect trinitrotoluene, but results were obtained using copper nanoparticles [45]. Application of a reduced graphene oxide–AuNP composite for modification of a GC electrode allowed selective determination of 4-nitrophenol with an LOD of 10 nM using DPV [124]. However, in the determination of methyl parathion it was shown that better sensitivity in square-wave voltammetric measurements was obtained for a gold atomic cluster–modified electrode than for an AuNP-modified electrode [57]. The effective enhancement of sensitivity and LOD in the determination of organophosphorus pesticide chlorpyrifos was reported for a GC electrode modified with AuNPs, and then with molecularly imprinted polyaminothiophenol membrane [119] (Figure 4). The application of an MIP membrane resulted in good sensitivity of chlorpyrifos detection over five other examined pesticides.

There different voltammetric enzymatic biosensors with application of gold nanostructures have been reported. They are usually electrodes with multilayer modifications of the sensing surface. Direct enzymatic sensing was reported, e.g., for glucose and phenol. In the case of glucose, a GC electrode was modified by electrochemical deposition of a nanocomposite of AuNPs and chitosan [126]. On the top of this layer, GOD and cytochrome c was placed by adsorption from aqueous solution. Such a biosensor showed a wide concentration range of linear response in DPV measurements, with an LOD of 0.15 mM, and was evaluated as a satisfactory point-of-care glucose monitor. In the case of a biosensor for phenol, AuNPs were electrodeposited onto a boron-doped diamond electrode; the working electrode was modified by cysteamine, and tyrosinase was attached using glutaraldehyde [107]. A satisfactory square-wave voltammetry (SWV) response was reported with an LOD of 70 nM, allowing, e.g., real environmental applications. A more complex system was reported for the indirect voltammetric enzymatic determination of sulfite [90]. In this case, enzyme polyphenol oxidase was immobilized onto the layer of AuNPs electrodeposited onto a GC electrode. The sensing of sulfite in DPV measurements with an LOD of 0.4 mM was based on the inhibitory effect of sulfite on the activity of polyphenol oxidase. The obtained LOD was competitive compared with numerous other sulfite biosensors reported in the literature that are based on the application of sulfite oxidase.

A large variety of designs employing AuNPs were reported for voltammetric immunosensors. In the simplest case, the AuNP loaded with paraoxon

Ab was immobilized on a GC electrode using Nafion membrane [85]. A much more complex system was developed for voltammetric immunosensing of another small molecule pharmaceutical, chloramphenicol [48]. In that design, the electrocatalysis of AuPd nanoparticles toward the oxidation of $NaBH_4$ was employed to produce a voltammetric analytical signal. The detection concept was based on the use of adamantine-labeled Ab to bind the antigen competitively on the CNT-modified surface of a GC electrode and the target species in solution. The immobilized adamantine bound β-cyclodextrin−functionalized AuPd nanoparticles to catalyze the oxidation of $NaBH_4$. The detection concentration range was reported for six orders of magnitude, with an LOD of 4.6 ng/L.

Several AuNP-based immunosensors were reported for the determination of proteins. In a design involving magnetic beads, which allow easy magnetic separation, signal amplification was attributed to AuNPs [40,132]. As shown in Figure 6(A), for the determination of IgG, Ab−magnetic beads and Ab−AuNP conjugates sandwich a target IgG analyte. Such a complex is attached to an ITO electrode modified with ferrocenyl-tethered dendrimers by an external magnet. A redox cycling involving generation of *p*-aminophenol greatly amplifies the DPV signal, allowing an extremely low LOD of 5 aM (1 pg/L) [132]. Several designs for a voltammetric immunoassay have been reported for a specific protein carcinoembryonic antigen, which was identified as a biomarker for breast cancer and ovarian carcinoma [47,50,59]. In a graphene-based immunosensor, gold nano-flowers and ssDNA were assembled onto a graphene surface on a gold electrode and then covered with thionine, of which the amino groups could conjugate gold nano-flowers [59] (Figure 2(B)). Antibodies were then immobilized for a sandwich-type immunoassay based on Ab-Au nano-flower−HRP as a label and H_2O_2 as the enzyme substrate. Gold nano-flowers with electrochemical activity provide a large surface area for the label of biomolecules, and an LOD of 10 ng/L was reported in DPV measurements. A significantly lower LOD value for immunosensing of the same antigen was reported for graphene−AuNP based with triple signal amplification, as mentioned previously [50]. The main contribution to a low LOD value of 0.12 ng/L is the result of ASV detection of the deposited silver nanoparticles. In another design for a voltammetric immunosensor for carcinoembryonic antigen, dendrimer-encapsulated AuNPs were used [47]. The application of dendrimer-AuNP as a sensor platform not only increases the amount of immobilized Ab and mediator, but also accelerates the electron transfer process for signal enhancement. In SWV determination the LOD was evaluated as 4.4 ng/L. The same SWV detection was used in the previously mentioned immunoassay for the detection of phosphorylation based on enzyme-linked electrocatalysis [134].

Several designs for voltammetric aptamer-based sensors and DNA genosensors are presented in detail in Section 6, including an aptamer sensor with DPV detection for carcinoembryonic antigen [139] or with square-wave ASV for leukemia cells [75], and a biosensor for the DPV detection of DNA using thionine-capped DNA−AuNP conjugates as signal tags [136]. Voltammetric aptasensors employing AuNPs have been reported for cocaine [140] and thrombin [125]. Both of these were used in DPV measurements. In the case of a cocaine aptasensor, ferrocene-appended poly-(ethyleneimine) and AuNPs were assembled layer by layer on the surface of an ITO electrode via electrostatic interaction, followed by the covalent attachment of cocaine aptamer fragments [140]. A design for an aptasensor for thrombin was based on modification of the GC electrode surface with an AuNP−thionine−graphene composite [125]. This served as a support for immobilization of the ethanethiol-substituted oligonucleotide probe, whereas thionine performed as an electrochemical redox probe, and an impressive LOD for the DPV detection of thrombin was reported at 93 pM.

In one of the earliest reported electrochemical DNA biosensors involving DNA, the probe DNA was immobilized on a silanized surface of ITO electrode with attached AuNPs [79]. This biosensor was employed for the DPV determination of pharmaceutical mifepristone, based on its intercalation into a DNA double helix with an LOD of 0.2 μM.

Gold nanoparticles were also employed in the construction of voltammetric cytosensors. The one developed for ultrasensitive detection of apoptotic cells, important in the etiology of leukemia, is based on ASV detection of Cd(II) [71]. As shown in Figure 1(A), a complex measuring system involves a electrochemical cytosensing interface combining CNTs and AuNPs, and a nanoprobe assembled of concanavalin A on CdTe QD-labeled silica nanoparticles. The binding of the nanoprobe to the cytosensing interface in the presence of determined cells was monitored by the dissolution of CdTe and detection of Cd(II) ions. The cytosensor for monitoring the carbohydrate expression on living cells with DPV detection is presented in Section 6.

7.3 Amperometry

In the case of electrochemical sensors employed in amperometric measurements at a given potential of the working electrode, AuNPs have found most applications in the construction of enzymatic biosensors and immunosensors. Numerous examples can be also found for chemical sensors of different analytes, as well as aptasensors and DNA genosensors.

An example of chemical sensors developed for the amperometric detection of inorganic ions and small molecules is nitrite, oxygen, or H_2O_2 sensors. In designing a sensor for the determination of nitrite, a GC electrode was modified with a composite of AuNPs and sulfonated graphene [103]. The amperometric response was satisfactorily selective for environmental applications, and the LOD was evaluated at 0.2 μM. Evidence of improvement in LOD for the amperometric detection of nitrite down to 10 nM was reported in the modification of a GC electrode with graphene−AuNP hybrids with immobilized hemoglobin [49]. An original paper-like electrochemical amperometric sensor employing AuNPs was described for oxygen determination [80]. A well-defined gold electrode array on cellulose membranes was produced by inkjet printing using AuNP suspension, and the sensor obtained with an ionic liquid electrolyte was employed for oxygen detection in gaseous phase. The immobilization of hemoglobin by self-assembling on n AuNP/poly(*p*-aminobenzene sulfonic acid)-modified Pt disk electrode allows a nonenzymatic H_2O_2 amperometric sensor to be obtained, exhibiting an LOD of 70 nM [102]. Because of the low working potential (30 mV), several examined electroactive species do not interfere.

Numerous amperometric sensors involving AuNPs were developed for the detection of other electroactive small molecules. For instance, a sensor for hydroxylamine was prepared as the arrays of AuNPs deposited on flexible film of single-walled CNTs [120]. The LOD of the amperometric response was reported to be 0.72 μM, but selectivity was not discussed. A similar design was reported for a uric acid sensor, in which sputtering of AuNPs onto vertically aligned CNTs on a copper support was carried out [74]. In this case, selectivity against other electroactive species present in real urine samples was demonstrated, and the LOD was 0.1 μM, which is comparable to or better than that reported earlier for other nonenzymatic sensors for uric acid. Similar values of LOD for amperometric response to uric acid and ascorbic acid were achieved for amperometric sensor, in which hexane-thiolate—protected AuNPs were entrapped into the sol—gel network on the surface of GC electrode [87]. The oxidation potentials of both analytes, however, are similar: 380 and 405 mV for ascorbic acid and uric acid, respectively.

Amperometric sensors developed for sulfur-containing analytes were based on the application of AuNP composites. The composite of AuNPs with palladium nanoparticles and Prussian blue was prepared in graphite paste in Nujol, and the obtained electrode exhibited larger sensitivity of the amperometric response to L-cysteine than carbon paste electrodes with Prussian blue or Prussian blue with AuNPs only, and an LOD of 0.18 μM was reached [67]. In the case of another sensor developed for biothiols, including cysteine, the colloid solution of NiO/PPy-Au nanocomposite was cast onto the GC electrode surface [52].

A more complex design was reported for an amperometric sensor for polychlorinated biphenyls (PCB), in which the response was attributed to specific inhibition of the charge transport when analyte is captured by the cavities of β-cyclodextrin attached to AuNPs and deposited onto an Si wafer with SO_2 coating (Figure 3(A)) [100]. The response to several PCB congeners was examined in the presence of different common organic environmental pollutants, and satisfactory selectivity with an LOD of 1 nM was achieved.

One can also find a design for an amperometric sensor for protein not involving immunochemical interactions. For this purpose, a carbon SPE was used. In such a disposable sensor for the determination of glycated hemoglobin in blood, the AuNPs were electrodeposited on SPE and covered by the conducting film of poly(terthiophene benzoic acid), to which aminophenyl boronic acid was attached, and served as a host to capture the analyte [112]. The amperometric response was based on the catalytic reduction of H_2O_2 by captured hemoglobin.

Gold nanoparticles and other nanostructures have found especially wide applications in the construction of amperometric enzymatic biosensors, both for effective immobilization of enzymes and electrocatalytic enhancement of the electrical signal. Often in this field, numerous developments are been reported for glucose sensors employing GOD as the model enzyme for enzymatic amperometric biosensors. In a multilayer construction of glucose sensor, oxidized GOD and AuNPs were associated by layer-by-layer covalent attachment on an aminated gold disk electrode [91]. The amperometric response was correlated to the number of deposited layer, and in optimum configuration the LOD was evaluated as 8 μM. In the design of a first-generation glucose sensor, a Pt electrode was modified with a composite film of 3-mercaptopropyl trimethoxysilane xerogel embedded with GOD, doped with alkanethiolate-protected AuNPs, and coated with an outer polyurethane layer [88]. The presence of AuNPs in xerogel resulted in a significant improvement in response time and sensitivity of response, whereas a polyurethane layer was employed to prevent GOD from leaching and to enhance the selectivity of response. The LOD value was evaluated as 23.2 μM, which is well below the limit of physiological concentrations. In another construction, GOD was immobilized in a composite material based on polyvinyl alcohol and partly prehydrolyzed tetraethyl orthosilicate on the surface of graphite electrodes [46]. Comparison of gold and silver nanoparticles in such applications indicates the advantage of AuNPs especially in terms of the concentration range of response. The LOD was evaluated to be 0.7 mM.

Two other gold nanostructures were also employed in the construction of amperometric glucose sensors. For instance, highly ordered Au nanowire arrays were modified with Pt nanoparticles, on which GOD was adsorbed [108]. It was showed that Pt nanoparticles enhances the electrocatalytic activity of AuNPs. As already mentioned, the same sensor can be used in potentiometric measurements; examples of obtained responses are shown in Figure 11. With the use of nanoporous gold placed on a GC electrode, different self-assembled thiolated monolayers were examined as a support for the immobilization of GOD after activation with carbodiimide and by *N*-hydroxy succinimide [141]. The use of nanoporous gold compared with planar polycrystalline gold provides essential improvement in the sensitivity of the biosensor response.

Different construction approaches can be found in the literature in the design of an HRP-based sensor for the detection of hydrogen peroxide. In the construction of a third-generation amperometric biosensor, a layer of silica gel was formed on the surface of a gold electrode, on which AuNPs were chemisorbed onto the thiol groups of the sol–gel, and then HRP was adsorbed [98]. Fast response to H_2O_2 was observed with an LOD of 2 μM. In another, similar design, suspensions of CNTs, AuNPs, and thionine were evaporated on the surface of a GC electrode and HRP was adsorbed, allowing a linear response to H_2O_2 with an LOD of 0.1 μM [94]. When a composite of AuNPs with Nafion nanofibers was deposited on the stainless-steel electrode, and HRP was adsorbed in the presence of poly(acrylic acid), a linear response to H_2O_2 was observed in the concentration range of 0.06–0.8 μM with an LOD of 38 nM [44]. A much wider dynamic range of response from 65 to 1600 μM, but with a poorer LOD of 5 μM, was reported for the much more complex design of an HRP-based H_2O_2 amperometric sensor [54]. In this case, a sensor based on the use of AuNP-tubular TiO_2 nanoclusters was prepared by modifying a GC electrode with a mixture of thiolated HRP, AuNPs linked to TiO_2 nanotubes by 12-phosphotungstate, hydrophilic liquid, and Nafion. A thiolated enzyme significantly improved the stability and direct electron transfer of HRP.

A much simpler procedure was used to prepare an enzymatic sensor for superoxide anion, in which AuNPs were electrodeposited on an ITO electrode and superoxide dismutase was adsorbed from aqueous solution [109]. In the same way, urease was adsorbed on the surface of carbon SPE covered with an electrodeposited layer of AuNPs, and this sensor was employed for the inhibitive determination of Hg(II), with a reported LOD of 58 nM [114].

In past years numerous immunosensors have been developed with application of AuNPs. Similar to enzymatic biosensors, the application of AuNPs is especially helpful in immobilizing antibodies to perform an immunoassay with a low limit of detection. In most cases reported below, AuNPs are deposited directly on the surface of the working electrode or entrapped in the layer of supporting material, and on the top of that antibody is adsorbed or covalently immobilized.

The detection of specific IgG anti-*Trypanosom cruzi* antibodies was developed for a microfluidic format with SPE [118], in which the surface of the working electrode was covered with electrodeposited AuNPs, and 3-mercaptopropionic acid was employed for the immobilization of *T. cruzi* proteins. Amperometric detection was based on the use of Ab labeled with HRP, which catalyzes the oxidation of 4-*tert*-butylcatechol in the presence of H_2O_2. Similarly, for microfluidic application in the detection of two cancer biomarker proteins in serum, an electrode array featuring eight carbon SPEs and glutathione-covered AuNPs was deposited onto an underlayer of polycation on which the capture antibody was immobilized [40]. Detection of the protein analyte was based on the use of the magnetic particle–Ab-HRP conjugates (Figure 6(B)). An application of the developed sensor was shown in the determination of prostate-specific antigen and interleukin-6. Another attempt to design an amperometric immunosensor to detect cancer biomarkers in microfluidic applications by the same group was based on the covalent attachment of Ab to the layer of glutathione-covered AuNPs by covering them with a thin layer of poly(diallyl dimethyl ammonium chloride) and activation with carbodiimide [142]. A microfluidic detection array was developed for the simultaneous detection of four protein biomarkers with an LOD in the range of 5–50 pg/L. The same strategy was used in design of an immunoassay for antibodies to peanut allergen Ara h2, based on the formation a synthetic peptide layer of the major IgE-binding epitope from Ara h2 attached to an AuNP film on a pyrolytic graphite electrode [84].

Two different amperometric immunosensors were developed for the detection of tumor marker α-fetoprotein, which involved AuNPs. One was based on the use of mixed self-assembled monolayers of two thiols on a gold-sputtered electrode and a layer of AuNPs, and then adsorption of Ab [92]. In the second approach, a gold electrode was coated with Nafion, soaked with thionine solution, and covered with an AuNP layer on which Ab was immobilized [93]. For both immunosensors similar values of LOD were reported: 2.4 μg/L [93] and 5.0 μg/L [92].

For detection of nitric oxide synthase, an AuNP encapsulated-conductive polymer was used, to which antibody was covalently bound by activating the carboxylic acid groups of the polymer [143]. A more complex fabrication process was reported for amperometric immunosensor for antibodies to *Plasmodium falciparum* histidine-rich protein-2 [115]. A carbon SPE was modified with alumina sol–gel, followed by electrochemical deposition of AuNPs. The detection was based on the use of enzyme-labeled antibody-ALP, employing as substrate 1-naphthyl phosphate, and product 1-naphthol was detected amperometrically.

The amperometric biosensor for the detection of the anticancer drugs antracycline, separated in a capillary electrophoretic microfluidic system, was developed with the use of AuNPs and was based on interaction with dsDNA [117]. To fabrication a biosensor, AuNPs were electrodeposited onto a carbon SPE and covered with thiophene-based polymer, which after carbodiimide activation served as a support for covalent immobilization of amine-terminated dsDNA. Finally, the sensor was conditioned in cardiolipin and covered with an outer layer of AuNPs. Such a sensor was employed in detection in the electrophoretic separation of mitoxantrone, idarubicin, doxorubicin, and daunomycin with LODs in the range of 1.2–5.5 fM.

7.4 Impedance Spectroscopy, Electrochemiluminescence, and Other Detection Methods

Besides the three most common electrochemical detections discussed previously, in which the modification of electrodes by AuNPs is advantageous in many cases, one can also find in the recent literature several other less common electrochemical methods of detection in which the application of gold nanostructures can be profitable. One of these is electrochemical impedance spectroscopy (EIS), is commonly employed, aside from cyclic voltammetry, in investigations of the mechanisms of electrode processes, especially in the case of modified or membrane electrodes.

Examples are, e.g., some works on the construction of aptasensors and immunosensors. From the point of view of procedures applied in the modification of electrode surface, they are simple. In designing an aptasensor for the tumor marker MUC1, AuNPs were electrodeposited onto a carbon SP, and the aptamer was immobilized by adsorption onto a modified working electrode [111]. In impedance measurements, an increase in charge transfer resistance after the immobilization of proteins to the sensor surface is an analytical signal. The LOD value in impedance spectroscopy was evaluated as 3.6 μg/L, although in DPV measurements in the same system, the LOD was four times lower. A similar modification was reported in the design of an aptasensor for bacterial typing on the detection of *Salmonella enteritidis* strain [68]. In this case, a thiolated aptamer was self-assembled onto an AuNP-modified carbon SPE. Binding of bacteria by aptamer resulted in a decrease in charge transfer, which is associated with a linear increase in charge transfer resistance (Figure 9(C)). The reported LOD for *S. enteritidis* was 600 CTU/mL. The impedimetric immunosensor was developed for the determination of chicken interferon-gamma, which regulates cell-mediated immunity against intracellular pathogen [97]. The design of this sensor consists of a GC electrode modified with polythionine, which is an electron transfer mediator, and an AuNP self-assembled layer on which Ab is adsorbed. Under optimized conditions the change in impedance was proportional to the logarithm of the analyte concentration with an LOD of 0.1 μg/L.

As partly electrochemical detection can be also considered electrochemiluminescence (ECL) with electrochemical production of luminescent species; some examples of such sensors and biosensors involving AuNPs can be found in the literature. A reported ECL H_2O_2 sensor was based on a gold nanocluster–silica nanoparticle nanocomposite [56]. The nanocomposite was synthesized by dispersing bovine serum albumin conjugates with gold nanoclusters in solution with the surface aminated silica cores, and the obtained nanocomposite was incorporated into a Nafion layer on the GC electrode surface. The anodic ECL originates from the formation of excited state gold nanoclusters, and it is decreased in the presence of H_2O_2. A linear response to analyte concentration was observed in the concentration range of 6.5–32.6 mM at a potential of 1.4 V. Several nanocomposites have been compared in studies of electrogenerated chemiluminescence of luminol [72]. It was found that CNT–graphene composite on a GC electrode is an ideal matrix for the deposition of AuNPs, and the effect of graphene on luminol ECL is more predominant than that of CNT. In the studies on the mechanism of luminol ECL, the adsorbed oxygen could be electrochemically reduced to $-OO^-$, which reacts with luminol to generate strong cathodic emission, and which is also increased when AuNPs are deposited onto a GC electrode modified with CNT−graphene composite. The developed ECL aptasensor for thrombin detection involving AuNPs and CdS QDs as an ECL emitter [38] is presented in detail in Section 6 (Figure 9(B)). The obtained semilogarithmic dependence of the ECL signal decrease versus concentration can be used for the determination of thrombin in an impressive wide concentration range from 100 aM to 100 fM.

In the recent literature one can also find an example of an ECL immunosensor, based on CdS QD in a nanocomposite film with CNTs and combined with AuNPs [144]. The multilayer structure of the immunosensor and the mechanism of ECL emission are shown in Figure 13. An AuNP−chitosan layer was used to immobilize antibody to target IgG; the principle of ECL detection is based on the increment of steric hindrance after the immunoreaction, which causes the decrease in ECL emission. The LOD was evaluated as 1 ng/L.

Among other electrochemical detection methods involving the modification of electrodes with AuNPs is the application of chronocoulometry and constant capacitance mode detection. In the development of a microfluidic system with aptamer-based assay for the determination of small molecules, it was indicated that chronocoulometric is more accurate for detecting the signal of an electrostatically trapped redox marker than other electrochemical methods [145]. The microfluidic system was developed using an Au-Ag dual-metal array, three-electrode system on a chip to carry out the simultaneous determination of small molecules. Sensing was based on the use of an $[Ru(NH_3)_6]^{3+}$ probe in measuring solution, whereas AuNPs were used to immobilize aptamer and transport them in a microfluidic system, and to enhance the sensitivity of detection. The operation of the developed system was showed in the determination of ATP and cocaine, with LOD values of 0.3 and 70 nM for ATP and cocaine, respectively. Chronocoulometric detection was employed in a DNA sensor for the determination of nuclear factor-κB [137], which is presented in Section 6.

For the field-effect AuNP-based glucose sensor with a capacitive electrolyte-insulator-semiconductor structure, constant capacitance mode detection was employed [100]. The design of the sensor is based on a hybrid system consisting of functionalized AuNPs and GOD associated with a surface of an Si-based field-effect electronic transducer. The binding of GOD to AuNPs was realized through a bifunctional coupling reagent. The capacitance of the system was kept constant by the appropriate electronic circuit and the voltage change was proportional to the analyte concentration. The measuring signal is a function of the change of charge status at the interface to the analyte. The co-immobilization of ferrocene on AuNPs resulted in additional enhancement of the biosensor's sensitivity.

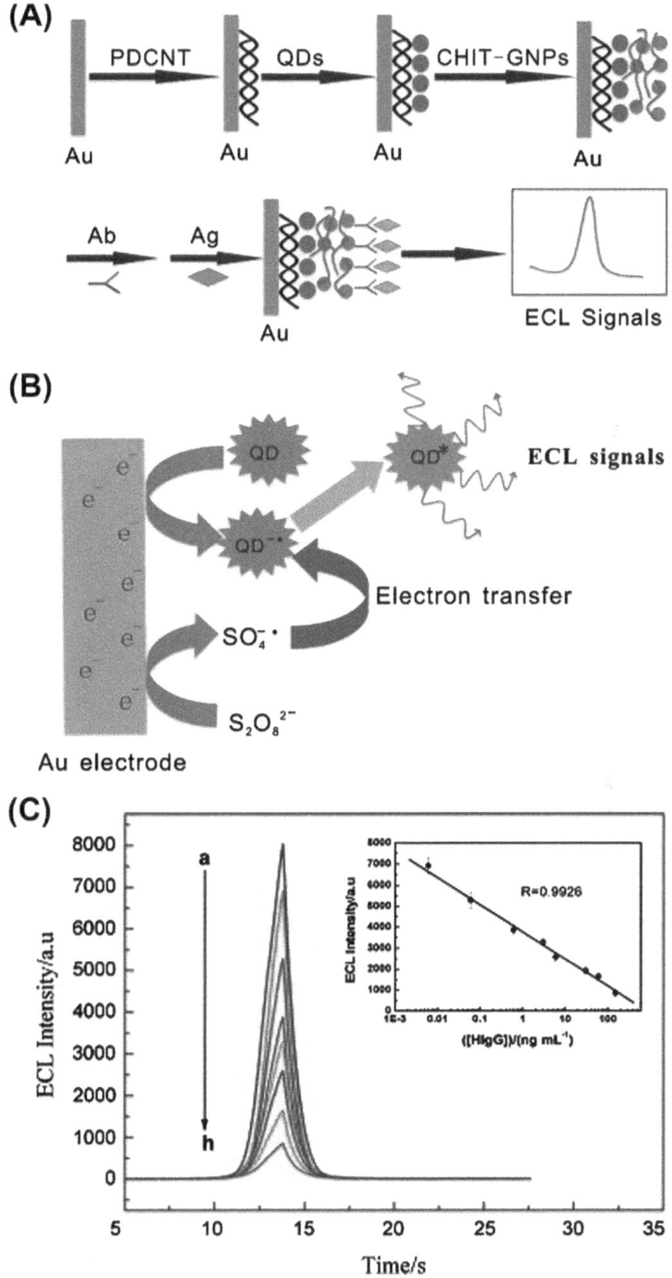

FIGURE 13 Construction and response of an electrochemiluminescence immunosensor based on a nanocomposite film of CdS QD-CNT combined with AuNP–chitosan [144]. (A) The fabrication steps of the ECL immunosensor. (B) ECL mechanisms of the CdS QDs. (C) ECL responses of the immunosensor with different concentrations of human IgG in the range of 0.006–150 ng/mL.

8. CONCLUSIONS

The intensive development of nanotechnology in recent decades is widely reflected in analytical applications. Increasing numbers of various nanostructures fabricated from different materials are the most significant starting point for analytical applications and many other areas of contemporary everyday life and industry [146]. On a large scale those materials gain increasing environmental application, e.g., in remediation strategies for detoxification of organic pollutants [147]. Numerous applications of nanomaterials are also connected to the problem of the ecotoxicology of those materials. Although their negative impact is already well documented for species used in regulatory testing and fresh water organisms, there is still a lack of data on higher organisms including humans [148]. Current issues in this field are transformations in the environment, including reactions with macrobiomolecules, redox reactions, or aggregation [149].

In the increasing application of AuNPs in chemical and biological sensing discussed in other chapters, electrochemical applications are a significant fraction of those applications, as documented by review papers mentioned previously [20,23,25]. An illustration of research activity in this field is the number of papers abstracted in the ISI Web of Knowledge database (Table 1). Approximately every 10th paper published about AuNPs concerns different aspects or applications in electrochemistry; in just 3 years, about 500 papers per year have been published. Those large numbers suggest that a review of

TABLE 1 Citation Data about the Present Status of Electroanalytical Applications of Gold Nanoparticles Based on Example Number of Published Papers in ISI Web of Knowledge (November 18, 2013)

Keyword(s)	Number of Cited Published Papers
Colloidal gold and electrochemical	661
Gold nanoparticles	104,269 (?) [a]
Gold nanoparticles and electrochemical	12,342 (?)
Gold nanoparticles and voltammetry	5,289 (?) [b]
Gold nanoparticles and amperometric	1,277 [c]
Gold nanoparticles and ion-selective electrode	147 [d]

[a] *Question mark indicates the approximate number given by the data-base system*
[b] *About 350 papers published annually in last 3 years*
[c] *About 200 papers published annually in last 3 years*
[d] *About 15 papers published annually in last 3 years*

scientific achievements in the field needs to be based on an arbitrary selection of cited papers and presented examples.

The electrochemical applications of AuNPs reported in this chapter demonstrate a significant contribution of this approach to the development of sensor technology. An important aspect of this approach is not only improvement in the functional parameters of novel electrochemical sensors, but also an essential contribution to their miniaturization and application in microfluidic systems. Further attempts can be expected, e.g., a comparison of different types of gold nanostructures for given applications, focusing more attention on an examination of the physical and chemical durability, and reproducibility of fabrication, because those factors are especially significant elements determining the possibilities of their routine applications.

LIST OF ABBREVIATIONS

Ab Antibody
ALP Alkaline phosphatase
ASV Anodic stripping voltammetry
AuNP Gold nanoparticles
CNT Carbon nanotubes
CV Cyclic voltammetry
DPV Differential pulse voltammetry
ECL Electrochemiluminescence
EIS Electrochemical impedance spectroscopy
GC Glassy carbon
GOD Glucose oxidase
HRP Horseradish peroxidase
ISE Ion-selective electrode
ITO Indium tin oxide
LOD Limit of detection
LSV Linear sweep voltammetry
MIP Molecularly imprinted polymer
PPy Polypyrrole
QD Quantum dot
SPE Screen-printed electrode
SWV Square wave voltammetry

REFERENCES

[1] [a] A.S. de Dios, M.E. Diaz-Garcia, Anal. Chim. Acta 666 (2010) 1–22;
[b] K. Scida, P.W. Stege, G. Haby, G.A. Messina, C.D. Garcia, Anal. Chim. Acta 691 (2011) 6–17;
[c] L. He, C. Toh, Anal. Chim. Acta 556 (2006) 1–15.

[2] M. Trojanowicz, Trends Anal. Chem. 25 (2006) 480–489.

[3] D.T. Pierce, J.X. Zhao (Eds.), Trace Analysis with Nanomaterials, Wiley-VCH, Weinheim, 2010.

[4] R. Sitko, B. Zawisza, E. Malicka, Trends Anal. Chem. 37 (2012) 22–31.
[5] K. Yang, B. Xing, Chem. Rev. 110 (2010) 5989–6008.
[6] M. Guihen, Trends Anal. Chem. 46 (2013) 1–14.
[7] A. Speltini, D. Merli, A. Profumo, Anal. Chim. Acta 783 (2013) 1–10.
[8] Y. Wang, J. Ouyang, W.R. Baeyens, J.R. Delanghe, Export Rev. Proteomics 4 (2007) 287–298.
[9] [a] Z. Zhu, V.M. Otello, R.W. Vachet, Analyst 134 (2009) 2183–2188;
[b] C. Lei, K. Qian, O. Noonau, A. Nouweus, C. Yu, Nanoscale 5 (2013) 12033–12042.
[10] X. Jiang, K. Huang, D. Deng, H. Xia, X. Hou, C. Zheng, Trends Anal. Chem. 39 (2012) 38–59.
[11] [a] J. Shi, Y. Zhu, X. Zhang, W.R.G. Baeyens, A.M. Garcia-Campana, Trends Anal. Chem. 23 (2004) 351–360;
[b] M.J. Ruedas-Rama, J.D. Walters, A. Orte, E.A.H. Hall, Anal. Chim. Acta 751 (2012) 1–23.
[12] F.W. Campbll, R.G. Compton, Anal. Bioanal. Chem. 396 (2010) 241–259.
[13] L. Rassaei, F. Marken, M. Sillanpaa, M. Amiri, C.M. Cirtin, M. Sillanpaa, Trends Anal. Chem. 30 (2011) 1704–1715.
[14] [a] A.J. Killard, M.R. Smyth, Electroanalysis 18 (2006) 319–326;
[b] W. Singaproh, W. Dungchai, P. Rattanarat, O. Chailapakul, Anal. Chim. Acta 690 (2011) 10–26.
[15] M.M.P. da Silva Neves, M.B.G. Garcia, C. Delerue-Matos, A.C. Garcia, Electroanalysis 23 (2011) 63–71.
[16] T. Yin, W. Qin, Trends Anal. Chem. 51 (2013) 79–86.
[17] S. Marin, A. Merkoci, Electroanalysis 22 (2010) 1–11.
[18] A. Merkoci, Electroanalysis 25 (2013) 15–27.
[19] D.W.M. Arrigan, Analyst 129 (2004) 1157–1165.
[20] B.J. Plownan, S.B. Bhargava, A.P. O'Mullane, Analyst 136 (2011) 5107–5119.
[21] M.C. Daniel, D. Astruc, Chem. Rev. 104 (2004) 293–346.
[22] M.T. Fernandez-Abedul, A. Costa-Garcia, Anal. Bioanal. Chem. 390 (2008) 293–298.
[23] J.M. Pingarron, P. Yanez-Sedeno, A. Gonzalez-Colrtes, Electrochim. Acta 53 (2008) 5848–5866.
[24] K. Saha, S.S. Agasti, C. Kim, X. Li, W.M. Rotello, Chem. Rev. 112 (2012) 2739–2779.
[25] S. Guo, E. Wang, Anal. Chim. Acta 598 (2007) 181–192.
[26] M. Faraday, Philos. Trans. 147 (1857) 145–181.
[27] M.A. Hayat, Colloidal Gold, Principles, Methods and Applications, Academic Press, New York, 1989.
[28] J.G. Stonehuerner, J. Zhao, J.P. O'Daly, A.L. Crumbliss, R.W. Henkens, Biosens. Bioelectron. 7 (1992) 421–428.
[29] A.L. Crumbliss, S.C. Perine, J. Stonehuerner, K.R. Tubergen, J. Zhao, R.W. Henkens, J.P. O'Daly, Biotechnol. Bioeng. 40 (1992) 483–490.
[30] D. Hernandez-Santos, M.B. Gonzalez-Garcia, A. Costa-Garcia, Electrochim. Acta 46 (2000) 607–615.
[31] C. Damaille, M. Brust, M. Tsionsky, A.J. Bard, Anal. Chem. 69 (1977) 2323–2328.
[32] M. Lahav, A.N. Shipway, I. Willner, J. Chem. Soc. Perkin Trans. 2 (1999) 1925–1931.
[33] Y. Xiao, F. Patolsky, E. Katz, J.F. Hainfeld, I. Willner, Science 299 (2003) 1877–1881.
[34] B.M. Quinn, P. Liljeroth, V. Ruiz, T. Laakson, K. Konturri, J. Am. Chem. Soc. 125 (2009) 6644–6645.
[35] J. Turkevitch, P.C. Stevenson, J. Hillier, Discuss. Faraday Soc. 11 (1951) 55–75.

[36] M. Brust, M. Walker, D. Bethell, D.J. Schiffrin, R.J. Whymann, J. Chem. Soc. Chem. Commun. (1994) 801–802.

[37] W. Woźnica, M.M. Wójcik, M. Wojciechowski, J. Mieczkowski, E. Bulska, K. Maksymiuk, A. Michalska, Anal. Chem. 84 (2012) 4437–4442.

[38] J. Wang, Y. Shan, W. Zhao, J. Xu, H. Chen, Anal. Chem. 83 (2010) 4004–4011.

[39] R. Kumar, A.K. Pandey, A.K. Tyagi, G.K. Dey, S.V. Ramagiri, J.R. Bellare, A. Goswami, J. Coll. Interface Sci. 337 (2009) 523–530.

[40] B.V. Chikkaveeraiah, V. Mani, V. Patel, J.S. Gutkind, J.F. Rusling, Biosens. Bioelectron. 26 (2011) 4477–4483.

[41] M. Du, T. Yang, K. Jiao, Talanta 81 (2010) 1022–1027.

[42] G. Maduraiveeran, R. Ramaraj, Electrochem. Commun. 9 (2007) 2051–2055.

[43] Y. Song, G. Muthuraman, J. Zen, Electrochem. Commun. 8 (2006) 1369–1374.

[44] A. Devadoss, H. Han, T. Song, Y. Kim, U. Paik, Analyst 138 (2013) 5025–5030.

[45] S. Hrapovic, E. Majid, Y. Liu, K. Male, J.H.T. Luong, Anal. Chem. 78 (2006) 5504–5512.

[46] U. Lad, G.M. Kale, R. Bryskova, Anal. Chem. 85 (2013) 6349–6355.

[47] B. Jeong, R. Akter, O.H. Han, C.K. Rhee, M.A. Rahman, Anal. Chem. 85 (2013) 1784–1791.

[48] L. Wang, J. Lei, R. Ma, H. Ju, Anal. Chem. 85 (2013) 6505–6510.

[49] J. Jiang, W. Fan, X. Du, Biosens. Bioelectron. 51 (2014) 343–348.

[50] D. Lin, J. Wu, M. Wang, F. Yan, H. Ju, Anal. Chem. 84 (2012) 3662–3668.

[51] L. Wang, X. Gan, Microchim. Acta 164 (2009) 231–237.

[52] D. Jia, Q. Ren, L. Sheng, F. Li, G. Xie, Y. Miao, Sens. Actuat. B 160 (2011) 168–173.

[53] X. Tian, C. Cheng, H. Yuan, J. Du, D. Xiao, S. Xie, M.M.F. Choi, Talanta 93 (2012) 79–85.

[54] X. Liu, J. Zhang, S. Liu, Q. Zhang, X. Liu, D.K.Y. Wong, Anal. Chem. 85 (2012) 4350–4356.

[55] J. Xie, Y. Zheng, J.Y. Ying, J. Am. Chem. Soc. 131 (2009) 888–889.

[56] Y. Wu, Y. Wu, J. Huang, T. Zhou, M. Rong, Y. Jiang, X. Chen, Analyst 138 (2013) 5563–5565.

[57] S. Anandhakumar, K. Dhanalakshimi, J. Mathiyarasu, Electrochem. Commun. 38 (2014) 15–18.

[58] J.P. Xie, Q.B. Zhang, J.Y. Lee, D.I.C. Wang, ACSNano 2 (2008) 2473–2480.

[59] B. Su, J. Tang, H. Yang, G. Chen, J. Hang, D. Tang, Electroanalysis 23 (2011) 832–841.

[60] K. Dawson, A. Wahl, R. Murphy, A. O'Riordan, J. Phys. Chem. 116 (2012) 14665–14673.

[61] H. Ciftci, U. Tamer, React. Func. Polym. 72 (2012) 127–132.

[62] G. Jagerszki, A. Takacs, I. Bitter, R.E. Gyurcsanyi, Angew. Chem. Int. Ed. 50 (2011) 1656–1659.

[63] X. Niu, C. Chen, Y. Teng, H. Zhao, M. Lan, Anal. Lett. 45 (2012) 764–773.

[64] M.H. Mashhadizadeh, H. Khani, Anal. Meth. 2 (2010) 24–31.

[65] M.H. Mashhadizadeh, S. Ramezani, S. Ebrahimi, Sens. Actuat. B 169 (2012) 305–311.

[66] C. Yang, Y. Chai, R. Yuan, W. Xu, T. Zhang, F. Jia, Talanta 97 (2012) 406–413.

[67] P.C. Pandey, A.K. Oandey, D.S. Chauhan, Electrochim. Acta 74 (2012) 23–31.

[68] M. Labib, A.S. Zamay, O.S. Kolovskaya, I.T. Reshetneva, G.S. Zamay, R.J. Kibbee, S.S. Sattar, T.N. Zamay, M.V. Berezovski, Anal. Chem. 84 (2012) 8114–8117.

[69] E. Jaworska, M. Wójcik, A. Kisiel, J. Mieczkowski, A. Michalska, Talanta 85 (2011) 1986–1989.

[70] Y. Lan, H. Luo, X. Ren, Y. Wang, L. Wang, Anal. Lett. 45 (2012) 1184–1196.

[71] J. Zhang, T. Zheng, F. Cheng, J. Zhang, J. Zhu, Anal. Chem. 83 (2011) 7902–7909.

[72] Y. Dong, Chu, Anal. Meth. 5 (2013) 5954–5960.

[73] B. Sebez, L. Su, B. Ogorevc, Y. Tong, X. Zhang, Electrochem. Commun. 25 (2012) 94–97.
[74] J. Wang, W. Zhang, J. Electroanal. Chem. 654 (2011) 79–84.
[75] K. Zhang, T. Tan, J. Fu, T. Zheng, J. Zhu, Analyst 138 (2013) 6323–6330.
[76] L. Ding, Q. Ji, R. Qian, W. Cheng, H. Ju, Anal. Chem. 82 (2010) 1292–1298.
[77] F. Melin, T. Meyer, S. Lankiang, S.K. Choi, R.B. Gennis, C. Blanck, M. Schmutz, P. Hellwig, Electrochem. Commun. 26 (2013) 105–108.
[78] Y. Park, S. Yoo, S. Park, Langmuir 24 (2008) 4370–4375.
[79] J. Xu, J. Zhu, Y. Zhu, K. Gu, H. Chen, Anal. Lett. 34 (2001) 503–512.
[80] C. Hu, X. Bai, Y. Wang, W. Jin, X. Zhang, S. Hu, Anal. Chem. 84 (2012) 3745–3750.
[81] D. Nakashima, F. Marken, M. Oyama, Electroanalysis 25 (2013) 975–982.
[82] Y. Yang, Z. Wang, M. Yang, M. Guo, Z. Wu, G. Shen, R. Yu, Sens. Actuat. B 114 (2006) 1–8.
[83] H. Xu, Shi, Y. Xian, L. Jin, Electrochim. Acta 10 (2008) 1839–1843.
[84] H. Liu, R. Malhotra, M.W. Peczuh, J.F. Rusling, Anal. Chem. 82 (2010) 5865–5871.
[85] S. Hu, J. Xie, Q. Xu, K. Rong, G. Shen, R. Yu, Talanta 61 (2003) 769–777.
[86] D.P. Tang, R. Yuan, Y.Q. Chai, X. Zhong, Y. Liu, J.Y. Dai, L.Y. Zhang, Anal. Biochem. 333 (2004) 345–350.
[87] S.S. Kumar, K. Kwak, D. Lee, Anal. Chem. 83 (2011) 3244–3247.
[88] M.H. Freeman, J.R. Hall, M.C. Leopold, Anal. Chem. 85 (2013) 4057–4065.
[89] D. Tang, R. Yuan, Y. Chai, Y. Fu, Electrochem. Commun. 7 (2005) 177–182.
[90] E.R. Sartori, F.C. Vicentini, O. Fatibello-Filho, Talanta 87 (2011) 235–242.
[91] W. Yang, J. Wang, S. Zhao, Y. Sun, C. Sun, Electrochem. Commun. 8 (2006) 665–672.
[92] Y.Y. Xu, C. Bian, S. Chen, S. Xia, Anal. Chim. Acta 561 (2006) 48–54.
[93] Y. Zhuo, R. Yuan, Y. Chai, D. Tang, Y. Zhang, N. Wang, X. Li, Q. Zhu, Electrochem. Commun. 7 (2005) 355–360.
[94] Z. Wang, M. Li, P. Su, Y. Zhang, Y. Shen, D. Han, A. Ivaska, L. Niu, Electrochem. Commun. 10 (2008) 306–310.
[95] L.B. Israel, N.N. Kariuki, L. Han, M.M. Maye, J. Luo, C. Zhong, J. Electroanal. Chem. 517 (2001) 69–76.
[96] B. Zanfrognini, C. Zanardi, F. Terzi, T. Aaritalo, A. Viinikanoja, J. Lukkkari, R. Seeber, J. Solid State Electrochem. 15 (2011) 2395–2400.
[97] H. Dai, G. Yang, F. Qi, X. Jiao, Anal. Meth. 5 (2013) 5684–5693.
[98] J. Jia, B. Wang, A. Wu, G. Cheng, Z. Li, S. Dong, Anal. Chem. 74 (2002) 2217–2223.
[99] D. Tang, R. Yuan, Y. Chai, Y. Liu, J. Dai, X. Zhong, Anal. Bioanal. Chem. 381 (2005) 674–680.
[100] J. Gun, M.J. Schöning, M.H. Abouzar, A. Poghossian, E. Katz, Electroanalysis 16 (2008) 1748–1753.
[101] Y. Yu, X. Chen, Y. Wei, J.Liu, X. Huang, Anal. Chem. 84 (2013) 9818–9824.
[102] F. Goo, R. Yuan, Y. Chai, M. Tang, S. Cao, S. Chen, Coll. Surf. A Physicochem. Eng. Aspects 295 (2007) 223–227.
[103] S. Li, G. Zho, R. Zhang, Y. Hou, L. Liu, H. Pang, Microchim. Acta 180 (2013) 821–827.
[104] H. Zhang, J.J. Xu, H.Y. Chen, J. Phyc. Chem. C 112 (2008) 13886–13892.
[105] E. Majid, S. Hrapovic, Y. Liu, K.B. Male, J.H.T. Luong, Anal. Chem. 78 (2006) 762–769.
[106] Y. Zheng, Z. Huang, C. Zhao, S. Weng, W. Zheng, X. Lin, Microchem. Acta 180 (2013) 537–544.
[107] B.C. Janegitz, R.A. Medeiros, R.C. Rocha-Filho, O. Fatibello-Filho, Diamond Rel. Mat. 25 (2012) 128–133.
[108] K. Xu, G. Xu, J. Lv, J. Cui, Y. Wu, J. Solid State Electrochem. 17 (2013) 2381–2389.

[109] L. Wang, W. Mao, D. Ni, J. Di, Y. Wu, Y. Tu, Electrochem. Commun. 10 (2008) 673–676.
[110] G. Dutta, H. Yang, Electrochem. Commun. 13 (2011) 1328–1331.
[111] A. Florea, Z. Taleat, C. Cristea, M. Mazloum-Ardakani, R. Sandulescu, Electrochem. Commun. 33 (2013) 127–130.
[112] D. Kimand, Y. Shim, Anal. Chem. 85 (2013) 6536–6543.
[113] O. Dominguez-Renedo, L. Ruiz-Espelt, N. Garcia-Astorgano, M.J. Arcos-Martinez, Talanta 76 (2008) 854–858.
[114] O. Dominguez-Renedo, M.A. Alonso-Lomillo, L. Ferreira-Goncalves, M.J. Arcos-Martinez, Talanta 79 (2009) 1306–1310.
[115] M.K. Sharma, G.S. Agarwal, V.K. Rao, S. Upadhyay, S. Merwyn, N. Gopalan, G.P. Rai, R. Vijayaraghavan, S. Prakash, Analyst 135 (2010) 608–614.
[116] A. Calvo-Perez, O. Dominguez-Renedo, M.A. Alonso-Lomillo, M.J. Arcos-Martinez, Electroanalysis 22 (2010) 2924–2930.
[117] P. Chandra, S.A. Zaidi, H. Noh, Y. Shim, Biosens. Bioelectron. 28 (2011) 326–332.
[118] S.V. Pereira, F.A. Bertolino, M.A. Fernandez-Baldo, G.A. Messina, E. Salinas, M.I. Sanz, J. Raba, Analyst 136 (2011) 4745–4751.
[119] C. Xie, H. Li, S. Li, J. Wu, Z. Zhang, Anal. Chem. 82 (2010) 241–249.
[120] M.N. Bui, X. Pham, K.W. Han, C.A. Li, E.K. Lee, H.J. Chang, G.H. Seong, Electrochem. Commun. 12 (2010) 250–253.
[121] N.F. Atta, R.A. Ahmed, H.M.A. Amin, A. Galal, Electroanalysis 24 (2012) 2135–2146.
[122] D. Martin-Yerga, M.B. Gonzalez-Garcia, A. Costa-Garcia, Sens. Actuat. B 165 (2012) 143–150.
[123] F. Xiao, Z. Mo, F. Zhao, B. Zeng, Electrochem. Commun. 10 (2008) 1740–1743.
[124] Y. Tang, R. Huang, C. Liu, S. Yang, Z. Lu, S. Luo, Anal. Meth. 5 (2013) 5508–5514.
[125] Z. Zhang, L. Luo, L. Zhu, Y. Ding, D. Deng, Z. Wang, Analyst 138 (2013) 5365–5370.
[126] Y. Song, H. Liu, Y. Wang, L. Wang, Anal. Meth. 5 (2013) 4165–4171.
[127] L. Qu, S. Xia, C. Bian, J. Sun, J. Han, Biosens. Bioelectron. 24 (2009) 3419–3424.
[128] B.K. Jena, S.J. Percival, B. Zhang, Anal. Chem. 82 (2010) 6737–6743.
[129] T. Gu, L. Bu, Z. Huang, Y. Liu, Z. Tang, Y. Liu, S. Huang, Q. Xie, S. Yao, X. Tu, X. Luo, S. Luo, Electrochem. Commun. 33 (2013) 43–46.
[130] K.Y. Chambimuni-Torres, Z. Dai, N. Rubinova, Y. Xiang, E. Pretsch, J. Wang, E. Bakker, J. Am. Chem. Soc. 128 (2006) 13676–13677.
[131] J. Wu, K.Y. Chumbimuni, M. Galik, C. Thammakhet, D.A. Haake, J. Wang, Anal. Chem. 81 (2009) 10007–10012.
[132] T. Selvaraju, J. Das, S.W. Han, H. Yang, Biosens. Bioelectron. 23 (2008) 932–938.
[133] J. Ishii, M. Chikae, M. Toyoshima, Y. Ukita, Y. Miura, Y. Takamura, Electrochem. Commun. 13 (2011) 830–833.
[134] J. Wang, Y. Cao, Y. Li, Z. Liang, G. Li, J. Electroanal. Chem. 656 (2011) 274–278.
[135] J. Wang, D. Xu, A.,N., Kawde, R. Polsky, Anal. Chem. 73 (2001) 5576–5581.
[136] W. Wang, L. Song, Q. Gao, H. Qi, C. Zhang, Electrochem. Commun. 34 (2013) 18–21.
[137] M. Shen, M. Yang, H. Li, Z. Liang, G. Li, Electrochim. Acta 60 (2012) 309–313.
[138] J. Szücs, R.E. Gyurcsanyi, Electroanalysis 24 (2012) 146–152.
[139] H. Shu, W. Wen, H. Xiong, X. Zhang, S. Wang, Electrochem. Commun. 37 (2013) 15–19.
[140] Y. Du, C. Chen, J. Yin, B. Li, M. Zhou, S. Dong, E. Wang, Anal. Chem. 82 (2010) 1556–1563.
[141] X. Xiao, H. Li, M. Wang, K. Zhang, P. Si, Analyst 139 (2014) 488–494.
[142] R. Malhotra, V. Patel, B.V. Chikkaveeraiah, B.S. Munge, S.C. Cheong, R.B. Zain, M.T. Abraham, D.K. Dey, J.S. Gutkind, J.F. Rusling, Anal. Chem. 84 (2012) 6249–6255.

[143] W.C.A. Koh, P. Chandra, D. Kim, Y. Shim, Anal. Chem. 83 (2011) 6177–6183.
[144] G. Jie, P. Lei, L. Wang, S. Zhang, Electrochem. Commun. 12 (2010) 22–26.
[145] Y. Du, C. Chen, M. Zhou, S. Dong, E. Wang, Anal. Chem. 83 (2011) 1523–1529.
[146] A. Lopez-Serrano, R.M. Olivas, J.S. Landaluze, C. Camara, Anal. Meth. 6 (2014) 38–56.
[147] G.E. Fryxell, G. Cao (Eds.), Environmental Applications of Nanomaterials. Synthesis, Sorbents and Sensors, World Scirentific, Singapore, 2007.
[148] R.D. Handy, R. Owen, E. Valsami-Jones, Ecotoxicology 17 (2008) 315–325.
[149] G.V. Lowry, K.G. Gregory, S.C. Apte, J.R. Lead, Environ. Sci. Technol. 46 (2012) 6893–6899.

Chapter 12

Spectroscopic Techniques Based on the Use of Gold Nanoparticles

María Jesús Almendral Parra[1,*] **and Sara Sánchez Paradinas**[1,2]
[1] *Departamento de Química Analítica, Nutrición y Bromatología, University of Salamanca, Plaza de la Merced s/n, Salamanca, Spain;* [2] *Institut für Physikalische Chemie und Elektrochemie, Leibniz Universität Hannover, Schneiderberg, Hannover, Germany*
**Corresponding author: E-mail: almendral@usal.es*

Chapter Outline

Gold Nanoparticles in Analytical Chemistry. http://dx.doi.org/10.1016/B978-0-444-63285-2.00012-2

1. INTRODUCTION: OPTICAL PROPERTIES OF GOLD NANOPARTICLES AS A STARTING POINT FOR USE IN SPECTROSCOPIC TECHNIQUES

In comparison with other nanoparticles (quantum dots (QDs), magnetic nanoparticles, lipid and polymeric nanoparticles, and carbon-based nanoparticles), metal nanoparticles are preferred owing to reasons such as the possibility of controlling, in a relatively simple way, their size, shape, composition, and structure during their synthesis, the tunability of their optical properties as a function of the previous characteristics, and the simple bioconjugation with other types of molecules.

Over the years, several methods for the synthesis of this type of nanoparticle have been developed and perfected, and special attention has been given to colloidal gold nanospheres. In this sense, different and very interesting reviews have been published [1–4].

When a material is exposed to light, a series of processes is initiated. First, the material is able to absorb light and that light can be reemitted, for example, in the form of fluorescent radiation. It can also be reflected (scattered) at the same frequency as the incident light (Mie or Rayleigh scattering). In other cases, even the electromagnetic field of the incident light can be enhanced, thereby increasing the spectroscopic properties of the material in question at surface level, giving rise to phenomena that can be encompassed within the term "surface-enhanced spectroscopy." In the case of gold nanoparticles (AuNPs), all these processes are enhanced in an important way owing to the interaction of the radiation with the free electrons at the surface of the metal particles. These phenomena are shown schematically in Figure 1.

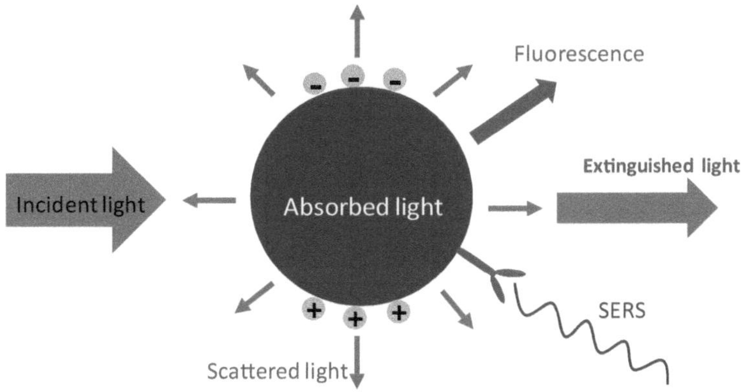

FIGURE 1 Optical processes derived from the interaction of light with a gold nanoparticle: absorption, scattering, surface-enhanced Raman scattering, and surface-enhanced fluorescence.

Along this chapter, we aim to address each of the processes brought about by the interaction of radiation and AuNPs in a more detailed way. We also discuss how these phenomena are used in spectroscopic techniques, offering examples of the most updated applications in fields such as biology and medicine.

1.1 Surface Plasmon Resonance

This phenomenon occurs mainly in nanoparticles known as Plasmon nanoparticles, made of noble metals such as gold and silver. Of all nanoparticles, AuNPs are those that have attracted the most attention owing to their unique optical properties, known as Surface Plasmon Resonance (SPR).

The research field addressing the light–metal interaction is called Plasmonics. In a pioneer work by G.A. Mie [5], and in some publications that appeared in the 1950s [6–9], the authors reported on the optical properties of colloidal solutions of metals. Huang and El-Sayed [10], basing their notions on all these theories, clearly explained what is known as SPR: "When a metal particle is exposed to light, the oscillating electromagnetic field of the light induces a collective coherent oscillation of the free electrons (conduction band electrons) of the metal. This electron oscillation around the particle surface causes a charge separation with respect to the ionic lattice, forming a dipole oscillation along the direction of the electric field of the light (Figure 2). The amplitude of the oscillation reaches maximum at a specific frequency, called Surface Plasmon Resonance (SPR)."

Since SPR is directly related to the free electrons of the conduction band, anything that affects the density of the electronic charge at the surface of the nanoparticle will strongly affect SPR, and hence the wavelength and intensity of the SPR band will depend on the type of metal, the size of the nanoparticle,

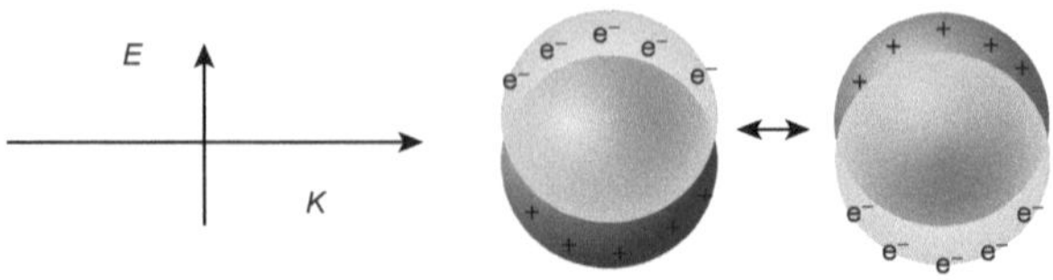

FIGURE 2 Schematic illustration of Surface Plasmon Resonance in plasmonic nanoparticles. *Reproduced from Ref. [10], Figure 1, p. 14.*

its shape, structure, composition, and the dielectric constant of the surrounding medium. All these parameters can be expressed in a simple equation (Eqn (1)), following Mie's theory [5], to explain SPR in quantitative terms:

$$C_{\text{ext}} = \frac{24\pi^2 R^3 \varepsilon_m^{3/2}}{\lambda} \frac{\varepsilon_i}{(\varepsilon_r + 2\varepsilon_m)^2 + \varepsilon_i^2} \tag{1}$$

where C_{ext} is the extinction cross-section, which is related to extinction coefficient by ε (M^{-1} cm^{-1}) $= 10^{-3}$ $N_0 \cdot C_{\text{ext}}$ (cm^2)/2.303; λ is the wavelength of the incident light; ε is the complex dielectric constant of the metal given by $\varepsilon = \varepsilon_r(\omega) + i\varepsilon_i(\omega)$; $\varepsilon_r(\omega)$ and $i\varepsilon_i(\omega)$ are the real part and imaginary part, respectively, of the dielectric function of the metal; and ε_m is the dielectric constant of the surrounding medium, which is related to the refractive index of the medium by $\varepsilon_m = n_m^2$. The real part of the dielectric constant of the metal determines the SPR position and the imaginary part determines the bandwidth. From this equation, one can predict the position and line shape of Plasmon absorption in spherical metal particles. When ε_i is small or does not change very much around the band, the extinction coefficient has a maximum value at a resonance frequency that satisfies $\varepsilon_r(\omega) = -2\varepsilon_m$.

AuNPs have attracted huge interest owing to their potential application in optical devices such as sensors that are used in the detection and capture of images. When AuNPs are dispersed in solution, this shows a different color owing to their small size and they exhibit a strong extinction band in the UV–Vis range of the spectrum. The SPR bands of nanoparticles made of noble metals are located in the visible region of the spectrum and this band is strongly dependent on the size, shape, composition, and crystallinity of the nanoparticles, and also on the spaces between the nanoparticles. For gold nanospheres, resonance occurs at approximately 520 nm, and hence solutions of this type of nanoparticle display a highly characteristic red color.

In the nanometric dimension, when they are exposed to electromagnetic radiation, the free electrons at the surface of NPs absorb the radiation and the color of the solution becomes red. The changes in color of a solution can be monitored via a UV–Vis spectrophotometer and can be used as colorimetric sensors for the detection of chemical species of different natures, such as antibiotics [11], glucose [12,13], melanin [14], etc. In comparison with other analytical techniques, spectroscopic analysis and colorimetric methods are sensitive, cheap, safe, easy to handle, and only require a simple pretreatment step. This, together with the possibilities offered by nanotechnology, opens a research field with important current applications in fields such as chemistry, biology, pharmacy, and medicine.

1.2 Surface-Enhanced Raman Scattering

Raman and Krishnan were the first to observe that certain molecules scatter a small fraction of light that is different from the energy of the incident beam. The strength of the Raman signal is weak since only about 1 in 10^6-10^{10} photons are scattered inelastically, limiting the application of the technique as a tool for analysis for many years. In the 1970s, researchers discovered an enhancement in Raman scattering for molecules placed near a roughened surface, which is now known as surface-enhanced Raman spectroscopy (SERS) [15].

SERS is a vibrational molecular spectroscopy that derives from an inelastic light scattering process. During the Raman scattering process, a laser photon is scattered by a sample molecule. The energy change of the irradiating photon is characteristic for a particular chemical bond in the molecule. Thus, SERS is a powerful biodiagnostic platform that affords a precise molecular spectral fingerprint.

In general terms, the SERS phenomena can be explained in terms of a combination of an electromagnetic mechanism that describes the movement of electrons at the surface of a metal substrate and a chemical mechanism related to charge transfer between the substrate and the molecules of the analyte in question. The electromagnetic mechanism is responsible for up to a 10^6-10^7 increase in Raman scattering, which occurs as the surface Plasmon becomes excited by the incident light and amplifies the electromagnetic field of the metal surface [16,17]. By contrast, the chemical mechanism provides up to 10^2 increases in Raman scattering, and it occurs when the molecule adsorbs strongly on the surface of the metal, which leads to changes of its polarizability. In SERS measurements, several factors affecting these two mechanisms need to be taken into consideration:

- The distance between the molecule and the SERS substrate, in the present case AuNPs. The closer the molecule to the surface, the stronger the SERS intensity.
- The location of the analyte on the substrate surface.
- The orientation and conformation of the molecule, the strongest enhancement taking place when the polarizability of the molecule is parallel to the electrical field vector of the enhanced light field.
- The strength of interaction between the molecule and the substrate. The stronger the interaction, the shorter the distance between the molecule and the substrate, and hence the stronger the SERS. When the interaction is

very strong, charge transfer between the molecule and metal substrate can take place, which is often used to explain the "chemical enhancement."

When Plasmon nanoparticles are excited by a laser light, the originating electromagnetic field around the metal nanoparticles enhances the Raman cross-section of nearby molecules. Thus, SERS becomes a spectroscopic method that provides specific information at molecular level, with a sensitivity comparable to that of methods that employ fluorescence.

SERS offers the possibility of Raman signals (usually very weak), enhanced by about 10^8-10^4, and the unique ability to recognize at molecular level an analyte present in a solution at a very low concentration, making it a highly appropriate technique for ultrasensitive chemical and biological analyses and environmental analysis. As well as its high sensitivity, another important aspect of SERS is that it offers a specific level of detection that can be attained by controlling the chemistry surrounding the metal surface. By adding a specific functional moiety to the surface of SERS, it is possible to detect a single species present in a complex mixture at nano- and femtomolar concentrations without the need to separate the interfering species physically.

All the above characteristics—the high selectivity and sensitivity of SERS, together with the vast amount of information offered by the Raman spectrum—mean that methods based on SERS offer a feasible alternative to more common optical sensing methods.

In this sense, SERS has been used for the detection of a large number of explosives used as bioterrorism agents [18]. In that work, Dasary et al. used AuNPs modified with cysteine for the sensitive and selective detection of TNT in aqueous solutions. SERS has also been used successfully for the detection of different inorganic, organic, and biological molecules. Generally, the larger the molecule, the more complex the SERS spectrum; and the more polarizable the molecule, the stronger the SERS signal.

Owing to the strong degree of enhancement, the SERS technique has been studied extensively over the last few decades, and such work has uncovered its many advantages, including its nondestructive nature, low detection limit, high sensitivity, and easy sample preparation. Thus, SERS has been widely applied in many different areas [17,19,20].

Because of the intrinsic characteristics of AuNPs, it seems clear that these nanoparticles could offer an ideal substrate for SERS, and this is indeed the case; they are currently being used and investigated as agents for SERS enhancement.

1.3 Size- and Shape-Based Colorimetric Variations

As reported above, photon confinement to a small particle size—nanosize—causes SPR, and this in turn enhances all the radiative and nonradiative processes. SPR induces a strong absorption of the incident light and hence can be measured using a UV–Vis absorption spectrometer. AuNPs show strong SPR bands in the visible region. The extinction spectra of AuNPs depend on particle size. Accordingly, the contributions of the absorption and scattering processes to the total light extinction will have different percentages, depending on that size. For AuNPs of 20 nm, total extinction is nearly all due to absorption [21] as shown in Figure 3(A). As size increases, so does scattering (Figure 3(B)). For AuNPs of 80 nm, the contributions of absorption and scattering to total extinction are equal, as depicted in Figure 3(C).

In the figure, it is also possible to see that as size increases the SRP band is displaced toward longer wavelengths and its intensity becomes higher. Thus AuNPs of 20 nm show a band at 520 nm, which red-shifts when particle size increases.

It can be proposed that in metallic nanoparticles with a diameter of <30 nm, the scattering processes are usually negligible. The particle absorbs energy through two fundamental mechanisms: collective excitation of free electrons and electronic transitions of bound electrons from occupied to empty bulk bands. In contrast, surface scattering is a mechanism for dissipating the energy already absorbed by the first process, *Plasmon excitation*. The optical response is dominated by scattering when the nanoparticles are 100 nm in diameter and larger.

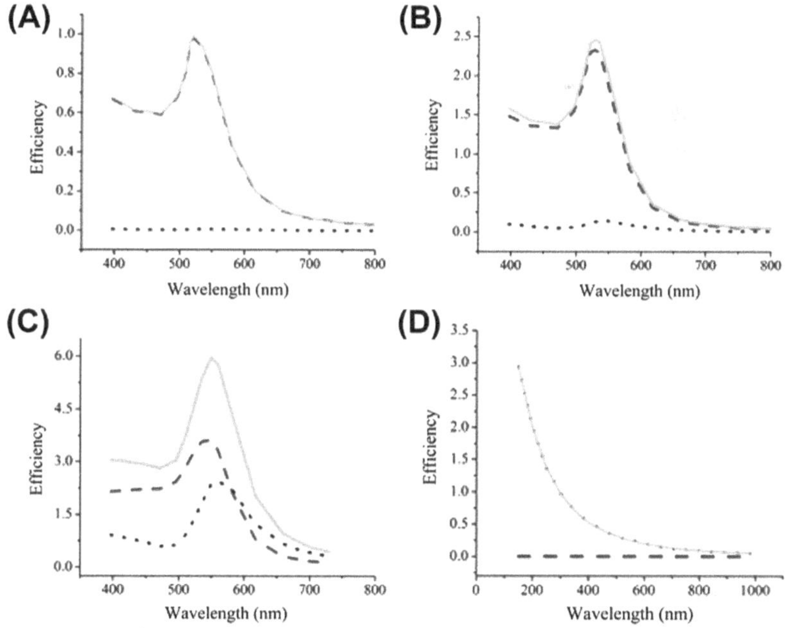

FIGURE 3 Changes in absorption and scattering due to changes in particle size. Gold nanoparticles of (A) 20 nm, (B) 40 nm, and (C) 80 nm. (D) Influence of the gold nanoparticle diameter on the ratio of the scattering-to-absorption cross-section. *Reprinted with permission from Ref. [21]. Copyright (2006) American Chemical Society.*

In sum, according to Figure 3(D), it may be concluded that when particle size increases, the scattering-to-absorption ratio increases. This is important when choosing AuNPs for some specific purpose. For bioimaging applications, it would be suitable to choose larger nanoparticles owing to their higher scattering efficiency. However, if one were interested in using AuNPs in photothermal therapy, the smallest nanoparticles should be chosen, with a greater absorption cross-section and greater efficiency in transforming the light absorbed into heat.

The optical properties of AuNPs are modified not only by their size but also by their shape and structure. As predicted by Gans theory in 1915 [22], as the shape of the nanoparticles changes from spherical to rod-like, the SPR band is split into two. The strongest band, referred to as the *longitudinal band*, is located in the NIR region corresponding to electron oscillations along the long axis. This band is displaced toward longer wavelengths, with increases in the length/width ratio, causing a color change from blue to red. The weak band, referred to as the *transverse band*, remains in the visible region at a wavelength similar to that of the nanospheres and does not undergo modifications with changes in size. This can be seen in Figure 4.

Because spherical particles are in principle completely symmetrical, there is only one dipolar Plasmon resonance, or, in other words, all the possible dipolar modes are degenerate. By elongating a spherical particle in one dimension, it is possible to form a second, lower energy dipolar resonance band in the longitudinal direction while the original transverse mode persists.

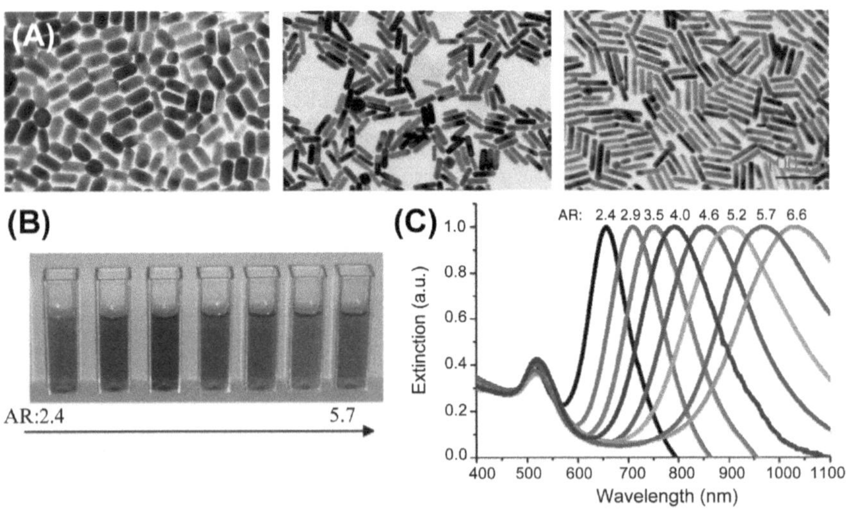

FIGURE 4 Gold nanorods of different aspect ratios (length/width). (A) TEM images, (B) color of the solutions, and (C) Surface Plasmon Resonance signals. *Reproduced from Ref. [10], Figure 3, p. 16.*

As the particles become more complex, many more nondegenerate dipolar modes arise and their optical absorption becomes more complex, with many bands over a broad spectral region [23].

By changing the synthesis conditions, anisotropic AuNPs with other different shapes including stars, tetrapods, triangles, cubes, etc. can be synthesized successfully [24–28]. These different-shaped nanostructures have vastly different absorption spectra and generate different colors, the result of multiple resonances in the complex structures and different resonance modes of electrons. Accordingly, the optical properties of AuNPs can be altered at will by manipulating the details of the structure in a controllable manner. These tunable and controllable optical properties are highly desirable for many applications related to the light absorption of metals, including SPR, SERS, sensing, imaging, therapy, and photocatalysis.

As an example, Mirkin et al. synthesized gold nanoprisms by using 4–6 nm citrate-capped gold dots as the seeds and a mixture of $HAuCl_4$, ascorbic acid, and CTAB as the growth solutions. This seed-mediated, surfactant-based system produced gold nanoprisms of high quality with a narrow size distribution and, at the same time, pseudospherical nanoparticles [25].

The development of different morphologies is particularly interesting because symmetry-decreased anisotropic particles exhibit new and unusual chemical and physical types of behavior, desirable in the field of nanomaterials research and of great relevance in spectroscopic techniques.

1.4 Enhanced Optical Properties of Gold Nanoparticles by Surface Functionalization

The spectral and optical properties of AuNPs, together with the possibility of a relatively easy functionalization with a broad variety of biological molecules and molecules of other types, make this type of nanoparticle a particularly useful tool for catalysis, sensing, and medicine. Owing to the strong affinity of the Au–S bond, AuNPs readily bind to thiols and disulfides. They also bind easily to amines. This allows their conjugation to peptides, proteins, oligonucleotides, DNA, etc. Even the absorption of biomolecules at the surface is possible with AuNPs capped with citrate molecules. Either by physical adsorption or by coupling through the Au–S dative bond, AuNPs are used in combination with biomacromolecules to afford structures known as bioconjugates, whereas the binding of these biomacromolecules to the surface of the nanoparticle is known as *functionalization*. This type of bioconjugate takes advantage of the properties offered, on one hand, by the conjugated molecules (specific coupling with a target) and, on the other, the metal core serves as an optical label.

As discussed along this chapter, because (as its name indicates) it is a surface phenomenon, the surface Plasmon on the gold surface will be strongly dependent on the state of the surface and its properties. Accordingly, the binding of ligands to a receptor immobilized at the surface of nanoparticles will elicit a measurable signal related to the concentration of the ligand, a phenomenon that will serve as a basis for the quantification of analytes in spectroscopic techniques.

Many of the methods aimed at the detection of certain analytes are based on the biospecific aggregation of conjugates, which is initiated by the binding of the target molecules in solution or by the binding of target molecules adsorbed on another conjugate. These changes in the aggregation state of AuNPs give rise to changes in color and hence in the absorption spectrum, which can be monitored with UV—Vis spectroscopy.

Some authors have proposed simple colorimetric methods using unmodified AuNPs to detect analytes [14]. Although these methods exhibit high selectivity for the analyte in question, they usually require time-consuming sample pretreatments, thus limiting their potential application. Surface-modified AuNPs with the appropriate modifier can provide multiple strong hydrogen-binding sites for analytes, inducing the aggregation of the AuNPs as a molecular linker directly, which results in an evident color change from wine-red to blue.

A well-designed chemical interaction between the analyte and the NP environment leads to a change in color (red to blue for AuNPs from well-spaced to aggregated ones), allowing visual detection of the target analyte.

For biomedical applications, the surface functionalization of NPs with biospecific molecules is a crucial step determining the properties of conjugates and the overall success of the technologies applied [29]. Below, we comment on this phenomenon and provide some examples in greater detail.

2. GOLD NANOPARTICLES AS THE BASIS OF SPECTROPHOTOMETRIC METHODS OF DETERMINATION

In any laboratory, there are well-established techniques and methods of quantitative chemical analysis that can be used to determine the unknown concentration of a given analyte. The operation of some of these methods such as high-pressure liquid chromatography, gas chromatography, and capillary electrophoresis (CE) is complex and requires expensive apparatus and costly maintenance. Electrochemistry methods, such as stripping voltammetry with a hanging mercury drop electrode, are not recommended due to the toxicity of the metal. Spectroscopic analysis methods are simple and sensitive. Colorimetric methods are generally of low cost since they do not require any expensive or complex instrumentation, and they are safe, easy to use, and generally need only a simple pretreatment step.

In chemistry, spectrophotometry is used for the quantitative measurement of the reflective and transmissive properties of a material as a function of wavelength. Thus, spectrophotometry deals with visible light, near-UV, and near-IR.

Nanoparticles have now entered all fields of science, providing platforms that some years ago were unheard of, and are now used in different techniques in widely varying disciplines. AuNPs are very attractive contrast agents because they can be visualized with a broad range of different methods. Among these, the most outstanding are those based on the interaction between NPs and light. AuNPs absorb and reflect (scatter) visible light strongly, giving rise to different optical phenomena, such as those described above.

The interaction with light can be exploited for the visualization of the actual particles themselves. AuNPs >20 nm can be visualized directly with phase-contrast light microscopy or in differential interference contrast (DIC) mode [30]. In dark-field microscopy, only the light reflected (scattered) by AuNPs is detected by light microscopy, such that particles >20 nm can be imaged.

Owing to the large light absorption cross-section of AuNPs, absorbed light leads to the heating of the particles and subsequently to the heating of the particle environment. This can be observed with two different techniques: photothermal imaging and photoacoustic imaging. The former makes use of DIC microscopy to record density fluctuations of the liquid environment around the particles (for example, local variations in the refraction index). The latter technique is based on the expansion of the liquid surrounding the nanoparticles due to a local heat-pulse coming from light absorption, which gives rise to a sound wave that can be detected with a microphone [31,32]. Traditional imaging techniques have some limitations, above all in biological applications, such as shallow penetration, low sensitivity, low specificity, and poor spatial resolution. Emergent modes, such as the molecular imaging modality, contribute to improving most of these limitations. However, they too have a drawback: i.e., low signal contrast. And it is precisely here where AuNPs provide advantages since they are proving to be promising signal amplifiers in these new modalities. Moreover, it has been reported that smaller AuNPs are able to emit fluorescence after excitation and hence can be visualized with fluorescence microscopy (to be addressed below in Section 3 of this chapter).

All the previously mentioned techniques (phase-contrast/interference contrast microscopy, dark-field microscopy, photothermal imaging, photoacoustic imaging, and fluorescence microscopy) offer sufficient sensitivity for detection at the single-particle level.

Many different nanoparticles have been used for different applications in the traditional disciplines of analytical chemistry, including spectroscopy, electronic detection, and separation, as well as in several sensor technologies. In the field of spectroscopy, nanoparticles have been used in SPR spectroscopy, Localized Surface Plasmon Resonance (LSPR) spectroscopy, fluorescence spectroscopy, SERS, mass spectrometry, nuclear magnetic resonance (NMR) spectroscopy, magnetic resonance imaging (MRI), and infrared (IR) spectroscopy. Here we focus on the use and improvements offered by NPs, in particular AuNPs, in the detection of chemical and biochemical analytes using several analytical techniques and sensor strategies. We shall not address the analysis with these techniques of AuNPs themselves because this was dealt with in a previous chapter in this book.

There are many ways in which the optical and electronic properties of AuNPs can be exploited to design analytical methods for the detection of chemicals and biomolecules. In this part of the chapter, we shall look at the colorimetric methods based on the aggregation (and consequent change in color) of AuNPs. These methods are currently drawing considerable attention in the sphere of biochemical applications owing to their simplicity, high sensitivity, and low cost. The change in the color of the solution provides an elegant platform for colorimetric detection based on absorption with nanoparticles as signal reporters.

2.1 Absorption-Based Colorimetric Sensing/Sensing Approaches Based on NP Aggregation (Colorimetric Assays)

Today, AuNPs are of great relevance in the development of new materials with optical properties never envisaged previously. From the point of view of analytical applications, AuNPs are potentially useful in colorimetric detection assays since they have a high molar absorption coefficient and a high sensitivity of their SPR band to changes in the state of dispersion. Changes in the refraction index of the local environment and the formation of aggregates cause changes in the spectral properties. The aggregation of dispersed AuNPs gives rise to a change in color from red to blue due to the coupling of the Plasmon, as reported in pioneer studies [33–35].

The aggregation of AuNPs of suitable size ($d > 3.5$ nm) induces the coupling of the surface Plasmon among the particles, leading to a change in color from red to blue at nanomolar concentrations. With the development of nanotechnology, new colorimetric methods have been exploited. AuNPs have been used extensively as colorimetric probes for detection and biodetection owing to their optical properties, dependent on size and the interparticle distance, and the subsequent change in the color of the solution. The change in color during the process of AuNP aggregation (or redispersion of an aggregate) provides a practical platform for colorimetric determination based on the absorption of any target analyte. Any analyte that causes the aggregation of AuNPs (and hence a change in color) can be detected. In this sense, we offer some examples based on the nature of the molecules detected: anions, metallic ions, DNA, genes, proteins, and cancerous cells. For each block we have chosen some representative examples since owing to the exhaustive bibliography and the huge variety of methods that can be used for analyte determination, it would be impossible to include them all. At the end of this section, we offer a table with some examples from each block; these should help to understand the enormous heterogeneity of this field of enquiry.

2.1.1 Detection of Anions

Anionic species are important in biological systems and often appear as polluting substances in the environment. The selective detection of anions in water or aqueous solution remains a tricky problem. The general methodology for the detection of hydrophilic anions is the creation of artificial host molecules with specific hydrogen-bonding sites such as N—H and C=O groups, which act as proton-donating and proton-accepting groups for anion binding, respectively. The low charge density, geometrical variety, and high hydration energy of anions [36] mean that their detection in aqueous media remains a challenge. Suitably modified AuNPs can be used as highly sensitive and selective colorimetric sensors for specific analytes based on aggregation-induced changes in the Plasmon band.

Many examples of surface Plasmon absorption change-based spectrophotometric detection of anions with AuNPs can be found. One of them is the method proposed by Kado et al. [37]. In that work, the authors prepared AuNPs modified with a thiol compound that had a phenylurea moiety. Based on the observed changes in the Plasmon intensity, they were able to detect anions as dihydrogen phosphate. The authors went further and prepared bifunctional AuNPs, modified with both the phenylurea derivative and another water-soluble thiol (L-cysteine), which is selective for anions, observing changes in the Plasmon band accompanied by increasing absorbance at a longer wavelength due to AuNP aggregation.

The determination of highly toxic anions, such as cyanide ions, is of great relevance from the environmental point of view since even though cyanide is produced in large quantities for use in industrial processes, it is extremely toxic for humans. Accordingly, its detection has attracted considerable attention in recent years and many cyanide sensors have been developed, including those based on nanotechnology. Kim et al. developed a competitive colorimetric assay for detection of cyanide in aqueous solution, using adenosine triphosphate-stabilized AuNPs as the reporter unit and a Cu^{2+}—phenanthroline complex as the receptor unit [38].

2.1.2 Detection of Metallic Ions/Inorganic Cations

Methodologies based on the use of AuNPs for the colorimetric detection of metallic ions do not require light-sensitive dye molecules, lengthy protocols, or sophisticated instrumentation, thereby overcoming the limitations inherent to more conventional methods.

A general method for the detection of metallic ions with AuNPs is the incorporation of chelating agents to the surface of the nanoparticle. Lin et al. [39] reported the use of 15-crown-5 functionalized AuNPs for the detection of K^+. In the presence of the cation, the colloidal solution changed from red (dispersed nanoparticles) to blue (aggregated nanoparticles), as depicted in the scheme shown in Figure 5.

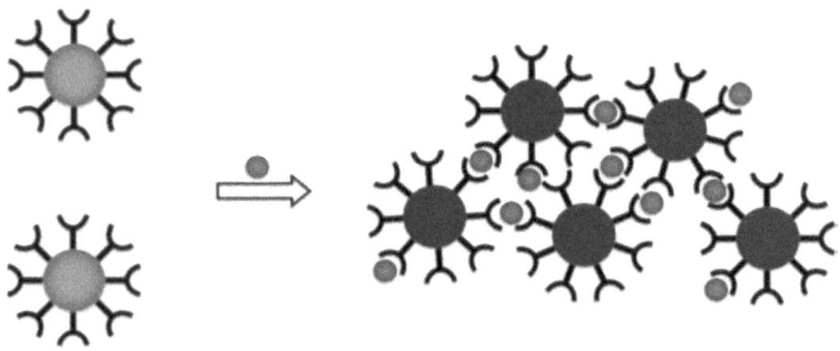

FIGURE 5 General schematic representation of the colorimetric determination of metallic ions using gold nanoparticles functionalized with chelating agents. *Reprinted with permission from Ref. [40]. Copyright (2012) American Chemical Society.*

In another very recent article [41], AuNPs modified with pyridine derivatives were used for the selective detection of Cu^{2+} and Ag^{+} in aqueous solution. In the presence of these ions, the nanoparticles aggregated, giving rise to the above described color change, which could be monitored with UV–Vis spectroscopy.

Regarding these, the number of possible examples is unlimited, including heavy metal ions such as Hg^{2+} [42], Pb^{2+} [43], and As^{3+} [44], among others, although the ones mentioned are sufficient to understand the general methodology and highlight the fact that the techniques developed with AuNPs offer advantages over traditional methods, such as not requiring the use of enzymatic reactions, light-sensitive dye molecules, lengthy protocols, or sophisticated instrumentation.

2.1.3 Detection of DNA/Genes/Proteins

The detection of DNA and important proteins is of great significance in the (bio)-analytical field. The polymerase chain reaction is the most widely used strategy for the detection of nucleic acids; in general, this consists of obtaining a large number of copies of the same fragment of DNA [45]. The technique allows the quantification of DNA specifically, sensitively, and efficiently [46]. However, despite the success of this technique in routine analyses, it requires special reagents and cost-effective equipment. Thus, in certain situations, such as under field conditions, it is necessary to have a tool that will allow an initial rapid and simple detection of DNA. Colloidal AuNPs provide a new platform for colorimetric DNA detection [47,48]. Over the years, several tests for the detection of DNAs have been developed, based on the change in solution color and the Plasmon resonance shift that occurs due to particle aggregation [33,49].

Nucleic acid detection tests are crucial tools for the detection of genetic mutations, and hence the early diagnosis of many diseases, including cancer.

AuNP-based colorimetric assays have been demonstrated to offer a highly competitive technology for oligonucleotide targets. AuNPs functionalized with thiolated DNA are used for the colorimetric detection of target oligonucleotides.

Li et al. [50] designed a bioassay for the detection of adenosine. They used two pieces of single-stranded DNA (ssDNA) of the adenosine aptamer attached to different AuNPs. Aptamers usually consist of short strands of oligonucleotides (DNA or RNA aptamers) or short peptide domains (peptide aptamers) that bind to a specific target molecule. In the presence of adenosine, the two pieces are reassembled into the intact aptamer tertiary structure and induce nanoparticle aggregation.

A novel colorimetric method to detect mutagenic DNA photodimers with the naked eye using AuNPs has been described by Kim and Chung [51]. Straight ssDNA is adsorbed physically on the AuNPs, and thus the stability of the nanoparticles in a high-ionic strength solution is preserved. It was found that UV-irradiated DNA was less adsorptive onto AuNPs because this DNA underwent a conformational change. Accordingly, the cumulative structural deformation in poly(dT)20 after prolonged UV irradiation resulted in an ineffective adsorption of the deformed oligonucleotides onto the AuNPs, leading to AuNP aggregation after the addition of salt (NaCl), observing a color change from red to purple.

One of the general methods to detect proteins is the use of AuNPs conjugated to metal ions, based on the well-known interactions between these ions and specific amino acids. Thus, nanoparticles modified with metal ions can cross-link to aggregate in the presence of specific amino acids, and hence this assay can be used to detect proteins.

As an example, on the basis of the strong coordination between myoglobin (Mb) and Cu^{2+}, Zhang et al. [52] demonstrated the qualitative and quantitative detection of Mb by iminodiacetic acid (IDA)-functionalized AuNPs. The IDA-modified AuNPs were treated with Cu^{2+} before exposure to Mb. The color changed from red to purple after the addition of Mb, with a corresponding red shift from 518 to 552 nm due to aggregation of the NPs caused by the cross-linking of IDA$-$$Cu^{2+}$$-$histidine (His) residues of Mb.

Willner et al. [53] employed an aptamer-functionalized AuNPs as a catalytic label for the amplified detection of thrombin (Thr), an extracellular serine protease that plays a crucial role in the blood coagulation cascade, thrombosis, and hemostasis. Citrate-stabilized AuNPs were functionalized with the thiolated aptamer (red color). Then, aptamer-functionalized AuNPs were reacted with Thr, resulting in the aggregation of the AuNPs due to the fact that Thr includes two binding sites for the aptamer (blue color).

2.1.4 Detection of Bacteria

The demand for biosensors for the detection of pathogens and pollutants in foodstuffs is increasing rapidly. The diagnosis of bacteria using the methods based on the phenomenon of the color change of nanoparticles has the potential to be rapid, easy, inexpensive, and able to maintain a high level of accuracy. The pioneer detection of bacterial DNA by probes based on AuNPs is reported in the work of Mirkin et al. These authors demonstrated that as little as 10 fmol of an oligonucleotide analyte could be detected by exploiting the agglomeration phenomenon of AuNPs [33]. Baptista et al. [54] functionalized AuNPs with a specific oligonucleotide complementary to a *Mycobacterium tuberculosis* sequence. Positive detection was indicated by retention of the red color of the AuNPs following the addition of a NaCl solution. In negative or control samples, no stabilization of the AuNPs occurred and the nanoparticles aggregated when NaCl was added. Su et al. [55] developed a method for the detection of E.coli O157:H7 using Au@Pt nanoparticles as substitutes for the natural enzymes. 4-mercaptophenylboronic acid (MPBA)-modified Au@Pt NPs were absorbed electrostatically on the surface of the bacteria, catalyzing the oxidation of typical organic substrates such as 3,3,5,5-tetramethylbenzidine (TMB) and giving rise to a color change that could be monitored by absorption.

2.1.5 Detection of Cancer Cells

As well as the detection of DNA and proteins, the detection of diseased cells is an important element in the biomedical and clinical field. Owing to the burgeoning needs of prompt and precise detection of cancer, existing techniques must be improved because they are often time-consuming and expensive. Accordingly, the contribution of colorimetric assays may be important for the direct detection of diseased cells.

Apart from analytical areas such as chemical sensing and (bio-)sensing, AuNPs have immense potential in medicine for cancer diagnosis and therapy since their size is similar to that of the biological molecules and structures. Furthermore, two of the characteristics of AuNPs commented on along this chapter, high-scattering cross-section and photostability, mean that they can be considered powerful tools in medical applications based on imaging techniques.

The specificity and selectivity required in diagnosis and therapy can be achieved with thanks to the bioconjugation capacity of AuNPs, described above. The synthesis and bioconjugation of AuNPs can be modified readily as a function of the aim pursued. This, together with the intense surface-Plasmon-enhanced scattering from Au nanoparticles, means that cancer cell imaging using AuNP bioconjugates is currently a hot-spot in science.

Many authors have addressed this avenue of enquiry. Accordingly, in this section, we have selected recent examples of two general types of colorimetric assays based on the use of AuNPs: namely, those which use aptamer-conjugated AuNPs and those using antibody-conjugated AuNPs.

Medley et al. [56] developed a colorimetric assay for the direct detection of cancer cells by using aptamer-conjugated AuNPs. The assay is based on the high selectivity and affinity of the aptamer for cancer cells. This aptamer was conjugated to AuNPs through a thiol group in its sequence, and in the presence of cancer target cells, the aptamer-conjugated AuNPs aggregated on them, changing the color from red to blue. Rapidly, and without any complex instrumentation, this assay was able to distinguish cancerous from noncancerous samples, combining the selectivity and affinity of aptamers and the spectroscopic advantages of AuNPs.

Many publications based not only on aggregation but also on the enhanced absorption and scattering of AuNPs have emerged over the years. The body of works addressing AuNPs conjugated with monoclonal antibodies, such as anti-epithelial growth factor receptor (EGFR), is also impressive [57]. In these cases, anti-EGFR antibody-modified AuNPs allow the specific detection of cancerous cells with EFGRs overexpressed on their surface, thereby differentiating the strong scattering in the target cells from the randomly distributed signals coming from healthy cells. Huang et al. [58] used gold nanorods conjugated with anti-EGFR antibodies as a novel contrast agent for cancer cell imaging using conventional dark-field microscopy. Antibody-conjugated nanorods bind to the cancer cells and show a strongly enhanced scattering in the NIR region (Figure 6). Gold nanospheres and gold nanorods currently offer a good alternative to fluorescent labels for cancer imaging.

As shown in the last example, AuNPs act as excellent sensors and novel contrast agents for optical detection due to their enhanced absorption and scattering.

As a summary of this part of the chapter, Table 1 is shown below with the examples discussed above of colorimetric assays using AuNPs for the detection of different types of analytes. We have added a few more representative examples to show the huge variety of assays developed to date involving the use AuNPs as signal reporters.

So far, a rational view of the main achievements reported in the relevant literature according to sensing approaches based on AuNPs aggregation has been discussed.

Recently, a few colorimetric probes based on the antiaggregation of AuNPs have been proposed with the aim of improving selectivity and lowering false-positive signals. These are blue-to-red colorimetric probes based on the morphological change of AuNPs from an aggregated to a dispersed state. Cross-linked AuNP aggregates are used to detect analytes that dissociate the cross-linkers and redisperse the AuNP aggregates. In this sense, Chen et al. [80] developed an assay for the detection of I^-, using a mercapto-functionalized thymine at the surface of the AuNPs able to coordinate Hg^{2+} and thus induce the aggregation of modified AuNPs. In the presence of I^-, aggregation decreased.

In another work, Aslan et al. [81] applied dextran-modified AuNPs for the detection of glucose. In their assay, dextran-modified AuNPs were aggregated by using Concanavalin A (Con A) as cross-linker, which has multiple binding

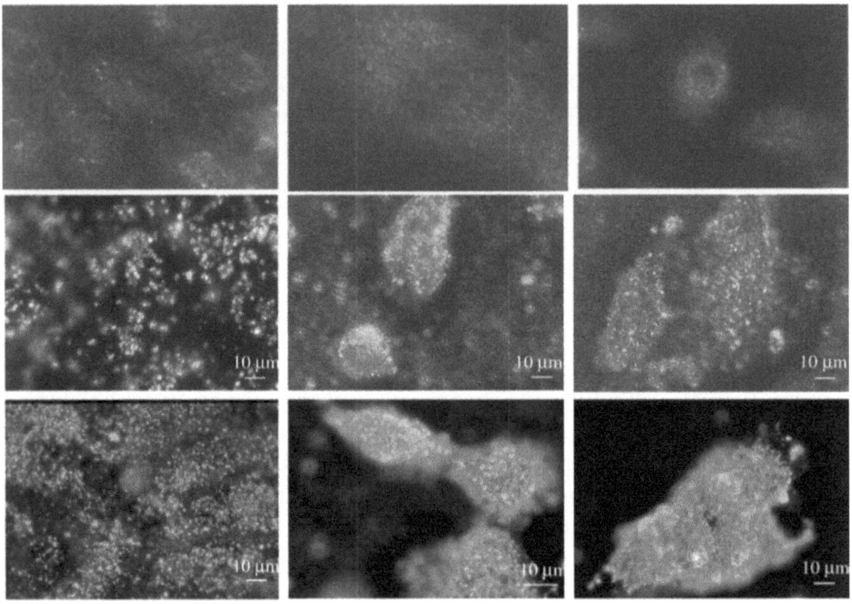

FIGURE 6 Light scattering images of Hacat normal cells (left column), HSC cancer cells (middle column), and HOC cancer cells (right column) without nanoparticles (first row). *(Reprinted with permission from Ref. [59]. Copyright (2005) American Chemical Society.)* Light scattering images after incubation with anti-epithelial growth factor receptor (anti-EGFR)/Au nanospheres (second row). Light scattering images after incubation with anti-EGFR/Au nanorods (third row). *Reprinted with permission from Ref. [58]. Copyright (2006) American Chemical Society.*

sites for dextran. Glucose binds competitively to Con A, and hence its addition dissociates the AuNP aggregates into dispersed AuNPs, leading to a purple-to-red color change. This type of competitive assay has also been used to detect Pb^{2+} [82], adenosine, and cocaine [83], and also to study protein—protein interactions [84].

2.2 Surface Plasmon of Gold Nanoparticles as Probes for the Detection of Analytes

SPR is an optical detection technique based on the measurement of changes in the surface Plasmon absorption band of a thin metal layer caused by interaction with a target analyte of interest. Typically, a polarized light hits a thin metal layer (usually gold) and, under certain conditions, free electrons at the surface absorb incident light photons and convert them into surface Plasmon waves. Changes in the plasmonic resonance signals at a thin metal film are strongly dependent on the refractive index of the medium (Figure 7). The general principle upon which SPR detection rests is that the wavelength of the

TABLE 1 Analyte Detection Using Gold Nanoparticle-Based Colorimetric Assays

Analyte	Sample Matrix	Detection Level	Approach/Functionalization	References
Anions				
CN^-	AS	10^{-5} M	Adenosine triphosphate-stabilized AuNPs	[38]
F^-	AS	mM	Thioglucose-capped gold nanoparticles	[60]
Dihydrogen phosphate	AS		Thiol-modified gold nanoparticles	[37]
Hypochlorite (OCl^-)	Water	μM	Citrate-capped AuNPs	[61]
Sulfate	AS	ppb	Positively-charged gold nanoparticles (cysteamine-AuNPs)	[62]
Cations				
K^+	AS		15-crown-5 functionalized AuNPs	[39]
Cu^{2+}, Ag^+	AS		Pyridine-functionalized AuNPs	[41]
Pb(II)	Water	100 ppt	Glutathione-modified AuNPs	[43]
Hg^{2+}, Pb^{2+}, and Cu^{2+}	Water		DNA-functionalized Au nanomaterials	[63]
Cr (VI)	AS	nM	1,4-Dithiothreitol-functionalized AuNPs	[64]
As^{3+}	Water	ppt	Label-free AuNPs	[44]

Continued

TABLE 1 Analyte Detection Using Gold Nanoparticle-Based Colorimetric Assays—cont'd

Analyte	Sample Matrix	Detection Level	Approach/Functionalization	References
Toxics				
Melamine	Milk products, eggs and feeds	Mg L^{-1}	Cysteamine-modified gold nanoparticles	[65]
Melamine	Milk		Unmodified gold nanoparticles	[66]
Melamine	Milk	ppm	Label-free gold nanoparticles	[67]
Hydrazine	Water	μM level	Detection based on the reduction of gold ions in solution by the analyte as a reducing agent	[68]
Pharmaceuticals				
Isoniazid (antibiotics)	Tablets and serum	μg mL^{-1}	Detection based on the reduction of gold ions in solution by the analyte as a reducing agent	[11]
Neomycin (antibiotics)	Eye- and ear-drops	28 ng mL^{-1}	Label-free gold nanoparticles	[69]
Dopamine		nM		[70]
Dihydralazine sulfate (DHZS)	Urine samples	μM	Dextran-capped AuNPs	[71]
Sugars				
Glucose	Human serum	μg mL^{-1}	Mannose-stabilized gold nanoparticles	[72]
Sugars	Commercial soft drinks	10 μM	Detection based on the reduction of gold ions in solution by the analyte as a reducing agent	
Glucose	Serum	μM	Detection based on the reduction of gold ions in solution by the analyte as a reducing agent	[13]

DNA/Peptides				
Telomere DNA		Pico-molar	Unmodified AuNPs	[73]
Adenosine			Aptamer-conjugated AuNPs	[50]
Phosphorylated peptides			Zinc(II) dipicolylamine-appended AuNPs	[74]
Protein	Milk		Starch-stabilized AuNPs	[75]
Cysteine	AS	10 nM	Unmodified AuNPs	[76]
Cysteine	Urine samples	μM	Carboxymethyl cellulose-functionalized AuNPs (CMC-AuNPs)	[77]
Bacteria				
Escherichia coli O157:H7			Mercaptoethylamine-modified AuNPs (MEA-AuNPs)	[78]
Cancer Cells				
CCRF-CEM cells and Ramos cells			Aptamer-conjugated AuNPs	[56]
Others				
Cationic surfactants	AS, hair conditioners		Negatively-charged AuNPs (cysteamine-AuNPs)	[79]

AS, aqueous samples; AuNPs, gold nanoparticles.

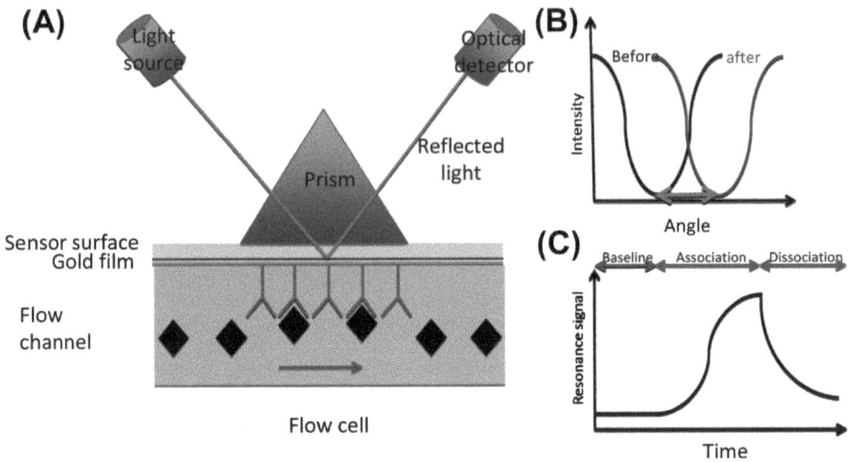

FIGURE 7 The general Surface Plasmon Resonance (SPR)-based detection system. The reflected light is measured from the metal–solution interface (A). The surface Plasmon of the metal film excited by the p-polarized light causes a strong decrease in reflectivity at a specific angle of the incident light, which depends on the refractive index of the medium. A solution containing analyte flows under the surface, and upon recognition and binding of the analyte to the Au surface, the refractive index changes, shifting the angle of minimum reflection (B). The binding and desorption of the analyte on the Au surface affords an SRP sensogram (C). The change in the refractive index upon analyte binding and the intensity of the signal increase. The signal decreases back to the baseline upon dissociation of the analyte from the receptor. *Reprinted with permission from Ref. [15]. Copyright (2011) American Chemical Society.*

SPR band is displaced owing to local dielectric changes caused by analyte adsorption at the metal surface. Improvements in these techniques using metal nanostructures, instead of thin films, have been explored, and SPR sensing has become a label-free and sensitive technique for the detection of molecular interactions and represents one of the simplest and most powerful applications of the optical properties of metal nanostructures.

The main limitation of SPR technology is the detection of analytes at very low concentrations in real samples and for the detection of small molecules. To overcome sensitivity problems, strategies using metal nanoparticles have been proposed.

AuNPs supported on metallic surfaces have been used for signal amplification of SPR. This arrangement allows the coupling of the localized surface Plasmon of metal NPs and the surface Plasmon wave of the metal film afford strong signal amplification for small-molecule detection. In this type of device, the surface Plasmon excited by the incident light causes a strong decrease in reflectivity at a specific angle of the incident light, which depends on the refractive index of the medium. Upon recognition and binding of the analyte to the AuNP surface, the refractive index changes, thus changing the angle and intensity of the reflected light, this being detectable by the instrument. Even the binding of a relatively small amount of analyte is sufficient to cause a detectable change in the Plasmon and hence the technique is considered to be highly sensitive, which is of great relevance for analytical and biomedical applications.

Examples of the use of AuNPs as SPR amplification probes include the detection of biomarkers, heavy metal ions, small biomolecules, and so on. Some of these examples are discussed below.

Sim et al. [85] developed a sandwich assay platform for the detection of protein biomarkers. They used functionalized nanorods to enhance the sensitivity of SPR measurements for the ultrasensitive detection of IgE proteins.

SPR sandwich assays have also been used for the detection of metal ions in aqueous solution. These are difficult to detect with conventional SPR because small molecules and ions usually show smaller refractive index changes. Using AuNPs, Wang et al. [86] detected Hg^{2+} in an aqueous solution by immobilizing a mercury-specific oligonucleotide probe on the Au film, which specifically interacted with the Hg^{2+} from the analyte solution. Then, AuNPs coated with the complementary part of the DNA strand were used to form the sandwich assay and amplify the signal.

Apart from sandwich assays, other configurations have been developed. When steric hindrance problems appear, owing to the small size of the gold probe with respect to the target molecule, it is possible to use dendrimer-encapsulated AuNPs instead of bare AuNPs. This was addressed by Fasconi et al. [87] for the detection of insulin. Direct modification of the metal film is also possible. AuNPs can be deposited at the surface forming a nanocomposite, such as the Au/Al_2O_3 nanocomposite film proposed by Gao et al. [88], or they can be cross-linked to a gold electrode, as developed by Riskin et al. [89] for the detection of explosives.

By combining the aptamer technique with the amplifying effect of AuNPs, Wang and Zhou [90] designed a sensitive SPR sensor for detecting adenosine. This sensor consists of the aptamer immobilized on SPR gold film with its ssDNA structure on one hand, and Au nanoparticle-tagged complementary ssDNA on the other. The hybridization between both ssDNA molecules leads to a large change in the SPR signal. However, in the presence of adenosine, the aptamer will change its structure from ssDNA to the tertiary structure and will not be able to hybridize with the Au nanoparticle-tagged complementary ssDNA. The number of aptamers with a tertiary structure is proportional to the concentration of adenosine.

In general, it has been observed that these methods work well for the detection of analytes of different types and the amplification strategies permit very low limits of detection to be attained. SPR spectral location is sensitive to the local environment, the changes in the local refractive index near the surface, and hence, the binding events on this surface. Willets and Van Duyne

[91] described the SPR wavelength shift ($\Delta\lambda_{max}$) of metal NPs with the following equation:

$$\Delta\lambda_{max} = m(n_{adsorbate} - n_{medium})\left(1 - e^{-\frac{2d}{l_d}}\right) \quad (2)$$

where m is the global RI sensitivity (nm/refractive index unit (RIU)), $n_{adsorbate}$ and n_{medium} are, respectively, the refractive indices of the adsorbate and the surrounding medium, d is the effective adsorbate layer thickness (nm), and l_d is the electromagnetic field decay length (nm).

A red shift of the SPR λ_{max} is usually observed upon binding of analytes to NPs, especially for biomolecules, since the RIs of the adsorbate (molecules and ions) are usually larger than the RI of the surrounding medium (buffer solution, air, or N_2).

There are numerous references in the literature to SPR sensors based on local binding events. Proteins are one of the most common analytes detected with metal NP-based SPR sensing owing to their large size and the highly specific recognition and binding between antibody and antigen. In most cases, NPs are conjugated with a specific antibody or antigen label. Then, the target antigen or antibody in solution will interact with the complementary antibody or antigen metal NP-bound receptor, leading to a local change in the refractive index and a shift in the SPR band. Modified metal NPs can be immobilized on a surface or suspended in solution.

An interesting example is that of the work of Dondapati et al., who detected streptavidin by the shift in the high-sensitivity SPR band of a single Au nanostar [92]. In that work, the nanostar was first functionalized with biotin and then exposed to solutions containing different concentrations of streptavidin. A red shift of 2.3 nm occurred in response to a 100 pM solution of streptavidin, with high specificity.

The techniques based on the SPR phenomenon take advantage of the plasmonic properties of metal NPs for sensing analytes based on their strong color, changes in the refractive index of the entire medium in the vicinity of the particles, local changes in the refractive index of the medium upon analyte binding, and Plasmon coupling induced by the presence of analyte. These methods are highly sensitive and are proving to be promising tools for the resolution of present and future analytical challenges.

2.3 SERS-Based Sensing

SERS is one of the most important applications of metallic nanoparticles. Au and Ag are the metals most commonly used for SERS because of their strong LSPR band in the visible range.

As pointed out above, SERS is based on the enhancement of Raman scattering of an analyte molecule close to or in contact with a roughened metal substrate surface. This enhancement, in turn, is due to an enhanced electromagnetic field at the metal surface due to increased absorption of the incident light.

Normal Raman scattering is a widely used spectroscopic technique for chemical analysis, detection, and imaging that provides molecular specificity or selectivity, allowing for the unique identification of samples; this differentiates it from SPR. The latter, itself, does not offer molecular specificity or selectivity. The major limitation of normal Raman spectroscopy is the very weak signal or low quantum yield (QY) of scattering (in the order of 10^{-7}). SERS combines the advantages of molecular specificity and a strong signal.

Owing to factors such as the distance from the surface, the orientation and conformation of the molecules, and the strength of the interaction between the molecule and substrate, among others, SERS spectra can be quite different from normal Raman spectra for the same molecule as regards peak intensity distribution and the particular vibrational modes detected. Therefore, the interpretation of SERS spectra must be made with caution [93].

Below, we comment on some of the most recent examples of the use of SERS with AuNPs for analyte detection.

Herrera et al. [94] evaluated the potential application of Au nanoparticles in the detection of explosives by SERS. The AuNPs were prepared by laser ablation and deposited on various substrates (such as Al sheets) and subsequently evaluated as SERS substrates with the objective of detecting explosives such as TNT. The results were promising for SERS-based studies and could lead to the development of microsensors.

Senapati et al. [95] reported for the first time a tryptophan-protected popcorn-shaped gold NPs probe for SERS detection of Hg(II) ions. The central sphere of these structures acts as an electron reservoir and the tips are responsible for the strong SERS signal observed for tryptophan. In the presence of Hg(II) ions, the Hg(II) interacts with the indole ring of tryptophan and forms a stable complex through intermolecular π interactions. As a result, tryptophan is released from the NPs, leading to a dramatic decrease in the SERS signal. This approach is rapid, easy, and highly selective for the recognition of Hg(II) ions at the 5 ppb level in aqueous solution. Moreover, it does not require any DNA or dye tag and Hg(II) detection takes <15 min to complete.

One approach for the detection of Fe^{3+} ions in water samples has been demonstrated using AuNRs and crystal violet (CV) [96]. When CV molecules are dispersed in an Au nanosensor solution, a strong enhancement of the signals from the CV molecule is observed and a number of peaks appear in the SERS spectra. The addition of Fe^{3+} ions leads to the suppression of the signal from the CV molecules. AuNRs act as sensitive nanosensors for the detection of Fe^{3+} ions in water.

The detection of several biomolecules such as peptides, proteins, and DNA is important in biomedical research, medical diagnostics, and proteomics. The

SERS technique has been widely applied in biochemical analyses for clinical diagnostics.

AuNPs-based SERS has been used for the detection of several proteins, such as human serum albumin (HSA) [97], and even prions—infectious agents that are very hard to detect [98]. Yoon et al. [99] used aptamer-conjugated AuNPs (as sensing probes) and magnetic beads (as the supporting substrate) for the rapid and reproducible detection of thrombin. It is a sandwich aptamer system between 15-mer thrombin-binding aptamers (TBA_{15}) immobilized onto the surface of magnetic beads, and then thrombin antigens and 29-mer thrombin-binding aptamer (TBA_{29})-conjugated AuNPs. Since the reaction progresses in solution, this SERS-based immunoassay technique can solve the diffusion-limited kinetic problems of solid substrates.

In an assay performed in 2010, Ray et al. [100] used Au nanopopcorn structures for the in situ monitoring of prostate cancer cells. Au nanopopcorn was multifunctionalized with anti-prostate-specific membrane antigen (PSMA) antibody and a Raman dye (RH6G) attached to A9 RNA anti-PSMA aptamers. The tips of the nanopopcorn provided SERS enhancements of several orders of magnitude. These multifunctional popcorn-shaped AuNPs bound to prostate cancer cell lines, LNcap, which express a high level of PSMA. In the presence of LNcap, the multifunctional popcorn-shaped AuNPs bound to PSMA receptors on the cancer cells and formed aggregates, providing a strong RH6G Raman signal.

Enzymatic bioassays for the detection of nicotinic acid (NAADP), a calcium secondary messenger that is responsible for intracellular calcium release, usually require sample pretreatment. Vitol et al. [101] used glass slides coated with AuNPs as an SERS substrate for the detection of NAADP in an easy and rapid way, at a concentration of 10 nM in aqueous solution.

3. APPLICATION OF GOLD NANOPARTICLES IN THE DEVELOPMENT OF FLUORESCENCE-BASED METHODOLOGIES

Nanobiotechnology is considered to be the integration of nanotechnology with biology, biochemistry, and medicine, and is now emerging as a very active field of research. In particular, the use of nanostructured materials in the life sciences, biology, and biomedicine has been extensively explored during the past decade [102], the development of new fluorescent labels starting from luminescent nanomaterials attracting special attention [103]. To date, researchers have developed several different types of luminescent nanomaterials, including semiconductor QDs, dye-doped nanoparticles, up-converting lanthanide-doped nanoparticles, and carbon nanodots [104]. In comparison with existing small-molecule dyes and fluorescent proteins, these luminescent nanoprobes exhibit highly improved photophysical properties, huge surface-to-volume ratios, easy surface tailorability, and color tunability.

Consequently, they are expected to complement or possibly even replace conventional fluorescent probes, and they offer great opportunities for advances in fields such as the development of biosensors, molecular imaging, optoelectronics, and nanomedicine.

In comparison with other techniques, fluorescence methods are extraordinarily attractive, with the advantages of being inexpensive, having high sensitivity and selectivity, speed, and simplicity. In analytical chemistry, fluorescence is one of the most popular methods of detection because of its high sensitivity. Nonetheless, there are still some limitations, namely photobleaching, the short lifetime of fluorophores and their limited brightness. Nanoparticles are ideal analytical tools in fluorescence-based methods to circumvent these disadvantages. Water-soluble metal NPs with different ligands and tunable emission colors have emerged as a new class of ultrasmall, biocompatible fluorophores for applications as biological labels or optoelectronic emitters. These new luminescent nanomaterials are typically composed of a few to roughly one hundred atoms, and are called metal nanoclusters (NCs) [105]. Metal NCs typically have diameters below 2 nm and properties between isolated atoms and larger nanoparticles or even bulk noble metals.

Photoluminescence from noble metals was first observed by Mooradian en 1969 [106], but for years failed to attract much attention due to the extremely low QY obtained. Recently, researchers have developed various approaches to synthesize water-soluble fluorescent metal NCs with highly enhanced QYs, in the range of 10^{-3} to 10^{-1}, simply by choosing suitable agents capable of stabilizing clusters against aggregation and avoiding polydisperse nanoparticle size distributions. Without devoting too much time to discussing the synthesis of this type of NC, since this would be out of the scope of this chapter, suffice it to mention that the most common routes of synthesis for obtaining NCs, in particular gold nanoclusters (AuNCs), include the use of thiol-containing small molecules, dendrimers, and polymers [107], among others. As an example, the preparation of Au nanodots within biocompatible poly(amidoamine) (PAMAM) dendrimer hosts affords nanodots with 100-fold greater fluorescence QYs than any of those prepared previously [108]. The strong size-dependent blue emission from these monodisperse Au nanodots was 41% in aqueous solution and 52% in methanol. The nanodots synthesized by Zheng et al. were easily prepared, water-soluble, and had high QY, providing insight into the molecular nature of small metal NCs. In future, they may find application as novel biocompatible fluorophores.

As explained in the first part of this chapter, the optical properties of metal nanoparticles are dominated by the collective oscillation of electrons at their surfaces, known as SPR. As size is reduced to the subnanometer scale, metal clusters become too small to support Plasmons, and a transition from metal to molecule ensues [102]. Currently, several mechanisms have been proposed to explain the emission properties of ultrasmall metal NCs, such as that proposed by Huang and Murray [109]. These authors suggested that the visible luminescence of monolayer-protected metal NCs would be associated with interband transitions between the filled *d* band and the *sp* conduction band. Other workers have suggested alternative sources for the fluorescence of metal NCs: the metal core, with its intrinsic quantization effects, and the particle surface, whose properties are governed by the chemical interactions between the metal core and the surface ligands. No general mechanism is currently known that is able to explain the emission from all metal NCs.

Photoluminescence from Au surfaces and nanoparticles has been reported to lie with the maximum ranging between 500 and 590 nm [110] and also with the maximum at 410–430 nm [111].

Luminescence in metals is a three-step process involving the photoexcitation of electro–hole pairs, the relaxation of excited electrons and holes, and the emission from the electron–hole recombination [103]. In gold, the 500–590 nm range of emission wavelength corresponds to electronic transitions between the *sp* band just below the Fermi level and the first *d* band (d_1), the second emission band, 410–430 nm, corresponding to the recombination of electrons with holes in the lower energy band, d_2.

As previously mentioned, fluorescent NPs are often more stable and resistant to photobleaching, and have higher luminescence intensity and longer lifetimes than organic dyes. One important characteristic is that their emission is narrow and symmetric and is tunable by size, shape, and composition [112].

AuNPs with a size <2 nm show strong fluorescence. Fluorescent surface-passivized AuNCs ranging in size from 28 atoms to smaller particles (<1.2 nm) show emission in the NIR, red, and blue, with increasingly higher energy emission with decreasing NC size. For NCs of this size, Plasmon absorption disappears and Mie's theory can no longer be applied [113]. These novel optical properties make fluorescent noble metal NCs ideal fluorophores for multicolor and multiplexing applications in biomedical engineering and molecular biotechnology [114].

3.1 Functionalized Gold Nanoparticles as Fluorophores

In nanobiotechnology, AuNPs are used in combination with recognition biomacromolecules bound to their surface by physical adsorption or common coupling through the dative Au–S bond with, for example, simple-chain nucleotides, peptide antibodies, carbohydrates, etc. Most common functionalization technologies are based on the use of thiolated intermediate linkers such as $HS(CH_2)_nR$ (R = COOH, OH or SO_3H; n = 11–22), or thiolated derivatives of oligonucleotides, proteins, immunoglobulins, avidin and antibodies, or linkers with phosphine, amine, or carboxylic groups, and even a new type of mixed matrices. Glutathione has been shown to be an excellent stabilizer for synthesizing AuNCs with visible luminescence by reducing Au^{3+} in the presence of glutathione with sodium borohydride ($NaBH_4$); the QY of these NCs was reported to be in the order of 10^{-3}. Instead of using the strong

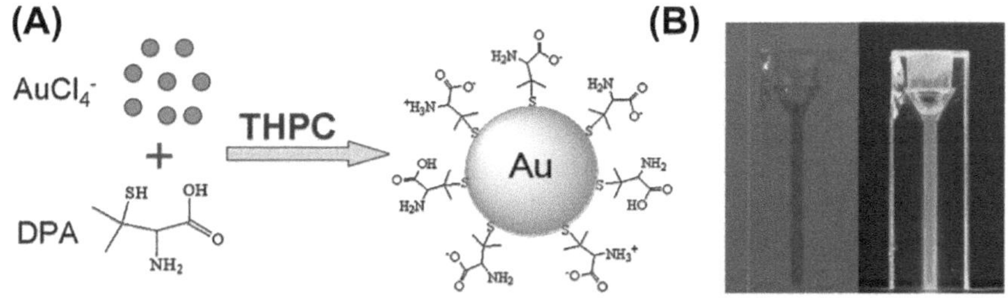

FIGURE 8 (A) Scheme for the synthesis of D-penicillamine (DPA)-stabilized fluorescent gold nanoclusters (AuNCs) using tetrakis(hydroxymethyl)phosphonium chloride (THPC) as the reducing agent. (B) Photographs of an aqueous solution of DPA-AuNCs in room light (left) and under a UV light source with excitation at 365 nm (right). *Reproduced from Ref. [102], Figure 1, p. 402.*

reducing agent $NaBH_4$, Nienhaus' group [107a] recently reported AuNC synthesis by using a mild reducing agent, tetrakis(hydroxymethyl)phosphonium chloride (THPC). In this approach, a zwitterionic thiolate ligand, D-penicillamine (DPA), was chosen as the stabilizer (Figure 8). The DPA-capped AuNCs obtained emitted bright fluorescence at 610 nm (QY: 1.3%) and displayed excellent stability in aqueous solvents across the physiologically relevant pH range.

At present, most biochemical applications based on fluorescent AuNPs use conjugated (biofunctionalized) nanostructures.

Typically, a fluorimetric detection technique is based on the principle that when a target analyte of interest is within close proximity to, or binds to biological receptors labeled with a fluorophore, a measurable change in the fluorescence emission intensity can be observed. This is due to quenching, Förster Resonance Energy Transfer (FRET), or surface-modified fluorescence (SMF). Below, we shall take a closer look at these phenomena, the use of AuNPs as mediators in these processes, and their usefulness in analyte detection using spectroscopic techniques.

3.2 Quenching Processes

Quenching is a process of decreasing the intensity of fluorescence emission. Two modalities are possible: a static mode, in which a covalently bonded complex of a quencher (an electron-accepting dye) and a fluorophore (an electron donor) absorbs light and returns to the ground state from an excited state or a dynamic mode, where the quencher molecule diffuses into a fluorophore in the excited state [115]. The mechanism accounting for this is that when a fluorophore tagged to a biomolecule is assembled on the quencher, fluorescence emission is reduced. The fluorescence emission intensity would increase again if a target analyte were to displace the quencher away from the fluorophore.

Most studies addressing fluorescence with AuNPs carried out to date have focused on the use of fluorescent ligands such as pyrenyl [116], polyoctylthiophenyl [117], fluorenyl [118], rhodamine B [119], and others. In some works, AuNPs have also been reported to exhibit an enhancement in fluorescence at appropriate fluorophore-to-metal distances on solid substrates [120]. However, the most widely observed phenomenon when an AuNP is tagged with a fluorophore is *fluorescence quenching*. This quenching occurs when the emission spectrum overlaps with the gold surface Plasmon band. Thus, AuNPs can serve as excellent fluorescence quenchers for FRET-based assays.

The interactions between the AuNPs and the fluorophore depends on various factors, including the size of the nanoparticles, the distance between the metal surface and the fluorophore moiety, and the spectral overlap between the fluorophore's emission and the nanoparticle's surface Plasmon absorption band; the result will be either an enhancement or a quenching of the fluorescence.

3.3 FRET Mechanisms

Förster resonance energy transfer (FRET) is also a mechanism that causes a change in fluorescence emission intensity. FRET refers to the energy transfer between two chromophores, a donor and an acceptor, via a nonradiative mechanism without the absorption or emission of photons. When both, donor and acceptor, are fluorescent, FRET is also termed as fluorescence resonance energy transfer. FRET spectroscopic techniques are based on a Coulomb interaction-mediated energy-transfer process, occurring between an excited donor fluorophore and an acceptor fluorophore via induced dipole–dipole interactions.

The quantum efficiency of the energy transfer is given by the Forster formula:

$$E = \frac{1}{1 + (R/R_0)^6} \tag{3}$$

where R is the distance between donor and acceptor, and R_0 is the distance at which 50% of the energy is transferred and is a function of multiple factors such as the refractive index of the medium, and donor and acceptor absorption [121].

Taking the previous equation into account, one of the main requirements for FRET occurrence is close proximity of the donor and acceptor so that energy can be transferred instead of being dissipated in the environment. The FRET technique works effectively when the distance between the two chromophores is <10 nm because of the huge decrease in energy transfer with distance ($1/R^6$) between the donor and acceptor (Eqn (3)). Another important requirement for FRET is an overlap between the donor emission and acceptor excitation spectra. Finally, the relative orientation of the dipole moments of the chromophores also influences FRET efficiency; the highest FRET occurs for the parallel dipole orientation. FRET-based assays have often been used to study receptor–ligand interactions and structural changes occurring in molecules.

3.4 Surface-Modified Fluorescence

SMF is another important mechanism that can be applied to measure the changes in fluorescence emission intensity caused by excited fluorophores. The governing principle of SMF is based on the concept that the fluorescence emitted by a fluorophore can be enhanced or quenched by modifying or controlling its surrounding electromagnetic environment. This can be done by conjugating the fluorophore with a metal particle or metal surface, in the present case AuNPs. Based on the distance (if <10 nm) between the coupled entities, i.e., fluorophore and metal particle (or surface), the latter absorbs energy and causes an enhancement or quenching of the fluorescence signal [115].

In light of the three mechanisms explained above, we now describe several representative examples of fluorescence methodologies based on the use of AuNPs for the detection of different types of analytes.

3.5 Fluorescent Gold Nanostructures for the Detection of Analytes: Representative Examples

Fluorescent AuNPs show great promise as optical reporters for sensitive detection owing to their good water-solubility, large Stokes shifts, low toxicity, and high emission rates.

Metal NCs have also been used as metal-ion sensors based on the quenching of their fluorescence. Ying et al. [122] developed a selective and sensitive sensor for the detection of Hg^{2+} using BSA-stabilized AuNCs. The sensing mechanism was based on the high-affinity metallophilic Hg^{2+}–Au^{+} interactions, which efficiently quenched the fluorescence of AuNCs. Another important metal ion from both the environmental and biological points of view is Cu^{2+}. Tu et al. [123] developed a novel NIR luminescent AuNPs assay for the detection of Cu^{2+} based on the quenching of luminescent AuNCs. In another assay, Ni^{2+} was detected using fluorescent bovine serum albumin-protected gold nanoparticles (BSA-AuNPs) [124]. The fluorescence of BSA-AuNPs was selectively quenched by nickel ions, a linear relationship being observed between the fluorescence intensity of BSA-AuNPs and the nickel concentration.

The detection of H_2O_2, one of the products of enzymatic reactions carried out by nearly all oxidases, enables the development of quantitative assays for the determination of the activity of the enzyme as well as several enzyme substrates such as glucose. Chang et al. [125] developed an assay system for the quantitative determination of H_2O_2 using fluorescent 11-mercaptoundecanoic acid-bound Au nanodots (11-MUA-AuNCs) based on luminescence quenching. In the presence of H_2O_2, the luminescence was reduced due to the oxidation of 11-MUA units bound to the cluster surface through Au–S bonds to form an organic disulfide product. H_2O_2 was detected over a wide dynamic range (100 nM–1.0 mM) with an LOD of 30 nM, making this method comparable with other optical H_2O_2 sensors. Further combination of luminescent AuNCs with glucose oxidase enabled the sensitive determination of glucose.

The general configuration for the detection of proteins with fluorescent metal NCs is the conjugation of selective receptor molecules (e.g., antibodies and aptamers) to the surface of metal NCs. The first application of fluorescent metal NCs for protein detection was reported in 2006 by the group of Leblanc [126]. They constructed a human IgG immunoassay with the polyclonal goat-derived antihuman IgG antibody electrostatically conjugated with PAMAM dendrimer-encapsulated AuNCs. This system detected IgG by linear fluorescence quenching in response to changes in the surface electrostatic properties of the AuNCs due to the formation of antigen–antibody immunocomplexes. More recent examples include the detection of Concanavalin A [127] and immunoglobulin G [128]. Recently, protein interactions have been investigated with ultrasmall AuNCs (diameter 3.2 nm) coated with dihydrolipoic acid [129]. This approach is based on the increase in the fluorescence intensity of AuNCs upon protein adsorption. The apparent binding affinities of four different proteins (HSA, apotransferrin, lysozyme, and apolipoprotein E4) to AuNCs were measured, yielding detection limits in the micromolar range. Time-resolved fluorescence studies further revealed significantly enhanced long-lifetime components of the AuNC luminescence decay curves upon protein association.

Fluorescent techniques have also been implemented for DNA detection by using fluorophore-labeled oligonucleotide probes attached to AuNPs by covalent chemical modification [130] or by physical adsorption [131].

Murray et al. reported an FRET-based assay for the detection of several metal ions [132]. In this work, the authors used a well-known cationic fluorophore, $[Ru(bpy)_3]^{3+}$, which interacts electrostatically with tiopronin-coated AuNPs, leading to the fluorescence quenching of the $[Ru(bpy)_3]^{3+}$. FRET quenching mechanisms with AuNPs have been used for the detection of metal ions such as Hg^{2+} and Cu^{2+} [133].

This type of assay with AuNPs based on FRET mechanisms has also been used for the detection of small molecules such as thiols, cholesterol, homocysteine, etc. [134].

Au nanoparticles were used as the energy acceptor in an assay developed by Peng et al. [135]. They employed up-converting phosphors (a kind of promising anti-Stokes fluorophore) in FRET-based methods for the determination of glucose in human serum samples. Rare-earth (Yb and Er) codoped

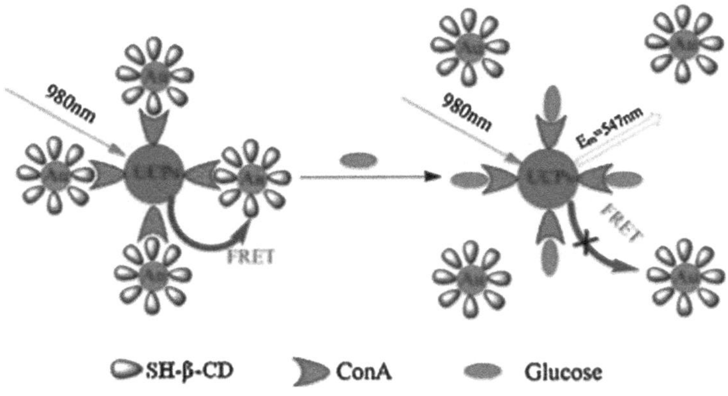

FIGURE 9 Schematic representation of the principle of the biosensor based on up-converting Forster Resonance Energy Transfer for the detection of glucose. *Reproduced from Ref. [135], Figure 1, p. 416.*

$NaYF_4$ nanocrystal, which is regarded as one of the most efficient up-converting materials, was adopted as the energy donor, and was labeled with Concanavalin A (Con A). AuNPs were used as the energy acceptor, which were tagged to thiolated β-cyclodextrins. The assay is based on the competition between thiolated β-cyclodextrins and glucose for Con A. The binding of Con A to dextran brings the AuNPs into close proximity with up-converting nanoparticles (UC-NPs), enabling the FRET process (Figure 9). In the presence of glucose, the UC-NPs detach from the AuNPs, suppressing the FRET process. This new biosensor is able to selectively respond to glucose in serum, with high sensitivity and a strong ability to resist background interference.

Either way, AuNPs influencing the fluorescence of fluorophores or their own fluorescence being influenced by other molecules, and whatever the mechanism (quenching or FRET) through which the change in fluorescence takes place, there is no doubt that AuNPs are promising tools for sensing applications.

4. GOLD NANOPARTICLES IN PHOSPHORESCENCE PROCESSES

The combination of the properties of AuNPs with the intrinsic functionalities of organic biomolecular systems can afford hybrid materials with new properties and functions, as mentioned above.

The role of AuNPs may be important in this context. Recent studies have shown that AuNPs can be readily taken up by human cells and that, under appropriate conditions, they are not toxic. This affords them an interesting potential for cell labeling at specific sites and for the nuclear delivery of diagnostic probes and therapeutics.

Although the effects of AuNPs on neighboring fluorescent molecules have been investigated in depth, few authors have focused on the influence of AuNPs on the luminescence properties of nearby phosphorescent molecules (phosphors), whose dipole strength is weakened by orders of magnitude as compared to that of fluorophores and hence phosphors exhibit radiative rates and rates of energy transfer to the AuNPs that are both reduced by orders of magnitude [136].

In this sense, the phosphorescent metal complexes ruthenium(II) tris(2,2′-bipyridyl) (Ru(bpy)3) and osmium(II) tris(2,2′-bipyridyl) (Os-(bpy)3) have been investigated, either adsorbed directly on an AuNP [137] or electrostatically bound very close to the nanoparticle's surface [132]. Both studies report luminescence quenching. In the case of the adsorbed phosphors, the quenching was explained in terms of electron transfer from the phosphors to the AuNP, while for the electrostatically bound phosphors, the decrease in luminescence was considered to be due to energy transfer.

4.1 Complex Formation of Functionalized Gold Nanoparticles with Eu(III)/Tb(III)

Another interesting use of AuNPs is when they are conjugated with phosphorescent elements. Ipe, Yoosaf, and Thomas [138] have described a procedure for the use of functionalized AuNPs as phosphorescent nanomaterials and sensors. They described a novel strategy for the design of phosphorescent nanomaterials (Figure 10) through a bottom-up approach, first by functionalizing a monolayer of 2,2′-bipyridine units on Au nanoparticles (Au-BT) and then further complexing them with lanthanide metal ions such as europium and terbium. Lanthanide complexes with coordinating ligands exhibit long-lived line-like emission and are widely used for the detection of biologically important cations and anions, as well as organic molecules (for example, phosphoglycerate).

They described the complex with a metal core of around 4 nm, with a shell of ~340 bipyridine ligands around the Au nanoparticle. The high local concentration of the chelating ligands (~5 M) around the Au nanoparticle makes these particles excellent ion sponges, and their complexation with Eu(III)/Tb(III) ions yields phosphorescent nanomaterials. Absorption spectral studies confirm a 1:3 complexing between Eu(III)/Tb(III) ions and bipyridines, functionalized on the surface of the Au nanoparticles. The red-emitting Au-BT: Eu(III) complex has a long lifetime of 0.36 ms with six line-like emission peaks, whereas the green-emitting Au-BT:Tb(III) complex has a lifetime of 0.7 ms with four line-like emission peaks. These phosphorescent nanomaterials composed of BT, for example, Eu(III) complexed with AuNPs, were further employed as sensors for metal cations. A dramatic decrease in luminescence was observed upon the addition of alkaline earth metal ions (Ca^{2+}, Mg^{2+}) and transition metal ions (Cu^{2+}, Zn^{2+}, Ni^{2+}), resulting from an

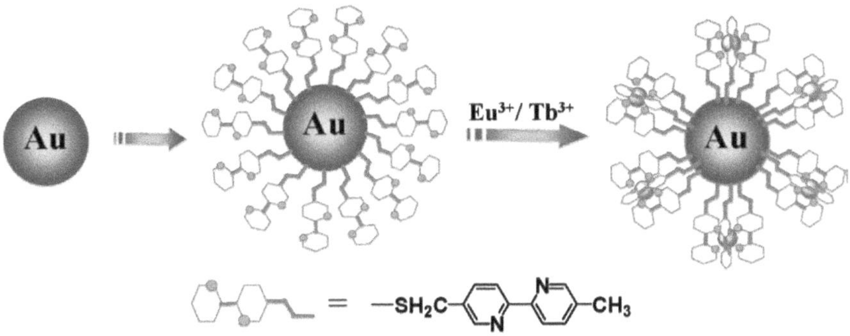

FIGURE 10 Design strategy of functionalization of gold nanoparticles with lanthanide complexes to create phosphorescent nanomaterials. *Reprinted with permission from Ref. [138]. Copyright (2006) American Chemical Society.*

isomorphous substitution of Eu(III) ions, whereas luminescence intensity was not influenced by the addition of Na^+ and K^+ ions. A direct interaction of bipyridine-capped Au nanoparticles with Cu^{2+} ions brings the nanohybrid systems closer, leading to the formation of three-dimensional superstructures. Strong interparticle Plasmon interactions were observed in these closely spaced Au nanoparticles.

5. APPLICATIONS OF GOLD NANOPARTICLES IN CHEMILUMINESCENCE REACTIONS

The chemiluminescence (CL) phenomenon of luminol was first reported by Albrecht [140] and since then CL has garnered extensive interest owing to its attractive features, such as high sensitivity, simple and inexpensive instrumentation, rapid analysis, and low background signals [141]. CL-related analytical techniques have been developed as important and powerful tools in diverse fields, such as biotechnology, food analysis, environmental analysis, pharmaceutical analysis, and clinical assays [142].

In general terms, CL is the emission of light (luminescence) as the result of a chemical reaction. CL is based on the production of electromagnetic radiation observed when a chemical reaction yields an electronically excited intermediate, which either luminesces (direct CL) or donates its energy to another molecule responsible for the emission (indirect or sensitized CL). CL is the electromagnetic radiation (UV, vis, or IR) produced by a chemical reaction, and several CL systems [e.g., luminol, peroxyoxalate and tris(2,2′-bipyridyl)ruthenium(II)] have been developed. CL and electrogenerated chemiluminescence (ECL) in aqueous-phase systems are generated mainly by redox reactions with suitable CL reagents and/or electrochemical reaction conditions. The reaction process involves the formation of intermediate radicals, giving rise to unstable products that decompose to form electronically excited molecules that deactivate with CL emission.

CL techniques have attractive advantages in comparison with other optical techniques such as spectrophotometry and fluorescence. Examples are the absence of unwanted strong background light, low noise signals, better sensitivity, and a broad linear dynamic range. Moreover, because there is no excitation source and no optical filters, the apparatus is simple, robust, cost-effective, and suitable for automation [143].

In contrast to bioluminescence systems, the CL emission processes generated during the oxidation of organic molecules afford a low QY. An enhancement of CL emission is therefore necessary for application in trace analyses in analytical chemistry. For such a purpose, catalysts such as transition metal ions and peroxidase have usually been used as signal enhancers [144]. Other compounds or reaction strategies that can act as signal enhancers have been proposed for the same aim (e.g., surfactants, hydroxylamine, and solvent preoxidation) [145].

In recent years, nanomaterials with unique redox catalytic properties have gained increasing attention as novel alternatives to catalyze redox CL reactions, providing amplified CL emission. Many researchers have indicated that the use of metal or semiconductor nanoparticles in CL reactions could provide new approaches to enhance the inherent sensitivity and expand new applications of this mode of detection. Apart from the general properties discussed along this book (e.g., optical, mechanical, thermal, size, and shape), transition-metal NPs show excellent properties for redox reactions: strong catalytic activity, which results in rapid detection; intense size- and shape-dependent catalytic action, which results in selective reaction; controllable charge-transfer events; and compatibility with biomaterials. All these properties can be employed to increase the kinetics and the selectivity of chemical reactions as well as to enhance the response signals of instrumental methods of analysis [146].

The study of CL was extended to NP systems from traditional molecular systems in order to improve sensitivity and stability, such improvement mainly resulting from the large surface area and special structure of nanomaterials. In these systems, metal NPs are able to participate in CL reactions as catalysts, reducing agents, luminophores, or energy acceptors. CL-functionalized nanoparticles, which enable multiple CL signal-generating molecules on a nanoparticle host and lead to excellent signal amplification, have become a new trend with innovative concepts in nanoscience and have drawn increasing attention [147].

Analytes can either enhance or inhibit the NP-induced enhanced CL signals, thus providing the basis for the development of a number of sensitive analytical methods. Also, NPs used as labels conjugated to biomolecules or linked to organic molecules to form complexes with the analyte can offer amplified methods by dissolution, followed by CL detection. Moreover, CL detection can be enhanced considerably when NPs are used as scaffolds carrying a number of labels (e.g., enzymes or even other NP conjugates or NPs coated with a shell of metal that is more soluble in a less acidic environment). Finally, highly selective determinations can be achieved through the coupling of CL detection with enzymatic reactions, bioaffinity or DNA-hybridization reactions.

In the next part of this chapter, recent trends in the analytical applications of AuNPs as amplifiers in CL-reaction strategies and ECL sensors are addressed. Among all the publications to date, representative examples in trace analysis, food chemistry, pharmaceuticals, and biochemical analysis are shown.

5.1 AuNPs as Catalysts in CL Reactions in the Luminol–H_2O_2 System

Based on their unique chemical and electrochemical properties, metal NPs (gold, silver, platinum, etc.) have an excellent ability to catalyze different redox reactions. In comparison with AuNP colorimetric methods, which often yield nanomolar LODs, the CL method can lower LODs to 0.1 pM.

AuNPs with diameter 38 nm were found to enhance the CL of the luminol–H_2O_2 system to an important extent [139]. In this work from 2005, Zhang et al. proposed a mechanism for the CL reaction of luminol–H_2O_2 catalyzed by AuNPs (Figure 11).

The enhancement mechanism of AuNPs on luminol CL is assumed to originate from the catalysis of AuNPs, which facilitates the radical generation and electron-transfer processes taking place on the surface of AuNPs.

The analytical usefulness of the catalytic effect of NPs on CL reaction systems stems from the fact that the CL intensity of the above reactions can be enhanced or inhibited by several analytes and this effect can be employed to develop CL methods of analysis. Giokas et al. [146] reviewed the use of nanoparticles in enhanced and amplified CL detection. For our purposes, we have summarized some of their examples, specifically those related to luminol–H_2O_2–AuNP CL methods, for the detection of analytes of different nature. They are shown in Table 2.

In these examples, the effect of Au-colloid NPs on the luminol–H_2O_2 system and the inhibitory effect of several organic compounds containing –OH, –NH_2, and –SH groups were studied, these methods proving to be potential analytical tools for detection. In all cases, it was suggested that the CL enhancement would be due to the ability of metal NPs to form surface radicals of OH· from the disruption of the O–O bond (H_2O_2) and to facilitate the electron-transfer process between the radical intermediates formed.

In all the above studies, sensitivity is the major benefit deriving from the use of AuNPs in CL reactions. However, routine samples include interfering elements that, in some cases, can seriously modify the response signal. Special attention must therefore be paid to issues such as restricting the access of

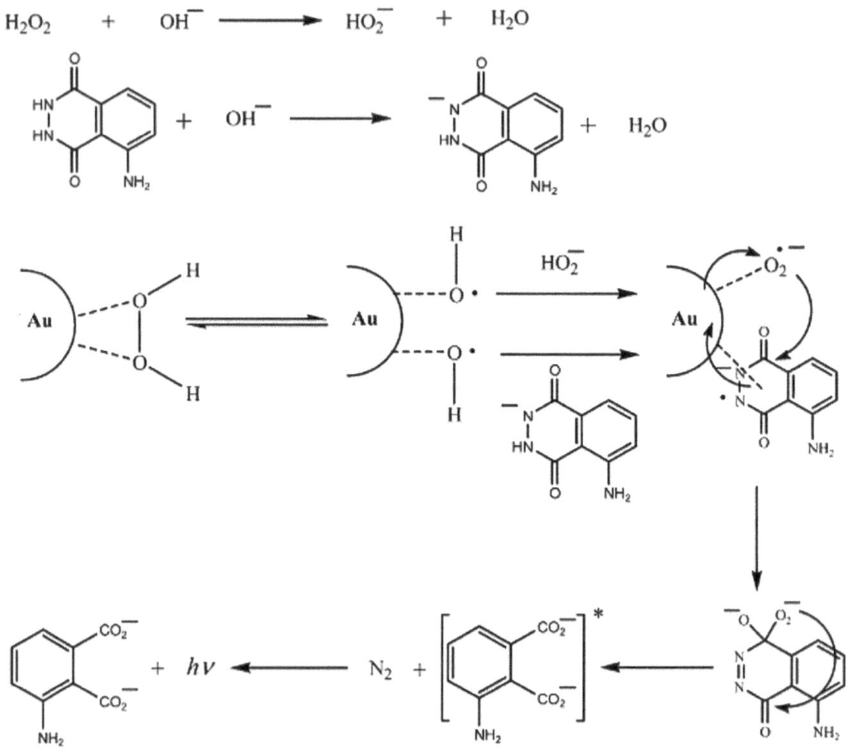

FIGURE 11 Mechanism proposed by Zhang et al. [139] for the luminol−H_2O_2−gold nanoparticles chemiluminescence system. *Reprinted with permission from Ref. [139]. Copyright (2005) American Chemical Society.*

interfering species to the reactor environment. Conventional separation procedures are the best choice to resolve selectivity barriers to CL detection, but it should not be forgotten that nanoparticles offer promising options for nanodesigns to improve selectivity.

5.2 Chemiluminescence Immunoassays

Metal NPs in chemiluminescence immunoassays (CLIAs) are mainly used as functionalized labels conjugated to antibodies or aptamers. In this way, they can enhance sensitivity or amplify detection; they are stable and they do not alter the biological properties of their conjugates. Solid-phase CLIA procedures involve noncompetitive or competitive sandwich-type reactions for the formation of specific immunocomplexes based on antigen−antibody or aptamer−bioaffinity systems (Table 3). Based on the catalytic action of AuNPs on the oxidation of luminol for enhanced CL-detection signal, several analytes have been detected, such as IgG [151] and CEA (carcinoembryonic antigen) [152].

TABLE 2 Analytes Determined by AuNPs-Catalyzed Chemiluminescence Methods Using Luminol–H_2O_2 Reagent

Analyte	LOD (M)	References
Catechol	1.7×10^{-9}	[139]
Ascorbic acid	2.0×10^{-9}	
Adrenalin	1.9×10^{-9}	
Noradrenalin	8.4×10^{-10}	
Dopamine	1.9×10^{-10}	
L-histidine	9.3×10^{-10}	
L-cysteine	6.4×10^{-10}	
Norfloxacin	1.0×10^{-8}	[148]
Ciprofloxacin	2.9×10^{-8}	
Lomefloxacin	2.0×10^{-8}	
Fleroxacin	2.4×10^{-8}	
Ofloxacin	2.2×10^{-8}	
Levofloxacin	2.2×10^{-8}	
Estrone	3.2×10^{-9}	[149]
Estradiol	7.7×10^{-9}	
Estriol	49×10^{-9}	
Glutathione	1.1×10^{-8}	[150]
6-mercaptopurine	1.1×10^{-9}	

5.3 CL-Hybridization Assays

After Mirkin's work on the detection of DNA through hybridization, DNA-functionalized AuNPs acquired widespread acceptance, specifically for single-nucleotide polymorphism methods of analysis. Aptamers and DNA enzymes, both containing ssDNA or RNA, are used to prepare bioconjugates of AuNPs with a specific ability to recognize proteins and small analyte molecules (Table 3). CL-hybridization assay (CLHA) protocols mainly comprise a sandwich-type noncompetitive hybridization reaction that forms biocomplexes between an immobilized molecule on a solid-support capture DNA probe, or a target (ss) DNA analyte and a labeled reporter DNA probe, or by a hybridization reaction between a target (ss) DNA analyte and a DNA probe-functionalized NP label [153].

TABLE 3 Analytical Applications of Chemiluminescence Immunoassays and Hybridization Assays

Method	Analyte	Range	LOD (3σ)	RSD (%)	Application	References
Chemiluminescence Immunoassays						
Amplified CLIA-detection MBs/ AuNPs label	hIgG	0.1–100 ng L^{-1}	3.1×10^{-12} 0.5 ng/mL			[31]
GFQIA-sensor FITC fluorescent agent	α-fetoprotein		0.17 nM		Patient serum samples	[37]
CL-imaging IA AuNP–protein (or chit.) composite(Lum. + H_2O_2 + PIP)	H_2O_2 rHu-IL-6	1.0×10^{-6}–1.0×10^{-4} M 2.0–312 pg/mL	2.0×10^{-7} 0.5 pg/mL	3.8% ($n = 6$ at 3.0×10^{-5} M) 4.4% ($n = 6$ at 39 pg/mL)	Serum samples	[36]
IA amplified CL-detection Ag on AuNPs (Ag^+-$K_2S_2O_8$-Mn^{2+})/ (H_3PO_4-luminol)	hIgG	0.02–50 ng/mL	0.005 ng/mL	3.5% ($n = 7$ at 0.5 ng/mL)	Human serum samples	[32]
IA amplified CL-detection Au-NPs label ($AuCl_4^-$–Lum.)	Goat-anti-human IgG	5.0 ng/mL to 10 μg/mL	1.5 ng/mL	4.2–4.6% ($n = 3$ at 5–10 ng/mL)	Immuno-serum	[33]

IA amplified CL-detection DC-AuNPs labels	AFP	0.008–0.3 ng/mL	5 pg/mL		Chemical serum samples Recovery: 96.2–103.0%	[34]
IA amplified CL-detection Multi. Ab-AuNP conjugates (Luminol-$AgNO_3$-AuNPs)	hIgG	25–5000 ng/mL	12.8 ng/mL	8.6% (at 100 ng/mL) 7.1% (at 500 ng/mL) 3.0% (at 1000 ng/mL)		[35]
Chemiluminescence Hybridization Assays						
HA amplified FI/CL-detection CuS. NP-labeled-DNA probe-functionalized AuNPs	Target DNA	2.0×10^{-14}–2.0×10^{-12} M	4.8×10^{-15} M	3.4% ($n = 11$ at 4.0×10^{-14} M)		[44]
HA amplified CL-detection NH_2OH extra gold loaded AuNPs Capture target DNA	CL det → target DNA nucleotide	0.01–1.0 fM >100 fM	10 M	4–11% (($n = 11$ at 0.075–1.0 fM))	105 ± 2.1% (at 0.075 fmol) 101.0 ± 5.0% (at 0.25 fmol) 111.0 ± 12.0% (at 1.0 fmol)	[43]

MBs, magnetic beads; GFQIA, gold fluorescence quenching immunoassay; AFP, α-fetoprotein; FITC, fluorescein isothiocyanate; DC-AuNPs, double-codified gold nanoparticles; rHu-IL-6, recombinant human interleukin-6.
Reproduced from Ref. [146], Table 1b, p. 1118.

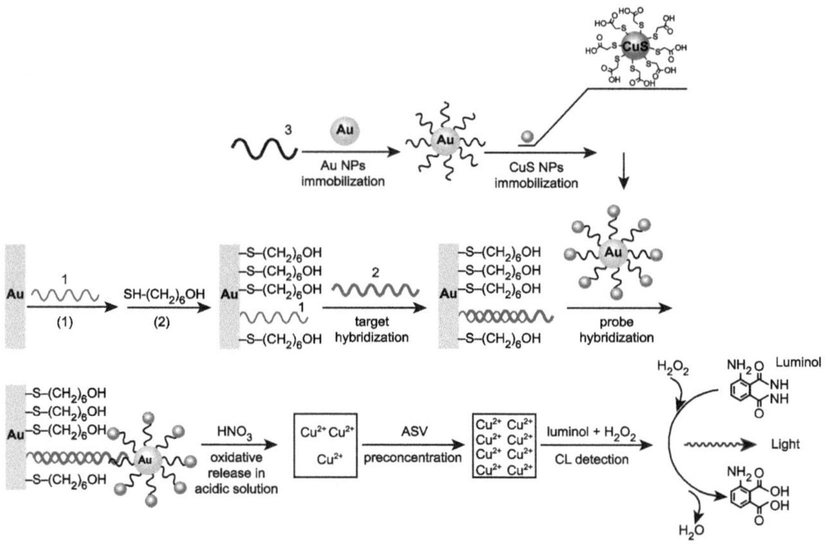

FIGURE 12 Chemiluminescence detection of DNA hybridization through two steps of amplification. Capture DNA sequence (1): 5′-TGG AAA ATC TCT AGC AGT CGT-$(CH_2)_6$-SH-3′, Target DNA sequence (2): 5′-ACT GCT AGA GAT TTT CCA CAC TGA CTA AAA GGG TCT GAG GGA-3′, Signal DNA probe sequence (3): 5′-SH-$(CH_2)_6$-ATG TCC CTC AGA CCC TTT-$(CH_2)_6$-NH_2-3′. *Reprinted with permission from Ref. [153d]. Copyright (2008) American Chemical Society.*

An example of this type of reaction can be found in the novel and sensitive flow injection CL assay for sequence-specific DNA detection based on signal amplification with NPs (Figure 12). The "sandwich-type" DNA biosensor was made with the thiol-functionalized capture DNA first immobilized on an Au electrode and then hybridized with one end of the target DNA, the other end of which was recognized by a signal DNA probe labeled with CuS NPs and AuNPs on the 3′- and 5′-ends, respectively. The hybridization events were monitored by the CL intensity of luminol$-H_2O_2-Cu^{2+}$ after the cupric ions had been released from the hybrids. The incorporation of AuNPs in this sensor design significantly enhanced sensitivity and selectivity because a single AuNP can be loaded with hundreds of signal DNA probe strands, which in this case were modified with CuS NPs.

5.4 Chemiluminescence-Resonance-Energy Transfer

Chemiluminescence-resonance-energy transfer (CRET), which involves non-radiative (dipole−dipole) energy transfer from a CL donor to a suitable acceptor molecule, is a powerful technique for probing very small changes in the distance between the CL donor and acceptor [154]. The CRET principle is similar to FRET, and CRET efficiency (E) can also be defined in the equation,

$E = 1/(1 + R^6/R_0^6)$, where R_0 is the distance leading to 50% of energy transfer from the donor to the acceptor [155]. In contrast to FRET, CRET occurs through the oxidation of a luminescent substrate without an excitation source. Since no external light source is used for excitation in CRET, the nonspecific signals caused by external light excitation often observed in fluorescence-based measurements can be minimized. This makes CRET an attractive light-measuring scheme in bioassays. Currently, CRET modes are used for assays of biomolecules and small molecules by intramolecular [156a] and intermolecular energy transfer [156].

Luminol and its derivatives are mostly used in CRET system as energy donors [157a] and small-molecule fluorophores (e.g., fluorescein and rhodamine [157b]) are used as energy acceptors. Compared with current analytical methods (e.g., colorimetric and electrochemical detection), CRET-based assays are characterized by their simplicity, high sensitivity and low cost, and low reagent and sample consumption when coupled with other analytical methods (e.g., CE, microchip CE, and droplet analysis). More importantly, CRET is usually used for the homogeneous analysis of complex samples without separation.

Although the CRET phenomenon was observed some time ago, current applications of conventional CRET are very scarce owing to its poor energy-transfer efficiency and limited number of possible energy acceptors. Recently, significant progress in CRET has been made after the introduction of certain nanomaterials (e.g., QDs, AuNPs, and grapheme) as energy acceptors. These nanomaterials are excellent CRET acceptors owing to their large Stokes shifts and high energy-transfer efficiency.

One example of a CRET strategy using AuNPs was reported by Huang and Ren [158]. Based on the fact that the CL spectrum of luminol ($\lambda_{max} = 425$ nm) partially overlaps with the visible absorption bands of AuNPs, they chose the highly sensitive CL reaction of luminol and hydrogen peroxide catalyzed by horse radish peroxidase (HRP).

By conjugating two antibodies (anti-AFP-1 and anti-AFP-2) to AuNPs and HRP, respectively (Figure 13), they developed a sandwich immunoassay for the detection of the alpha fetoprotein (AFP) cancer marker. The sandwich-type immunoreactions between the AFP (antigen) and the two different antibodies bridged the donors (luminol) and acceptors (AuNPs), which led to the occurrence of CRET from luminol to AuNPs upon the CL reaction. The quenching of the CL signal depended linearly on the AFP concentration, within a range of concentration from 5 to 70 ng mL^{-1}, and the detection limit of AFP was 2.5 ng mL^{-1}. The method was successfully applied for the determination of AFP levels in sera from cancer patients, and the results were in good agreement with ELISA assays.

The set of applications shown above points the great potential of AuNP--assisted CL reactions in many, different fields of contemporary analytical science.

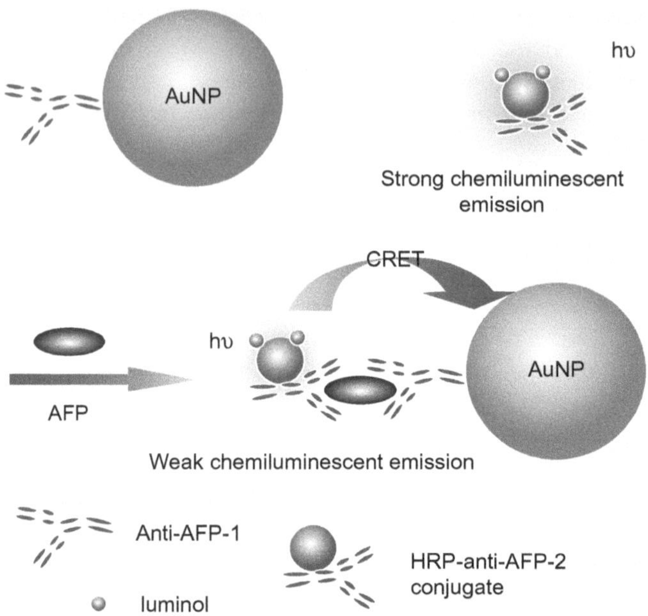

FIGURE 13 Schematic illustration for the chemiluminescence-resonance-energy transfer (CRET) process between anti-alpha fetoprotein (AFP)-1—horse radish peroxidase (HRP)-catalyzed luminol (donor) and anti-AFP-1—gold nanoparticles (AuNPs) (acceptor). *Reproduced from Ref. [158], Scheme 1, p. 117.*

6. GENERAL CONCLUSIONS TO THIS CHAPTER

In this chapter, we have explored the optical properties of AuNPs as a starting point for use in spectroscopic techniques.

The optical processes derived from the interaction of light with an AuNP, such as absorption, scattering, SERS, surface-enhanced fluorescence, and phosphorescence are studied and documented with examples. Of all plasmonic particles, AuNPs are those that have received the most attention owing to their unique properties, such as SPR. Since SPR is directly related to the free electrons in the conduction band, everything that affects the electronic charge densities at the surface of the NP will also strongly affect SPR. Accordingly, the wavelength and the intensity of the SPR band will depend on the size of the NP, its shape, and the composition and dielectric constant of the surrounding medium. AuNPs can be used as colorimetric sensors for the detection of chemical species of different nature by simple colorimetric methods. This, together with the possibilities offered by nanotechnology, opens a field of enquiry with important applications in chemistry, biology, pharmacy, and medicine. The spectral and optical properties of AuNPs, together with the possibility of relatively simple functionalization with a broad variety of biological and other types of molecules, mean that they can be applied in different fields. Owing to the strong affinity of the Au−S bond, AuNPs bind strongly to thiols and disulfides. They also bind easily to amine groups, which allows conjugation with peptides, proteins, oligonucleotides, DNA, etc. This type of bioconjugate takes advantage of the possibilities offered, on one hand, by the conjugated molecules (specific coupling with a target) and, on the other, the metal core serves as an optical label.

The surface Plasmon on the gold surface depends strongly on the state of the surface and its properties. Thus, the binding of ligands to a receptor immobilized on the surface of nanoparticles will elicit a measureable signal related to the concentration of the ligand, a phenomenon that serves as a basis for the quantification of analytes in spectroscopic techniques.

In light of the above, different methodologies have been developed for the use of AuNPs as a basis of spectrophotometric methods for the determination of anions, metallic ions/inorganic ions, DNA/genes/proteins, the detection of bacteria, and the detection of cancer cells described in this chapter. The importance of the surface Plasmon of AuNPs as probes for the detection of analytes and their use based on SERS techniques of enhancement of the Raman scattering of an analyte molecule close to a roughened metal substrate surface should also be taken into account. The application of AuNPs in the development of fluorescence-based methodologies is also addressed in depth. The importance of functionalized AuNPs as fluorophores or in processes of quenching, FRET or SMF, and the use of AuNPs as mediators in these processes and in the determination of a large number of analytes are of great interest, and here we describe some of the most representative examples.

The importance and future possibilities of the application of AuNPs in phosphorescence processes is also stressed, although the applications described here are few since not many have appeared in the literature. Some phosphorescent metal complexes of ruthenium (II) and osmium (II), either adsorbed directly on an AuNP or bound electrostatically very close to the nanoparticle surface have been explored as well as the complex formation with Eu(III)/Tb(III) of functionalized AuNPs.

Likewise, the applications of AuNPs in CL reactions have been described. The use of CL-functionalized nanoparticles, which enrich multiple CL signal-generating molecules on a nanoparticle host and afford excellent signal amplification, has become a new trend with innovative concepts in nanoscience and has drawn increasing attention. The use of AuNPs in CL reactions in the luminol−H_2O_2 system, their application as a basis in procedures of CLIAs or in CL-hybridization assays (CLHAs) and in processes of CRET point to the great potential of AuNPs-assisted CL reactions in many different fields of analytical science.

The different sections of this chapter show the huge possibilities of the application of AuNPs as nanoplatforms in different spectroscopic techniques, providing an attractive and applicable scaffold for the determination of analytes of interest. Currently, nanotechnology is a hot topic in which AuNPs play a major role, having unprecedented potential for signal amplification, detection, and quantification in the field of analytical chemistry.

ACKNOWLEDGMENTS

We would like to thank Dr Bernardo Moreno Cordero and Dr Manuel Fuentes García for their critical comments on the manuscript.

REFERENCES

[1] P. Zhao, N. Li, D. Astruc, Coord. Chem. Rev. 257 (2013) 638–665.
[2] M. Grzelczak, J. Perez-Juste, P. Mulvaney, L.M. Liz-Marzán, Chem. Soc. Rev. 37 (2008) 1783–1791.
[3] T.K. Sau, C.J. Murphy, J. Am. Chem. Soc. 126 (28) (2004) 8648–8649.
[4] M.C. Daniel, D. Astruc, Chem. Rev. 104 (2004) 293–346.
[5] G. Mie, Ann. Phys. 25 (1908) 377–445.
[6] M. Kerker, The Scattering of Light and Other Electromagnetic Radiation, Academic Press, New York, 1969.
[7] G.C. Papavassiliou, Prog. Solid State Chem. 12 (1979) 185–271.
[8] C.F. Bohren, D.R. Huffman, Absorption and Scattering of Light by Small Particles, Wiley, New York, 1983.
[9] U. Kreibig, M. Vollmer, Optical Properties of Metal Clusters, Springer, 1995.
[10] X. Huang, M.A. El-Sayed, J. Adv. Res. 1 (2010) 13–28.
[11] B. Zargar, A. Hatamie, Spectrochimica Acta Part A: Mol. Biomol. Spectrosc. 106 (2013) 185–189.
[12] G. Palazzo, L. Facchini, A. Mallardi, Sens. Actuators, B 161 (2012) 366–371.
[13] X.W. Shen, C.Z. Huang, Y.F. Li, Talanta 72 (2007) 1432–1437.
[14] H. Su, H. Fan, S. Ai, N. Wu, H. Fan, P. Bian, J. Liu, Talanta 85 (2011) 1338–1343.
[15] F.P. Zamborini, L. Bao, R. Dasari, Anal. Chem. 84 (2012) 541–576.
[16] L. Xia, J. Wang, S. Tong, G. Liu, J. Li, H. Zhang, Vib. Spectrosc. 47 (2008) 124–128.
[17] S. Hong, X. Li, J. Nanomaterials (2013). Article ID 790323, 9 pp.
[18] S.S.R. Dasary, A.K. Singh, D. Senapati, H. Yu, P.C. Ray, J. Am. Chem. Soc. 131 (2009) 13806–13812.
[19] S. Nie, S.R. Emory, Science 275 (1997) 1102–1106.
[20] R.A. Alvarez-Puebla, L.M. Liz-Marzán, Small 6 (2010) 604–610.
[21] P.K. Jain, K.S. Lee, I.H. El-Sayed, M.A. El-Sayed, J. Phys. Chem. B 110 (14) (2006) 7238–7248.
[22] R. Gans, Ann. Phys. 47 (1915) 270–284.
[23] J.Z. Zhang, C. Noguez, Plasmonics 3 (2008) 127–150.
[24] C.J. Murphy, T.K. San, A.M. Gole, C.J. Orendorff, J.X. Gao, L. Gou, S.E. Hunyadi, T. Li, J. Phys. Chem. B 109 (2005) 13857–13870.
[25] J.E. Millstone, S.J. Hurst, G.S. Métraux, J.I. Cutler, C.A. Mirkin, Small 5 (2009) 646–664.
[26] N.R. Jana, Small 1 (2005) 875–882.
[27] M. Liu, P. Guyot-Sionnest, J. Phys. Chem. B 109 (47) (2005) 22192–22200.
[28] L.M. Liz-Marzan, Mater. Today 7 (2) (2004) 26–31.
[29] N.G. Khlebtsov, L.A. Dykman, J. Quant, Spectrosc. Radiat. Transfer 111 (2010) 1–35.
[30] W. Sun, L. Xiao, N. Fang, Methods Mol. Biol. 931 (2013) 169–186.

[31] C. Bao, N. Beziere, P. del Pino, B. Pelaz, G. Estrada, F. Tian, V. Ntziachristos, J.M. de la Fuente, D. Cui, Small 9 (1) (2013) 68–74.
[32] S. Ha, A. Carson, A. Agarwal, N.A. Kotov, K. Kim, Biomed. Opt. Express 2 (3) (2011) 645–657.
[33] R. Elghanian, J.J. Storhoff, R.C. Mucic, R.L. Letsinger, C.A. Mirkin, Science 277 (1997) 1078–1081.
[34] J.J. Storhoff, R. Elghanian, R.C. Mucic, C.A. Mirkin, R.L. Letsinger, J. Am. Chem. Soc. 120 (1959) 1998.
[35] J.J. Storhoff, A.A. Lazarides, R.C. Mucic, C.A. Mirkin, R.L. Letsinger, G.C. Schatz, J. Am. Chem. Soc. 122 (2000) 4640.
[36] P.D. Beer, P.A. Gale, Angew. Chem. Int. Ed. 40 (2001) 486.
[37] S. Kado, A. Furui, Y. Akiyama, Y. Nakahara, K. Kimura, Anal. Sci. 25 (2009) 261–265.
[38] M.H. Kim, S. Kim, H.H. Jang, S. Yi, S.H. Seo, M.S. Han, Tetrahedron Lett. 51 (2010) 4712–4716.
[39] S.Y. Lin, S.W. Liu, C.M. Lin, C.H. Chen, Anal. Chem. 74 (2) (2002) 330–335.
[40] K. Saha, S.S. Agasti, C. Kim, X. Li, V.M. Rotello, Chem. Rev. 112 (5) (2012) 2739–2779.
[41] A. Alizadeh, M.M. Khodaei, Z. Hamidi, M. bin Shamsuddin, Sens. Actuators, B 190 (2014) 782–791.
[42] X. Xu, J. Wang, K. Jiao, X. Yang, Biosens. Bioelectron. 24 (10) (2009) 3153–3158.
[43] L. Beqa, A.K. Singh, S.A. Khan, D. Senapati, S.R. Arumugam, P.C. Ray, ACS Appl. Mater. Interfaces 3 (2011) 668–673.
[44] J.R. Kalluri, T. Arbneshi, S.A. Khan, A. Neely, P. Candice, B. Varisli, M. Washington, S. McAfee, B. Robinson, S. Banerjee, A.K. Singh, D. Senapati, P.C. Ray, Angew. Chem. Int. Ed. 48 (2009) 9668–9671.
[45] K.B. Mullis, F.A. Faloona, Methods Enzymol. 155 (1987) 335–350.
[46] M.J. Espy, J.R. Uhl, L.M. Sloan, S.P. Buckwalter, M.F. Jones, E.A. Vetter, et al., Clin. Microbiol. Rev. 19 (2002) 165–256.
[47] C.A. Mirkin, R.L. Letsinger, R.C. Mucic, J.J. Storhoff, Nature 382 (1996) 607–609.
[48] H.G. Park, Biotechnol. Bioprocess Eng. 8 (2003) 221–226.
[49] N.L. Rosi, C.A. Mirkin, Chem. Rev. 105 (2005) 1547–1562.
[50] F. Li, J. Zhang, X. Cao, L. Wang, D. Li, S. Song, B. Ye, C. Fan, Analyst 134 (2009) 1355–1360.
[51] J.H. Kim, B.H. Chung, Biosens. Bioelectron. 26 (2011) 2805–2809.
[52] X. Zhang, X. Kong, W. Fan, X. Du, Langmuir 27 (10) (2011) 6504–6510.
[53] V. Pavlov, Y. Xiao, B. Shlyahovsky, I. Willner, J. Am. Chem. Soc. 126 (2004) 11768–11769.
[54] P.V. Baptista, M. Koziol-Montewka, J. Paluch-Oles, G. Doria, R. Franco, Clin. Chem. 52 (2006) 1433–1434.
[55] H. Su, H. Zhao, F. Qiao, L. Chen, R. Duana, S. Ai, Analyst 138 (2013) 3026–3031.
[56] C.D. Medley, J.E. Smith, Z. Tang, Y. Wu, S. Bamrungsap, W. Tan, Anal. Chem. 80 (2008) 1067–1072.
[57] K. Sokolov, M. Follen, J. Aaron, I. Pavlova, A. Malpica, R. Lotan, R. Richards-Kortum, Cancer Res. 63 (2003) 1999–2004.
[58] X. Huang, I.H. El-Sayed, M.A. El-Sayed, J. Am. Chem. Soc. 128 (6) (2006) 2115–2120.
[59] I.H. El-Sayed, X. Huang, M.A. El-Sayed, Surface plasmon resonance scattering and absorption of anti-EGFR antibody conjugated gold nanoparticles in cancer diagnostics: applications in oral cancer, Nano Lett. 5 (5) (2005) 829–834.

[60] S. Watanabe, H. Seguchi, K. Yoshida, K. Kifune, T. Tadaki, H. Shiozaki, Tetrahedron Lett. 46 (2005) 8827.
[61] J.X. Wang, X. Yang, Analyst 137 (2012) 2806–2812.
[62] M. Zhang, Y.Q. Liu, B.C. Ye, Analyst 136 (2011) 4558–4562.
[63] Y.W. Lin, C.C. Huang, H.T. Chang, Analyst 136 (2011) 863–871.
[64] F. Tan, X. Liu, X. Quan, J. Chen, X. Li, H. Zhao, Anal. Methods 3 (2011) 343–347.
[65] X. Liang, H. Wei, Z. Cui, J. Deng, Z. Zhang, X. You, X.E. Zhang, Analyst 136 (2011) 179–183.
[66] H. Chi, B. Liu, G. Guan, Z. Zhang, M. Han, Analyst 135 (2010) 1070–1075.
[67] L. Guo, J. Zhong, J. Wu, F. Fu, G. Chen, X. Zheng, S. Lin, Talanta 82 (2010) 1654–1658.
[68] B. Zargar, A. Hatamie, Sens. Actuators, B 182 (2013) 706–710.
[69] V.V. Apyari, S.G. Dmitrienko, V.V. Arkhipova, A.G. Atnagulov, M.V. Gorbunova, Y.A. Zolotov, Spectrochimica Acta Part A: Mol. Biomol. Spectrosc. 115 (2013) 416–420.
[70] Y. Zhang, B. Li, X. Chen, Microchim. Acta 168 (2010) 107–113.
[71] Y. Wang, L. Zhan, C.Z. Huang, Anal. Methods 2 (2010) 1982–1988.
[72] K.R. Lim, J.M. Park, H.N. Choi, W.Y. Lee, Microchem. J. 106 (2013) 154–159.
[73] Y. Qi, L. Li, B. Li, Spectrochimica Acta, Part A 74 (2009) 127–131.
[74] S. Zhang, J. Wang, L. Han, C. Li, W. Wang, Z. Yuan, Sens. Actuators, B 147 (2010) 687–690.
[75] S. Vantasin, P. Pienpinijtham, K. Wongravee, C. Thammacharoen, S. Ekgasit, Sens. Actuators, B 177 (2013) 131–137.
[76] L. Li, B. Li, Analyst 134 (2009) 1361–1365.
[77] X. Wei, L. Qi, J. Tan, R. Liu, F. Wang, Anal. Chim. Acta 671 (2010) 80–84.
[78] H. Su, Q. Ma, K. Shang, T. Liu, H. Yin, S. Ai, Sens. Actuators, B 161 (2012) 298–303.
[79] C.L. Kuong, W.Y. Chen, Y.C. Chen, Anal. Bioanal. Chem. 387 (2007) 2091–2099.
[80] L. Chen, W. Lu, X. kun Wang, L. Chen, Sens. Actuators, B 182 (2013) 482–488.
[81] K. Aslan, J.R. Lakowicz, C.D. Geddes, Anal. Biochem. 330 (2004) 145–155.
[82] J. Liu, Y. Lu, J. Am. Chem. Soc. 127 (2005) 12677–12683.
[83] J. Liu, Y. Lu, Angew. Chem. Int. Ed. 45 (2005) 90–94.
[84] C.S. Tsai, T.B. Yu, C.T. Chen, Chem. Commun. (Camb.) 34 (2005) 4273–4275.
[85] H. Rim Sim, A.W. Wark, H. Jin Lee, Analyst 135 (2010) 2528–2532.
[86] L. Wang, T. Li, Y. Du, C. Chen, B. Li, M. Zhou, S. Dong, Biosens. Bioelectron. 25 (2010) 2622–2626.
[87] M. Frasconi, C. Tortolini, F. Botrè, F. Mazzei, Anal. Chem. 82 (2010) 7335–7342.
[88] S. Gao, N. Koshizaki, H. Tokuhisa, E. Koyama, T. Sasaki, J.K. Kim, J. Ryu, D.S. Kim, Y. Shimizu, Adv. Funct. Mater. 20 (2010) 78–86.
[89] M. Riskin, R. Tel-Vered, I. Willner, Adv. Mater. 22 (2010) 1387–1391.
[90] J.L. Wang, H.S. Zhou, Anal. Chem. 80 (2008) 7174–7178.
[91] K.A. Willets, R.P. Van Duyne, Annu. Rev. Phys. Chem. 58 (2007) 267–297.
[92] S.K. Dondapati, T.K. Sau, C. Hrelescu, T.A. Klar, F.D. Stefani, J. Feldmann, ACS Nano 4 (2010) 6318–6322.
[93] L. Seballos, N. Richards, D.J. Stevens, M. Patel, L. Kapitzky, S. Lokey, G. Millhauser, J.Z. Zhang, Chem. Phys. Lett. 447 (2007) 335–339.
[94] G.M. Herrera, A.C. Padilla, S.P. Hernandez-Rivera, Nanomaterials 3 (2013) 158–172.
[95] T. Senapati, D. Senapati, A.K. Singh, Z. Fan, R. Kanchanapally, P.C. Ray, Chem. Commun. (Camb.) 47 (37) (2011) 10326–10328.
[96] S. Thatai, P. Khurana, S. Prasad, D. Kumar, Microchem. J. 113 (2014) 77–82.
[97] Z.H. Lin, I.C. Chen, H.T. Chang, Chem. Commun. 47 (2011) 7116–7118.

[98] R.A. Alvarez-Puebla, A. Agarwal, P. Manna, B.P. Khanal, P. Aldeanueva-Potel, E. Carbó-Argibay, N. Pazos-Pérez, L. Vigderman, E.R. Zubarev, N.A. Kotov, L.M. Liz-Marzán, Proc. Natl. Acad. Sci. USA 108 (2011) 8157–8161.
[99] J. Yoon, N. Choi, J. Ko, K. Kim, S. Lee, J. Choo, Biosens. Bioelectron. 47 (2013) 62–67.
[100] W. Lu, A.K. Singh, S.A. Khan, D. Senapati, H. Yu, P.C. Ray, J. Am, Chem. Soc. 132 (51) (2010) 18103–18114.
[101] E.A. Vitol, E. Brailoiu, Z. Orynbayeva, N.J. Dun, G. Friedman, Y. Gogotsi, Anal. Chem. 82 (2010) 6770–6774.
[102] L. Shang, S. Dong, G.U. Nienhaus, Nano Today 6 (2011) 401–418.
[103] M.A. Noginov, G. Zhu, V.I. Gavrilenko, Opt. Express 15 (2007) 15648–15655.
[104] [104a] J.P. Wilcoxon, J.E. Martin, F. Parsapour, B. Wiedenman, D.F. Kelley, J. Chem. Phys. 108 (1998) 9137–9143;
[104b] A. Lin, D.H. Son, H.H. Ahn, G.H. Song, W.T. Han, Opt. Express 15 (2007) 6374;
[104c] H. Wang, T.B. Huff, D.A. Zweifel, W. He, P.S. Low, A. Wei, J.X. Cheng, Proc. Natl. Acad. Sci. USA 102 (2005) 15752;
[104d] D. Philip, Spectrochim. Acta Part A 71 (2008) 80–85.
[105] J. Zheng, P.R. Nicovich, R.M. Dickson, Ann. Rev. Phys. Chem. 58 (2007) 409–431.
[106] A. Mooradian, Phys. Rev. Lett. 22 (1969) 185–187.
[107] [107a] L. Shang, R.M. Dörlich, S. Brandholt, R. Schneider, V. Trouillet, M. Bruns, D. Gerthsen, G.U. Nienhaus, Nanoscale 3 (2011) 2009–2014;
[107b] Y.C. Jao, M.K. Chen, S.Y. Lin, Chem. Commun. 46 (2010) 2626–2628;
[107c] B. Santiago-González, M.J. Rodríguez, C. Blanco, J. Rivas, M.A. Lòpez-Quintela, J.M.G. Martinho, Nano Lett. 10 (2010) 4217–4221.
[108] J. Zheng, J.T. Petty, R.M. Dickson, J. Am. Chem. Soc. 125 (2003) 7780–7781.
[109] T. Huang, R. Murray, J. Phys. Chem. B 105 (2001) 12498–12502.
[110] M.R. Beversluis, A. Bouhelier, L. Novotny, Phys. Rev. B 68 (2003) 115433.
[111] E. Dulkeith, T. Niedereichholz, T.A. Klar, J. Feldmann, G. von Plessen, D.I. Gittins, K.S. Mayya, F. Caruso, Phys. Rev. B 70 (2004) 205424.
[112] W.R. Algar, K. Susumu, J.B. Delehanty, I.L. Medintz, Anal. Chem. 83 (2011) 8826–8837.
[113] [113a] M.M. Alvarez, J.T. Khoury, T.G. Schaaff, M.N. S hafigullin, I. Vezmar, R.L. Whetten, J. Phys. Chem. B 101 (1997) 3706–3712;
[113b] P. Apell, R. Monreal, S. Lundqvist, Phys. Scr. 38 (1988) 174–179.
[114] C.A.J. Lin, C.H. Lee, J.T. Hsieh, H.H. Wang, J.K. Li, J.L. Shen, W.H. Chan, H.I. Yeh, W.H. Chang, J. Med. Biol. Eng. 29 (2009).
[115] V.K.K. Upadhyayula, Anal. Chim. Acta 715 (2012) 1–18.
[116] K.G. Thomas, P.V. Kamat, J. Am. Chem. Soc. 122 (2000) 2655–2656.
[117] P. Xu, H. Yanagi, Chem. Mater. 11 (1999) 2626.
[118] B. Dubertret, M. Calame, A. Libchaber, J. Nat, Biotechnol. 19 (2001) 365–370.
[119] [119a] H.H. Cai, H. Wang, J. Wang, W. Wei, P.H. Yang, J. Cai, Dyes Pigm. 92 (2012) 778–782;
[119b] J. Chen, A. Zheng, A. Chen, Y. Gao, C. He, X. Kai, G. Wu, Y. Chen, Anal. Chim. Acta 599 (2007) 134–142.
[120] J.R. Lakowicz, Anal. Biochem. 337 (2005) 171–194.
[121] P.C. Ray, G.K. Darbha, A. Ray, J. Walker, W. Hardy, Plasmonics 2 (2007) 173–183.
[122] J. Xie, Y. Zheng, J.Y. Ying, Chem. Commun. 46 (2010) 961–963.
[123] X. Tu, W. Chen, X. Guo, Nanotechnology 22 (2011) 095701, 7 pp.
[124] H.Z. Zheng, L. Liu, Z.J. Zhang, Y.M. Huang, D.B. Zhou, J.Y. Hao, Y.H. Lu, S.M. Chen, Spectrochim. Acta A Mol. Biomol. Spectrosc. 71 (5) (2009) 1795–1798.

[125] Y.C. Shiang, C.C. Huang, H.T. Chang, Chem. Commun. (2009) 3437–3439.

[126] R.C. Triulzi, M. Micic, S. Giordani, M. Serry, W.A. Chiou, R.M. Leblanc, Chem. Commun. (2006) 5068–5070.

[127] C.C. Huang, C.T. Chen, Y.C. Shiang, Z.H. Lin, H.T. Chang, Anal. Chem. 81 (2009) 875–882.

[128] Y.C. Shiang, C.A. Lin, C.C. Huang, H.T. Chang, Analyst 136 (2011) 1177–1182.

[129] L. Shang, G.U. Nienhaus, Mater. Today 16 (2013) 58–66.

[130] D.J. Maxwell, J.R. Taylor, S. Nie, J. Am. Chem. Soc. 124 (2002) 9606–9612.

[131] H. Li, L. Rothberg, Anal. Chem. 76 (2004) 5414–5417.

[132] T. Huang, R.W. Murray, Langmuir 18 (2002) 7077–7081.

[133] [133a] C.C. Huang, H.T. Chang, Anal. Chem. 78 (2006) 8332–8338;
[133b] X.R. He, H.B. Liu, Y.L. Li, S. Wang, Y.J. Li, N. Wang, J.C. Xiao, X.H. Xu, D.B. Zhu, Adv. Mater. 17 (2005) 2811.

[134] [134a] S.J. Chen, H.T. Chang, Anal. Chem. 76 (2004) 3727;
[134b] N. Zhang, Y.Y. Liu, L.L. Tong, K.H. Xu, L.H. Zhuo, B. Tang, Analyst 133 (2008) 1176;
[134c] J.H. Lin, C.W. Chang, W.L. Tseng, Analyst 135 (2010) 104.

[135] J. Peng, Y. Wang, J. Wang, X. Zhou, Z. Liu, Biosens. Bioelectron. 28 (2011) 414–420.

[136] T. Soller, M. Ringler, M. Wunderlich, T.A. Klar, J. Feldmann, Nano Lett. 7 (7) (2007) 1941–1946.

[137] W.R. Glomm, S.J. Moses, M.K. Brennaman, J.M. Papanikolas, S. Franzen, J. Phys. Chem. B 109 (2005) 804–810.

[138] B.I. Ipe, K. Yoosaf, A.G. Thomas, J. Am. Chem. Soc. 128 (2006) 1907–1913.

[139] Z.F. Zhang, H. Cui, C.Z. Lai, L.J. Liu, Anal. Chem. 77 (10) (2005) 3324–3329.

[140] H.O.Z. Albrecht, Phys. Chem. 136 (1928) 321–330.

[141] W. Chen, L. Hong, A.L. Liu, J.Q. Liu, X.H. Lin, X.H. Xia, Talanta 99 (2012) 643–648.

[142] [142a] A. Roda, P. Pasini, M. Mirasoli, E. Michelini, M. Guardigli, Trends Biotechnol. 22 (2004) 295–303;
[142b] D. Christodouleas, C. Fotakis, A. Economou, K. Papadopoulos, M. Timotheou-Potamia, A. Calokerinos, Anal. Lett. 44 (2011) 176–215;
[142c] M.J. Pringle, Adv. Clin. Chem. 30 (1993) 89–183;
[142d] J. Du, Curr. Pharm. Anal. 6 (2010) 30–38;
[142e] L. Zhao, L. Sun, X. Chu, Trends Anal. Chem. 28 (2009) 404–415;
[142f] P. Fletcher, K.N. Andrew, A.C. Calokerinos, S. Forbes, P.J. Worsfold, Luminescence 16 (2011) 1–23.

[143] Q. Li, L. Zhang, J. Li, C. Lu, Trends Anal. Chem. 30 (2011) 401–413.

[144] [144a] M. Yang, R. Liu, Y. Xu, Y. Zou, X. Song, Biosens. Bioelectron. 22 (2007) 871–876;
[144b] W. Som-Aum, J. Threeprom, H. Li, J.M. Lin, Talanta 71 (2007) 2062–2068;
[144c] A. Waseem, M. Yaqoob, A. Nabi, Luminescence 21 (2006) 174–178;
[144d] S. Bi, Y. Yan, X. Yang, S. Zhang, Chem. Eur. J. 15 (2009) 4704–4709.

[145] [145a] A. Safavi, M.A. Karimi, Anal. Chim. Acta 468 (2002) 53;
[145b] N.P. Evmiridis, Analyst (Cambridge, UK) 112 (1987) 825;
[145c] G.Z. Tsogas, D.L. Giokas, A.G. Vlessidis, N.P. Evmiridis, J. Chromatogr. A 1107 (2006) 208;
[145d] G.Z. Tsogas, D.L. Giokas, A.G. Vlessidis, N.P. Evmiridis, Anal. Chim. Acta 565 (2006) 56;
[145e] D.L. Giokas, A.G. Vlessidis, N.P. Evmiridis, Anal. Chim. Acta 589 (2007) 59.

[146] D.L. Giokas, A.G. Vlessidis, G.Z. Tsogas, N.P. Evmiridis, Trends Anal. Chem. 29 (2010) 1113–1126.

[147] J. Jiang, Y. He, X. Yu, J. Zhao, H. Cui, Anal. Chim. Acta 791 (2013).
[148] L. Wang, P. Yang, Y. Li, H. Chen, M. Li, F. Luo, Talanta 72 (2007) 1066–1072.
[149] Y. Li, P. Yang, P. Wang, L. Wang, Anal. Bioanal. Chem. 387 (2007) 585–592.
[150] P. Yang, Y. Chen, Q. Zhu, F. Wang, L. Wang, Y. Li, Microchim. Acta 163 (2008) 263–269.
[151] Z.P. Li, C.H. Liu, Y.S. Fan, Y.C. Wang, X.R. Duan, Anal. Biochem. 359 (2006) 247–252.
[152] M. Hao, Z. Ma, Sensors 12 (2012) 17320–17329.
[153] [153a] Z.-P. Li, C.-H. Liu, Y.-S. Fan, X.-R. Duan, Anal. Bioanal. Chem. 387 (2007) 613–618;
[153b] C. Ding, H. Zhong, S. Zhang, Biosens. Bioelectron. 23 (2008) 1314–1318;
[153c] A. Fan, C. Lau, J. Lu, Analyst (Cambridge, UK) 134 (2009) 497–503;
[153d] S. Zhang, H. Zhong, C. Ding, Anal. Chem. 80 (2008) 7206–7212.
[154] X. Huang, J. Ren, Trends Anal. Chem. 40 (2012) 77–89.
[155] [155a] T. Förster, Ann. Phys. 2 (1948) 55–75;
[155b] L. Stryer, R.P. Hangland, Proc. Natl. Acad. Sci. USA 58 (1967) 719–726.
[156] [156a] H. Xu, C.M. Liu, Y. He, H.W. Tang, Q. Wu, J. Lumin. 130 (2010) 1872–1879;
[156b] S. Zhao, J. Liu, Y. Huang, Y. Liu, Chem. Commun. 48 (2012) 699–701.
[157] [157a] J. Han, J. Jose, E. Mei, K. Burgess, Angew. Chem. Int. Ed. Engl. 46 (2007) 1684–1687;
[157b] B.M.L. Grayeski, P.A. Moritzen, Langmuir 13 (1997) 2675–2680.
[158] X. Huang, J. Ren, Anal. Chim. Acta 686 (2011) 115–120.

Chapter 13

Gold Nanoparticles as (Bio)Chemical Sensors

Miguel Peixoto de Almeida,[1] Eulália Pereira,[1] Pedro Baptista,[3] Inês Gomes,[1,4] Sara Figueiredo,[3] Leonor Soares[1,2] and Ricardo Franco[2]*

[1] *REQUIMTE, Departamento de Química e Bioquímica, Faculdade de Ciências, Universidade do Porto, Porto, Portugal;* [2] *REQUIMTE, Departamento de Química, Faculdade de Ciências e Tecnologia, Universidade Nova de Lisboa, Caparica, Portugal;* [3] *CIGMH, Departamento de Ciências da Vida, Faculdade de Ciências e Tecnologia, Universidade Nova de Lisboa, Caparica, Portugal;* [4] *Instituto de Medicina Molecular, Faculdade de Medicina da Universidade de Lisboa, Lisboa, Portugal*

**Corresponding author: E-mail: ricardo.franco@fct.unl.pt*

Chapter Outline

Gold Nanoparticles in Analytical Chemistry. http://dx.doi.org/10.1016/B978-0-444-63285-2.00013-4

1. INTRODUCTION

The focus of this chapter is on several sensing strategies and major recent advances in the use of gold nanoparticles (AuNPs) for detecting a variety of targets. These include chemical and biological entities—namely, metal ions, organic molecules, proteins, enzymes, antigens, nucleic acids (DNA and RNA), and microorganisms.

The unique chemical, optical, and electrical properties of AuNPs have led to their intensive use as (bio)chemical sensors with enhanced performance [1–3]. These proprieties include tunable surface plasmon resonances, distance-dependent fluorescence quenching or enhancement, high electrical conductivity, and exceptional light-scattering properties (Figure 1). Furthermore, AuNPs provide a suitable microenvironment for biomolecule immobilization, retaining their biological activity. Strategies for biomolecule binding to AuNPs include the formation of capping layers of dual-functionality molecules capable of binding strongly to the gold surface, and with a suitable chemical reactive group at the other end for further functionality. This type of capping agent provides a convenient, flexible, and simple "biologically friendly" environment for adsorption of various recognition elements to AuNP surfaces (Figure 1).

The need to detect and characterize biomolecules suitable for molecular diagnostics has prompted intense research and development toward the design of biosensors. Technological advances have focused on the selective recognition of relevant biomolecules with high accuracy, ideally as simple, rapid, and low-cost point-of-care bioanalytical devices. From the broad spectrum of biochemical sensors, strong preference has been devoted to those with a signal transduction that is a colorimetric change of the medium. This is mainly a result of the relative ease of interpretation of data derived from intuitive assessment of results. These have been extremely useful for the development of biosensors for analysis of active biomolecules relevant for human health and condition monitoring and diagnostics, food safety analysis, and environmental monitoring.

In a simplified vision, AuNP-based sensors are typically constituted by a recognition element coupled to AuNPs, the latter being the transducers that, on

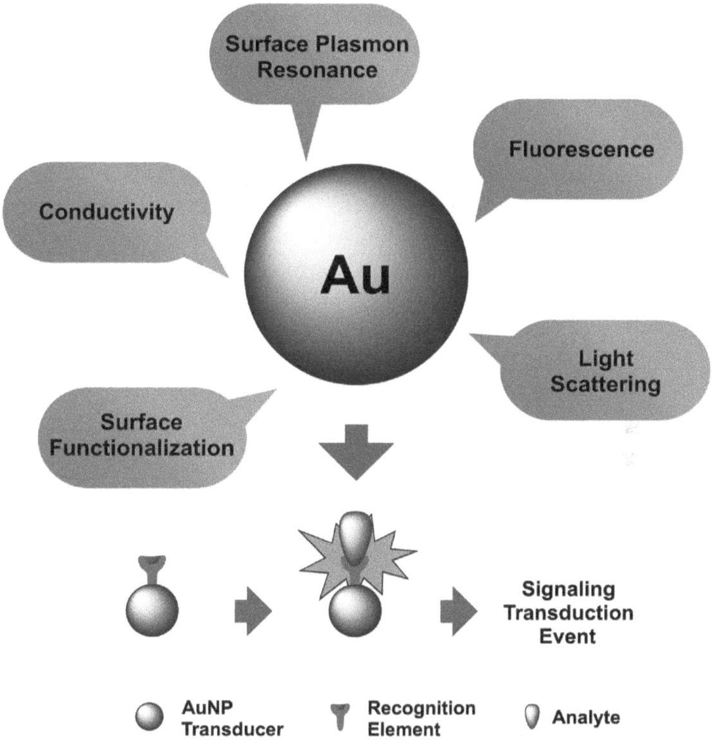

FIGURE 1 Relevant properties of gold nanoparticles (AuNPs) for sensing, and schematics of a standard AuNP-based sensor.

analyte binding, generate a signaling transduction event (Figure 1). The (bio) recognition element provides the sensors' specificity via biomolecules capable of interacting with the analyte of interest. Because these biomolecules may be of various types, biochemical sensors may be classified according to recognition element such as (1) immunosensors for detection of antigens, antibodies, and proteins; (2) nucleic acids sensors (DNA/RNA); and (3) enzymatic sensors, among others. (Bio)sensors can also be classified according to the relevant physicochemical property of the AuNP transducer, such as, for instance, optical (absorbance, luminescence, scattering, plasmon resonance), electrical/electrochemical (voltage, potentiometric, semiconductor sensors), magnetic, thermometric, and so on.

This chapter is divided in sections according to the detection technique used in a particular assay method. Section 2 deals with colorimetric sensing of surface plasmon resonance (SPR) changes, with the main detection technique being ultraviolet (UV)/visible spectroscopy. As a result of ease of use and the common availability of this type of spectroscopy, colorimetric-based sensing methods are, without question, the most widespread and used techniques. Section 3 describes the biobarcode assay and recent developments of this important method. Section 4 presents sensing based on photoluminescence phenomena. Fluorescence allows extremely high sensitivity in sensing, although selectivity might constitute a problem because many species present fluorescence and may cause interference. Section 5 explores detection schemes based on electrical and electrochemical measurements—by far the longest standing and widespread application of AuNPs in sensing. Last, Sections 6 and 7 refer to sensing systems that use the strong light-scattering properties of AuNP for sensing based on optoplasmonic or Raman effects, respectively.

2. SENSORS BASED ON COLORIMETRIC SENSING

Traditional reagents used for biochemical colorimetric assessment and analysis are reactive species that act as substrates and/or by-products of reaction, or as tags that label a recognition molecule, such as proteins, antibodies, and nucleic acids. These color-producing reagents are usually dyes (e.g., Coomassie Brilliant Blue G-250) that, once associated with a protein, induce color change. We focus on nanotechnology-based colorimetric assays and, although the previously mentioned dyes are of extreme relevance, they fall outside the scope of this chapter (for an in-depth review please refer to Sapan et al. [4] and references therein).

AuNPs have been considered ideal platforms for colorimetric sensing because they are extremely versatile, easily synthesized in a wide range of sizes and shapes, and can be modified with several chemical functional groups [2]. But above all, their unique optical properties, such as a high extinction coefficient leading to intense colors and/or exceptionally intense scattering, have been pivotal for the development of colorimetric biosensing strategies [5]. Optical behavior is strongly dependent on particle size, and solutions containing small nanoparticles (<40 nm in diameter) are red whereas larger particles (or aggregates) produce blue or purple solutions. Typically, for sensing applications, spherical AuNPs with a diameter ranging between 13 nm and 20 nm are used that show a characteristic intense SPR band at around 520 nm. In general, AuNP-based biosensors exploit the aggregation of nanoparticles that cause a shift in the absorption band to a higher wavelength, leading to a color change of the colloidal solution that may be detected visibly. In general, straightforward UV/visible spectrophotometry is enough for data assessment and quantitative analysis, in which the result is often presented as a ratio between the absorbance at 520 nm (corresponding to nonaggregated particles) and at a longer wavelength (near 600 nm—aggregated nanoparticles; Figure 2).

Spherical AuNPs with a diameter of 13–20 nm are usually synthesized by citrate reduction of a gold salt, and can be used as synthesized (labile citrate capping) or after surface modification with functional moieties. These functional moieties are generally thiolated at one end, providing a quasicovalent binding to the gold surface, and at the other end present functionalities

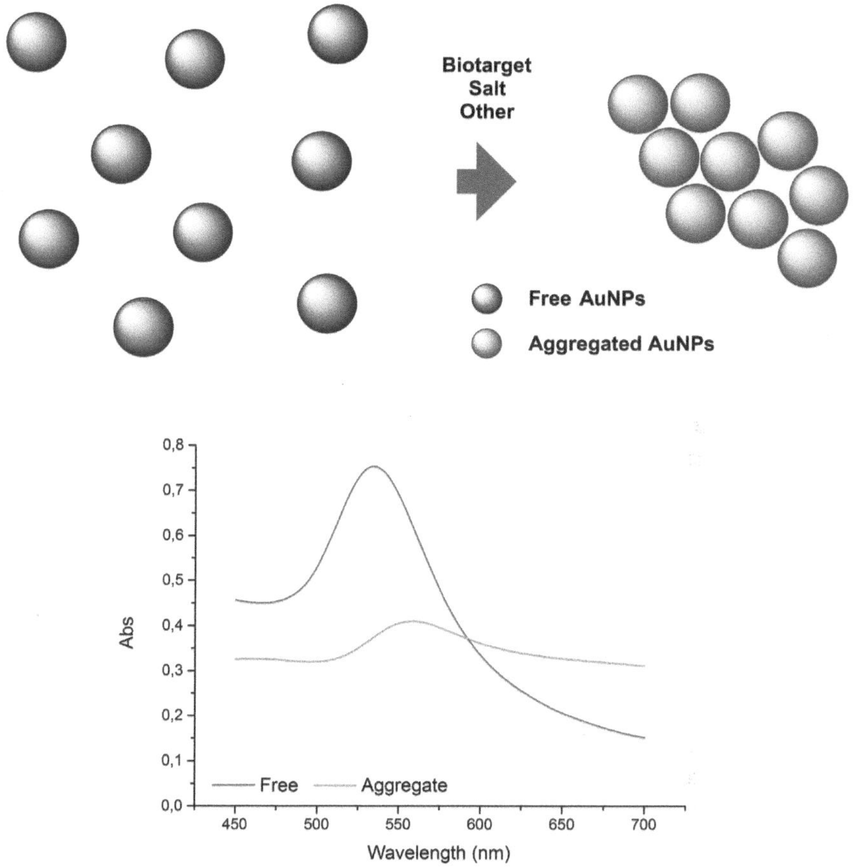

FIGURE 2 Schematic representation of colorimetric assays based on gold nanoparticle (AuNP) aggregation. A solution of dispersed AuNPs presents a red color, which changes to blue on AuNP aggregation.

that are responsible for target/analyte recognition. The analyte itself may promote or prevent aggregation of the AuNPs and has been associated mainly with two distinct mechanisms—interparticle cross-linking and noncross-linking aggregation—caused by loss of colloidal stability. Potential applications include identification of nucleic acid mutations associated with disease, detection of gene expression and pathogens, detection of small molecules or ions, and detection of proteins, such as in immunodetection and affinity binding.

2.1 Detection of Small Molecules

Several AuNP colorimetric sensing strategies have been proposed that target detection and/or quantitation of small molecules, and metal ions in particular.

Metal ions are usually associated with organic molecules. Only traces of metal ions can be found in the human body, and rare transition metals interact with biomolecules, originating adverse responses. DNA molecules and protein monomers have been used extensively as the recognition moiety for colorimetric biosensors of small molecules based on AuNPs (for additional insights, please refer to Lin et al. [6,7] and references therein).

2.1.1 Metal Ions

Heavy metal ions are highly toxic and cause long-term damage to living organisms, mainly as a result of significant oxidative damage. Because of their capability to interact with biomolecules, these ions can tune the structure and conformation of these molecules by a rather specific mechanism. As such, these molecules and their optical properties can be used to recognize the ion of interest and can be used in a biosensor. Lee et al. [8] proposed a system for the detection of Hg^{2+} based on the increased binding affinity of T—T mismatches to the Hg^{2+} cation, using this ion as a bridge between the bases, stabilizing them. In their system, single-strand DNA (ssDNA) sequences with an engineered T—T mismatch were conjugated to two AuNPs, which after hybridization yielded a blue solution. Denaturing of the DNA strands by heat led to dispersion of AuNPs and the solution turned red. Complexation with Hg^{2+} results in stabilization of the T—T mismatch, leading to an increase in the melting temperature (T_m) correlated directly to the concentration of the ion [8]. Xue et al. [9] adapted the method of Lee et al. [8] into a system capable of performing at room temperature by controlling the number of T—T mismatches, thus modulating the strands' T_m. Cross-linking of two DNA-functionalized AuNPs by a proper oligonucleotide linker was possible only in the presence of Hg^{2+}, leading to DNA/NP aggregates [9]. Later, Lee and Mirkin [10] adapted the same principle to a chip-based technique.

Although Hg^{2+} can induce aggregation of nanoparticles, other metals can be used as inductors of disassembling. Such is the case for the platform proposed by Liu and Lu [11] based on DNAzyme for the detection of Pb^{2+}. DNAzymes (also called *deoxyribozymes*) are able to cleave a DNA substrate strand in the presence of one of its cofactors (highly specific for Pb^{2+}, Cu^{2+}, and Zn^{2+}). The proposed sensor was based on oligonucleotide-functionalized AuNPs and a linker capable of hybridizing specifically at each end, inducing AuNP aggregation. In presence of Pb^{2+}, the DNAzyme 17E cleaves the substrate and "dissolves" the aggregates, in an event associated with a color change that can be monitored by UV/visible spectroscopy. The method allows the determination of metal ion concentration in a range suitable for environmental monitoring.

A simple and real-time analysis of lanthanides was proposed by Lisowski et al. [12] using malonamide-functionalized AuNPs, the cross-linking of which was induced by the bonding between malonamide and the metal, resulting in plasmon coupling and a concomitant red-to-blue change in the solution.

Furthermore, a titration was performed showing linear correlation between the absorbance and the concentration of the studied lanthanide at a limit of detection of 50 nM.

Several other methods have been proposed for the detection of heavy metals and small ions (e.g., Li^{+}, K^{+}, Mg^{2+}, PO_4^{3-}, UO_2^{2+}, and so on; Table 1).

2.1.2 Aptamers for Colorimetric Sensors

The combination of aptamers with AuNPs has attracted considerable interest and led to advances in biosensing. Aptamers are short, ssDNA/RNA molecules that can bind various molecular targets such as small molecules, proteins, and nucleic acids. For each application, an appropriate aptamer is required and must be prepared through chemical synthesis. These molecules can assume a variety of shapes because they tend to form single-strand loops; thus, their specificity is determined by their tertiary structure [18]. When aptamers bind to their targets, they adopt specific conformations, such as G-quadruplex and hairpins. Furthermore, aptamers are robust at high temperatures and are easily modified. Because of these advantages, aptamers can be used in bioassays in two distinct ways. First, the aptamer is linked to the AuNP, stabilizing the system. After the aptamer binds to its target, it is released from the AuNPs, leading to aggregation. Second, the aptamer is attached to the AuNP, functioning as a cross-linking agent, and leading to aggregation of the aptamer-modified AuNPs (blue solution). After its binding to the target, the particles

TABLE 1 Recent Progress in Colorimetric Metal Chemosensors

Element	Limit of Detection	Method	References
Ln^{3+}	50 nM	Cross-linking	[12]
Hg^{2+}	3 μM (naked eye)	Cross-linking	[9]
Hg^{2+}	10 nM	On chip	[10]
Hg^{2+}	100 nM	Cross-linking	[13]
Pb^{2+}	100 nM	DNAzyme	[11]
Pb^{2+}	0.025 nM	POEGMA-AuNP on glass	[14]
Mg^{2+}	800 nM	Mg^{2+} triggers dispersion or aggregation	[15]
PO_4^{3-}	11 ppm	AuNPs functionalized with MPTP-Zn	[16]
UO_2^{2+}	50 nM	DNAzyme-AuNP	[17]

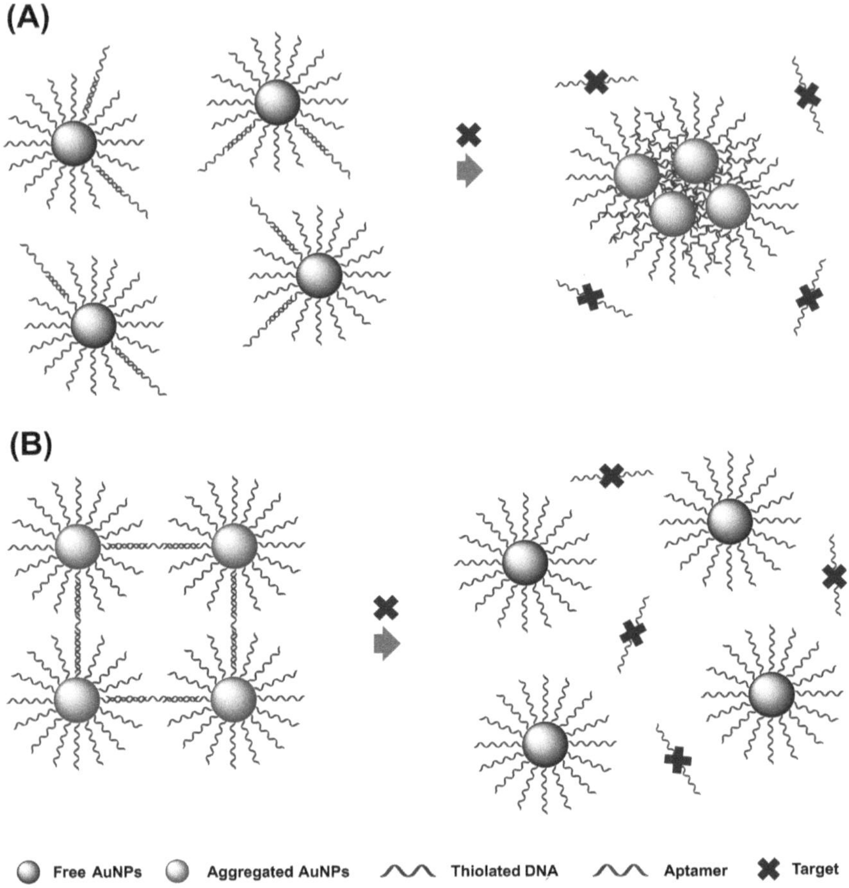

FIGURE 3 (A, B) Schematic illustrations representing sensors based in aptamers using gold nanoparticles (AuNPs), in which the release of the aptamer induces aggregation of AuNPs (A) and the release of the aptamer induces disaggregation of AuNPs (B).

disassemble, leading to a change in the color from blue to red (Figure 3). Aptamers have been widely used in the detection of small molecules such as adenosine triphosphate (ATP), Pb^{2+}, K^{+}, and cocaine.

2.2 DNA/RNA Applications

A remarkable colorimetric detection platform for high-throughput DNA/RNA screening was developed by Mirkin et al. [19] based on the cross-linking of two ssDNA-functionalized AuNPs (Au nanoprobes) on hybridization to a specific target sequence. Cross-linking of the nanoprobes brings the AuNPs into close vicinity, resulting in a red-shift of the surface plasmon, with concomitant color change of the solution from red to blue [19,20]. This concept was the basis for a plethora of detection strategies relying on the modulation of hybridization and control over the extent and kinetics of cross-linking (for additional information, see Mirkin et al. [19] and references therein).

As described before, AuNPs are extremely sensitive to changes in the dielectric constant of the medium and aggregate on an increase of the ionic strength of the medium (e.g., after salt addition). Functionalization of AuNPs with oligonucleotides stabilizes the suspension. Larguinho and Baptista [21] made use of this property to develop a noncross-linking method in which hybridization of the gold nanoprobe to the complementary target prevents salt-induced aggregation. This approach was applied successfully to the detection of pathogens (e.g., *Mycobacterium tuberculosis* [22]), characterization of gene expression in cancer (e.g., chimeric BCR-ABL messenger RNA responsible for chronic myeloid leukemia [23]), and detection of point mutations [24].

Another approach for the cross-linking method involves the use of AuNPs modified with peptide nucleic acids (PNAs). PNAs are DNA analogues in which the phosphate backbone is replaced by a polyamide backbone, providing neutrality and high rigidity. Kim et al. [25] used PNAs bonded to a glass slide to act as a capture probe for specific DNA. Then, 4-nm, positively charged AuNPs (coated with cetyltrimethylammonium bromide) were added and gold/DNA supramolecular adducts formed through electrostatic interactions. Detection of AuNPs bound to the DNA was amplified using a metal enhancement solution, allowing visualization of DNA spots by the naked eye (in the presence of a noncomplementary DNA, the spot is almost invisible) [25].

Although a large number of methodologies and techniques have been presented for the detection of DNA/RNA through AuNPs, for pathogen detection and genetic profiling, most of them have used synthetic molecules as targets, and only a few have been applied for the detection of DNA/RNA directly in clinical samples.

2.3 Applications to Protein/Antigen Detection

Sensitive methods for protein or antigen detection are considered fundamental in clinical diagnosis. Recently, Kim et al. [26] proposed an aptamer—AuNP conjugate for the detection of a specific protein target, using lysozyme as a representative example. The proposed assay was based solely on visual recognition, avoiding the need for any type of instrumentation. Furthermore, the authors claim that the platform exhibited the highest sensitivity reported for colorimetric detection systems of lysozyme and that it was optimized for fast output because the target can be used without pretreatment [26].

Other examples include the colorimetric detection of protein kinase activity via biotin-labeled ATP used to monitor the phosphorylation of peptide-modified AuNPs [27]. These phosphorylated AuNPs with a biotin label are further reacted with another avidin-capped AuNP, leading to aggregation and concomitant color change. As a result of the costs associated with the label, Yao et al. proposed a label-free detection method using a peptide harboring the phosphorylation site *Ser*, which acts as substrate for cyclic adenosine monophosphate-dependent protein kinase (PKA). The substrate contains, beside the phosphorylation site, additional positively charged residues and a cysteine residue at the N-terminal for anchoring on the surface of citrate-capped AuNPs. When bonded to the AuNPs, the substrate peptide is capable of inducing AuNP aggregation by binding to adjacent negatively charged particles. Carboxypeptidase (CPY) is capable of discriminating phosphorylated and nonphosphorylated peptides, and is thus able to recognize kinase catalysis. Treatment of nonphosphorylated substrate peptide with CPY results in cleavage of the peptide and a consequent decrease in the aggregation of AuNPs [28]. The simplicity of colorimetric detection minimizes the costs associated with the equipment. Furthermore, the colorimetric sensors described combine cutting-edge technologies with nanotechnology, endowed with high sensitivity and selectivity.

3. GOLD NANOPARTICLE-BASED BIOBARCODE ASSAY

A decade after the first description of the biobarcode assay by Mirkin's group [29], there is already a substantial amount of published work based on this method. The nonbiological core components of this technique are magnetic microparticles (MMPs), probes, and AuNPs combined in various ways and with different types of assay-specific biomolecules. The work that established the principles of the method was developed during the early 2000s, with prostate-specific antigen (PSA) as the target of interest. Two main groups of biomaterials were prepared: (1) polyamine particles with a diameter of 1 μm with magnetic iron oxide cores, functionalized with PSA monoclonal antibodies (mAbs) and (2) AuNPs functionalized heavily with oligonucleotides (ssDNA) that hybridize with the biobarcode DNA (bcDNA), and PSA polyclonal antibodies (pAbs). The target molecule functions as a bridge between the monoclonal and the polyclonal antibodies—in other words, between one MMP–mAb assembly and one or more (bcDNA)AuNP–pAb assemblies. Each one of the MMP–mAb structures can capture several (bcDNA) AuNP–pAb units; therefore, a large quantity of bcDNA is present for each PSA biomolecule. When the reaction mixture is exposed to a magnetic field, all the MMPs–mAbs are projected to the vessel's wall almost instantly. Only the (bcDNA)AuNPs–pAbs that are linked to the attracted MMP–mAb assemblies, as a result of the presence of the PSA, are separated, which means that if there is no target analyte in the mixture, no bcDNA will be present in the following steps. By rejecting the reaction mixture and using the aggregate structures attracted by the magnet in additional steps, these are washed to dehybridize the bcDNA from the complementary strand anchored on the

AuNP. Again, the magnet is used to remove the aggregates, but in this case keeping the bcDNA-rich solution, which can then be identified easily by standard DNA detection methods. The intermediate assembly that will ensure PSA detection is presented in Figure 4(A), which depicts an MMP surrounded by an AuNP (barcode DNA carriers) in the presence of PSA (analyte and linking agent) [29,30].

Taking advantage of the AuNP probe carrying a large number of oligonucleotides per protein binding event, this work showed there is substantial amplification of the PSA signal, which can be detected at a 30-aM concentration, a five order-of-magnitude improvement over clinically accepted conventional assays for detecting the same target, with sensitivity limits of ~3 pmol. It is important to mention that some properties, such as particle size, can be tuned to improve sensitivity and that it is possible, with this method, to detect simultaneously distinct analytes by creating distinct identifiable barcodes for nearly every target of interest [29].

In the following years, the main principles of this bcDNA method were the base for new applications proposed by Mirkin's group and others. Less than 1 year later, the same group proposed a similar method for DNA detection [30]. In this case, DNA strands complementary to the analyte were used to functionalize the nanoparticles. The supported DNA fragments are different between the MMP and the AuNPs, and only complementary to a portion of the target DNA, so the

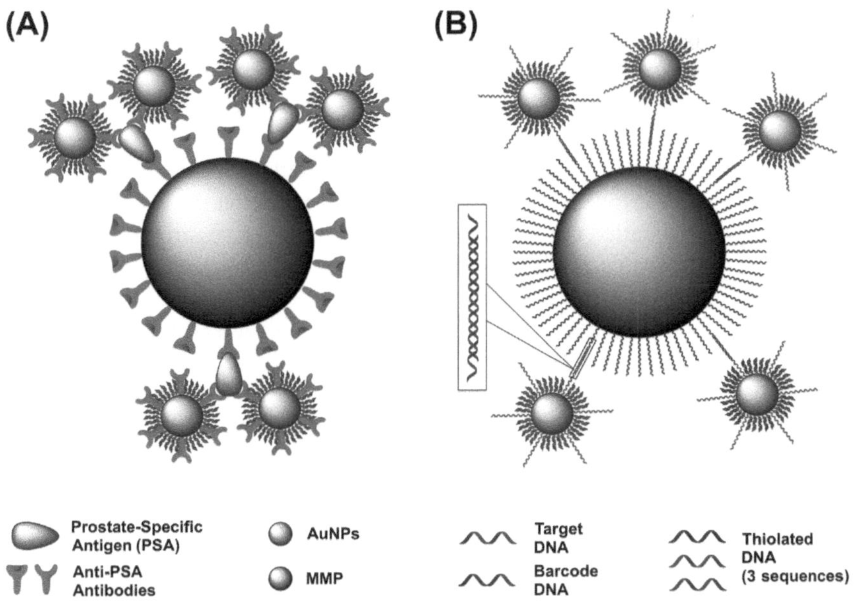

FIGURE 4 (A, B) Biobarcode analysis essential intermediates for immuno-based detection (A) and for DNA detection (B). MMP, magnetic microparticle.

target acts as bridge between the two types of particles (Figure 4(B)). This technique allows specific detection of small amounts of a particular DNA sequence, without the need to amplify the target DNA by polymerase chain reaction (PCR), using instead signal amplification by bcDNA detection. The method has high selectivity with a sensitivity that is equivalent to many PCR-based approaches, but without the need for enzymatic amplification [30].

A year later, a new approach also used AuNPs, not as bcDNA carriers but as a latter intervenient, for direct bcDNA detection [31]. This work used again two distinct antibodies for antigen detection (interleukin-2) acting as a bridge. AuNPs were substituted with amine-modified porous silica microparticles, and ssDNA complementary to the bcDNA was anchored together with the first antibody. The following steps are similar to the ones described earlier. This time, the use of AuNPs takes place after the bcDNA release from the probes, enabling detection of cytokines at attomolar concentrations. Two groups of AuNPs were prepared with different ssDNA complementary to the bcDNA and these probes—when in the presence of bcDNA—are linked, undergoing aggregation. Aggregation causes a color change, with different intensities depending on the quantity of bcDNA in the media. Further details about the principle behind this colorimetric assay are presented in Section 6.1.

A recent review [32] states that the capture moiety most commonly used in recent works is the antibody, but also points to some of its disadvantages. The antibody-based biobarcode assay has been developed successfully for the detection of several cancer-specific biomarkers and viral proteins. However, when the antibody-based detection is compared with DNA biomarkers, it is easy to notice that proteins are less stable and more sensitive to thermal and chemical degradation. Also, for many protein targets, two distinct antibodies are not available for two distinct epitopes, preventing the antigen from functioning as a bridge.

One of the possible techniques for detecting bcDNA, besides the classic ones, was also developed by the group behind the biobarcode method. This technique is often called the *scanometric method* and it also relies on the use of AuNPs [33]. With this method several components are involved—namely, a DNA chip to which a DNA fragment is attached (A), AuNPs functionalized with other DNA fragments (B), and finally the target DNA fragment (C). Sequences A and B are not complementary to each other, but are both complementary to different portions of sequence C, so when target DNA is present (e.g., bcDNA), the AuNP is attached to the chip via hybridization of C with A and B. In contrast to the use of fluorophore probes, AuNPs alter substantially the melting profiles of the targets from an array substrate, allowing the distinction of an oligonucleotide sequence from targets with single-nucleotide mismatches, with a selectivity that is more than three times greater than for fluorophore-labeled targets. Additional signal amplification is also possible using nanoparticle-catalyzed reduction of silver(I), leading to an increase of sensitivity that surpasses the analogous fluorophore system by two orders of magnitude.

4. SENSORS BASED ON FLUORESCENCE-RELATED PHENOMENA

Fluorescent markers have been widely used for labeling of cellular components, and also as an imaging tool in clinical diagnosis. In addition to the very high sensitivity of fluorescence detection, it is possible to use different fluorophores for the colocalization of molecules that are in close proximity. However, the resolution of the image is often hampered by the resolution of the fluorescence microscope used during the analysis.

The search for new fluorescence reporters led to the introduction of dyes combined with nanomaterials. These nanomaterials present improved photophysical behavior, such as, for instance, enhancement or quenching of the signal. AuNPs are commonly associated with fluorescence quenching when fluorophores are attached to their surface. Actually, AuNPs serve as excellent fluorescence quenchers for fluorescence resonance energy transfer-based assays. On this basis, fluorescence sensors using AuNPs have become popular, for which both "switch on" and "switch off" sensing mechanisms have been proposed for fluorescence-based assays. Therefore, two distinct approaches have been widely reported: (1) molecular beacons that depend on functionalization of AuNPs with fluorescently labeled oligonucleotide strands and (2) AuNPs functionalized with oligonucleotide strands that hybridized to another oligonucleotide probe that is fluorescently labeled (illustrated in Figure 5(A) and (B), respectively) [1,34].

4.1 Förster Resonance Energy Transfer-Based Detection

Förster resonance energy transfer (FRET) is a distance-dependent mechanism of nonradiative energy transfer between two chromophores, in which energy from the molecule excited by irradiation (donor) is transferred to another

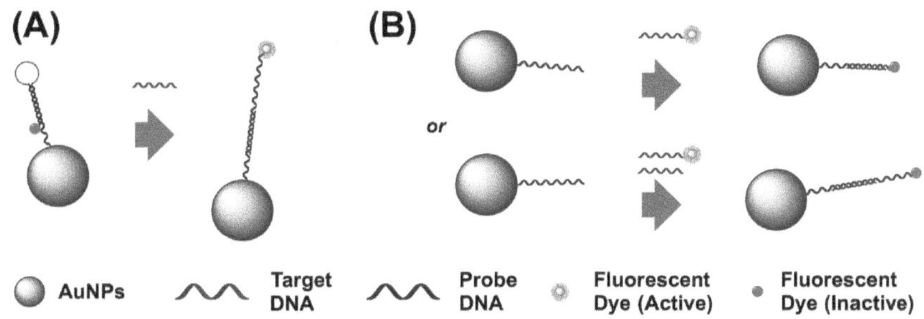

FIGURE 5 (A, B) Schematic representation of fluorescent-based sensors using gold nanobeacons (A) and probes (B) that take advantage of the interdistance between the fluorophore and the quencher.

molecule (acceptor) that then emits lower energy light. The efficiency of energy transfer is highly dependent on the distance between fluorophores. In fact, energy transfer efficiency between donor and acceptor molecules depends on the inverse sixth power of the intermolecular distance. Hence, FRET can be used to determine whether two molecules are within a certain distance from each other, and can be considered an important technique that gathers information on a variety of biological phenomena. A possible application is determination of the colocalization of proteins and other molecules. In theory, energy transfer measurements can be used to estimate the distances between regions within a certain molecule; therefore, FRET could be considered an effective molecular ruler. However, for resonance energy transfer to occur, the donor and acceptor molecules need also to have overlapping emission and absorption spectra.

Besides being applied in optical microscopy techniques, FRET has also been used as a spectroscopic technique in several areas, as in structure analysis, nucleic acid analysis, and signal transduction, to name a few. More recently, use of nanoparticle-based systems has led to the use of FRET within biosensors. The versatility of modification of the surface chemistry of nanomaterials—in particular, AuNPs—encouraged the development of probes for the recognition of peptides, nucleic acids, or small molecules. As it is known, several factors can influence metal-induced modulation of fluorescence, such as the physical properties of the nanoparticles or the properties of the fluorophores [35,36].

4.2 Fluorophores and AuNPs

Ray et al. [37] presented a rather remarkable example reporting an Au-based FRET assay to monitor the cleavage of DNA by nucleases. Mainly, AuNPs functionalized with an oligonucleotide were hybridized to a complementary Cy3-labeled oligonucleotide. The hybridization induced quenching of the dye by AuNPs and the assembly was then treated with a nuclease that cleaves the DNA strand, inducing the recovery of the fluorescence.

Because the impact of biosensors is particularly relevant in clinical applications, the development of a rapid detection system for malaria diagnostics had high impact in the area. Guirgis et al. [38] developed an immunoassay based on the fluorescence quenching of Cy3B by AuNPs and, as a proof-of-concept, researchers successfully detected recombinant *Plasmodium falciparum* heat shock protein 70 antigen, with a limit of detection (LOD) of 2.4 μg/mL. Gold nanosystems have also been used for forensic analysis, in both fingerprint and metabolites identification [39]. Russell et al. proposed a platform of anticotinine–AuNP conjugates for the detection of cotinine in the fingerprints of smokers. This assembly was detected using a fluorescently tagged secondary antibody fragment, and the fluorescence images obtained enabled both the identification of an individual and simultaneous determination of cotinine present in the sweat deposited in the fingerprint. It is noteworthy that the presence of AuNP conjugates induced an enhancement in the fluorescence signal, mainly as a result of the multiple binding sites for the secondary antibody. These results underscore the importance of the multivalency effect presented by AuNPs for increased detection sensitivity. For additional insights on other platforms using AuNPs as sensors, and based on fluorescence-related phenomena, see Table 2.

One of the most promising fluorescent systems relies on the quenching effect of AuNPs on fluorophore-labeled hairpin ssDNA, a system referred to as *gold nanobeacons*. In its more common form, molecular beacons act like switches that, when closed, hold in close proximity the fluorophore and the quencher, thus resulting in fluorescence quenching ("off" signal). When the target binds to the beacon, it undergoes conformational transition that switches on fluorescence as the fluorophore is pulled apart from the quencher ("on" signal) (Figure 5(A)).

Using this principle, Maxwell et al. [40] have shown that AuNPs functionalized with ss-oligonucleotide, with a fluorophore at the 3′-end, can assemble into an archlike conformation, inducing fluorescence quenching. On binding with target DNA, the conformation opens and the fluorophore is separated from the AuNP, switching on the fluorescence [40]. These models can detect single-base mismatches [45] and can be used to follow transcription of a given gene in real time [46]. A similar approach used nanoflares (i.e., fluorescently labeled siRNA that is released from AuNPs on target recognition) [47].

4.3 Quantum Dots and AuNPs

As nanoparticles evolved within biosensing, core−shell NPs have attracted considerable attention because they are able to improve the performance of biosensors by enhancing their LOD. Among these, quantum dots (QDs) have become popular because this type of NPs has unique optical properties, high quantum yields (typically presenting molar extinction coefficients significantly greater than organic dyes), a tunable emission spectrum, and outstanding photostability. AuNPs and QDs have been combined as an FRET donor−acceptor couple, and are used in methods to detect nucleic acids, proteins, and antigens. Because the size of the QD is correlated to its optical properties, several colors may be used, allowing for multiplexing and thus potentiating the simultaneous analysis of more than one type of biomolecule. This is a clear advantage over traditional fluorophores because there are almost no limitations for response wavelengths and emission peak discrimination provided by QDs.

In one interesting report, the fluorescence intensity and lifetime of the fluorophore could be used in sensing applications based on the distance dependence of interaction between the donor and the quencher. A "DNA

TABLE 2 Recent Progress in Fluorescent and AuNP-Based Sensors

Element to Detect	Fluorophore	LOD/Linear Detection Range	Approach	Reference
DNA–single-base mismatch detection	Fluorescein or tetramethylrhodamine	ND/ND	AuNP as quencher	[40]
Glucose		Buffer, 0.043 μM; serum, 0.065 μM/0.4–10 μM	Up-converting fluorescence	[41]
Protein detection–thrombin	—	ND/ND	AuNPs as quencher and aptamer as probe	[42]
Malaria–recombinant *Plasmodium falciparum* Hsp70	Cy3B	2.4 μg/mL/8.2–23.8 μg/mL	AuNPs as quenchers and anti-Hsp70 monoclonal antibody as probe	[38]
Detection of DNA cleavage by nucleases	Cy3	950 fM Cy3-ss-DNA/ND	AuNP-based FRET; fluorescence enhancement after cleavage	[35]
Alpha-fetoprotein (serum biomarker of hepatocellular carcinoma)	Fluorescein isothiocyanate	0.17 nM/ND	AuNP fluorescence quenching immunoassay; AuNPs coated with monoclonal antibody	[43]
Pb^{2+}	Fluorescein	5 nM/up to 10^3	DNAzyme-based fluorescence sensors; fluorescence enhancement resulting from enzymatic cleavage of substrates immobilized on the AuNPs	[44]

AuNP, gold nanoparticle; Hsp, heat shock protein; LOD, limit of detection; ND, not determined.

origami" was used to tune interparticle distance, thus allowing control of the enhancement or quenching of the fluorescence signal [48]. More recently, several reports have been presented using AuNPs combined with QDs for FRET-based detection of nucleic acid sequences [49] and proteins [50]. For example, Kim et al. [51] published an interesting study on a multiplexed, high-throughput, chip-based system in which the fluorescence of a QD immobilized onto a glass slide is quenched by AuNPs. Quenching is modulated by the presence of proteases that cleave the linker peptide between the donor and the acceptor, making this a quantitative and highly sensitive assay for protease activity [51].

5. ELECTRICAL AND ELECTROCHEMICAL SENSING

AuNPs exhibit high surface area-to-volume ratios, excellent conductivity, and catalytic properties that make them ideal materials for use in electrocatalytic and electrochemical sensing. Because of their large surface area, AuNPs are used for biomolecular recognition at the interface with electrochemical signal transductions, and for amplification of the resulting electrical response. In electrochemical applications, AuNPs have the important role of conducting electrons directly from immobilized biomolecules to an electrode surface, and they have the ability to increase electrochemical transduction in the absence of electron mediators. AuNPs can decrease the overpotentials of electroanalytical reactions and maintain the reversibility of redox reactions [2,52]. AuNPs have been used to enhance electrochemical detection of small molecules, such as epidopamine, glucose, uric acid, ascorbic acid, and nitrite. Also, AuNP-based electrodes have been used to detect toxic ions such as chromium, mercury, antimony, and arsenic; pesticides such as atrazine, paraoxon ethyl, and carbofuran; drugs such as paracetamol, atenolol, prednisolone, and ethamsylate; and antituberculosis and antitumor drugs. Electrochemical sensors with AuNPs have been used in the detection of plant based-toxins, with aflatoxin being the most relevant [2,52].

5.1 AuNP-Based Electrochemical Biosensors

To mediate electrochemical reactions and amplify recognition signals for biological processes, AuNPs can connect enzymes to electrode surfaces by using different strategies such as: (1) direct attachment by the use of cysteine (2) via thiol linkers, or (3) by covalent bonds. In DNA detection, AuNPs offer attractive properties to act as DNA hybridization tags in the developing of electrochemical genosensor and conductometric detection schemes [1,2,53].

5.1.1 DNA Detection

The most important strategies used to integrate AuNPs in DNA detection systems are (1) AuNP dissolution by an acid, (2) direct detection of AuNPs anchored onto the surface of the DNA genosensor, (3) conductometric detection with silver, (4) enhancement of AuNPs anchored to the DNA genosensor surface by using silver or Au, (5) AuNPs as carriers of other AuNPs, and (6) AuNPs as carriers of other electroactive labels (Figure 6) [2,53,54].

AuNPs bound to a DNA molecule can be detected by dissolving AuNPs tags oxidatively in HBr/Br_2 solution followed by accumulation and stripping analysis of the resulting gold ions solution. Ozsoz et al. [55] developed a sandwich-type electrochemical genosensor in which the labeled target was captured by probe strands immobilized onto a graphite electrode and hybridization was detected electrochemically by gold oxidation. The response is greatly enhanced as a result of the large electrode surface area and the availability of many oxidizable gold atoms in each nanoparticle label, with a detection limit for PCR amplicons as low as 0.78 fmol [55].

Castañeda et al. [56] reported two AuNP-based genosensor designs for the detection of DNA hybridization. Both assay formats were based on a magnetically induced direct electrochemical detection of the AuNP tags on a magnetic graphite–epoxy composite electrode. The first assay is based on the hybridization between two single-strand, biotin-modified DNA probes: a

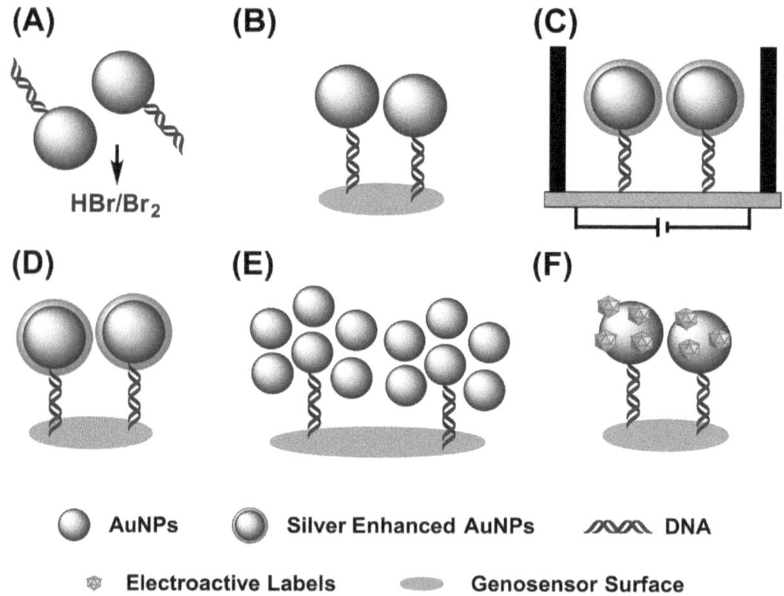

FIGURE 6 Schematic representation of the different strategies used for integration of gold nanoparticles (AuNPs) into DNA sensing systems. (A) Dissolution of AuNPs by HBr/Br_2 followed by detection of Au(III). (B) Direct detection of AuNPs anchored onto the surface of the genosensor. (C) Conductometric detection. (D) Enhancement with silver or gold followed by detection. (E) AuNPs as carriers of other AuNPs. (F) AuNPs as carriers of other electroactive labels. *Adapted with permission from Ref. [53]. Copyright 2007, Wiley Publications 2007.*

capture DNA probe and a target DNA related to the BRCA1 breast cancer gene. The second assay is based on hybridization between three DNA strands: a biotin-modified capture DNA probe, a target DNA related to the cystic fibrosis gene, and a DNA signaling probe modified with AuNPs via biotin—streptavidin conjugation [56].

In electrochemical sensors, signal amplification by silver accumulation on AuNPs is a very successful technique. Mirkin's group developed an electronic DNA detection sensor with high selectivity and sensitivity [57]. The sensor used a microelectrode array with 20-μm gaps, within which short oligonucleotides were immobilized. The target DNA functionalized AuNPs were subsequently hybridized to the short oligonucleotides, and subsequently silver was deposited onto the AuNPs to enhance conductivity. By measuring the resistance values across the electrode gaps, a sensitivity of 500 fmol was achieved, with a point mutation selectivity factor of 100,000:1 in target DNA [57]. Wang et al. [58] reported a novel silver-enhanced electrochemical stripping sensor for DNA with 83-fold enhancement of the signal response. Silver was precipitated on AuNPs after DNA hybridization, then dissolved in HNO_3 and measured subsequently by anodic stripping voltammetry with a detection limit in the femtomolar range [58]. Rochelet-Dequaire et al. [59] developed and evaluated a new efficient protocol for the sensitive quantification of a 35-bp human cytomegalovirus nucleic acid target. In the assay, the hybridization of the target adsorbed on the bottom of microwells with an oligonucleotide-modified AuNP detection probe was monitored by anodic stripping detection of the chemically oxidized gold label at a screen-printed microband electrode [59].

AuNPs can also be carriers for other AuNPs or other electroactive labels enhancing DNA detection. Kawde and Wang [60] described the strategy for amplifying particle-based electrical DNA detection based on oligonucleotides functionalized with polymeric beads carrying AuNPs tags. The gold-tagged beads were prepared by binding biotinylated metal AuNPs to streptavidin-coated polystyrene spheres. The amplified electrical transduction allows detection of DNA targets down to the 300-amol level, and offers great promise for ultrasensitive detection of other biorecognition events [60].

Another signal amplification strategy is to attach electroactive molecules onto the AuNPs labels. Wang et al. [61] demonstrated that AuNP—streptavidin conjugates covered with 6-ferrocenylhexanethiol were attached to a biotinylated DNA detection probe of a sandwich DNA complex, with a detection level down to 2.0 pM.

5.1.2 Electrochemical Immunosensors

Signal enhancement associated with the use of nanomaterial-amplifying labels and with the formation of AuNP—antibody—antigen conjugates provides the basis for immunosensors and immunoassays. Immunosensors for antigens present similar strategies as genosensors [54]. Because the antigen can be recognized by different antibodies at the same time, the sandwich-type assays are mostly used in electrochemical immunosensors. For the sandwich-type immunosensors, antibody 1 is immobilized on the electrode and antibody 2 is conjugated with AuNPs. The high affinity of the antibodies is usually used during the amplification of the electrochemical signal. Cui et al. [62] developed an amplified immunosensor of the AuNP–colloidal carbon nanosphere hybrid material by using a sandwich-type assay to detect human immunoglobulin (IgG). This approach provided a linear response range between 0.01 ng/mL and 250 ng/mL, and a detection limit of 5.6 pg/mL. The method is versatile, offers enhanced performance, and can be extended easily to other protein detection and DNA analyses [62].

Mao et al. [63] described a multiplex electrochemical immunoassay based on the use of AuNP probes and immunochromatographic strips (ISs) for rabbit IgG and human IgM detection. The AuNP-based sandwich immunoreactions were performed on the IS, and the captured AuNP labels on the test zones were determined by highly sensitive stripping voltammetric measurement of the dissolved gold(III) ions with a carbon paste electrode. The detection limits are 1.0 ng/mL and 1.5 ng/mL, with a linear range of 2.5–250 ng/mL for quantitative detection of rabbit IgG and human IgM, respectively [63].

Ho et al. [64] reported an AuNP-based electrochemical sandwich immunosensor for enolase antigen, a potential diagnostic marker for lung cancer. The immunosensor operates through physisorption of the anti-ENO1 monoclonal antibody on polyethylene glycol-modified disposable screen-printed electrodes, with a detection limit as low as 11.9 fg/mL [64].

Amperometric immunosensors are widely used for electrochemical-based sensing. Tang et al. [65] developed an amperometric immunosensor for ultrasensitive determination of carcinoembryonic antigen (CEA), using as labels thionine-doped magnetic AuNPs and HRP-encapsulated nanogold hollow microspheres. Under optimized conditions, the linear range of the proposed immunoassay without horseradish peroxidase (HRP) as an enhancer was 1.2–125 ng/mL CEA, whereas the assay sensitivity with HRP as enhancer could be increased further to 0.01 ng/mL, with a linear range from 0.01 to 160 ng/mL CEA [65].

Usually, potentiometric, impedimetric, and conductometric immunosensors are used in the label-free electrochemical immunosensors. Zhang et al. [66] devised a sandwich-type impedance immunosensor to detect human IgG using nanogold-labeled secondary antibodies for signal amplification on a primary antibody-modified electrode with high resistance. Liu et al. [67] developed a conductometric immunoassay for a hepatitis B surface antigen-based bioelectrocatalytic reaction on a microcomb-type electrode by using double-codified AuNPs as labels.

AuNPs also have an important role in electrochemiluminescence (ECL) sensors, because they amplify the chemiluminescence signal, as reported by Jie et al [68]. This group found that, after applying a dense monolayer of silica-coated AuNPs on the surface of QDs, the ECL signal was about 17 times higher than with pure QDs [68]. On a typical ECL immunoassay, the human IgG analyte is first captured by the primary antibody, followed by sandwich with a secondary antibody labeled with AuNPs. Li et al. [69] reported an ECL immunoassay based on the sandwich-type immunoreaction coupled with the chemiluminescent reaction of the gold(III)−luminol−H_2O_2 system. The gold(III) dissolved from rabbit−antigoat IgG AuNP labels served as an analyte in the ECL reaction for the indirect measurement of goat−antihuman IgG [69].

In conclusion, AuNPs have an important role in electrochemical immunosensors, with high detection sensitivity, selectivity, and specificity for clinical diagnosis. Table 3 summarizes the main characteristics of several AuNP-based immunosensors described in recent literature.

5.2 AuNPs in Quartz Crystal Microbalance-Based Sensing

Quartz crystal microbalance (QCM) is an ultrasensitive piezoelectric device that determines mass arising from analyte binding by measuring the change in frequency on a quartz crystal resonator. The QCM technique has been widely used in the detection of oligonucleotides, antigens, and proteins using AuNPs. The use of AuNPs in QCM-based sensing systems allows enhancement of detection sensitivity. This is achieved, on the one hand, because of the high surface area of AuNPs and, on the other hand, by mass amplification, with the latter improving the QCM frequency signal.

5.2.1 DNA Detection

In the QCM system, AuNPs have been used mostly through thiol-modified DNA coupling. Using DNA-capped AuNPs, they act as mass amplifiers, promoting the enhancement of the immobilization capacity and the detection limit for DNA sensing. Lin et al. [75] demonstrated that the immobilization of ~12-nm-diameter AuNPs onto a gold-coated QCM resulted in an easier attachment of thiol containing a single DNA strand, and an increased capacity for nucleic acid detection, allowing a higher sensitivity by amplifying frequency changes and increased capacity for DNA detection.

AuNPs can equally amplify QCM-D (QCM with dissipation) DNA sensor signals using sandwich hybridization. Sandwich-based approaches using ssDNA-modified AuNPs can improve the LOD of the technique significantly, in which one end of the target oligonucleotide hybridizes with the immobilized ssDNA (recognition element) while the opposite end hybridizes with an oligonucleotide bound to AuNPs (mass enhancer) [76]. Zheng et al. [77] developed a reusable aptamer-based AuNP-enhanced QCM-D biosensor for detection of adenosine molecules. This aptasensor showed a linear response to the increase of the adenosine concentration in the range of 0−2 μM, with a detection limit of 65 nM. The sensor showed high sensitivity, selectivity, stability, and reproducibility [77].

TABLE 3 Recent Progress in AuNP-Based Immunosensors with Detection by Electrochemical and Chemiluminescent Methods

Detection Principle	Biological Target	LOD/linear Detection Range	Reference
AuNP electrochemical immunosensor	Human IgG	0.01–250 mg/mL/5.6 pg/mL	[62]
AuNP electrochemical stripping	Human IgM, rabbit IgG	1.5 ng/mL, 1.0 ng/mL/2.5–250 ng/mL	[63]
	Human IgG, goat IgG	0.5 ng/mL/5–5000 and 5–400 ng/mL	[70]
	Goat IgG	0.5 ng/L/0.5–100 ng/mL	[71]
AuNP electrochemical sandwich immunosensor	Enolase antigen	11.9 fg/mL/10^{-8}–10^{-12} g/mL	[64]
AuNP amperometric immunosensor	Carcinoembryonic antigen (CEA)	ND/1.2–125 ng/mL, ND/0.01–160 ng/mL (when using HRP as enhancer)	[65]
Gold(III) from AuNP stripping catalyzes luminol chemiluminescence	Goat–antihuman IgG	1.5 ng/mL/5–10,000 ng/mL	[69]
	Human IgG	0.5 ng/mL/0.1–100 ng/mL	[72]
	ApxIV antibody in pig serum	1:40,000 (serum dilution)	[73]
Catalytic activity of irregularly shaped AuNPs on luminol CL	IgG in human plasma	17 pM (~2.7 ng/mL)/0.05–10 nM	[74]

AuNP, gold nanoparticle; Ig, immunoglobulin; LOD, limit of detection; ND, not determined.

5.2.2 Detection of Antigens and Proteins

AuNPs on QCM have been used for immunosensing of proteins in sandwich-type assay formats. The use of AuNP labels in sandwich assays can increase the surface stress on the immunocomplex mass, allowing an important increase in sensitivity for these electromechanical assays.

AuNPs, as signal amplifiers, were used to detect C-reactive protein (CRP) in a flow-type, indirect-competitive QCM immunosensor system. The sensor chip was constructed with 2 mg/mL CRP as the coating antigen, and anti-rat CRP biotinylated antibody conjugated with streptavidin-coated AuNPs was added, allowing a frequency shift of 139.8 ± 0.3 Hz. Compared with the frequency shift of 91.1 ± 1.3 Hz found with the addition of the unmodified antibody only, the shift increased 53.4% on AuNP binding, which resulted in sensitivity improvement of the immunosensor [78]. In a similar method format, although not requiring antibody biotinylation, Ding et al. [79] described a highly sensitive method to detect CRP by using a QCM immunosensor based on 20 nm AuNPs conjugated with anti-CRP antibodies. CRP was recognized by the anti-CRP antibody, which was immobilized on the gold electrode of the QCM and the sensitivity was enhanced further by binding of the AuNP conjugated anti-CRP antibody. A linear response could be obtained in the 20 ng/mL−30 μg/mL range [79].

Lyu et al. [80] proposed a highly sensitive QCM microgravimetric lectin Con A biosensor based on carbohydrate (mannose)-stabilized AuNPs as a signal amplifier. Mannose-stabilized AuNPs formed a sandwich-type complex with the Con A target bound specifically to a mannose-modified gold electrode to give an amplified frequency response [80].

Chen et al. [81] developed an aptamer−functionalized AuNPs sandwich QCM-D biosensor for sensitive detection of human α-thrombin (Figure 7).

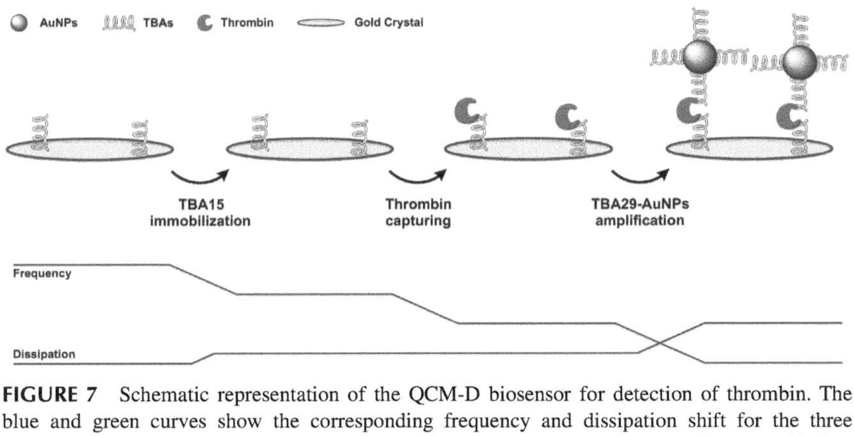

FIGURE 7 Schematic representation of the QCM-D biosensor for detection of thrombin. The blue and green curves show the corresponding frequency and dissipation shift for the three depicted immobilization steps. Aptamers are 15-mer and 29-mer thiolated thrombin binding aptamers (TBA15 and TBA29) [81].

Thrombin was determined using AuNPs functionalized with an appropriate aptamer (TBA 15 and TBA29) to enhance both frequency and dissipation signals. This aptamer-based QCM-D biosensor showed a low detection limit of 0.1 nM and provided a promising approach for sensitive detection of proteins with two binding sites [81].

6. AUNP-BASED SURFACE PLASMON RESONANCE SENSORS

SPR can be defined simply as a method that enables the detection of (bio) chemical modifications at the surface of a thin noble metal film. SPR-based sensors are one of the most important types in the class of label-free sensors, and they are capable of excellent results in terms of specificity and sensitivity. SPR sensors can be used in several fields—from medical diagnostics and research to environmental monitoring. In fact, SPR sensors are capable of sensing proteins, nucleic acids, or other biomolecules, with the advantage of real-time measurement [82]. These sensors can overcome the limitations of other methods such as enzyme-linked immunosorbent assays or PCR, which are usually expensive and time-consuming.

Surface plasmons are electron cloud oscillations that occur at specific material interfaces as a consequence of the interaction of incident light and changes at or near these interfaces (e.g., metal–dielectric interface), such as those induced by the presence of an analyte, which results in optical signal shifts [82]. The use of AuNPs in these sensors can be as signal-enhancers, or AuNPs can form part of the sensing surface itself, and their optical behavior will depend on its size, shape, and surrounding environment [83].

6.1 Sensors Based on Change in Localized SPR of AuNPs

AuNPs have been widely used as an efficient localized surface plasmon resonance (LSPR) probe because of their strongly enhanced SPR at optical frequencies that make them excellent scatterers and absorbers of visible light. LSPR sensors are based on sensing changes in the surface refractive index and transducing these changes into wavelength shifts of the LSPR extinction maximum as light is shined onto the patterned sensing platforms. In contrast to SPR long decay lengths (>100 nm), LSPR decay lengths are considerably shorter (5–10 nm), and LSPR is less sensitive to interference. In LSPR, the electric field component of the incident light interacts with the electrons of each metallic particle, causing the collective electron density to oscillate at a frequency that is resonant with the incoming light. The versatile nature of LSPRs can be attributed to their strong dependence on nanoparticle composition, shape, size, interparticle spacing, as well as dielectric properties and those of the local environment (i.e., substrate, solvent, and surface-confined molecules).

LSPR sensors are one of the most powerful technologies for ultrasensitive, real-time, and multiplex sensing [84]. In particular, nanobiosensors based on LSPR have great advantages for the detection of biomolecules, providing high levels of sensitivity, without labeling requirements, and ease implementation for the development of low-cost microfluidic systems. Mechanisms of LSPR wavelength shifts are essentially based on analyte−surface binding interactions. These include the adsorption of small molecules [85], ligand−receptor binding [86], protein adsorption on self-assembled monolayers (SAMS), DNA and RNA hybridization [87], and protein−DNA interactions [88].

LSPR sensors using AuNPs can be prepared according to two approaches—namely, by fabrication of AuNP patterned surfaces resorting to lithography and/or by the deposition of a colloidal gold solution onto a surface. All spectroscopy measurements associated with the detection can be performed in air or under analyte solution flow, providing a more efficient binding monitoring. Because it is localized, their sensing features are much more effective when performed at a single particle level using darkfield microscopy, rather than as part of an ensemble. Moreover, the corresponding wavelength shifts are found to be proportional to the amount of analytes, and LOD of the systems is determined by the surface-confined binding constant between the capture ligand on the surface and the analytes in solution.

Chemical LSPR-based sensors for selective detection of metal ions, such as Hg^{2+}, Pb^{2+}, Cu^{2+}, and Ag^{+}, have demonstrated to be very useful for the determination of these toxic metal ions in water samples and industrial products. Depending on the AuNP surface modification, sensor response can be specific to certain chemical entities, either by aggregation or simply because of changes in the vicinity. LSPR-based biosensors have been studied in different assay formats, such as immunoassays, DNA−DNA and DNA−-protein interactions, protein adsorption on SAMs. This localized approach leads to significant improvements in sensitivity [87,89]. Recently, the use of anisotropic AuNPs— mostly rods, plates, and stars—is enhancing even more the sensitivity of LSPR sensors (Figure 8) as a result of the presence of "hot spots" (usually located on the tips and edges) [90].

6.2 AuNP-Mediated SPR Signal Amplification

The involvement of AuNPs in SPR sensors as labels is particularly relevant because of the improved sensitivity; however, this converts a label-free technique into a labeling-dependent one. The sensitivity improvement is related to a coupling interaction between the surface plasmon characteristic from the "traditional" assembling of the sensor and the AuNP localized surface plasmons, but also as a result of a greater amount of analyte deposited on the sensor [82].

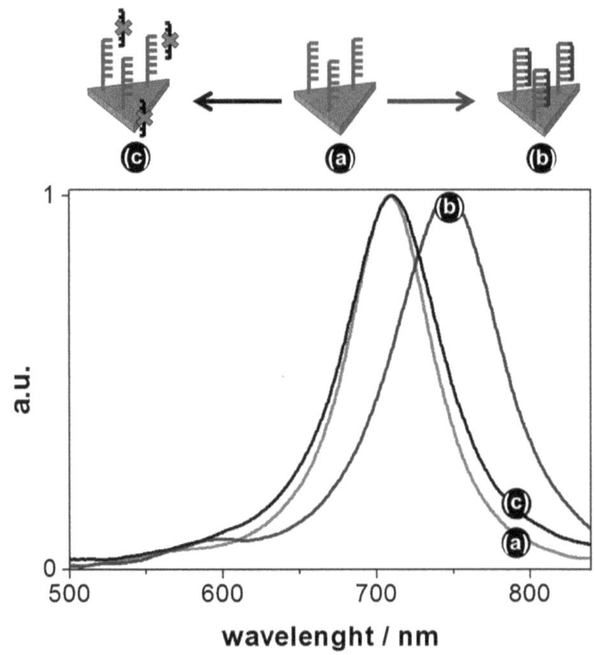

FIGURE 8 Schematic representation of a localized surface plasmon resonance biosensor based on triangular gold nanoparticles (AuNPs) for DNA hybridization detection. The red trace (a) corresponds to the spectrum of the triangular AuNPs with immobilized thiolated oligonucleotide probes. After addition of a fully complementary target, the spectrum red-shifts by 35 nm (blue trace (b)), whereas no appreciable spectrum shift is observed in the presence of a noncomplementary target (black trace (c)).

Usually a "sandwich" format is used for this purpose, in which the analyte is trapped between two components: a surface-bound receptor and a second detecting agent, which is linked to an AuNP (in fact, AuNPs are functionalized with this secondary detecting agent). These types of assemblies are presented in Figure 9 [82]. The introduction of the AuNP complex near the analyte causes much larger optical shifts in the system than when the analyte is added alone. The work of Natan's group during the late 1990s describes and proves the efficiency of this method. Lyon et al. [91] presented a method in which a gold film-immobilized antibody is bound to an AuNP–antibody complex via the antigen, as shown in Figure 9(A). Great signal amplification was observed, comparing the AuNP–antibody conjugate results with the addition of a free antibody solution (six orders of magnitude more concentrated), resulting in a 25-fold larger signal and allowing picomolar detection of human IgG by this type of particle enhancement. Also, Lyon et al. [91] reported a quasilinear relationship between particle coverage and plasmon angle shift, pointing to a direct correlation between plasmon shift and antigen concentration.

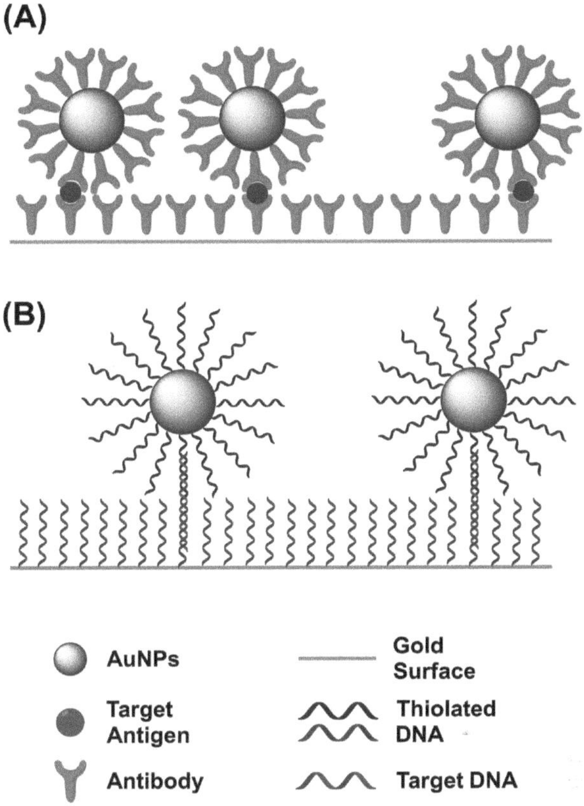

FIGURE 9 (A, B) Gold nanoparticle-enhanced surface plasmon resonance sensor for immuno-based detection (A) and DNA detection (B) [91,92].

He et al. [92] described a method for oligonucleotide detection based on DNA hybridization in conjunction with AuNP tagging. A direct 10-fold increase in angle shift was achieved, corresponding to a more than 1000-fold improvement in sensitivity for the target oligonucleotide compared with the unamplified binding event. This advance results in a limit of quantification of ~10 pM for 24-mer oligonucleotides. This type of SPR sensor assembly is presented in Figure 9(B), with the target oligonucleotide acting as a linker between the smaller oligonucleotide attached to the sensor surface and the AuNP–oligonucleotide probe.

6.3 Sensors Based on AuNPs Plasmon Resonance Scattering

AuNP plasmon resonance scattering is another distinctive property that leads to another role of AuNPs in analytical chemistry—namely, from the perspective of imaging techniques [83]. El Sayed et al. [93] were able to obtain SPR light-scattering images, which reveal that AuNPs can be useful as biosensors in living cells—namely, with application in (oral) cancer diagnostics. In this work, AuNPs were conjugated to anti-EGFR monoclonal antibodies and incubated with malignant and nonmalignant cell lines. AuNP biosensors bind to cancerous cells with 6 × larger affinity than to normal ones, showing a sharper and red-shifted SPR absorption band, that allows a clear distinction between both types of cells observed in light-scattering images.

Outside of the imaging techniques field and the immunological approach, scattering phenomena inspired Du et al. [94] to create a (linear) light-scattering assay for DNA hybridization in a homogeneous solution by using AuNPs as the labels of oligonucleotide probes. With this assay, target DNA can be detected at a concentration as low as 0.1 pm (no washing steps are needed), with the possibility of discrimination between perfectly matched target DNA and targets with single-base pair mismatches. These conclusions resulted from a study of a 30-base target DNA fragment of the human p53 gene (exon 8) using two distinct 15-base probes, complementary to two different ends of the target DNA, with both probes presenting a AuNP tail.

7. SURFACE-ENHANCED RAMAN SCATTERING-BASED SENSING

Raman spectroscopy is vibrational spectroscopy based on the inelastic scattering of incident radiation by the analyte. Because of the absence of water interference, and taking into consideration the low number of Raman-active symmetrical molecule vibrational modes, Raman spectroscopy is rather appropriate for biomolecule detection. Special care should be taken to avoid fluorescence effects. In fact, the fluorescence phenomenon is orders of magnitude more intense than Raman, masking Raman signals. Fluorescence can be avoided by careful choice of the Raman-active species and by selecting an appropriate laser line. The sensitivity and selectivity of Raman spectroscopy can be especially enhanced by coupling radiation scattering with an electronic transition, producing a resonance Raman signal. Another sensitivity enhancement possibility is surface-enhanced Raman spectroscopy (SERS), which allows detection of analytes at extremely low concentrations.

In SERS, the Raman signal of the probe molecule is amplified through excitation of LSPR of the substrate, usually in the form of metal particles or a roughened metal film. According to this "plasmonic theory" of SERS, the optical signals observed originate primarily from the metal and not the molecules [95]. Namely, the enhancement is particularly strong at hot spots, often nanoscale crevices and interstices in or between nanostructures. In recent years, significant improvements in nanofabrication techniques paved the way for the controlled design of reliable and effective SERS nanostructures. These SERS-active surfaces have been the theme of excellent reviews [96,97] and are not dealt with further here, because our focus is on colloidal AuNPs. Great efforts have also been devoted to develop new nanoparticle assembly strategies in suspension with improved control over hot spot architecture and cluster structure [98].

The fundamental metrics for SERS activity is the enhancement factor (EF), which quantifies the increase in signal intensity (counts/s/mW) per molecule. Commonly, EFs range from 10^3 to 10^7, with a few highly enhancing substrates achieving EFs greater than 10^8 [99].

7.1 SERS Based on Spherical AuNPs in Solution

The use of AuNPs in label development has additional advantages. First, gold is a strongly enhancing surface at longer wavelength excitation, and therefore is less susceptible to interference by native fluorescence. Typically, 633-nm (visible) or 785-nm (infrared) lasers are used for the excitation of AuNP assemblies, whereas shorter wavelengths are the rule for the more common silver colloids. Second, each nanoparticle is coated with a large number of Raman reporter molecules (10^3–10^4). As such, the response of an individual binding event is amplified markedly. Third, gold surfaces are readily modified by thiols and disulfides, a chemistry that provides a versatile platform for label construction.

A large number of SERS-based detection schemes based on AuNPs rely on hot spots that are generated on AuNP aggregation induced by the presence of the analyte. Because AuNPs carry Raman-active dyes, the enhanced signal obtained can be correlated to analyte concentration. This is the case for a system that uses 4-mercaptopyridine-modified AuNPs, which aggregate in solution by increasing melamine concentration (Figure 10). This simple melamine sensor in solution allowed an LOD as low as 0.1 ppb of melamine, with an excellent linear detection range of 0.5–100 ppb [100].

Wang et al. [101] developed a single-step SERS immunoassay based on plasmonic coupling enhancement via a sandwiched antibody–antigen assembly. The SERS nanoparticles were AuNPs and gold nanorods decorated with antibody half-fragments and Raman-active dyes. On target antigen binding, AuNPs aggregate and generate a SERS signal. This antigen-controlled assembly of SERS AuNPs is realized with a strong SERS signal achieved via plasmonic coupling, creating an immunoassay platform for rapid, sensitive, multiplexed quantification of proteins.

Ruan et al. [102] reported the development of a AuNP-based SERS assay that allows rapid and ultrasensitive analysis of alkaline phosphatase (ALP) at concentrations as low as 4×10^{-15} M. They demonstrated that 5-bromo-4-chloro-3-indolyl phosphate (BCIP), a substrate of ALP, becomes highly SERS active on exposure to an active alkaline phosphatase enzyme. In fact, the enzymatic hydrolysis of BCIP led to the formation of indigo dye derivatives, which were found to be highly SERS active.

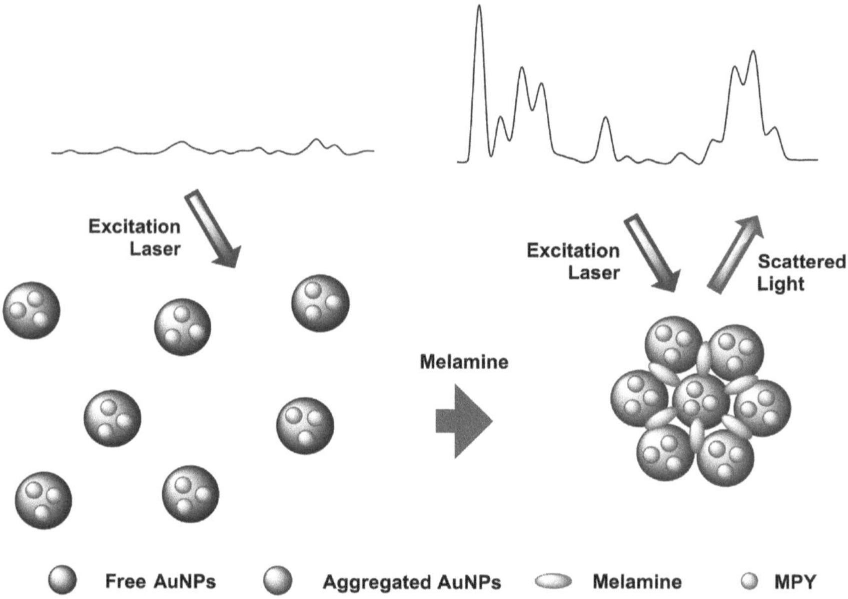

FIGURE 10 Schematic diagram of the indirect surface-enhanced Raman spectroscopy (SERS) method of measuring melamine using 4-mercaptopyridine (MPY)-modified gold nanoparticles (AuNPs). *Reproduced from Ref. [100] with kind permission from Springer Science and Business Media.*

One of the most interesting recent application of SERS induced by AuNPs aggregation is for monitoring changes in local pH of the cellular compartments of living NIH/3T3 cells [103]. Kneipp et al. [103] used a mobile SERS nanosensor made from gold nanoaggregates, and 4-mercaptobenzoic acid (pMBA) attached as a reporter, to follow the dynamics of local pH in individual live cells at subendosomal resolution in a timeline of cellular processes.

Jung et al. [104] also used AuNP aggregation inside cells to elicit a SERS signal selectively at cancer cells. This theranostic approach allowed for simultaneous Raman imaging/diagnosis and photothermal therapy. Spherical 10 nm AuNPs were functionalized with pH-responsive ligands and presented Raman probes on the surface. AuNP aggregation provides hot spots for the SERS probe, with the EF reaching 1.3×10^4, and shifts the absorption to the infrared region, optimizing deep-tissue penetration for excitation provided by a 785-nm laser. This shift was exploited successfully for low-threshold photothermal therapy (Figure 11).

A AuNP-based SERS immunoassay has been developed for the detection of two waterborne pathogens—*Cryptosporidium parvum* and *Giardia lambia* [105]—coupled with x−y Raman mapping. Ideal Raman labels were optimized by conjugating 40 nm AuNPs with rhodamine B isothiocyanate and monoclonal IgM antibodies, for *Cryptosporidium parvum*, and malachite green isothiocyanate and monoclonal IgG1 antibodies, for *Giardia lambia*

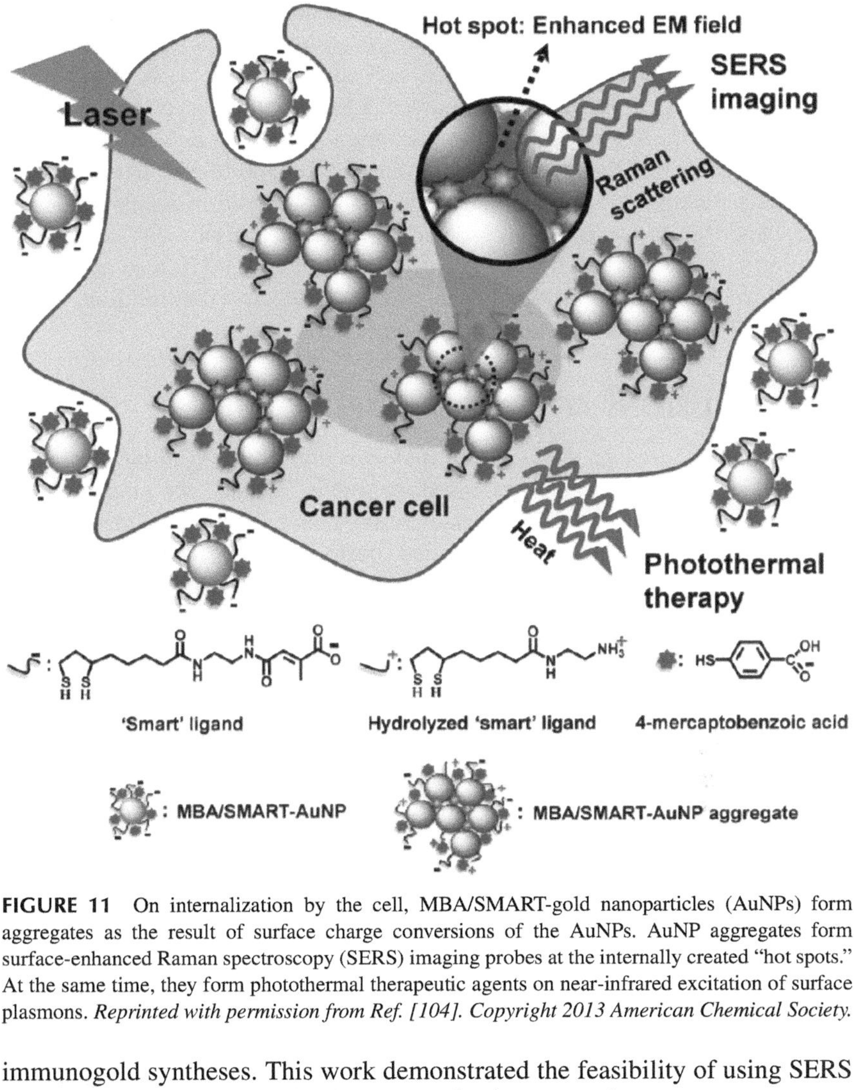

FIGURE 11 On internalization by the cell, MBA/SMART-gold nanoparticles (AuNPs) form aggregates as the result of surface charge conversions of the AuNPs. AuNP aggregates form surface-enhanced Raman spectroscopy (SERS) imaging probes at the internally created "hot spots." At the same time, they form photothermal therapeutic agents on near-infrared excitation of surface plasmons. *Reprinted with permission from Ref. [104]. Copyright 2013 American Chemical Society.*

immunogold syntheses. This work demonstrated the feasibility of using SERS labeling for sensitive multipathogen monitoring strategies.

7.2 SERS Based on Anisotropic AuNPs in Solution

In recent years, AuNPs with novel morphologies and consequently new plasmonic properties became available. These new plasmonic properties can include the necessary hot spots for sensitive SERS detection. The presence of these hot spots in the individual NPs circumvents the necessity for NP aggregation or its deposition onto nanostructured surfaces. Ideally, a single nanoconstruct can be used both for capture and detection of biomolecules.

Pereira's group explored star-shaped, gold-coated magnetic NPs functionalized with a nickel(II)-nitrilotriacetate group to capture selectively a histidine-tagged protein from a crude cell extract [106]. Labeling with a Raman-active dye (Astra Blue) allowed SERS detection of these nanoconstructs [106]. A similar approach coupled with online SERS detection was undertaken by Bhana et al. [107] that could capture and detect cancer cells magnetically in whole blood under a flow condition using oval iron oxide–AuNPs. An online SERS detection was then used with a limit of detection of 1–2 cells/mL blood. Bhana et al. [107] used no Raman-active dye labeling, relying instead on the intensity of the 1496/cm vibration line, which increased linearly with the number of cancer cells present.

8. CONCLUSIONS AND FUTURE PERSPECTIVES

This chapter focused on some of the major recent advances in biosensing of chemical and biological analytes based on AuNPs. In fact, the unique chemical, optical, and electrical properties of AuNPs have led to their intensive use as biochemical sensors with enhanced performance. Properties that are explored intensely in biosensing devices include tunable SPRs, distance-dependent fluorescence quenching or enhancement, high electrical conductivity, and exceptional light-scattering properties.

Colorimetric detection is especially interesting from the application point of view, as this approach minimizes costs associated with equipment. Furthermore, colorimetric sensors combine cutting-edge analytical tools with nanotechnology endowed with high sensitivity and selectivity. With this type of detection, the highly dispersive properties of silver NPs are used advantageously in silver enhancement of AuNP-based detection. The biobarcode detection method deserved a full section in this chapter. In fact, despite its rather complex workout, this technique has allowed extremely sensitive multiplex detection of various antigens and DNA targets.

The use of photoluminescence-based biosensors has allowed extremely low LODs and easy multiplexing associated with different fluorophores. AuNPs have pushed the detection limit of these systems further, thanks to interesting fluorescence quenching/enhancing properties. In fact, AuNPs can act as fluorophore enhancers or quenchers, participate in FRET-based detection systems, and be active in different detection geometries, including the use of QDs as fluorophores, potentiating the advantages of both nanosystems.

Applications of the electrical properties of AuNP represent by far the most abundant and diverse applications of AuNPs in biosensing. Although sensitivity of detection is important, this technique is more relevant for the construction of flexible biosensors, detecting a plethora of different analytes. QCM applications are also described as an emerging sensing technique that makes use of the high density of AuNPs for piezoelectric-based detection.

Future trends in nanobiosensing will rely strongly on plasmonic-based sensors because they are extremely flexible platforms that can be adapted easily for point-of-care devices as a result of ongoing developments in micro- and nanoelectronics. These methodologies, discussed in Sections 6 and 7, rely on detection at the surface of AuNPs, avoiding the need for AuNP aggregation or other more or less complicated detection geometries. High sensitivity and selectivity can also be achieved with plasmonic detection using, for instance, anisotropic AuNPs, which allow detection down to a single molecule of the analyte.

Nanobiosensors have yet to fulfill their potential in terms of sensitivity and limits of detection, and strict validation is still required for most of these novel platforms before they can exert their full impact in the field. Nevertheless, nanobiosensors have gone where traditional techniques and systems have not been before in terms of sensitivity and specificity, all while being made portable and ready to use at the point of care or in fully automated platforms. Understanding the impact of these newly achieved LODs and their translation into screening and diagnostics potential, still requires comprehensive investigation. The future of nanobiosensors based on AuNPs will reshape the way we perceive diagnostics and monitoring, doing it at reduced costs and providing for widespread application of personalized medicine, localized environmental control, and personal food stuff control.

LIST OF ABBREVIATIONS

AuNP Gold nanoparticle
ELISA Enzyme-linked immunosorbent assay
FRET Förster resonance energy transfer
HRP Horseradish peroxidase
LOD Limit of detection
(L)SPR (Localized) surface plasmon resonance
NP Nanoparticle
PCR Polymerase chain reaction
POC Point-of-care
QCM Quartz crystal microbalance
QD Quantum dot
SERS Surface enhanced Raman scattering/spectroscopy

ACKNOWLEDGMENTS

The authors acknowledge Fundação para a Ciência e a Tecnologia (FCT-MEC) through grants COMPETE PTDC/CVT/110732/2009 (to M.P.A. and E.P.), PTDC/CTM/NAN/112241/2009 (to R.F.), PTDC/BBB-NAN/1812/2012 (to P.B.), PEst-C/EQB/LA0006/2011 & 2013 (to Associate Lab REQUIMTE), PEst-OE/SAU/UI0009/2011 & 2013 (to CIGMH), postdoctoral fellowship SFRH/BPD/63850/2009 (to I.G.) and PhD fellowship SFRH/BDE/51100/2010 (to L.S.), and a joint venture with STAB-Vida Lda. The LungCard Project (FP7-SME action) is acknowledged for a postdoctoral fellowship to S.F.

REFERENCES

[1] P.V. Baptista, G. Doria, P. Quaresma, M. Cavadas, C.S. Neves, I. Gomes, P. Eaton, E. Pereira, R. Franco, Chapter 11-Nanoparticles in molecular diagnostics, in: V. Antonio (Ed.), Progress in Molecular Biology and Translational Science, vol. 104, Academic Press, 2011, pp. 427–488.

[2] K. Saha, S.S. Agasti, C. Kim, X. Li, V.M. Rotello, Gold nanoparticles in chemical and biological sensing, Chem. Rev. 112 (2012) 2739–2779.

[3] R. Wilson, The use of gold nanoparticles in diagnostics and detection, Chem. Soc. Rev. 37 (2008) 2028–2045.

[4] C.V. Sapan, R.L. Lundblad, N.C. Price, Colorimetric protein assay techniques, Biotechnol. Appl. Biochem. 29 (1999) 99–108.

[5] P. Baptista, E. Pereira, P. Eaton, G. Doria, A. Miranda, I. Gomes, P. Quaresma, R. Franco, Gold NPs for the development of clinical diagnosis methods, Anal. Bioanal. Chem. 391 (2008) 943–950.

[6] Y.-W. Lin, C.-C. Huang, H.-T. Chang, Gold nanoparticle probes for the detection of mercury, lead and copper ions, Analyst 136 (2011) 863–871.

[7] D. Liu, Z. Wang, X. Jiang, Gold NPs for the colorimetric and fluorescent detection of ions and small organic molecules, Nanoscale 3 (2011) 1421–1433.

[8] J.-S. Lee, M.S. Han, C.A. Mirkin, Colorimetric detection of mercuric ion (Hg^{2+}) in aqueous Media using DNA-functionalized gold nanoparticles, Angew. Chem. Int. Ed. 46 (2007) 4093–4096.

[9] X. Xue, F. Wang, X. Liu, One-step, room temperature, colorimetric detection of mercury (Hg^{2+}) using DNA/Nanoparticle conjugates, J. Am. Chem. Soc. 130 (2008) 3244–3245.

[10] J.-S. Lee, C.A. Mirkin, Chip-based scanometric detection of mercuric ion using DNA-functionalized gold nanoparticles, Anal. Chem. 80 (2008) 6805–6808.

[11] J. Liu, Y. Lu, A colorimetric lead biosensor using DNAzyme-directed assembly of gold nanoparticles, J. Am. Chem. Soc. 125 (2003) 6642–6643.

[12] C.E. Lisowski, J.E. Hutchison, Malonamide-functionalized gold nanoparticles for selective, colorimetric sensing of trivalent lanthanide ions, Anal. Chem. 81 (2009) 10246–10253.

[13] C.-C. Huang, H.-T. Chang, Parameters for selective colorimetric sensing of mercury(II) in aqueous solutions using mercaptopropionic acid-modified gold nanoparticles, Chem. Commun. (2007) 1215–1217.

[14] A.R. Ferhan, L. Guo, X. Zhou, P. Chen, S. Hong, D.-H. Kim, Solid-phase colorimetric sensor based on gold nanoparticle-loaded polymer brushes: lead detection as a case study, Anal. Chem. 85 (2013) 4094–4099.

[15] X. Zhuang, D. Wang, L. Yang, P. Yu, W. Jiang, L. Mao, Cysteine-modulated colorimetric sensing of extracellular Mg^{2+} in rat brain based on the strong chelation interaction between dithiothreitol and Mg^{2+}, Analyst 138 (2013) 3046–3052.

[16] G. He, L. Zhao, K. Chen, Y. Liu, H. Zhu, Highly selective and sensitive gold nanoparticle-based colorimetric assay for PO_4^{3-} in aqueous solution, Talanta 106 (2013) 73–78.

[17] J.H. Lee, Z. Wang, J. Liu, Y. Lu, Highly sensitive and selective colorimetric sensors for Uranyl (UO_2^{2+}): development and comparison of labeled and label-free DNAzyme-gold nanoparticle systems, J. Am. Chem. Soc. 130 (2008) 14217–14226.

[18] K.-M. Song, S. Lee, C. Ban, Aptamers and their biological applications, Sensors 12 (2012) 612–631.

[19] C.A. Mirkin, R.L. Letsinger, R.C. Mucic, J.J. Storhoff, A DNA-based method for rationally assembling nanoparticles into macroscopic materials, Nature 382 (1996) 607–609.

[20] R. Elghanian, J.J. Storhoff, R.C. Mucic, R.L. Letsinger, C.A. Mirkin, Selective colorimetric detection of polynucleotides based on the distance-dependent optical properties of gold nanoparticles, Science 277 (1997) 1078–1081.

[21] M. Larguinho, P.V. Baptista, Gold and silver nanoparticles for clinical diagnostics—from genomics to proteomics, J. Proteomics 75 (2012) 2811–2823.

[22] P.V. Baptista, M. Koziol-Montewka, J. Paluch-Oles, G. Doria, R. Franco, Gold-nanoparticle-probe-based assay for rapid and direct detection of *Mycobacterium tuberculosis* DNA in clinical samples, Clin. Chem. 52 (2006) 1433–1434.

[23] J. Conde, J. de la Fuente, P. Baptista, RNA quantification using gold nanoprobes -application to cancer diagnostics, J. Nanobiotechnol. 8 (2010) 5.

[24] G. Doria, R. Franco, P. Baptista, Nanodiagnostics: fast colorimetric method for single nucleotide polymorphism/mutation detection, IET Nanobiotechnol. 1 (2007) 53–57.

[25] S.K. Kim, H. Cho, J. Jeong, J.N. Kwon, Y. Jung, B.H. Chung, Label-free and naked eye detection of PNA/DNA hybridization using enhancement of gold nanoparticles, Chem. Commun. 46 (2010) 3315–3317.

[26] B.-H. Kim, I.S. Yoon, J.-S. Lee, Masking nanoparticle surfaces for sensitive and selective colorimetric detection of proteins, Anal. Chem. 85 (2013) 10542–10548.

[27] Z. Wang, R. Lévy, D.G. Fernig, M. Brust, Kinase-catalyzed modification of gold nanoparticles: a new approach to colorimetric kinase activity screening, J. Am. Chem. Soc. 128 (2006) 2214–2215.

[28] J. Zhou, X. Xu, X. Liu, H. Li, Z. Nie, M. Qing, Y. Huang, S. Yao, A gold nanoparticles colorimetric assay for label-free detection of protein kinase activity based on phosphorylation protection against exopeptidase cleavage, Biosens. Bioelectron. 53 (2014) 295–300.

[29] J.-M. Nam, C.S. Thaxton, C.A. Mirkin, Nanoparticle-based bio-bar codes for the ultrasensitive detection of proteins, Science 301 (2003) 1884–1886.

[30] J.-M. Nam, S.I. Stoeva, C.A. Mirkin, Bio-bar-code-based DNA detection with PCR-like sensitivity, J. Am. Chem. Soc. 126 (2004) 5932–5933.

[31] J.-M. Nam, A.R. Wise, J.T. Groves, Colorimetric bio-barcode amplification assay for cytokines, Anal Chem. 77 (2005) 6985–6988.

[32] N.M. Adams, S.R. Jackson, F.R. Haselton, D.W. Wright, Design, synthesis, and characterization of nucleic-acid-functionalized gold surfaces for biomarker detection, Langmuir 28 (2011) 1068–1082.

[33] T.A. Taton, C.A. Mirkin, R.L. Letsinger, Scanometric DNA array detection with nanoparticle probes, Science 289 (2000) 1757–1760.

[34] G. Doria, J. Conde, B. Veigas, L. Giestas, C. Almeida, M. Assunção, J. Rosa, P.V. Baptista, Noble metal nanoparticles for biosensing applications, Sensors 12 (2012) 1657–1687.

[35] P. Ray, G. Darbha, A. Ray, J. Walker, W. Hardy, Gold nanoparticle based FRET for DNA detection, Plasmonics 2 (2007) 173–183.

[36] J.P. Rosa, J.C. Lima, P.V. Baptista, Experimental photophysical characterization of fluorophores in the vicinity of gold nanoparticles, Nanotechnology 22 (2011) 415202.

[37] P.C. Ray, A. Fortner, G.K. Darbha, Gold nanoparticle based FRET asssay for the detection of DNA cleavage, J. Phys. Chem. B 110 (2006) 20745–20748.

[38] B.S. Guirgis, C. Sá e Cunha, I. Gomes, M. Cavadas, I. Silva, G. Doria, G. Blatch, P. Baptista, E. Pereira, H.E. Azzazy, M. Mota, M. Prudêncio, R. Franco, Gold nanoparticle-based fluorescence immunoassay for malaria antigen detection, Anal. Bioanal. Chem. 402 (2012) 1019–1027.

[39] R. Leggett, E.E. Lee-Smith, S.M. Jickells, D.A. Russell, "Intelligent" fingerprinting: simultaneous identification of drug metabolites and individuals by using antibody-functionalized nanoparticles, Angew. Chem. Int. Ed. 46 (2007) 4100–4103.

[40] D.J. Maxwell, J.R. Taylor, S. Nie, Self-assembled nanoparticle probes for recognition and detection of biomolecules, J. Am. Chem. Soc. 124 (2002) 9606–9612.

[41] J. Peng, Y. Wang, J. Wang, X. Zhou, Z. Liu, A new biosensor for glucose determination in serum based on up-converting fluorescence resonance energy transfer, Biosens, Bioelectron. 28 (2011) 414–420.

[42] W. Wang, C. Chen, M. Qian, X.S. Zhao, Aptamer biosensor for protein detection using gold nanoparticles, Anal. Biochem. 373 (2008) 213–219.

[43] L. Ao, F. Gao, B. Pan, R. He, D. Cui, Fluoroimmunoassay for antigen based on fluorescence quenching signal of gold nanoparticles, Anal. Chem. 78 (2006) 1104–1106.

[44] J.H. Kim, S.H. Han, B.H. Chung, Improving Pb^{2+} detection using DNAzyme-based fluorescence sensors by pairing fluorescence donors with gold nanoparticles, Biosens. Bioelectron. 26 (2011) 2125–2129.

[45] V. Beni, K. Hayes, T.M. Lerga, C.K. O'Sullivan, Development of a gold nano-particle-based fluorescent molecular beacon for detection of cystic fibrosis associated mutation, Biosens. Bioelectron. 26 (2010) 307–313.

[46] J. Rosa, J. Conde, J.M. de la Fuente, J.C. Lima, P.V. Baptista, Gold-nanobeacons for real-time monitoring of RNA synthesis, Biosens. Bioelectron. 36 (2012) 161–167.

[47] A.E. Prigodich, D.S. Seferos, M.D. Massich, D.A. Giljohann, B.C. Lane, C.A. Mirkin, Nano-flares for mRNA regulation and detection, ACS Nano 3 (2009) 2147–2152.

[48] S.H. Ko, K. Du, J.A. Liddle, Quantum-dot fluorescence lifetime engineering with DNA origami constructs, Angew. Chem. Int. Ed. 52 (2013) 1193–1197.

[49] Q. Wang, H. Wang, C. Lin, J. Sharma, S. Zou, Y. Liu, Photonic interaction between quantum dots and gold nanoparticles in discrete nanostructures through DNA directed self-assembly, Chem. Commun. 46 (2010) 240–242.

[50] Y.-P. Kim, S. Park, E. Oh, Y.-H. Oh, H.-S. Kim, On-chip detection of protein glycosylation based on energy transfer between nanoparticles, Biosens. Bioelectron. 24 (2009) 1189–1194.

[51] Y.-P. Kim, Y.-H. Oh, E. Oh, S. Ko, M.-K. Han, H.-S. Kim, Energy transfer-based multiplexed assay of proteases by using gold nanoparticle and quantum dot conjugates on a surface, Anal. Chem. 80 (2008) 4634–4641.

[52] V.K.K. Upadhyayula, Functionalized gold nanoparticle supported sensory mechanisms applied in detection of chemical and biological threat agents: a review, Anal. Chim. Acta 715 (2012) 1–18.

[53] M.T. Castañeda, S. Alegret, A. Merkoçi, Electrochemical sensing of DNA using gold nanoparticles, Electroanalysis 19 (2007) 743–753.

[54] F. Lu, T.L. Doane, J.-J. Zhu, C. Burda, Gold NPs for diagnostic sensing and therapy, Inorg. Chim. Acta 393 (2012) 142–153.

[55] M. Ozsoz, A. Erdem, K. Kerman, D. Ozkan, B. Tugrul, N. Topcuoglu, H. Ekren, M. Taylan, Electrochemical genosensor based on colloidal gold nanoparticles for the detection of factor V Leiden mutation using disposable pencil graphite electrodes, Anal. Chem. 75 (2003) 2181–2187.

[56] M.T. Castañeda, A. Merkoçi, M. Pumera, S. Alegret, Electrochemical genosensors for biomedical applications based on gold nanoparticles, Biosens. Bioelectron. 22 (2007) 1961–1967.

[57] S.-J. Park, T.A. Taton, C.A. Mirkin, Array-based electrical detection of DNA with nano-particle probes, Science 295 (2002) 1503–1506.

[58] J. Wang, R. Polsky, D. Xu, Silver-enhanced colloidal gold electrochemical stripping detection of DNA hybridization, Langmuir 17 (2001) 5739–5741.
[59] M. Rochelet-Dequaire, B. Limoges, P. Brossier, Subfemtomolar electrochemical detection of target DNA by catalytic enlargement of the hybridized gold nanoparticle labels, Analyst 131 (2006) 923–929.
[60] A.-N. Kawde, J. Wang, Amplified electrical transduction of DNA hybridization based on polymeric beads loaded with multiple gold nanoparticle tags, Electroanalysis 16 (2004) 101–107.
[61] J. Wang, J. Li, A.J. Baca, J. Hu, F. Zhou, W. Yan, D.-W. Pang, Amplified voltammetric detection of DNA hybridization via oxidation of ferrocene caps on gold nanoparticle/streptavidin conjugates, Anal. Chem. 75 (2003) 3941–3945.
[62] R. Cui, C. Liu, J. Shen, D. Gao, J.-J. Zhu, H.-Y. Chen, Gold Nanoparticle–Colloidal carbon nanosphere hybrid material: preparation, characterization, and application for an amplified electrochemical immunoassay, Adv. Funct. Mater. 18 (2008) 2197–2204.
[63] X. Mao, M. Baloda, A.S. Gurung, Y. Lin, G. Liu, Multiplex electrochemical immunoassay using gold nanoparticle probes and immunochromatographic strips, Electrochem. Commun. 10 (2008) 1636–1640.
[64] J.-a.A. Ho, H.-C. Chang, N.-Y. Shih, L.-C. Wu, Y.-F. Chang, C.-C. Chen, C. Chou, Diagnostic detection of human lung cancer-associated antigen using a gold nanoparticle-based electrochemical immunosensor, Anal. Chem. 82 (2010) 5944–5950.
[65] D. Tang, R. Yuan, Y. Chai, Ultrasensitive electrochemical immunosensor for clinical immunoassay using thionine-doped magnetic gold nanospheres as labels and horseradish peroxidase as enhancer, Anal. Chem. 80 (2008) 1582–1588.
[66] Y. Zhang, H. Wang, J. Nie, Y. Zhang, G. Shen, R. Yu, Individually addressable microelectrode arrays fabricated with gold-coated pencil graphite particles for multiplexed and high sensitive impedance immunoassays, Biosens. Bioelectron. 25 (2009) 34–40.
[67] H. Liu, Y. Yang, P. Chen, Z. Zhong, Enhanced conductometric immunoassay for hepatitis B surface antigen using double-codified nanogold particles as labels, Biochem. Eng. J. 45 (2009) 107–112.
[68] G.-F. Jie, P. Liu, S.-S. Zhang, Highly enhanced electrochemiluminescence of novel gold/silica/CdSe-CdS nanostructures for ultrasensitive immunoassay of protein tumor marker, Chem. Commun. 46 (2010) 1323–1325.
[69] Z.-P. Li, Y.-C. Wang, C.-H. Liu, Y.-K. Li, Development of chemiluminescence detection of gold nanoparticles in biological conjugates for immunoassay, Anal. Chim. Acta 551 (2005) 85–91.
[70] C. Leng, G. Lai, F. Yan, H. Ju, Gold nanoparticle as an electrochemical label for inherently crosstalk-free multiplexed immunoassay on a disposable chip, Anal. Chim. Acta 666 (2010) 97–101.
[71] M. Dequaire, C. Degrand, B. Limoges, An electrochemical metalloimmunoassay based on a colloidal gold label, Anal. Chem. 72 (2000) 5521–5528.
[72] A. Fan, C. Lau, J. Lu, Magnetic bead-based chemiluminescent metal immunoassay with a colloidal gold label, Anal. Chem. 77 (2005) 3238–3242.
[73] D. Hu, H. Han, R. Zhou, F. Dong, W. Bei, F. Jia, H. Chen, Gold(III) enhanced chemiluminescence immunoassay for detection of antibody against ApxIV of *Actinobacillus pleuropneumoniae*, Analyst 133 (2008) 768–773.
[74] Z. Wang, J. Hu, Y. Jin, X. Yao, J. Li, In situ amplified chemiluminescent detection of DNA and immunoassay of IgG using special-shaped gold nanoparticles as label, Clin. Chem. 52 (2006) 1958–1961.

[75] L. Lin, H.-q. Zhao, J.-r. Li, J.-a. Tang, M.-x. Duan, L. Jiang, Study on colloidal Au-enhanced DNA sensing by quartz Crystal microbalance, Biochem. Biophys. Res. Commun. 274 (2000) 817–820.

[76] X.C. Zhou, S.J. O'Shea, S.F.Y. Li, Amplified microgravimetric gene sensor using Au nanoparticle modified oligonucleotides, Chem. Commun. (2000) 953–954.

[77] B. Zheng, S. Cheng, W. Liu, M.H.-W. Lam, H. Liang, Small organic molecules detection based on aptamer-modified gold nanoparticles-enhanced quartz crystal microbalance with dissipation biosensor, Anal. Biochem. 438 (2013) 144–149.

[78] N. Kim, D.K. Kim, Y.J. Cho, Gold nanoparticle-based signal augmentation of quartz crystal microbalance immunosensor measuring C-reactive protein, Curr. Appl. Phys. 10 (2010) 1227–1230.

[79] P. Ding, R. Liu, S. Liu, X. Mao, R. Hu, G. Li, Reusable gold nanoparticle enhanced QCM immunosensor for detecting C-reactive protein, Sens. Actuators, B 188 (2013) 1277–1283.

[80] Y.-K. Lyu, K.-R. Lim, B.Y. Lee, K.S. Kim, W.-Y. Lee, Microgravimetric lectin biosensor based on signal amplification using carbohydrate-stabilized gold nanoparticles, Chem. Commun. (2008) 4771–4773.

[81] Q. Chen, W. Tang, D. Wang, X. Wu, N. Li, F. Liu, Amplified QCM-D biosensor for protein based on aptamer-functionalized gold nanoparticles, Biosens. Bioelectron. 26 (2010) 575–579.

[82] E.E. Bedford, J. Spadavecchia, C.-M. Pradier, F.X. Gu, Surface plasmon resonance biosensors incorporating gold nanoparticles, Macromol. Biosci. 12 (2012) 724–739.

[83] S. Eustis, M.A. El-Sayed, Why gold nanoparticles are more precious than pretty gold: Noble metal surface plasmon resonance and its enhancement of the radiative and nonradiative properties of nanocrystals of different shapes, Chem. Soc. Rev. 35 (2006) 209–217.

[84] K. Jia, J.L. Bijeon, P.M. Adam, R.E. Ionescu, Sensitive localized surface plasmon resonance multiplexing protocols, Anal. Chem. 84 (2012) 8020–8027.

[85] R.E. Messersmith, G.J. Nusz, S.M. Reed, Using the localized surface plasmon resonance of gold nanoparticles to monitor lipid membrane assembly and protein binding, J. Phys. Chem. C 117 (2013) 26725–26733.

[86] X. Liu, Q. Zhang, Y. Tu, W. Zhao, H. Gai, Single gold nanoparticle localized surface plasmon resonance spectral imaging for quantifying binding constant of carbohydrate–protein interaction, Anal. Chem. 85 (2013) 11851–11857.

[87] T. Schneider, N. Jahr, J. Jatschka, A. Csaki, O. Stranik, W. Fritzsche, Localized surface plasmon resonance (LSPR) study of DNA hybridization at single nanoparticle transducers, J. Nanopart Res. 15 (2013) 1531.

[88] K.-M. Lo, C.-Y. Lai, H.-M. Chan, D.-L. Ma, H.-W. Li, Monitoring of DNA–protein interaction with single gold nanoparticles by localized scattering plasmon resonance spectroscopy, Methods 64 (2013) 331–337.

[89] D.-K. Kim, T.J. Park, E. Tamiya, S.Y. Lee, Label-free detection of leptin antibody-antigen interaction by using LSPR-based optical biosensor, J. Nanosci. Nanotechnol. 11 (2011) 4188–4193.

[90] M.M. Kathryn, H. Feng, L. Seunghyun, N. Peter, H.H. Jason, A single molecule immunoassay by localized surface plasmon resonance, Nanotechnology 21 (2010) 255503.

[91] L.A. Lyon, M.D. Musick, M.J. Natan, Colloidal Au-enhanced surface plasmon resonance immunosensing, Anal. Chem. 70 (1998) 5177–5183.

[92] L. He, M.D. Musick, S.R. Nicewarner, F.G. Salinas, S.J. Benkovic, M.J. Natan, C.D. Keating, Colloidal Au-enhanced surface plasmon resonance for ultrasensitive detection of DNA hybridization, J. Am. Chem. Soc. 122 (2000) 9071–9077.

[93] I.H. El-Sayed, X. Huang, M.A. El-Sayed, Surface plasmon resonance scattering and absorption of anti-EGFR antibody conjugated gold nanoparticles in cancer diagnostics: applications in oral cancer, Nano Lett. 5 (2005) 829–834.
[94] B.-A. Du, Z.-P. Li, C.-H. Liu, One-step homogeneous detection of DNA hybridization with gold nanoparticle probes by using a linear light-scattering technique, Angew. Chem. Int. Ed. 45 (2006) 8022–8025.
[95] M. Moskovits, Persistent misconceptions regarding SERS, Phys. Chem. Chem. Phys. 15 (2013) 5301–5311.
[96] J.A. Dieringer, A.D. McFarland, N.C. Shah, D.A. Stuart, A.V. Whitney, C.R. Yonzon, M.A. Young, X. Zhang, R.P. Van Duyne, Introductory lecture surface enhanced Raman spectroscopy: new materials, concepts, characterization tools, and applications, Faraday Discuss. 132 (2006) 9–26.
[97] B. Sharma, M. Fernanda Cardinal, S.L. Kleinman, N.G. Greeneltch, R.R. Frontiera, M.G. Blaber, G.C. Schatz, R.P. Van Duyne, High-performance SERS substrates: advances and challenges, MRS Bull. 38 (2013) 615–624.
[98] L. Guerrini, D. Graham, Molecularly-mediated assemblies of plasmonic nanoparticles for surface-enhanced Raman spectroscopy applications, Chem. Soc. Rev. 41 (2012) 7085–7107.
[99] L. Rodríguez-Lorenzo, R.n.A. Álvarez-Puebla, I. Pastoriza-Santos, S. Mazzucco, O. Stéphan, M. Kociak, L.M. Liz-Marzán, F.J. García de Abajo, Zeptomol detection through controlled ultrasensitive surface-enhanced Raman scattering, J. Am. Chem. Soc. 131 (2009) 4616–4618.
[100] T. Lou, Y. Wang, J. Li, H. Peng, H. Xiong, L. Chen, Rapid detection of melamine with 4-mercaptopyridine-modified gold nanoparticles by surface-enhanced Raman scattering, Anal. Bioanal. Chem. 401 (2011) 333–338.
[101] Y. Wang, L.-J. Tang, J.-H. Jiang, Surface-enhanced raman spectroscopy-based, homogeneous, multiplexed immunoassay with antibody-fragments-decorated gold nanoparticles, Anal. Chem. 85 (2013) 9213–9220.
[102] C. Ruan, W. Wang, B. Gu, Detection of alkaline phosphatase using surface-enhanced Raman spectroscopy, Anal. Chem. 78 (2006) 3379–3384.
[103] J. Kneipp, H. Kneipp, B. Wittig, K. Kneipp, Following the dynamics of pH in endosomes of live cells with SERS nanosensors, J. Phys. Chem. C 114 (2010) 7421–7426.
[104] S. Jung, J. Nam, S. Hwang, J. Park, J. Hur, K. Im, N. Park, S. Kim, Theragnostic pH-sensitive gold nanoparticles for the selective surface enhanced Raman scattering and photothermal cancer therapy, Anal. Chem. 85 (2013) 7674–7681.
[105] K.L. Rule, P.J. Vikesland, Surface-enhanced resonance Raman spectroscopy for the rapid detection of *Cryptosporidium parvum* and *Giardia lamblia*, Environ. Sci. Technol. 43 (2009) 1147–1152.
[106] P. Quaresma, I. Osorio, G. Doria, P.A. Carvalho, A. Pereira, J. Langer, J.P. Araujo, I. Pastoriza-Santos, L.M. Liz-Marzan, R. Franco, P.V. Baptista, E. Pereira, Star-shaped magnetite@gold nanoparticles for protein magnetic separation and SERS detection, RSC Adv. 4 (2014) 3659–3667.
[107] S. Bhana, E. Chaffin, Y. Wang, S.R. Mishra, X. Huang, Capture and detection of cancer cells in whole blood with magnetic–optical nanoovals, Nanomedicine 9 (5) (2014) 593–606.

Chapter 14

Lateral Flow Biosensors Based on Gold Nanoparticles

Lourdes Rivas,[1,2] Alfredo de la Escosura-Muñiz,[1] Josefina Pons[2] and Arben Merkoçi[1,3,*]
[1] *Institut Catala de Nanociencia i Nanotecnologia (ICN2), Bellaterra (Barcelona), Spain;* [2] *Inorganic Chemistry Unit, Chemistry Department, Science Faculty, Autonomous University of Barcelona, Barcelona, Spain;* [3] *Institucio Catalana de Recerca i Estudis Avançats (ICREA), Barcelona, Spain*
**Corresponding author: E-mail: arben.merkoci@icn.cat*

Chapter Outline

1. GENERAL INTRODUCTION: HISTORICAL PERSPECTIVE

The concept of rapid body fluid tests is documented in the history since 1000 years ago. In ancient China, a saliva-based diagnostic called “the rice test” determined the guilt of an accused person [1]. The earliest urine-based

Gold Nanoparticles in Analytical Chemistry. http://dx.doi.org/10.1016/B978-0-444-63285-2.00014-6

tests for determining the pregnancy was documented on Egyptian papyrus of 1350−1200 BC [2,3]: if the urine of the woman was able to provoke (or not) the growth of wheat, then it was possible to give a pregnancy diagnosis. In the Middle Ages and Renaissance, pregnancy and some diseases could be diagnosed by the color of the urine. In the nineteen century, physicians still practiced the examination of urine in a more rational way including scientific approaches, but it was not until early 1900 that the term "hormones" was introduced by Ernest Starling [4]. After this study, scientists began to explore in detail a particular hormone that can be found only in pregnant woman, called human chorionic gonadotropin (hCG) [5,6]. In this context, many advances have been achieved until the development of devices for rapid diagnostic methods, with lateral flow (LF) devices being one of the most important.

The basic principles of the LF date to the mid 1950s, starting with the work of Plotz and Singer using the latex agglutination test [7]. At the same time, other applications in immunoassays with special importance for LF were developed. The radioimmunoassay (RIA) by Berson and Yalow [8], based on the detection of radioactive compounds, although being a very sensitive and specific technique due to the special equipment and precautions required, was replaced by a safer and nonhazardous technique such as the enzymatic immunoassay (EIA) [9,10]. In 1980, Leuvering et al. [11] reported the sol particle immunoassay (SPIA) in which silver and gold colloids were used as labels for sandwich immunoassays obtaining limits of detection comparable to RIA and superior in comparison to EIA. This work became pioneer in the use of nanoparticles as labels in immunoassays.

The main application for LF technology is the human pregnancy test developed in the mid 1970s. The "early pregnancy test" (e.p.t.™) developed by Warner-Chilcott was a complex system that involved the use of different tubes and reagents, and the result was obtained about 2 h making the determination of the end point difficult and ambiguous. Since the launching of this home-diagnostic method, lateral flow assay (LFA) gained more importance in the biosensors field and continued progressing becoming completely established in the late 1980s by the introduction of several patents related to LF devices for pregnancy test [12−14]. In 1988, Unipath Ltd. launched the first one-step rapid LF assay for home pregnancy test known as Clearblue™. This test was specifically qualitative: if two blue lines develop, the woman is pregnant and the result is visualized within few minutes. In a pregnant woman, levels of hCG rapidly rise and can be detected in urine from 9 to 10 days after conception [15] and Clearblue™ is able to detect hCG protein in urine. Although pregnancy tests have been conceived for providing qualitative visual results (yes/no response), it is possible to obtain semiquantitative results to estimate the pregnancy stage. The pregnancy test market has also evolved for generating new and more accurate devices, and nowadays it is possible to find a digital pregnancy test that gives semiquantitative results. This is a new generation of devices where LF technology is integrated with electronics in the same platform.

2. LF TECHNOLOGY AND ITS FORMAT ASSAYS

2.1 Operation Principles

The LFAs are based on chromatographic principle in which test sample flows through a solid substrate via capillary action. Each strip is composed of dry prestored reagents and different porous membranes that have specific functions.

Figure 1 shows a typical scheme of LF strip. As prerequisite for the proper functioning of LFA, all the components of the strips must be previously treated for making them hydrophilic and overlapped between them over an adhesive backing card to guarantee the continuous flow of the sample. The fluid sample applied at the end of the strip (at the sample pad) migrates through the next region, the conjugate pad, where nano/microparticles conjugates are immobilized. This conjugate is composed by the biological components of the assay (e.g., antibodies), which are attached to the particles and previously have been deposited in the conjugated pad. In this step, the analyte interacts with the biological component of the conjugated particle forming a complex. There is a wide variety of particles used as labels (see Section 2.2.2), but the most commonly used are latex microparticles and gold nanoparticles. The sample continues migrating by capillary forces through the detection pad, composed by a porous membrane in which the biological complementary part of the assay has been previously deposited. The capture lines, test and control line, are

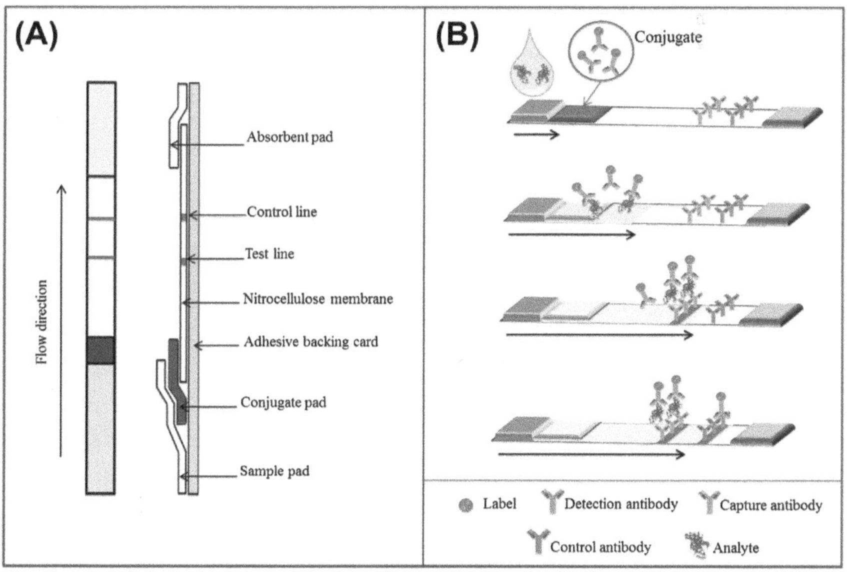

FIGURE 1 (A) Components of a lateral flow strip. (B) Scheme of a typical lateral flow assay (LFA).

constituted by antibodies, antigens, or proteins, and their function is to capture the analyte and the excess of conjugated particle, respectively. The excess of reagents is absorbed by the wicking or absorbent pad located at the end of the LF strip. Visual results can be interpreted as the presence or absence of analyte in sample and can be detected by the naked eye or using a strip reader.

According to the type of format used, the results can be interpreted in opposite ways. In Figure 2, details of two main LFA formats such as the sandwich (direct) and competitive assay are shown. For the detection of molecules with a high molecular weight, with two or more antigenic sites (e.g., big proteins), the sandwich assay is used (Figure 2(A)). In this case, the presence of analyte shows two colored zones: test and control lines. On the other hand, if the analyte is a low-molecular-weight molecule with one antigenic site (hapten) for binding with antibody (e.g., pesticides or toxins), the competitive assay is used. In this type of assay, the analyte and the analyte are immobilized at the membrane compete for binding with the detector antibody. The presence of analyte is indicated by the absence of a test line; therefore, only one line at the control zone is observed (see Figure 2(B)).

In both cases, the control line is always visible and independent of the format used, allowing adequate migration of the sample through the strip to be controlled.

2.2 Components

2.2.1 Assay Membranes

The function and the materials that constitute the different components of LFA are summarized in Table 1 and discussed next.

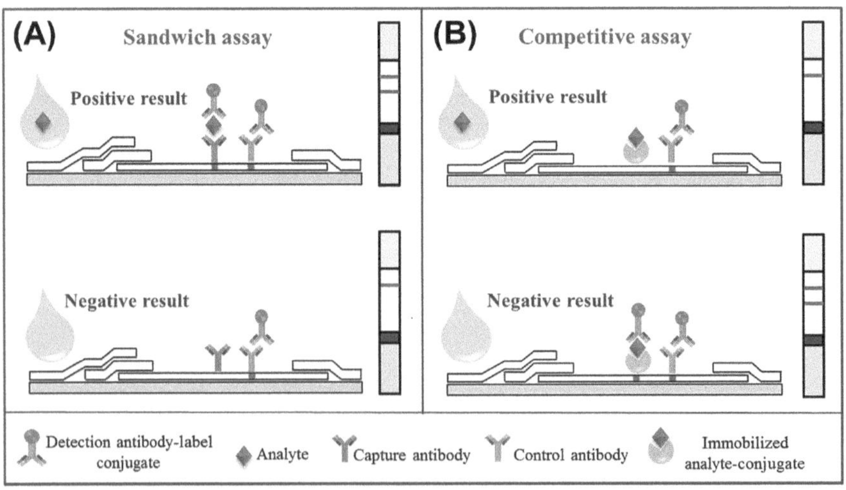

FIGURE 2 Typical formats of lateral flow assays (LFAs): (A) direct and (B) competitive assays.

TABLE 1 Components of Lateral Flow Assay and Their Functions

Components	Material	Function
Sample pad	Cellulose	Distributes the sample in a controlled way for releasing the analyte with high efficiency
Conjugation pad	Glass fiber	Retains the conjugate between its fibers until the test is running. Coupling of analyte and conjugate takes place in this pad
Detection pad	Nitrocellulose	Binds irreversibly the capture reagents and facilitates the visualization of test and control lines
Absorbent pad	Cellulose	Drains off the excess of fluid of the detection pad

1. Sample pad

   Its purpose is to receive and then distribute the sample in a controlled and uniform way, releasing the analyte with high efficiency [16]. Additionally, it controls the rate of flow that enters to the conjugation pad, preventing flooding of device. The most commonly used sample pads are porous materials such as cellulose. Cellulose has large bed volumes (>25 $\mu L/cm^2$) and is extensively used for testing various kinds of liquid samples (e.g., urine). Due to adsorption capability, loading of blocking agents, pH and ionic strength modifiers, and viscosity enhancers is easy to be performed.

   When pretreated by impregnation with proteins, buffers, and detergents, the sample pad can modulate the flow properties, increasing the sample viscosity, preventing nonspecific adsorptions of both the conjugate and analyte, increasing the reaction time in the conjugate pad, and thereby modulating the interaction with the sample.

2. Conjugate pad

   The function of this pad is to keep the dried conjugated particles and hold them stable in the pad until the test is run and to uniformly transfer the signaling reagent (e.g., gold nanoparticles conjugates) and sample onto the membrane [16]. When the sample flows through the conjugate pad, the signaling reagent solubilizes and recognizes the analyte, and the formed conjugate moves through the membrane. Assay sensitivity can be severely affected by the conjugate deposition, drying, inadequate conjugate mixing, and releasing from the pad. Ideally, the conjugate pad must have low nonspecific binding of signaling reagent or analyte to the pad, allowing a suitable formation of the immunocomplex at the test line. In addition, it must ensure a constant flow and bed volume so as to avoid problems of uneven signals that can alter the sensitivity of assay. Glass fibers, cellulose filters, surface-treated (hydrophilic) polyester or polypropylene filters, and rayon are the most-often used materials for a conjugate pad.

To ensure an optimal release of the sample, the pad must be pretreated with a pH-adjusted buffer containing proteins and surfactants, followed by a suitable drying.

The deposition of conjugated particles can be achieved manually (batch mode) or by using a dispensing machine that delivers in a continuous mode solutions of reagents including suspension of nanoparticles modified with antibodies.

3. Detection pad

Its main purpose is to bind irreversibly the capture reagents at the test and control lines and facilitate the visualization of colored bands indicating the presence or absence of analyte in the sample [16].

Nitrocellulose (NC) is the most widely used material for detection membranes in LF technology and other point-of-care (POC) devices. Depending on the application, other materials, such as nylon and polyethersulfone, can be used as a detection membrane.

NC is a versatile polymer that has been used for years due to its unique biomolecule-adsorbing properties. For this reason, the use of NC has been extended to powerful blotting techniques for identifying the presence of biomolecules such as proteins (Western blot), DNA (Southern blot), and RNA (Northern blot) [17–19]. Based on these applications, during the past 25 years, NC has gained importance as a suitable porous support in LF technology.

Detection membrane is the most important part of a lateral flow assay because of the binding and detection that occur within it. Therefore, the adsorption of biomolecules is an essential aspect for a good LFA performance. Even though the binding protein to NC has been reported and used for powerful and existing techniques, the binding mechanism is still not sufficiently clear [18]. Different mechanisms have been proposed for explaining the binding mechanism. According to one of the theories, proteins are thermodynamically driven to adsorb to NC by noncovalent interactions [17]. Others claim that an initial attraction of protein to NC involves the interaction between the dipoles of the nitro groups on the NC and the carbonyl groups in peptide bonds of protein [20]. Either way, different forces (hydrogen bonding, hydrophobic, and electrostatic interactions) are involved in the adsorption process of protein to NC; as recommended, LF manufacturers must consider all these interactions, materials, and assay reagents for optimizing the binding of the proteins to NC [21].

Membranes made of NC are hydrophobic by nature. However, membranes can easily be moistened due to the presence of surfactants added during membrane production [16]. Type and amount of surfactant is usually recommended by the manufacturer but can also be adapted according to the application. Once the membrane is processed and ready to be used, depositing the capture reagents at test and control lines can be carried out using either a contact or a noncontact dispensing system. The best option is a noncontact dispenser, typically an inkjet system, because the membrane does not sustain damages that can impede a normal flow of the sample. Nevertheless, low-contact dispenser systems are available [22]. In this case, the nozzles can dispense the sample with a minimal contact pressure, avoiding any mechanical stress on the membrane.

Flow rate is difficult to be measured since decay occurs exponentially while the fluid front moves through the membrane [16]. A common parameter to evaluate the NC membranes is the capillary flow time, which is the time required for a liquid to move along and fill completely a 4-cm-long section of membrane. The capillary flow time and the sensitivity of the assay increase when the pore size decreases, since the reagents spend more time on the membrane and the formation of immunocomplex between the analyte and the label is more effective at the test line.

Another parameter is the thickness of the membrane since it has a direct influence on bed volume and signal visibility. Bed volume is the volume of liquid required to saturate a given area of membrane and is related to the porosity and the thickness. As the thickness increases, the bed volume also increases. However, this consideration may be not important when an absorbent pad is used because the total volume of the sample is governed mainly by the bed volume of the absorbent pad [16]. When capture reagents are dispensed onto the membrane, the liquid penetrates downward into it and moves laterally. In a thin membrane, spreading of the reagents will be occurring since its insufficient depth provokes the diffusion of the liquid (including nanoparticles used as labels) through the whole membrane, causing signals (diffused/undefined detection and control lines) to be difficult to visualize.

4. Absorbent pad

Located at the end of the strip, its main function is to drain the fluid from the membrane. Using an absorbent pad, the volume of the sample can be increased, with the background signal reduced and the sensitivity of the assay improved.

Most of absorbent pads are made of high-density cellulose and no treatments are required. However, once the experiment is achieved and a visual record is needed, it is recommendable to remove the absorbent pad because it can act as reservoir once the liquid has evaporated and it might release the excess of label reagent, causing false positives [23].

2.2.2 Labels

Immunoassay conjugate consists of an antibody with an appropriate label whose function is to generate a signal that correlates with the immunocomplex formed and its antigen, allowing the detection by using techniques such as colorimetry, fluorescence, and luminescence [24].

Most of the commercial LFAs use colored latex beads, carbon nanoparticles, and gold nanoparticles, which allow visual detection without the use of readers if the test provides a qualitative response. On the other hand, if LFA is intended to be quantitative, the use of a reading system is required depending on the type of label used, such as charged couple devices (CCDs) cameras for AuNPs or colored beads; readers based on light-emitting diode (LED) technology for fluorescent labels; and magnetic readers such as the Magnetic Assay Reader (MAR™) for superparamagnetic labels.

Applications of traditional labels will be briefly summarized in this section, while gold nanoparticles will be discussed in detail in Section 3.

1. Latex beads

   Latex beads are spherical microparticles composed of an amorphous polymer (usually polystyrene). These beads have been used as the first labels in LF technology due to their high binding protein capacity [24]. Latex beads of various sizes and with controlled surface chemistries are able to incorporate colored dyes, fluorescent dyes, and magnetic or paramagnetic components have been already developed and are available in the market. The sensitivity of LFA that uses colored latex beads is, in principle, lower compared with that achieved with gold nanoparticle labels because their smaller size (typically 20–40 nm) and, consequently, their higher surface area–to–volume ratio allow them to bind more antibodies onto their surface.

2. Liposomes

   Liposomes are colloidal vesicles formed by a lipid bilayer membrane capable to encapsulate part of the aqueous media in which they are dispersed [25]. They can encapsulate (i.e., visual and fluorescent dyes or enzymes) inside their cores, improving the sensitivity of LFA. It is possible to add different functional groups to the lipid surface for coupling with biomolecules or chemicals. Different studies with interest, for example, for DNA [26], allergenic agents [27], and other analytes have been reported. However, liposomes have some disadvantages related to their high production cost, short shelf-life, and propensity to leak and fuse encapsulated molecules.

3. Carbon nanoparticles

   Carbon nanoparticles are the most inexpensive label for LFA and provide an excellent color contrast to the white NC membrane. These amorphous labels are suitable for physical adsorption because of their hydrophobic features and have been used for DNA [28,29] and small molecule (e.g., chemical contaminants [30,31]) detection. The main limitations of carbon nanoparticles are related to their low density, which make difficult their visualization by electron microscopy (transmission or scanning); their irregular sizes and shapes; and the requirement of surfactant addition to stabilize the carbon dispersions [32].

4. Fluorescent labels

Organic and inorganic fluorophores can also be used as labels. A typical organic fluorophore for labeling antibodies is fluorescein isothiocyanate (FITC), which is reactive to the primary amines of antibodies producing a fluorescent conjugate [33]. LFA using fluorescent labels to detect, for example, influenza virus has been shown to be more sensitive that colorimetric assays [34].

Organic fluorescent dyes (e.g., $[Ru(bipy)_3]^{2+}$) have been used for coating of silica nanoparticles, and their subsequent use in LFA have demonstrated to be remarkably more sensitive than gold nanoparticles for nucleic acid detection [35].

Inorganic fluorophores such as quantum dots (QDs) and lanthanides also have been reported for LF applications. QDs are semiconductor nanoparticles in size range of 1−10 nm composed of GaN, GaP, ZnO, ZnS, CdS, or CdSe [36]. As a consequence of their small sizes, they have unique properties such as the high quantum yield and molar extinction coefficients, broad absorption spectra, and resistance to photobleaching and chemical degradation [37]. Uses of QDs in biosensors have grown in the past years because of their advantages over common fluorophores. QDs can be conjugated to different biomolecules, especially antibodies. However, the main limitation of QDs is related to their low colloidal stability in biological media [36].

QDs have been used as labels in LFA in competitive assays for detection of a primary metabolite biomarker for pesticides [38], toxin detection [39], and in sandwich formats for cancer biomarkers [40].

Lanthanides (e.g., Eu(III)) also have unique fluorescent properties that can be applied for their use as sensitive labels for prostate-specific antigen (PSA) detection in LFA, increasing 300-fold the sensitivity of assay [41].

2.3 LF Assay Advantages and Limitations

POC testing (POCT) is defined as a rapid diagnostic test performed at or close to the site of patient care. According to the World Health Organization (WHO), a POC device must be fit in the ASSURED criteria: *a*ffordable, *s*ensitive, *s*pecific, *u*ser-friendly, *r*apid and robust, *e*quipment free, and *d*eliverable to those who need it [42]. There is a wide variety of POC devices for different applications, with the pregnancy test being one of the most often used [24].

LF is a well-established technology that has many advantages such as (1) rapid qualitative or semiquantitative tests; (2) being easy to use (results can be interpreted by untrained personnel); (3) use for in-field applications in remote areas where high-tech equipment is not available; (4) prolonged shelf-life (about 1 year) and often without refrigeration; (5) small volumes required; (6) relatively low-cost production; (7) good sensitivity and specificity; and (8) possibility of integration with electronics and electrochemical systems.

Despite the mentioned advantages, LF technology has also limitations, especially related to reproducibility and sensitivity issues, such as (1) restricted volume around the microliter level; (2) higher volumes that could deteriorate the strip and delay the assay time; (3) pretreatment required for nonliquid samples and filter must be added to the strip; (4) integration with electronic/electrochemical systems still a challenge; (5) less sensitive than other immunoassay techniques, such as ELISA; and (6) multiple detection is challenging due to some cross-reactivity issues and the fact that the setup for dispensing on membrane requires spatial differentiation.

Nevertheless, LF technology has a wide range of applications in different fields, and it is the most used POC test because fulfills the requirements for this class of device. Many efforts are currently being conducted in order to improve either the used materials or the assay platforms with the purpose of solving sensitivity issues beside others. In this context, new strategies to enhance the sensitivity of LFA will be discussed in Section 4.1.

3. GOLD NANOPARTICLE-BASED LF BIOSENSORS

AuNPs have been used for many years in biomedical applications, and their unique chemical and physical properties make them attractive for areas such as diagnostic, therapy, and immunology [43–47].

In 1971, Faulk and Taylor [48] published the first report about the use of AuNPs as labels for immunoelectron microscopy. Nine years later, Leuvering et al. [11] presented the "sol particle immunoassay" (SPIA), the first colorimetric report using AuNPs as labels for hCG detection. Nowadays, AuNPs are the most used labels for LFA due to their signaling properties combined with their biocompatibility and low or missing toxicity overall for in vitro applications.

There are a wide range of synthetic methods for obtaining AuNPs, most of them based on chemical reduction of chloroauric acid, $HAuCl_4$ [44,49]. Reduction of Au(III) to Au(0) in aqueous media in presence of citrate as mild reducing agent and stabilizer is the most used method for synthesizing AuNPs, as pioneered by Turkevich and co-workers [50]. Well-dispersed and size-controlled AuNPs are obtained by controlling the ratio citrate:Au [51]. Noncontrolled sizes and uneven shapes of AuNPs can affect the performance of LFA.

AuNPs surface can be functionalized by biomolecules (e.g., antibodies and nucleic acids) due to their affinity for thiol-containing ligands. Antibodies can be attached (conjugated) to the AuNPs surface through hydrophobic and ionic interactions or through chemisorption of thiol groups present in their structure [52].

Three are the most reported methods for binding proteins to AuNPs [53]: (1) gold nanoparticles surrounded by a layer of negative charges can be conjugated firmly to any protein positively charged due to the ionic binding; (2) amino acid-containing proteins bind to the AuNPs through hydrophobic binding; and (3) gold and sulfur bonding. AuNPs can be conjugated to antibodies by the cysteine residues in the protein.

Control of pH and a suitable concentration of antibody must be determined for avoiding flocculation of AuNPs [54]. The pH of an AuNP dispersion must be adjusted before the conjugation at slightly above the isoelectric point (pK_i) of the antibody. Aggregation of AuNPs conjugated will occur if the pH is below the pK_i. By contrary, if the pH is above the pK_i of the antibody, a limited adsorption will occur because of the charge repulsion between the AuNPs and the antibody [55].

The optimum size of AuNPs for using as labels in LFA is around 20–40 nm: below 10 nm, AuNPs do not have bright and intense color, and above 40 nm, AuNPs are prone to flocculation.

Lou et al. [56] studied the influence of the nanoparticle size for lateral flow applications. They prepared conjugates with four different AuNPs sizes (14, 16, 35, and 38 nm), and they found that different-sized AuNPs had different conjugation efficiencies under different pH values and concentrations of antibody. Moreover, they proved that 16-nm AuNPs remained stable at the lowest concentration of antibody, but less sensitivity was observed than with 38-nm AuNPs in real samples.

In the following section, some of the most important applications of AuNP-based LF for different analytes such as protein, DNA, etc. are described and accordingly discussed. Table 2 gives an overview of the applications for different fields of AuNP-based LF technology.

3.1 Protein Detection

LF technology is well diversified in several sectors, with pregnancy self-testing being the largest segment on the LF market, representing a great number of companies that commercialize these tests [55]. In addition, a high number of approaches for the detection of proteins of interest in different fields have been reported in the past years.

A study related to the detection of thrombin (a serine protease related to the coagulation system) and aptamer-based LFA was reported by Xu et al. [72]. Aptamers are synthetic nucleic acids able to recognize a wide range of analytes with the same or even higher specificity than do antibodies. The strategy developed by Xu and co-workers for the detection of thrombin introduced the use of aptamers instead of antibodies in an LFA, achieving a limit of detection (LoD) of 2.5 nM thrombin. Comparing the aptamer-based LFA with the traditional antibody-based LFA, the authors found more sensitive assay using the synthetic nucleic acids.

TABLE 2 Overview of Gold-Based Lateral Flow Assay (LFA) Applications in Different Fields

Type of Analyte	Specific Target	Detection	Cut-off/LoD	References
Proteins	CEA	Instrumental	5 ng/mL	[57]
	PSA	Visual	1 ng/mL	[58]
	cTnI	Instrumental	0.01 ng/mL	[59]
	HIgG		200 pg/mL	[60]
	RIgG and HIgM		1.0 ng/mL and 1.5 ng/mL	[61]
Infectious agents	*Legionella*	Visual	0.5 CFU/mL	[62]
	HIV-1		10^{11} copies/mL	[63]
	Escherichia coli O157:H7		0.4 nM	[64]
Mycotoxins	ZEA, DON	Visual	1500 and 100 μg/kg	[65]
	Type B fumonisins	Instrumental	199 μg/kg	[66]
	OTA		0.18 ng/mL	[67]
Heavy metals	Pb^{2+}	Visual	0.5 μM	[68]
	Hg^{2+}	Instrumental	0.1 nM	[69]
	Cd^{2+}		0.1 nM	[70,71]

Cancer biomarker detection using LFA also has been reported. One of the examples includes carcinoembryonic antigen (CEA), which acts as biomarker for several cancers, including colon, breast, ovarian, and lung. Zeng and co-workers [57] developed an LFA for detecting CEA in human serum. They could detect CEA in a 10-min assay with an LoD of 5 ng/mL. Although this value was higher than the LoD provided by a commercial ELISA kit for CEA detection, the assay time was considerably shorter (around 90 min) and simpler, opening the way to rapid and simple diagnostic kits.

PSA is a glycoprotein synthesized in the prostate gland and the presence in high concentrations in male serum is related to prostate cancer and other prostate disorders. Fernández-Sánchez et al. [73] reported the sensitive detection of free and total PSA (f-PSA and t-PSA, respectively). They developed an integrated AuNP-based LF strip with an electrochemical system by using impedimetric detection, achieving an LoD of 3 ng/mL PSA. One year later, the same group [58] reported a sensitive detection of f-PSA and t-PSA in one-step LFA. The authors achieved a series of optimizations including a careful selection of materials and experimental conditions, for achieving a visual LoD of 1 ng/mL for both biomarkers.

Another prostatic cancer biomarker is prostatic acid phosphatase (PAP), which is combined with a PSA test for more accurate cancer diagnostics. For this reason, Fang and co-workers [74] developed a semiquantitative barcode LFA for PAP detection in human serum samples. Based on the stepwise capture of analyte, they achieved an LoD of 0.25 ng/mL in 30 min.

Choi and co-workers [59] reported an enhanced LFA for the detection of troponin I, a biomarker related to the acute myocardial infarction (AMI). They detected up to 0.01 ng/mL troponin I in serum, which is lower than the cutoff value of 0.1 ng/mL for the diagnosis of myocardial infarction. Following a similar strategy, Zhu et al. [75] reported the simultaneous detection of two biomarkers (troponin I and myoglobin) related to AMI. They used an AuNP doubly labeled complex, in which biotinylated single-stranded DNA (ssDNA) was used as a linkage to integrate two AuNPs of different sizes (13 and 41 nm). With this enhanced strategy, very low LoDs of 1 ng/mL and 1 μg/mL were reached for troponin I and myoglobin, respectively.

3.2 Nucleic Acids

Nucleic acid detection is of outstanding importance in molecular diagnosis due to the interest in human and veterinary research [76,77]. When a small amount of DNA is available, it is required a previous amplification that can be achieved by using the polymerase chain reaction (PCR) or isothermal amplification techniques. During the past years, a new generation of integrated and sophisticated miniaturized POC nucleic acid devices is emerging [78]. An easy and low-cost alternative approach is the use of nucleic acid lateral flow (NALF) assay, which uses nucleic acid hybridization events to capture and detect synthetic DNA sequences or amplified nucleic acid products [79]. Usually, the capture of nucleic acids can be performed in two main formats: (1) antibody-dependent or (2) antibody-independent manner (see Figure 3). The former one, named nucleic acid lateral flow immunoassay (NALFIA), uses an antibody that is located at the test line and a labeled amplicon or a complementary sequence to the amplicon. When an antibody-independent format is used, the attachment of the oligonucleotide probe can be carried out through biotin-streptavidin linkage or by using UV irradiation for covalently binding it to the NC surface.

Piepenburg et al. [80] developed an isothermal amplification method (recombinase polymerase amplification (RPA)) that allows them to detect a few copies of pathogen genomic DNA, using an NALFIA format (see Figure 3(A)). In this work, a double-labeled (biotin/carboxyfluorescein-FAM)

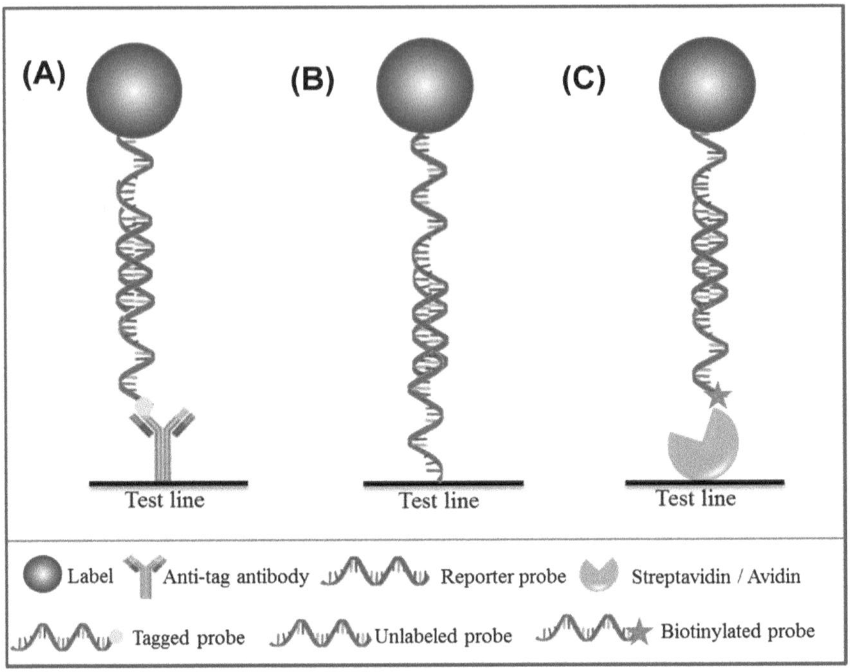

FIGURE 3 General strategies for the capture probe immobilization in nucleic acid lateral flow (NALF) assays. (A) Antibody-dependent format. An anti-tag antibody is attached at test line and recognizes a tagged probe which hybridizes with a reporter labeled probe. (B) Antibody-independent format without molecular receptor. An unlabeled probe is directly attached at test line. Detection is produced when a reporter labeled probe hybridize to the other probe. (C) Antibody-independent format using a molecular receptor. The most common strategy takes advantage of the biotin/streptavidin linkage for attaching the biotinylated probe on the test line.

RPA amplicon was captured by an anti-biotin antibody immobilized at test line. When the flow is passing through the membrane, the visualization of the lines is produced by the binding between an anti–FAM-AuNP conjugate and the amplicon tail labeled with FAM.

A similar approach was reported by He and co-workers [81], who developed an NALFIA for detection of the mutation in the *R156H* gene of keratin. The gene was amplified by isothermal strand-displacement polymerase reaction (ISDPR) technique using primers labeled with digoxin and biotin and finally detected using streptavidin receptors and anti-digoxigenin-AuNP labels, obtaining an LoD of 1 fM of *R156H*-mutant DNA.

Detection and quantification of microRNA (miRNA) can be achieved with NALFIA devices as reported by Hou and co-workers [82]. AuNPs were conjugated to a thiol-DNA that acts as detection probe, whereas a biotin ssDNA serves as the capture probe. Detection is produced when the anti-avidin antibody immobilized at test line recognizes the avidin–biotin–AuNP sample complex. The authors reported a perfect match between the capture probe with tested miRNA and detection probe, instead of the common sandwich in which the tested probe hybridizes with the capture and detection probes. Finally, they used the silver enhancement method to catalytically growth silver on the AuNPs and in this way increase the sensitivity of the assay, reaching an LoD of 5 attomol, opening the way to further application of this method for the detection of other short oligonucleotides.

Detection of single DNA mutations, also known as a single oligonucleotide polymorphism (SNP), can be achieved with high sensitivity and specificity by using NALFIA devices as reported by Xiao et al. [83]. They used AuNPs as label, the circular strand displacement reaction for amplifying the target, and the ligation reaction performed by T4 DNA ligase (the most commonly used ligase, an enzyme that facilitates the joining of DNA strands), which provides high specificity recognizing single-base mismatches. With this strategy, the authors reported an LoD of 0.01 fM for wild-type target DNA with high specificity, and the performance of the device using human genomic DNA clinical samples was shown to be excellent.

On the other hand, in the antibody-independent NALF, the capture reagent is an oligonucleotide probe that can be labeled (e.g., biotin) or not and is located at test line instead of antibody (see Figure 3(B) and (C)). An unlabeled oligonucleotide probe can be covalently attached to nylon and NC membranes by cross-linking using UV irradiation. In 2007, Aveyard and co-workers [84] reported the first NALF device with these features for the detection of amplified PCR products. They reported a sandwich assay between the target probe that was complementary to the AuNP-labeled probe and the capture probe attached to the NC. Also, they studied the effect of different AuNP sizes in the performance of the assay, and they found that sensitivities of devices with AuNPs of 80 nm were higher than AuNPs of 40 nm, which are the most common size used for LF purposes and showed the possibility to detect unpurified PCR products using an antibody-independent format.

The most common format in NALF devices uses biotin-labeled oligonucleotide probes attached to the NC surface via biotin-streptavidin linkage (see Figure 3(C)). Following this format, Mao and co-workers [85] developed a sensitive and rapid NALF for DNA detection with an LoD of 50 pM of target. They used an enzymatic-signal enhancement and applied the device for detection of human genomic DNA obtaining an LoD of 1.25 fM, a value considerably lower compared with the obtained with alternative techniques such as microarray or surface plasmon resonance–based analysis.

Lie and co-workers [86] developed an NALF for the detection of amplified DNA sequences by isothermal strand-displacement polymerase reaction (ISDPR). The sensitivity of this device based on this isothermal amplification is comparable or even better than that obtained with colorimetric or fluorescence-based analysis. Following this approach, they were able to detected human genome at levels of 25 ng/mL in clinical samples.

3.2.1 Detection of Infectious Agents

LFAs of nucleic acids with interest for POC infectious disease diagnostics also have been reported. Infectious diseases cause millions of death every year, and most of them are caused by late detection and lack of suitable treatment overall in countries with less resources. The application of LFA should be of great interest to detect an infectious disease in the early stage [87–90].

Although there are five mayor types of infectious agents (bacteria, viruses, fungi, protozoa, and helminthes [91]), in this section the described examples will include bacteria and viruses detection based on the use of NALF and NALFIA devices.

Horng et al. [62] developed an NALFIA device for the detection of *Legionella* bacteria from environmental cooling tower samples using multiplex-nested PCR as amplification method of 16S rDNA and *dnaJ* DNA fragments. In multidetection and specific AuNP-based NALFIA, they could identify the presence of *Legionella* species and, more specifically, *Legionella pneumophila* if one or two test lines were colored, respectively. The authors reported an LoD of 0.5 colony-forming-unit (CFU)/mL and 100% sensitivity and specificity for *L. pneumophila* compared with traditional detection methods.

For *Vibrio cholerae* detection, a foodborne pathogen, Chua and co-workers [92] developed an NALFIA that consisted of an antibody-dependent format by using the PCR technique and double-labeled primers (biotin/fluorescein and digoxigenin/fluorescein). Microspheres coated with streptavidin, anti-digoxigenin, and anti-mouse IgG antibody were deposited on the strip, which acts as test and internal control of PCR amplification and migration control lines, respectively. The DNA target, biotin-labeled DNA, was captured by the streptavidin located at the test line. The internal control, digoxigenin-labeled DNA, was captured by the anti-digoxigenin antibody, and the migration control line was visualized through the binding between AuNP–mouse anti-fluorescein antibody conjugate and an anti-mouse antibody. Using this strategy, they could verify the efficiency of the DNA amplification and detect up to 5 ng of target DNA (see Figure 4(A)).

Recently, Liu et al. [93] developed an ultrasensitive NALFIA for *Salmonella* detection, a common foodborne pathogen; it was based on the hybridization between AuNP-DNA probes complementary to the 16S ribosomal RNA and DNA of the bacteria. An LoD of 5 fmol was found for the synthetic ssDNA. For the *Salmonella* cultured samples, nucleic acids from 10^7 cells were detected in 30 min by the naked eye. Using silver enhancement of AuNPs, the LoD was improved up to 10^4 cells in an additional 10 min.

The catalytic deposition of gold(III) on AuNPs is an amplification strategy used for the enhancement of the NALF signal. Two representative works have applied this strategy for enhanced detection of infectious agents. In the first one, a semiquantitative detection of amplified HIV-1 RNA using an

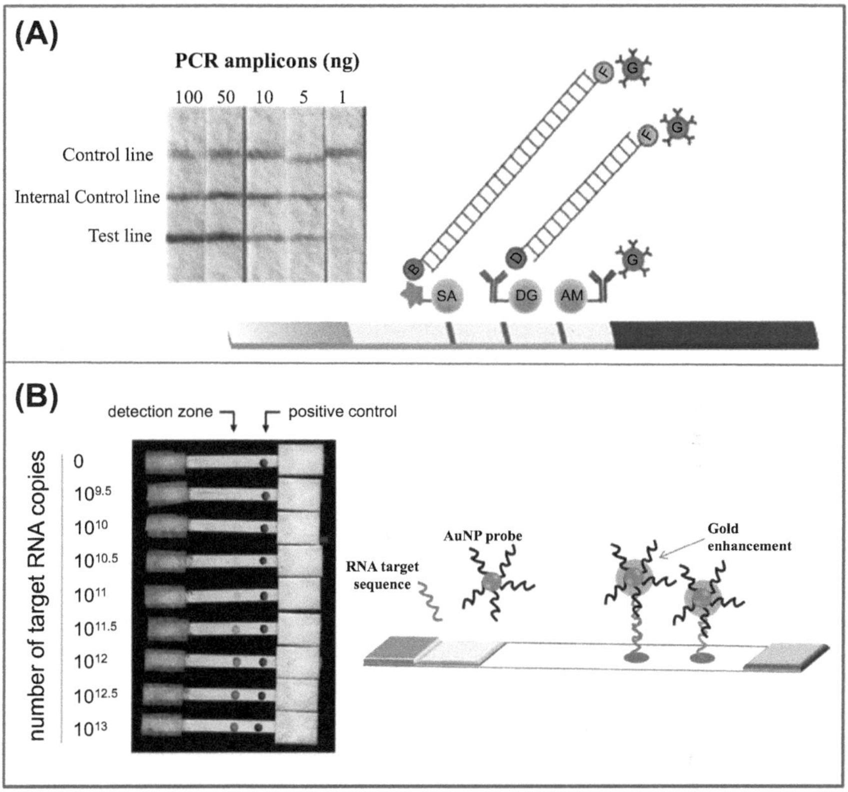

FIGURE 4 NALFIA and NALF assays for detection of infectious agents. (A) Scheme of NALFIA for visualization of PCR amplicons by using capture reagents coated onto carrier beads and immobilized onto a glass fibre membrane and picture of strips for a PCR amplicon range of 1–100 ng. (B) Left: Image of NALF devices for a series of in vitro transcribed RNA from 9.5 to 13 log10 copies. Right: NALF design for HIV-1 detection. NALFIA, nucleic acid lateral flow immunoassay; PCR, polymerase chain reaction. *Adapted from Refs [63,92] with permission.*

NALF device using was developed by Rohrman et al. [63]. Two complementary probes were covalently attached to the NC surface by UV irradiation, and the RNA target sequence was isothermally amplified by nucleic acid sequence–based amplification (NASBA) technique (see Figure 4(B)). The authors reported an LoD of 10^{11} copies/mL of amplified HIV RNA using gold enhancement that matches the clinical range for viral HIV-1 loads. In the second one, detection of *Escherichia coli* O157:H7 in an NALF device using locked nucleic acid (LNA)–AuNP conjugates was reported by Rastogi and co-workers [64]. With the enhancement protocol, the sensitivity of the assay was improved by 3-fold when LNA–AuNP conjugates were used in an NALF device instead of DNA–AuNP conjugates.

3.3 Mycotoxins

A toxin is a poisonous substance produced by living cells and organisms. Among the different classes of toxins, mycotoxins are natural secondary metabolites produced by different species of fungi that mainly belong to *Aspergillus*, *Penicillium*, and *Fusarium* genera [94]. Mycotoxins represent a health risk problem for human and animals because they are capable of causing diseases. First-screening methods for mycotoxin detection as LFAs have been commercially developed and reviewed in the past few years [95–97].

Because mycotoxins are very small compounds, a suitable LFA for their detection is the competitive format as stated in Section 1. Therefore, some works related to the detection of mycotoxins using AuNPs as labels in a competitive LFA format will be discussed in this section.

Aflatoxins are carcinogen mycotoxins that are produced by *Aspergillus flavus, Aspergillus nomius*, and *Aspergillus parasiticus*, species of fungi. Delmulle et al. [98] developed a qualitative LFA for the detection of aflatoxin B_1 (AFB_1) in porcine feed. Rabbit anti-mouse antibodies and AFB_1-bovine serum albumin conjugate were immobilized at the control line and test line, respectively. When the liquid sample passes through the membrane, AuNP-labeled monoclonal anti-AFB_1 antibodies are saturated by the presence of the toxin, resulting in a nonvisible test line and visible control line, which indicate the proper performance of the LFA. Using this LFA format, they obtained a visual LoD of 5 μg/kg AFB_1 in a simple and rapid test.

A modified competitive format assay was proposed by Moon and co-workers [99] for the detection of aflatoxin B_1. The authors reported the immobilization of polyclonal anti-AFB_1 on the NC membrane and AuNP–AFB_1–BSA conjugate as detection reagent. In the absence of AFB_1, the conjugate can bind to the anti-AFB_1 located at detection zone, giving rise to the visible line. On the contrary, if the AFB_1 is present in the sample, it inhibits the binding of the conjugate with antibodies avoiding the visualization of test zone. With this modified competitive LFA format, the authors reported a visual LoD of 10 μg/mL AFB_1 and examined the cross-reactivity to similar compounds such as ochratoxin-A (OTA), finding that the device was only specific to AFB_1. One year later, the same group [134] improved the modified competitive format, and semiquantification was possible using a smartphone for acquisition and processing of the image. With this technology, it is possible to discriminate between the blank from contaminated samples on the base of peak and area values, provided by Android application, being able to quantify up to 5 μg/kg AFB_1.

Zearalenone (ZEA) and deoxynivalenol (DON) are secondary metabolites produced by fungi *Fusarium* species, which can be found in grains. Simultaneous detection of these mycotoxins was first reported by Kolosova and co-workers [65]. They developed an LFA in a traditional competitive format, and results were visually evaluated by using AuNP of 40 nm. In the absence of toxins in the sample, three lines were observed (migration control and test lines). For positive samples, only the control line is visualized due to the presence of both mycotoxins. Finally, if two lines appeared (control and one of the test line), the sample was positive for ZEA or DON. Cut-off levels observed for both mycotoxins were 1500 and 100 μg/kg for DON and ZEA, respectively. In addition, spiked wheat samples were analyzed with the LF strips resulting in good agreement with the amount spiked, proving the reliability of the assay.

Detection of DON using AuNPs as labels was also reported by Kolosova et al. [100]. According to experimental results, they could detect DON at low and high concentration levels between 250–500 and 1000–2000 μg/kg, respectively. Spiked wheat and pig feed samples demonstrated accurate and reproducible results. Moreover, LFA were tested in real field conditions, proving the applicability of the devices.

Fumonisins belong to the same family of ZEA and DON mycotoxins. They occur mainly in maize, wheat, and cereals. From the group B fumonisins, fumonisin B1 (FB1) is considered as a possible carcinogen to humans by the International Agency for Research on Cancer [101]. A semiquantitative LFA was developed for detection of total type B fumonisins in maize by Molinelli et al. [66]. They reported an LoD of 199 μg/kg in a test range up to 4000 μg/kg, with no cross-reactivity toward other mycotoxins, and the applicability of the device was proved by the screening of 23 naturally contaminated maize samples.

Ochratoxins are produced by fungi species such as *Aspergillus ochraceus, Aspergillus carbonarius, Aspergillus niger* and *Penicillium verrucosum.* The most relevant toxin of this group is OTA, one of the most abundant food-contaminating mycotoxins. A semiquantitative LFA was developed by Anfossi et al. [102] for the detection of OTA in cereals. They reported the use of ovalbumin as an AuNP stabilizer and as a membrane blocker so as to reduce nonspecific adsorptions. Additionally, the use of polyethylene glycol instead of organic solvents for OTA extraction allowed avoidance of the matrix effects caused for the cereals used. Under these conditions, they reported an LoD of 1.5 μg/kg OTA in samples.

Recently, semiquantitative detection of OTA in wine and grape must, which represent a major source of OTA dietary intake, was described by Anfossi and co-workers [103]. The reported LFA allowed OTA detection in wine in 5 min with an LoD of 1 μg/L with a simple sample pretreatment that only implied its dilution with sodium bicarbonate, polyethylene glycol, and ethanol.

Another interesting strategy for OTA detection involving aptamer-based LFA was developed by Wang and co-workers [67]. The principle of the assay consisted of the competition for the aptamer (labeled with AuNPs) between OTA and complementary DNA probes located at the test line. The results indicated that the developed aptamer-based LFA was more sensitive than the conventional immunoassays due to their high specificity to a wide range of targets including small compounds. Visual LoD was 1 ng/mL and semiquantitative LoD was 0.18 ng/mL, obtained with a strip reader (see Figure 5).

3.4 Pesticides

The term *pesticide* refers to substances used to control or destroy plant or animal pests and includes herbicides, fungicides, and insecticides [104]. The health and environmental risks represented by these substances make necessary the control of their use by setting the amount of pesticide in agricultural products. Nowadays, nanomaterials are bringing new opportunities for pesticide detection because the analysis can be achieved in situ by using electrochemical and optical techniques [105,106].

For screening and rapid detection of pesticides, different LFAs in a competitive format due to the low molecular weight of these substances have been developed.

Wang et al. [107] developed a semiquantitative LFA for detection of the carbaryl that belongs to the class of carbamates, and it is used as an insecticide. They could obtain an LoD of 100 μg/L, higher than conventional methods as membrane-based ELISA, but the LFA provided semiquantitative carbaryl detection in one step, avoiding the use of carcinogenic and mutagenic substrates.

Atrazine is an organic compound used as a herbicide that was detected in water by Shim and co-workers [108] using an AuNP-based LFA in a competitive format. The authors reported a visual limit of detection of 3 ng/mL, and specificity tests showed a high selectivity to atrazine without cross-reactivity with other pesticides. Additionally, the results for the spiked water samples were in good agreement with the obtained with direct competitive ELISA and high-performance liquid chromatography (HPLC). An improved detection of atrazine by using AuNP-based LFA in a competitive format was presented by Kaur et al. [109]. In this work, they prepared a novel and very stable tracer by synthesizing derivatives of atrazine, which served as hapten molecules, that were first conjugated to a carrier protein (e.g., BSA) and finally to AuNPs, forming the hapten–protein–AuNPs conjugate. As the typical competitive assays, the atrazine and the conjugate compete with the available sites in the anti-atrazine antibody. The authors reported limits of detection below 0.1 ng/mL atrazine in water samples.

Triazophos belongs to the class of organophosphates pesticides and are widely used as insecticides and acaricides. Gui and co-workers [110] developed a competitive AuNP-based LFA for detecting residues of this compound in a short time, obtaining an LoD of 5 ng/mL in water and soil.

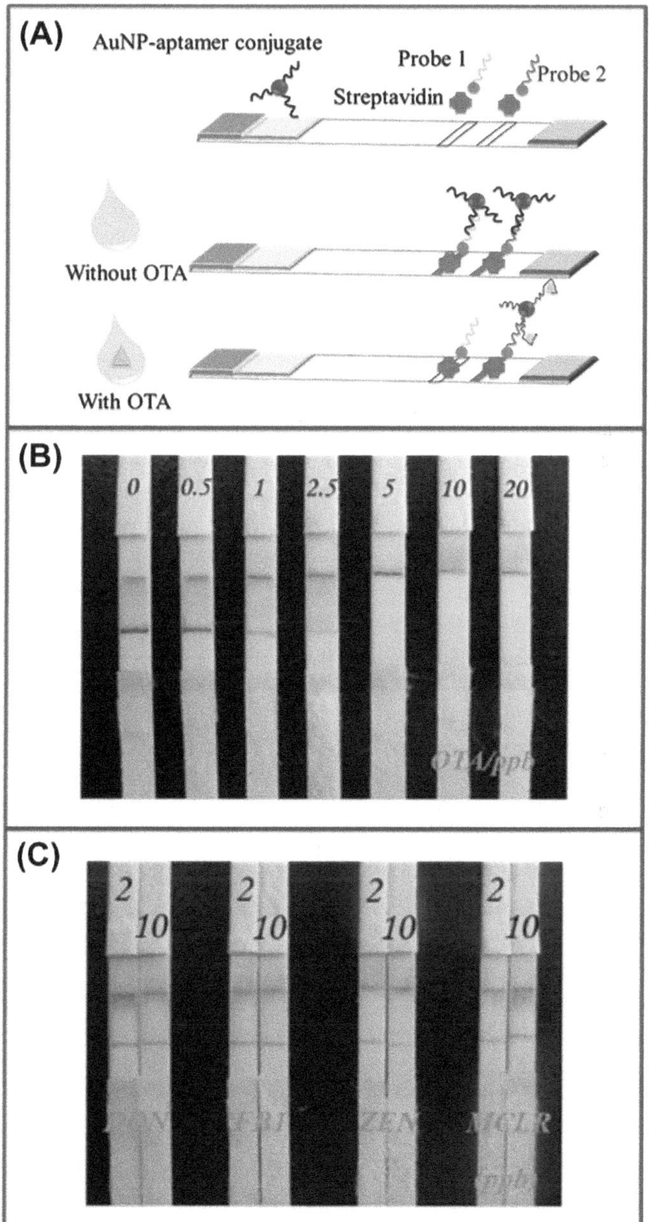

FIGURE 5 Detection of ochratoxin-A in an aptamer-based lateral flow assay (LFA). (A) Scheme of competitive aptamer-based LFA. (B) Picture of the strips for different OTA concentrations. (C) Specificity test toward several mycotoxins. *Adapted from Ref. [67] with permission.*

3.5 Heavy Metals

Detection of metal ions is of vital importance in environmental monitoring due to the potential hazardous effects that can affect human health. Heavy metals such as mercury, cadmium, and lead are released into the environment due to industrial activities, causing contamination to water and soil. Biological uptake of heavy metal ions through the food chain can cause chronic poisoning. Other heavy metal ions such as chromium, zinc, or copper are necessary for metabolic functions, but their presence at high levels can cause damage to liver, pancreas, and stomach, seriously affecting human health [111].

Traditional analytical techniques for metal detection such as inductively coupled plasma mass spectroscopy (IP-OES) or atomic absorption spectroscopy (AAS) are used for metal detection due to their reliable results, but these analyses are not compatible with the field applications due to the use of complex instrumentation. In recent years, the use of biosensors for metal ion detection has been growing because of their remarkably selectivity through different metals, which makes them sensitive and easy-to-use devices. In this context, Mazumdar et al. [68] described an LFA for lead detection in paints using AuNP–DNAzyme conjugates. A deoxyribozyme or DNAzyme is a DNA molecule that is able to catalyze a reaction in the presence of a particular metal ion. In this approach, the authors prepared an enzyme–substrate–AuNP complex composed by a modified 8-17 DNAzyme with biotin at one tail and thiol group in the other one, in order to be conjugated to the AuNPs and then, spotted at the conjugation pad. In the absence of Pb^{2+}, the substrate remains uncleaved and a red line at the control zone is observed due to the binding with the biotin modified tail of the complex. On the other hand, if Pb^{2+} is present on the sample, the cleavage of the substrate is produced and can migrate and be captured by a complementary DNA sequence located at the test line. With this strategy, the authors reported an LF device with useful industrial applications and with an LoD of 0.5 μM Pb^{2+}.

Following a similar approach, Fang and co-workers [112] reported the detection of Cu^{2+} with high sensitivity and selectivity. In a sample containing Cu^{2+}, the cleavage of the substrate is produced and, then, the cleaved product is analyzed in the NALF device. Two complementary DNA sequences are deposited on the strip to form the test and control lines. AuNPs are modified with a first probe, which is complementary to the cleaved piece at 5′ end, forming a DNA-AuNPs conjugate that migrates through the membrane. The test line is visualized when the second DNA probe located in this zone hybridizes with the cleaved piece at the 3′ end. Finally, the excess of DNA–AuNP conjugate migrates until the control zone and the hybridization between the third DNA probe and the conjugate is produced. The intensity of color of the test line is proportionally related to the concentration of Cu^{2+} in the sample, obtaining a limit of detection of 10 nM. In the absence of copper ion, the cleavage in the substrate is not produced and only the control line is visualized, indicating that the biosensor works properly. In addition, the authors used sample solutions containing different metal ions and the results showed high specificity toward Cu^{2+}.

He et al. [69] reported the detection of Hg^{2+} in aqueous solutions based in the thymine–Hg^{2+}–thymine coordination chemistry by using thymine-rich hairpin DNA probes. Briefly, in the presence of Hg^{2+} and the thymine-rich DNA probe, the thiolated hairpin DNA probe immobilized on the surface of AuNPs opens to form DNA duplexes (thymine–Hg^{2+}–thymine). While these DNA duplex moves through the membrane, the test line is visualized due to the recognition between the anti-digoxin and the digoxin labels on the duplexes. Finally, the control line is visualized through the hybridization between the thiolated hairpin DNA and a biotinylated control probe. Using this strategy, the authors reported a limit of detection of 0.1 nM and showed that the presence of other metal ions in samples do not affect the detection of Hg^{2+}. In a similar and less sensitive approach, Yang and co-workers [113] described the detection of Hg^{2+} using a thiolated DNA probe conjugated to AuNPs, which hybridizes with a second biotinylated probe, forming a DNA duplex due to the strong thymine–Hg^{2+}–thymine coordination. This complex is visualized because of the binding between the streptavidin located at test line and the biotinylated DNA duplex. A third DNA probe, complementary to the thiolated one, produced the visualization of control line. In the absence of a mercury ion, only the control line is observed. By using this strategy, the authors reported an LoD of 6 nM for Hg^{2+} and a high selectivity toward this metal ion.

Detection of Cr^{3+} ion in water and serum samples by using a gold-based LFA was reported by Liu et al. [114]. The principle of the assay consisted in a competitive format, where the AuNPs were conjugated to antibodies against isothiocyanobenzyl-EDTA (iEDTA)–chelated Cr^{3+}. In the absence of Cr^{3+}, the conjugated nanoparticle can be captured by the coating antigen and the goat anti-EDTA deposited at the test and control line, respectively. Therefore, two lines can be observed on the strip. In contrast, in the presence of Cr^{3+}, the nanoparticle-conjugated reacts first with the Cr^{3+} in the sample and in consequence just the control line is observed. The authors reported an LoD of 4.8 ng/mL and tested samples with different metal ions showing a high specificity toward Cr^{3+} ion.

Determination of Cd^{2+} in drinking and tap water using a competitive LFA was recently introduced by López-Marzo and co-workers [70,71]. The detection principle of Cd^{2+} ion is based on the competition between the Cd–EDTA complexes with the Cd–EDTA–BSA–AuNP conjugate for the same binding site of a purified monoclonal antibody, which specifically recognizes Cd–EDTA complexes but no free Cd^{2+} ions. This affinity is associated with certain residues such as histidine-H96 and tryptophan, on the heavy chain variable regions of the antibody, which are important as antigen

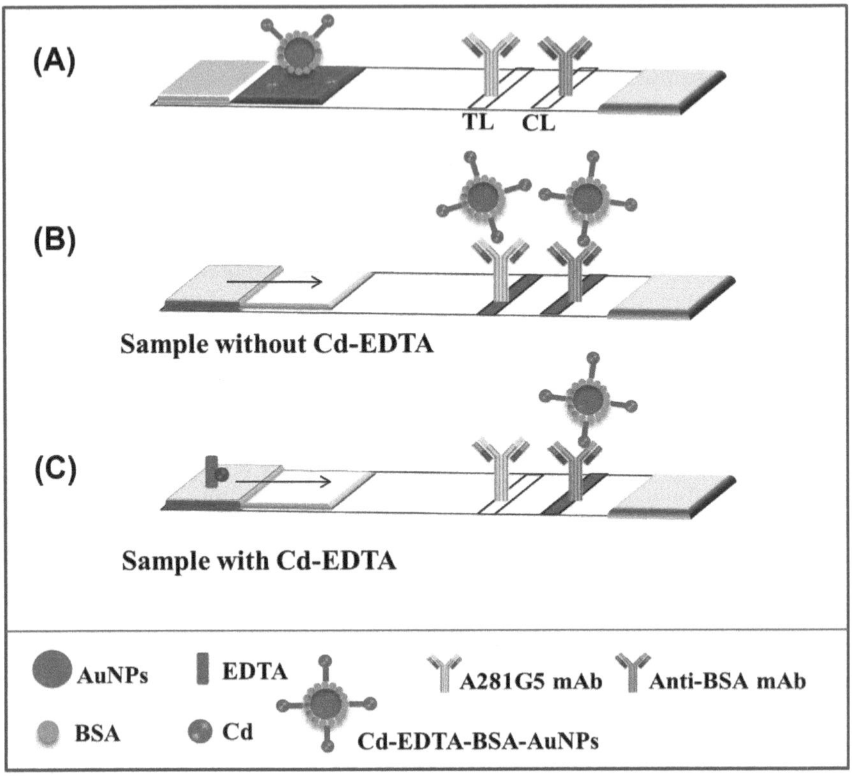

FIGURE 6 Cadmium analysis in waters through lateral flow assay (LFA)-based immunodetection of Cd–EDTA. (A) Configuration of the LF strip and its state after adding the sample without (B) and with (C) Cd–EDTA. *Adapted from Ref. [70] with permission.*

recognition sites. A very low limit of detection of 0.1 nM was found, which constitutes the lowest value reported for paper-based biosensors (see Figure 6). The authors also tested the specificity of the assay using different metal ions with Cd^{2+} ion, and the results showed several degrees of interference that can be reduced optimizing the concentration of reagents such as EDTA and OVA or BSA.

4. NEW TRENDS IN LF TECHNOLOGY

4.1 Amplification Strategies

In the past years, many efforts have been focused in order to improve the sensitivity of LFAs, being of special interest to those that exploit the benefits offered by the nanotechnology.

Most of the different strategies used for developing more sensitive devices require additional assay steps. The immunogold silver-staining (IGSS) is a technique used for enhancing the signal, which is based on the reduction of silver ions over the surface of AuNPs, increasing their size and thereby improving the sensitivity of the assay. Shyu et al. [115] described an AuNP-based LFA for determining ricin, reporting a limit of detection of 50 ng/mL in a standard assay, but when they used the silver-enhancement as signal amplifier method, the LoD decreased considerably to 100 pg/mL. Cho and co-workers [116], reported the detection of cardiac troponin I in a cross-flow chromatographic assay using the IGGS as signal amplifier, and this method increased the sensitivity by 51-fold compared with the conventional gold-based LFA.

Another method for enhancing the sensitivity of LFA by in situ increasing the size of AuNP is the gold-mediated amplification. Rastogi and co-workers [64] developed a device for detection of DNA specific to *E. coli* O157:H7 by using locked nucleic acids (LNA) and this amplification method. To achieve autocatalytic deposition of gold over the AuNP surface and subsequent growth of AuNP, an aliquot of amplifier solution composed by $HAuCl_4$ and formaldehyde in a suitable ratio was dropped directly over the membrane once the conventional assay was developed. Using this strategy, the authors reported 0.4 nM as LoD.

Signal amplification in an LFA for human IgG detection using gold nanoparticles as carriers of enzymatic labels has been recently reported by Parolo et al. [60]. In this work, they described the use of AuNP as direct labels giving a less sensitive detection range, but when the nanoparticles were coupled with an antibody modified by horseradish peroxidase (HRP), they could act also as carriers labels. Three different HRP substrates were tested and added to the strips after the direct detection with AuNP. The 3,3,5,5-tetramethylbenzidine gave the best result for LFA applications (see Figure 7), enhancing the LoD up to 1 order of magnitude (200 pg/mL) compared with the traditional assay (5 ng/mL).

Although the mentioned methods greatly increase the sensitivity of LFA, they required two steps: detection and signal amplification, which implies more reagents, additional procedures, and longer detection times.

Another interesting method for intensifying the signal involves the use of gold nanoparticles of different sizes in the same lateral flow strip. Choi et al. [59] studied the effect produced in the LFA sensitivity using gold nanoparticles of different sizes for troponin I detection. The assay consisted in the recognition between 40-mn AuNP nanoparticles and 10-nm AuNP primarily accumulated at the test line. Briefly, smaller AuNPs were conjugated to anti-troponin I antibody and blocked by BSA, bigger AuNPs were conjugated to anti-BSA antibody, and both groups were retained in different conjugate pads. When the sample flows through the membrane, the smaller nanoparticles recognize the analyte and flow fast, arriving first to the test zone. In contrast,

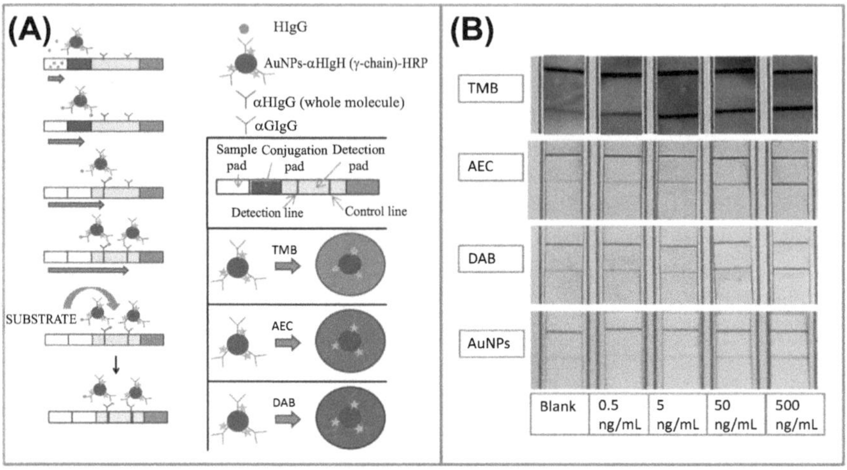

FIGURE 7 LFA signal amplification using AuNPs as carriers of enzymes. (A) Principle of the enhanced assay using HRP enzyme and three different substrates: 3,3,5,5-tetramethylbenzidine, AEC, and DAB. (B) Pictures of LFA strips for different HIgG concentrations using the three different substrates. HRP, horseradish peroxidase; LAF, lateral flow assay. *Adapted from Ref. [60] with permission.*

bigger nanoparticles flow slower due to their size, they recognize the conjugated of 10 nm AuNP and an increase in sensitivity of around 100-fold is consequently achieved (see Figure 8).

Other methods for signal amplification using a composite formed by the immobilization of AuNP over a Fe_2O_3 nanoparticles core was reported by Tang and co-workers [117]. They applied this strategy for fast screening of aflatoxin B2 in a competitive LFA and found an LoD of 0.9 ng/mL, which was 3-fold lower than the value found for conventional LFA using AuNP as label.

Inspired by this work, Liu and co-workers [118] reported the use of Fe_3O_4 nanoparticles aggregates as labels for detection of pesticide residues. The authors achieved a controlled aggregation of magnetic nanoparticles by cross-linking the polyethylene glycol-coated Fe_3O_4 NP bearing surface carbonyl groups with poly-L-lysine, followed by the conjugation with anti-pesticide antibodies. Employing this approach, they reported an LoD of 1.7 ng/mL which was 40-fold lower that the assay with Fe_3O_4 NPs.

An interesting and simple strategy based on changes of architecture of gold-based LFA was also recently presented by Parolo et al. [119] The authors studied the effect on the sensitivity of the assay by simply changing the dimensions of the conjugate and sample pad. Increasing the size of sample pad and keeping constant the conjugate pad size, no clear positive effect on sensitivity was observed. However, by increasing the size of both the conjugate and sample pad, an 8-fold improvement in the limit of quantification was achieved (see Figure 9).

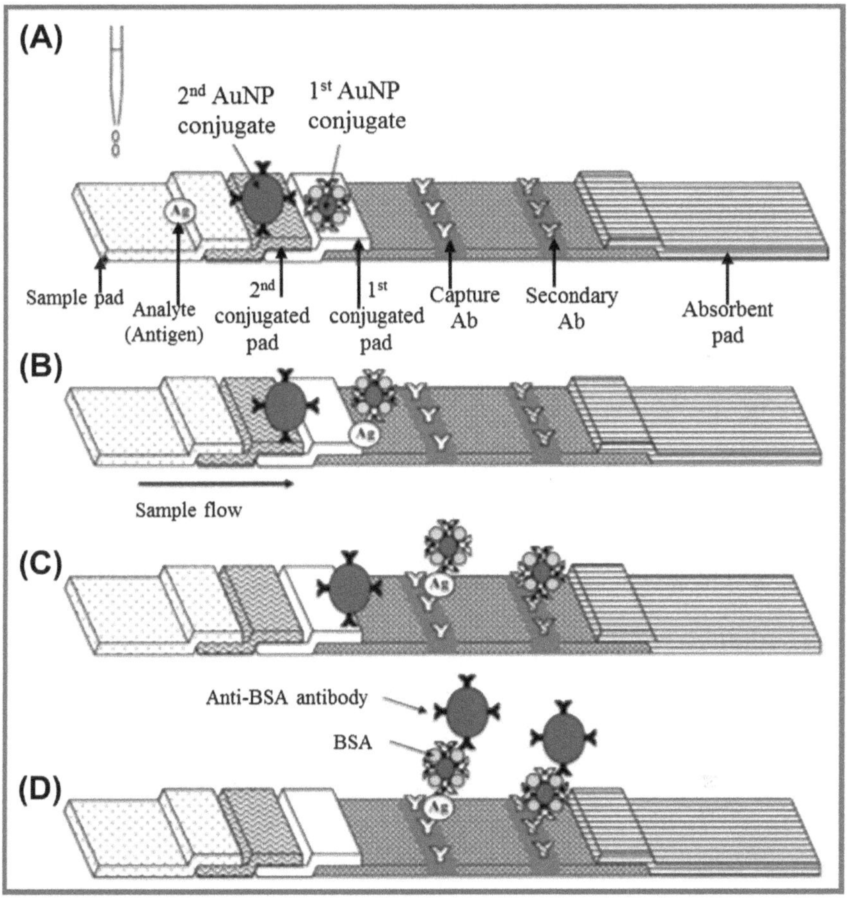

FIGURE 8 Schematic representation of the Lateral flow assay (LFA) signal amplification using dual AuNP conjugate label. (A) Components of the strip and sample insertion; (B) Sample flow; (C) First immunoreaction; (D) Final result. *Adapted from ref. [59] with permission.*

4.2 Multianalyte Detection

In most diseases, the detection of more than one biomarker is usually required for offering an accurate diagnostic. In consequence, the need to detect multiple analytes in a single device is becoming a growing topic in paper-based biosensors field. Multianalyte LF assays are important because they allow the simultaneous detection of different analytes in a single assay with short detection times and low cost.

Nevertheless beside advantages the possible cross-reactivities between antigens and antibodies in the sample that could affect the results of the LF assay should be carefully considered.

Commercial multianalyte LFA for different applications such as (1) the triple line QuickVue® developed by Quidel Corporation for detection of

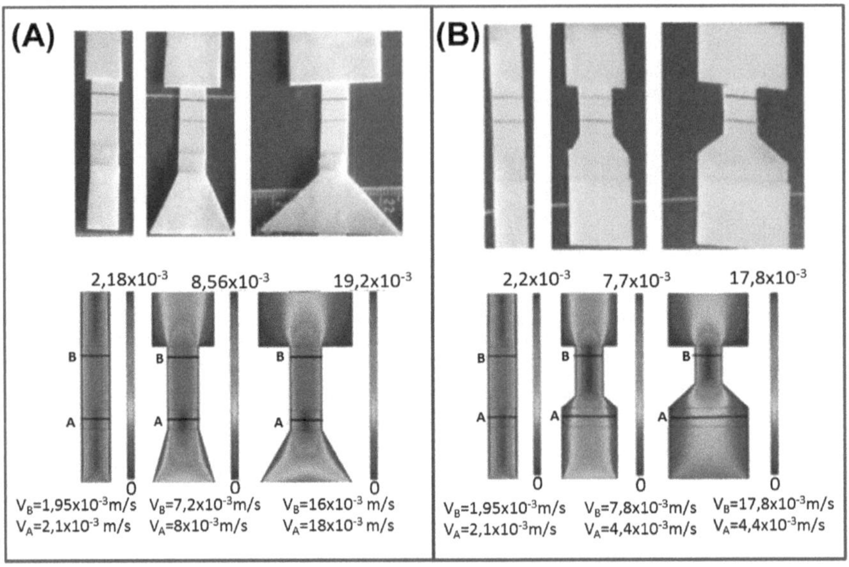

FIGURE 9 Lateral flow assay (LFA) signal amplification by changing the standard LF strip geometries. (A) Above: Pictures of LFA with different sample pad architectures for 60 ng/mL of human IgG (HIgG). Bottom: Flow speed simulations for sample pads with different sizes. (B) Above: Pictures of LFA with different sample and conjugation pads designs for 60 ng/mL HIgG. Bottom: Flow speed simulations for sample and conjugation pads with different designs. *Adapted from Ref. [119] by permission of The Royal Society of Chemistry.*

influenza A or B virus [120]; (2) a multiplexed LFA for HIV/syphilis by CTK Biotech [121]; (3) the dual detection of *Cryptosporidium parvum* and/or *Giardia duodenalis* (pathogens related to animal diseases) developed by Megacor Diagnostik [122]; and (4) the sensitive detection of *E. coli* Shiga toxin one and two commercialized by Meridian Bioscience have been reported [123].

Recently, Yonekita et al. [124], described a novel method by using AuNP-labeled antimicrobial peptides as recognition molecules for bacteria in LFA. In addition, the authors developed a multianalyte LFA for detection of *E. coli* strains O157, O26, and O111 up to 10^4 CFU/mL in beef.

Even though the detection of toxins is vital for avoiding the effects produced to the human health, few works have been reported for their multidetection. Unisensor [125] developed the first multiplex dipstick for detection of four main classes of mycotoxins: zearalenone (ZEN), T-2 and HT-2 toxins, deoxynivalenol (DON), and fumonisins (FB_1/FB_2). This device allows the detection in just less than 30 min with high specificity and sensitivity.

4.3 Integration of Novel Materials

Magnetic particles have been used for years as well-established labels in bioassays because they are not affected by reagents, are stable over time and can be manipulated by applying a magnetic field. In recent years, the use of superparamagnetic nanoparticles (SPMNPs) has become an attractive alternative to the traditional labels for LFA, because of their quantitative measurement that can be achieved over the entire thickness of NC membrane, instead of the eye-naked detection, which uses color intensity as signal only if the labels are located 10 μm coming from the top layer of membrane [16]. With the use of these new labels, it is consequently possible to obtain better sensitivities in LFA.

In this context, Wang et al. [126] presented a study about the size and magnetite content of SPMNP and their use as label in LFA using hCG as model protein. The authors found that smaller sizes of SPMNP between 111 and 133 nm are related to shorter times of detection. In addition, while higher was the content of magnetite higher was the signal intensity, which was stable over time, allowing a recheck of the results.

An application of magnetic LFA was developed by Xu and co-workers for detection of cardiac troponin I [127]. They could detect up to 0.01 ng/mL of analyte (1000-fold lower than ELISA test) in a wide range of five orders of magnitude, just in 15 min.

A study of magnetic LFA was described by Zheng et al. [128] for rapid detection of parvalbumin. The authors reported limit of detection of 5 μg/mL and 0.046 μg/mL parvalbumin for visual and quantitative detection. The results were obtained in less than 20 min and were in agreement with Western blot assay (93% of accuracy).

Other materials to replace the traditional supports, such as NC in LFA, have been also studied. In this sense, cotton threads have been used for Zhou et al. [129] as support for transporting and mixing liquids, applied to the detection of C-reactive protein and demonstrating the possibility of a quantitative and multiplex assay. The authors reported the AuNP-based immunochromatographic assay thread (ICAT), using cotton threads and nylon fibers bundles as membrane and pads, respectively. Briefly, the assay consisted in placing the sample onto the nylon fiber bundle where detection antibody–AuNP conjugate, as well the capture and control antibodies located in the cotton thread at test and control zone, respectively were vacuum dried. Immediately after this step, a buffer solution for allowing the flow of the sample and the conjugate through the cotton threads and subsequent detection was added (see Figure 10). With this strategy, it was possible to detect up to 377 pM CRP, value comparable to other reported LFA for CRP but with the advantage of the possible simultaneous detection of several analytes (CRP, leptin, and osteopontin) in a reduced space and the possibility to customize ICAT à la carte thanks to the modularity of this assay. Currently, studies related to long-term stability and the type of sample are being investigated by this group.

Even though the NC is the most common material for immobilizing antibodies in LFA, the research about the use of cellulose substrates for developing bioactive paper biosensors has highly increased in the past years. Usually the functionalization of cellulose is achieved under harsh conditions, but Credou

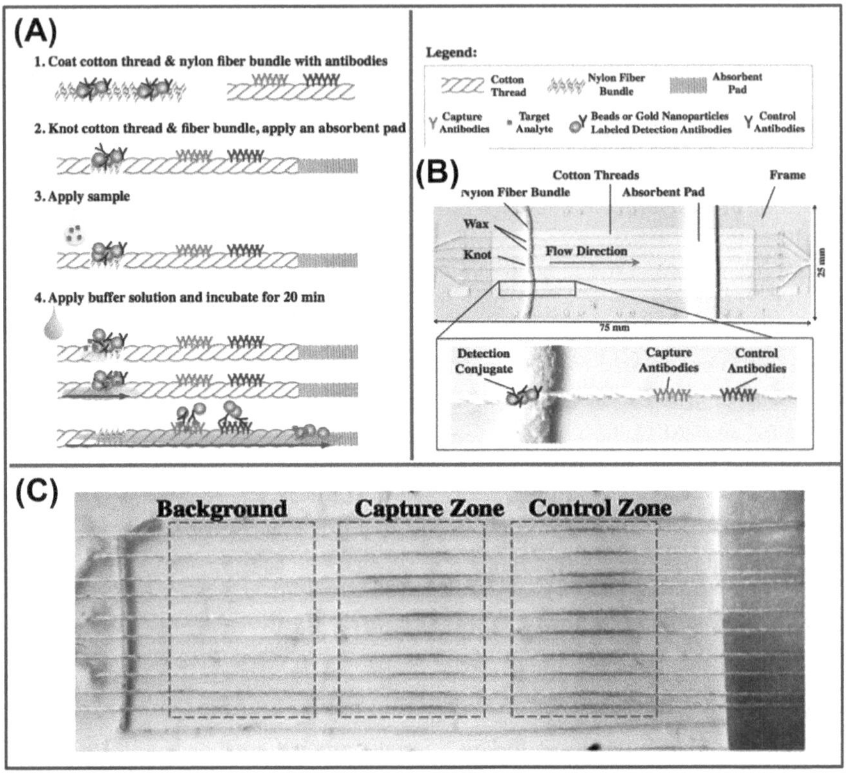

FIGURE 10 Immunochromatographic assay thread (ICAT). (A) Schematic representation of the ICAT and steps for detection. (B) Photograph of an ICAT cartridge with six threads and configuration of ICAT. (C) Photograph of a completed ICAT after detection. Red color is due to AuNPs located at the capture and control zone. Unbound AuNPs are absorbed by the wicking pad, on the right. *Adapted from Ref. [129] with permission.*

and co-workers [130] recently developed a method for functionalizing cellulose based in the aryldiazonium chemistry. All the functionalization procedures were done in mild conditions: in water, in a single step and at room temperature. The authors applied the biofunctionalized cellulose as detection pad in an AuNP-based LFA, which resulted to be less efficient than NC but more than pristine cellulose. Nevertheless, the authors developed a new method for modifying cellulose without damaging the activity of biomolecules and they successfully proved the use of biofunctionalized sheets in LF technology.

4.4 Integration with Electrochemistry

Electrochemical detection in conjunction with colorimetric methods enables to quantify the results of an assay, taking advantage of the red-ox properties of some of the tags used in LFA.

Detection of PSA has been achieved in an LFA coupled to electrochemistry using core–shell quantum dots of CdSe@ZnS [131]. In this work, a sandwich LFA format was achieved and the captured labels at test zone were detected by dissolving cadmium contained in the NP which was subsequently detected by stripping voltammetry in a disposable screen-printed electrode. The authors reported a reproducible coupled assay with a low detection limit of 0.02 ng/mL PSA.

Mao et al. [61] developed a simultaneous visual detection of rabbit IgG (RIgG) and human IgM (HIgM) as protein models in an LFA and subsequent quantitative detection of dissolved Au(III) ions by stripping voltammetry. LoD achieved with this methodology were 1.0 ng/mL and 1.5 ng/mL RIgG and HIgM, respectively.

Sensitive detection of trichloropyridinol, a metabolite biomarker of exposure to organophosphorus insecticides was achieved in an LFA integrated with electrochemistry [132]. The enzyme-linked immunoreaction was developed in the LF strip and all the captured HRP-labeled antibodies located at the test zone were measured in a screen-printed electrode by using square wave voltammetry, allowing detection of up to 0.1 ng/mL trichloropyridinol.

Du et al. [133] developed a sensitive integrated LFA electrochemical detection for the determination of organophosphorus pesticides. This work is based on parallel measurements of post exposure and baseline acetylcholinesterase enzyme activity. The proposed device was composed by the LFA and a screen-printed electrode located under the test zone and the signal was enhanced using carbon nanotubes allowing the electrochemical detection at low potentials. Authors reported a limit of detection of 0.02 nM of organophosphorus pesticide.

5. CONCLUSIONS AND FUTURE PERSPECTIVES

Lateral flow assays are excellent tools that fulfill the requirements of an ASSURED device. Incorporation of nanotechnology to these devices brings a wide number of advantages for improving their sensitivity.

LFA technology has been quickly established with a good market position due to two main reasons: (1) potentiality of satisfying the necessity of rapid diagnostic devices with moderate sensitivity for being used outside of lab settings, and (2) low cost analysis and easy production. Even though the cost LFA equipment fabrication, it is considerably reasonable compared with other immunoassay technologies and the final costs of mass producing lateral flow strips are significantly low. In fact, LF technology is still improving with regard to the recent market needs for more quantitative and rapid diagnostic devices making possible the extension of LFA to other areas, e.g., agriculture, veterinary, biodefense, food, environmental control, and nucleic acid testing.

Gold nanoparticles are the most popular label for LFA, due to their biocompatibility, easy and inexpensive preparation, stability, and strong red

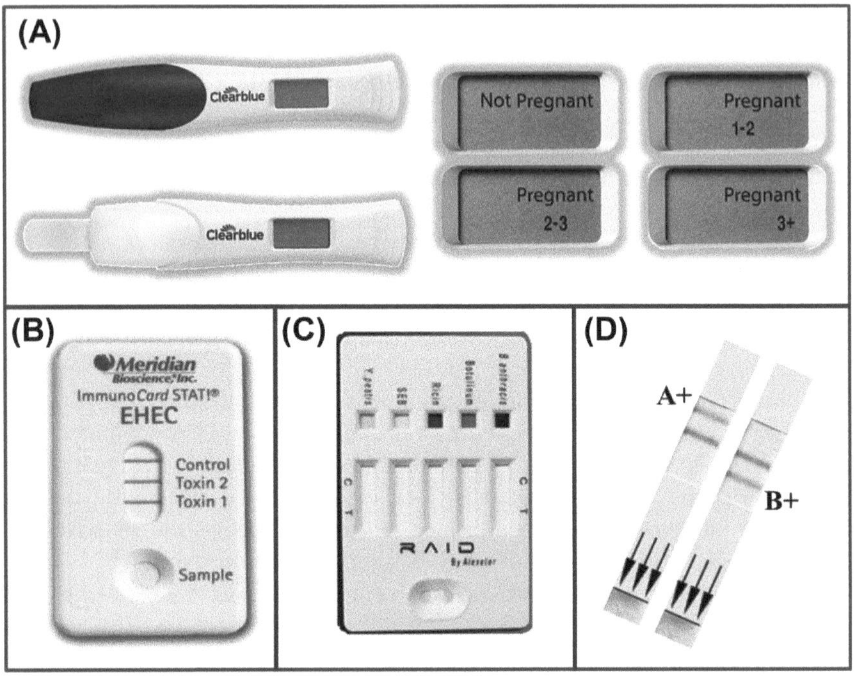

FIGURE 11 Commercial lateral flow assay (LFA) devices. (A) Clearblue™ Advance Pregnancy Test with Week Estimator. Photo reproduced with kind permission from SPD Swiss Precision Diagnostics GmbH, owner of the trademark ClearblueTM © 2014. All rights reserved. (B) ImmunoCard STAT! (R) for *Escherichia coli* Shiga toxin 1 and 2 detection. Meridian Bioscience, Inc. All rights reserved. (C) RAID five for biological threat screening. Alexeter Technlogies. All rights reserved. (D) QuickVue® for Influenza A and B detection. Quidel Corporation. All rights reserved.

color coming from their surface plasmon absorption. A wide variety of analytes have been tested by using LF technology, including different fields such as medical diagnostics, environment, and food. Some of these devices can be found in the market (see Figure 11).

Although the LFA is an old technology, it is under an extensive research due to the new opportunities coming from the use of nanotechnologies and nanomaterials. The main challenge in the field consists in improving the sensitivity of the assays so as to reach the requested low detection limits for real samples applications.

Gold nanoparticle–based LFA can be further improved by developing of more sensitive detection systems. Beside possible improvements of current optical detectors, the integration of AuNP-electrochemical detection systems may bring significant results. This can be related to the possibility of the use of electrochemical detection systems that use electrocatalytic properties of AuNP well known for their higher sensitivity in conventional batch systems (e.g., screen-printed based electrodes already reported in biosensing of DNA, proteins and cells using AuNP). This would imply careful studies and further modification of AuNP-related electrocatalytic reactions so as to adapt these to LF designs and the composition of the LF pads/platforms.

Future trends beside the use of gold nanoparticles seem to be focused on the development of new labels such as other catalytic nanoparticles, new nano/micromaterials for LF platform modification, as well as of new architectures and strategies that would allow for faster and higher sensitive detection. Such improvements, some of which are already reported as proof of concept, are expected to bring these devices to real analytical applications in the near future for a broader range of analytes.

ACKNOWLEDGMENTS

We acknowledge the support from MINECO (Spain) under project MAT2011-25870 and POC4PETS FP7 EU project.

REFERENCES

[1] I.D. Mandel, Ann. N.Y. Acad. Sci. 64 (1993) 1–10.
[2] A Timeline of Pregnancy Testing. http://history.nih.gov/exhibits/thinblueline/timeline.html.
[3] P. Ghalioungui, S. Khalil, A.R. Ammar, Med. Hist. 7 (1963) 241–246.
[4] E.H. Starling, Lancet 166 (1905) 339–341.
[5] G.H. Ettinger, G.L.M. Smith, E.W. McHenry, Can. Med. Assoc. J. 24 (1931) 491–495.
[6] L. Wide, C.A. Gemzell, Acta Endocrinol. 35 (1960) 261–267.
[7] C.M. Plotz, J.M. Singer, Am. J. Med. 21 (1956) 888–892.
[8] B.S.A. Berson, R.S.J. Yalow, Clin. Invest. 38 (1959) 1996–2016.
[9] E. Engvall, P. Perlman, Immunochemistry 8 (1971) 871–874.
[10] B.K. Van Weemen, A.H. Schuurs, FEBS Lett. 15 (1971) 232–236.
[11] J.H. Leuvering, P.J. Thal, M. van der Waart, A.H.J. Schuurs, Immunoassay 1 (1980) 77–91.
[12] K. May, M.E. Prior, I. Richards, Immunoassays and Devices therefore, International Patent WO 88/08534 (1988).
[13] R.W. Rosenstein, Solid Phase Assay, European Patent EP 0 284 232 (1988).
[14] R. Campbell, D.B. Wagner, J. O'Connel, Solid-Phase Assay with Visual Readout, US Patent 4,855,240 (1987).
[15] A.J. Wilcox, D.D. Baird, C.R.N. Weinberg, Engl. J. Med. 340 (1999) 1796–1799.
[16] Millipore Corporation, Rapid Lateral Flow Test Strips: Considerations for Product Development, 2002.
[17] C.J. Van Oss, R.J. Good, M.K. Chaudhury, J. Chromatogr. 391 (1987) 53–65.
[18] J.L. Tonkinson, B.A. Stillman, Front. Biosci. 7 (2002) c1–12.
[19] M.W. Nicholas, K. Nelson, J. Invest. Dermatol. 133 (2013) 1–3.
[20] R.C. Wong, H.Y. Tse (Eds.), Drugs of Abuse: Body Fluid Testing, Humana Press, New Jersey, 2005.
[21] K.D. Jones, IVD Technol. 5 (1999) 32–34.
[22] Imagene Technology. http://www.imagenetechnology.com/.
[23] A.K. Yetisen, M.S. Akram, C.R. Lowe, Lab Chip 13 (2013) 2210–2251.

[24] D.G. Wild (Ed.), The Immunoassay Handbook, Fourth ed., Elsevier Ltd., Great Britain, 2013.
[25] D.D. Lasic, D. Papahadjopoulos (Eds.), Medical Applications of Liposomes, Elsiever Science, Amsterdam, 1998.
[26] K.A. Edwards, A.J. Baeumner, Anal. Bioanal. Chem. 386 (2006) 1335–1343.
[27] H.W. Wen, W. Borejsza-Wysocki, T.R. DeCory, R.A. Durst, Anal. Bioanal. Chem. 382 (2005) 1217–1226.
[28] P. Noguera, G.A. Posthuma-Trumpie, M. van Tuil, F.J. van der Wal, A. de Boer, A.P.H.A. Moers, A. van Amerongen, Anal. Bioanal. Chem. 399 (2011) 831–838.
[29] D.P. Kalogianni, L.M. Boutsika, P.G. Kouremenou, T.K. Christopoulos, P.C. Ioannou, Anal. Bioanal. Chem. 400 (2011) 1145–1152.
[30] L. Vanrell, A. Gonzalez-Techera, B.D. Hammock, G. Gonzalez-Sapienza, Anal. Chem. 85 (2013) 1177–1182.
[31] C. Suárez-Pantaleón, J. Wichers, A. Abad-Somovilla, A. van Amerongen, A. Abad-Fuentes, Biosens. Bioelectron. 42 (2013) 170–176.
[32] G.A. Posthuma-Trumpie, J.H. Wichers, M. Koets, L.B.J.M. Berendsen, A. van Amerongen, Anal. Bioanal. Chem. 402 (2012) 593–600.
[33] K. Wilson, J.M. Walker (Eds.), Principles and Techniques of Biochemistry and Molecular Biology, seventh ed., Cambridge University Press, Cambridge, 2010.
[34] G.P. Leonardi, A.M. Wilson, A.R. Zuretti, J. Virol. Methods 189 (2013) 379–382.
[35] Y. Wang, S.R. Nugen, Biomed. Microdevices 15 (2013) 751–758.
[36] N. Hildebrandt, ACS Nano 5 (2011) 5286–5290.
[37] F.A. Esteve-Turrillas, A. Abad-Fuentes, Biosens. Bioelectron. 41 (2013) 12–29.
[38] Z. Zou, D. Du, J. Wang, J.N. Smith, C. Timchalk, Y. Li, Y. Lin, Anal. Chem. 82 (2010) 5125–5133.
[39] L. Wang, W. Chen, W. Ma, L. Liu, W. Ma, Y. Zhao, Y. Zhu, L. Xu, H. Kuang, C. Xu, Chem. Commun. 47 (2011) 1574–1576.
[40] Q. Yang, X. Gong, T. Song, J. Yang, S. Zhu, Y. Li, Y. Cui, Y. Li, B. Zhang, J. Chang, Biosens. Bioelectron. 30 (2011) 145–150.
[41] E. Juntunen, T. Myyryläinen, T. Salminen, T. Soukka, K. Pettersson, Anal. Biochem. 428 (2012) 31–38.
[42] R.W. Peeling, K.K. Holmes, D. Mabey, A. Ronald, Sex. Transm. Infect. 82 (2006) v1–6.
[43] R. Wilson, Chem. Soc. Rev. 37 (2008) 2028–2045.
[44] E. Boisselier, D. Astruc, Chem. Soc. Rev. 38 (2009) 1759–1782.
[45] D.A. Giljohann, D.S. Seferos, W.L. Daniel, M.D. Massich, P.C. Patel, C.A. Mirkin, Angew. Chem. Int. Ed. Engl. 49 (2010) 3280–3294.
[46] L. Dykman, N. Khlebtsov, Chem. Soc. Rev. 41 (2012) 2256–2282.
[47] E.C. Dreaden, A.M. Alkilany, X. Huang, C.J. Murphy, M.A. El-Sayed, Chem. Soc. Rev. 41 (2012) 2740–2779.
[48] F.W. Faulk, G.M. Taylor, Immunochemistry 8 (1971) 1081–1083.
[49] P. Zhao, N. Li, D. Astruc, Coord. Chem. Rev. 257 (2013) 638–665.
[50] J. Turkevich, P.C. Stevenson, J. Hillie, Discuss. Faraday Soc. (1951) 55–75.
[51] G. Frens, Nat. Phys. Sci. 241 (1973) 20–22.
[52] G. Hermanson, Microparticles and Nanoparticles. Bioconjugate Techniques, third ed., Academic Press, San Diego, 2008.
[53] J. Carney, H. Braven, J. Seal, E. Whitworth, IVD Technol. 12 (2006) 41–49.
[54] A. Ambrosi, M.T. Castañeda, A.J. Killard, M.R. Smyth, S. Alegret, A. Merkoçi, Anal. Chem. 79 (2007) 5232–5240.

[55] R.C. Wong, H.Y. Tse (Eds.), Lateral Flow Immunoassay, Humana Press, New Jersey, 2009.
[56] S. Lou, J. Ye, K. Li, A. Wu, Analyst 137 (2012) 1174–1181.
[57] Q. Zeng, X. Mao, H. Xu, S. Wang, Am. J. Biomed. Sci. 1 (2009) 70–79.
[58] F.C. Fernández-Sánchez, C.J. McNeil, K. Rawson, O. Nilsson, H.Y. Leung, V. Gnanapragasam, J. Immunol. Methods 307 (2005) 1–12.
[59] D.H. Choi, S.K. Lee, Y.K. Oh, B.W. Bae, S.D. Lee, S. Kim, Y.B. Shin, M.G. Kim, Biosens. Bioelectron. 25 (2010) 1999–2002.
[60] C. Parolo, A. de la Escosura-Muñiz, A. Merkoçi, Biosens. Bioelectron. 40 (2013) 412–416.
[61] X. Mao, M. Baloda, A.S. Gurung, Y. Lin, G. Liu, Electrochem. Commun. 10 (2008) 1636–1640.
[62] Y.T. Horng, P.C. Soo, B.J. Shen, Y.L. Hung, K.Y. Lo, H.P. Su, J.R. Wei, S.C. Hsieh, P.R. Hsueh, H.C. Lai, Water Res. 40 (2006) 2221–2229.
[63] B.A. Rohrman, V. Leautaud, E. Molyneux, R.R. Richards-Kortum, PLoS One 7 (2012) 1–5.
[64] S.K. Rastogi, C.M. Gibson, J.R. Branen, D.E. Aston, A.L. Branen, P.J. Hrdlicka, Chem. Commun. 48 (2012) 7714–7716.
[65] A.Y. Kolosova, S. De Saeger, L. Sibanda, R. Verheijen, C. Van Petghem, Anal. Bioanal. Chem. 389 (2007) 2103–2107.
[66] A. Molinelli, K. Grossalber, R. Krska, Anal. Bioanal. Chem. 395 (2009) 1309–1316.
[67] L. Wang, W. Ma, W. Chen, L. Liu, W. Ma, Y. Zhu, L. Xu, H. Kuang, C. Xu, Biosens. Bioelectron. 26 (2011) 3059–3062.
[68] D. Mazumdar, J. Liu, G. Lu, J. Zhou, Y. Lu, Chem. Commun. 46 (2010) 1416–1418.
[69] Y. He, X. Zhang, K. Zeng, S. Zhang, M. Baloda, A.S. Gurung, G. Liu, Biosens. Bioelectron. 26 (2011) 4464–4470.
[70] A.M. López, J. Pons, D.A. Blake, A. Merkoçi, Biosens. Bioelectron. 47 (2013) 190–198.
[71] A.M. López Marzo, J. Pons, D.A. Blake, A. Merkoçi, Anal. Chem. 85 (2013) 3532–3538.
[72] H. Xu, X. Mao, Q. Zeng, S. Wang, A.N. Kawde, G. Liu, Anal. Chem. 81 (2009) 669–675.
[73] C. Fernández-Sánchez, C.J. McNeil, K. Rawson, O. Nilsson, Anal. Chem. 76 (2004) 5649–5656.
[74] C. Fang, Z. Chen, L. Li, J. Xia, J. Pharm. Biomed. Anal. 56 (2011) 1035–1040.
[75] J. Zhu, N. Zou, D. Zhu, J. Wang, Q. Jin, J. Zhao, H. Mao, Clin. Chem. 57 (2011) 1732–1738.
[76] P. Craw, W. Balachandran, Lab Chip 12 (2012) 2469–2486.
[77] C. Parolo, A. Merkoçi, Chem. Soc. Rev. 42 (2013) 450–457.
[78] F. Ahmad, S.A. Hashsham, Anal. Chim. Acta 733 (2012) 1–15.
[79] J. Seal, H. Braven, P. Wallace, IVD Technol. 12 (2006) 41–51.
[80] O. Piepenburg, C.H. Williams, D.L. Stemple, N. A, PLoS Biol. 4 (2006) 1115–1121.
[81] Y. He, K. Zeng, S. Zhang, A. S.Gurung, M. Baloda, X. Zhang, G. Liu, Biosens. Bioelectron. 31 (2012) 310–315.
[82] S. Hou, Y. Hsiao, M. Lin, C. Yen, C. Chang, Talanta 99 (2012) 375–379.
[83] Z. Xiao, P. Lie, Z. Fang, L. Yu, J. Chen, J. Liu, Chem. Commun. 48 (2012) 8547–8549.
[84] J. Aveyard, M. Mehrabi, A. Cossins, H. Braven, R. Wilson, Chem. Commun. 41 (2007) 4251–4253.
[85] X. Mao, Y. Ma, A. Zhang, L. Zhang, L. Zeng, G. Liu, Anal. Chem. 81 (2009) 1660–1668.
[86] P. Lie, J. Liu, Z. Fang, B. Dun, L. Zeng, Chem. Commun. 48 (2012) 236–238.
[87] World Health Organization Mortality and Global Health Estimates. Cause-Specific Mortality, 2000–2011. http://apps.who.int/gho/data/node.main.CODWORLD?lang=en.

[88] F.S. Rodrigues Ribeiro Teles, L.A. Pires de Távora Tavira, L.J. Pina da Fonseca, Crit. Rev. Clin. Lab. Sci. 47 (2010) 139–169.

[89] W.G. Lee, Y.G. Kim, B.G. Chung, U. Demirci, A. Khademhosseini, Adv. Drug Deliv. Rev. 62 (2010) 449–457.

[90] A. Niemz, T.M. Ferguson, D.S. Boyle, Trends Biotechnol. 29 (2011) 240–250.

[91] N. Shetty, J.W. Tang, J. Andrews, Infectious Disease: Pathogenesis, Prevention and Case Studies, Wiley-Blackwell, Malasia, 2009.

[92] A. Chua, C.Y. Yean, M. Ravichandran, B. Lim, P. Lalitha, Biosens. Bioelectron. 26 (2011) 3825–3831.

[93] C. Liu, C. Yeung, P. Chen, M. Yeh, S. Hou, Food Chem. 141 (2013) 2526–2532.

[94] J.W. Bennett, M. Klich, Clin. Microbiol. Rev. 16 (2003) 497–516.

[95] R. Krska, A. Molinelli, Anal. Bioanal. Chem. 393 (2009) 67–71.

[96] B. Ngom, Y. Guo, X. Wang, D. Bi, Anal. Bioanal. Chem. 397 (2010) 1113–1135.

[97] L. Anfossi, C. Baggiani, C. Giovannoli, G.D. Arco, G. Gianfranco, Anal. Bioanal. Chem. 405 (2013) 467–480.

[98] B.S. Delmulle, S.M.D.G. De Saeger, L. Sibanda, I. Barna-Vetroand, C.H.J. Van Peteghem, Agric. Food Chem. 53 (2005) 3364–3368.

[99] J. Moon, G. Kim, S. Lee, Materials 5 (2012) 634–643.

[100] A.Y. Kolosova, L. Sibanda, F. Dumoulin, J. Lewis, E. D-uveiller, C. Van Peteghem, S. De Saeger, Anal. Chim. Acta 616 (2008) 235–244.

[101] International Agency for Research on Cancer, Some Traditional Herbal Medicines, Some Mycotoxins, Naphthalene and Styrene, in: IARC Monographs on the Evaluation of Carcinogenic Risks to Humans, vol. 82, 2002.

[102] L. Anfossi, G. D'Arco, C. Baggiani, C. Giovannoli, G. Giraudi, Food Control 22 (2011) 1965–1970.

[103] L. Anfossi, C. Giovannoli, G. Giraudi, F. Biagioli, C. Passini, C.J. Baggiani, Agric. Food Chem. 60 (2012) 11491–11497.

[104] U.S. Environmental Protection Agency. http://www.epa.gov/pesticides/about/index.htm.

[105] G. Aragay, F. Pino, A. Merkoçi, Chem. Rev. 112 (2012) 5317–5338.

[106] C. Wang, C. Yu, Rev. Anal. Chem. 32 (2013) 1–14.

[107] S. Wang, C. Zhang, J. Wang, Y. Zhang, Anal. Chim. Acta 546 (2005) 161–166.

[108] W.B. Shim, Z.Y. Yang, J.Y. Kim, J.G. Choi, J.H. Je, S.J. Kang, A.Y. Kolosova, S.A. Eremin, D.H. Chung, J. Agric. Food Chem. 54 (2006) 9728–9734.

[109] J. Kaur, K.V. Singh, R. Boro, K.R. Thampi, M. Raje, G.C. Varshney, C.R. Suri, Environ. Sci. Technol. 41 (2007) 5028–5036.

[110] W.J. Gui, S.T. Wang, Y.R. Guo, G.N. Zhu, Anal. Biochem. 377 (2008) 202–208.

[111] P.G. Georgopoulos, A. Roy, M.J. Yonone-Lioy, R.E. Opiekun, P.J. Lioy, J. Toxicol. Environ. Heal. Part B Crit. Rev. 4 (2001) 341–394.

[112] Z. Fang, J. Huang, P. Lie, Z. Xiao, C. Ouyang, Q. Wu, Y. Wu, G. Liu, L. Zeng, Chem. Commun. 46 (2010) 9043–9045.

[113] F. Yang, J. Duan, M. Li, Z. Wang, Z. Guo, Anal. Sci. 28 (2012) 333–338.

[114] X. Liu, J.J. Xiang, Y. Tang, X.L. Zhang, Q.Q. Fu, J.H. Zou, Y. Lin, Anal. Chim. Acta 745 (2012) 99–105.

[115] R.H. Shyu, H.F. Shyu, H.W. Liu, S.S. Tang, Toxicon 40 (2002) 255–258.

[116] I.H. Cho, S.M. Seo, E.H. Paek, S.H. Paek, J. Chromatogr. B 878 (2010) 271–277.

[117] D. Tang, J.C. Sauceda, Z. Lin, S. Ott, E. Basova, I. Goryacheva, S. Biselli, L. Lin, R. Niessner, D. Knopp, Biosens. Bioelectron. 25 (2009) 514–518.

[118] C. Liu, Q. Jia, C. Yang, R. Qiao, L. Jing, L. Wang, C. Xu, M. Gao, Anal. Chem. 83 (2011) 6778–6784.
[119] C. Parolo, M. Medina-Sánchez, A. de la Escosura-Muñiz, A. Merkoçi, Lab Chip 13 (2013) 386–390.
[120] Quidel Corporation. http://www.quidel.com/.
[121] CTK Biotech. http://www.ctkbiotech.com/.
[122] Diagnostik Megacor. http://www.megacor.at/.
[123] Meridian Bioscience. http://www.meridianbioscience.com/.
[124] T. Yonekita, R. Ohtsuki, E. Hojo, N. Morishita, T. Matsumoto, T. Aizawa, F.J. Morimatsu, Microbiol. Methods 93 (2013) 251–256.
[125] Unisensor. http://www.unisensor.be/.
[126] Y. Wang, H. Xu, M. Wei, H. Gu, Q. Xu, W. Zhu, Mater. Sci. Eng. C 29 (2009) 714–718.
[127] Q. Xu, H. Xu, H. Gu, J. Li, Y. Wang, M. Wei, Mater. Sci. Eng. C 29 (2009) 702–707.
[128] C. Zheng, X. Wang, Y. Lu, Y. Liu, Food Control 26 (2012) 446–452.
[129] G. Zhou, X. Mao, D. Juncker, Anal. Chem. 84 (2012) 7739–7743.
[130] J. Credou, H. Volland, J. Dano, T.J. Berthelot, Mater. Chem. B 1 (2013) 3277–3286.
[131] Y.Y. Lin, J. Wang, G. Liu, H. Wu, C.M. Wai, Y. Lin, Biosens. Bioelectron. 23 (2008) 1659–1665.
[132] L. Wang, D. Lu, J. Wang, D. Du, Z. Zou, H. Wang, J.N. Smith, C. Timchalk, F. Liu, Y. Lin, Biosens. Bioelectron. 26 (2011) 2835–2840.
[133] D. Du, J. Wang, L. Wang, D. Lu, Y. Lin, Anal. Chem. 84 (2012) 1380–1385.
[134] S. Lee, G. Kim, J. Moon, Sensors 13 (2013) 5109–5116.

Index

Note: Page numbers followed by "f" and "t" denote figures and tables, respectively.

A

B

C

F

N

O

P

T

U

V

X

Z

FSC
www.fsc.org
MIX
Papier | Fördert gute Waldnutzung
FSC® C083411

Kolibri 360 GmbH
Ferdinand-Jühlke-Straße 7
99095 Erfurt, Deutschland
produktsicherheit@kolibri360.de